GENERAL PHYSICS WORKBOOK

GENERAL PHYSICS WORKBOOK

PHYSICS PROBLEMS AND HOW TO SOLVE THEM

FOSTER STRONG
California Institute of Technology

W. H. FREEMAN AND COMPANY
San Francisco

Printed in the United States of America

International Standard Book Number: 0-7167-0339-4

9 8 7 6 5 4 3 2 1

CONTENTS

PREFACE

This Workbook is intended to give the student of elementary physics three dividends for the work he invests in it: an experience in solving a wide range of problems that are intimately related both to an understanding of physics and to its use, a problem-solving equipment that includes methods and techniques not discussed in the average textbook, and a solid confidence that physics problems can be solved.

To meet these objectives the Workbook concentrates on the development of thought processes and basic understanding rather than on the collection and display of information. It does not, to any appreciable extent, infringe on the role of the usual elementary physics textbook, which is required to serve as an encyclopedia of known facts and theories. Thus the Workbook and a textbook complement each other and should be used together.

Elementary textbooks are organized into topics and categories of topics. Since no two modern textbooks agree on the sequence of this organization, it would be impossible to construct a problem workbook that would mesh with more than one textbook in the sequence of topics or in their relative weight of treatment. The Workbook avoids this difficulty by concentrating on the fundamental principles underlying all topics; selected topics appear not in any standard sequence but only as appropriate illustrations of the operation of fundamental principles.

Physics, in contrast to information about physics, consists of motion and forces and momentum and energy and of the operational and conservation laws that control and predict their behaviour. These fundamental components of physics are independent of both time and topic; the motion of a planet being measured by Tycho

Brahe and the motion of a proton in the Cosmotron at Brookhaven are physically alike, the thrust exerted by Samson on the gate pillars at Gaza is physically like the electrostatic force between two neighboring protons, the momentum of a punt on the river Cam being watched by Newton is physically like the momentum an electron carries into a relativistic collision in a bubble chamber, the energy that made Watt's steam engine possible is physically the same as the energy being lost by a photon that is moving downward in a gravitational field. By concentrating on the universal aspects of all such physical phenomena the Workbook is free to cut across all topic lines and across all distinctions between "classical" and "modern" physics.

This approach to elementary physics has always appealed to me; apparently it appeals to users of physics as well. Rarely has a visit with any former student failed to include a strongly worded comment that what he valued from his elementary physics course was not the incidental topical knowledge he had picked up but the feeling of respect he had acquired for the universality of the fundamental laws and the confidence he had developed that *any* problem could be solved once he had reduced it to basic elements. All these alumni have made the point that their involvement in any particular topical area frequently changes, thus requiring the continued acquisition of new information, but that the fundamental lines of attack by which this information is absorbed, assessed, related, and used remain unchanged.

It is possible that these alumni took in as much from their elementary physics course as they did not only because the emphasis was on fundamental thinking, but also because the approach to this thinking was through the real world. Since World War II elementary physics courses and their texts have become more scientific, and this has been a healthy development. Unfortunately in many instances being "scientific" meant treating anything that was labeled engineering or was of current practical use as irrelevant and of no "scientific" interest. But students in elementary physics courses—even future theoretical physicists—still live in the real world and are still curious about the practical aspects of their environment. Hence it should not be surprising that the interest of students should become more deeply engaged, and their motivation for working at learning should be stronger, when they can see a clear relation between their study and the world they live in.

A majority of the problems in this Workbook are related to the real world that students see; either the problems simulate a real situation, or they are component sub-problems of a larger real problem. All problems are solvable with the date available, with many providing an opportunity for exercising judgment—which also duplicates the real world. All problems are provided with answers; I have never been in favor of playing games with students by withholding answers. A student working at a problem is trying to learn; even if in desperation he tries working backward from the answer he still learns something.

January 1972 *Foster Strong*

ACKNOWLEDGMENTS

Permission from Robert B. Leighton and Rochus E. Vogt to use some of the problems in their book *Exercises in Introductory Physics* is gratefully acknowledged.

In addition to this specific acknowledgment, a more general acknowledgment is due. No instructor operates in a vacuum free of a strong influence and a constant input from those with whom he is professionally involved. Therefore the author wishes to record his debt to his colleagues, who have been stimulating, challenging, and always generously helpful, and to all his students over the years who have taught him what fun teaching can be.

1

PREPARATION FOR PROBLEM SOLVING

HOW TO SOLVE PHYSICS PROBLEMS

The easiest way to solve complex problems is to break them down into simple components. A practical test of simplicity is to count the number of broad physical principles, or laws, that must be considered in preparing a solution. A linear problem in statics or the question of current flowing in a single-loop, direct-current circuit are about as simple as problems can get; they usually contain only one relationship, $\Sigma F = 0$ or $V = IR$, solvable by means of elementary algebra. A problem about the energy available in a certain process may seem like a simple application of the Law of Conservation of Energy. Before you can do the energy inventory, however, you may first have to do a force analysis, determining a normal force in order to evaluate work done against friction. Obviously such a problem is not of first-order simplicity, but can be separated into simple subproblems. Here again, algebra and trigonometry may be all the mathematics needed to make the necessary calculations.

Rarely is the *physics* of a physics problem more complex than was hinted at in the energy problem. However, many, if not most, of the interesting and illuminating problems are problems of process: a force changes with the distance it moves; energy is transferred at a specified rate; a charge on a capacitor decays with time. In such problems the major challenge often lies in the *mathematics* required, rather than in the physics. Newton met such challenges by inventing the calculus; we are spared the necessity of inventing it, but we are often required to know how to use it. No one can hope to acquire proficiency over a significant range of physics problems who cannot do simple differentiation and integration. It is hoped that users of this book will come to it equipped with a reasonable competence in calculus, or will seek to gain such competence concurrently with using this book.

Formal Steps

1. When a problem consists of only one physical relationship, the main burden on the solver of the problem is to recognize this relationship.

a. "A resistance R is connected in series with a battery and an ammeter of negligible resistance. If the voltage at the battery terminals is 7.5 volts and the ammeter reads 0.01 ampere, what is the value of R?" Here the challenge is to recognize that the situation is completely describable by Ohm's Law, $V = IR$. Then, $R = V/I = 7.5/0.01 = 7.5 \times 10^2$ ohms.

b. "A car moving at 30 mi hr^{-1} is brought skidding to a stop in 40 ft. Assuming the acceleration during the skid was constant, what was it?" Here the challenge is to recognize that one of the equations developed for motion with uniform acceleration applies: $v_t^2 = v_0^2 + 2aS$. A secondary challenge exists in the necessity to make the units consistent: note that 30 mi hr^{-1} and 40 ft are not consistent. Then

$$a = \frac{v_t^2 - v_0^2}{2S} = \frac{0 - 44^2}{2 \times 40} = -24 \text{ ft sec}^{-2}.$$

c. "A mole of gas is confined in a cylinder with a movable piston; the cylinder is in thermal contact with a reservoir maintained at T°K. When the piston is slowly pulled out so as to double the volume of the gas, what happens to the pressure in the gas?" The main challenge here is to recognize that the gas law equation $PV = NRT$ includes all the elements of the problem. A secondary challenge is to recognize that the given information implies that the temperature of the gas remains constant, so that NRT remains constant. Then, $P_1V_1 = NRT = P_2V_2$; $P_2 = (V_1/V_2)P_1 = (V_1/2V_1)P_1 = \frac{1}{2}P_1$.*

To meet challenges such as these one must have a prior knowledge of the principles and equations of physics. It is true that the working of physics problems reinforces and illuminates and adds zest to the study of physical principles. However, in this case the egg very definitely comes before the chicken; there is no point

*Note that in the examples the symbols V and v appear; in (a) the V stands for potential difference in volts, in (b) the v stands for speed, in (c) the V stands for volume. The literature of science is filled with such confusing duplications; there simply are not enough symbols to assign a unique one to each aspect of the physical world. Fortunately, the context almost always makes it quite clear just what the symbol represents.

in your attempting to solve physics problems until you have first studied physics.

This book is not the text for that study. This is an auxiliary book, designed on the assumption that the user has available a standard text on elementary physics. However, some text-like material has been inserted to give background and context for the problems.

The problems discussed in this section are simple, but not trivial. Even one-line "plug-in" problems offer a necessary training in the tools of the trade; they help to consolidate your easy familiarity with the basic relationships, and they provide a practice arena for developing confidence and accuracy in the handling of numbers, units, and dimensions. In spite of the admitted value of such problems, they are in the minority in this book. Almost any basic text you use will have a good supply of them.

2. For the simple problems already discussed, the steps proceeding to a solution are almost always obvious and uncomplicated. In more complex problems, it is often not so easy to see a clear sequence of steps. Many times it is difficult even to see where to start. For such problems you need a formal, organized plan of attack that will carry you forward even though you do not see where you are going. For this purpose the following plan is strongly recommended.

a. Read the problem entirely through, rather casually. This first reading should set the problem in context and, most important, should provide initial momentum; even a casual reading starts your mind to moving toward contact with what you know.

b. Now read the problem through a second time, this time thoughtfully. You still do not pay attention to the details of the data given, but you do attempt to assess the kinds of information available. In this reading your purpose is to determine the basic pattern of the problem, and to identify if you can the simpler elements that make up the basic structure. Once you recognize these elements you automatically bring to the front of your mind the fundamental laws and equations developed for that area of physics that you have studied in a textbook.*

c. Following your decision about the basic structure of the problem, you are ready to make an inventory of the information available, *in terms consistent with this basic structure*. Thus if you decided (or at least guessed) that the problem concerned the conversion of energy from one form to another, and

*Unfortunately, in the usual textbook, and in this book also, you do not get much practice in facing a completely new problem and being forced to decide, from scratch, what fields of physics are going to be used in its solution. In the usual text, problems following chapters on kinematics are clearly going to be exercises in the use of the equations of motion, while problems following chapters on thermodynamics are almost certain to call on the gas laws and the equations of kinetic theory. In real life, problems do not confront the engineer or scientist in this orderly, prelabeled manner. Therefore, to provide some practice with problems not identified by category, a section at the end of this book contains a random variety of problems whose position has no relation to the areas of physics pertinent to the problems.

was therefore a Conservation of Energy problem, you would tabulate the information in energy terms. You would not include in this tabulation any item of information that did not belong, either by itself or in combination, in an energy inventory. The stated or implied unknown, the end product of the problem, would also be included in the inventory.

The making of this inventory serves several purposes:

- It forces you to extract the analytical data from all the wordage of the problem statement, and to put that data in an organized array.

- The process of organizing the data almost always suggests the next step. This may be a substitution in a standard formula, it may be the geometrical solution of a vector diagram, it may be the setting up of a differential or integral equation, it may be a search in a handbook for additional information. The point to be made here is the extreme desirability of maintaining momentum; anything at all you can see to do that is consistent with the problem is more productive than staring at a blank paper with a blank mind.

- Restricting the inventory to "terms consistent with the basic structure" reduces the chance of your being confused or trapped by extraneous information. This danger is not present to any great extent in made-up problems such as appear in this book, but is a constant menace in real problems. In the real world, trivial or useless or inapplicable or conflicting information usually outweighs what is applicable; one of the most penetrating measures of the competence of an engineer or scientist is his ability to separate a few pertinent nuggets from a confused and incompatible mess of information.

This inventory of the data may take different forms; its utility will be greatly enhanced by choosing the form that is most effective for a particular problem and its data. In problems in which the basic data is scalar in nature, and the solution usually requires only the manipulation of algebraic equations, a simple tabular array of data is sufficient to organize the information and to call attention to the need, if any, for making units internally consistent. In problems dealing with directions, an appropriate vector diagram is an essential part of the inventory. Where relative velocities are involved, a velocity vector diagram not only organizes that information but often reveals a preferred frame of reference or coordinate system. In equilibrium and dynamical problems, the free-body diagram is another form of vector diagram. When the data either expresses or implies relationships between certain variables of the problem, a graph of this relationship might be a central feature of the inventory; the v-t graph in motion problems and the P-V graph in thermodynamic problems are obvious examples.

d. By the time the basic structure has been identified and the inventory made, usually sufficient momentum has been generated so that the final step, the calculation leading to the answer, is fairly straightforward and obvious. In this step, the unknown should be made the target as soon as possible. The equation leading to the solution should start $x = \ldots$, where x is the principal unknown or quantity sought. Keeping the variables separated in this way, with the target isolated on the left-hand side of the equation, always contributes to the clarity of your development and a greater understanding of what you are doing.

The computation should be carried out in literal, or symbolic, form rather than with numbers. All of the algebraic manipulations should be done with symbols, and the answer should first be expressed in symbols. The importance of this procedure cannot be emphasized too strongly. If the given data includes numerical values, *only in the final step* should these numbers be substituted to obtain a numerical answer.

There are three reasons for this, all interrelated and all about equally important. It is an easily demonstrated fact that you work with greater clarity and better organization when you deal with symbols rather than with numbers. A problem work sheet done in symbols looks better, is easier to follow and to check, in some subtle fashion gives you more confidence in what you are doing; a work sheet done in figures often loses all semblance of order, and too often looks like a random scrambling for some lucky combination that will match the published answer. A symbolic solution carries with it a constant picture of the physics involved, something that a mess of numbers can never do. And finally, an answer in symbolic terms permits easy dimensional checking. In order to encourage the habit of working with symbols, the majority of problems in this book are written in terms of symbols only.

e. Even after the answer has been computed, it must be checked dimensionally and for reasonableness. For example, if you were attempting to determine a grating space, which has the dimension L, and the answer came out with the dimension L^{-1}, not only would you know that you had made an error, but you would also suspect that the error was one of inversion. Thus dimensional checks often give a clue to the kind of error that must be located and corrected.

Dimensional analysis will not reveal errors of magnitude; for these, aside from careful and accurate work, the problem solver must depend on a sense of reasonableness, or appropriateness, in the answer. For every numerical answer the question should be asked, "Is this a reasonable answer for this problem?" If, for example, you were working on a synchrotron problem and you came out with an electron travel time of 17.42 hr or a track radius of 0.174 cm, you ought to be able to recognize that these answers aren't very reasonable. Those who give attention to the reasonableness of the answer soon develop an ability to guess the answers to most problems to within an order of magnitude.

General Suggestions

1. In made-up problems the only numbers you can trust are those that are given to you in the statement of the problem. You cannot trust, to the same degree, numbers you yourself have derived. This means that throughout the problem you work as closely to the original data as you can. There are many multi-part problems in which an answer calculated for one part can be fed into a subsequent calculation

for a quick answer to another part. Computers can do this, but you should resist that temptation except for checking purposes.

The reason for this is to avoid compounding errors. Since you are not a computer there is always the possibility of your making an error; when you carry this error into another calculation, you double the number of errors. What is more serious—you usually quadruple the difficulty of locating the error. Until you have developed considerable skill in problem solving, and have sharply reduced your error probability, you will find the apparent extra work required in operating with original data much less than the work required in tracing compounded errors.

2. In any equation the units used must be consistent. You cannot mix a speed of m sec^{-1} and an acceleration of ft sec^{-2}; you cannot plug a value of electric charge in esu units into an MKSC form of Coulomb's Law; you cannot add watts to horsepower and get anything but trouble.

To avoid this trouble, in the inventory stage of your analysis of the problem you should note carefully the units of the data. If there are inconsistencies, you should decide at once what system of units you want to use, and convert all inconsistent items to this chosen system.

In problems with symbolic data, unless there is some clear indication to the contrary you can assume that the symbols represent compatible units.

3. *Don't be a nit-picker.* Every teacher has had experience with students who complain, "The problem didn't *say* the electron was injected parallel to the plates, so how could I work the problem?" This was in spite of a diagram showing a collimating anode, and in spite of the context in which the problem appeared. Even without these, the student should have recognized that the problem made more sense, and was more "real," when the electron injection was parallel to the plates.

The point to be made here is that there are very few problems, either in this book or in any standard textbook, in which every i is dotted and every t is crossed. There will be no problems which attempt to trap you by omitting essential data; to explain everything in detail, however, and to cover every contingency, would require more wordage than you would be willing to wade through. A valid problem will contain all the necessary data that is special to it, but the general conditions of the problem will most often be left to your common sense, to your willingness to look up the meaning of words you do not understand, and to your general knowledge of the real world you live in.

DIMENSIONS AND UNITS

Fundamental Dimensions

It is an amazing fact that all of science and technology is built of only four ingredients—mass, length, time, and electric charge. These are called the fundamental dimensions.

They are called "fundamental" because they exist independently of any definition. (As an exercise in philosophical futility you might try, for example, to define the concept of length without using the concept of length in the definition.) The fundamental dimensions are the primordial facts of the universe we live in; everything we do to try to understand that universe, or to manipulate it for our purposes, involves an arrangement of these fundamental dimensions.

1. Mass—symbol M

Theoretically, scientists can distinguish between two kinds of mass—inertial and gravitational. Experimentally, scientists have gone to great lengths to prove that these two masses are identical. Newton assumed they were indistinguishable, and so will we in this book.

Mass is not related to size, or to volume occupied. Neither is mass the same as weight, although masses can be, and usually are, compared by weight.

2. Length—symbol L
3. Time—symbol T
4. Electric charge—symbol Q

No attempt has been made to define these four dimensions. (As was indicated above, to attempt to do so would only lead to circular statements.) They are intrinsic properties; two, mass and charge, are properties of the matter we find in our universe; length is a property of the space in our universe; time is a property of the change, the dynamics, of our universe.*

Dimensional Analysis

For most of us, our awareness of these fundamental dimensions will come from their use in dimensional analysis. Dimensional analysis serves two purposes: it forces us to balance our equations dimensionally, which helps guide our thinking and prevents errors; and because equations must balance dimensionally, we are able to guess at relationships that might be difficult to discover by other means.

In nonscientific statements, categories can be so broad that almost anything can be related to almost anything else. In scientific statements we have no such latitude; every term in an equation-statement must carry the same dimensional identity.

The necessity for dimensional consistency helps in setting up relationships to solve problems. Suppose you wanted to use an equation of uniformly accelerated motion that reads $S = v_0t + \frac{1}{2}a(?)$; you got as far as the $\frac{1}{2}a$ and then couldn't remember whether the complete term was $\frac{1}{2}at$ or $\frac{1}{2}at^2$. You could, of course, look it up,

*Occasionally, for convenience in dimensional analysis, a "dimension" will be assigned to temperature or to luminous intensity. However, these are not fundamental dimensions, since they can be defined.

but you could resolve the uncertainty more quickly by making a mental dimensional check, thus: S has the dimension L, v the dimension LT^{-1}, and a the dimension LT^{-2}, so the equation reads $L = LT^{-1}T + LT^{-2}(?)$. The only way the equation can balance dimensionally, so that it reads $L = L + L$, is for the (?) to be T^2. Hence the correct equation was $S = v_0t + \frac{1}{2}at^2$.

Dimensional analysis is also useful in predicting unknown relationships. Suppose you were concerned about the force of the wind on a camper you proposed to drive from Palm Springs to Indio during a "Santa Ana" wind. You would feel reasonably certain that the speed of the wind and the area exposed to the wind would be key factors. Since that was all you could think of, you guessed that the force F must be some function of v and A, where A was the area. You had no reason to be sure that F would be linear in either v or A, so you tentatively proposed that $F \propto A^a v^b$, where a and b were unknown exponents. The equation had to balance dimensionally, so you then had to write $[F] = [A^a] \times [v^b]$.* The dimensions of F are MLT^{-2}, so your dimensional equation was $MLT^{-2} = (L^2)^a (LT^{-1})^b$. As soon as you wrote this you realized something was wrong; no matter what a and b might be, there would be no M on the right side of the equation to balance the M on the left. This forced you to reconsider, and you guessed that the density of the air, which has the dimensions ML^{-3}, might be a factor. So you rewrote your equation thus: $F \propto A^a v^b \rho^c$. Now your dimensional equation became $MLT^{-2} = (L^2)^a (LT^{-1})^b (ML^{-3})^c$. Equating the exponents of each dimension separately and solving the algebra, you found that $a = 1$, $b = 2$, and $c = 1$. So by dimensional analysis alone, without any recourse to physical theory, you deduced that the force of the wind, F, was $\propto \rho A v^2$, or $F = k\rho A v^2$, where k is a dimensionless number whose value must be determined from experiment.

Later in this book you will encounter equations like $y = y_0\, e^{-1/2\, \gamma t} \cos(\omega t - \theta)$; an equation describing the position of a damped harmonic oscillator as a function of the time t. What is the dimension of γ?

Since an exponent must be a pure number and hence dimensionless, $[\gamma t] = M^0L^0T^0Q^0 = 1$; then $[\gamma]T = 1$, or $[\gamma] = T^{-1}$.

If a beam brings a number N_0 of charged particles per second to the face of an absorber of thickness t, the number N getting through per second is evaluated by $N = N_0\, e^{-\mu t}$. What is the dimension of μ? (*Answer*: L^{-1})

The professional manipulator of scientific equations and relationships constantly keeps in mind a running check of dimensional consistency. Problems such as those in this book offer a good training ground for the development of this very useful habit.

In the two examples just given you will note that t appears both as a measure of time and a measure of length. This illustrates the mild confusion you will encounter in any area of knowledge where there has developed an extensive use of equations and analytical statements as a shorthand method of description. In most areas there just aren't enough unique symbols available, so that the symbols we do have must often do multiple duty. The only solution for this difficulty is to keep wide awake and constantly aware of the context of the data you are working with.

*Enclosing a single term in brackets means "the dimensions of."

Units

Dimensional Units

Nature provided us with the fundamental dimensions; the units we use in evaluating these dimensions we ourselves provide, to suit our convenience.

The distance from some particular "here" to some particular "there" is independent of us; it exists just as it is, regardless of whether we measure it by how many strides we take in going from here to there, or by how many strides our small son takes, or by how many chairs we can line up between here and there. Obviously, some methods of counting off a distance are going to be more convenient, and more widely acceptable, than others. However, it should be understood that all such methods are arbitrary; we choose a method of measurement for personal reasons, not for natural reasons.

The systems of measurement units* that will confront most of you are the MKSC and the English fps systems. Electric charge—the C of the MKSC system—is much more difficult to measure than is current, the amount of charge passing along a conductor in unit time. Because of this, a working substitute for the MKSC system has been developed as the MKSA system—a system in which the current A measured in amperes is treated as the fundamental dimension of electricity. These systems are tabulated below.

Also included in the tabulation is the cgs system. Most of the development of

Comparison of Systems of Units

System	Length	Mass	Time	Charge	Current
MKSC	meter (m)	kilogram (kg)	second (sec)*	coulomb (C)	—
MKSA	meter (m)	kilogram (kg)	second (sec)	—	ampere (A)
fps (English)	foot (ft)	pound (lb)	second (sec)	—	—
cgs	centimeter (cm)	gram (gm)*	second (sec)	esu or emu	

*The current official tabulation of international units symbolizes seconds by s and grams by g. However, to avoid confusion with other s and g symbols, the author prefers to use the older notation.

*In chronological development, length and time were the first fundamental dimensions recognized, then followed charge, and finally mass. In the development of dimensional analysis it has been customary to list them in the order mass, length, time, and charge. In establishing systems of units the order usually is length, mass, time, and charge. There is no logical explanation for these inconsistencies; science just grew that way, and the customs are now imbedded.

electrical theory took place in cgs units, and they are still used in specialized areas of physics. Very little use will be made of cgs units in this book, however. It seems better to encourage the growth of an easy self-confidence in one system of primary units; if necessary, conversion to other units can be made easily by the use of conversion tables, Appendix B.

Quantitative measurement of a physical entity consists of a number and a unit identification. The fundamental, or primary, units are listed above; derived units will appear as needed in the text.

There are two things to be said about numbers: they should be as neatly expressed as possible, and there should be some indication of their accuracy. The first requirement suggests the use of unit exponents and powers of ten; the second requirement leads to a discussion of significant figures.

Unit Exponents

Most physical measurements you will deal with in problems will represent combinations of units. Thus speed will be distance divided by time: $v = s/t$, density will be mass divided by volume: $\rho = m/V$. In standard units speed could be written

$$v = \frac{\text{mi}}{\text{hr}} \text{ or mi/hr}$$

(read miles per hour) and density could be

$$\rho = \frac{\text{gm}}{\text{cm}^3} \text{ or gm/cm}^3$$

(read grams per cubic centimeter). You will see these forms in many textbooks. However, usage in scientific journals and reports suggests a strong professional preference for unit exponents—a clearer notation and one that permits easy dimensional checking. With unit exponents v is thus expressed as

$$\text{mi hr}^{-1} \text{ and } \rho \text{ as gm cm}^{-3}.$$

Powers of Ten

There are two parts to any number—the actual figures, and the position of the decimal point. It is correct to say that the speed of light in free space $c = 299{,}792{,}000$ m sec^{-1}. It is also correct, and also much neater and more easily understood, to say $c = 2.99792 \times 10^8$ m sec^{-1}. It is correct to say that at a certain place on the earth's surface the earth's magnetic field $B = 0.000057$ webers m^{-2}; it is correct, and better, to say that $B = 5.7 \times 10^{-5}$ webers m^{-2}. In other words, keep numbers neat and under tight control, and let the exponent of 10 tell where the real decimal point lies.

A natural byproduct of the use of powers of ten is the phrase "order of magnitude." Some one will say of a tentative answer, "Well, at least that's the right order of magnitude." This means that if the tentative answer is $A.B \times 10^n$, where A, B, and n are numbers, the 10^n is correct, and $A.B$ must lie somewhere between 0.1 and 9.9. When an experimentalist claims he has improved the accuracy of an experiment by "two orders of magnitude," he is saying that he estimates the current probable error is only about 0.01, or 10^{-2}, of what it was before the improvement.

Significant Figures

No measurement is perfect; there is a limit to the accuracy of even such elaborately tested values as the speed of light, c. That limit is indicated by the numbers given in a measurement; to say that $c = 2.99792 \times 10^8$ m sec^{-1} is to indicate that what comes after the final 2 is not known. To say that $B = 5.7 \times 10^{-5}$ webers m^{-2} is to say that the measurement establishing that figure was accurate to only two figures. Such accurately known numbers are called "significant figures." Thus there are six significant figures for the value of c, but only two significant figures for B.

The accuracy of a calculation is limited to the accuracy of the least accurate number used in the calculation. To avoid unnecessary work (and the added possibility of error) it is desirable to balance equations for significant figures just as it is necessary to balance them dimensionally. For example, in a calculation involving c and other numbers of only three-figure accuracy, for c you would use the value 3.00×10^8 m sec^{-1}.

In problems designed for learning, it is expected that most calculations will be made on a slide rule, which has a readable accuracy over most of its range of only three figures. For that reason, data for problems in this book will not usually be given to more than three significant figures. Actually, this is not the artificial restriction it may seem; most of the data of the real world is uncertain beyond the third or fourth significant figure.

2

MOTION

Motion, the change of position in space during an interval of time, is basic to almost all segments of physics. The experimental physicist, the theoretical physicist, the technologist, all work with motion, measuring it, accounting for or predicting it, putting it to use, or controlling it. It is therefore quite appropriate to start any analytical work in physics with the study of motion.

A second justification for beginning with motion lies in its simplicity. Many of the important aspects of physics are complex and require a background of the simpler aspects for understanding. Motion is one of the simplest. It was defined above as the change of spatial position with time; it thus comprises only the dimensions L and T. And as we shall see shortly, the relationships between these two dimensions of motion are very simple.

Motion is a function only of spatial coordinates and of time. What it is that is moving—a baseball, a shock wave from a fast airplane, an electron in a TV picture tube, an electromagnetic wave reflected from a space satellite—is not pertinent to the study of motion as motion.

MOTION AT MUNDANE SPEEDS

Straight-line Motion

Let us begin our study by considering motion along a straight line. Although this may seem to be a rather limited and special case, its study does in fact lead to the most general treatment. Even a complicated curved motion can be (and in computerized computations must be) analyzed as a continuous array of a very large number of very short, straight-line motions.

For our straight line, let us use the x-axis of the standard three-dimensional coordinate system. Then our changes in spatial position will be changes in the x-location of the object whose motion we are considering. If our object is at x_1 at a time t_1 and is at x_2 at some later time t_2, then the change of spatial position, usually called the displacement, is $x_2 - x_1$ and the time interval is $t_2 - t_1$. Since x_1 and x_2 are lengths measured from some reference point on the x-axis, $x_2 - x_1$ is also a length;* t_1 and t_2 are successive † clock readings.

All of us have an intuitive concept of speed; we will now use our displacement and time interval to make an analytical definition of speed, thus:

$$v_{av} = \frac{x_2 - x_1}{t_2 - t_1}, \tag{2-1}$$

where v_{av} is the average speed over the time interval $t_2 - t_1$. If we change notation by saying $x_2 - x_1 = \Delta x$ and $t_2 - t_1 = \Delta t$, we simplify our statement, and at the same time make it more compatible with the development into calculus notation that will come later. With this new notation,

$$v_{av} = \frac{\Delta x}{\Delta t}, \tag{2-2}$$

where $\Delta t > 0$. The dimensions of v are L/T, or LT^{-1}.

Our straight-line motion along the x-axis can thus be defined by a relationship between x and t. If we consider t to be an independent variable and x the dependent variable, we can graph an x-t relationship as shown in Figure 1. There are, of

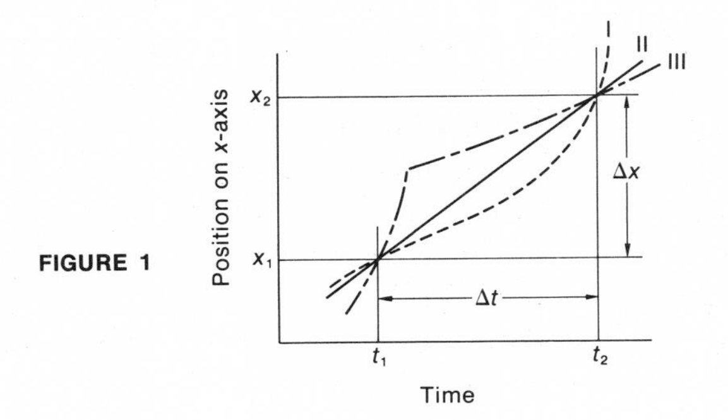

FIGURE 1

*Note that the displacement $x_2 - x_1$ is not affected by the location of the reference point from which x_1 and x_2 were measured. For if you measure from a reference point x_0 different from the first reference point, your new x_1' becomes $x_1 + x_0$ and your new x_2' becomes $x_2 + x_0$; however, your new displacement $x_2' - x_1' = (x_2 + x_0) - (x_1 + x_0) = x_2 - x_1$ is the same displacement as before.

†Normally, $t_2 > t_1$; i.e., t_2 is later than t_1. In our real world, "Time's arrow" points in only one direction. This is not true of displacement, however; motion toward the origin and motion away from the origin are equally possible. Thus $x_2 - x_1$ can be either positive or negative.

course, an infinite number of ways to go from x_1 to x_2 in the time $t_2 - t_1$; only three are shown.

Constant Speed

In Figure 1 clearly $v_{av} = \Delta x/\Delta t$ has no general usefulness for motions I and III, since for any other equal Δt the displacement will not, unless accidentally, be numerically equal to $x_2 - x_1$. For II, the straight-line motion however, $v_{av} = \Delta x/\Delta t$ is the slope of the straight line, and has the same numerical value for all possible Δt. For example, II might be the motion of a car being driven at a steady pace of 30 mi hr^{-1} along a straight road; in one hour it traversed 30 mi, in $\frac{1}{2}$ hr it covered 15 mi, in any 10-min period it went 5 mi, in any 1-min period it went $\frac{1}{2}$ mi, etc. Note that $\Delta x/\Delta t$ for all of these periods was the same:

$$v_{av} = \frac{\Delta x}{\Delta t} = \frac{30}{1} = \frac{15}{1/2} = \frac{5}{1/6} = \frac{1/2}{1/60}.$$

Hence a straight line on a displacement-time graph with linear scales indicates a constant speed. In this kind of motion, the displacement is linear with time.

For such cases we can replace v_{av} by v_c, where v_c is a constant, so that our definition becomes

$$v_c = \frac{\Delta x}{\Delta t}. \tag{2-3}$$

Since Δx and Δt are finite, this equation can be manipulated algebraically to produce

$$\Delta x = v_c \times \Delta t \tag{2-4}$$

and

$$\Delta t = \frac{\Delta x}{v_c}. \tag{2-5}$$

In many texts the straight-line displacement will be indicated by S or D and the time interval by t, so that equation 2-4, for example, may appear as

$$S \text{ (or } D) = vt. \tag{2-6}$$

Since displacement is the difference between two measured positions in space, the most general—and therefore the most useful—form of equation 2-4 would be

$$x_t - x_o = v_x (t - t_o)$$

or

$$x_t = x_o + v_x (t - t_o). \tag{2-7}$$

PROBLEM 1
How long after a solar flare erupts do we observe it?

SOLUTION
As time is the unknown quantity to be calculated, we would use the form shown in equation 2-5. The Δx is the distance from the sun to the earth, 1.0 A.U. (astronomical unit), or 1.49×10^{11} m, and $v = 3.00 \times 10^{8}$ m sec^{-1}, the speed of light in free space. Hence,

$$\Delta t = \frac{1.49 \times 10^{11}}{3.00 \times 10^{8}}$$

$$= 4.97 \times 10^{2} \text{ sec}$$

$$= 8 \text{ min } 17 \text{ sec.}$$

PROBLEM 2
A "light year" is a distance—the distance light travels in one year.

a. How far is that in kilometers?
b. The average radius of the earth's orbit is 1 A.U.; how many A.U. are in a light year?

PROBLEM 3
You are driving down a main street at 35 mi hr^{-1}, and you glance sideways at a disturbance among the people on the sidewalk. In the one second you took your eye off the road, how far did you travel toward the delivery truck double-parked down the street ahead of you?

PROBLEM 4
A certain soccer player can run 100 yds in 9.5 sec. How long should it take him to run 100 m?

PROBLEM 5
You see a lightning flash and hear the corresponding thunder t sec later. If x is the distance in miles between you and the position of the flash, what is the formula for x in terms of t? (What assumption did you have to make?)

PROBLEM 6
A motor yacht uses fuel at the rate of 1.7 gal hr^{-1} when cruising at 10 knots. If the fuel tank holds 23 gal,
a. how long can the yacht cruise at 10 knots on a tank filling,
b. how far did it travel in that time?

PROBLEM 7
If an electromagnetic pulse, which travels through the atmosphere with the speed of light, is partly reflected from an object back to the source of the pulse, the time between the emission of the pulse and the arrival of the reflection determines the distance between source and reflecting object. (This is the basic principle of radar.) If the

range R, the source-object distance, is wanted in nautical miles, and the round-trip time t can be measured in μsec, what is the formula for a scale connecting R and t?

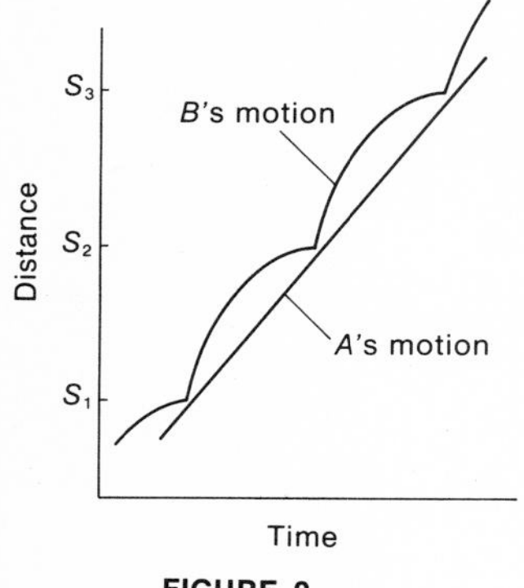

FIGURE 2

PROBLEM 8

Traffic lights along a main street are at S_1, S_2, and S_3, an equal number of blocks apart. The signals are timed to turn green to permit a steady flow of traffic at 30 mi hr^{-1}. (See Figure 2.) Driver A is aware of this, and drives accordingly. Driver B is an impatient driver who "burns rubber" when the light turns green, and then has to brake down at the next light.

a. What is A's speed?
b. What is B's average speed?
c. Estimate B's maximum speed.
d. How much time between S_1 and S_3 did B save, relative to A?

PROBLEM 9

In a lecture hall you sit 30 ft from the lecturer, but sound from him that has traveled by reflection as far as 140 ft is still audible. How "rounded," in terms of duration in time, are the sounds from the lecturer's speech?

PROBLEM 10

On a train on the main line of the Pennsylvania RR between Philadelphia and New York, a mother traveling with two precocious children asks them to determine the train's speed. She gives one child her watch, which has a split-second hand. To the other child she points out the towers evenly spaced along the right-of-way that carry the electric power for the trains. After some practice the children determine, with good consistency, that there are 16 towers per mile, and that they are passing 20 towers every 90 sec. Assuming that their reaction times remained constant during the measurements, what was the speed of the train?

PROBLEM 11

Shortly after making the above speed determination, the children note that it took exactly 18 sec to go by a string of 24 connected, identical freight cars standing on a passing track alongside the main line. Since these were smart children, they were able to tell their mother the correct overall length of a freight car. What was it?

PROBLEM 12

With the intent of providing a compatible environment for a traveling exhibit of abstract modern art a museum curator decides to paint the exhibition hall with narrow stripes of deep purple flecked with yellow. The room is rectangular, 50 ft long by 30 ft wide by 10 ft high, with plane walls and ceiling. The purple stripes are to run in direct straight lines from the center of the ceiling down to the four corners at floor level; since a floor corner can be approached in straight lines from the ceiling center along either an end wall or a side wall, eight stripes must be painted. Two junior members of the museum staff are given the job, A to do the stripes coming down the end wall and B the ones that come down the side wall. Both start at the center of the ceiling at the same instant, and both paint at the rate of 4.0 ft of stripe per minute. Which finishes his chore first, and by how many minutes? *Suggestion*: To visualize the situation, make a "doll-house cutout" plan of the exhibition room.

PROBLEM 13

At the end of a 31-min taxi trip across town, the meter indicates a charge corresponding to a trip of 7.75 mi.

a. What was the average speed for the journey?
b. If the driver said that he averaged about 23 mi hr^{-1} when he was moving, how much time during the trip had he spent standing still?

PROBLEM 14

Driving on the freeway at a constant maximum legal speed of 65 mi hr^{-1}, you can get from Pasadena to Newport Harbour in 1 hr, 10 min. On a day when your engine was knocking and you felt like driving at only 55 mi hr^{-1}, how much longer did the trip take you?

PROBLEM 15

A car is moving east on Main Street, at a steady 45 mi hr^{-1}. At a certain instant it is reported to be 3 blocks west of Central Avenue. Where will it be 1.0 min later? (In this town, the east-west measurement of a block is 440 ft.)

PROBLEM 16

At the instant the car in problem 15 is reported, a police car is at Central Avenue and Highland Street, which is 10 blocks north of Main Street. (North and south measurement of a block is 220 ft.) The police car immediately goes east on Highland 7 blocks to Cooper Avenue, and then south on Cooper to Main. If the police car is to intercept the other car at Main and Cooper, what is the minimum steady speed it must be driven? (Assume no time lost in the turn at Highland and Cooper.)

PROBLEM 17

In a 1500-m free-style swimming race held in a standard "Olympic" pool (50.0 m in length), the two leading swimmers reach the end of the pool simultaneously, 17 min, 26.5 sec after the start. One of these swimmers is the winner, having swum the full 1500 m in this time; the other swimmer has been "lapped" by the winner and still has two lengths of the pool to go. Assuming the second swimmer continues at his average rate, what will be his finishing time?

PROBLEM 18

The lead swimmer laps the second swimmer just as they reach the end of the pool, but instead of finishing as in problem 17, he gets the warning gun, which means he still has two lengths to go. He ultimately finished in the same time as in problem 17. Assuming both swimmers continue to swim at the same constant rate, what will be the finishing time of the second swimmer?

Constant Acceleration

Constant speed in a straight line does not occur too often (except in physics problems). No restrictions were placed on equation 2-1 as a definition of speed. Since it is a general equation, it would be helpful if we could enlarge its range of usefulness to include speeds not constant.

We get a hint on how to proceed (and at the same time make a tangential approach to calculus) by noting that for all possible curves in Figure 1, any v_{av} we might calculate would be a function of both t_1 and t_2 and of x_1 and x_2—two sets of variables. We usually have enough trouble when the quantity we are interested in is a function of one variable; obviously it would help if we could avoid dealing with more than one.

Let us consider a linear motion in the y direction whose graph (Figure 3) looks

FIGURE 3

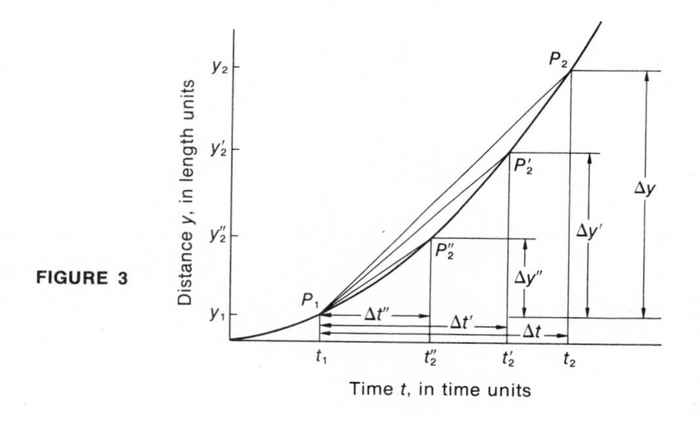

something like 1 in Figure 1. In accordance with our definition in equation 2-1 we can say.

$$v_{av_{t_2 - t_1}} = \frac{y_2 - y_1}{t_2 - t_1} = \frac{\Delta y}{\Delta t}$$

$$v_{av_{t_2' - t_1}} = \frac{y_2' - y_1}{t_2' - t_1} = \frac{\Delta y'}{\Delta t'}$$

$$v_{av_{t_2'' - t_1}} = \frac{y_2'' - y_1}{t_2'' - t_1} = \frac{\Delta y''}{\Delta t''}, \text{ etc.}$$

always leaving t_1 and y_1 as reference points, but considering shorter and shorter displacements and corresponding time intervals.

Obviously v_{av} changes each time, both in numerical value and in the direction of the secant P_1P_2. It should now be easy to see that as $t_2 \longrightarrow t_1$, $P_2 \longrightarrow P_1$, the direction of P_1P_2 approaches the direction of the tangent to the curve at P_1, and the value of $(y_2 - y_1)/(t_2 - t_1)$ approaches as a limit the value of the slope of the tangent at P_1. Writing this in conventional form:

$$\lim_{t_2 \to t_1} v_{av_{t_2 - t_1}} = \lim_{\Delta t \to 0} \left.\frac{\Delta y}{\Delta t}\right|_{t_1}.$$

(This notation reads as follows: "The limit, as t_2 approaches t_1 of v_{av} over the interval $t_2 - t_1$, equals the limit, as Δt approaches zero, of the ratio $\Delta y/\Delta t$ evaluated at t_1.")

As $t_2 \longrightarrow t_1$, $v_{av_{t_2-t_1}}$ becomes an average speed over a shorter and shorter interval until its value reaches as a limit the instantaneous value of v at t_1. Hence

$$\lim_{t_2 \to t_1} v_{av_{t_2 - t_1}} = v_{t_1},$$

where v_{t_1} denotes the instantaneous speed at t_1. Similarly, from analytic geometry,

$$\lim_{\Delta t \to 0} \left.\frac{\Delta y}{\Delta t}\right|_{t_1} = \left.\frac{dy}{dt}\right|_{t_1} = \text{slope of curve at } t_1.$$

Putting all this together, we have arrived at a new and more generally useful definition:

$$v_y = \frac{dy}{dt}, \tag{2-8}$$

where v_y, the instantaneous speed in the y-direction, and the slope dy/dt are evaluated at the same instant. Obviously for any y-t curve not a straight line, dy/dt, and hence v_y, varies from point to point on the curve, and the speed is not constant.

In developing Figure 3, we happened to choose the y-direction for the direction of our straight-line motion. We could just as easily have chosen the x- or z-direction, in which case our definition would have been

$$v_x = \frac{dx}{dt} \tag{2-9}$$

or

$$v_z = \frac{dz}{dt}. \tag{2-10}$$

We can include all of these possibilities in a generalized definition based on a generalized displacement coordinate s, by saying

$$v_s = \frac{ds}{dt}. \tag{2-11}$$

By developing the concept of instantaneous speed we accomplished what we set out to do—reduce the number of variables we needed to consider. We have thus increased the range of usefulness of our definition of speed, but at a price. When we restricted ourselves to constant speed, Δx and Δt were finite values and we needed only the simplest kind of algebra to make our computations. However, when dealing with ds/dt, the limit of a ratio of infinitesimals, we usually need the methods of calculus.

The Special Case of Free Fall

Fortunately, there is one class of linear motions in which the speed is not constant but for which the computations are simple. This type of motion is characterized by a speed that changes at a constant rate. An electron moving nonrelativistically under the single influence of a constant, uniform electric field exhibits this motion. The most easily observed example is that of free fall in a constant gravitational field.

Let us study the motion of a body falling freely without air friction or drag in a constant gravitational field. This theoretical case of free fall can be approximated in the real world (for short distances and low speeds) by dropping a streamlined object from a hovering helicopter. Photographs taken from the ground at exactly timed intervals are measured to produce the data in the table below. (In the first clear picture of the object it has fallen free of the helicopter, hence the zeros of time and distance do not coincide with the beginning of the motion.)

When the y-t data in the table below are plotted and the best curve is drawn through the points, Figure 4 results. Then, in accordance with the development that

Observational data		Computed data
Time (sec)	y (meters)*	$v_y = \frac{dy}{dt}$ m sec^{-1}
0	0	
0.5	3.2	8.8
1.0	8.7	13.6
1.5	16.7	18.5
2.0	27.1	23.2
3.0	55.4	32.9
4.0	93.4	42.3
5.0	139.9	51.5
6.0	196.7	60.4

*The vertically downward displacement is recorded and graphed as positive, i.e., the coordinate system used is

$$\begin{array}{c} - \\ + \\ + \end{array}$$

This is inconsistent with what was said earlier about normal coordinate systems; in this case it is believed it contributes to clarity to avoid negative values.

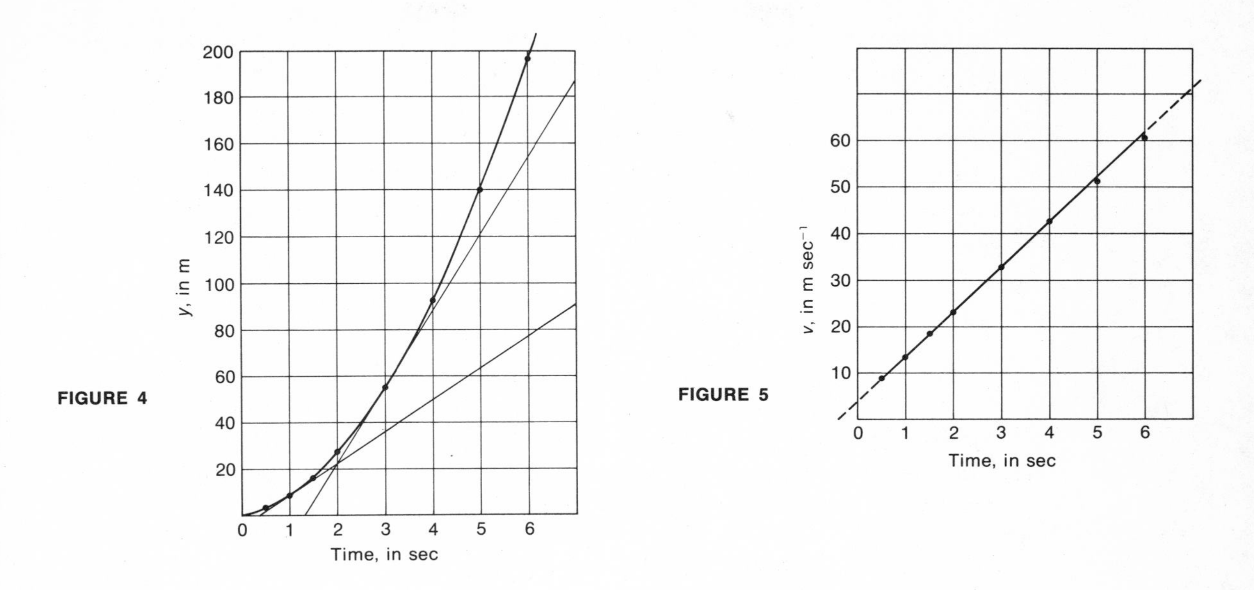

followed Figure 3, tangents to the curve are drawn and the slope evaluated. (To avoid showing too many lines, only the tangents at $t = 1.0$ sec and $t = 3.0$ sec are shown in Figure 4.)

The measured values of the slopes of the tangents are entered in this table as the computed value of v. For example, the slope of the tangent of $t = 3.0$ sec is found by dividing the maximum y distance, 187 m, by the corresponding t distance, 5.68 sec; hence $v_{t=3} = 187/5.68 = 32.9$ m sec^{-1}. By similar means all the values of v shown in this table were determined.

No obvious and correct relationship between y and t can be deduced by inspection of the curve of Figure 4. One might guess that the curve was essentially parabolic, so that a relationship $y = A + Bt + Ct^2$ might be appropriate. However, even if this was a correct guess, the determination of the constants A, B, and C would be laborious. Fortunately, there is a simple way.

The values of v and t in the table are plotted in Figure 5. To our surprise and delight the best curve through most of the points seems to be a straight line. We are delighted because we know how to get meaning out of a straight line. The general equation for a straight line is $y = b + mx$, where y is the dependent variable, x is the independent variable, b is the y-intercept (the value of y at $x = 0$), and m is the slope.

Looking carefully at Figure 5, we see that the straight line represents the relationship

$$v_t = 3.9 + 9.8t,$$

where v_t is the instantaneous speed at the instant t, 9.8 is the slope, and 3.9 is the value of v_0 (v_t at $t = 0$).

The value of the slope of the line in Figure 5 was found by dividing a value of an elapsed v by the corresponding value for the elapsed time: $\Delta v/\Delta t$. Both Δv and

Δt were finite. Just as we earlier defined $v_{av} = \Delta x/\Delta t$, we can now define a new concept, which we will call acceleration, by

$$a_{av} = \frac{\Delta v}{\Delta t}. \tag{2-12}$$

Again applying the method of a limit of a ratio of infinitesimals as we did for the definition of instantaneous speed, we can now define the instantaneous acceleration by

$$a = \lim_{\Delta t \to 0} \frac{\Delta v}{\Delta t} = \frac{dv}{dt}. \tag{2-13}$$

Thus, a is the instantaneous slope of the v-t graph. The dimensions of a are $(LT^{-1})/T = LT^{-2}$.

In the free fall plotted in Figure 5 the acceleration is obviously constant up to $t = 4.0$ sec; dv/dt has the same value for all $t < 4.0$ sec. Thus linear motion with constant linear acceleration exhibits a relationship between v and t that results in a straight line on a v-t graph. The relationship must then be expressible by the straight-line equation

$$v_t = v_0 + at. \tag{2-14}$$

This is the prime, independent equation governing linear motion with constant linear acceleration. It is important for efficiency in solving motion problems that both sides of this coin be recognized and appreciated: if a linear motion can be plotted as a straight line in a normal v-t graph, then you know that the acceleration was constant; if, on the other hand, you know that the acceleration is constant, you *know* that the resulting motion appears as a straight line on a normal v-t graph, and is described by equation 2-14.

The v-t Graph

We have already discussed v, t, and a, and how a v-t graph displays their relationship. It will be useful to see if we can extract further value from such a graph. The graph in Figure 6 shows a small gray trapezoid with the width Δt and the height v_{av}, the average value of v over the specific interval Δt. The area of the trapezoid is obviously $v_{av}\Delta t$.

Does that look familiar? It should. If we rearrange equation 2-2, we have

$$v_{av}\Delta t = \Delta x.$$

Thus the gray area must represent the distance Δx traversed in time Δt.

Suppose we divide the time t into a large number of small Δt's, each the base of its own gray area. The sum of these areas would make up the total area under the v-t curve from $t = 0$ to $t = t$. Also the sum of these areas must make up the sum of the distances traversed during all the Δt's, or during t. Hence we come to a most

FIGURE 6

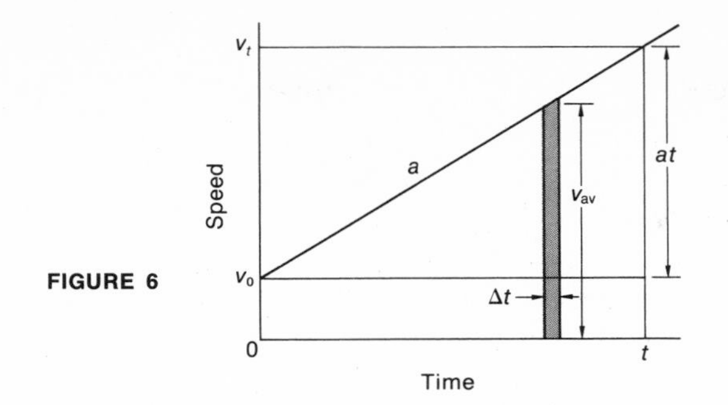

important aspect of the v-t graph. *The area under a v-t graph between any limits t_1 and t_2 gives exactly the distance traversed during the time interval $t_2 - t_1$.**

With this "area equals distance" statement in mind, let us now look at Figure 6 again. The area under the curve between $t = 0$ and $t = t$ is made up of the rectangle of altitude v_0 and base t, and of the triangle of altitude at and base t. Then the total area $= v_0 t + \frac{1}{2}at^2$. But the area also must equal the total distance S traversed in time t. Hence, finally,

$$S = v_0 t + \tfrac{1}{2}at^2. \tag{2-15}$$

The area $v_0 t$ represents the distance traveled in time t at a constant speed v_0; the area $\frac{1}{2}at^2$ represents the extra distance traveled in time t because of the constant acceleration a.

Throughout the section on straight-line motion we have been dealing with motion in only one direction. Because this singularity underlies the whole development, it has been possible to be free and casual about the coordinates under discussion—sometimes x, sometimes y, and sometimes the generalized coordinates. However, at this point it should be emphasized that the generalized equation in linear motion,

$$v_t = v_0 + at,$$

is shorthand for these three independent equations governing linear motion with constant linear acceleration in three-dimensional x-y-z space:

$$\begin{aligned} v_{x,t} &= v_{x,0} + a_x t, \\ v_{y,t} &= v_{y,0} + a_y t, \\ v_{z,t} &= v_{z,0} + a_z t; \end{aligned} \tag{2-16}$$

and the generalized statement

$$S = v_0 t + \tfrac{1}{2}at^2$$

represents

$$S_x = x_t - x_0 = v_{x,0}\, t + \tfrac{1}{2}a_x t^2, \text{ etc.} \tag{2-17}$$

*Although this statement was developed from the straight-line graph of Figure 6, it is equally valid for *any* v-t graph. The summation of $v_{av} \times \Delta t$ trapezoids can be carried out in exactly similar fashion under any v-t curve.

By eliminating t between equations 2-16 and 2-17 we can obtain another useful relationship applying to linear motion with constant acceleration:

$$v_t^2 = v_0^2 + 2aS \tag{2-18}$$

or

$$v_{x,t}^2 = v_{x,0}^2 + 2a_x(x_t - x_0), \text{ etc.} \tag{2-19}$$

It should be noted that when $a = 0$, equation 2-14 reduces to $v_t = v_0 =$ constant as it should, and equation 2-17 reduces to equation 2-7 as it should.

By eliminating a between equations 2-14 and 2-15, we obtain an additional equation:

$$S = \frac{v_t + v_0}{2}\, t = v_{\text{av}} t \quad \text{when } a = \text{constant.} \tag{2-20}$$

Caution: $v_{\text{av}} = (v_t + v_0)/2$ *only* when the v-t curve is a straight line, i.e., only when the linear acceleration is constant. The statement $v_{\text{av}} = \Delta S/\Delta t$ is always true; it is the definition of v_{av}. However, even this true statement must be used with judgment. To illustrate, consider the New York City subway shuttle train that runs under 42nd Street from Grand Central Terminal to Times Square—a total travel distance of about 3/4 mi. On a typical schedule a train will arrive at Times Square 2 min after leaving Grand Central, will wait at Times Square for 5 min while passengers are unloading and loading, and will then return to Grand Central in 2 min.

For the trip from Grand Central to Times Square, or the reverse, $\Delta x = 3/4$ mi and $\Delta t = 2$ min. Hence $v_{\text{av}} = (3/4)/(2/60) = 22.5$ mi hr^{-1}, a sensible value that could be used by a schedule maker. But what about the visitor to Grand Central who looks at the shuttle platform every 9 min and sees the same train standing there every time? For him $\Delta x = 0$, hence he would say $v_{\text{av}} = 0$. Although true by substitution in the formula, it is a meaningless statement, and of no use to anyone. Thus when motion is not continuous in one direction, any statement about average speed should be made with extreme care.

To solve problems of linear motion involving either constant speed or constant acceleration, we have available definitions given by equation 2-2, 2-11, and 2-13, and by operational equations 2-14, 2-15, 2-18, and 2-20. Given sufficient time, paper, and independent data, any such problem can be solved by grinding through the algebra. However, anyone who wishes to solve such problems easily (and in a professional manner) should recognize the usefulness of the v-t graph. Everything to be considered in linear motion is on the graph—the independent and dependent variables t and v, the acceleration given by the slope, and the displacement given by the area under the curve. The graph summarizes the information and the relationships clearly; it makes many solutions obvious by inspection, and other solutions easy.

GRAPHICAL SOLUTION TO PROBLEM 14

Since the distance traversed is the same for both trips, the area under the solid line must equal the area under the dashed line (Figure 7). Then the two gray areas must be equal. Hence,

$$(65-55)\,\frac{7}{6}=55\;\Delta t;$$

$$\Delta t=\frac{70}{330}=\frac{7}{33}\text{ hr}=13\text{ min.}$$

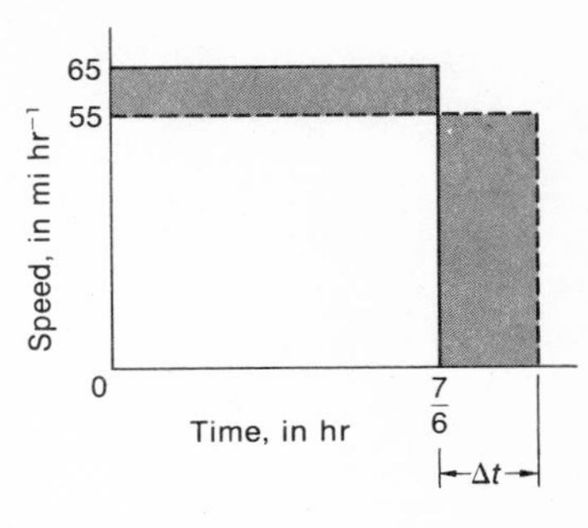

FIGURE 7

PROBLEM 19

If in Figure 5 the straight line is extrapolated back, it intercepts the t-axis at $t=-0.4$ sec. Does this time have any meaning?

PROBLEM 20

a. What is the physical meaning of the value of $v_0=3.9$ m sec^{-1} in Figure 5?

b. In Figure 4 the slope of the curve of the origin does not appear to be zero; what is its value?

PROBLEM 21

In Figure 5, beyond $t=4.0$ sec the observed speeds seem to be less than predicted by the straight line. Assuming Figure 5 was accurately plotted from correct data, why does the actual v-t curve turn downward as speeds increase? At what speed does the linear relationship between v and t break down?

PROBLEM 22

In Figure 5 the extrapolated dashed portion of the straight line intercepts the time axis at $t=-0.4$ sec. What is the physical meaning of the area under the line between $t=-0.4$ sec and $t=0$?

PROBLEM 23

By evaluating areas under a v-t curve, determine the equation which measures the distance traversed in the time interval t_2-t_1, $t_1\neq 0$, for constant linear acceleration. (This is a useful alternate for equation 2-14; many times you might be interested in a portion of the motion not based on $t=0$, and it would be cumbersome, and involve unnecessary work, to refer always to the motion from $t=0$.)

PROBLEM 24

Once, when traveling at 30 mi hr^{-1}, the taxi of problem 13 skidded to a stop to avoid a collision. The passenger, looking back, estimated the skid marks were 60 ft long. What was the average acceleration during the skidding?

PROBLEM 25

For a TV commercial, an advertising agency arranges to film a race between a famous track star and a new amphibious vehicle. Both start from rest at the same starting line; the star reaches the finish line 100 yd away at the same time as the vehicle, which had been driven at constant acceleration. If the track star's time was 9.4 sec,

a. what was the acceleration of the vehicle?
b. What was its speed as it passed the finish line?
c. Can you tell, by looking at the appropriate v-t graph, that the vehicle's speed at finish must be approximately twice the sprinter's speed?

PROBLEM 26

The motion of a London Underground train between Oxford Circus and Tottenham Court Road is graphed in Figure 8.

a. How long did the train take to get up to running speed?
b. How long was the braking period?
c. What was the maximum acceleration felt by the passengers, and when did this occur?
d. What is the approximate distance between the two stations?

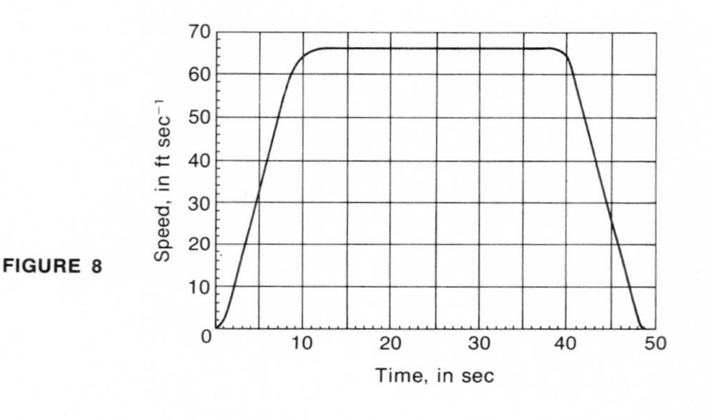

FIGURE 8

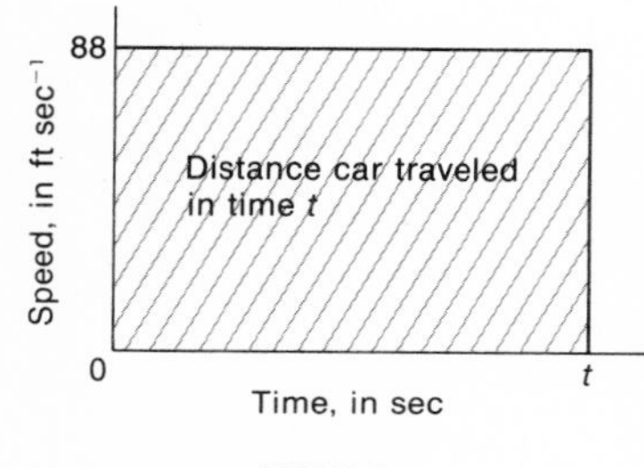

FIGURE 9

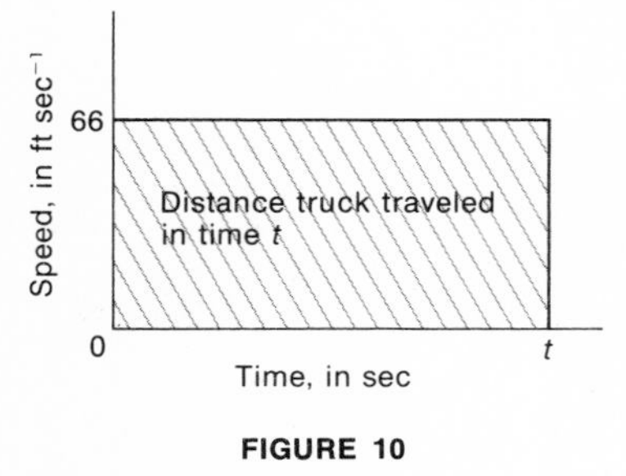

FIGURE 10

PROBLEM 27

The minimum time required for a car to pass safely another vehicle on the highway is usually measured from the time the front of the passing car is abreast of the rear of the other vehicle, until the rear of the passing car is one of its lengths fully ahead of the passed vehicle. If a car 16 ft long and traveling at 60 mi hr^{-1} passes a truck-trailer combination 45 ft long and moving at 45 mi hr^{-1},

a. how long did it take to pass safely?
b. How far did the car travel in this time?

SOLUTION

If we consider only the car, its v-t graph is as shown in Figure 9; in time t it will have traveled a distance equal to the hatched area. If now we consider only the truck, in the same time t it has traveled the distance equal to the hatched area as shown in Figure 10. Both these motions are shown on one graph, in Figure 11. The upper area, hatched but not crosshatched, is the excess distance, relative to the truck, that the car traveled in time t; this distance is $(88\text{-}66)t$.

If we make this excess distance equal to the required passing distance, then t must be the time required for passing. Hence,

$$(88\text{-}66)t = 2 \times 16 + 45$$

$$22t = 77$$

$$t = 3.5 \text{ sec.}$$

Then, the total distance traveled by the car during the passing time equals the total hatched area shown in the first graph (Figure 9). Since now we know that $t = 3.5$ sec, then $S_{\text{car}} = 88 \times 3.5 = 308$ ft. (The 3.5-sec value is derived, not original, data. Unfortunately, in this problem there is no simple way to avoid using derived data.)

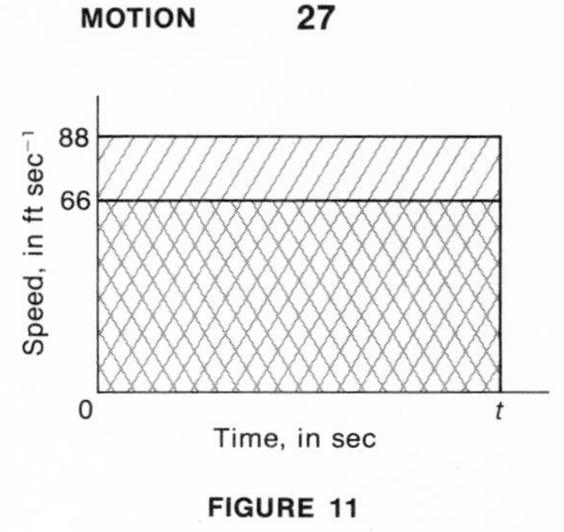

FIGURE 11

PROBLEM 28

A fast train running at speed v_1 rounds a hill onto a straightaway; the engineer observes, at distance d ahead, a slower train on the same track going in the same direction at speed v_2. The engineer instantly applies his brakes, which give a constant acceleration of $-a$ to the fast train. Determine the formula for the minimum value of d such that there shall be no rear-end collision.

SOLUTION

This is a standard problem, one found, with variations in wording, in many elementary physics textbooks. There are at least three methods of solution; all three of course require that in order to avoid a collision the speed of the fast train shall decrease to the speed of the slow train by the time the fast train catches up with the slow train.

1. Straightforward computation. $S_1 = v_1 t - \frac{1}{2}at^2$, the distance the fast train traveled in time t, $t = (v_1 - v_2)/a$, the time it took for the fast train to change its speed from v_1 to v_2, and $S_2 = v_2 t$, the distance the slow train traveled in time t. For no collision, d must $= S_1 - S_2$.

2. Relative speeds. At $t = 0$, speed of fast train relative to slow train $= v_1 - v_2$. At $t = (v_1 - v_2)/a$, the speed of the fast train relative to the slow train should $= 0$ for no collision; $d = [0 - (v_1 - v_2)^2]/2(-a)$.

3. The v-t graph, with d obtainable by inspection (Figure 12).

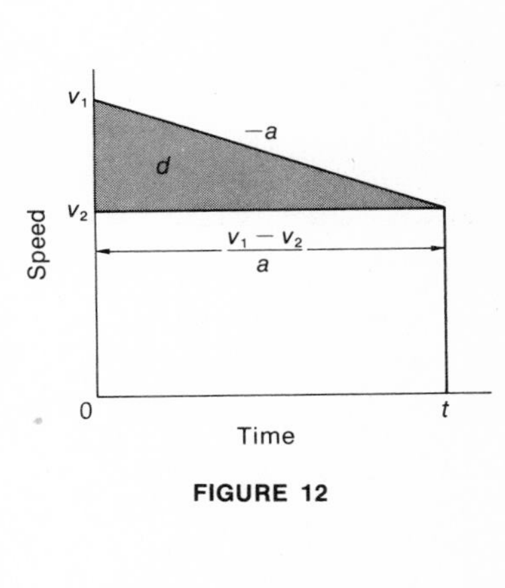

FIGURE 12

3a. The v-t graph further simplified by consideration from the point of view of a coordinate system moving at constant speed with the slow train. In this coordinate system, the original speed of the fast train is the relative speed $v_1 - v_2$ and the final speed is 0, the relative acceleration of the fast train is still $-a$, and the relative distance traveled by the fast train is d. Hence for this coordinate system the v-t graph is that shown in Figure 13, which is just the upper portion of the graph in Figure 12. This is a standard illustration of what happens when one shifts a coordinate system.

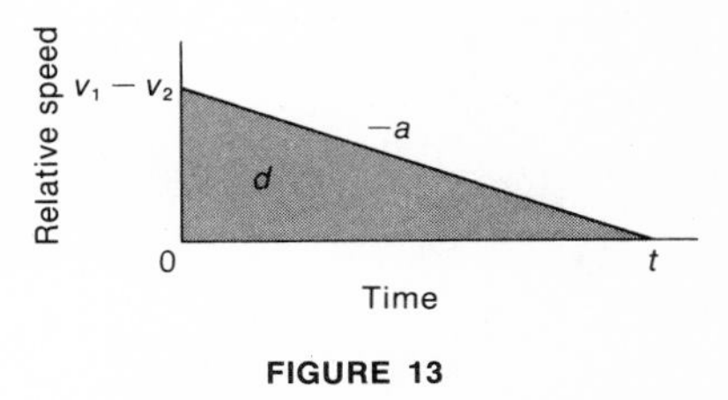

FIGURE 13

Try all three methods of solution, and determine for yourself which is the easiest and most informative.

PROBLEM 29

The children in problem 10 noted later that as the train approached the Princeton station, they passed two towers in 6.8 sec, and the next two towers in 10.2 sec. By assuming that the train was slowing down at a constant rate, they were able to compute the acceleration. What was it?

PROBLEM 30

A sophomore driving along a straight, level road with no other traffic on it comes upon a "mileage test" section and decides to check the maximum acceleration of his car. As he passes the "0" mark he pushes the accelerator pedal so as to hold the acceleration constant; at the same time a freshman riding with him starts a stopwatch he just happens to have with him. Exactly 4.0 sec later the freshman notes the speedometer reads 30 mi hr^{-1}. In his excitement he drops the watch, but thinking quickly he notes that as they pass the 0.2-mi post the speedometer reads 75 mi hr^{-1}. Assuming the acceleration was constant, what was it?

PROBLEM 31

A freshman, inexperienced with suburban traffic officers, has just received a ticket for speeding. Thereafter, when he comes upon one of the "mileage test" sections on a level stretch of highway, he decides to check his speedometer reading. As he passes the "0" start of the marked section he presses his accelerator so as to hold his car at constant acceleration. He notices that he passes the 0.10-mi post 16 sec after starting the test, and 8.0 sec later he passes the 0.20-mi post.

a. What should his speedometer have read at the 0.2-mi post?
b. What was his acceleration?

PROBLEM 32

A car waiting at a signal is passed by a truck that goes through the red light at 30 mi hr^{-1} without stopping. At 1.0 sec after the truck has passed, the driver of the car assumes the light has changed, and starts up with an acceleration of 6.0 ft sec^{-2}. At 2.0 sec after the car starts a motorcycle officer waiting behind the car starts with a constant acceleration A_0. He plans to catch both offenders just as the car comes abreast of the truck. What constant acceleration does he need to accomplish this?

PROBLEM 33

At Edwards Air Force Base, both rocket and jet motors can be tested by being mounted on sleds that run freely down the long, straight, horizontal test track. A rocket motor, started from rest, accelerated constantly until its fuel was exhausted, and then ran at constant speed. It was observed that the rocket fuel became exhausted as the rocket passed the midpoint of the measured test distance. Then a jet motor, started from rest, was run down the track with a constant acceleration for the entire test distance. It was observed that both rocket and jet motor covered the test distance in exactly the same time. What was the ratio of the acceleration of the jet motor to that of the rocket motor?

PROBLEM 34

A passenger seated facing forward in a bus can be accelerated forward by a push from the seat back, but can be accelerated backward only by friction between his clothing and the seat. On a particular bus, seat friction can supply only -0.20 g acceleration; for any greater negative acceleration the passenger slides forward on the seat. (Assume he is engrossed in reading a magazine and is not braced against sliding.) However, acceleration from a push in back can rise to 0.50 g before the passenger feels discomfort. If the maximum speed of the bus is 30 mi hr^{-1} and it is not driven beyond the limit of passenger comfort,

a. how long does it take the bus to reach running speed after a stop?
b. How far ahead must the driver see a clear road if he is to drive at maximum speed?
c. What is the least time he can drive between bus stops 880 ft apart?

PROBLEM 35

At a certain instant on a straight, horizontal road a motorcycle officer is abreast of a car he is chasing, and both vehicles are at the same speed v. At that instant the car accelerates with a constant acceleration a_c, and the policeman accelerates with a constant acceleration a_p. However, $a_c > a_p$, so the policeman falls behind. At t seconds after the car and policeman were abreast the policeman fires a bullet with a muzzle speed v_m at the car; T sec later it hits the car. Develop the quadratic equation for T, but do not spend time solving for T. For simplicity, assume the bullet had a straight horizontal trajectory. Suggestion: A v-t diagram is recommended.

PROBLEM 36

In freight train sorting yards, cars or strings of unconnected cars are pushed over a "hump" and allowed to coast freely down a uniform incline, where they can be switched to different make-up tracks. The front of the lead car of such a string passes a switch point while moving at a speed of 5.0 mi hr^{-1}; 15 sec later the rear of the next-to-last car passes the same point at 10 mi hr^{-1}.

a. How many cars were in the string?
b. If the acceleration of a car rolling freely down an incline is given by $a = g \sin \theta$, where θ is the angle of the incline with the horizontal, what was the slope of the track?
c. What assumption did you have to make in working part (a)?

PROBLEM 37

The axle of the wheels nearest the end of a freight car on a level track is 8.0 ft horizontally from the center of the separation between two adjacent cars. With the string of cars in problem 36, how much time did a switchman have to switch the last car to a different track?

The Gravitational Acceleration g

Although Figure 5 was developed for a case of free fall, the concept of acceleration that followed, and its symbol a, was general in character. However, it is customary to use the symbol g in problems concerned with a constant gravitational acceleration. It is also customary to use g to indicate only a numerical value and its units; the direction of the acceleration is indicated by a plus or minus sign, appropriate to the coordinate system in use. Thus, to describe motion depending on gravitational acceleration and in the normal coordinate system in which the upward direction is positive, equation 2-16 would be written

$$v_{y,t} = v_{y,0} - gt; \tag{2-21}$$

equation 2-17 would appear as

$$S_y = v_{y,0}t - \tfrac{1}{2}gt^2; \tag{2-22}$$

and equation 2-19 would be

$$v_{y,t}^2 = v_{y,0}^2 - 2g(y_t - y_0). \tag{2-23}$$

PROBLEM 38

Refer to the free-fall discussion beginning on page 20. Assume the position on the helicopter from which the object was released was 1.0 km above the ground.

a. How long did the object take to fall?

b. How fast was it going when it hit the ground? *Note:* In view of the drooping v-t curve in Figure 4, it would be improper to treat this as free fall with constant acceleration. First, compute the theoretical answers based on constant acceleration; then, by studying the departure of the real curve from the theoretical straight line, try to guess how the real answers depart from your theoretical answers.

FIGURE 14

PROBLEM 39

A simple way to test reaction time without a stopwatch is as shown. A new dollar bill is held vertically at the upper edge by the tester (Figure 14). The testee holds his thumb and forefinger open at the bottom edge of the bill. When the bill is released at some instant that cannot be anticipated by the testee, the testee attempts to close his fingers and grasp the bill as soon as possible. Develop a formula relating the distance d in inches from the bottom edge where the bill was grasped, and reaction time.

PROBLEM 40

Paratroopers are dropped from a transport helicopter at 1.0-sec intervals. Neglecting air resistance, how far are the first four paratroopers spread out below him as the fifth paratrooper leaves the helicopter?

PROBLEM 41

a. To throw a 2-oz tennis ball straight up so that it rises 25 ft., you must get it to leave your hand at what speed?
b. What speed would be required to project a 5-oz baseball to the same height?

PROBLEM 42

A pebble is dropped in a well; by stopwatch it is found that 4.2 sec elapse after release of the pebble before the splash is heard. What is the depth to the water surface?

PROBLEM 43

A pedestrian crossing the Golden Gate Bridge on a walkway 258 ft above water level looks over the railing and sees a garbage scow being towed at 8.0 knots about to pass underneath the bridge. He drops an orange over the railing just as the bow of the scow passes directly beneath him. The orange *just* misses the stern of the scow. If air resistance can be neglected, how long was the scow?

PROBLEM 44

Suppose the pedestrian repeated the orange-dropping experiment of problem 43, this time with another scow moving at the same speed, but from a part of the bridge whose height above the water he did not know. However, he did note that again the orange just missed the stern of the scow, and that he saw the splash 3.9 sec after he released the orange. How long was the second scow?

PROBLEM 45

In attempting to hit a pitched baseball, a batter hits the ball almost straight up in the air. If the catcher catches the ball 5.0 sec later,

a. what was the minimum vertical component of the speed of the ball as it left the bat?
b. How far did the ball rise?

SOLUTION

Let us consider the motion of the ball in a $\overset{+}{\underset{-}{+}}$ coordinate system. At $t = 0$ the ball is given by the bat some upward speed $v_{y,0}$. Following that, and neglecting air drag, the ball is subject to a downward, and hence negative, acceleration until it is caught. The v-t graph of its motion would be as shown in Figure 15.

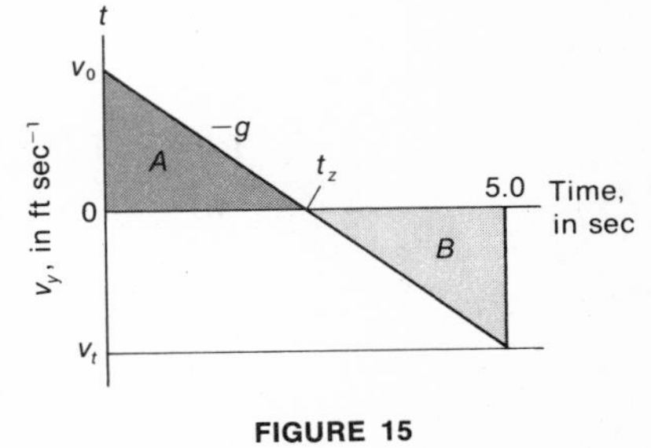

FIGURE 15

What is the meaning of t_z? Since the graph of the ball's motion crosses the time axis at that point, it represents the time for $v_y = 0$. It must, then, represent the time at which the ball reached zenith, the top of its path, and came momentarily to rest before falling back down.

What does the area A represent? It is a distance, the distance traversed between the time the ball started up, $t = 0$, and t_z, when it reached its zenith. In other words, A is the total distance the ball rose.

What happens after t_z? The ball falls from rest with an acceleration $-g$. How far does it fall? Obviously as far as it went up. (It is reasonable to assume that the ball was caught at about the same height at which it was struck by the bat.) Then area B must equal area A in magnitude.

Analytically, it would be more precise to say

$$A + B = 0, \text{ or } B = -A,$$

since at the end of 5.0 sec, the *net* vertical displacement is zero. Is B negative and A positive? The area A is an area determined by the multiplication of a positive v by a positive t; hence positive. The area B is determined by the multiplication of a negative v by a positive t; hence negative.

Since the triangles A and B are similar, and the areas are equal, the time interval $0 - t_z$ must equal the interval $t_z - 5$. Hence $t_z = 2.5$ sec. Also, by the same reasoning $v_t = -v_0$.

At $t = 5.0$ sec,

$$y = 0 = v_0 \times 5 - \tfrac{1}{2} \times 32 \times 5^2,$$

from which

$$v_0 = 80 \text{ ft sec}^{-1},$$

and

$$y_{\max} = A = \tfrac{1}{2} g t_z^2 = \tfrac{1}{2} \times 32 \times 2.5^2 = 100 \text{ ft}.$$

It should be clear from the graph of Figure 15 that the motion of an object projected upward is symmetrical about its zenith, providing there is no drag. and the motion does not extend beyond the limit of constant gravitational acceleration.

PROBLEM 46

We simplified problem 45 by neglecting the "drag" from the air. Without knowing the exact value of this drag, make the reasonable assumption that in some way it is a function of the ball's speed and that it produces an acceleration in a direction opposite to that of the speed.

a. Draw a v-t graph that takes this drag into account.
b. Is the motion symmetrical about the zenith?
c. Will the ball rise as high as it did without drag?
d. Did v_t (the speed with which the ball returned to the starting level) $= -v_0$ as before?
e. Was the time of rise $t_z >$, $=$, or $<$ the time of fall?

PROBLEM 47

A juggler tossing billiard balls can handle two balls per second; i.e., it takes at least 0.50 sec for him to catch a ball with one hand and start it upward again with the other hand. If he is working with 5 balls, to what minimum height must he throw them?

PROBLEM 48

Traffic control during the Tournament of Roses parade in Pasadena is directed from a dirigible flying back and forth above the parade route at a ground speed three times the speed of advance of the parade. For simplicity, consider the dirigible's turnaround time to be negligible. If the length of the parade is 4 mi, how far has the head of the parade advanced during the circuit of the dirigible from the rear to the head of the parade and back to the rear again? (Try reading the answer by inspection from the v-t graph.)

PROBLEM 49

A student standing on the express platform of the Concorde Metro (subway) station in Paris notes that the first 3 cars of an arriving train pass him in 3.0 sec and the next 3 cars pass him in 5.0 sec. The cars are each 20 m long, and the braking acceleration of the train is constant. When the train comes to a stop, the student finds himself opposite the rear section of the last car of the train. How long was the train?

PROBLEM 50

Cars of average length L, whose braking acceleration with brakes jammed full on $= -a$, have drivers whose reaction time in which to recognize an emergency and jam on brakes is t_0. They are traveling in line on a freeway lane, all at a steady speed v. If the cars are being driven so that rear-end collisions in event of emergency are just barely avoidable,

a. what should v be for maximum flow of traffic?
b. What is this maximum flow, in terms of a, t_0, and L?

Note: Two practical situations should be distinguished: (1) The space between cars at any moment is the total stopping distance available; and (2) the total stopping distance can include the distance the car ahead travels after *its* brakes are jammed on. Assume L, t, a, and v are in compatible units.

PROBLEM 51

A sounding rocket is fired vertically upward. The rocket motor burns for t_R sec and gives the rocket a constant upward acceleration of a_R relative to the ground during that time. Neglecting effects of wind, yaw, and the rotation of the earth, you can assume the rocket case lands alongside the place from which it was fired.

a. By inspection of the v-t graph for the flight, prove that regardless of the value of a_R or t_R, the rocket always takes longer to go up than to come down.
b. What was the maximum speed reached on the upward flight?
c. How high did the rocket rise?
d. How fast was the rocket case moving when it hit the ground?
e. As an extension of (a), determine the value of the excess time Δt of the upward flight over the falling flight.

Varying Acceleration

When we developed the

$$\lim_{\Delta t \to 0} \frac{\Delta y}{\Delta t},$$

and found it equal to the slope of a y-t curve, we were flirting with the origins of the differential calculus. And when we summed up all the infinitesimal $v \times \Delta t$ areas under the v-t graph, to equate this sum to the total distance traversed, we were flirting with the origins of the integral calculus. However, up to this point we have managed to avoid getting involved in anything more complicated than simple algebra.

In the general case, and particularly in studying many of the most interesting kinds of motion, we will not find either constant speed or constant acceleration. We will be restricted to the use of instantaneous values only, and for this we will need the methods of calculus.

Let us confirm some of the equations we have already developed, as a sort of warm-up review of calculus. In equation 2-13 we defined instantaneous acceleration by $a = dv/dt$, from which we get $dv = a\ dt$.

Integrating dt between the limits 0 and t, and dv between the corresponding limits of v,

$$\int_{v_0}^{v_t} dv = \int_0^t a\ dt.$$

For the special case of $a =$ constant, a can come out from under the integral sign, and after integrating we have

$$v_t - v_0 = at,$$

which confirms equation 2-14.

The instantaneous value of v was defined in equation 2-8 by

$$v = \frac{dy}{dt}.$$

Since we also have the instantaneous value of v at any t given by $v_t = v_0 + at$ for the special case of $a =$ constant,

$$\frac{dy}{dt} = v_0 + at,$$

and

$$\int_{y_0}^{y_t} dy = \int_0^t (v_0 + at)\ dt.$$

Integrating results in

$$y_t - y_0 = v_0 t + \tfrac{1}{2}at^2,$$

which confirms equation 2-17.

Before going further into motions requiring calculus for solution, we will expand our notation to bring it into line with professional usage. In this new notation a dot over a symbol indicates differentiation with respect to time, two dots indicate two successive differentiations with respect to time:

$$v = \frac{ds}{dt} \equiv \dot{S};$$

$$a = \frac{dv}{dt} \equiv \dot{v} \equiv \frac{d(\dot{S})}{dt} \equiv \frac{d^2S}{dt^2} \equiv \ddot{S}.$$

When values are known, or when established relationships are expressed, it is customary to use the older notation:

"The muzzle speed of the bullet $v_0 = 9.0 \times 10^2$ ft sec^{-1}."
"$v_t = v_0 + at$."
"For a freely falling object in the normal coordinate system, $a = -g$."
"$a = (v_t^2 - v_0^2)/2S$," etc.

When the values are unknown or appear as a variable in a differential equation, it is customary to use the dot notation:

$$\ddot{y} = -g - p\,\dot{y}.$$

$$\ddot{x} = -\frac{k}{m}x\text{ , etc.}$$

PROBLEM 52

A body is constrained to move in the x direction. Its linear acceleration is proportional to, and in the opposite direction from, its displacement from $x = 0$. Determine its position as a function of time.

SOLUTION

The data given in problem 52 conform to the last sample equation shown. We are told that the acceleration $\ddot{x} = cx$, where c is a constant of proportionality and x is a displacement. Since the displacement and the acceleration are in opposite directions, we also need a minus sign. Hence the statement of the information given is

$$\ddot{x} = -cx.$$

For a solution to exist, x must be expressible as some function of t such that x and d^2x/dt^2 possess the same root form. There are three such functions:

$$x = \sin \omega t, \qquad \ddot{x} = -\omega^2 \sin \omega t = -\omega^2 x;$$

$$x = \cos \omega t, \qquad \ddot{x} = -\omega^2 \cos \omega t = -\omega^2 x;$$

$$x = e^{\alpha t}, \qquad \ddot{x} = \alpha^2 e^{\alpha t} = \alpha^2 x.$$

The use of $x = e^{\alpha t}$ will lead to a solution involving complex num bers. Let us postpone that approach and work instead with the trigonometric functions.

The trick in solving differential equations of the type $\ddot{x} = -cx$ is to assume a solution (using of course, the best judgment you can) and then see if, and under what conditions, the assumed solution actually is a solution.

Since both $x = \sin \omega t$ and $x = \cos \omega t$ meet the fundamental require that $\ddot{x}$ and x have the same root form, let us use both of them for maximum generality; let us assume that

$$x = A \sin \omega t + B \cos \omega t$$

is a solution of $\ddot{x} = -cx$.
Differentiating

$$\dot{x} = \omega A \cos \omega t - \omega B \sin \omega t$$

and

$$\begin{aligned} \ddot{x} &= -\omega^2 A \sin \omega t - \omega^2 B \cos \omega t \\ &= -\omega^2 (A \sin \omega t + B \cos \omega t) \\ &= -\omega^2 x. \end{aligned}$$

Obviously, $\ddot{x} = -\omega^2 x$ and $\ddot{x} = -cx$ are a match if $\omega^2 = c$.
Hence $x = A \sin \omega t + B \cos \omega t$ is a solution of the initial problem when $\omega = \sqrt{c}$.

This is a general solution; to make it a particular solution for our particular problem we must identify the constants A and B. To do this we need to know the initial conditions to which any particular solution must conform.

Suppose for our initial conditions we say that when $t = 0$, $x = x_0$ and $\dot{x} = 0$.* Substituting $t = 0$ and $\dot{x} = 0$ in the equation above for $\dot{x}$, we obtain

$$\begin{aligned} 0 &= \omega A \cos 0 - \omega B \sin 0 \\ &= \omega A. \end{aligned}$$

Hence $A = 0$. Substituting $t = 0$ and $x = x_0$ in the above equations for x (with $A = 0$),

$$x_0 = B \cos 0 = B.$$

So finally, the particular solution for our particular problem with the stated particular initial conditions is

$$x = x_0 \cos \omega t, \text{ where } \omega = \sqrt{c}.$$

---

*This condition is met when the body is held at rest at a distance x_0 from the origin, and then released.

Graphed, our solution is as shown in Figure 16. The equation and its graph tell us that an object whose acceleration is proportional to its displacement from some reference point, and in the opposite direction, oscillates between $+x_0$ and $-x_0$ according to the cosine function of time when timing is started when $x = \pm x_0$.

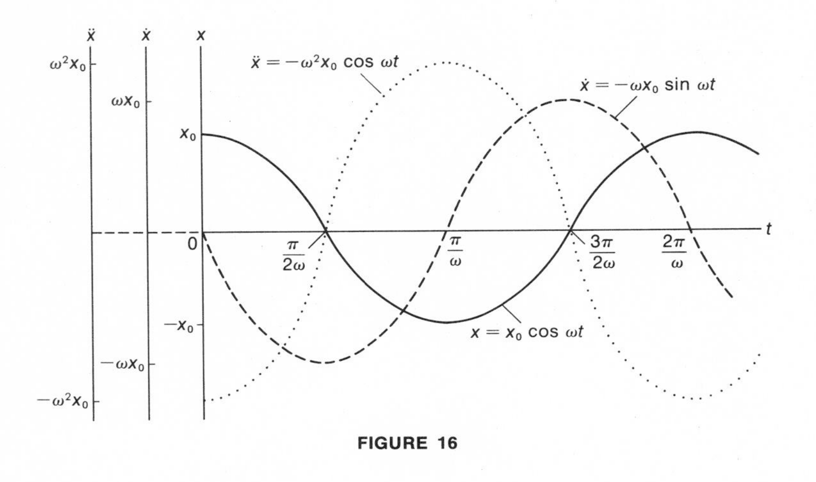

FIGURE 16

Obviously, by the same method, if $t = 0$ when $x = 0$ and $\dot{x} = v_0$, then $x = (v_0/\omega) \sin \omega t$, where $\omega = \sqrt{c}$. Thus the body oscillates according to the sine function of time if timing is started when the body is at the origin. This motion can be initiated by giving a body at rest at the origin an impulse to start it in motion with a speed v_0. As we shall see later in more detail, these motions occur frequently and are called simple harmonic motion.

PROBLEM 53

In problem 46 we attempted to move from the idealized world to the real world by a qualitative approach. In this problem we take a further step toward reality by a quantitative approach.

A ball is propelled vertically upward with an initial speed v_0. In addition to a constant gravitational acceleration downward, the ball's motion is affected by a "drag" acceleration numerically equal to γv^2, and in a direction opposite to that of v. (The symbol γ stands for an experimentally determined coefficient whose value depends both upon the form of the moving object and the range of speeds involved.)

a. What are the dimensions of γ?
b. How much time did the ball take to rise to its zenith?
c. How high did the ball rise?
d. How fast was the ball moving when it returned to the starting level?
e. How long did it take to fall?

SOLUTION

If we adopt the coordinate system

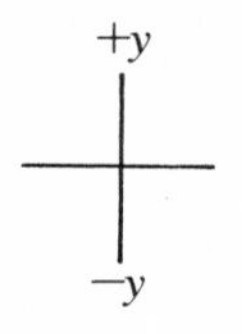

for the up trip the information given about the ball's acceleration can be stated in the equation:

$$\ddot{y} = -g - \gamma\dot{y}^2$$

$$\ddot{y} = \frac{d\dot{y}}{dt} = -(g + \gamma\dot{y}^2).$$

(The gravitational acceleration is obviously negative; since on the up trip the opposing drag is downward, it is negative also.)

Rearranging results in

$$\frac{d\dot{y}}{g + \gamma\dot{y}^2} = -dt.$$

When $t = 0$, $\dot{y} = v_0$; when $t = t$, $\dot{y} = \dot{y}$. Putting in these limits of integration, we have

$$\int_{v_0}^{\dot{y}} \frac{d\dot{y}}{g + \gamma\dot{y}^2} = -\int_0^t dt$$

$$\left[\frac{1}{\sqrt{\gamma g}} \tan^{-1} \sqrt{\frac{\gamma}{g}}\,\dot{y}\right]_{v_0}^{\dot{y}} = -t$$

$$\tan^{-1} \sqrt{\frac{\gamma}{g}}\, v_0 - \tan^{-1} \sqrt{\frac{\gamma}{g}}\,\dot{y} = \sqrt{\gamma g}\, t$$

$$\tan^{-1} \sqrt{\frac{\gamma}{g}}\,\dot{y} = \tan^{-1} \sqrt{\frac{\gamma}{g}}\, v_0 - \sqrt{\gamma g}\, t$$

$$\dot{y} = \sqrt{\frac{g}{\gamma}} \tan\left(\tan^{-1} \sqrt{\frac{\gamma}{g}}\, v_0 - \sqrt{\gamma g}\, t\right)$$

$$= \frac{v_0 - \sqrt{g/\gamma}\tan\sqrt{\gamma g}\, t}{1 + \sqrt{\gamma/g}\, v_0 \tan\sqrt{\gamma g}\, t}.$$

b. When $\dot{y} = 0$, $t = t_z$; then

$$v_0 - \sqrt{\frac{g}{\gamma}} \tan\sqrt{\gamma g}\, t_z = 0$$

$$t_z = \frac{1}{\sqrt{\gamma g}} \tan^{-1} \sqrt{\frac{\gamma}{g}}\, v_0.$$

c. From above,

$$\dot{y} = \frac{dy}{dt} = \frac{v_0 - \sqrt{g/\gamma}\tan\sqrt{\gamma g}\, t}{1 + \sqrt{\gamma/g}\, v_0 \tan\sqrt{\gamma g}\, t}.$$

When $y = 0$, $t = 0$; when $y = y$, $t = t$. Then, integrating,

$$\int_0^y dy = \int_0^t \frac{v_0 - \sqrt{g/\gamma} \tan \sqrt{\gamma g}\, t}{1 + \sqrt{\gamma/g}\, v_0 \tan \sqrt{\gamma g}\, t}\, dt$$

$$y = \left[\frac{v_0}{1 + (\gamma v_0^2/g)} \left(t + \frac{\sqrt{\gamma/g}\, v_0}{\sqrt{\gamma g}} \ln \left(\cos \sqrt{\gamma g}\, t + \sqrt{\frac{\gamma}{g}}\, v_0 \sin \sqrt{\gamma g}\, t\right)\right)\right.$$

$$\left. - \frac{\sqrt{g/\gamma}}{1 + (\gamma v_0^2/g)} \left(\sqrt{\frac{\gamma}{g}}\, v_0 t - \frac{1}{\sqrt{\gamma g}} \ln \left(\cos \sqrt{\gamma g}\, t + \sqrt{\frac{\gamma}{g}}\, v_0 \sin \sqrt{\gamma g}\, t\right)\right)\right]_0^t$$

(from any standard table of integrals)

$$= \frac{1}{\gamma} \ln \left(\cos \sqrt{\gamma g}\, t + \sqrt{\frac{\gamma}{g}}\, v_0 \sin \sqrt{\gamma g}\, t\right).$$

When

$$t = t_z = \frac{1}{\sqrt{\gamma g}} \tan^{-1} \sqrt{\frac{\gamma}{g}}\, v_0, \quad y = H.$$

Then,

$$H = \frac{1}{\gamma} \left\{\ln \left[\frac{1}{\sqrt{1 + \gamma v_0^2/g}} + \sqrt{\frac{\gamma}{g}}\, v_0 \times \frac{\sqrt{\gamma/g}\, v_0}{\sqrt{1 + (\gamma v_0^2/g)}}\right]\right\}$$

$$= \frac{1}{2\gamma} \ln \left(1 + \frac{\gamma v_0^2}{g}\right).$$

Alternate Solution to Question (c).

In solving question (b) for t_z, at which time $y = H$, we had gotten as far as $\dot{y} = dy/dt$. It then seemed reasonable to integrate again and solve for y as a function of t, in order to get to $y = H$. If H alone is all that is wanted, however, there is a shorter way:

$$\ddot{y} = \frac{d\dot{y}}{dt} = \frac{d\dot{y}}{dy}\frac{dy}{dt} = \dot{y}\, \frac{d\dot{y}}{dy}.$$

Going back to our original statement about $\ddot{y}$, we have

$$\ddot{y} = \frac{\dot{y}\, d\dot{y}}{dy} = -\,(g + \gamma \dot{y}^2).$$

Rearranging results in

$$\frac{\dot{y}\, d\dot{y}}{g + \gamma \dot{y}^2} = -\, dy.$$

Integrating $\dot{y}$ between the limits of v_0 and $\dot{y}$, and the corresponding limits for y, we have

$$\int_{v_0}^{\dot{y}} \frac{\dot{y}\,d\dot{y}}{g+\gamma\dot{y}^2} = -\int_0^y dy$$

$$\left[\frac{1}{2\gamma}\ln(g+\gamma\dot{y}^2)\right]_{v_0}^{\dot{y}} = -y$$

$$y = \frac{1}{2\gamma}\ln\left(\frac{g+\gamma v_0^2}{g+\gamma\dot{y}^2}\right)$$

$$= \frac{1}{2\gamma}\ln\left[\frac{1+(\gamma v_0^2/g)}{1+(\gamma\dot{y}^2/g)}\right].$$

When $\dot{y}=0$, $y=H$.
Hence

$$H = \frac{1}{2\gamma}\ln\left(1+\frac{\gamma v_0^2}{g}\right),$$

as before.

Whenever you reach a computation that can be checked, it is a good practice to make the check. Such a check point occurs with our computation for H. We know from work we have already done that under idealized conditions, when $\gamma = 0$, H should $= v_0^2/2g$. It would therefore be useful to see if our above computation for H reduces to $v_0^2/2g$ when γ goes to 0. Considering the above equation for H,

$$\lim_{\gamma\to 0} H = \frac{\ln 1}{0} = \frac{0}{0},$$

which is indeterminate. Proceeding to the next step, by taking derivatives separately of the numerator and denominator—see any calculus text on how to resolve the indeterminate 0/0—

$$\lim_{\gamma\to 0} H = \lim_{\gamma\to 0} \frac{(v_0^2/g)/[1+(\gamma v_0^2/g)]}{2}$$

$$= \frac{v_0^2}{2g}.$$

---

d. For the down trip, since in our coordinate system the speed is negative, the drag acceleration will be positive. Then,

$$\ddot{y} = -g + \gamma\dot{y}^2 = -(g-\gamma\dot{y}^2).$$

Using the second form developed above, because of its direct relationship between $\dot{y}$ and y,

$$\ddot{y} = \dot{y}\frac{d\dot{y}}{dy} = -(g-\gamma\dot{y}^2)$$

$$\int_0^{\dot{y}} \frac{\dot{y}\,dy}{(g-\gamma\dot{y}^2)} = \int_H^y -dy$$

$$\left[-\frac{1}{2\gamma}\ln(g-\gamma\dot{y}^2)\right]_0^{\dot{y}}=-(y-H)$$

$$1-\frac{\gamma}{g}\dot{y}^2=e^{2\gamma(y-H)}$$

$$\dot{y}=\sqrt{\frac{g}{\gamma}}\left[1-e^{2\gamma(y-H)}\right]^{1/2}$$

$$=\sqrt{\frac{g}{\gamma}}\left[1-\frac{e^{2\gamma y}}{1+(\gamma v_0^2/g)}\right]^{1/2}$$

after putting in the value for H. When $y=0$, $\dot{y}=\dot{y}_D$, the speed on return to the starting level. Then,

$$\dot{y}_D=\sqrt{\frac{g}{\gamma}}\left[1-\frac{1}{(1+\gamma v_0^2/g)}\right]^{1/2}$$

$$=\frac{-v_0}{\sqrt{1+(\gamma v_0^2/g)}}.$$

(Because of the square root, the numerator was $\pm v_0$; the minus sign was chosen to make the answer consistent with the coordinate system.) The speed of arrival back at the starting level is obviously less than the speed with which it started up from that level. Does this seem reasonable, and if so, why?

e. Integrating the above equation for $\dot{y}$ to get a relation between y and t looks more difficult than starting fresh from the definition of $\ddot{y}=d\dot{y}/dt$;

$$\ddot{y}=\frac{d\dot{y}}{dt}=-(g-\gamma\dot{y}^2)$$

$$\int_0^{\dot{y}}\frac{d\dot{y}}{g-\gamma\dot{y}^2}=-\int_0^{\tau}dt,$$

where τ is time measured from the zenith. Then,

$$\left[\frac{1}{2\sqrt{\gamma g}}\ln\frac{g+\sqrt{\gamma g}\,\dot{y}}{g-\sqrt{\gamma g}\,\dot{y}}\right]_0^{\dot{y}}=-\tau;$$

$$\tau=-\frac{1}{2\sqrt{\gamma g}}\ln\frac{1+\sqrt{\gamma/g}\,\dot{y}}{1-\sqrt{\gamma/g}\,\dot{y}}.$$

When

$$\dot{y}=\frac{-v_0}{\sqrt{1+(\gamma v_0^2/g)}},\qquad \tau=\tau_D,$$

the time of fall. Hence

$$\tau_D=-\frac{1}{2\sqrt{\gamma g}}\ln\frac{1+\sqrt{\gamma/g}[(-v_0)/\sqrt{1+(\gamma v_0^2/g)}]}{1-\sqrt{\gamma/g}[(-v_0)/\sqrt{1+(\gamma v_0^2/g)}]}$$

$$=\frac{1}{2\sqrt{\gamma g}}\ln\frac{\sqrt{1+(g/\gamma v_0^2)}+1}{\sqrt{1+(g/\gamma v_0^2)}-1}.$$

PROBLEM 54

For the ball of problem 53, $v_0 = 80$ ft sec^{-1}, and $\gamma = 5.0 \times 10^{-4}$ ft^{-1}. Plot v against t for the entire motion of the ball. Compare your graph of this problem with your graph of problem 46.

PROBLEM 55

If the falling object of page 20 experienced an upward acceleration γv^2 from air resistance, where $\gamma = 2.95 \times 10^{-4}$ m^{-1}, what are the real answers to problem 38? How close did you come in question (b) of that problem to these real answers?

PROBLEM 56

A child's toy balloon filled with helium escapes from the child's hand in quiet air. Because the balloon's buoyancy exceeds its weight, it is subject to an upward acceleration of 0.1 g; its motion is opposed by a drag acceleration equal to $0.01\,v$,* when its speed v is in cm sec^{-1}

a. How fast was the balloon rising at the end of 10 sec?
b. How high was it at that time?
c. What was the balloon's terminal speed, i.e., the speed at which its acceleration has dropped to zero value?
d. In this problem the numerical value of the drag coefficient is 0.01; what is its dimension?

*For very low speeds, the drag acceleration of an object moving through a viscous medium like the atmosphere is proportional to the speed, instead of to v^2 as in the previous problems. (This is Stokes' Law.)

PROBLEM 57

A yachtsman who has just bought a new power boat is experimenting with its "run" in the water. He finds that when he is moving through the water at 5.0 mi hr^{-1} and shuts off the power, the boat's speed is reduced to 2.0 mi hr^{-1} in 9.0 sec. He assumes (correctly) that the boat's deceleration is proportional to its speed, at least for speeds less than 5.0 mi hr^{-1}. Later, he approaches his dock at 4.0 mi hr^{-1} and wishes to reach the dock moving at only 0.5 ft sec^{-1}.

a. How far out from the dock should he shut off the power?
b. How much time did he spend in coasting in to the dock?

Digression—Calculation of an Average of a Function. Before we leave these illustrations of the use of calculus in motion, let us consider the warning given earlier about the determination of v_{av}.

PROBLEM 58

For the harmonic oscillator discussed in problem 52, what is the average speed over the following intervals:

a. The first half-cycle, from $t = 0$ to $t = \pi/\omega$?
b. The second quarter-cycle, from $t = \pi/2\omega$ to $t = \pi/\omega$?
c. The third quarter-cycle, from $t = \pi/\omega$ to $t = 3\pi/2\omega$?

d. The half-cycle made up of the above two quarter-cycles, i.e., from $t = \pi/2\omega$ to $t = 3\pi/2\omega$?

SOLUTION

Since by definition, $v_{av} = \Delta S/\Delta t$, and since in this problem our Δt's are specified, we need only to determine the corresponding ΔS's. There is an obvious way and a less obvious way.

The obvious way. From the graph in problem 52 or from the equations for x if you would rather compute and subtract, the ΔS's are easily obtained for each time interval, as follows:

a. For the time interval between 0 and π/ω,

$$\Delta x = x_{\pi/\omega} - x_0 = - x_0 - (+ x_0) = - 2x_0.$$

Hence,

$$v_{av} = \frac{-2x_0}{\pi/\omega} = - \frac{2x_0\omega}{\pi}.$$

b. For the time interval $\pi/2\omega$ to π/ω,

$$\Delta x = x_{\pi/\omega} - x_{\pi/2\omega} = - x_0 - 0 = - x_0.$$

Hence,

$$v_{av} = \frac{-x_0}{\pi/\omega - \pi/2\omega} = - \frac{2x_0\omega}{\pi}.$$

c. For the time interval π/ω to $3\pi/2\omega$,

$$\Delta x = 0 - (-x_0) = x_0$$

and

$$v_{av} = \frac{x_0}{3\pi/2\omega - \pi/\omega} = \frac{2x_0\omega}{\pi}.$$

d. For the time interval $\pi/2\omega$ to $3\pi/2\omega$,

$$\Delta x = 0 - 0 = 0$$

and

$$v_{av} = \frac{0}{3\pi/2\omega - \pi/2\omega} = 0.$$

As was pointed out before, the average speed computed by the definition for a complete round trip = 0, and has no useful meaning. It *is* a round trip: at $\pi/2\omega$ the object is at $x = 0$; it then goes to $-x_0$, and then back to $x = 0$ at $3\pi/2\omega$.

The less obvious way, by integration. The first way was easy, because we knew the displacement explicitly. Much of the time, however, we will only know v as a function of t. From our definition of $v = ds/dt$, $\Delta S = \int v\, dt$. We then have

$$v_{av} = \frac{\Delta S}{\Delta t} = \frac{1}{\Delta t} \int v\, dt.$$

For the present problem, from above we have

$$v = \dot{x} = - \omega x_0 \sin \omega t.$$

a. For the time interval 0 to π/ω,

$$\bar{\dot{x}} = \frac{1}{\pi/\omega} \int_0^{\pi/\omega} -\omega x_0 \sin \omega t \, dt^*$$

$$= \omega/\pi \left[x_0 \cos \omega t \right]_0^{\pi/\omega} = -\frac{2x_0\omega}{\pi}.$$

b. For the time interval $\pi/2\omega$ to π/ω,

$$\bar{\dot{x}} = \frac{1}{\pi/2\omega} \int_{\pi/2\omega}^{\pi/\omega} -\omega_0 x_0 \sin \omega t \, dt$$

$$= 2\omega/\pi \left[x_0 \cos \omega t \right]_{\pi/2\omega}^{\pi/\omega} = -\frac{2x_0\omega}{\pi}.$$

c. For the time interval π/ω to $3\pi/2\omega$,

$$\bar{\dot{x}} = \frac{1}{\pi/2\omega} \int_{\pi/\omega}^{3\pi/2\omega} -\omega \, x_0 \sin \omega t \, dt$$

$$= \frac{2\omega}{\pi} \left[x_0 \cos \omega t \right]_{\pi/\omega}^{3\pi/2\omega} = \frac{+2x_0\omega}{\pi}.$$

d. For the time interval $\pi/2\omega$ to $3\pi/2\omega$,

$$\bar{\dot{x}} = \frac{1}{\pi/\omega} \int_{\pi/2\omega}^{3\pi/2\omega} -\omega \, x_0 \sin \omega t \, dt$$

$$= \frac{\omega}{\pi} \left[x_0 \cos \omega t \right]_{\pi/2\omega}^{3\pi/2\omega} = 0.$$

Fortunately the two sets of answers agree.

It is useful to look at the geometrical interpretation of the second method. We should remember that

$$\int_0^{\pi/\omega} - \omega x_0 \sin \omega t \, dt$$

is the area between the curve of $\dot{x} = -\omega x_0 \sin \omega t$ and the t-axis between the limits of 0 and π/ω – the gray area shown in Figure 17; $\bar{\dot{x}}$ is such that this area is equalled by the rectangular area between the dashed line and the t-axis, of width π/ω and altitude $-\bar{\dot{x}}$.

General definition of an average. Although the concept of an average was developed for motion, it has general application and can be applied to any function that can be integrated. Speaking generally,

*A bar over a symbol is a standard notation for the average value of that symbol; thus, $\bar{\dot{x}} = \dot{x}_{av}$.

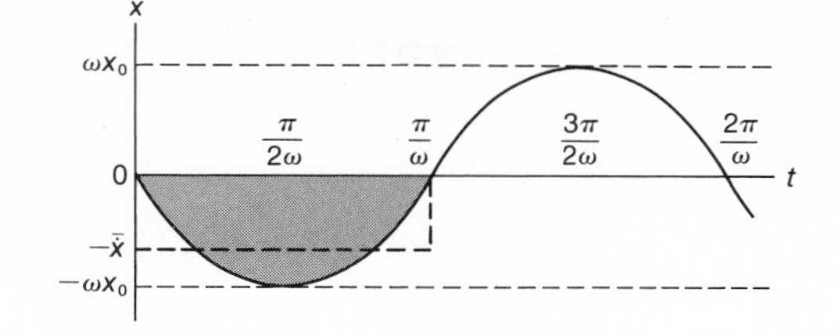

FIGURE 17

if p is an integrable function of some variable q, i.e., if $p = f(q)$, then for the interval between q_1 and q_2,

$$\bar{p} = \frac{1}{q_2 - q_1} \int_{q_1}^{q_2} p \, dq. \tag{2-24}$$

PROBLEM 59

A linear speed v is described by $v = 3.9 + 9.8\,t$, when v is in m sec^{-1} and t is in seconds.

a. What is the average speed between $t = 1.0$ sec and $t = 4.0$ sec?
b. Does your answer agree with what you can calculate from Figure 5?

PROBLEM 60

If the instantaneous value of the alternating current flowing in a circuit is described by $I = I_0 \sin \omega t$, what is the average value of the current

a. over the half-cycle from $t = 0$ to $t = \pi/\omega$;
b. over an entire cycle?

PROBLEM 61

Many useful physical values—the kinetic or potential energy of an oscillator, the power being delivered in an alternating current circuit, the energy transported by a wave, for examples—can be described by an equation of the form $U = A_0^2 \sin^2 bt$ or $U = A_0^2 \cos^2 bt$, A_0 and b being independent of the time t.

a. What is the average value of U over a cycle? (A cycle would run from $bt = 0$ to $bt = 2\pi$, or more generally, from $bt = 2n\pi$ to $2(n+1)\pi$.)
b. How does this average value of U compare with U's maximum value?

PROBLEM 62

In an electrical circuit containing only a pure resistance R, an alternating voltage $V = V_0 \sin \omega t$ will cause a current $I = I_0 \sin \omega t$ to flow through the resistor. An alternating current (AC) ammeter in the circuit will not read I_0 but some I_{eff}, such that I_{eff}^2 equals the average value of I^2 over a complete cycle. (Similarly, an AC voltmeter will read some V_{eff}, such that V_{eff}^2 equals the average of V^2 over a cycle.) What is I_{eff}, in terms of I_0 (or V_{eff} in terms of V_0)?

PROBLEM 63
The harmonic oscillator of problem 52 possesses at any time t a kinetic energy equal to $\frac{1}{2}m\dot{x}^2 = \frac{1}{2}mx_0^2\omega^2 \sin^2 \omega t$, and a potential energy equal to $\frac{1}{2}mcx^2 = \frac{1}{2}mcx_0^2 \cos^2 \omega t$. When averaged over a complete cycle, how does the average of the kinetic energy compare with the average of the potential energy?

PROBLEM 64
In problem 63 the kinetic and potential energies of an oscillator were expressed as functions of the time t. They can also be expressed as functions of x, the displacement of the oscillator from its equilibrium position; thus the potential energy of an undamped oscillator at x is given by $\frac{1}{2}kx^2$, and the corresponding kinetic energy at x equals $\frac{1}{2}k(A^2 - x^2)$.

a. What is the average potential energy of such an oscillator during its excursion from A to $-A$?
b. What is the average kinetic energy over the same excursion?
c. How do these averages compare?
d. Is your result for question (c) inconsistent with your answer to problem 63?

MOTION IN TWO DIRECTIONS

Trajectories

After motion in one direction the obvious next step for study is motion in two directions, or motion in a plane. If we describe this motion in terms of an orthogonal coordinate system—the usual x-y Cartesian coordinates, for example—we can consider motion in a plane to be made up of two separate linear motions, each perpendicular to, and independent of, the other. Thus motion in a plane will consist of a combination of linear motions with which we know how to deal.

To give this statement an experimental basis, let us repeat the dropping of an object from a helicopter (discussed on page 20). This time, however, instead of

Time, sec	Vertical y coordinate, m	Horizontal x coordinate, m
0	0	0
0.5	−3.2	13.4
1.0	−8.8	26.8
1.5	−16.9	40.2
2.0	−27.4	53.6
3.0	−55.7	80.4
4.0	−93.9	107.1
5.0	−141.2	133.8
6.0	−198.9	160.6

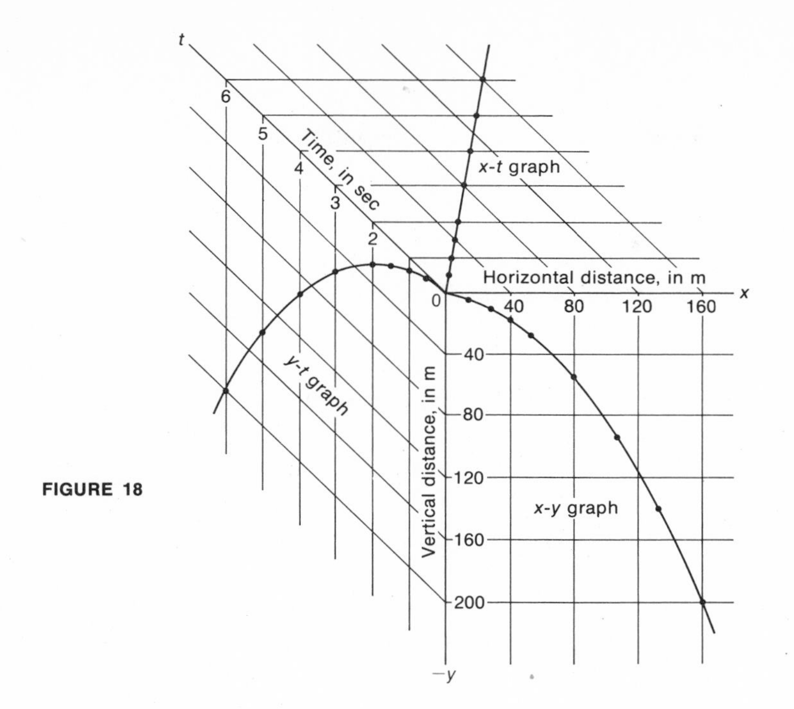

FIGURE 18

hovering, the helicopter is moving on a straight, horizontal course at 60 mi hr^{-1}. This time the tracking data are as shown in the table on page 46. The x-direction is the direction of the helicopter's motion, and the object moves in the x-y plane. Again we will consider only that portion of the motion in which drag and variation of gravitation is negligible.

The data are graphed in Figure 18. This is a three-dimensional graph; the orthogonal axes are $+x$, $-y$, and $+t$. (In the earlier discussion of the drop from a stationary helicopter, y was treated as positive downward, so as to concentrate attention on the relationships developing. In the present discussion we return to the normal coordinate system, in which up is treated as positive.)

The graph in the x-y plane shows the motion as it was actually photographed, in a vertical plane which contains the helicopter's forward motion. The y-t graph can be seen to be an exact repeat of the previous y-t graph of Figure 4: the forward motion of the helicopter has not affected the free-fall motion of the object. The x-t graph shows that the uniform forward motion imported to the object by the helicopter has not been affected by the fact that the object is falling. Hence the independent equations of the motion are

$$y = -\tfrac{1}{2}gt^2 \qquad \text{(from equation 2-22, with } v_{y,0} = 0\text{)}$$

and

$$x = v_{x,0}t \qquad \text{(from equation 2-7).}$$

By eliminating t between these two equations, we can determine that the path in the x-y plane traversed by the object is given by

$$y = -\frac{g}{2(v_{x,0})^2} x^2, \tag{2-25}$$

which is a parabola.

In trajectories in which horizontal motion is combined with vertical motion, the range R—the horizontal displacement from release point to point of impact—is

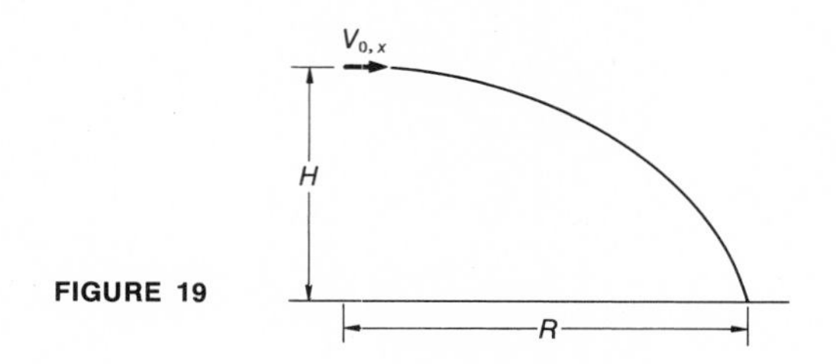

FIGURE 19

of major importance (Figure 19). For the object dropped from a horizontally moving helicopter which is H above level ground, equation 2-25 becomes

$$-H = -\frac{1}{2}\frac{g}{(v_{x,0})^2} R^2,$$

or the range

$$R = v_{x,0}\sqrt{\frac{2H}{g}}.$$

The more usual trajectory is the path of an object that is dispatched forward and upward from some initial point and returns to some impact point not necessarily on the same level as the initial point. Such a trajectory is described in problem 65, which is of a standard textbook type.

PROBLEM 65

A shell is fired with a muzzle speed v_0 from a gun barrel which is pointed upward at an angle θ above the horizontal. (Effects from the air and the rotation of the earth are to be assumed negligible.)

a. How far away on the same horizontal plane does the shell land?
b. What should θ be for maximum range?

SOLUTION

To take advantage of the previous discussion about the independence of vertical and horizontal motion, we need first to determine what these motions are. Heretofore, our initial speed in a chosen direction, our v_0, has been explicitly provided. In the current problem, however, our v_0 is not in either of the independent vertical or horizontal directions we are interested in, but at an angle to them.

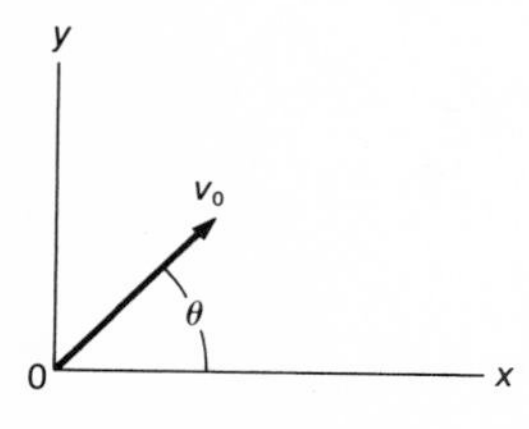

FIGURE 20

In Figure 20, the projection of v_0 on the x-axis is $v_0 \cos\theta$. If an

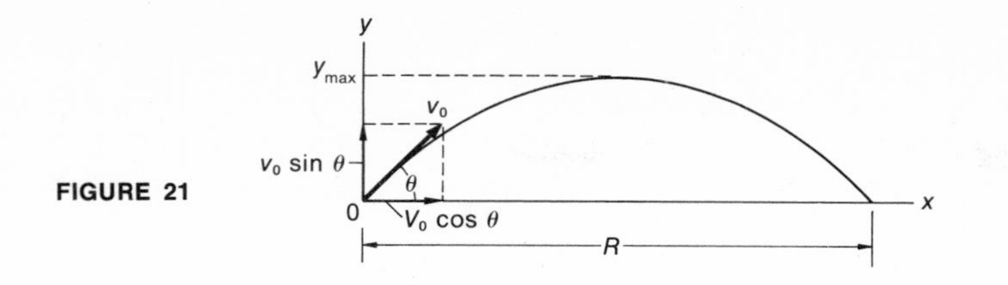

FIGURE 21

observer traveling in the x-direction with a speed $v_0 \cos \theta$ left the origin at the same instant the shell was fired there, he would at all times be directly under the shell, i.e., his progress in the x-direction and that of the shell would be alike. Thus $v_{0,x} = v_0 \cos \theta$; it is called the component of v_0 in the x-direction. Similarly, $v_{0,y}$, the component of v_0 in the y-direction, $= v_0 \sin \theta$. (See Figure 21.)

Returning now to the motion of our shell, for the horizontal motion we have

$$x_t = R = v_{0_x} t = v_0 \cos \theta \times t,$$

where t is the time of flight. In the same time t the net vertical motion is zero, hence

$$y_t = 0 = v_{0_y} t - \tfrac{1}{2} g t^2 = v_0 \sin \theta \times t - \tfrac{1}{2} g t^2.$$

Solving this for t,

$$t = \frac{2 v_0 \sin \theta}{g}.$$

Putting that value into the above equation for R, the solution for question (a) is:

$$R = v_0 \cos \theta \times \frac{2 v_0 \sin \theta}{g} = \frac{v_0^2 \times 2 \sin \theta \cos \theta}{g} = \frac{v_0^2 \sin 2\theta}{g}. \qquad (2\text{-}26)$$

Since v_0 and g are constants for the problem in question (b), R can be increased or decreased only by changing θ, and hence $\sin 2\theta$. For maximum R, it is clear that $\sin 2\theta$ must be a maximum. Since the maximum value of the sine is 1,

$$R_{max} = \frac{v_0^2}{g}.$$

For this maximum range,

$$2\theta = \sin^{-1} 1 = 90°,$$

$$\theta = 45°.$$

PROBLEM 66

An artillery piece which consistently shoots its shells with the same muzzle speed has a maximum range of R. To hit a target which is 2/3 R from the gun and on the same level, at what elevation angle(s) should the gun be pointed?

PROBLEM 67

Do the results of problem 66 suggest any general relationship for the two elevation angles which will produce the same range from a particular gun and shell combination? State the relationship and prove it.

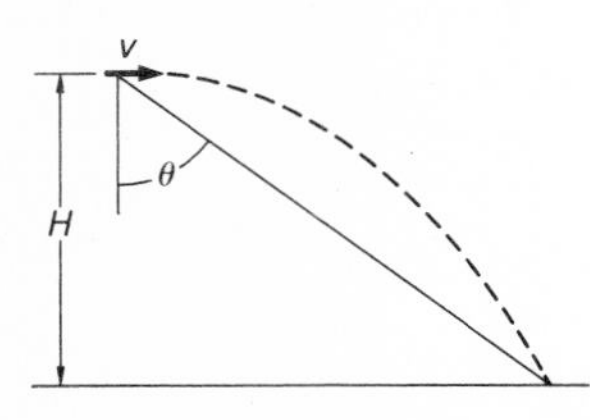

FIGURE 22

PROBLEM 68

A simplified form of bombsight, for bombing from an airplane in level flight, can be made from a single telescope. (In this simple form, air effects are neglected.) A sighting angle, θ with the vertical, can be determined as a function of the height of release H above the target and the horizontal speed v of the plane relative to the ground (Figure 22). With the telescope set at this angle and the airplane moving in a vertical plane that passes through the target, when the target appears in the line of sight the bomb is released, and theoretically a hit occurs. Determine the formula for θ.

PROBLEM 69

A bomber drops a bomb while flying on a steady, horizontal course 8.25 miles above ground at 6.00×10^2 mi hr^{-1}. Neglect any effect of the air on the bomb.

a. At what angle below the horizontal was the bombardier sighting on the target when he released the bomb, if it made a perfect hit?
b. If the bombsight was perfectly adjusted, and the bombardier thought he pressed the release button at the exact instant the target passed the cross hair of the bombsight, but the bomb fell 110 ft beyond the target, what was the reaction time of the bombardier?

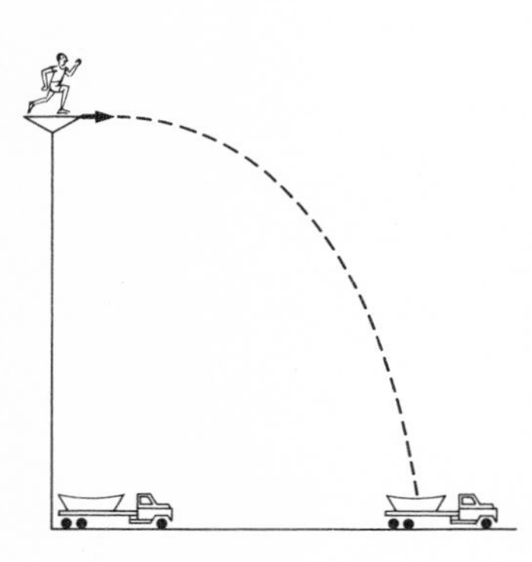

FIGURE 23

PROBLEM 70

A stunt performer is to run and dive off a tall platform and land in a net in the back of a truck below (Figure 23). Originally the truck is directly under the platform; it starts forward with a constant acceleration A at the same instant the performer leaves the platform. If the platform is H above the net in the truck, what horizontal speed V must the performer have as he leaves the platform?

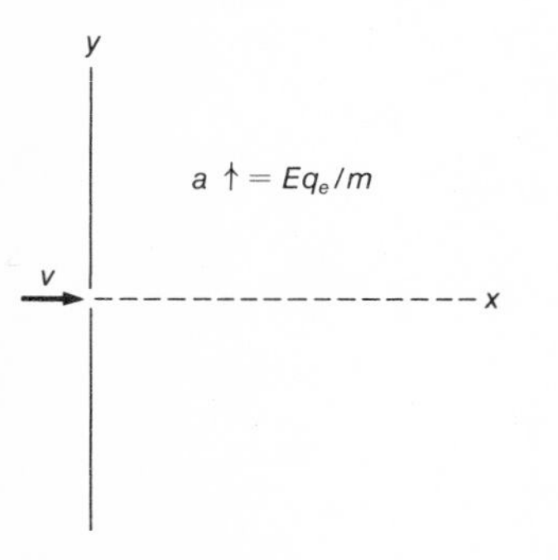

FIGURE 24

PROBLEM 71

An electron is injected, with a horizontal speed v, into a region in which a uniform electric field gives an upward acceleration of Eq_e/m to the electron. Coordinate axes are as shown in Figure 24.

a. What is the equation of the path of the electron for $x > 0$?
b. At $x = d$, what is the y coordinate of the electron; what is v_y?
c. If the accelerating field is zero beyond $x = d$, what is the equation of the path of the electron for $x > d$?

PROBLEM 72

An electron gun shoots a stream of electrons along a horizontal path that lies midway between two parallel, horizontal, deflection plates. The electrons enter the region between the plates with a horizontal speed of 2.3×10^6 m sec^{-1}; after leaving the plate region they hit a vertical fluorescent screen, creating a spot of light. (See Figure 25.) The plates are conducting, and between them an electric field can be established that gives to the electrons a vertical acceleration of 1.5×10^{15} m sec^{-2}. Assume that this accelerating field is uniform over the region between the plates, and zero outside that region.

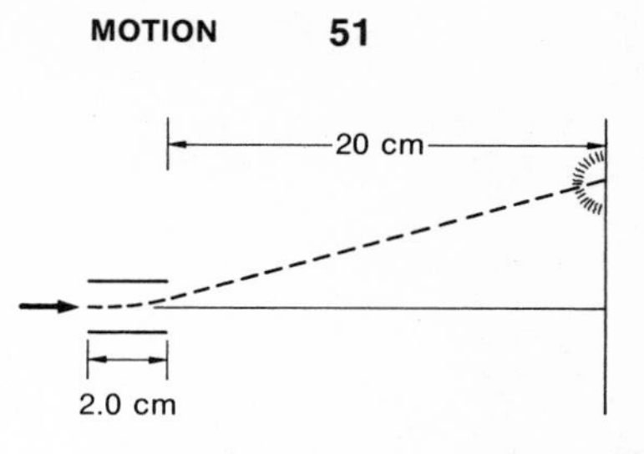

FIGURE 25

a. When the field is turned on, and then reversed in direction, how far does the spot of light move?
b. By how much did the acceleration of gravity affect your answer?
Note: Although considerably simplified compared to a real apparatus, this problem does illustrate the principle of the cathode ray oscilloscope.

PROBLEM 73
a. If in problem 71 the field E is doubled, how is y affected?
b. Does this suggest that a linear measurement Δy could be used to measure a change in the electric field E?

PROBLEM 74
At a county fair, a stunt flyer is to fly over the fair grounds at a constant elevation of 400 ft and drop packages of fire-extinguishing chemicals on bonfires that have been spaced 165 ft apart in a straight line. He released the first package when he was moving at 75 mi hr^{-1}, but immediately "revved up" his engine (to make more noise and thereby increase the excitement) so that thereafter he was accelerating horizontally at a rate of 10 ft sec^{-2}. Assuming that he was successful in hitting his targets, and that air resistance can be neglected,

a. how far back from the first fire was he when he made the first release?
b. What were the time intervals between the first and second release, and between the second and third release?
c. How far horizontally was he beyond the third fire when the third package landed there?

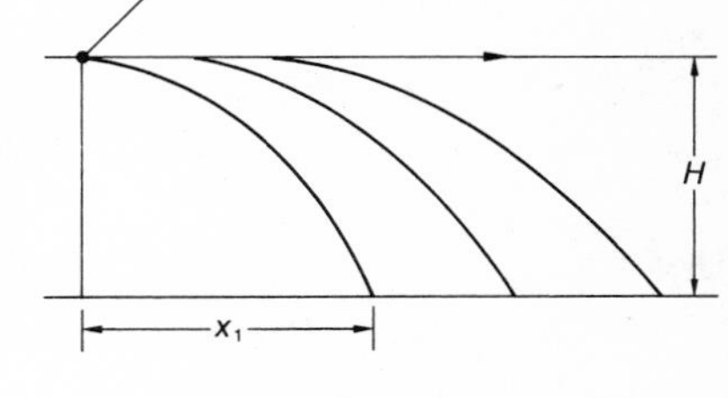

FIGURE 26

SOLUTION
A sketch of the drop pattern, Figure 26, ought to be helpful. Since all packages drop the same distance, the time of fall will be the same for all, and its determination is a useful start. Since for the packages $v_{y,0} = 0$, then

$$t_y = \sqrt{\frac{2\,(-H)}{-g}} = \sqrt{\frac{2 \times 400}{32}} = 5.0 \text{ sec}$$

and

$$v_{x,0} = 75 \text{ mi hr}^{-1} = 110 \text{ ft sec}^{-1}.$$

a. Then $x_1 = v_{x,0}\, t_y = 110 \times 5 = 550$ ft.
b. One could determine the value of x_2 by algebra alone, and then set $x_2 - x_1 = 165$; thus,

$$v_{x,t_1} = v_{x,0} + a\, t_1,$$

the forward speed of the plane at the time of the second drop, and

$$x_2 = v_{x,0}t_1 + \tfrac{1}{2}at_1^2 + v_{x,t_1} \times 5,$$

the horizontal distance the plane moved during t_1 plus the horizontal distance the package traveled after release. Putting in the above value for v_{x,t_1},

$$x_2 = v_{x,0}t_1 + \tfrac{1}{2}at_1^2 + (v_{x,0} + at_1)\ 5.$$

Then

$$x_2 - x_1 = (v_{x,0} + 5a)\ t_1 + \tfrac{1}{2}at_1^2 = 165.$$

Putting in numbers, we get

$$5t_1^2 + 160\ t_1 = 165,$$

from which $t_1 = 1.0$ sec.

By similar algebra, either using $x_3 - x_2 = 165$ or $x_3 - x_1 = 330$, you can determine that $t_2 = 1.94$ sec. Thus the time interval between the second and third drops is $t_2 - t_1 = 0.94$ sec.

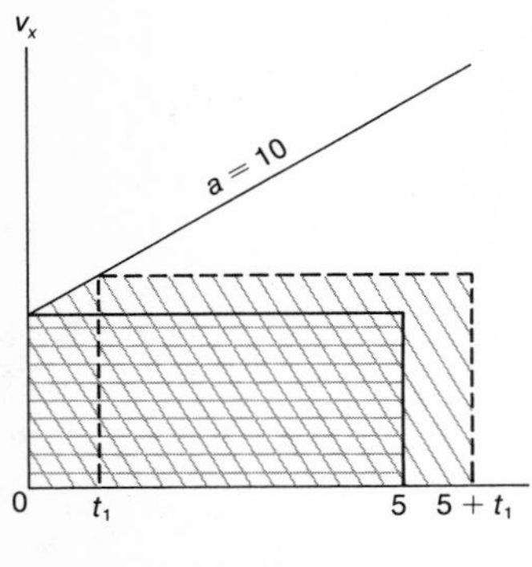

FIGURE 27

It is probably more illuminating to get this same result from a v-t graph. (See Figure 27.) The area of the solid rectangle—with horizontal hatching—gives directly the horizontal distance traveled by the first package from the first release point, the x_1 of our algebraic discussion. The diagonally hatched area under the heavy dashed outline gives directly the horizontal distance traveled by the second package from the first release point to its landing point. This area includes two areas: that from 0 to t_1, which is the horizontal distance traveled by the accelerating plane between the first and second release points, and the area from t_1 to $t_1 + 5$, which is the horizontal distance traveled by the second package after release, during its 5.0 sec of fall. This entire crosshatched area is the x_2 of our algebraic treatment. The difference between the x_2 area and the x_1 area—the area diagonally hatched but not crosshatched—must equal 165 ft.

Reading off the graph this difference in areas, we have

$$\tfrac{1}{2}at_1^2 + 5at_1 + 110t_1 = 165,$$

which gives $5t_1^2 + 160\ t_1 = 165$ as before. Similarly, you can determine that

$$\tfrac{1}{2}at_2^2 + 5at_2 + 110t_2 = 330.$$

(Once you form the habit of using v-t graphs in motion problems, you will find you can get information from them more quickly and more clearly than you can from algebraic equations.)

c. The answer can be done in your head. After the third drop was made, the plane would have remained at all times during the drop directly above the third package if it had not continued to accelerate. You will recall from page 000 that the extra distance traveled because of the acceleration $= \tfrac{1}{2}at^2$. In this instance t is the time of fall (5.0 sec), so that

$$\Delta X = \tfrac{1}{2} \times 10 \times 5^2 = 125 \text{ ft}.$$

PROBLEM 75
A shore battery (at sea level) spots an enemy ship moving directly away from it with speed u. The battery fires a shell of muzzle velocity v at elevation angle θ, thereby scoring a direct hit. Find the distance of the ship from the battery at the instant the shell is fired.

PROBLEM 76
Shortly after lifting off from the ground, a jet air transport is moving in a straight line with an air speed of 270 mi hr^{-1} and a rate of increase of altitude of 40 ft sec^{-1}. Just as the plane passes over a runway marker at an altitude of 600 ft, a small chunk of metal falls off the plane. Neglecting air resistance, how far from the marker does the metal hit the ground?

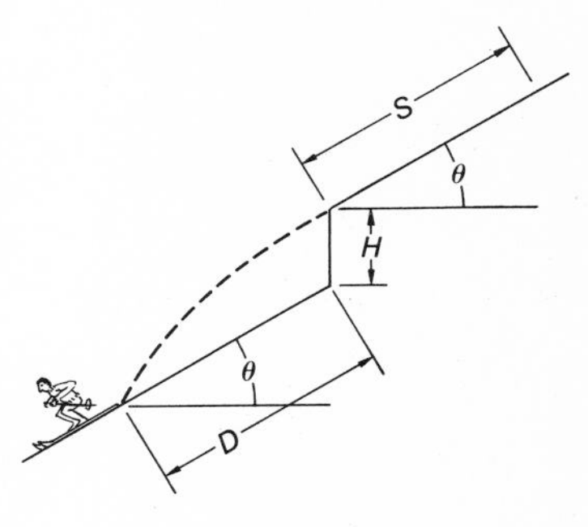

FIGURE 28

PROBLEM 77
After starting from rest straight down a uniform slope of snow at angle θ with the horizontal, S m along his track a skier reaches the edge of a vertical cliff H m high (Figure 28). Below the cliff the snow continues at the same angle θ. How far down the slope from the base of the cliff did the skier land? (Assume negligible friction between ski and snow.)

PROBLEM 78
a. At what angle above the horizontal should a gun be fired to hit a target S distance from the gun, at the same elevation, in the shortest possible time?
b. What is that time?
Assume the muzzle speed of the shells is v_0 and the air effects are negligible. *Reminder:* $\sin \theta = [(1 - \cos 2\theta)/2]^{1/2}$.

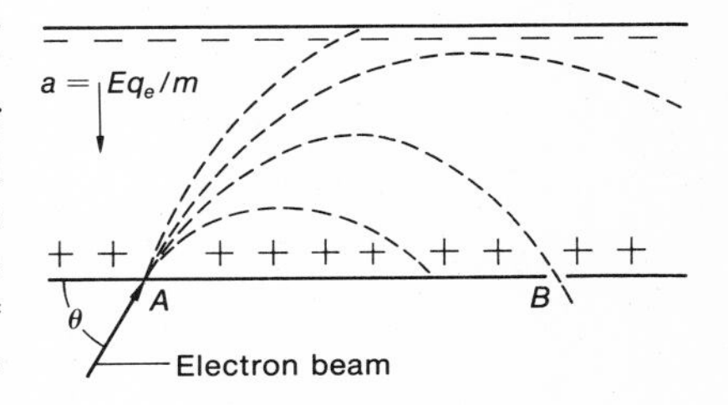

FIGURE 29

PROBLEM 79
One form of a "velocity selector" for charged particles consists of two parallel charged plates, as shown in Figure 29. A collimated beam of electrons of various speeds is introduced at A, at an angle θ with the lower plate. When the plates are charged as shown, the electrons between the plates are accelerated downward with $a = Eq_e/m$, the resulting trajectories depending on their initial speed at A.

a. Where should an exit slit B be placed so as to collect only electrons entering A with a particular speed v?
b. If the slit is so placed, what is the speed of the electrons exiting at B, in terms of v? *Suggestion:* place the origin of an x-y coordinate system at A.

PROBLEM 80
At what angle above the horizontal should a gun be fired to hit a target that is at a horizontal distance S from the gun, and at an elevation H above the gun? The muzzle speed of the shell is v_0; neglect air effects.

PROBLEM 81

A basketball player is a horizontal distance D away from a basket that is H above the level at which he releases a ball when he attempts a shot.

a. Develop the formula connecting D, H, the speed v with which he releases the ball in shooting, and the angle θ from the horizontal at which he shoots, that will enable him to score a basket.
b. The correct formula developed in (a) is a quadratic in D, which yields two values for D for any given H. For a successful shot, which value is appropriate? *Hint:* Visualize the location of the roots on the trajectory.

PROBLEM 82

The axis of a telescopic gun-sight points directly at the target, whereas the axis of the gun barrel on which the sight is mounted must point above this line if a bullet from the gun is to hit the target. If bullets leave the gun with a muzzle speed of 1400 ft $\sec^{-1}$, what should be the angle between sight and barrel for a hit on a target that is 3500 ft away horizontal measurement and

a. at the same elevation;
b. at an elevation 400 ft higher than the gun?

PROBLEM 83

In a football game between the Los Angeles Rams and the New York Giants, it is the Rams' ball, third down, on the Giants' 44-yd line. When the ball is snapped, the Rams' left end cuts diagonally across to his right, beyond the line of scrimmage, until he reaches the Giants' 30-yd line, at a point 15 yd in from the right sideline; at this point he turns sharply and runs at a speed of 9 yd $\sec^{-1}$ directly toward the Giant goal line. At the exact instant the end starts straight downfield the Ram quarterback (who has reached a point on the 50-yd line that is 15 yd in from the same right sideline) throws a pass parallel to the sideline. (The Ram quarterback's important passes are always thrown at an elevation of 45° above the horizontal.) The end caught the pass and was tackled immediately, but momentum carried him 4.0 yd beyond where he caught the pass before he fell.

a. What did the field judge, who was nearest to the point of catch, do?
b. Justify quantitatively your answer to (a). (Professional quarterbacks solve trajectory problems like this in their heads, in tenths of seconds; you ought to be able to solve it in ten minutes!)

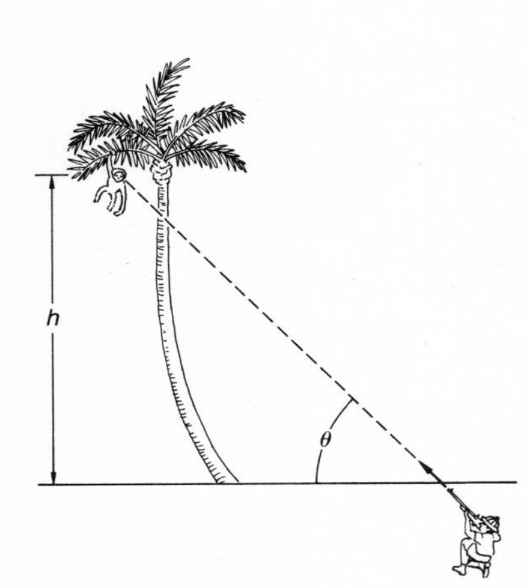

FIGURE 30

PROBLEM 84

Probably the most common demonstration experiment on trajectories is the "monkey and the hunter," illustrated in Figure 30. The hunter aims a rifle at an elevation angle θ directly at a monkey hanging from a limb of a tree; the monkey is h above the ground, which is essentially level between the tree and the hunter.* Just as the

*This condition obtains when the demonstration experiment is conducted entirely on the top of a lecture table.

hunter fires a bullet with a muzzle speed v_0, the monkey releases his hold and drops to the ground to avoid the bullet. In the experiment the bullet "always" hits the monkey. The "always" is not always true; there is a minimum value of $v_0 \sin\theta$ below which a hit cannot be made. Determine this minimum value.

PROBLEM 85

In problem 84 the ground was level, so that the distance h the monkey dropped was $D \tan\theta$, where D was the horizontal distance between hunter and monkey. Suppose now the terrain is different (Figure 31); the hunter is still a horizontal distance D from the monkey, but there is a steep hillside below the tree, so that the monkey's height above ground is greater than $D \tan\theta$*. Again, the bullet does not "always" hit the monkey. This time there is a minimum value of $v_0 \cos\theta$ below which a hit cannot be made. Determine this minimum value.

*This condition obtains when the "monkey" is allowed to drop beyond the edge of the lecture table on which the "gun" is located.

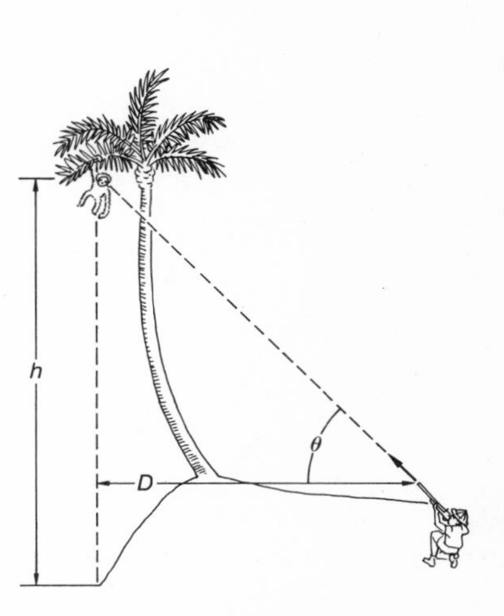

FIGURE 31

PROBLEM 86

A pilot of a military aircraft operating near a hostile gun emplacement needs to know the umbrella of coverage of that gun, i.e., the envelope of all of the possible trajectories from that gun, assuming the gun can be trained at all elevation angles from horizontal to vertical. If the center of a Cartesian coordinate system is established at the gun (Figure 32), what is the equation of the envelope of its trajectories, if the gun is known to eject shells with a muzzle speed of v_0? (For simplicity, assume no air resistance.)

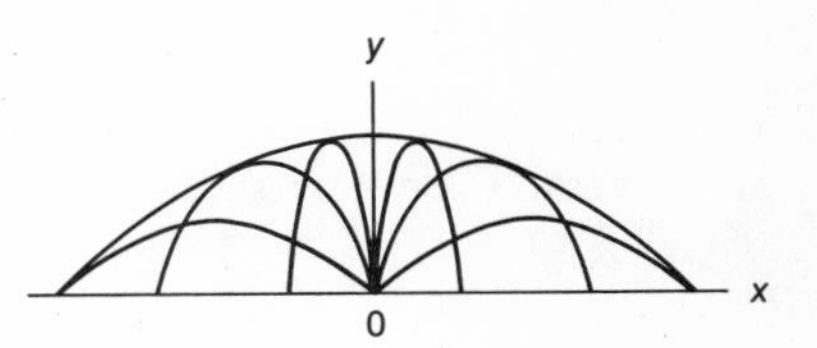

FIGURE 32

SOLUTION

For the vertical motion of any trajectory, we have

$$y = (v_0 \sin\theta)t - \tfrac{1}{2}gt^2;$$

for the horizontal motion,

$$x = (v_0 \cos\theta)t.$$

Eliminating t between these,

$$y = v_0 \sin\theta \times \frac{x}{v_0 \cos\theta} - \tfrac{1}{2}g\left(\frac{x}{v_0 \cos\theta}\right)^2$$

$$= x \tan\theta - \frac{gx^2}{2v_0^2 \cos^2\theta} \qquad (2\text{-}27)$$

for $0 < \theta < \pi/2$. For $\theta = \pi/2$, $y = v_0^2/2g$, from application of equation 2-23.

Every point in the envelope must correspond to a point on one of the infinite number of possible trajectories; in addition, at that point the slope of the envelope and the slope of the trajectory must be equal. (This arises from the definition of an envelope—that it be tangent to all of the family of curves it encloses.) Hence we next need the parametric equation for the slope, dy/dx. Differentiating the above equation for y,

$$\frac{dy}{dx} = \tan\theta - \frac{gx}{v_0^2 \cos^2\theta}. \tag{2-28}$$

For this slope equation to apply to every point on the envelope, its dependence upon θ must be removed. To accomplish this elimination and to assure that every y of the equation of the envelope corresponds to some value of y from equation 2-27, the values of the y points we are going to use from equation 2-27 must be independent of θ, and must depend only on x. In other words, $dy/d\theta$ must equal zero for the points on the envelope.

Differentiating equation 2-27, and equating to zero,

$$\frac{dy}{d\theta} = \frac{x}{\cos^2\theta} - \frac{gx^2 \sin\theta}{v_0^2 \cos^3\theta} = 0,$$

from which

$$\tan\theta = \frac{v_0^2}{gx}$$

and

$$\frac{1}{\cos^2\theta} = 1 + \tan^2\theta = 1 + \frac{v_0^4}{g^2x^2}.$$

Substituting in equation 2-28, we get

$$\frac{dy}{dx} = \frac{v_0^2}{gx} - \frac{gx}{v_0^2}\left(1 + \frac{v_0^4}{g^2x^2}\right) = -\frac{gx}{v_0^2},$$

the differential equation for the envelope. Integrating results in

$$y = -\frac{gx^2}{2v_0^2} + C.$$

Evaluating C from the boundary conditions that when $x = 0$, $y = v_0^2/2g$ (or when $y = 0$, $x = x_{\max} = v_0^2/g$),

$$C = \frac{v_0^2}{2g}.$$

Hence, finally, the equation of the envelope is

$$y = \frac{v_0^2}{2g} - \frac{gx^2}{2v_0^2}. \tag{2-29}$$

As it stands, this gives only the equation of the envelope in a vertical plane through the y-axis. Obviously the complete umbrella for the gun, assuming its ability to fire in any compass direction, would be obtained by rotating equation 2-29 about the y-axis.

Motion about an Axis

So far we have considered only those aspects of linear motion in which successive displacements did not change in direction (except possibly by 180°). If we had to, we could continue to analyze all possible motions we would be interested in by describing them in linear terms. To illustrate with a simple case, motion in a circle of radius R in the x-y plane, about a fixed z-axis, can be described by the following instantaneous linear relationships for the point P in Figure 33:

$$x = R \cos \theta$$

and

$$y = R \sin \theta,$$

where θ is measured counterclockwise from the $+x$-axis (the usual convention for measuring θ).

Since point P is moving in a circle, its motion is always instantaneously tangent to the circle; hence its speed v is in the direction of the tangent as shown. From Figure 33 it can be seen that

$$v_x = - v \sin \theta$$

and

$$v_y = v \cos \theta.$$

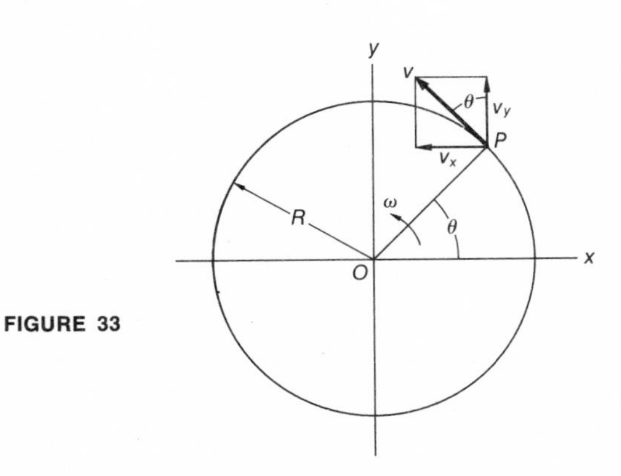

FIGURE 33

To put off a little longer the use of calculus, we note that P travels through a small arc ΔS on the circle in the same small time Δt that the radius OP rotates through a small angle $\Delta\theta$. Then v, the instantaneous tangential speed at P, equals $\Delta S/\Delta t$, and ω, the instantaneous angular speed of the rotating radius, equals $\Delta\theta/\Delta t$.

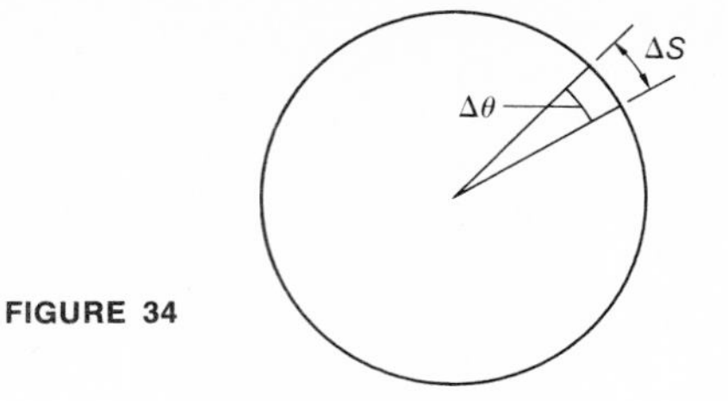

FIGURE 34

Then, equating the Δts, $\Delta t = \Delta S/v = \Delta\theta/\omega$. But by geometry, $\Delta S = R\Delta\theta$* (Figure 34), hence

$$\frac{R\,\Delta\theta}{v} = \frac{\Delta\theta}{\omega},$$

which gives us

$$v = R\,\omega. \tag{2-30}$$

This relationship holds for motion in a circle, whether v and ω are instantaneous values that are changing with time or are constants. It should be noted that v is perpendicular to R. Substituting from equation 2-30 in the equations for v_x and v_y,

$$v_x = -R\omega \sin\theta;$$

and (2-31)

$$v_y = R\,\omega\cos\theta.$$

This result could have been obtained more quickly by the use of calculus. Going back to $x = R\cos\theta$ and $y = R\sin\theta$ and differentiating these equations with respect to time, we have

$$\dot{x} = v_x = -R\sin\theta \times \frac{d\theta}{dt} = -R\,\dot{\theta}\sin\theta = -R\,\omega\sin\theta;$$

and

$$\dot{y} = v_y = R\cos\theta \times \frac{d\theta}{dt} = R\,\dot{\theta}\cos\theta = R\,\omega\cos\theta.$$

Continuing with calculus, by differentiating $\dot{x}$ and $\dot{y}$ with respect to time we obtain the accelerations

$$\ddot{x} = a_x = -R\,\frac{d\dot{\theta}}{dt}\sin\theta - R\,\dot{\theta}\cos\theta\,\frac{d\dot{\theta}}{dt} = -R\,\ddot{\theta}\sin\theta - R\,\dot{\theta}^2\cos\theta\dagger$$

$$= -R\,\alpha\sin\theta - R\,\omega^2\cos\theta, \tag{2-32}$$

where $\alpha = \dfrac{d^2\theta}{dt^2} = \ddot{\theta}$.

*Actually this is the fundamental definition of the angle $\Delta\theta$.

†As was stated earlier for v_x and $\dot{x}$, a_x and $\ddot{x}$, when values are known, or established relationships are expressed, it is customary to indicate angular speed by ω and angular acceleration by α. When these terms appear as unknowns in a differential equation, $\dot{\theta}$ and $\ddot{\theta}$ are usually used.

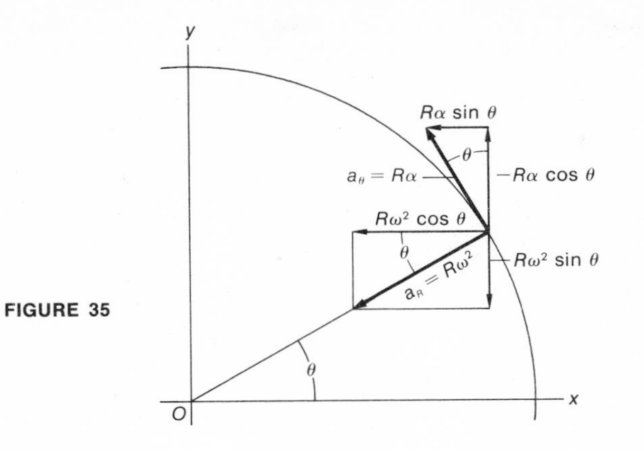

FIGURE 35

Also,

$$\ddot{y} = a_y = R\,\ddot{\theta}\cos\theta - R\,\dot{\theta}^2\sin\theta = R\,\alpha\cos\theta - R\,\omega^2\sin\theta. \tag{2-33}$$

Both a_x and a_y have two components, one dependent on the angular speed ω and the other dependent on the angular acceleration α. These components are shown separately in Figure 35. From the diagram it can be seen that the total acceleration a is made up of an acceleration a_R inward along the radius OP, where

$$a_R = \sqrt{(R\,\omega^2\sin\theta)^2 + (R\,\omega^2\cos\theta)^2} = R\,\omega^2, \tag{2-34}$$

and an acceleration a_θ tangentially in the direction of motion with

$$a_\theta = \sqrt{R\,\alpha\sin\theta)^2 + (R\,\alpha\cos\theta)^2} = R\,\alpha. \tag{2-35}$$

In most cases where circular motion is to be considered, ω will be constant, and hence α will $= 0$. In such cases, for point P the tangential speed $v = R\,\omega$ is constant; the instantaneous acceleration of P is then only $a_R = R\,\omega^2 = v^2/R$, directed centripetally.

Note that this is an important result to remember: An object moving in a circle of radius R about a fixed axis perpendicular to the plane of the circle, at a constant angular speed ω about that axis, has an instantaneous linear speed $v = R\,\omega$ that is always tangent to the circle of motion, and an instantaneous linear acceleration $a_R = R\,\omega^2 = v^2/R$, directed inward along a radius and hence is always perpendicular to the instantaneous tangential speed.

The converse of this is also true, and should be remembered. If an object moving at speed v is accelerated in a direction at right angles to its velocity, the value of the object's speed will not change but the direction of its motion will change. If the acceleration remains constant in value but changes direction so as always to be directed perpendicular to the instantaneous velocity—as in the case of a charged particle moving in a plane that is perpendicular to a constant magnetic field—then the change of direction of motion will be uniform and continuous, and the object

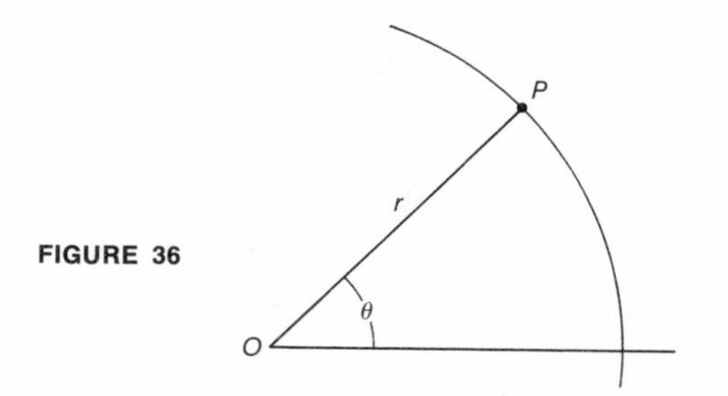

FIGURE 36

will move in the plane in a circle of radius R. R can be evaluated from $R = v^2/a_R$ when v and a_R are known; v can be computed if a_R is known and R is measured.

Although we started this section by describing rotational motion in linear coordinates, in general, rotational motion can best be discussed in rotational coordinates. For a position P in a plane, these coordinates consist of a distance r to P from some perpendicular axis of reference through O, and an angle θ measured to the line OP from some chosen reference line (Figure 36). A description of motion in these coordinates consists of appropriate statements about changes in r and θ.

When the distance r remains constant, as it does in pure circular motion, then the entire motion can be described in terms of angular displacement and a rate of change of angular displacement. But we have already learned how to handle displacements and their rates of change; that is how we developed the equations governing linear motion. Except for a little terminology, we do not need to learn anything new! We can transfer, by a point-to-point correspondence, what we have developed for linear motion and have it apply to circular motion. This correspondence appears clearly in the following tabulation:

Concept	Linear	Rotational	Geometrical connection for r constant, i.e., for circular motion
Displacement	ΔS, or S, or $S_2 - S_1$	$\Delta\theta$, or θ, or $\theta_2 - \theta_1$	$\Delta S = r\,\Delta\theta$
Speed	$v = \dfrac{dS}{dt} = \dot{S}$	$\omega = \dfrac{d\theta}{dt} = \dot{\theta}$	$v = r\omega$
Acceleration	$a = \dfrac{dv}{dt} = \dfrac{d^2S}{dt^2} = \ddot{S}$	$\alpha = \dfrac{d\omega}{dt} = \dfrac{d^2\theta}{dt^2} = \ddot{\theta}$	$a = r\alpha$*

*Note that this is the tangential component only.

For the special, but widely useful, case of constant acceleration:

Linear (a = constant)	Rotational (α = constant)
$v_t = v_0 + at$	$\omega_t = \omega_0 + \alpha t$
$S = v_0 t + \frac{1}{2}at^2$	$\theta = \omega_0 t + \frac{1}{2}\alpha t^2$
$v_t^2 = v_0^2 + 2aS$	$\omega_t^2 = \omega_0^2 + 2\alpha\theta$
$v_{av} = \dfrac{v_0 + v_t}{2}$	$\omega_{av} = \dfrac{\omega_0 + \omega_t}{2}$
$S = v_{av}\, t = \dfrac{v_0 + v_t}{2}\, t$	$\theta = \omega_{av}\, t = \dfrac{\omega_0 + \omega_t}{2}\, t$

When α is not constant, i.e., when $\ddot{\theta}$ is a function of time or a function of θ, the pertinent physical information must be stated in the form of a differential equation—exactly as we discussed earlier, beginning on page 35. The solutions then follow exactly the same course as discussed there. In fact, it should be emphasized that all of the methods we developed for the analysis of linear motion, including the illuminating use of v-t graphs, apply in corresponding fashion to the analysis of rotational motion.

It should also be pointed out that in many cases a complicated motion can be separated into linear and circular motions. To describe in one clear statement the motion of a mark on a tire on a car that is accelerating up a hill might appear difficult; when viewed as a pure linear motion of the tire axis plus a pure rotational motion of the mark about that axis, the description becomes obvious and simple.

PROBLEM 87

Making the reasonable approximation that the earth's orbit around the sun is circular,

a. what is the linear speed of the center of the earth in its orbital motion;
b. what is the earth's acceleration toward the sun?

PROBLEM 88

A flywheel rotating at 2400 rpm is braked uniformly to a stop in 34 sec.

a. What was its angular acceleration during the braking period?
b. Through how many revolutions did it turn during the braking period?

PROBLEM 89

A bicycle moving at 12 mi hr^{-1} is brought uniformly to a stop, without skidding, in 44 ft. The bicycle wheels are 28 inches in outside diameter.

a. How many revoltuions did the wheels turn during the braking period?
b. What was the angular acceleration of the wheels?

PROBLEM 90

A "24-hour" or "synchronous" satellite of the earth is one that remains constantly, directly above the same point on the earth's equator; i.e., it revolves around the earth's axis exactly in time with the earth. The orbit of such a satellite lies in the equatorial plane, is circular, and has a constant radius from the earth's center of 4.22×10^7 m.

a. What is the linear speed of the satellite?
b. What is the earth's gravitational acceleration at the orbital distance from the earth's center? (Effect of the moon and sun can be neglected.)

PROBLEM 91

An automobile's speedometer reading is directly related to the average speed of rotation of the rear wheels. A certain speedometer is correctly calibrated for use on a car with stiff American tires, which when properly inflated have an effective load-bearing radius of 12 inches. If flexible European tires having the same nominal size but designed to flex under load one inch more than American tires are now placed on the car, what is the percentage error of the speedometer reading?

PROBLEM 92

A uniformly rotating disc has a point of light at its center and one near its outer edge. A camera aimed perpendicular to the disc takes a picture with the shutter speed set at 1/25 sec; on an enlargement the blur from the outer light point is found to subtend an angle of $9.^\circ 40$ at the center.

a. The owner of the camera believes his shutter timing is accurate to within 1%; what does he compute as the rotational speed of the disc?
b. The owner of the disc knows its rotational speed is $33\frac{1}{3}$ rpm to within 1%; what can he tell the camera owner about the timing error of his camera shutter?

PROBLEM 93

In a fireworks display a rotating wheel, 6.0 ft in diameter and completely covered with burning "sparklers," slips off its axle and starts falling under gravity. A careful observer noted that during the first 2.3 sec of fall there was always some sparkler that seemed instantaneously motionless, but that beyond that time all sparklers moved all the time. How fast was the wheel turning when it slipped off its axle?

PROBLEM 94
What was the firing elevation of an artillery piece, if at zenith the trajectory of the shell it fired was congruent with a circle whose center was directly below the zenith and on the same level as the gun? (Neglect air friction.)

PROBLEM 95
A small object is moving with constant angular speed ω in the x-y plane in a circle of radius R about a perpendicular axis through O (Figure 37). With timing $t = 0$ started when the object is at $\theta = 0$, sketch the graphs for x–t, y–t, v_x–t, v_y–t, a_x–t, and a_y–t for one revolution about O.

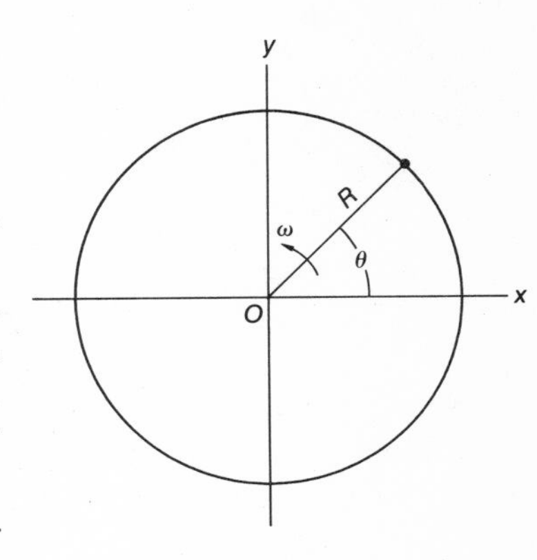

FIGURE 37

PROBLEM 96
Relative to the earth's axis,

a. a man standing at the earth's equator is moving with i. what speed, ii. what acceleration?
b. Where in the United States is a man standing when his speed and acceleration are one-third of the values found in question (a)?

PROBLEM 97
In the evening, a jet plane is flying due west at latitude 40° N, maintaining a constant altitude H. If H is small compared to the radius of the earth, how fast relative to the ground must the plane be flown so that the pilot has a chance of seeing the sun rise in the west? *Hint:* For the sun to "rise" in the west, the sun must appear to be moving west to east relative to the pilot, which means that the pilot must be moving west faster than the earth's surface beneath him is moving east.

PROBLEM 98
In a certain machine, two steel plates are separated by a hardened steel cylindrical roller (Figure 38). In operation, the plates move back and forth horizontally, perpendicular to the axis of the roller, and the roller rolls freely between the plates without slipping on either one. At a particular instant plate A is moving with a speed of 18 cm sec^{-1} to the right and an acceleration of 32 cm sec^{-2} to the left, and plate B is moving with a speed of 6.0 cm sec^{-1} to the right and an acceleration of 8.0 cm sec^{-2} to the left. Determine the following, at that instant, for the roller:

a. Its angular speed.
b. Its angular acceleration.
c. The linear speed of its axis.
d. The linear acceleration of its axis.

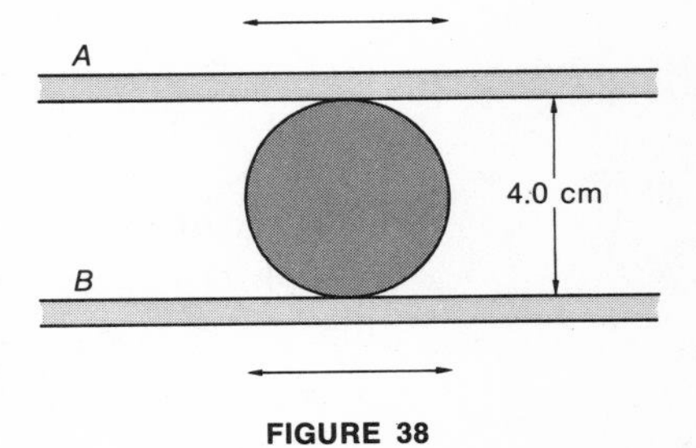

FIGURE 38

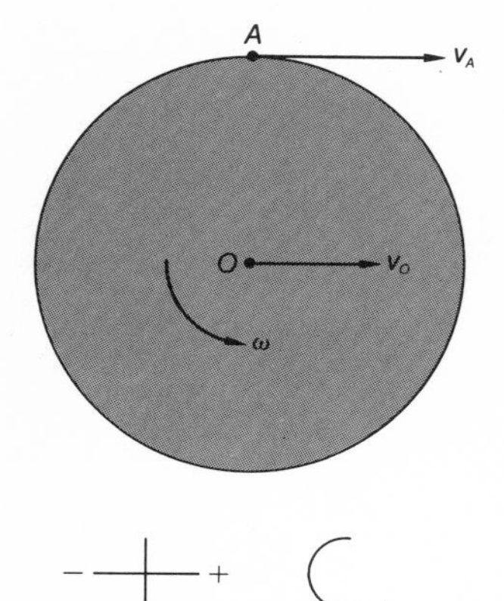

FIGURE 39

START OF SOLUTION

Because of the statement of "no slipping," point A on the roller has the same instantaneous linear motion as plate A. But the motion of point A is compounded of the motion of the axis through O and the motion due to rotation about a horizontal axis through O (Figure 39). In equation form, this information is

$$v_A = v_O - r\omega.$$

PROBLEM 99

The motion of a particular particle can be described by the equations $x = L \cos \omega t$ and $y = -L/3 \sin \omega t$, L being a known distance.

a. Determine the equation for, and sketch, the orbit of motion; indicate on the sketch the direction of motion.
b. Compute the instantaneous speed of the particle.
c. Compute the instantaneous acceleration of the particle.
d. Investigate whether the acceleration is centripetal, i.e., toward the origin.
e. At what points on the orbit are the speed and the acceleration perpendicular to each other?

Vectors and Relative Velocities—Newtonian Relativity*

We could have obtained the centripetal acceleration that goes with circular motion by a trigonometric method. This would give us more insight into the necessity of considering direction as well as scalar values of motion.

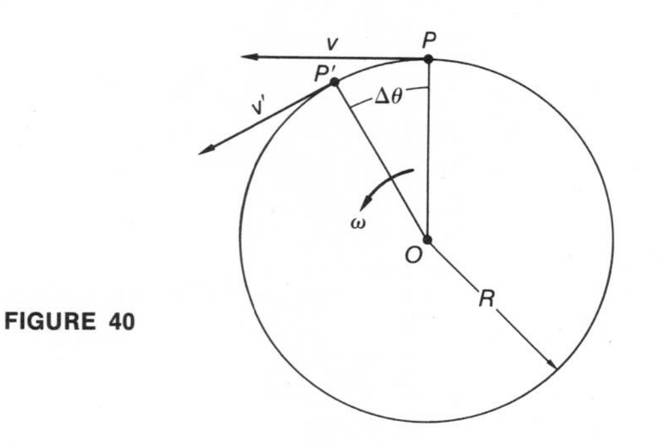

FIGURE 40

Consider a small object moving in a plane in a circle of radius R about an axis through O (Figure 40). At one instant the object is at P; Δt later it is at P'. If the object is moving with uniform motion, i.e., if ω is constant, then the value of the tangential speed at P equals the value of the tangential speed v' at P'. However, if we compare these two speeds, we can see that in the direction of v, v' is less than v; in the direction perpendicular to v there has now appeared some speed that did not exist before. These changes are represented by Δv, which supplies exactly the

*Also called Galilean Relativity.

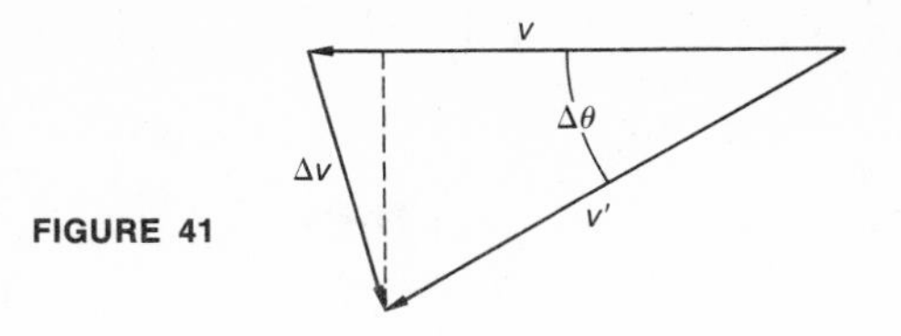

FIGURE 41

proper decrease in one direction, and proper increase in the other direction, to account for the change from v to v' (Figure 41).

Evaluating Δv by the cosine formula (and remembering that the numerical values of v and v' are equal), we have

$$\Delta v = \sqrt{2}\, v \sqrt{1 - \cos \Delta\theta}.$$

Let us now divide by Δt, and recall that by definition

$$\lim_{\Delta t \to 0} \frac{\Delta v}{\Delta t}$$

is the acceleration:

$$\lim_{\Delta t \to 0} \frac{\Delta v}{\Delta t} = a = \lim_{\Delta t \to 0} \frac{\sqrt{2}\, v \sqrt{1 - \cos \Delta\theta}}{\Delta t}.$$

For

$$\Delta\theta << 1,\ \sqrt{1 - \cos \Delta\theta} \approx \sqrt{\frac{(\Delta\theta)^2}{2}}.$$

Hence,

$$a = \lim_{\Delta t \to 0} \frac{\sqrt{2}\, v}{\sqrt{2}} \frac{\Delta\theta}{\Delta t} = v \lim_{\Delta t \to 0} \frac{\Delta\theta}{\Delta t} = v\omega.$$

Since $v = R\omega$ for the uniform circular motion we have been considering,

$$a = R\omega^2 = \frac{v^2}{R}, \tag{2-36}$$

as before. Further, as $\Delta t \to 0$, $\Delta\theta \to 0$; as $\Delta\theta \to 0$ the interior angles between Δv and v or between Δv and v' approach $\pi/2$ and at the limit equal $\pi/2$. Hence at the limit, Δv is perpendicular to v. The change in v, and hence the acceleration, is perpendicular to the tangent and is therefore centripetal, as before.

This approach to circular motion tells us nothing new, except for one very important indication—that *change of direction of motion* requires an acceleration just as much as does change of the numerical measure of motion. Heretofore the pertinence of direction has been hidden because we limited our discussion to linear motion; we selected some particular direction and then stayed with it. Even when we discussed trajectories we reduced a two-dimensional motion to separate, independent linear motions. We strained a bit at this limitation in Problems 96 through 99; it is now time to remove this limitation and develop a generalized approach to motion, one that will allow us to consider direction as easily as we consider numbers.

Vectors and vectorial space have been invented for this purpose. A vector is a directed line of limited length; its direction in space is the direction of the physical entity it represents, and its length represents the numerical value of that entity in the terms of some chosen scale. Only those physical entities that require a direction for complete description are represented vectorially. The amount of milk spilled on a highway after a collision between a tank truck and a speeding car, the volume of a weather balloon, the temperature of a spring day, the age of the universe—these are completely specified by numbers and units and are called scalars. As we have already seen, displacement and motion—and as we shall see later, forces and momentum, torques and angular momentum—have a direction as well as a number, and hence must be represented by vectors.

When you go from the railroad station 5 mi out Cactus Boulevard to the entrance to the airport, your displacement is represented by the vector $\vec{A}$ in Figure 42, in which each scale division represents a mile.* This vector can be described as 5 mi northeast of the station, 5 mi beyond the station and at right angles to the track, or 5 mi from the RR station on an azimuth, (or bearing) of 45°.

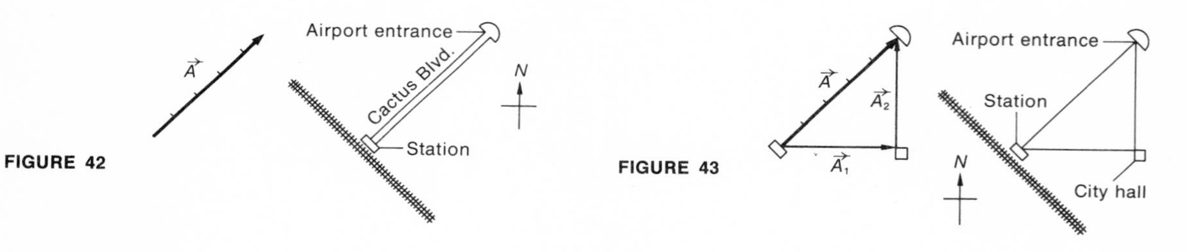

FIGURE 42

FIGURE 43

These three different identifications for the same vector illustrate the important point that a vector is independent of the coordinate system used to describe it. A vector represents an independent physical entity; a coordinate system is chosen to suit your convenience.

You could have gotten from the station to the airport by going east to City Hall, and then north to the airport entrance. For this displacement the vectors would be as shown in Figure 43. Note that when you add $\vec{A}_2$ to $\vec{A}_1$ you get $\vec{A}$; no matter how you go from the station to the airport, when you get to the airport your displacement from the station is still just 5 mi northeast. This addition of displacements is written in the usual equation form: $\vec{A}_1 + \vec{A}_2 = \vec{A}$.

Let us extend this addition of vectors to motion. To do this, we first must make one change in terminology. When we were not concerned with direction (because

*In all material that is printed it is standard practice to identify vectors in boldface type, thus: **A**. In writing, either at your desk or at the blackboard, this is not possible. Hence in writing it is customary to indicate a vector by a short arrow over the letter identifying the vector, thus: $\vec{A}$. Since this book is intended mainly for the use of people who are working either at their desk or at the blackboard, and are confined to the use of handwritten symbols, the arrow designation will be used for vectors.

all motion was in the same direction), we called the scalar v, defined as time rate of change of displacement, by the name "speed." Now that we have entered the realm of vectors where direction must be considered, and the displacement S is a vector, we must define a new term $\vec{v} = d\vec{S}/dt$; this new term we call velocity. Hereafter, when direction as well as the scalar speed are to be considered, we will call the combination "velocity." If "velocity" is called for in a problem, the answer must include a statement of direction as well as a numerical value of speed.*

First degree equations consisting of terms that are scalar values of vectors can be changed to vector equations by the simple process of putting → over all terms. Thus the scalar equation developed for motion with constant acceleration,

$$v_t = v_0 + at,$$

becomes

$$\vec{v}_t = \vec{v}_0 + \vec{a}t \tag{2-37}$$

and equation 2-15, also developed for constant acceleration, becomes $\vec{S} = \vec{v}_0 t + \frac{1}{2}\vec{a}t^2$. (The attentive reader will have a strong suspicion—and will be right—that these vector equations are a shorthand equivalent of the equations 2-16 and 2-17.)

As a further example of the addition of vectors: at $t = 0$, a velocity $\vec{v}_0$ of an object relative to the ground is observed to be

$\vec{v}_0$

G O

At t sec later the object's velocity $\vec{v}_t$ is observed to be

O

$\vec{v}_t$

G

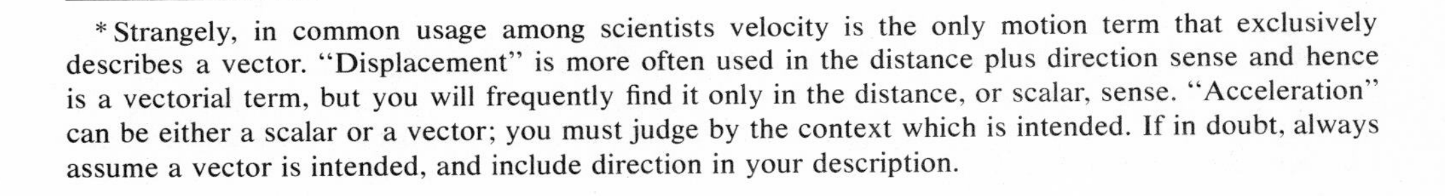
* Strangely, in common usage among scientists velocity is the only motion term that exclusively describes a vector. "Displacement" is more often used in the distance plus direction sense and hence is a vectorial term, but you will frequently find it only in the distance, or scalar, sense. "Acceleration" can be either a scalar or a vector; you must judge by the context which is intended. If in doubt, always assume a vector is intended, and include direction in your description.

Putting $\vec{v}_0$ and $\vec{v}_t$ together, it is clear that in the time t some change in velocity $\overrightarrow{\Delta v}$ had to be added to $\vec{v}_0$ to produce $\vec{v}_t$, thus:

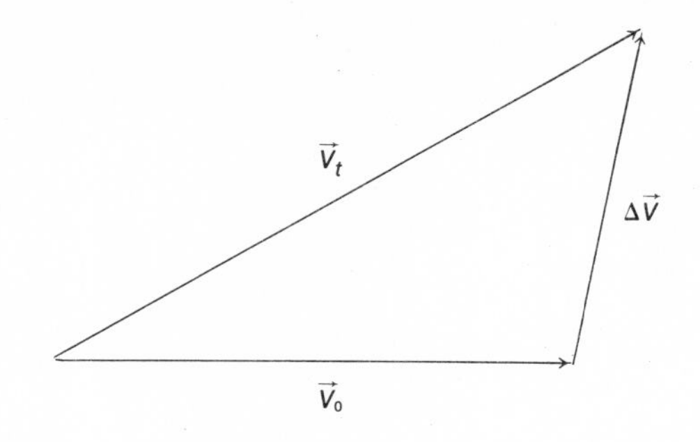

Without further information we cannot say any more about $\overrightarrow{\Delta v}$; it could be a sum of a number of smaller $\overrightarrow{\Delta v}$'s, each of which took place during a small portion of the time t, thus:

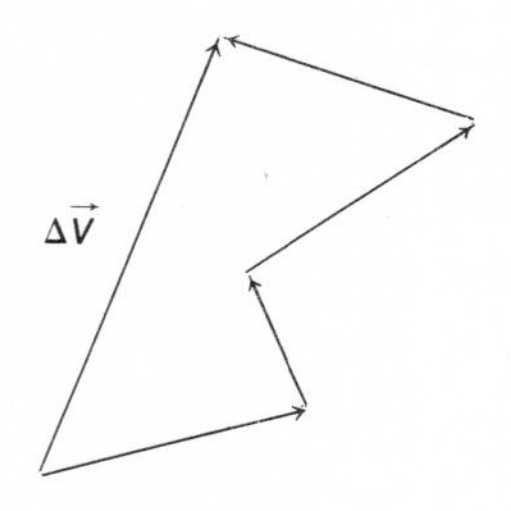

However, if it is further specified that during the time t the acceleration was constant both in direction and magnitude, the segmented possibility just shown has to be ruled out. For constant acceleration, the vector $\overrightarrow{\Delta v}$ can thus be replaced by $\vec{a}t$:

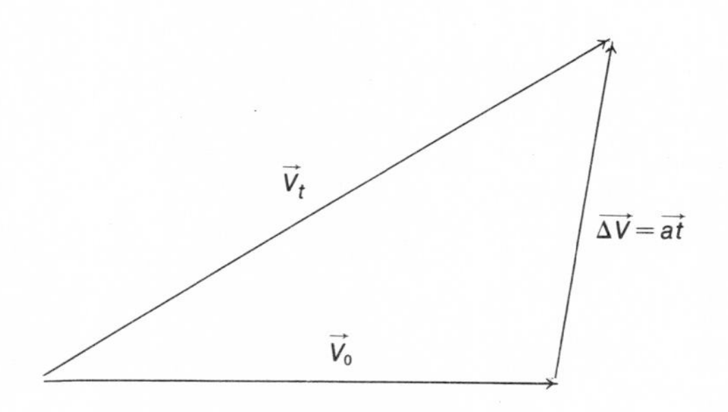

This is the vector diagram for the vector equation $\vec{v}_t = \vec{v}_0 + \vec{a}t$, given above.*

The addition of vectors is particularly useful in the determination of relative velocities. Of course a velocity itself is relative, in fact all vectors are relative—

*There is no inconsistency between this development and the one on page 19. Here, the time t, and of course Δv, are finite; there, Δt, and of course Δv, were small and were made to approach zero.

relative to the coordinate system in which they are observed and measured. For example, the displacement discussed earlier was relative to the railroad station. However, the term "relative velocity" usually implies a kind of "relative squared" relationship; a relative velocity usually refers to a velocity which is measured relative to a coordinate system which itself is moving relative to some "fixed" coordinate system.

As a simple case, consider a highway patrol officer pursuing at 80 mi hr^{-1} a car that is traveling down a straight road at 70 mi hr^{-1}. The velocity vector for the patrol officer relative to the ground would look like this:

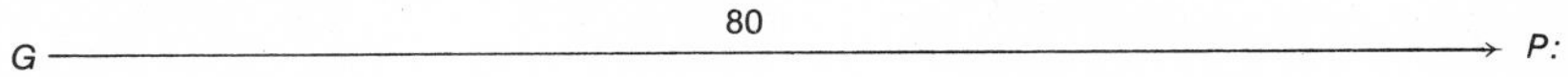

$$G \xrightarrow{\quad 80 \quad} P:$$

the velocity vector for the car, like this:

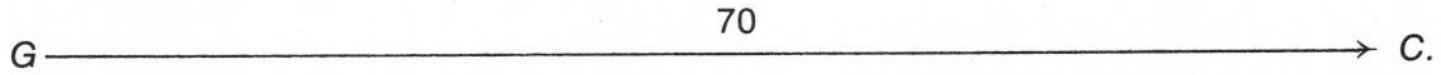

$$G \xrightarrow{\quad 70 \quad} C.$$

Put together, both relative to ground G, they look like this:

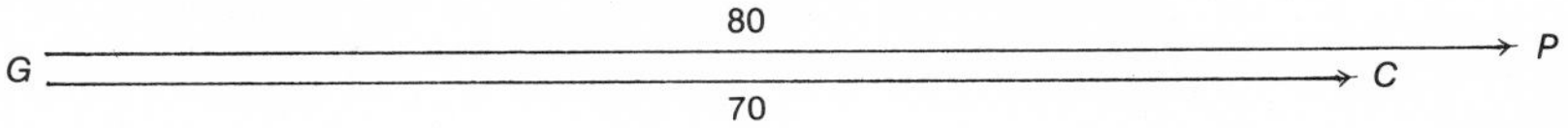

$$G \xrightarrow{\quad 80 \quad} P$$
$$G \xrightarrow[\quad 70 \quad]{} C$$

The velocity vector of the patrolman relative to the car then looks like this:

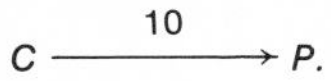

$$C \xrightarrow{\ 10\ } P.$$

In other words, the patrolman is approaching the car at a rate of 10 mi hr^{-1}; that is what the driver sees in his rear-view mirror.

The patrolman on the other hand sees the car backing toward him at 10 mi hr^{-1}, thus:

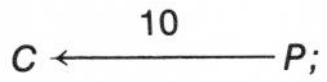

$$C \xleftarrow{\ 10\ } P;$$

this is the vector for the car relative to the patrolman.

The entire technique for dealing easily and clearly with relative velocity problems is illustrated in this simple example.

- The given vectors are treated *exactly as given.*
- Each vector is drawn to the same scale, with the arrow point representing the object moving and the tail of the vector representing what the object is moving relative to; i.e., in the vector

$$G \longrightarrow P,$$

the patrolman P is moving relative to the ground G;
- The heads and tails of the given vectors, together with their appropriate labelling, determine points in vector space (in this case velocity vector space) thus:

$\dot{G}$ $\qquad\qquad\qquad\qquad\qquad\qquad\qquad\qquad$ $\dot{C}$ $\qquad$ $\dot{P}$

• Vector arrows can be drawn between any of these points in vector space for any relative velocity desired. The arrows must be drawn toward the point representing the thing moving, and drawn away from the point representing what the thing moving is moving relative to. For example,

G ←——— 80 ——— P

indicates that, relative to the patrolman, the ground is moving backward at 80 mi hr^{-1} (which, of course, is exactly what the patrolman sees when he looks down at the ground).

PROBLEM 100

A man cycling north on a straight road observes from the speedometer on his bicycle that he is maintaining a steady speed of about 8.7 mi hr^{-1}. Because the man is an experienced sailor he is trained to evaluate wind velocities, and he estimates (correctly) that the wind he feels is blowing about 10 mi hr^{-1} from 30° E of N.

a. What is the wind velocity as measured by a stationary observer?
b. When the cyclist speeds up to 15 mi hr^{-1}, what is the velocity of the wind he feels?

SOLUTION

This is a standard type of relative velocity problem, similar to those found in almost any textbook.

a. We are given two vectors: the cyclist relative to ground vector, pointing north and 8.7 units long, and the wind relative to cyclist vector, pointing 30° *W* of *S* and 10 units long. Thus, as shown in Figure 44, *C* is the cyclist point, *G* is the ground point, and *W* is the wind point. Obviously the desired wind relative to ground vector completes the triangle.

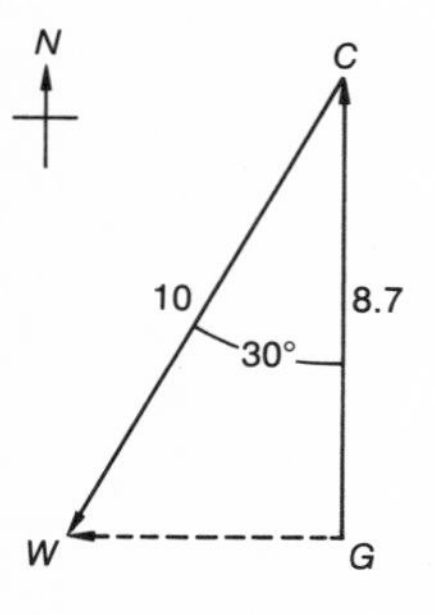

FIGURE 44

The hard way to get the value of $\vec{v}_{W-G}$ would be as follows: using the fact that you know two sides of a triangle and the included angle, by the cosine formula you could compute $v_{W-G} = 5$. Then using that value in the sine formula, you could determine that $\angle CGW$ is 90°. Putting these together, $\vec{v}_{W-G} = 5$ mi hr^{-1} from due east.

The easy way, and the way it is hoped most of you will see fairly quickly, is to recognize that 8.7 is very approximately 10 cos 30°, and hence CGW is a 30-60-90 triangle. Thus $v_{W-G} = 10 \sin 30 = 5$, and $\vec{v}_{W-G} = 5$ mi hr^{-1} from due east—all done by inspection and in your head.*

b. For this part we have the wind relative to ground vector just

*I might comment at this point that if you are at all interested in physics or engineering problems, there are some things you should have memorized by now; among these would certainly be the values of the functions of 30° and 45°.

determined in part (a) and a new cyclist relative to ground vector; thus the new wind relative to cyclist vector has a value

$$v_{W-C} = \sqrt{15^2 + 5^2} = 15.8 \text{ mi hr}^{-1},$$

and

$$\alpha = \tan^{-1} \tfrac{5}{15} = 18\overset{\circ}{.}4.$$

Hence (Figure 45) the new $\vec{v}_{W-C} = 15.8$ mi hr^{-1} from 18°.4 E of N.

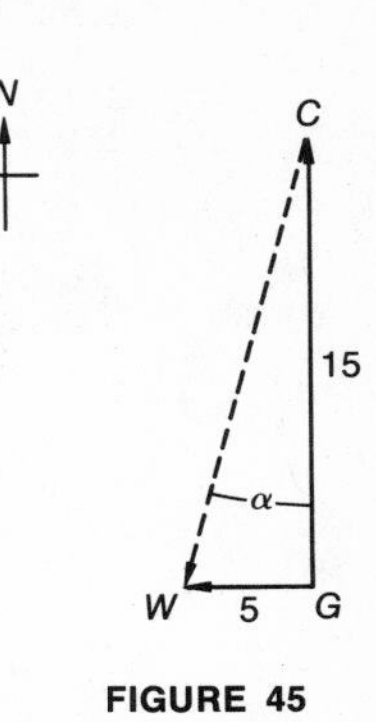

FIGURE 45

PROBLEM 101

The gas from a burning rocket motor escapes through the nozzle with an efflux speed of 5720 ft sec^{-1} relative to the nozzle. Relative to the space through which the rocket is traveling, how fast is the gas that has just been ejected moving,

a. when the rocket's speed is 100 mi hr^{-1},
b. when the rocket's speed is 1000 mi hr^{-1}?

PROBLEM 102

How fast is the rocket of problem 101 moving at the instant the gas just exhausted lies motionless in space?

PROBLEM 103

A sailboat on a large lake is moving at 6 knots northeast, under the influence of a west wind. (A wind is identified by the direction from which it comes.) The skipper notices that a wind vane at the top of the mast points at right angles to his direction of motion. What is the magnitude of the true wind?

PROBLEM 104

A ship is proceeding due east at a speed of 20.0 knots. At noon the officer on the bridge notes that the apparent wind is from 30° W of S and that the ship's anemometer registers 12.0 knots. What true wind speed and direction did the officer report to the weather stations on land?

PROBLEM 105

The normal nonstop flight path for jet planes between New York and San Francisco runs approximately east and west, and is about 2600 miles long. A passenger jet plane that cruises at an average air speed of 600 mi hr^{-1} takes about three-quarters of an hour longer in actual travel time to fly from New York to San Francisco than it takes to fly from San Francisco to New York, both flights being approximately at the same altitude. From this information, what can you say about the wind at the jet's cruising altitude?

PROBLEM 106

If, in addition to the information in problem 105, you are told that the prevailing wind over the flight path blows on the average from 30° S of W, what now can you say about the wind at the cruising altitude? *Comment:* Theoretically this is a perfectly feasible problem. However, the answer is a small difference of two large numbers, which is always of dubious validity when dealing with only two- or three-figure data.

PROBLEM 107

During the movement of a passenger elevator, a screw in the overhead grill comes loose and falls to the floor of the elevator, which is 2.9 m below the grill. Compute the times of fall for the screw when the elevator is

a. accelerating upward at a rate of 1/7 g,
b. accelerating downward at a rate of 1/7 g,
c. moving at constant speed.

Hint: Acceleration, as a vector, can be treated similarly to velocity vectors. For example, the acceleration vector relative to the ground of a freely falling object O would look like this,

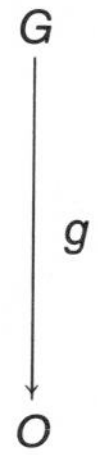

G

g

O

with its length equal to g.

PROBLEM 108

While the train of problem 10 was still running at the speed determined in that problem, the children observed a freight train going by them in the opposite direction on the other track. The child with the watch reported the train passed them in 17.5 sec; the other child reported the passing train contained 40 freight cars (see problem 11), a motor unit he estimated was about 1.5 freight cars long, and a caboose that seemed about half the length of a freight car. After a little figuring the children correctly told their mother the speed of the freight train. What was it?

PROBLEM 109

The meteorologist at airport A informs a pilot there, who wishes to fly to airport B which is 300 statute miles due north of A, that a steady east wind is blowing at a speed of 30 knots all the way between A and B. The pilot then makes out a flight plan based on an air speed for his plane of 150 knots.

a. What compass heading does he plan to fly in order to proceed from A to B in a single straight line?
b. What flight time between the two airports does he estimate?

PROBLEM 110

After concluding his business at B, the pilot in the above problem now wishes to fly to C, which is 300 statute miles due east of A. He is informed that the same steady 30-knot east wind is still blowing in that part of the country. He again wishes to fly at an air speed of 150 knots between airports.

a. How should he head to reach C from B in the shortest time?
b. How long will it take him to fly from B to C?

PROBLEM 111

Just as he is about to leave B for C, the pilot in problem 110 is informed that the airport at C will be closed in 2.0 hr, due to the expected arrival of the Presidential jet.

a. What minimum air speed must he maintain after leaving B, in order to be permitted to land at C?
b. In what compass direction should he head his plane?

PROBLEM 112

On finally returning home to A from C, the pilot in the above problems assumes the same east wind is still blowing, and so heads due west from C at an air speed of 150 knots. However, to his surprise, 87 min after leaving C he finds himself 30.5 statute miles due north of A.

a. On leaving C the pilot had radioed his ETA (estimated time of arrival) to his wife at A. What was it?
b. Assuming the wind was constant after the pilot left C, what was its speed and direction?
c. When the pilot discovered his error, he corrected it immediately and headed straight for A, still flying at an air speed of 150 knots. What compass heading did he fly on this final leg?
d. How much time over his ETA did he run?

PROBLEM 113

A physics student riding in a car notices that when the car is being accelerated at 60 mi hr^{-1} min^{-1}, streaks on the side windows made by falling rain change angle from 45° to 30° with the horizontal, in 15 sec. Assuming that the rain actually is falling vertically downward, he computes in his head the speed of the falling rain. What is that speed?

PROBLEM 114

Ship A is traveling at a constant 15 knots on a steady eastward course; ship B is moving at a constant 26 knots. At $t = 0$, a watch officer on A takes a sight on B and notes that B is 6.0 nautical miles due south of A. Later B was observed to pass behind A, and its distance of closest approach to A was measured to be 3.0 nautical miles.

a. Assuming B maintained a steady course, what was that course?
b. How far apart will the two ships be at $t = 1.0$ hr?

Note: The distance of closest approach of B to A is the length of the radius of a circle that is centered at A and is tangent to the line of motion of B relative to A.

PROBLEM 115

During training exercises, when a destroyer is running at a speed of 28 knots on a course 30° N of E, the watch officer notes, by the smoke, that the wind seems to be coming from 30° off the port bow. Later, when the destroyer is running at the same speed on a course 30° S of E, the smoke indicates the wind is coming from 60° off the port bow. Assuming the wind was constant throughout the period of observation, what was its velocity?

PROBLEM 116

At $t = 0$, an object I is at $x_0 = -A$, $y_0 = 0$, and is moving with a constant speed $+v_x$; a similar object II is at $y_0 = +B$ and $x_0 = 0$, and is moving with constant speed $-v_y$.

a. At $t = t$, where is I? Where is II?
b. What relationship must hold between A, B, v_x and v_y if the objects, no matter how small their dimensions, are to collide?
c. If this relationship in part (b) does not hold, how far apart are I and II at any time t?

PROBLEM 117

For the objects in problem 116, question (c),

a. at what time were they nearest each other?
b. What was their minimum separation?
c. Does this minimum separation reduce to zero when $Av_y = Bv_x$?

PROBLEM 118

The starboard gunner in a bomber flying on a horizontal course due north observes an enemy fighter flying due west at the same altitude. The bomber's ground speed is 600 mi hr^{-1}, and the fighter's ground speed is assumed to be 800 mi hr^{-1}. At the instant the fighter is observed to be directly northeast of the bomber, the gunner fires a shell whose muzzle velocity has a horizontal component of 1760 ft sec^{-1}. To make a hit, in what compass direction should the gun be aimed?

PROBLEM 119

Instead of using the fighter data given in problem 118, the gunner in that problem learns from the radar operator that the fighter's velocity *relative to the bomber* is 1200 mi hr^{-1} directly toward the bomber along a line bearing 60° E of N.

a. What is the fighter's velocity relative to the ground?
b. Now in what direction should the gun be aimed to make a hit?

PROBLEM 120

Light from a distant star takes a finite time to traverse the length of a telescope; during this time the earth has moved a finite distance in orbit. Because of this movement of the earth, the telescope must be canted slightly forward in the direction of orbital motion; this effect is called "aberration." (The motion of the telescope due to the rotation of the earth is about two orders of magnitude less than the orbital speed, and can be neglected.) To observe a star whose geometrical line of position is perpendicular to the earth's orbital motion, what must be the telescope's angle of tilt from the true geometrical direction of the star, in terms of the orbital speed v and the speed of light c?

PROBLEM 121

A small power boat enters a long narrow bay in a fog so heavy that the maximum safe speed, relative to the ground, is 8.0 ft sec^{-1}. During a brief break in the fog the boat's helmsman sights a mooring buoy 0.50 mi away, on a line bearing 34° W of N from him. He knows from Coast Guard tables that for the next half hour he will be in a tidal current running 22° W of S at a speed of 3.0 ft sec^{-1}. He steers so that the boat moves directly toward the buoy at maximum safe speed relative to the ground.

a. How fast is the boat moving relative to the water?
b. In what direction is the boat pointing?

PROBLEM 122

A man standing on the bank of a river 1.0 mi wide wishes to reach a point directly opposite him on the other bank. He can do this in two ways.

- He can swim somewhat upstream (relative to the water) so that his resultant motion is straight across.
- He can swim directly toward the opposite bank relative to the water, and then walk up along the bank from the point downstream to which the current carried him.

If he can swim 2.5 mi hr^{-1} and walk 4.0 mi hr^{-1} and if the current flows at 2.0 mi hr^{-1}, which is the faster way to cross, and by how much? (Assume current flows at same rate across the entire width of river.)

PROBLEM 123

A river of constant width flows due south with a uniform speed v_R relative to the ground. A man on the west bank desires to reach a point on the opposite bank directly east from where he starts. To cross the river he has a motor boat that runs at a constant speed v_B relative to the water. As $v_B < v_R$, he cannot go directly across the stream, but must land somewhere downstream and walk up the river bank to his objective. If he walks on land with a constant speed v_M relative to the ground, in what compass direction should he point the boat so as to reach his destination in the least time?

In problems 122 and 123 the simplifying assumption was made that the velocity of the river was constant across its width. This assumption served to place the emphasis in those problems where it should have been—on the treatment of relative velocities

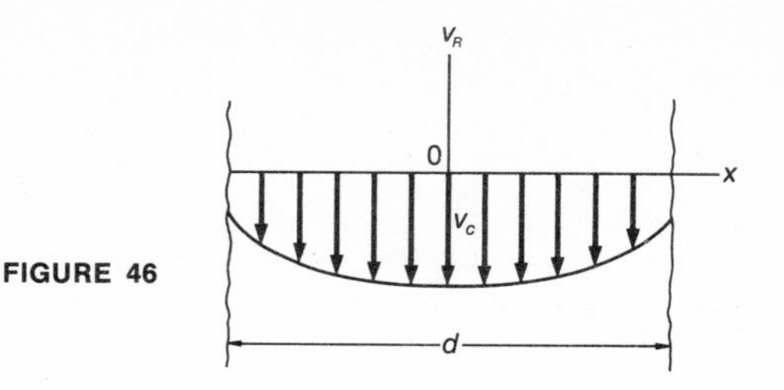

FIGURE 46

—but for most rivers the assumption is unrealistic. Stream-gauging measurements indicate the surface velocities of a straight river usually look something like the diagram shown in Figure 46. The velocity vectors are parallel to the banks, and the velocity profile looks like a parabola. One could guess that, for coordinates centered as shown, the river velocity at any x could be described by $v_{R,x} = -v_c + Bx^2$, where B must be evaluated for a particular river.

PROBLEM 124

The velocity profile of the surface water of a real river is parabolic, with the speed at the center equal to v_c and the speed at the banks equal to v_b. (See footnote on page 166.) At what constant angle to the current should a boat be pointed so that, when driven at a speed v relative to the water, the boat will land at a point on the opposite bank directly across from where it started? $v > v_c > v_b$.

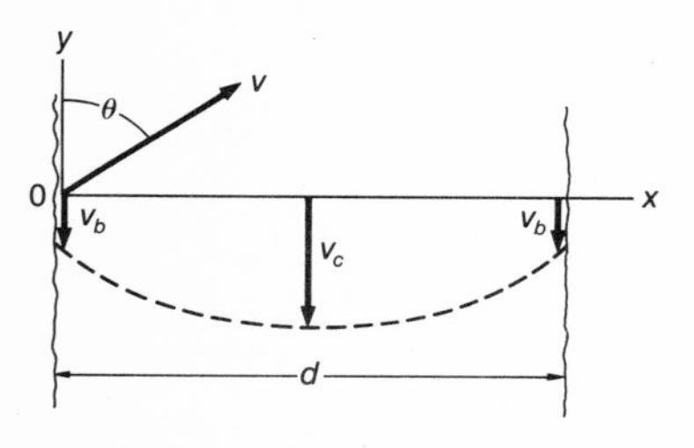

FIGURE 47

START OF SOLUTION

In looking ahead to how the solution might develop, it could occur to you that you would be more comfortable if the center of the coordinate system were placed at the boat's point of departure, as shown in Figure 47. Transferring the river velocity parabola to the same origin, $v_{R,x} = -v_c + B(x - d/2)^2$. Since for $x = 0$ and $x = d$, v_R must $= -v_b$, B must $= 4(v_c - v_b)/d^2$. Then, for this river,

$$v_{R,x} = -v_c + \frac{4(v_c - v_b)}{d^2}\left(x - \frac{d}{2}\right)^2$$

For the boat,

$$\dot{x} = v \sin\theta$$

and

$$\begin{aligned}\dot{y}_x &= v\cos\theta + v_{R,x}\\ &= v\cos\theta - v_c + \frac{4(v_c - v_b)}{d^2}\left(x - \frac{d}{2}\right)^2.\end{aligned}$$

Integrating,

$$\int_0^{y_x} dy = \int_0^t \left[v\cos\theta - v_c + \frac{4(v_c - v_b)}{d^2}\left(x - \frac{d}{2}\right)^2\right] dt$$

From $\dot{x} = dx/dt = v \sin\theta$ we get $dt = dx/(v\sin\theta)$. Changing variables from t to x in the above equation (the simplicity gained from our choice of coordinate origin shows up here in the integration limits) we have

$$\int_0^{y_x} dy = \int_0^x \left[\cot\theta - \frac{v_c}{v\sin\theta} + \frac{4(v_c - v_b)}{d^2 v \sin\theta}(x - \frac{d}{2})^2\right] dx.$$

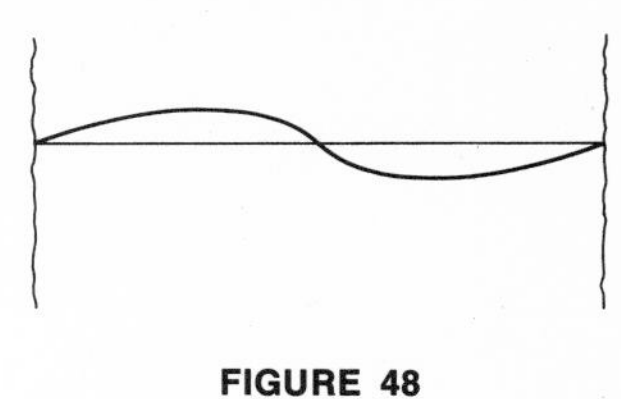

FIGURE 48

When this is integrated, you will have y_x as a function of x. To meet the requirement of the problem, y_x should $= 0$ when $x = d$; making this substitution should lead to the evaluation of the necessary θ. For further interest, you might put this value of θ back into the equation for y_x, and evaluate y_x for various values of x. From this you could plot the course of the boat across the river; you should get something like the drawing in Figure 48.

PROBLEM 125

An electron moves in the x-y plane, in the presence of an electric field that provides the electron with a constant acceleration vector in the x-y plane. At $t = 0.000$, the electron is observed to be moving parallel to the x-axis with a speed of 1.00×10^4 m sec^{-1}. At $t = 0.015$ sec the electron is observed to be moving in a direction at 30° with the x-axis, with a speed of 1.00×10^4 m sec^{-1}.

a. What is the value of the acceleration vector, in magnitude and direction?

b. At $t = 0.030$ sec what will be the velocity of the electron?

Suggestion: As a starting point for this problem, review page 68.

PROBLEM 126

Two competing speedboats have scored exactly equal times in straight races on a calm lake. A wind makes racing on the lake unsafe, and it is decided to determine the winner by speed runs made between two bridges D miles apart, which span both a river and a canal parallel to the river. The river flows at m mi hr^{-1}, the canal at $2m$ mi hr^{-1}. The time for each boat is to be the average of two runs, one with the current and one against. The drivers are given choice of river or canal. Driver A chooses the canal, driver B the river. Determine which driver knew his physics better by calculating the ratio of the average times for the two drivers.

PROBLEM 127

A motor yacht is cruising on a steady course at 10 knots off a strange coast at night. The skipper takes bearing sights on a distant light (Figure 49). From the flashing pattern he identifies the light as a Coast Guard beacon at the tip of a certain rocky headland. By taking several bearings one right after another, he determines that when the bearing angle is 45° to starboard of his course, it is increasing at the rate of 4° min^{-1}. If he maintains course,

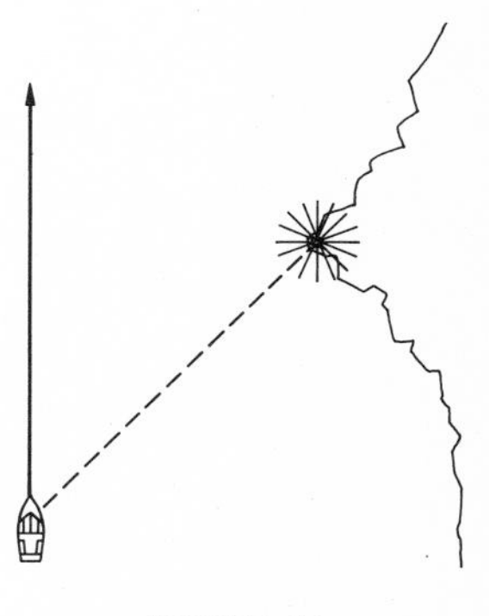

FIGURE 49

a. what will be the yacht's closest approach to the headland?

b. How soon after taking the 45° bearing will the light be broad abeam?

Hint: Make a space diagram showing the yacht, the light, the bearing line, and the course of the yacht. Set up the trigonometry for this diagram, and apply calculus to it.

PROBLEM 128

A Ferris wheel rotating at a uniform rate makes one rotation in 12 sec; its seats are located 16 ft from the axis of rotation.

a. When a passenger on the wheel is at the top of the arc, what is his velocity relative to the ground?
b. What is his acceleration?

PROBLEM 129

A "wheel of fortune," found at various merchandise booths at fairs and carnivals, consists of a perfectly balanced (presumably!) wheel rotating in a vertical plane about a fixed, frictionless horizontal axis. Around the rim of the wheel are usually 100 equally spaced pegs, numbered from 1 to 100; a flexible "flapper" mechanism can be set to strike the pegs as the wheel rotates (Figure 50). Such a wheel, of 1.0 m diameter, is set rotating at 50 rpm. At the start, peg No. 1 is directly above the axis, i.e., it is at "12 o'clock." What is the velocity of peg No. 1 as a function of time in seconds

FIGURE 50

a. if the wheel is allowed to rotate freely;
b. if the "flapper" mechanism is engaged and uniformly slows the wheel to a stop in 15 sec?

PROBLEM 130

A fly is riding on the rim of a 10-inch phonograph record that is rotating at $33\frac{1}{3}$ rpm. On the ceiling of a tall room, directly above the record player, another fly is at rest, watching the rotating fly. What is the velocity of the record fly, as seen by the ceiling fly, as a function of time in seconds? Start time at $t = 0$ when the record fly is moving due north.

PROBLEM 131

The locomotive driving wheel of a train going at a speed V rolls without slipping on a horizontal track. Relative to the ground,

a. what is the instantaneous velocity, and
b. what is the instantaneous acceleration vector for the following points on the rim of the wheel? i. The point instantaneously in contact with the track. ii. The point at the top of the wheel, directly above the center of the axle. iii. The point at the front of the wheel on a level with the center of the axle.
c. At any instant, what points on the circumference of the wheel have a forward velocity?

Hint: The motion of any point on the wheel will be the vector sum of the motion of the axle plus the rotational motion of that point about the axle.

PROBLEM 132

A wheel of radius R is rolling in a straight line without slipping on a plane surface; the plane of the wheel is vertical. For the instant when the axis of the wheel is moving with a speed v relative to the surface, calculate the instantaneous velocity of any point on the rim of the wheel, relative to the surface.

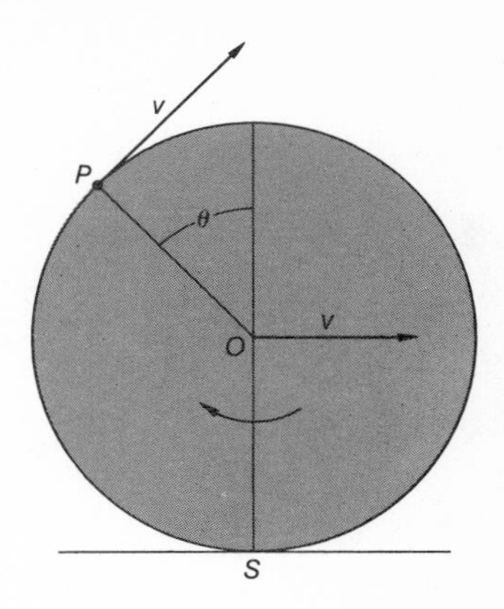

FIGURE 51

SOLUTION

This is an important problem, in that its solution will provide information that can be used later.

Point P is a randomly selected point on the rim (Figure 51). Relative to the axle it has a speed v in the direction of the tangent at P (or $\perp$ to the radius vector OP), thus:

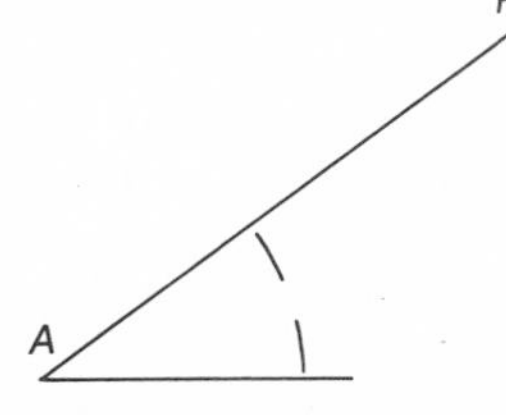

The axle is moving relative to the surface thus:

Putting these together and solving for the scalar value of v_{P-S} by the cosine formula,

$$\vec{v}_{P-S} = v\sqrt{2(1+\cos\theta)},$$

forward and at an angle $\frac{1}{2}\theta$ above the surface—as can be seen from the diagram in Figure 52.

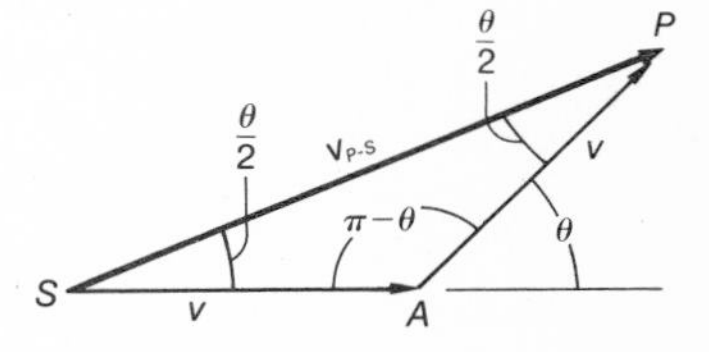

FIGURE 52

(For points on the right-hand side of the wheel, θ, measured as shown in Figure 52, will be minus, and for these points the vector velocity is downward.)

Since P was a randomly selected point, the evaluation of $\vec{v}_{P-S}$ applies to *all* points on the rim, θ being the only variable.

The calculated value of $\vec{v}_{P-S}$ was all that was asked for. However, this result is a minor part of what can be gained from this problem. Let us repeat the diagram of the wheel with the vector $\vec{v}_{P-S}$ imposed on it (Figure 53). Drawing the line SP, we have the triangle SOP, whose interior angles at S and $P = \frac{1}{2}\theta$. It should not now be too difficult to see that the angle between the vector $\vec{v}_{P-S}$ and the line SP is 90°; the velocity vector $\vec{v}_{P-S}$ is *perpendicular to the radius vector from the point in instantaneous contact with the surface.*

This means that the total instantaneous motion of point P is one of pure rotation about an axis through point S, which is instantaneously at rest relative to the surface.

Since P was a randomly chosen point on the rim, this conclusion applies to all points on the rim. With a little more complication in the geometry, the same conclusion can be reached for *any* point on the wheel. Thus we reach the important and useful statement: when a rigid, round body rolls without slipping on a surface, the entire mo-

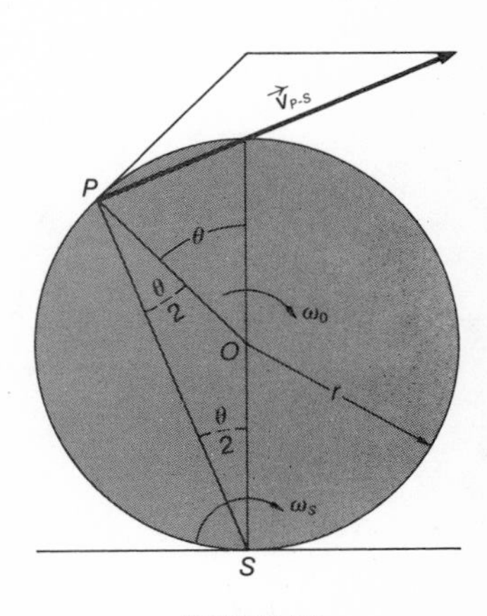

FIGURE 53

tion of the body can be treated as pure rotation about a transverse axis through the instantaneously at-rest point of contact with the surface.

This is not all we can get out of this problem. Since the tangential speed of a point on the rim is v, the rotational speed of the wheel about the axle is $\omega_0 = v/r$. Looking at Figure 53 we can see that the rotational speed of point P about the transverse axis through S is

$$\omega_s = \frac{v\sqrt{2(1+\cos\theta)}}{\text{length } SP}.$$

But by the cosine formula, length SP is

$$r\sqrt{2(1+\cos\theta)}.$$

Hence,

$$\omega_s = \frac{v\sqrt{2(1+\cos\theta)}}{r\sqrt{2(1+\cos\theta)}} = \frac{v}{r} = \omega_0.$$

If v is not constant (so the axle is moving with some linear acceleration) the wheel has an angular acceleration about its axis of α_0. By the same method as above, it can be determined that the wheel as a whole has an angular acceleration α_s about the transverse axis through S, and further, that $\alpha_s = \alpha_0$.

These results should be kept in mind; they will be found to have many useful applications in rotational dynamics.

PROBLEM 133

A carousel 60 ft in diameter is rotating at a speed of 6.0 rpm. A boy riding at the outer rim of the carousel tosses his fare to the attendant who is at the center of the rotating platform. The speed of the toss has a horizontal component of 30 ft sec^{-1}.

a. In what direction relative to his radius vector did the boy throw the money?

b. What time elapsed between the throw and the catch by the attendant?

c. If the attendant tossed a piece of money in change back at the same horizontal speed as the boy's throw, in what direction, relative to the radius vector to the boy, did the attendant throw?

PROBLEM 134

For the flies of problem 130, what is the velocity of the ceiling fly as observed by the fly on the record? (Is there any relation between this problem and the motion of the stars as observed from a point on the earth?)

PROBLEM 135
A fenderless bicycle is being ridden along a horizontal muddy road at a constant velocity v_0. As the bicycle proceeds, soupy mud is thrown from the wheels. (Ignore what is happening to the rider.) If the wheels have a radius R, what is the maximum height above the road that is reached by the flying mud?

PROBLEM 136
A differential pulley is as shown in Figure 54; a continuous flexible cable or link chain connects the double-sheave upper pulley with a load pulley. If one pulls the chain at P down with a speed v_P, how fast is the load rising?

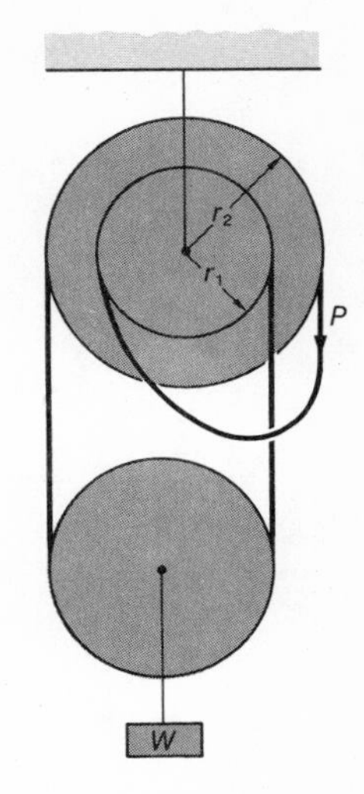

FIGURE 54

What we have been doing in the last group of problems, which were solved by the vector arrow method, is deducing from observations made directly in one coordinate system how the same phenomenon would have been observed in another coordinate system. It is now time to put this kind of transformation into a more analytical form.

Let us consider two coordinate systems: the x-y-z system (usually the system in which we or our laboratory are at rest), and the x'-y'-z' system in which the primed observer is at rest. Most of the situations we will be interested in come under one of two categories: in one, the coordinate axes of the two systems are parallel but the primed system is moving linearly relative to the unprimed system which is at rest; in the other, the z and z' axes are colinear but the primed system is rotating about its z' axis relative to the unprimed system at rest.

Relative Linear Motion

The primed system, axes parallel to those of the unprimed system, is moving in the $+x$ (and $+x'$) direction with a speed u relative to the unprimed system (Figure 55). (To the primed observer, the unprimed system is moving relative to his system with a speed $-u$.) The y and y' axes coincide at $t = 0$.

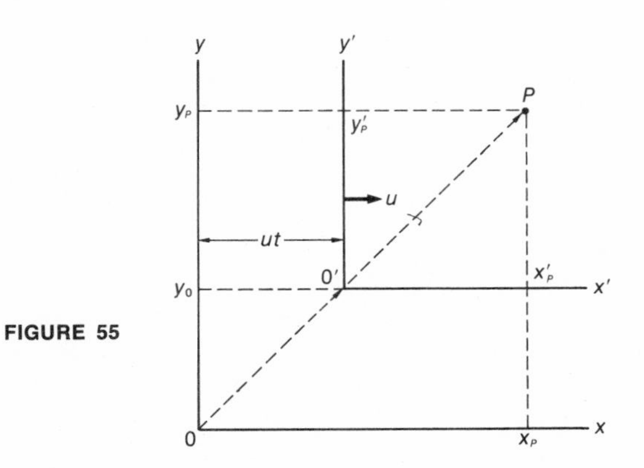

FIGURE 55

For this situation the transformation equations relating the primed and unprimed observations of the same phenomenon at P at the same instant of time are as follows.

$$x_P' = x_P - ut, \qquad y_P' = y_P - y_0, \qquad z_P' = z_P; \tag{2-38}$$

$$v_{P,x'}' = v_{P,x} - u, \qquad v_{P,y'}' = v_{P,y} \qquad v_{P,z'}' = v_{P,z}; \tag{2-39}$$

$$a_{P,x'}' = a_{P,x} \qquad a_{P,y'}' = a_{P,y} \qquad a_{P,z'}' = a_{P,z}. \tag{2-40}$$

These can be summarized by the following vector equations:

$$\vec{r}' = \vec{r} - \vec{r}_0; \qquad \vec{v}_P' = \vec{v}_P - \vec{u}; \qquad \vec{a}_P' = \vec{a}_P. \tag{2-41}$$

Relative Rotational Motion

The x'-y' axes rotate counterclockwise about the z' axis with an angular speed ω relative to the x-y axes (Figure 56). For generality the z' axis should be considered displaced from the z axis; however to prevent the transformation equations from becoming too cumbersome, the z' axis will be treated as colinear with the z axis. The x'-y' axes coincide with the x-y axes at $t = 0$.

For this situation simultaneous observations are related by the following equations. (Subscripts to indicate that the observations are made at P have been eliminated for clarity.)

$$x' = x \cos \omega t + y \sin \omega t,$$

$$y' = y \cos \omega t - x \sin \omega t,$$

$$z' = z;$$

$$v_{x'}' = (v_x + \omega y) \cos \omega t + (v_y - \omega x) \sin \omega t,$$

$$v_{y'}' = (v_y - \omega x) \cos \omega t - (v_x + \omega y) \sin \omega t,$$

$$v_{z'}' = v_z;$$

$$a_{x'}' = (a_x + 2\omega v_y - \omega^2 x) \cos \omega t + (a_y - 2\omega v_x - \omega^2 y) \sin \omega t,$$

$$a_{y'}' = (a_y - 2\omega v_x - \omega^2 y) \cos \omega t - (a_x + 2\omega v_y - \omega^2 x) \sin \omega t,$$

$$a_{z'}' = a_z.$$

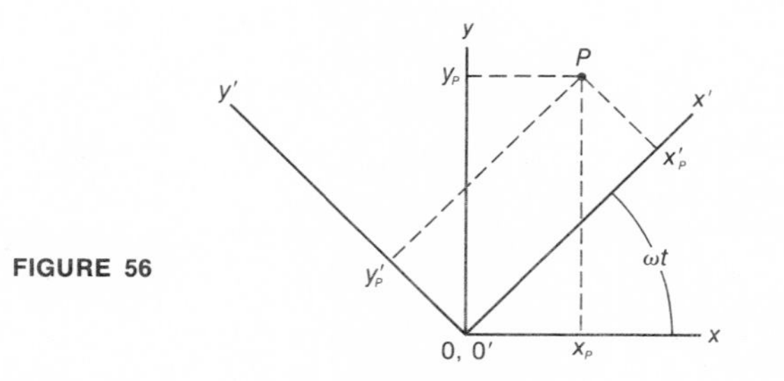

FIGURE 56

These can be summarized by the following vector equations.

$$\vec{r}' = \vec{r} + \vec{z}_0, \ (\vec{z}_0 \text{ is a constant}) \tag{2-42}$$

$$\vec{v}' = \vec{v} - \vec{\omega} \times \vec{r}, \tag{2-43}$$

$$\vec{a}' = \vec{a} - 2\,\vec{\omega} \times \vec{v} - \omega^2\,\vec{r}. \tag{2-44}$$

By this time we have accepted the vector nature of linear displacement, linear velocity, and linear acceleration, but why in the above equations is ω treated as a vector? The answer is that some way must be found to indicate that direction also enters into rotation; you recognize this connection instinctively every time you tip the bicycle you are riding from a vertical plane in order to turn a corner. From this recognition it is only a small step further to recognize that the only unique direction for all parts of a rotating body is the direction of the axis of rotation. We make this recognition official by saying that $\vec{\omega}$ is a vector lying in the direction of the axis of rotation (either colinear with it or parallel to it, as suits your convenience); that it points in the direction of the advance of a right-handed screw, turning in the same sense as the actual rotation; and that it is of a length proportional to the scalar value of ω. Thus the $\vec{\omega}$ for the rotation of the earth is a vector along the earth's axis (or parallel to the axis), pointing toward the North Pole, and of a length representing $2\pi/(24 \times 3600) = 7.27 \times 10^{-5}$ radians $\sec^{-1}$. (In this calculation the time unit is the sidereal second.)

Although the vector nature of rotation was introduced by way of angular velocity, it applies to the other aspects of rotation as well. Angular displacement $\vec{\Delta\theta}$ and angular acceleration $\vec{\alpha}$ also are vectors. So, just as we did before for linear motion, for rotational motion with constant acceleration we can rewrite our equations thus:

$$\begin{aligned} \vec{\omega}_t &= \vec{\omega}_0 + \vec{\alpha}t; \\ \vec{\Delta\theta} &= \vec{\omega}_0 t + \tfrac{1}{2}\vec{\alpha}t^2. \end{aligned} \tag{2-45}$$

The jump in this discussion from the Cartesian equations to the vector equations is not obvious. To comprehend the vector equations, and particularly the meaning of the vector cross-products $\vec{\omega} \times \vec{r}$ and $\vec{\omega} \times \vec{v}$, and to see clearly the vectors themselves, the reader is strongly urged at this point to study Appendix F and Appendix G before going further in the problems. The section in Appendix F on the vector cross-product should also prove useful for later developments.

To acquire some feeling of reality for the vector equations, let us check them with two examples whose answers we already know. For the first example, consider an object at P at rest in the unprimed system; then both $\vec{v}$ and $\vec{a} = 0$. Hence the primed observer must see $\vec{v}' = -\,\vec{\omega} \times \vec{r}$ and $\vec{a}' = -\omega^2\,\vec{r}$. Physically, this is what to expect: if the primed system is rotating counterclockwise with a speed ω relative to the unprimed system, an object at rest in the unprimed system is, to the primed observer, rotating clockwise in a circle of radius r, and has an angular speed $-\omega$, a linear tangential speed $-\omega r$, and a centripetal acceleration $\omega^2 r$. Now consider the converse of that example: this time an object is observed by the unprimed observer to be moving counterclockwise in a circle of radius r with an angular speed ω. Therefore, in the unprimed system the object has an instantaneous speed $v = \omega r$ that is perpendicular to $\vec{r}$ and in the counterclockwise direction—hence, $\vec{v} = \vec{\omega} \times \vec{r}$—and a centripetal acceleration $\omega^2 r$, or $\vec{a} = -\omega^2 \vec{r}$. If we substitute these in the appropriate equations given earlier in this paragraph,

$$\vec{v}' = \vec{\omega} \times \vec{r} - \vec{\omega} \times \vec{r}' = 0,$$

and

$$\vec{a}' = -\omega^2 \vec{r} - 2\,\vec{\omega} \times (\vec{\omega} \times \vec{r}) - \omega^2 \vec{r} = -2\omega^2 \vec{r} + 2\,\omega^2 \vec{r} = 0.$$

This is what a primed observer rotating counterclockwise should see—in his system the object is at rest.

In equations 2-42, 2-43, and 2-44, what the primed observer sees is expressed in terms of what the unprimed observer sees. But normally the primed observer doesn't know what the unprimed observer sees; he only knows what *he* sees. An astronomer on earth is a primed observer in a rotating coordinate system, and his observations are primed observations. From these, can he deduce what an observer at rest relative to the distant stars would observe?

The answer is, he can.* By algebraic manipulation of equations 2-42, 2-43, and 2-44, we can arrive at

$$v_x' = v_x \cos \omega t + v_y \sin \omega t + \omega y',$$

$$v_y' = v_y \cos \omega t - v_x \sin \omega t - \omega x';$$

$$a_x' = a_x \cos \omega t + a_y \sin \omega t + 2\omega v_y' + \omega^2 x',$$

$$a_y' = a_y \cos \omega t - a_x \sin \omega t - 2\omega v_x' + \omega^2 y'.$$

These can be summarized by

$$\vec{v}' = \vec{v} - \vec{\omega} \times \vec{r}'; \tag{2-46}$$

$$\vec{a}' = \vec{a} - 2\vec{\omega} \times \vec{v}' + \omega^2 \vec{r}'. \tag{2-47}$$

Rearranging, we get what the at-rest observer must see in terms of what the rotating observer did see:

$$\vec{v} = \vec{v}' + \vec{\omega} \times \vec{r}'; \tag{2-48}$$

$$\vec{a} = \vec{a}' + 2\vec{\omega} \times \vec{v}' - \omega^2 \vec{r}'. \tag{2-49}$$

In either $\vec{v}' = \vec{v} - \vec{\omega} \times \vec{r}$ or $\vec{v} = \vec{v}' + \vec{\omega} \times \vec{r}'$, it is not difficult to visualize that what happens in one system must appear in the other system to have a tangential velocity

*For three-figure accuracy he must include in $\vec{\omega}$ the revolution of the earth about an axis through the center of mass of the earth-moon system; for four-figure accuracy he must also include the revolution of the center of mass of the earth-moon system about the sun.

—the $\vec{\omega} \times \vec{r}$ term—imposed on it. Similarly, in $\vec{a}' = \vec{a} - 2\,\vec{\omega} \times \vec{v} - \omega^2\vec{r}$ or in $\vec{a} = \vec{a}' + 2\vec{\omega} \times \vec{v}' - \omega^2\vec{r}'$, the radial acceleration—the $\omega^2\vec{r}$ term—appears reasonable in view of the relative rotation. What is not expected is the $2\vec{\omega} \times \vec{v}$ term.

This term is called the Coriolis acceleration. It exists because any v not parallel to the axis of rotation seems in the primed system (or any v', in the unprimed system) to carry the moving object across radial lines, so that the object appears to follow a spiral path. This "spiralling" effect from the Coriolis acceleration should be clearly visible in the solution to problems 142 and 143.

So far, in an attempt at maximum clarity, we have restricted ourselves to a two-dimensional situation; only projections on the x-y and x'-y' planes could be compared, so it was assumed there was no motion in the z or z' directions. To include z motion and pass to the general, three-dimensional case, we need to make only one small change in the acceleration equations. For a complete description of three-dimensional relative rotational motion, we then have:

$$\vec{v}' = \vec{v} - \vec{\omega} \times \vec{r}, \qquad (2\text{-}50)$$

$$\vec{v} = \vec{v}' + \vec{\omega} \times \vec{r}'; \qquad (2\text{-}51)$$

$$\vec{a}' = \vec{a} - 2\vec{\omega} \times \vec{v} + \vec{\omega} \times (\vec{\omega} \times \vec{r}), \qquad (2\text{-}52)$$

$$\vec{a} = \vec{a}' + 2\vec{\omega} \times \vec{v}' + \vec{\omega} \times (\vec{\omega} \times \vec{r}'). \qquad (2\text{-}53)$$

It is hoped that this discussion of rotating coordinates makes clear the difference between appearance and reality. Real motions are the result of real forces, and are observable in inertial coordinate systems. Often these real motions are observed from accelerated, noninertial coordinate systems; the differences between the "accelerated" observations and the real motions are not the result of the application of forces and are not explained by Newton's Laws of Motion. A body moving at constant speed in a straight line in an inertial system—and therefore unaccelerated in that system—will be observed as accelerated by an observer in a rotating system. It should be clearly understood that the body is not accelerated in the physical sense that acceleration is related to the application of force; it appears accelerated to the rotating observer only because he is moving at a changing rate relative to the body. This apparent acceleration seen by the accelerated observer is the basis for the postulation of "pseudo" forces discussed in some textbooks; this postulation is unfortunate, because it adds unnecessary complications and tends to obscure the fact that in reality there are no such forces and that whatever is observed in addition to the real motion is solely the effect of the observer's relative acceleration.

PROBLEM 137

A bar lies parallel to the x'-axis, and in the primed system its length is measured as L'. What will be its length L as measured by an observer in the unprimed system, if the x-axis is parallel to the x'-axis, and

a. the origin of the primed coordinate system is displaced a distance

x_0 from the origin of the unprimed system, but the systems do not move relative to each other;
b. the primed system moves with a speed u in the x direction relative to the unprimed system;
c. the primed system moves with an acceleration a in the x direction relative to the unprimed system?

PROBLEM 138

The primed system is moving in the x direction with a speed u relative to the unprimed system; the x'-y' and x-y planes are parallel. A bar of length L', as measured in the primed system, lies in the $x'y'$ plane oriented at 45° with the x'-axis and moves in the direction of its length with a speed v'. What are the bar's length and velocity, as observed in the unprimed system?

PROBLEM 139

By considering the cyclist as an observer moving with the bicycle, solve problem 100 by the use of equation 2-41. Which method of solution seems to you more direct and comprehensible—the one you used here, or the one you used in problem 100?

PROBLEM 140

On a training flight out of Luke AFB a cadet is practicing estimating plane ground speed and course by visual tracking of ground objects. After tracking a gray house with red trim he reports the plane is moving 420 mi hr^{-1} on a course 30° E of N. Unfortunately, what he thought was a gray house was a caboose of the Santa Fe's Super C freight train, which at that time was moving eastward along a straight stretch of desert track at 60 mi hr^{-1}. What was the plane's correct speed and course?

PROBLEM 141

In a system at rest, an object moves with a velocity of 40 m sec^{-1} in a direction 30° E of N. If you are asked to write instructions for the movement of a coordinate system in which the same motion will be measured at a scalar value of 20 m sec^{-1}, what choice of directions for the movement are open to you?

PROBLEM 142

Sketch the path of the return throw, part (c) of problem 133,

a. as seen by a man standing on the ground outside the carousel;
b. as seen by a man riding on the carousel.

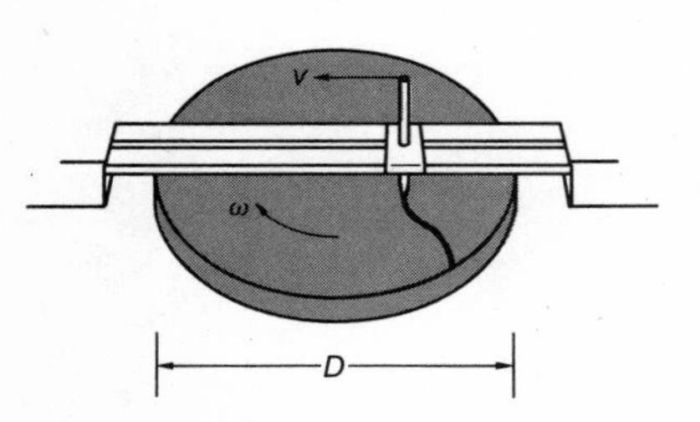

FIGURE 57

PROBLEM 143

A standard method for demonstrating Coriolis plus radial acceleration is to place a disc of polar coordinate paper on a horizontal turntable (Figure 57). Above the turntable is centered a fixed channel which can carry a marking pen across the disc at a constant linear velocity. (Thus $\vec{a} = 0$.)

a. If v is such that the pen moves the diameter of the disc in the time

the disc turns through 90° (which means $v = 2\omega D/\pi$), sketch the trace the pen makes.
b. Do the same for $v = \omega D/\pi$.
c. Do the same for $v = \omega D/2\pi$.
d. Is v' constant? If not, why not?
e. Should your curves show constant curvature?
f. If they don't, where is the greatest radius of curvature—at the edge of the disc or at the center?
Note: If you do this problem well, you will understand why a rotating observer doesn't "see it like it is."

PROBLEM 144
Occasionally on the earth's surface a local area of low atmospheric pressure is completely surrounded by higher atmospheric pressures, i.e., the neighboring isobars are closed curves. Since air flows "downhill" from high to low atmospheric pressure, air outside the low area acquires a velocity toward the low area. Relative to the earth, which is rotating underneath the air, the air then acquires a Coriolis acceleration perpendicular to this velocity. This creates a "cyclone" effect, in which the air relative to the earth moves *around* the low area instead of directly toward it. How does such a cyclone rotate (clockwise or counterclockwise) in the Northern hemisphere, as viewed

a. from an airplane above it;
b. as viewed by a spectator on the ground below it?

PROBLEM 145
Normally, a belt of high atmospheric pressure stretches across the United States at a latitude of about 30° N; both north and south of this belt the atmospheric pressures are lower. From this information, and with a knowledge of Coriolis acceleration, what can you deduce about the direction, relative to the earth's surface, of the prevailing winds

a. north of 30° N latitude, and
b. south of 30° N latitude?

PROBLEM 146
At a point on the surface of the earth at latitude θ, what is the angle between a plumb-bob line and the radial direction to the center of the earth?

SOLUTION
Even on an earth that was not rotating, a body at the surface of the earth—the plumb bob in this case—would experience a gravitational acceleration directed toward the center of the earth. Let us call this acceleration $\vec{g}_p$; it corresponds to $\vec{a}$, and its scalar value is very nearly constant everywhere on the earth's surface. Since the observation on the plumb bob is made in the primed system, the rotating earth, we would find equation 2-53 useful. By placing our center of coordinates at the center of the earth, the $\vec{r}'$ in that equation becomes $\vec{R}$, and $\vec{a}'$ becomes g_θ, the gravitational acceleration actually ob-

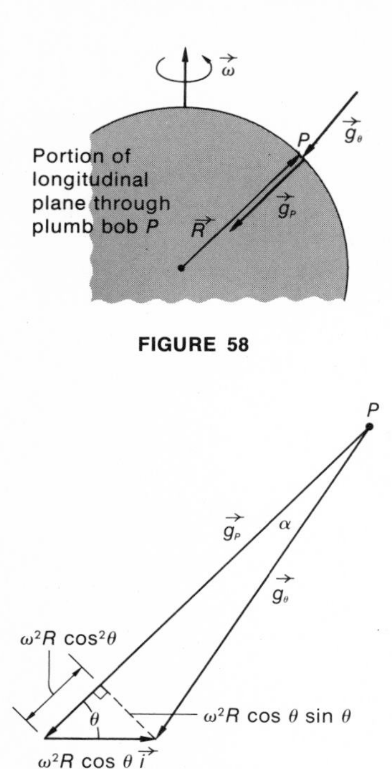

FIGURE 58

FIGURE 59

served at latitude θ. These vectors lie in the earth's longitudinal plane through the plumb bob P, and are shown in Figure 58.

Since the plumb bob is not moving relative to the earth's surface, $v' = 0$. Then equation 2-53, rearranged, becomes $\vec{g_\theta} = \vec{g_p} - \vec{\omega} \times (\vec{\omega} \times \vec{R}) = \vec{g_p} + \omega^2 R \cos\theta\, \vec{i}$, where $\vec{i}$ is a unit vector perpendicular to the earth's axis and pointing radially outward from that axis through the plumb bob. This vector equation is illustrated by the closed vector diagram in Figure 59.

In Figure 59, α is the angle between $\vec{g_\theta}$ and $\vec{g_p}$ that we were asked to find. From the diagram it can be determined that

$$\alpha = \tan^{-1} \frac{\omega^2 R \cos\theta \sin\theta}{g_p - \omega^2 R \cos^2\theta}.$$

The maximum value of α, which occurs at $\theta = 44°54'$, is $0°06'$.

Since α is so small, g_θ and g_p are essentially parallel, and it is reasonably accurate to say that

$$g_\theta = g_p - \omega^2 R \cos^2 \theta.$$

The direction of $\vec{g_\theta}$ is the direction of the local vertical: the horizon and other horizontal surfaces (the bed of a properly installed billiard table or the surface of a pool of mercury) are perpendicular to $\vec{g_\theta}$; structures erected by plumb line and leveled surveying instruments rise along the $\vec{g_\theta}$ line; the local North and South lie in the longitudinal plane and are perpendicular to $\vec{g_\theta}$.

PROBLEM 147

If $g_p = 983.21$ cm sec^{-2}, what would g_θ be at the equator, and at 45° latitude?

PROBLEM 148

Why is the earth an oblate spheroid?

PROBLEM 149

At what points on the earth's surface would a horizontal plane be perpendicular to the radius vector from the center of the earth?

PROBLEM 150

If the helicopter of problem 38 was completely at rest relative to the earth when it dropped the test object and there was no crosswind, and you can assume no air resistance,

a. how far from vertically beneath the release point did the object land if the test took place at the equator?

b. How far, if the test was conducted at a site near the center of New Zealand?

START OF SOLUTION

An object released from rest at a height h above the surface of the earth, $h << R$, will appear to the earthbound observer to experience three accelerations: the centrifugal acceleration $\omega^2 R \cos\theta$, the true gravitational acceleration g_p in a longitudinal plane, and the Coriolis acceleration $-2\omega \times v'$. (Since the body is falling, v' is not zero.) It should be evident from Figure 59 that the North-South components of the centrifugal acceleration and of g_p just cancel each other, so that in falling, the object does not deviate north or south of the local vertical. However, because of the Coriolis acceleration the object does not stay in its original longitudinal plane. v', the velocity acquired in falling, is directed toward the center of the earth (ignoring α as negligible); then $-2\omega \times v'$ is perpendicular to the longitudinal plane and is directed eastward in both the Northern and Southern hemispheres. Its scalar value is $2\omega v' \cos\theta$; since v' is a function of t, the Coriolis acceleration also varies with t, and the total eastward departure from the vertical for a falling body must be found by integration.

PROBLEM 151

The situation at the New Zealand site in problem 150 is reversed; the test object is projected vertically upward from the ground at such initial speed that it just reached a height above ground of 1.0 km. (The helicopter has been moved out of the way.)

a. How far was the zenith of the trajectory from directly above the projection point?
b. What was the velocity of this zenith point relative to the ground?

PROBLEM 152

How far from the projection point did the test object of problem 151 finally land? *Hint:* Put together questions (a) and (b) of problem 151 and question (b) of problem 150.

RELATIVE MOTION AT RELATIVISTIC SPEEDS

The title of this section sounds redundant. However, the term "relativistic" is widely accepted to refer to speeds that are a significant fraction of the speed of light. For calculations having no more than four-figure accuracy, this means speeds in excess of about 1.5 percent of the speed of light, or speeds $> 4.5 \times 10^6$ m sec^{-1}.

Speeds this great are not within the realm of our direct experience. In fact, engineering, chemistry, and most of physics can be carried on quite accurately using only the relations developed in the previous section for Newtonian, or Galilean, Relativity. These relations are valid, within any reasonable requirement for accuracy, for all operations at speeds normally encountered. However, in high-energy physics where speeds are greater than 4.5×10^6 m sec^{-1}, Newtonian Relativity is inadequate. For these high speeds, the methods of Einstein's Special Relativity are necessary. This theory is valuable on two counts: it provides a set of working equations which enables a physicist to deal accurately with high-speed phenomena; more important even than that, it contains an adjusting factor for speed which reduces it, in its application to motion, to Newtonian Relativity for low speeds. The Special Theory of Relativity is thus a general and comprehensive theory which includes as a lower limit for motion all that we have developed so far for mundane speeds.

It is customary to start the discussion of the Special Theory of Relativity with an extension of a situation previously discussed under Newtonian Relativity: a primed system, axes parallel to those of an at-rest unprimed system, moves relative to the unprimed system with a speed u in the x direction. Under Special Relativity the transformation equation corresponding to equation 2-38 now becomes

$$x' = \frac{x - ut}{\sqrt{1 - (u^2/c^2)}} \quad (c \text{ is the speed of light in free space}). \tag{2-54}$$

If the coordinate systems move relatively only in the x direction, $y' = y$ and $z' = z$ as before.

Under Newtonian Relativity we could shift our viewpoint back and forth between primed and unprimed coordinate systems rather easily and casually; if $x' = x - ut$ as in equation 135-1, then $x = x' + ut$, etc. Under Special Relativity we are required to be more careful about where our measurements are taken. In equation 2-54, both the displacement coordinate x and the time coordinate t must be measured in the unprimed system. If our measurements are made in the primed system, then we have

$$x = \frac{x' + ut'}{\sqrt{1 - (u^2/c^2)}}. \tag{2-55}$$

The ratio u/c and the term

$$\frac{1}{\sqrt{1 - (u^2/c^2)}}$$

occur so often in the formulae of Special Relativity that they are designated by special symbols, thus:

$$\frac{u}{c} = \beta \text{ and } \frac{1}{\sqrt{1 - (u^2/c^2)}} = \frac{1}{\sqrt{1 - \beta^2}} = \gamma.$$

With these symbols, equations 2-54 and 2-55 become

$$x' = \gamma\,(x - ut) \qquad \text{and } x = \gamma\,(x' + ut'). \tag{2-56}$$

One of the central features of Special Relativity is that in these equations, $t \neq t'$; in coordinate systems that move relative to each other, time does not seem to run the same, clocks do not seem to "tick" at the same rate. This is expressed analytically in the relation

$$\Delta t' = \sqrt{1 - \beta^2}\,\Delta t. \tag{2-57}$$

Here $\Delta t'$ is the time interval, as timed by a clock in the primed system, of an episode that occurred *at rest in the primed system*;* Δt is the time interval, as timed

*The interval $\Delta t'$ is called the "proper" time interval; "proper" time is measured by a clock at rest relative to what it is timing.

by a clock in the unprimed system, of the same episode. (Since the episode was at rest in the primed system the unprimed observation Δt was made on a moving episode.) Equation 2-57 is the famous time dilation, or "slow clock" formula; it states that an observer at rest relative to an episode will record a shorter time interval for the episode than will an observer who had to time it while it was moving relative to him.

Since $\Delta t' \neq \Delta t$, you would not expect dx/dt to equal dx/dt' and you would be correct. Thus the differentiation with respect to time of equations 2-54 and 2-55 to establish relationships between v_x and v'_x is more complicated than it was earlier. Under Special Relativity these relations are

$$v'_x = \frac{v_x - u}{1 - \beta(v_x/c)} \quad \text{and} \quad v_x = \frac{v'_x + u}{1 + \beta(v'_x/c)},$$
$$v'_y = \frac{v_y\sqrt{1-\beta^2}}{1 - \beta(v_x/c)} \quad \text{and} \quad v_y = \frac{v'_y\sqrt{1-\beta^2}}{1 + \beta(v'_x/c)}, \tag{2-58}$$
$$v'_z = \frac{v_z\sqrt{1-\beta^2}}{1 - \beta(v_x/c)} \quad \text{and} \quad v_z = \frac{v'_z\sqrt{1-\beta^2}}{1 + \beta(v'_x/c)}.$$

Since $y' = y$ and $z' = z$, it is somewhat startling to discover that $v'_y \neq v_y$, etc. This is, of course, a consequence of the fact that $dt' \neq dt$.

Let us return to equation 2-54 and consider a slender bar at rest in the primed system, parallel to the x-axis, with its end points at x'_B and x'_A. The unprimed observers will observe that the bar's end-joints are, at some time t in their system, at x_B and x_A. Since the bar is at rest in the primed system, the rest length of the bar is $x'_B - x'_A$. But by the operation of equation 2-54,

$$x'_B - x'_A = \frac{x_B - x_A}{\sqrt{1-\beta^2}}.$$

Then

$$x_B - x_A = \sqrt{1-\beta^2}\,(x'_B - x'_A)$$
$$\ell_{\text{moving}} = \sqrt{1-\beta^2}\,\ell_0, \tag{2-59}$$

where ℓ_0 is the proper rest length of the bar, and ℓ_{moving} is the length measured by an observer while the bar moved relative to him.* Equation 2-59 is the famous Lorentz-Fitzgerald contraction formula; it states that an observer measuring the length of an object moving relative to him will record a shorter scale length than will a measurer who is at rest relative to the object.

Equations 2-54 and 2-58 are useful and necessary in dealing with the motions of high-speed atomic particles, and the adjusting factor $\gamma = 1/(1-\beta^2)^{1/2}$ will appear in several places later in the text. In the application of equations 2-54 and 2-58 the main difficulty will arise in trying to keep straight which is the moving coordinate system, what is the proper time or length, etc.

*The length ℓ_0 could be called the "proper" length, in consistency with what was defined earlier as "proper" time.

PROBLEM 153

In a major postulate underlying the Special Theory of Relativity, the speed of light in free space is asserted to be a universal constant; even though various observers may move at constant velocities relative to each other, all will measure c to have the same value. In view of this postulate, is the answer to problem 120—the problem on aberration—correct?

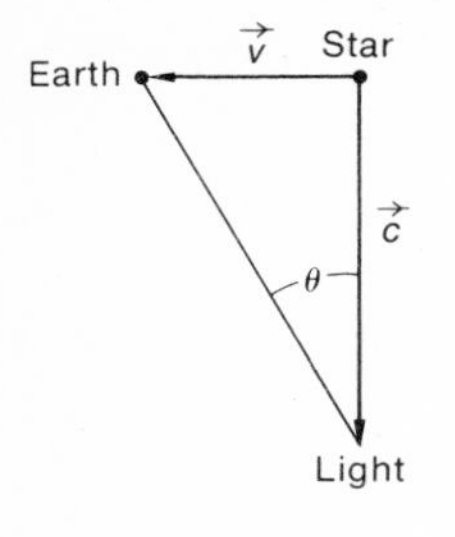

FIGURE 60

SOLUTION

In the original development of the aberration formula, the velocity of the star's light was assumed to be in the geometrical direction perpendicular to the observer's motion, as shown in Figure 60. In the light of the above postulate, let us look at the diagram in Figure 60 more critically.

By the postulate, the observer at the end of the telescope must observe light moving with a speed c. But the light he is observing is light that is moving down the axis of the telescope. Hence the correct vector diagram must be as shown in Figure 61.

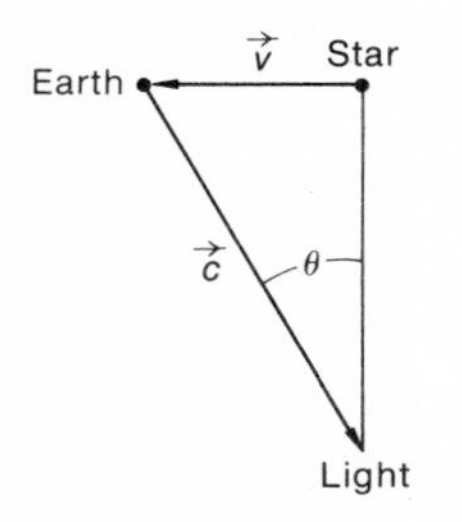

FIGURE 61

Then, $\theta = \sin^{-1}(v/c)$. This is the correct formula for the aberration angle θ. Since the angle is so small, there is no observable difference between $\sin\theta$ and $\tan\theta$.

A consequence of the Special Relativity equations is that c is the ultimate speed—there can be no speed greater than c. Had we known this earlier we could have seen then that our first solution to the aberration problem was incorrect; a look at the first vector diagram indicates a speed of $(v^2 + c^2)^{1/2}$ for the light coming down the axis of the telescope, which is forbidden.

PROBLEM 154

a. Repeat step (b) of problem 137 as a general case, in which u can have any value $< c$.
b. If the proper length L' in that problem is 1.0 m and $u = 0.5\,c$, how long does the meter stick appear to be in the unprimed system?

PROBLEM 155

A slender bar of proper length L' is at rest in the x'-y' plane in a primed system at an angle θ' with the x'-axis; the primed system moves with a speed u in the x direction relative to the unprimed system.

a. How does the bar seem oriented to the unprimed observers?
b. How long does it appear to be in the unprimed system?
(Do your answers check out reasonably at the limits of $u = 0$ and $u = c$?)

PROBLEM 156

If θ' in problem 155 $= 45°$, what must u be for the rod to appear, to unprimed observers, to lie at 60° with the x-axis?

PROBLEM 157

If u is an appreciable fraction of c, what is the velocity of the bar in problem 138 as measured by unprimed observers; $v' << c$?

PROBLEM 158

A visitor from outer space travels very fast, just above a major highway leading out of Chicago; various small shops face this highway. As the visitor passes a shop whose front is two-thirds as high as it is wide, he glances at it out of the corner of his right eye and notes that its facade seems square. How fast was he traveling relative to the ground?

PROBLEM 159

As the space visitor in problem 158 approaches the high-energy accelerator at Batavia, Illinois, something goes wrong at the accelerator; as a result a beam of protons, with a speed of 0.5 c relative to the ground, escapes the beam switchyard and moves straight at the space capsule toward a head-on collision. Relative to the space capsule, how fast are the protons moving? For the space capsule's speed, use the published answer to problem 158.

PROBLEM 160

To escape the oncoming proton beam of the above problem, the space visitor turns on his ion drive, which shoots ions out the back of his capsule with a speed of 0.1 c relative to the capsule. How fast do the ions appear initially to be moving to observers on the ground below?

PROBLEM 161

It has been calculated (*Physics Today, 21:10-41*) that a 100-kiloton spaceship could be accelerated to a speed of 10^4 km sec^{-1} by the explosion of thermonuclear bombs. When the spaceship has traveled for exactly one hour at this speed by its own clock, how long has it been at mission speed according to a terrestrial observer who is tracking it?

PROBLEM 162

In one laboratory, muons at rest are observed to have an average life of 2.2×10^{-6} sec before decaying. In another laboratory muons ejected from an accelerator target at high speeds are observed to have an average lifetime of 7.1×10^{-6} sec. What was the speed of the latter muons?

3

FORCES AND TORQUES

The motions we discussed in the previous section did not exist spontaneously—they were caused. The description of this causal operation is contained in Newton's Second Law of Motion, and is one of the great landmarks of science.

NEWTON'S SECOND LAW FOR LINEAR MOTION

Before we tackle Newton's Second Law, however, we need to introduce a new concept—the concept of inertial mass. Our experience helps us understand this concept: early in life our muscles learned they had to "do" more to hurl a stone at a certain speed than they had to "do" in hurling a tennis ball, of the same size as the stone, at the same speed. Or, if we used the same muscular effort on both a baseball and a solid lead ball, we soon discovered that the baseball left our hands with much greater speed, and went much farther, than did the lead ball. Or, to illustrate this more dramatically, we would rather catch a tennis ball than a baseball when the two are moving at about the same speed, and we would certainly prefer to catch a baseball rather than a lead ball even if the baseball is moving faster.

Note that these differences are related to changes of motion of different objects. Aside from appearance, there are no uncomfortable differences among a tennis ball, a baseball, and a lead ball when they are lying on the ground. It is only when we are throwing—changing motion positively—or catching—changing motion negatively—that we are physically aware of a difference.

To take care of these differences among objects we have the concept of inertial mass, a fundamental property of matter that was discussed in Chapter 1. To combine this with motion, Newton defined a new term which included the concept of inertial mass and the concept of motion. He called this new term the "quantity of motion;" this same thing we now call momentum. For linear motion momentum is symbolized by $\vec{p}$, and has as components the scalar m, for inertial mass, and the vector $\vec{v}$ representing motion; its dimensions are MLT^{-1}. Until the twentieth century it was assumed that $\vec{p}$ was completely described by the product $m\vec{v}$.

It is now accepted that the complete definition of $\vec{p}$ is given by $\vec{p} = m\vec{v}/[(1 - v^2/c^2)^{1/2}] = \gamma m\vec{v}$, and when v/c is not negligible, this relativistic form of p must be used. However, the science of dynamics evolved over a long period of time during which $\vec{p}$ was defined as $m\vec{v}$; for almost all useful purposes today it is still sufficiently accurate to describe it so. The enormously productive development of dynamics, from the establishment of units to comprehensive methods of analysis of physical systems, was carried on under the simpler, nonrelativistic definition of $\vec{p}$, and we will follow the same path.

The subjective differences in muscular effort which we discussed earlier can now be summarized analytically in Newton's Second Law—force is proportional to the time rate of change of linear momentum. In equation form,

$$\text{Force } \vec{F} = k\frac{d\vec{p}}{dt} = k\frac{d(\gamma m\vec{v})}{dt} \text{ always}$$

$$= k\frac{d(m\vec{v})}{dt} \text{ when } \gamma \simeq 1. \tag{3-1}$$

The symbol k stands for a dimensionless constant of proportionality.

Equation 3-1 is a double-headed equation. It can serve as a definition: if m is established independently, then the equation defines F and determines its dimensionality. If F is established independently, then m is defined and its dimensions determined.

Equation 3-1 can also serve as an operational equation; if F is known, the rate of change of linear momentum also becomes known. If the rate of change of linear momentum can be measured, then the net force F that is acting can be deduced. Those various operational applications have made Newton's Second Law the cornerstone of modern technology.

Force Defined

Absolute Units

If we set $k = 1$, which seems like a reasonable thing to do, then equation 3-1 defines what are called absolute units of force, inertial mass having already been established arbitrarily. In the MKS system, in which the arbitrary unit of inertial mass is 1 kilogram (kg), a unit rate of change of momentum would be (1 kg × 1 m sec^{-1}) sec^{-1}, so that

$$\text{unit } F = \frac{dp}{dt} = 1 \text{ kg m sec}^{-2}.$$

This unit of force is called a newton (N).

In the cgs system, in which the arbitrary unit of inertial mass is 1 gm,

$$\text{unit } F = \frac{dp}{dt} = (1 \text{ gm} \times 1 \text{ cm sec}^{-1}) \text{ sec}^{-1} = 1 \text{ gm cm sec}^{-2},$$

and this unit of force is called a dyne.

By a simple comparison of the right-hand side of the two equations, it can be determined that

$$1 \text{ newton} = 10^5 \text{ dynes}.$$

In the English system of units, where the arbitrary unit of inertial mass is 1 pound (lb),

$$\text{unit } F = \frac{dp}{dt} = (1 \text{ lb} \times 1 \text{ ft sec}^{-1}) \text{ sec}^{-1} = 1 \text{ lb ft sec}^{-2},$$

and this unit of force is called a poundal (lbal).

Absolute units as so defined are used almost exclusively in scientific computations, and their use in engineering calculations is increasing. However, much of engineering calculation is still carried on in so-called "practical" units.

Practical Units

Practical units of force developed, particularly in the English system, from a desire of engineers to have the *number* for the inertial mass of an object agree with the *number* for the weight of the object. (It is much easier to weigh an object than it is to determine its mass by comparison with a mass standard.) This equality of numbers can be accomplished by a suitable juggling of the constant k in equation 3-1.

As we shall see later, in absolute units the weight of an object of inertial mass m is mg. It is obvious the number value of this absolute weight force is g times the number value for the mass. However, if we define a new set of force units, in which the new force unit $= 1/g \times$ the corresponding value of the absolute force unit, we have the weight of the above object in new force units $= 1/g \times mg = m$.

We have achieved our desire of having the numbers for the force and the numbers for the mass agree. This amounts to making k in equation 3-1 $= 1/g$, so that, in practical units, Newton's Second Law for linear motion is

$$\vec{F}_{\text{pract}} = \frac{1}{g}\frac{d\vec{p}}{dt}.$$

Since g varies in value over the surface of the earth (problem 146), $k = 1/g$ would vary. To avoid this, a standardized value of g has been established by international agreement: $g_s = 980.665$ cm sec^{-2}, or $= 32.174$ ft sec^{-2}.

Thus the practical units of force are

$$F_{\text{pract}} = \frac{1}{9.80665} \times 1 \text{ kg} \times g \text{ m sec}^{-2}$$

$$= 1 \text{ kilogram weight (kgwt) } (\simeq 9.8 \text{ N}),$$

or

$$F_{\text{pract}} = \frac{1}{32.175} \times 1 \text{ lb} \times g \text{ ft sec}^{-2}$$

$$= 1 \text{ poundweight (lbwt) } (\simeq 32 \text{ lbal}).$$

The important thing to remember about practical units is that weight and mass are numerically equal. If you are told that a body "weighs" 19 kgwt, you know immediately that its mass is 19 kg. Similarly, if you know that the inertial mass of a body is 29 lb, you know it will "weigh" 29 lbwt.*

Occasionally in this book, in order to keep in harmony with the practical world, forces will be given in practical units. However, momentum and energy, which we will come to later, can only be handled easily in absolute units. Dynamics, which is the subject of this section, is more clearly and accurately done with absolute units, and most problems will be stated in such units.

Linear Dynamics

The one general and universally correct form of Newton's Second Law in linear form (in absolute units with $k = 1$) is

$$\vec{F} = \frac{d\vec{p}}{dt} = \frac{d(\gamma m\vec{v})}{dt}.$$

However, for mundane speeds this can be expressed in more workable form. Nonrelativistically, $p = mv$; hence nonrelativistically,

$$\vec{F} = \frac{d(m\vec{v})}{dt}.$$

*The convenience of this arrangement in common commercial and engineering considerations far outweighs the slight inaccuracies that result. Thus a given mass "weighs" the same in practical units everywhere on the globe, whereas its scientific weight in mg will vary slightly with both latitude and altitude.

If m is constant, as it is in a majority of scientific examples and in most engineering examples, this equation reduces still further to

$$\vec{F} = m \frac{d\vec{v}}{dt} = m\vec{a}. \tag{3-2}$$

In the above equation, m is the inertial mass of a particular closed system, a is the vector acceleration of the system as a whole, and F is the net *external* force applied to the system.

What constitutes "a particular closed system" is at your choice, made with regard to the particular problem you are attempting to study, and also with some due concern for clarity and ease of solution. In many cases the system you will isolate, and study by itself, will be a single body; in other cases it might be a collection of interconnected bodies; it might be more complex, such as a rocket and its train of exhausted propellant.

The phrase "vector acceleration of the system as a whole" is an awkward and not very informative reference; it would have been much simpler to have said "the vector acceleration of the center of mass of the system." However, it will be more appropriate to introduce, and to define, the "center of mass" later in this section; for now let it be sufficient to say that there is one unique point belonging to a closed system, and this point moves in accordance with equation 3-2 as if all the mass of the system were concentrated at that point, and all the external forces acting on the system acted at that point.

The net external force, F, acting on the system, may in fact be a vector sum of a number of forces. For that reason, to remind readers that it is a net sum, and to encourage a habit of consciously and accurately making an inventory of all the external forces, hereafter the left-hand side of equation 3-2 will be shown as ΣF. The Greek letter sigma (Σ) is the mathematical symbol for "the sum of." Thus the operating form of equation 3-2 is

$$\Sigma\vec{F} = m\vec{a}. \tag{3-3}$$

In the normal Cartesian coordinate system, this vector form is replaced by its components in the axial directions:

$$\begin{aligned} \Sigma F_x &= ma_x, \\ \Sigma F_y &= ma_y, \\ \Sigma F_z &= ma_z. \end{aligned} \tag{3-4}$$

Aside from equations 3-4, the most valuable tool to aid in solving dynamics problems is the free-body diagram. This diagram consists of an isolated drawing of the chosen closed system, the vector forces that are acting on the system, and a chosen coordinate system. At this point two things should be said, and said strenuously, about the coordinate system: one, once having chosen a coordinate system, stay with that system throughout the operation—don't change horses in midstream and don't change a coordinate system in the middle of a problem; two, don't be afraid of a minus sign; in fact, if you deal honestly with your coordinate system, the algebraic signs that come out with the answer may be as informative as the number you arrive at.

PROBLEM 163
A body of mass m is dropped in a free-fall apparatus in the laboratory, and its motion recorded. After the record is analyzed, it is computed that the body fell with a constant acceleration g. What force acted on the body?

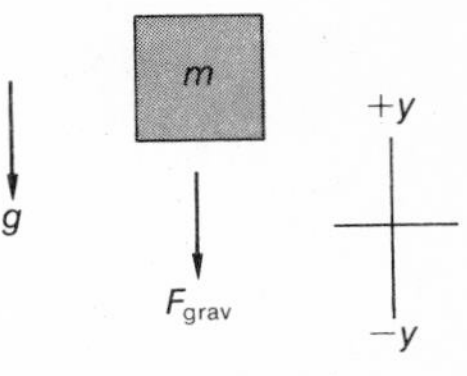

FIGURE 62

SOLUTION
In the coordinate system chosen, we know that $a_y = -g$. (See Figure 62.)

Hence, $\Sigma F_y = m(-g) = -mg$. If we assume that there is only one force acting, the force of gravitation, then

$$\begin{aligned} F_{\text{grav}} &= -mg \\ &= mg \text{ vertically downward.} \end{aligned}$$

This is the term that was referred to earlier as the weight of the body. The complete free-body diagram of the falling body would then look like Figure 63.

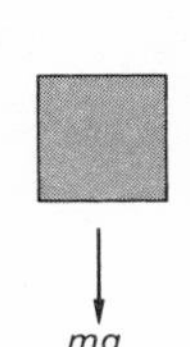

FIGURE 63

PROBLEM 164
The body of problem 163 is held in your hand, preparatory to making another drop. What force did your hand have to supply?

SOLUTION
From problem 163 we know that there is a force of mg acting downward on m; from the fact that the body is being held at rest, we know $a_y = 0$. (See Figure 64.)

Hence, $\Sigma F_y = F - mg = ma_y = m \times 0 = 0$, or $F = mg$ upward.

In some uses of the term, the F that was just solved for is called the weight of the body. In this context, weight is that force that the environment must supply to a body offsetting the gravitational force on the body, in order to maintain the body at rest in the environment. There should be no confusion over these different uses of the term "weight"; they represent the "equal but opposite" aspects of Newton's Third Law.*

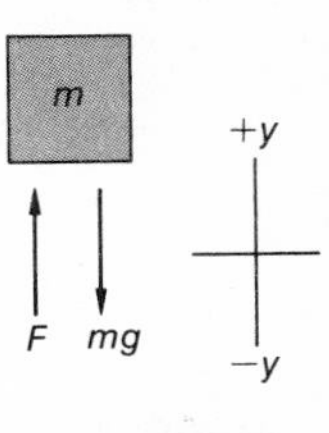

FIGURE 64

*Newton's Third Law concerns itself with action and equal reaction; if you push on B with a force F, automatically B pushes back on you with a force F. If you lean on a wall, the wall pushes back on you with equal force.

PROBLEM 165
In a supermarket aisle a woman pushing a shopping cart slips on a grape on the floor and falls. As she goes down, she automatically shoves the cart away from her, and it comes down the aisle straight toward you at 8 ft sec^{-1}. You reach out and bring it uniformly to a halt in 1.5 ft. If the cart and contents together weighed 80 lbwt, what force did you use in stopping the cart?

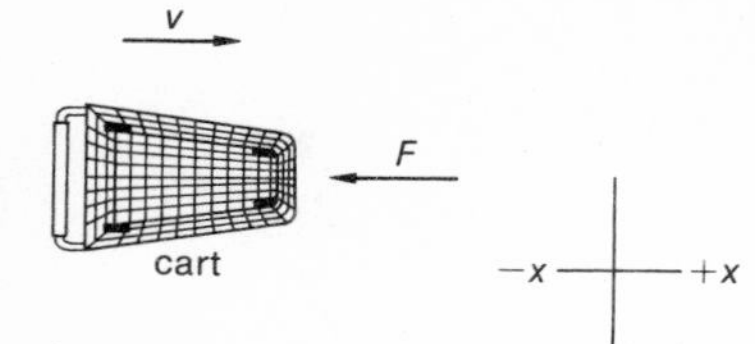

FIGURE 65

SOLUTION
See Figure 65.

$$\Sigma F_x = -F = ma_x = m\frac{(-v^2)}{2s},$$

$$F = 80^* \times \frac{8^2}{2 \times 1.5} = 1700 \text{ lbal},$$

$$F_{\text{pract}} = \frac{1}{32} \times 1700 = 53 \text{ lbwt}.$$

*Remember, if it weighed 80 lbwt, its mass was 80 lb.

PROBLEM 166
You are driving along a level street in your Mercedes-Benz 280 SE, towing a trailer loaded with sand and gravel you plan to use to build an addition to your patio. The trailer and load together weigh 1800 lbwt. You are going at 30 mi hr^{-1} and a boy chasing a ball darts in front of you. You come down on the brake pedal, and manage to stop uniformly in 50 ft. What horizontal force did the trailer hitch have to withstand? (Luckily for you, the trailer stayed straight behind the car during the braking!)

PROBLEM 167
A boy is walking along a level, snow-covered road at a speed v_0 ft sec^{-1} pulling a box sled of mass m lb. Soon snow begins to fall, filling the box at the rate of m lb min^{-1}. If the boy continues to walk at speed v_0, and the frictional force between sled and road remains independent of the additional load, what additional horizontal force did the boy have to supply during the snowfall?

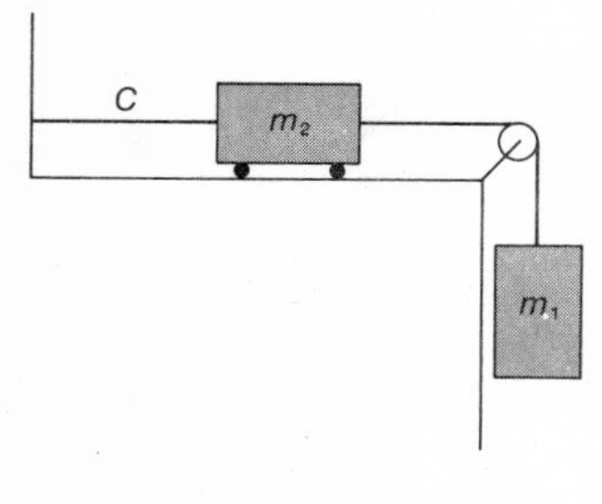

FIGURE 66

PROBLEM 168
Later, the snowfall in problem 167 stopped, and the skies cleared. When the sun began to shine, $m/3$ lb min^{-1} of snow melted and leaked out of the sled. How did this leakage affect the horizontal force the boy had to supply while continuing to move at v_0?

PROBLEM 169
A system (m_1, m_2) is held motionless by a cord C, as shown in Figure 66. When the cord is burned, what is the acceleration of m_1?

SOLUTION
Let us do this by separate free-body diagrams for m_1 and m_2 (Figure 67). Only the forces in the direction of motion of each body are shown. T is the unknown tension in the flexible but inextensible cord connecting m_1 and m_2; assuming no friction and negligible mass in the pulley, the tension T is constant throughout the cord.

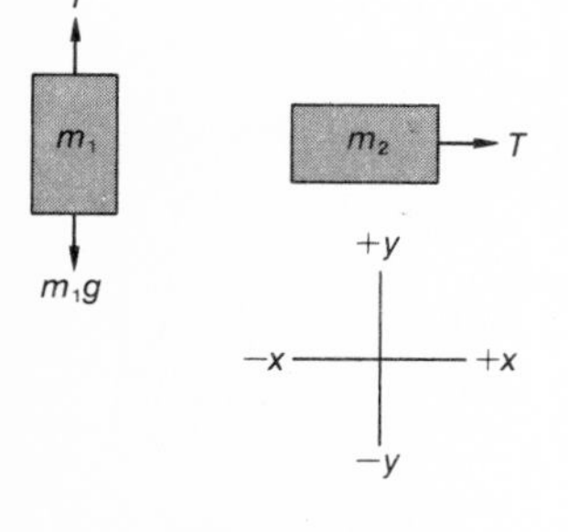

FIGURE 67

$$\text{On } m_1: \quad \Sigma F_y = T - m_1 g = m_1 a_{1y}.$$

$$\text{On } m_2: \quad \Sigma F_x = T \quad = m_2 a_{2x}.$$

This is as far as Newton's Second Law will take us, but it isn't enough. Note that we have three unknowns, but only two independent equations. What Newton can't supply, but we can, is the additional knowledge that m_1 and m_2 are constrained to move alike, i.e., the numerical values of the accelerations are equal. Since one acceleration is opposite in sign from the other (if m_1 moves negatively m_2 must move positively, or vice versa, in the coordinate system chosen), the statement of this constraint is

$$a_{2x} = -a_{1y}.$$

Since we were asked to find a_1, let us use this information to eliminate a_2, and we then have

$$T - m_1 g = m_1 a_{1y},$$
$$T = -m_2 a_{1y}.$$

Subtracting the bottom equation from the upper to eliminate the unknown and unwanted T, we have

$$-m_1 g = (m_1 + m_2)\, a_{1y}.$$

So finally,

$$a_{1y} = -\frac{m_1}{m_1 + m_2}\, g.$$

Some readers may wonder why m_1 and m_2 can't be treated together as a system, with a system mass of $m_1 + m_2$ and a force in the direction of motion of $-m_1 g$. This approach would lead immediately to the penultimate equation in the solution to problem 169. The answer is that there are a very few, very elementary, two-body situations where such a shortcut approach would produce the correct answer. However, most interesting dynamical problems could not be handled in this way—they must be done, if done by Newton's Second Law, through separate free-body diagrams, separate Second Law statements, and constraint statements. (For example, see problems 182, 184, 186.) Therefore it is good practice to use the more general approach even on simple, by-inspection problems. This builds both confidence and the basis for intuition. Ultimately, the more you know soundly and analytically, the more shortcuts you can take safely. At the beginning learning stage, shortcuts are not safe.

As another illustration of this method, let us consider a problem given by Professor Leighton, of Caltech, to his freshman physics class several years ago.

PROBLEM 170
What horizontal force must be constantly applied to M, as shown in Figure 68, so that m_1 and m_2 do not move *relative to M?*

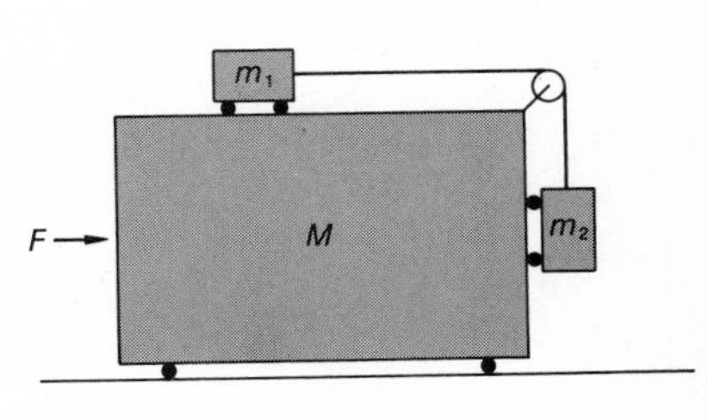

FIGURE 68

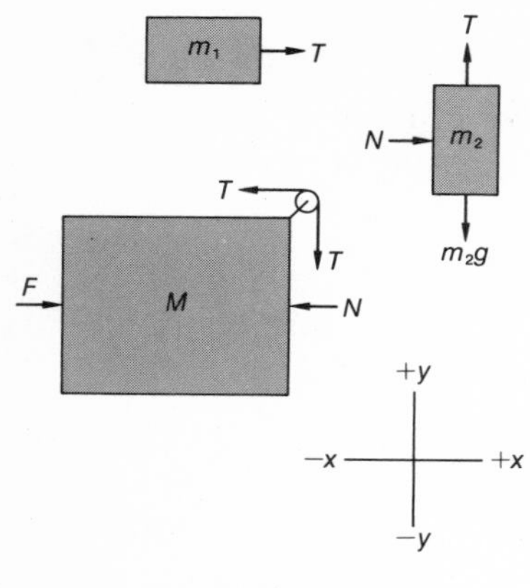

FIGURE 69

SOLUTION

Again we start with free-body diagrams (Figure 69) and the appropriate Second Law equations. The symbol N stands for some unknown normal force by which M pushes m_2 ahead of it.

$$\text{On } m_1\text{:}\quad \Sigma F_x = T = m_1 a_{1x}.$$

$$\text{On } m_2\text{:}\quad \Sigma F_x = N = m_2 a_{2x},$$

$$\Sigma F_y = T - m_2 g = m_2 a_{2y}.$$

$$\text{On } M\text{:}\quad \Sigma F_x = F - N - T = M a_{Mx}.$$

We must now add the constraint equations, which carry out the instruction that m_1 and m_2 do not move relative to M. For this to be true,

$$a_{2y} = 0 \quad \text{and} \quad a_{1x} = a_{2x} = a_{Mx}.$$

Adding the first, second, and fourth of the above equations after appropriate substitutions, gives us

$$F = (m_1 + m_2 + M) a_{1x};$$

and adding the first and third equations, after substitutions, give us

$$a_{1x} = \frac{m_2}{m_1} g.$$

Hence, finally,

$$F = \frac{(m_1 + m_2 + M) m_2}{m_1} g.$$

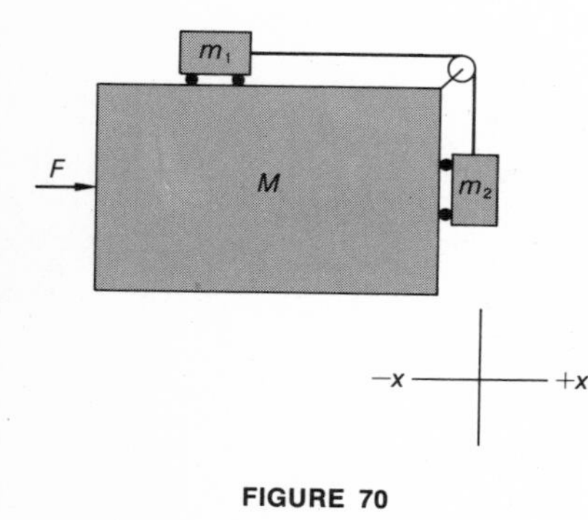

FIGURE 70

Note that again the work could have been shortened by considering m_1, m_2, and M as a system; for such a free-body diagram (Figure 70) we can immediately write $\Sigma F_x = F = (m_1 + m_2 + M) a_x$. However, in order to determine the acceleration, m_1 and m_2 must still be treated separately.

In summary, the one sure-fire method for solving dynamical problems by Newton's Second Law is to follow this schedule:

- Make a careful free-body diagram for each mass, making sure that every force acting on that body is shown.*

- Select a coordinate system.

- Write Newton's Second Law equations for each mass, in accordance with the chosen coordinate system.

- Express the appropriate geometrical relationships, or the design requirements, in constraint equations. The number of unknown quantities in the Second Law

*For clarity, forces not contributing to motion have been eliminated from the diagrams in the examples.

equations, minus the number of the Second Law equations, tells you how many independent constraint equations you need. In other words, altogether you need as many independent equations as you have unknowns. Be careful that these equations are really independent; in some situations it is easy to get trapped into making a statement that is simply a transposed copy, or a rearrangement, of a previous statement.

- Solve the algebra for the desired results.

It is worth saying something about the philosophical purpose of the constraint equations. Each mass is cognizant *only* of the forces acting on *it*, and it will move independently in response to those forces alone. Where those forces come from, or what the rest of the world is like, is of no concern to an individual mass. Only you, as an external observer, know where the forces come from, and only you know whether there are any geometrical relationships between any one mass and the world outside that mass. This external knowledge of yours is what must be expressed in the constraint equations.

PROBLEM 171
A sky diver free-falling before he opens his parachute reaches a terminal speed of about 75 mi hr^{-1}. What is the drag force on him from the atmosphere at that speed?

PROBLEM 172
An elevator starts upward from rest with an acceleration of 0.25 g. As it nears a floor at which it is to stop, the power is shut off and the elevator coasts to a stop. If your normal weight is 160 lbwt, what was your apparent weight* during

a. the start-up period;
b. the stopping period?

*Apparent weight is the force felt by the supporting surface.

PROBLEM 173
In problem 169, the system existed in an inertial environment. Now let us mount the assembly in an elevator. When the elevator was accelerating upward at $[m_2/(m_1 + m_2)]\, g$, and m_1 was 4.0 ft above the elevator floor, the cord C was burned. How long after that did m_1 hit the floor? (Obviously, for this experiment m_2 is more than 4.0 ft back of the pulley when the cord is burned.)

PROBLEM 174
An astronaut takes with him to the moon a calibrated spring balance and a kit of small tools. He knows that on earth the kit weighs 2.94 lbwt on the balance. On the moon he picks up a large lump of material and finds that on the spring balance it also weighs 2.94 lbwt. This

surprises him, so he again weighs his kit of tools and finds that on the moon the kit only weighs 0.50 lbwt.

a. What is the value of g on the moon?
b. What is the mass of the lump of material the astronaut picked up?

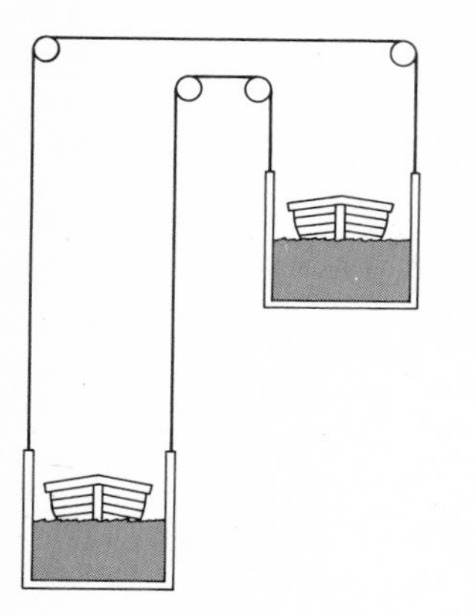

FIGURE 71

PROBLEM 175
Assume that in problem 170, $m_1 = 400$ gm, $m_2 = 100$ gm, and $M = 1500$ gm.

a. What was the acceleration of M?
b. What was the value of the force F?
c. What was the tension in the cord connecting m_1 and m_2?
d. What horizontal force existed between m_2 and M?

PROBLEM 176
On a canal in Belgium, barges are moved from one canal level to another by an *ascenseur* as shown in Figure 71, instead of being transferred by locks. Huge, identical, rectangular metal boxes, longer and wider than the largest barges on the canal, are partly filled with water in which the barges can float. If the water level in both boxes is the same, the system is balanced; to transfer barges all that is needed is to slightly overfill the upper box. Normally movement due to this imbalance is controlled by brakes. At a time when each box was balanced at a weight of 3.5×10^5 kgwt, just after the upper box was overfilled with 1.0×10^3 kgwt of water, the brakes failed. With what acceleration did the boxes then move? (Assume no friction and negligible mass in the support system.)

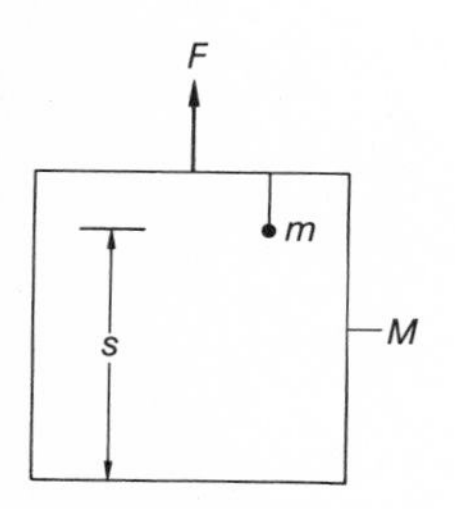

FIGURE 72

PROBLEM 177
An elevator of mass M has a mass m hanging from its ceiling (Figure 72). The elevator is being accelerated upward by a constant force F, $F > (M + m)g$. The mass m hangs a distance s above the elevator floor.

a. What is the acceleration of the elevator?
b. What is the tension in the string supporting m?
c. If the string supporting m breaks, what is the acceleration of the elevator immediately thereafter?
d. How long does it take m to reach the floor of the elevator?

PROBLEM 178
A 20-ton tank is crossing a sluggish river on a floating pontoon bridge 100 ft long. The bridge is held against each bank by ropes with a breaking strength of 1.0 tonwt each. As the tank enters the bridge at a speed of 7.5 mi hr^{-1} it gets orders to speed up, and uniformly accelerates so that it leaves the bridge at a speed of 30 mi hr^{-1}. At a minimum, how many ropes are needed at each end of the bridge?

PROBLEM 179
A small object of mass m moves in the x-y plane under the influence of a force lying in that plane. The force is described by $\vec{F} = -k\,\vec{r}$, where $\vec{r}$ is the radius vector from the origin to the mass, and k is a

constant. If at $t = 0$ the object is at $x = x_0$ and $y = 0$, and is moving with a velocity v_0 in the $+y$ direction, what is the path of the motion of the object for $t > 0$?

PROBLEM 180

Two iceboats, each of total mass with passengers of 200 kg, are at rest side by side on the smooth ice surface of a lake, both pointing in the same direction. A gust of wind hits the boats, and applies a steady force of 200 N against the sail of each boat. One boat's sail blows to pieces after that boat had gone 10 m. If the gust lasted for only 10 sec, how far apart are the boats 10 sec after the wind stopped, if neither boat altered course?

PROBLEM 181

A space ship approaches a landing on a strange, atmosphereless planet by backing down, using its rockets to slow its fall. The rockets are turned full on at the instant the ship's radar indicates it is 4.5 km above the planet's surface and moving toward the surface at a speed of 200 m sec^{-1}. At the same instant a radar target is dropped from the ship; radar tracking of the falling target indicates it is accelerating relative to the ship at a rate of 7.0 m sec^{-2}, and it is observed to hit the planet surface 20 sec after release from the ship.

a. What was the space ship's acceleration capability in free space?
b. Assuming 4.5 km $<<$ radius of the planet, what was the value of g on that planet?
c. How long did it take the rockets to bring the space ship to a stop?
d. When the ship stopped, how far above the planet's surface was it?

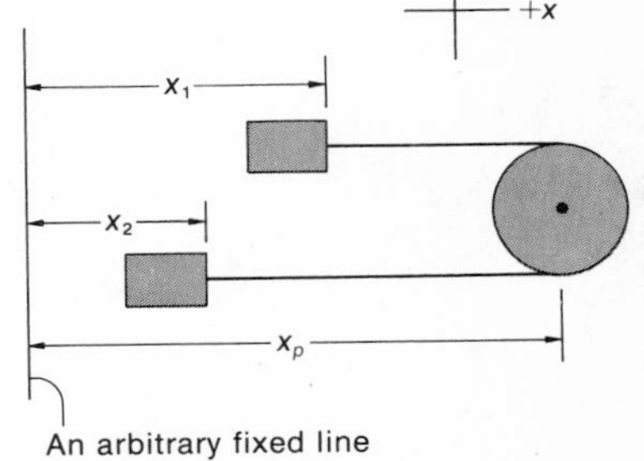

FIGURE 73 An arbitrary fixed line on the horizontal surface

So far constraint equations have been easy to formulate because we have dealt with, at most, two bodies simply connected. However, in general, the problems you will have to deal with are not that simple, and it will be useful to develop a general method for finding geometrical relationships even for complicated arrays.

These geometrical constraints always appear as algebraic relationships of the accelerations of the various bodies. Since, in turn, the accelerations are themselves functions of spatial position, it seems pertinent to start our general development with a look at the spatial coordinates directly.

As an example, let us consider the formulation of the constraint equation for the next problem—problem 182. If we establish the usual coordinate system, spatially the problem is as shown in Figure 73.

During movement of the assembly, x_p, x_1, and x_2 are all changing with time. However, one thing is constant—the length of the cord. If we call that length L, then $L = (x_p - x_1) + (x_p - x_2)$ (disregarding the constant wraparound on the pulley).

Differentiating this equation twice with time (which in a first-degree equation means simply replacing isolated constants by 0 and displacement variables by accelerations), we have

$$0 = a_p - a_1 + a_p - a_2$$

or

$$a_p = \frac{a_1 + a_2}{2},$$

which is the constraint equation we need for problem 182.

As a further example, consider problem 183. Here the space picture is a little more complicated (Figure 74). As it stands, the total length of cord L is given by $L = -y_3 - (y_2 - y_p) - y_1$. However, if L_1 is the length of the upper cord, $L_1 = -y_p - y_1$. Putting these together to eliminate y_p, $L + L_1 = (y_3 + y_2 + 2y_1)$. After differentiating, $a_1 = -(a_2 + a_3)/2$, which is the constraint equation needed in problem 183.

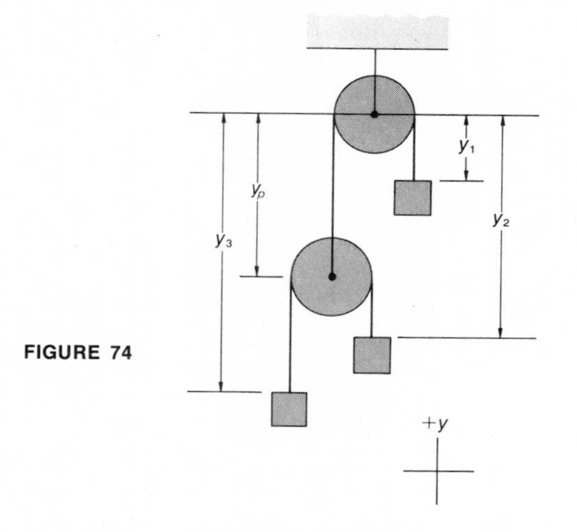

FIGURE 74

Two comments are appropriate. With the coordinate system chosen as usual, the displacements had to be identified as negative. For that particular problem we could just as easily have made downward positive. However, it was left in the usual form to encourage readers to get accustomed to dealing with negative coordinates. The other comment is to point out that this problem is, to a large degree, just a 90° rotation of the previous problem; since the pulley and mass m_1 move alike except for a minus sign, the constraint equation developed for problem 182 could have been used immediately for problem 183. The recognition of such similarities and shortcuts is a basic ingredient of a well developed intuition.

PROBLEM 182

Two masses, $m_1 = 6.0$ kg and $m_2 = 2.0$ kg, are lying on a smooth, horizontal surface and are connected by a flexible, inextensible cord over a pulley which also lies on the surface (Figure 75). The mass of the pulley is negligible. If a horizontal force of 15 N is applied to the axle of the pulley as shown, what will be the acceleration of the pulley?

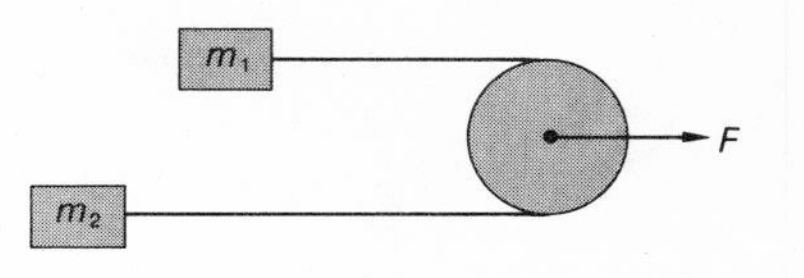

FIGURE 75

PROBLEM 183

In the system shown in Figure 76, the cords are flexible and inextensible, and the pulleys have no friction and negligible mass.

a. After release from rest, what is the acceleration of m_1?

b. What is the tension in the cord supporting the upper pulley?

c. Is there any set of values for m_2 and m_3 such that m_1 will not move even when the system is released?

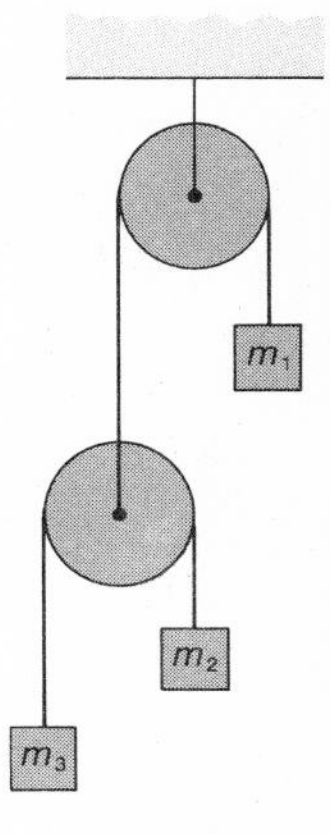

FIGURE 76

PROBLEM 184

A painter weighing 180 lbwt and working from a "bosun's" chair hung down the side of a tall building desires to move in a hurry (Figure 77). He pulls down on the fall rope with such a force that he presses against the chair only with a force of 100 lbwt. The chair itself weighs 30.0 lbwt.

a. What is the acceleration of painter and chair?

b. What is the total force supported by the pulley?

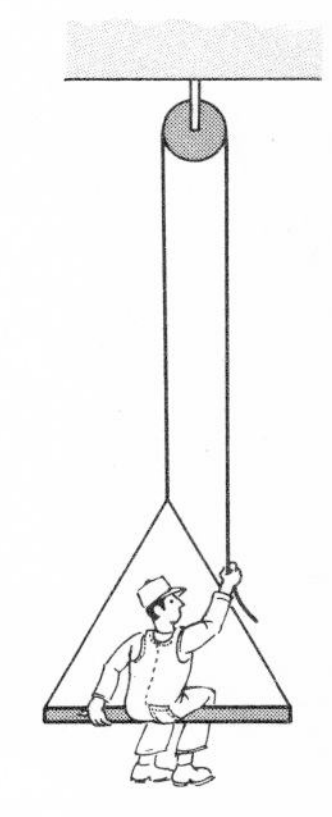

FIGURE 77

PROBLEM 185

When the system of masses pictured in Figure 78 is released from rest, it is observed that m_1 does not move.

a. What must then be the value of m_4, in terms of m_1, m_2, and m_3?

b. What is the upper limit for the ratio m_1/m_2 for the problem to be at all possible?

PROBLEM 186

A simplification of a particular kind of interlock is the system shown in Figure 79; all surfaces are frictionless. m has a mass of 150 gm and $M = 1650$ gm. When m is released 4.0 ft above the base of M, how long does it take for it to reach the base?

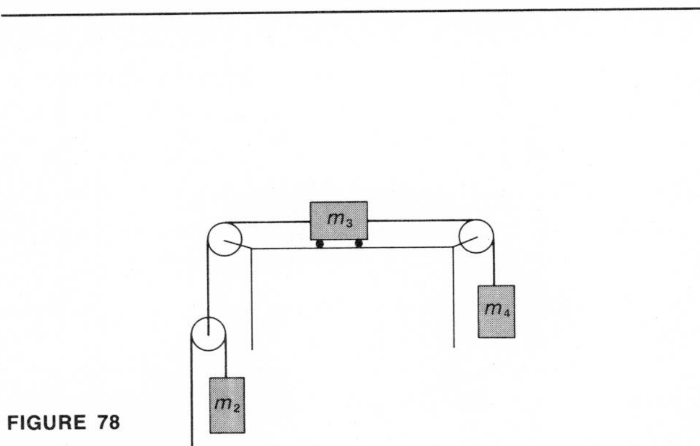

FIGURE 78

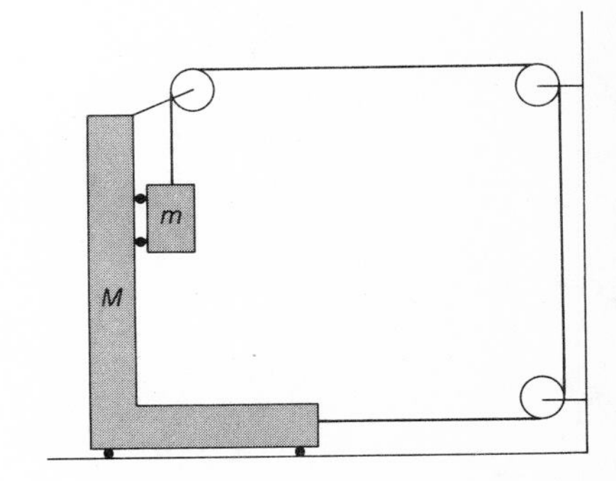

FIGURE 79

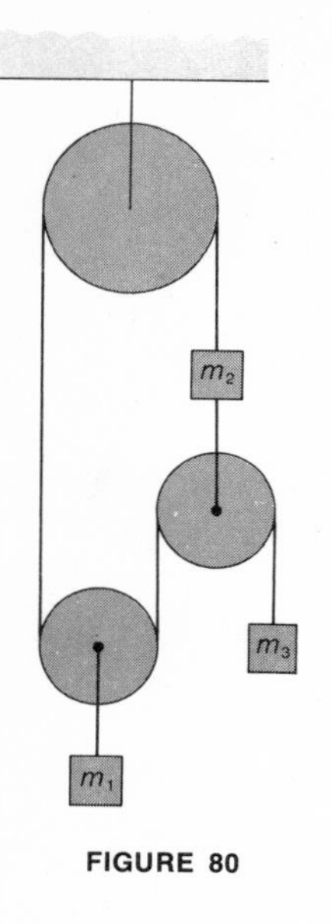

FIGURE 80

PROBLEM 187
The "freshman's delight" pulley system is as shown in Figure 80; the masses of the pulleys and cords are negligible, their axles are frictionless, and the cords are inextensible and perfectly flexible. (In other words, an "ideal" system.) After release from rest,

a. which mass will fall with an acceleration greater than g;
b. what is the acceleration of m_1;
c. what relationship among the masses will allow m_3 to accelerate upward?

Note: Question (a) should be answered by inspection; the other parts will require a complete solution of the dynamic equations.

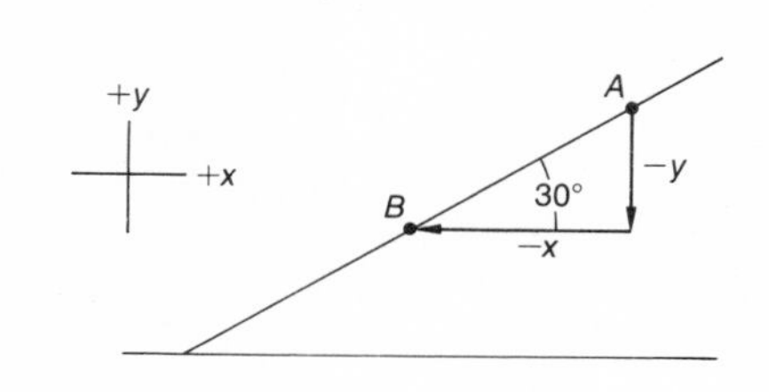

FIGURE 81

Not all displacement relations are nicely linear, or at convenient right angles, as they have been in the previous problems. Consider, for example, problem 188. Here the motion of the young man must follow the surface of the plane. If he was at A at $t = 0$, at some time later he will be at B (Figure 81). In that time his displacement in the y-direction was $-y$ and in the x-direction was $-x$. These are related by the equation

$$-y = -x \tan 30°.$$

Thus the constraint needed for that problem would be

$$a_y = \frac{a_x}{\sqrt{3}}.$$

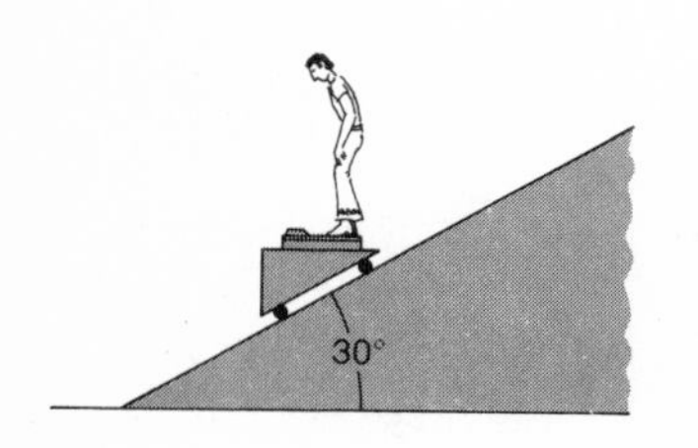

FIGURE 82

PROBLEM 188
A precocious young man arranges a scale on a small platform with rollers, so that it will roll down a 30° incline without friction. He stands on the scale, reading it as he moves down the incline (Figure 82). If he normally weighs 160 lbwt, what did the scale read during his descent?

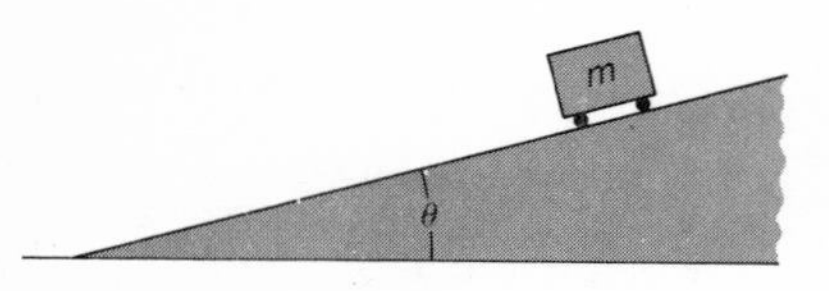

FIGURE 83

PROBLEM 189
An object of mass m is released from rest on a smooth incline that is θ above the horizontal (Figure 83).

a. What is the acceleration of the object relative to the fixed incline?
b. What are the vertical and horizontal components of this acceleration?

c. What is the normal force between the object and the incline? *Suggestion:* Make your coordinate system perpendicular to and parallel to the incline, and resolve the weight mg along those axes.

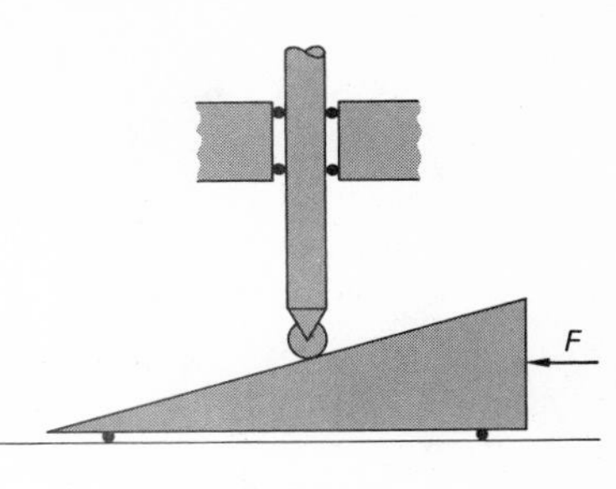

FIGURE 84

PROBLEM 190

A section of a cam arrangement is as shown in Figure 84. The vertical follower rod, whose effective mass is 400 gm, runs through frictionless vertical guides, and ends in a frictionless bearing that rests on the movable wedge. The wedge has a wedge angle of 11°.3 and a mass of 200 gm, and runs horizontally on frictionless bearings. In order to give the follower rod an upward acceleration of 2.0 g, what horizontal force must be applied to the wedge?

PROBLEM 191

A mass m is free to slide on the frictionless, inclined face of a wedge of mass M and wedge angle θ. When M is given a horizontal acceleration as shown in Figure 85, it is noted that for a particular value of the acceleration a_0, m does not move relative to the wedge.

a. Determine a_0.
b. What horizontal force F is required to produce a_0?

FIGURE 85

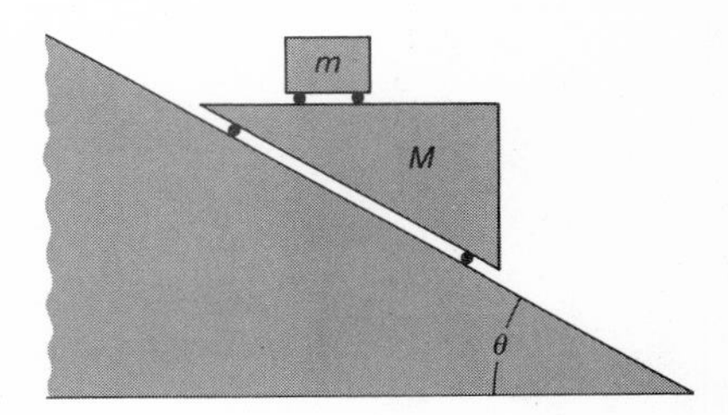

FIGURE 86

PROBLEM 192

After release (Figure 86), what is the vector acceleration of m?

PROBLEM 193

The essentials of a triggering element in an automatic filling machine are as shown in Figure 87. A small cube slides down the smooth face of a motionless wedge; when it reaches the restraining lip at the bottom, a circuit is closed, and the wedge is immediately given a horizontal acceleration a by an external force. If this acceleration is such that the cube slides back up the wedge in exactly the same time it took to slide down, what is the value of a?
Note: For your constraint equation you should get $a_x = a - a_y \cot\theta$, where the a_x and a_y refer to the cube.

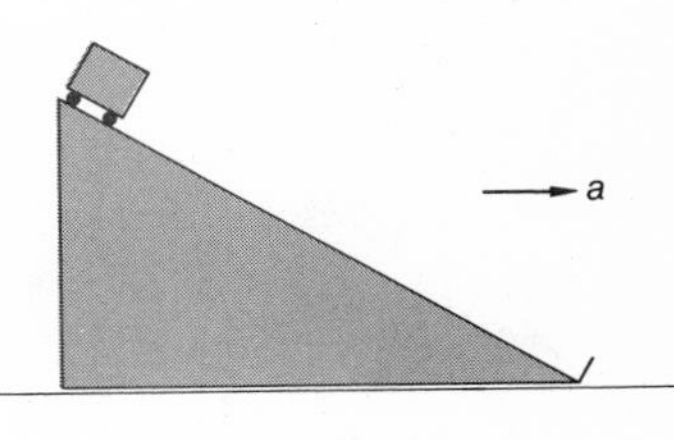

FIGURE 87

PROBLEM 194

What horizontal force was required to give the wedge the necessary acceleration in problem 193? Take the masses of the cube and wedge in that problem to be m and M respectively.

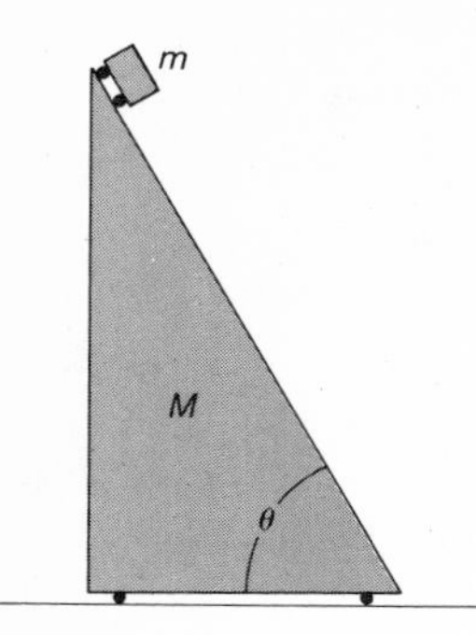

FIGURE 88

PROBLEM 195

In designing some automatic equipment you have the basic design shown in Figure 88 to work with. When the car, mass m, is at the top of the incline, mass M, the catches holding m and M at rest are released simultaneously. It is required that the car shall reach the bottom of the incline just as the incline itself has moved sideways half the width of its base. It is also required that the *vertical* component of the car's acceleration shall be 0.36 g downward.

a. What should M be, in terms of m?
b. What should the angle θ be?

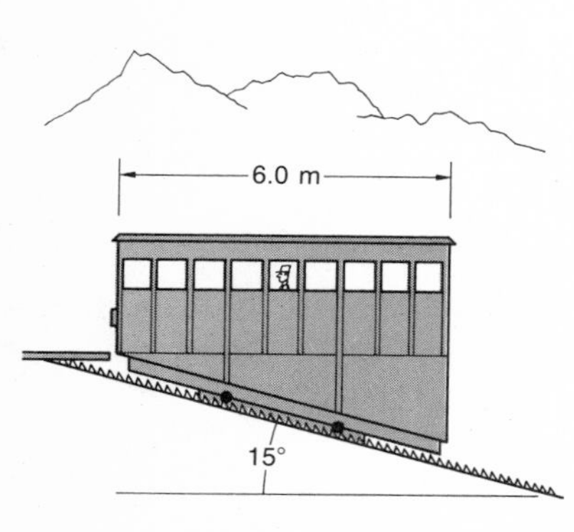

FIGURE 89

PROBLEM 196

On many of the cogwheel mountain railroads of Europe, the gradient is relatively constant; on such roads, for passenger comfort the cars are built with horizontal floors, as shown in Figure 89. Such a car, of mass 800 kg, on a 15° gradient, is stopped at an upper station platform, with the 70 kgwt attendant standing in the aisle in the exact middle of the car. The latch holding the brake lever gives way from fatigue, and the car is released to roll freely down the track. At the instant the latch broke the attendant realized what had happened and started running, at constant acceleration relative to the car, toward the open upper end of the car. When he escaped from the end of the car he found he had no horizontal speed relative to the ground.

a. How far vertically below the platform did he land?
b. With what vertical speed did he hit the ground?
c. How long did it take him to escape from the car?

Suggestion: Before starting the analytical work necessary for the solutions to questions (b) and (c), try to see the "obvious by inspection" answer to question (a).

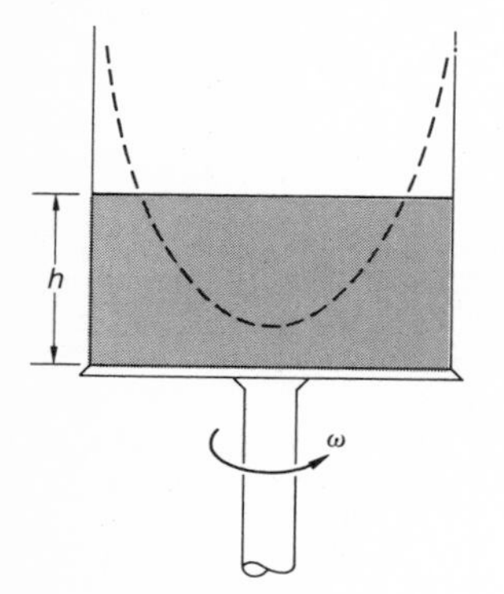

FIGURE 90

PROBLEM 197

A vertical cylindrical can, of radius R, centrally mounted on a rotator, is filled at rest with a liquid to a depth h (Figure 90). Then the can is rotated about its vertical axis with an angular speed ω. The y-axis is coincident with the axis of the can, and the origin of coordinates is at the base of the can.

a. Develop the equation describing the intersection of the liquid surface with the x-y plane, when dynamic equilibrium is attained at some ω, in terms of the variables h, R, and ω.
b. Determine the maximum value for ω for which the bottom of the can will be completely covered with liquid.

START OF SOLUTION

A point mass at the surface of a liquid is acted on by two forces: by its weight and by a buoyant force outward from the surface and perpendicular to it (Figure 91). The tangent to the surface is given by the ratio dy/dx. The buoyant force, when there is dynamical equilibrium, must support the mass in offset to its weight, and must also supply the necessary centripetal acceleration to keep the mass rotating about the central axis with an angular speed ω. Applying this to the diagram, it can be seen that

$$F \cos \theta \text{ must equal } mg$$

and

$$F \sin \theta \text{ must equal } mx\omega^2.$$

Dividing the lower equation by the upper, we have

$$\tan \theta = \frac{x\omega^2}{g}.$$

But $\tan \theta$ also equals dy/dx, so that, putting these together,

$$\frac{dy}{dx} = \frac{x\omega^2}{g}.$$

When this is integrated, the constant of integration must be evaluated from the fact that the total volume of liquid under the surface

$$\int_0^R y\, dx$$

must equal $\pi R^2 h$, the total volume we started with.

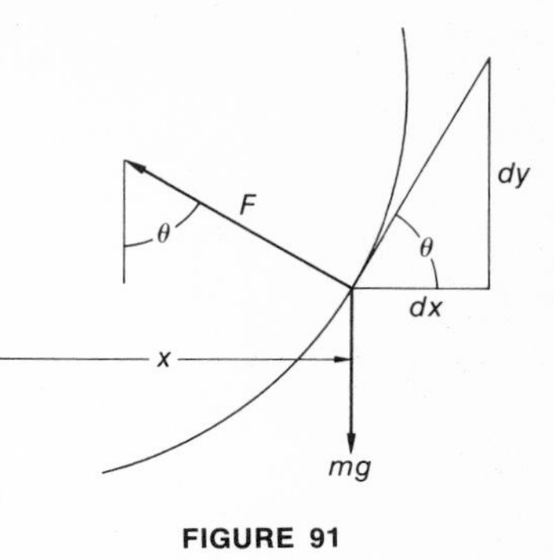

FIGURE 91

PROBLEM 198

A large bowl is formed in the shape of a paraboloid of revolution (Figure 92), i.e., the intersection of the bowl with any vertical plane through the vertical axis of symmetry is a parabola, $y = Bx^2$. At a point where the radius of the bowl is 20 cm, the bowl is 10 cm deep.

a. Find the angular speed at which the bowl must be rotated so that a small steel ball is in dynamical equilibrium *anywhere* on the inner surface of the bowl.

b. If the ball has a radius r not negligible compared to the bowl dimensions, investigate whether the angular speed required for dynamical equilibrium is still independent of the position of the ball in the bowl.

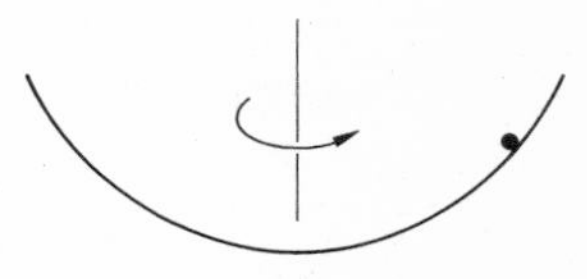

FIGURE 92

PROBLEM 199

In 1985, as a result of a series of accidents in launching space ships, the rate of rotation of the earth was increased (without any increase in the earth's orbital speed). In 1986 a test mass was found to weigh, at the equator, only 0.990 as much as it had weighed there in 1984 on the identical spring balance. How many sidereal days were there in the year 1986?

PROBLEM 200

An Air Force fighter-bomber, of 40 metric tons mass, takes off from a level air field at Great Falls, Montana, latitude 47°.5 N. Before lifting off the plane reaches a velocity of 300 km hr^{-1} due north. In order to keep the plane tracking exactly along a north-south line painted on the runway,

a. with what force must the ground push on the tires in a direction perpendicular to the plane's motion;

b. in which direction is this force? (Neglect any influence from the wind.)
Hint: A force is required to produce an acceleration that will exactly cancel the Coriolis acceleration.

Some Natural Forces

Gravitation

We used the gravitational force near the surface of the earth when we talked about the weight mg of a body of mass m. In a much more generalized form this force appears in the operational statement of Newton's Law of Gravitation:

$$F_{\text{grav},m} = \frac{GmM}{r^2}. \tag{3-5}$$

This states that two point masses m and M, which are r apart, are attracted toward each other along the line of r. The constant G is the universal constant of gravitation—universal because its value seems to remain constant throughout the observable universe. The more useful vector form of this law is

$$\vec{F}_m = -\frac{GmM}{r^3}\vec{r}. \tag{3-6}$$

In this form, F_m is the gravitational force on m in the presence of M, r is the displacement vector **from** M **to** m. Since r is measured from M, and the attractive force on m is toward M, the minus sign is necessary.

This is an appropriate place at which to add to our vocabulary the concept of "field." In equation 3-5 the gravitational force on any particular m is given by GmM/r^2, which obviously varies for different values of m. However, the force on a unit mass is simply GM/r^2, which is just a function of M and r. As we decrease m, keeping r constant, equation 3-5 tells us that the gravitational force resulting from the presence of M decreases, but that GM/r^2, the gravitational force on unit mass, stays constant. We can then assume that GM/r^2 is a property of the mass M and the distance from M. We call this property the gravitational field of M.

Thus "field," being a general property, is more physically fundamental than any particular force. Since gravitational field is force per unit mass, the gravitational force on an object at any point in space can be found simply by multiplying the gravitational field at that point by the mass of the object. Field is a vector, so that the direction of the force is provided by the direction of the field. The dimensions of the gravitational field are LT^{-2}. By comparison with what we have already said about weight, it should be obvious that the gravitational field of the earth, at a distance r from its center, is simply $\vec{g}_r$.

For any bodies other than point masses, or particles, there is the obvious question of where you measure $\vec{r}$ from, or to. For extended bodies, which consist of an enormous number of "point masses," the application of equation 3-5 or 3-6 requires

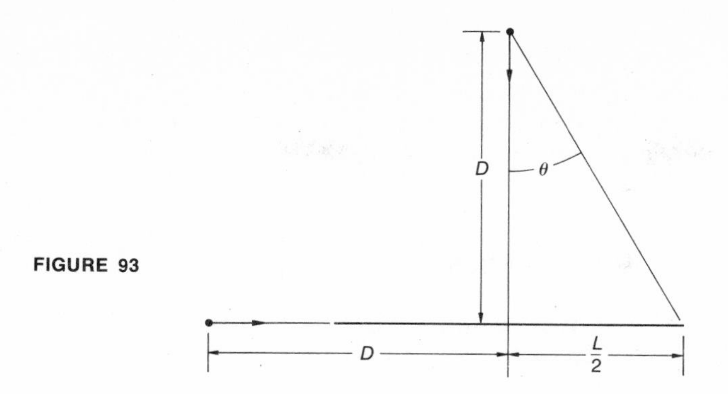

FIGURE 93

an integration over the volumes of m and M. For irregular shapes this can become a real challenge. Fortunately, we are usually interested in the gravitational force in only two kinds of cases: those in which the bodies are so small compared to the distances between them that, to the observational limit of accuracy, they can be treated as point masses; and those in which the attracting bodies are external to each other and are uniform spheres, either solid or hollow. For uniform spheres the gravitational force is exactly measured from the center of the sphere, as if all the mass of the sphere were concentrated there. The sun and planets, or the planetary electrons and the nucleus in a Bohr atom, are examples of the first case; a space vehicle in the earth's gravitational field is a combination of the two cases; and the Cavendish experiment to determine the value of G is an example of the second case.

With the exception of point masses and uniform bodies of spherical symmetry, the value of F computed by equation 3-6 for any two bodies is not unique and single valued but depends on the orientation of the bodies. To illustrate, say you want to find the gravitational field of a slender, uniform bar, of mass M and length L, at a point that is D from the center of the bar (Figure 93). If the point is on a line perpendicular to the bar and passing through its center, the gravitational field at that point is

$$\frac{GM}{D^2}\cos\theta \equiv \frac{GM}{D\sqrt{L^2/4 + D^2}},$$

directed toward the center of the bar. You will note that this is less than GM/D^2.

On the other hand, if the point is in line with the bar, the field is still directed toward the center of the bar but it now equals

$$\frac{GM}{(D^2 - L^2/4)}$$

which is greater than GM/D^2. Both of the values just calculated approach GM/D^2 for a D very large compared to L, which is what you should expect; the farther you get from the bar the more it looks to you like a point mass.

You are not likely, very many times in your life, to be interested in the gravitational field of a long, slender rod (although you will be interested in the electric field of a long, straight wire uniformly charged). But you might easily be concerned about the gravitational field of a large flat plate.

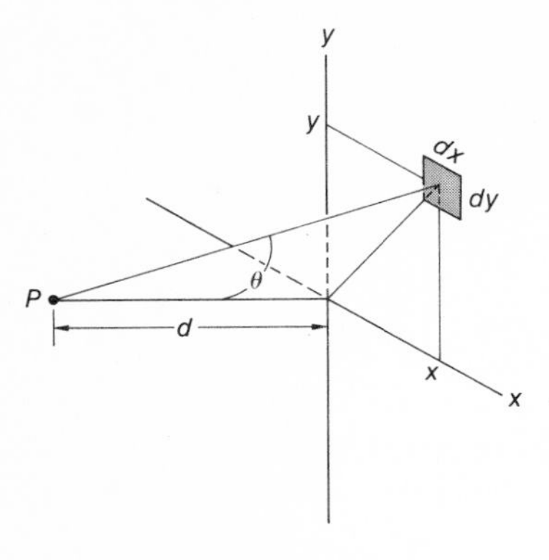

FIGURE 94

PROBLEM 201

What is the gravitational field at point P, a distance d away from a large, uniform flat plate, due to the plate alone? The dimensions of the plate are very large compared to d, and the thickness t of the plate is small compared to d.

SKELETON SOLUTION

Probably your first thought would be to set up a Cartesian coordinate system, as shown in Figure 94. From the small element whose volume is $dx\, dy\, t$ you would have, for the field at P,

$$\Delta g_{P,\perp} = \frac{G\ \rho\ dx\ dy\ t}{(d^2 + x^2 + y^2)} \cos\theta$$

$$= \frac{G\ \sigma\ dx\ dy}{(d^2 + x^2 + y^2)} \frac{d}{\sqrt{d^2 + x^2 + y^2}}$$

From symmetry, components of the field parallel to the plate at P cancel out in the summation over the plate, so that the only component that adds continuously during the summation is the one that is perpendicular to the plate—hence the $\perp$ subscript and the $\cos\theta$ term. In the final equation, ρt has been replaced by σ, where σ is called the surface density — kg m^{-2} in MKS units. There is no physical significance to this substitution; it is simply that in situations where the surface area but not the thickness is physically pertinent, the surface density σ is more convenient.

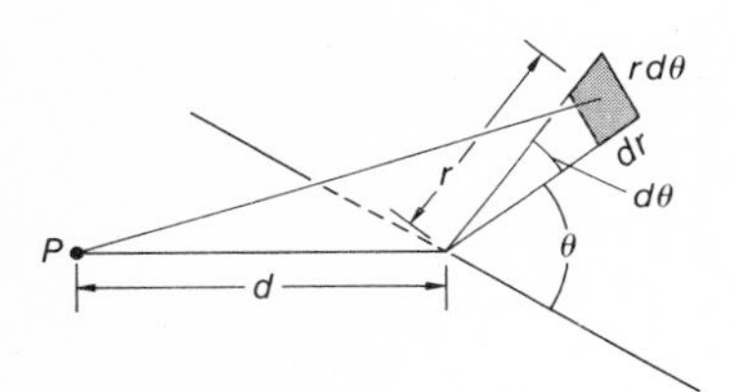

FIGURE 95

When you attempt to determine g_P by integrating this equation, you run into difficulties; the first integration goes through all right but it is not easy to find the second integrand in the usual published tables of integrals. So you turn to polar coordinates in the hope that they will give you a more manageable integrand. With the coordinates shown in Figure 95 you have

$$\Delta g_{P,\perp} = \frac{G\ \sigma\ r\ dr\ d\theta}{(r^2 + d^2)} \frac{d}{(r^2 + d^2)^{1/2}}.$$

You find that this integrates easily: with θ running from 0 to 2π and r from 0 to R,

$$g_{P,} = 2\pi G\ \sigma\ d\left(\frac{1}{d} - \frac{1}{\sqrt{R^2 + d^2}}\right).$$

For $R >> d$, which expresses the given condition that the plate is very large compared to d, this reduces to $g_{P,\perp} = 2\pi G\ \sigma$.

The interesting thing about this result is that no dimension appears explicitly in the answer. This means that near a very large, flat, uniform plate the gravitational field is everywhere perpendicular to the plate and of constant value $2\pi G\ \sigma$.

PROBLEM 202

a. At the center of a hemispherical shell of uniform surface density, what is the gravitational field due to the shell alone?

b. In consideration of the answer to (a) what is the gravitational field at the center of a complete spherical shell of uniform surface density, due to the shell alone? For an extension of this result to *anywhere* inside a spherical shell, see problem 211.

Digression: The Center of Mass. A conclusion that can be drawn from what we have just been investigating is that for most bodies near each other there is no unique value of the gravitational force between them; there is no unique point from which R can be measured for the force GmM/R^2. (Point masses and bodies with spherical symmetry are exceptions to this statement.) There is, however, one point in space applicable to any particular extended mass or assembly of mass points that is unique. Such a point is called the center of mass (c.m.).

If we look at a group of n mass points in the x-y plane, the center of mass for that group will be defined as that point whose x-coordinate is given by

$$x_{\text{c.m.}} = \frac{m_1x_1 + m_2x_2 + \cdots + m_nx_n}{m_1 + m_2 + \cdots + m_n}$$

$$= \frac{\sum_{i=1}^{n} m_ix_i}{\sum_{i=1}^{n} m_i} = \frac{\sum_{i=1}^{n} m_ix_i}{M}, \tag{3-7}$$

where M is the total mass of the group. Similarly,

$$y_{\text{c.m.}} = \frac{\sum_{i=1}^{n} m_iy_i}{M} \quad \text{and} \quad z_{\text{c.m.}} = \frac{\sum_{i=1}^{n} m_iz_i}{M}.$$

Fortunately, because of the properties of the space we operate in, any other orientation of the coordinate axes would not change the location of the center of mass relative to the other masses. Thus the center of mass is a unique point for any particular group of masses. As an example, consider the array of 4 masses shown in Figure 96. For these,

$$x_{\text{c.m.}} = \frac{m \times 3 + 2m(-1) + 3m \times 1 + 4m(-3)}{m + 2m + 3m + 4m}$$

$$= -\,0.8\ \text{m},$$

and

$$y_{\text{c.m.}} = \frac{m \times 2 + 2m \times 3 + 3m(-1) + 4m(-1)}{m + 2m + 3m + 4m}$$

$$= +\,0.1\ \text{m}.$$

When we have a continuous distribution of mass, instead of a group of discrete mass points, the summation must be made by integration. To illustrate, let us determine the center of mass of a uniform bar of mass m and length L (Figure 97).

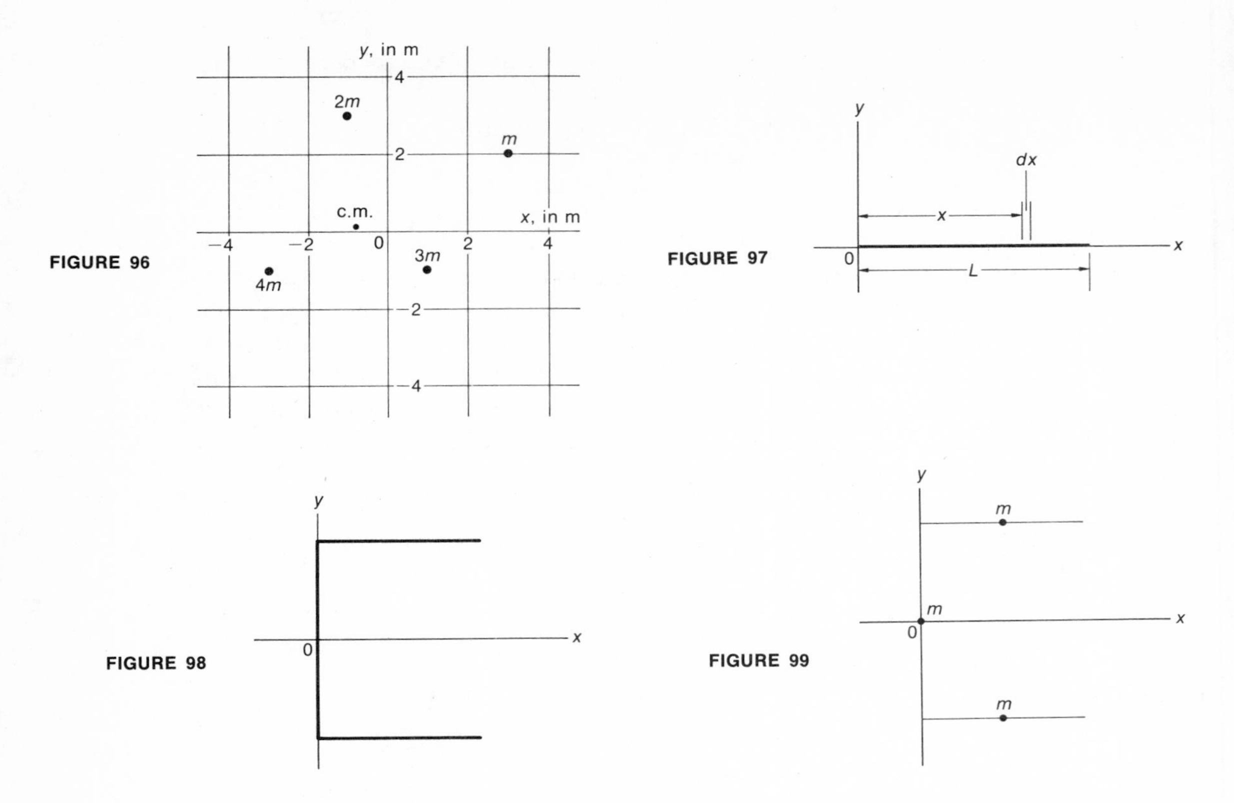

FIGURE 96

FIGURE 97

FIGURE 98

FIGURE 99

With the bar placed along the x-axis as shown, if λ is the mass of the bar per unit length, i.e., $\lambda = m/L$, then for $m_i x_i$ we have $(\lambda\, dx)x$, so that

$$x_{\text{c.m.}} = \int_0^L \frac{(\lambda\, x)\, dx}{m} = \left[\frac{\lambda x^2}{2m}\right]_0^L = \frac{L}{2}.$$

It should be quite obvious that $y_{\text{c.m.}} = 0$, so that the center of mass of a straight, uniform bar lies at the geometric center of the bar.

This result is far from startling; your intuition would have told you that the center of mass for a symmetrical body of uniform density would probably be at the geometric center. However, it's always comforting to see intuition and mathematical rigor confirm each other.

Let us now go from one bar to three identical bars, arranged as shown in Figure 98. With coordinates placed as indicated, you could safely say that $y_{\text{c.m.}} = 0$. If we set up the appropriate integration in the x-direction, as we did for the single bar, we can see that we would get the same result if we replaced the bars by equal point masses at the location of the c.m. of the bars, as done in Figure 99. From this array, and hence for the bar array, we can compute

$$x_{\text{c.m.}} = \frac{m(L/2) + m(L/2) + m \times o}{3m}$$

$$= \frac{L}{3}.$$

Thanks to an early Greek mathematician named Pappus, in many problems concerned with extended masses we can avoid integration. One of Pappus' theorems states that if we revolve an open curve about a line not cutting the curve, the surface area of the figure so generated equals the length of the curve multiplied by the distance the c.m. of the curve traveled during the generating revolution.

With this theorem in mind, let us look again at the same three-bar array of Figure 98. If we revolve the array completely about the axis A-A (Figure 100), the c.m. moves $2\pi x_{\text{c.m.}}$ and the surface generated equals $2 \times \pi L^2 + L \times 2\pi L$. Hence, $3L \times 2\pi x_{\text{c.m.}} = 4\pi L^2$ and $x_{\text{c.m.}} = \frac{2}{3}L$, which corresponds to the location previously computed.

Instead of a three-bar array, let us consider a uniform solid square of sides L (Figure 101). Another theorem of Pappus is an exact repetition of the theorem previously stated, except that the dimensions are increased by one; in this second theorem, the length of a curve is replaced by area, and the surface generated is replaced by volume generated. Using this second theorem for the solid square, the area $= L^2$, and the volume generated $= \pi L^2 \times L$. Hence, $\pi L^3 = L^2 \times 2\pi x_{\text{c.m.}}$ or $x_{\text{c.m.}} = L/2$. Again, not a very surprising result.

If we differentiate equation 3-7 twice with respect to time, we get, rearranged,

$$M\ddot{x}_{\text{c.m.}} = \sum_{i=1}^{n} m_i \ddot{x}_i. \tag{3-9}$$

By Newton's Second Law, $\Sigma m_i \ddot{x}_i$ is the sum of all the forces acting on all the masses, including internal and external forces. But by Newton's Third Law, the sum of all the internal forces must equal 0; equal and opposite internal forces cancel each other in a summation. Thus on the right-hand side of the equation we are left with just the x-components of the external forces:

$$M\ddot{x}_{\text{c.m.}} = \Sigma m_i \ddot{x}_i, = \Sigma F_{x,\text{ext}}. \tag{3-10}$$

Also, of course,

$$M\ddot{y}_{\text{c.m.}} = \Sigma m_i \ddot{y}_i, = \Sigma F_{y,\text{ext}};$$

$$M\ddot{z}_{\text{c.m.}} = \Sigma m_i \ddot{z}_i, = \Sigma F_{z,\text{ext}}.$$

Equation 3-10 is a powerful and very useful statement. It says that for a collection of masses, or for an extended massive object, the center of mass moves as if

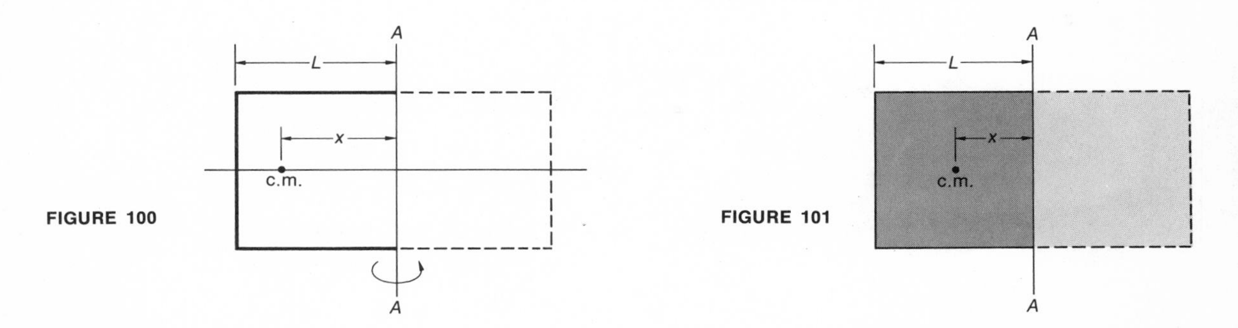

FIGURE 100

FIGURE 101

all the external forces acted at that point on the total mass concentrated at that point. You will find many occasions to take advantage of this particular aspect of the center of mass.

The Center of Gravity. An immediate application of equation 3-10 can be made when the $F_{i,\text{ext}}$ are due to a gravitational field, i.e., when $F_{i,\text{ext}} = m_i g$. If $\vec{g}$ is constant over the entire Σm_i (as would be the case, for instance, if the body represented by Σm_i was a relatively small body in the gravitational field of the earth outside the earth), then

$$\Sigma F_{i,\text{ext}} = m_1 g + m_2 g + \cdot \cdot \cdot + m_n g = g \, \Sigma m_i = Mg.$$

Going back to equation 3-10, we then have $M\ddot{y}_{\text{c.m.}} = Mg$. Thus in a uniform gravitational field the gravitational forces on the elements of an extended body act as if concentrated at the center of mass of the body. When used in this way, the center of mass is usually called the center of gravity. *Caution:* the center of mass and the center of gravity are identical *only* in the presence of a constant uniform gravitational field.

PROBLEM 203

You measure g at the surface of the earth, and measure G by a Cavendish-type experiment, and from surveying data you arrive at a value for the radius of the earth. Then what is the mass of the earth?

PROBLEM 204

a. What is the acceleration of the moon toward the earth, due to the gravitational attraction of the earth?

b. What is the centripetal acceleration of the moon in its orbit about the earth?

Historical note: it was a comparison of (a) and (b) that Newton used in checking the formula for his Law of Gravitation.

PROBLEM 205

a. Considering only the earth-moon system, at what point is an astronaut on his way from the earth to the moon really weightless, i.e., subject to no net gravitational force?

b. Relative to the earth-moon system, what is the curvature of his trajectory at that point?

c. Relative to an observer at rest with the sun, what is the curvature of the astronaut's trajectory at the point found in question (a)?

PROBLEM 206

In problem 90 you computed the value of g at the altitude of a synchronous satellite from data on the satellite's motion. In view of equation 3-5 do you now confirm that value?

PROBLEM 207

When astronauts returning from the moon to earth pass through the "weightless" point determined in problem 205, they are then in free fall toward earth.

a. Assuming that their thrust out of moon orbit was so weak that their space capsule barely got through the neutral point, how fast would the capsule be moving when it arrived at the upper limit of the earth's atmosphere, about 130 km above the earth's surface?
b. With what speed would the capsule reach the earth's atmosphere if it passed through the neutral point with a speed of about 700 m sec^{-1}? (The data given are reasonably valid for Apollo 12.)
c. If you neglected the effect of the moon's gravitational field in the region between the neutral point and the earth, how were your answers to questions (a) and (b) affected?

PROBLEM 208

To an observer on a star outside our galaxy, the Milky Way appears almost like an extensive sheet of fairly uniform density; particularly is this view valid in the area of the galaxy out where the solar system is located. If a star is observed over a long period of time to be approaching this galactic plane with an acceleration of about 16 km yr^{-2}, what is the apparent surface density of the galaxy?

PROBLEM 209

Your laboratory partner didn't trust your measurement of G in problem 203, so he decided to measure the mass of the earth in a different way. He got hold of a strong but sensitive equal arm balance and rigged it as shown in Figure 102, with a uniform solid sphere of known mass m in the lower pan and sufficient weights in the upper pan for exact balance. Then he brought a uniform solid sphere of mass M to a point d below m; to restore balance he had to add a small mass w to the upper pan. If the upper pan is R from the center of the earth, what was the mass of the earth?

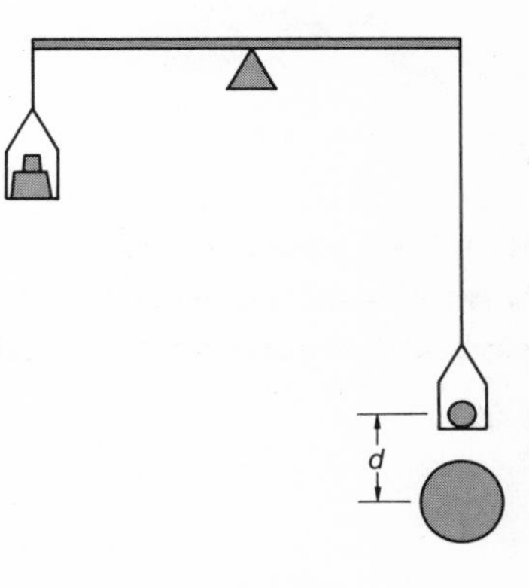

FIGURE 102

PROBLEM 210

In an airplane flying due east along the equator at a certain altitude with a ground speed v, an observer weighs a mass m on an accurate spring balance and records a weight W_{W-E}. On a return trip, flying due west along the equator at the same altitude and at the same ground speed, he again weighs the same mass on the same balance, and records a weight W_{E-W}. He notes that there is a difference between these two weights of ΔW, and then proceeds to compute m in terms of ΔW, v, and the rotational speed of the earth. What result did he get?

PROBLEM 211

In establishing his Theorem No. 30, Book I, Newton proved that inside a spherical shell of uniform surface density the gravitational field of the shell is zero; a point mass placed anywhere inside such a shell would feel no net gravitational force from the mass of the surrounding shell. Keep this theorem in mind, and assume that the earth is uniform and homogenous.

a. Determine g_r, the earth's gravitational field at a distance r from its center, $r < R$, where R is the radius of the earth.
b. Express g_r in terms of g_R, the earth's gravitational acceleration at its surface.
c. Plot g_r as a function of r, from $r = 0$ to $r > R$.

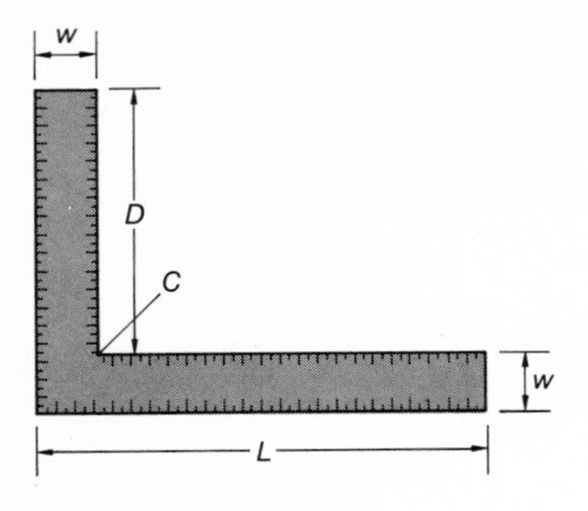

FIGURE 103

PROBLEM 212
How much lighter, by percent, will 2.0 mi of oil-well drill stem be when hanging freely in the well than when it lay stacked on the ground at the wellhead before use?

PROBLEM 213
A carpenter's square is made from a uniform sheet of steel, as shown in Figure 103.

a. Where is the c.m. of the square? *Hint:* Consider the square as made up of two rectangles, and treat each rectangle as a point mass concentrated at the rectangle's geometrical center. For normal values of w, the c.m. will not lie inside the square. There is nothing in the definition of the c.m. that requires that it be inside the system it refers to.
b. What should be the relation between L and D so that the c.m. is at the inside corner C?

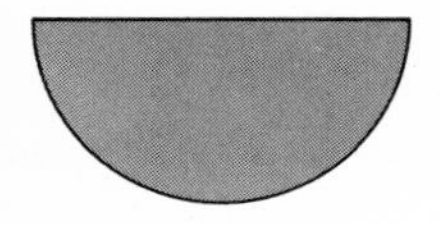

FIGURE 104

PROBLEM 214
A semicircle of radius R is cut from a sheet of plywood (Figure 104). Where is its c.m.?

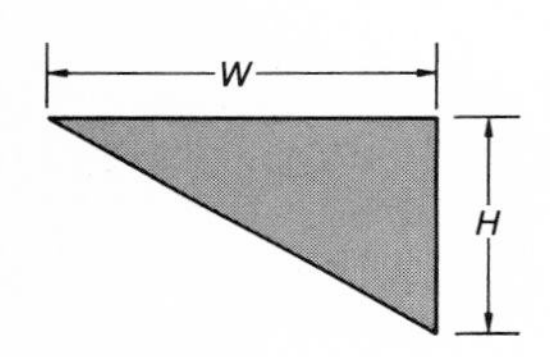

FIGURE 105

PROBLEM 215
A right triangle is cut from plywood to serve as a shelf base (Figure 105). Where is its c.m.?

PROBLEM 216
A uniform piece of wire is bent into a plane arc of a circle of radius R and subtended angle α (Figure 106). Where is the c.m. of the arc?

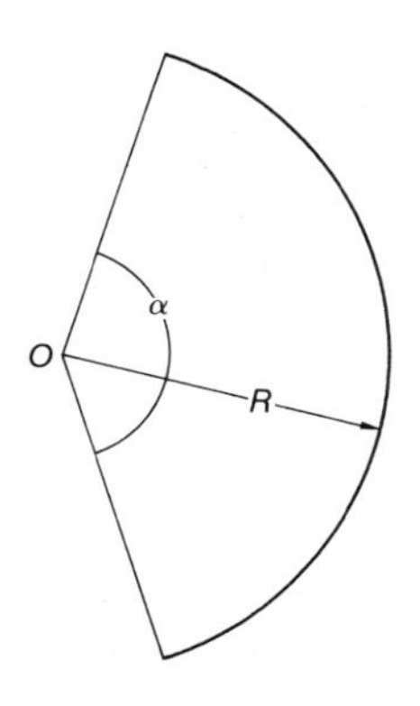

FIGURE 106

PROBLEM 217
To make a cam, a shaft hole of radius r is drilled through a solid disc of steel of radius R (Figure 107). The axis of the hole was parallel to the axis of the disc, and r from it. Where is the c.m. of the cam? *Note:* A hole can be treated as negative mass; to make a hole you add to a solid piece a hole-sized disc of negative mass. *Suggestion:* Put your center of coordinates at the center of the disc.

PROBLEM 218
Extremely delicate weighing is carried on in a ground-floor room of a laboratory. In an attempt to damp out vibrations a very large, steel

vault door is obtained from a bank building being demolished and is installed as a platform for the most sensitive balance.

a. As a result, what is the approximate percentage change in the gravitational field at the balance?
b. How does this change affect the weighing if the balance is i. a supersensitive beam balance, or ii. an electrostatic balance (such as is used in the Millikan oil-drop experiment)?

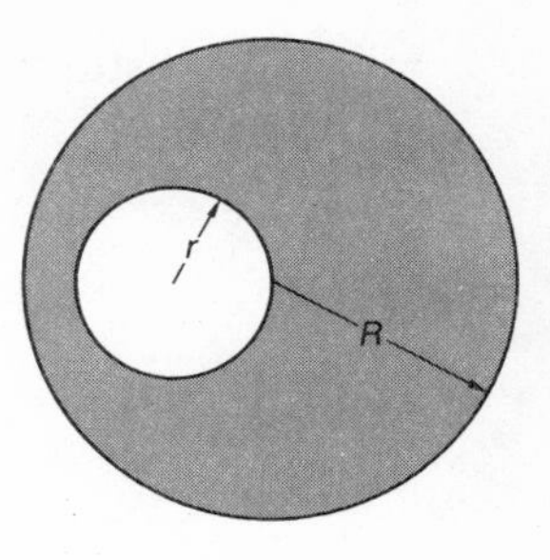

FIGURE 107

PROBLEM 219
It is the c.m. of the earth-moon system that is in orbit about the sun, and celestial observations made from the earth show, over a period of time, slight but measurable variations due to the earth's motion about this c.m. The "wobble" calculated from these variations indicates the c.m. of the earth-moon system is 2.9×10^3 mi from the center of the earth.

a. What is the mass of the moon in terms of the mass of the earth?
b. In view of the answer to problem 203, what is the mass of the moon?
c. What would be the value of the moon's gravitational acceleration at its surface?

PROBLEM 220
Where is the c.m. of the earth-sun system? Should you include the mass of the moon in your calculation?

In a simple planetary system—this could be Jupiter moving about the sun or a satellite moving about the earth—the gravitational force on the less massive body m—Jupitor or the satellite—is just the centripetal force necessary to keep that body in an orbit. Of course that same force, but opposite in direction, acts on the more massive body M to keep *it* in an orbit also. However, usually $M >> m$; then the c.m. of the system is so close to M that, to any reasonable accuracy, M can be regarded as stationary.

To keep the math simple, let us assume that the orbit of m about a relatively stationary M is circular, of radius r. We then have, from above,

$$\frac{GmM}{r^2} = mr\omega^2.$$

Here ω is the angular speed of m in its orbit about M. If T_m is the time in sec for one revolution in this orbit,* then $\omega = 2\pi/T$ rad sec^{-1}.

*Hereafter T will be called the period of revolution, or simply the period.

Substituting this, we have

$$\frac{GmM}{r^2} = mr\left(\frac{2\pi}{T_m}\right)^2$$

or

$$T_m^2 = \frac{4\pi^2}{GM} r^3.* \qquad (3\text{-}11)$$

Equation 3-11 is Kepler's Third Law, modified by Newton's Second Law. It is an unexpected relationship—it indicates that the period of m does not depend on the value of m. It is a useful relationship, because of the ease with which T_m can be measured.

If M is of the same order as m, instead of $M >> m$ as assumed above, then the c.m. of the system is not near M, and M cannot be treated as stationary. Let us consider a system in which M and m are separated by a distance a, and the only force acting is the internal gravitational force between M and m (Figure 108). Since there are no external forces the c.m. does not move, but M and m are dynamically balanced by the gravitational force, and move in opposition about the c.m. This motion is governed by

$$\frac{GmM}{(R+r)^2} = mR\,\omega^2 \; (= Mr\,\omega^2).$$

The c.m. is so located that $mR = Mr$. If we add MR to both sides of this equality, we have $(m+M)R = M(r+R)$, or $R = M(r+R)/(M+m)$.

Putting this value of R in the dynamical equation above we get

$$\frac{GmM}{(R+r)^2} = \frac{mM(R+r)}{(m+M)}\,\omega^2.$$

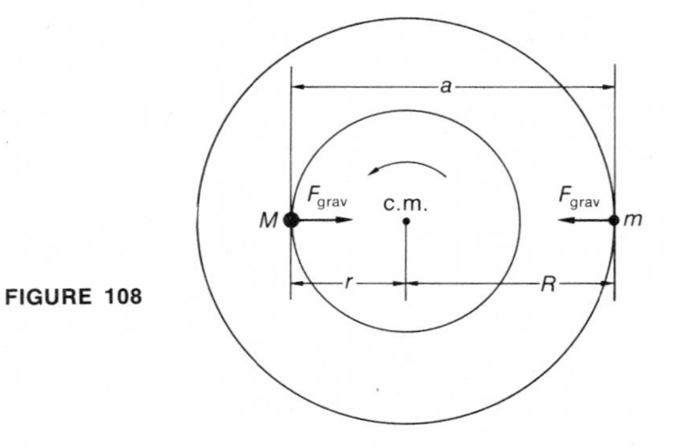

FIGURE 108

*Because of the assumption made earlier "to keep the math simple," this formula apparently would apply only to circular orbits. Actually, it applies with equal validity to elliptical orbits; for elliptical orbits, r represents the semi-major axis of the ellipse. In circular orbits, M is at the center of the circle; in elliptical orbits M is at one of the foci of the ellipse.

Since $\omega = 2\pi/T$, we finally get

$$T^2 = \frac{4\pi^2}{G(M+m)}(R+r)^3 = \frac{4\pi^2}{G(M+m)}a^3. \tag{3-12}$$

If T is expressed in years, a in A. U.,* and $(M+m)$ in sun masses, then the operating form of equation 3-12 is

$$T^2 = \frac{a^3}{(M+m)}. \tag{3-13}$$

Equation 3-12 is equation 3-11 adapted for use with binary star systems. Although developed from circular motion, equation 3-12 also is valid for elliptical orbits providing a is treated as the semi-major axis of the orbit of m relative to M.

* One astronomical unit (abbreviated A. U.) = radius of the earth's orbit = 92.9×10^6 mi = 1.49×10^{11} m.

PROBLEM 221

The moon has a certain period and orbit relative to the earth; the earth has a different period and orbit relative to the sun. By putting the data for these periods and orbits together, you can determine the mass of the sun in terms of the mass of the earth.

a. What is the mass of the sun, in earth masses?

b. In view of the answer to problem 203, what is the mass of the sun in kg?

PROBLEM 222

One of the moons of Mars travels in an orbit of radius 5.83×10^3 mi and makes a revolution in 7 hr and 39 min. The corresponding data about the earth's moon are 2.39×10^5 mi and 27 days, 8 hr. Calculate the mass of Mars in terms of earth masses.

PROBLEM 223

For an eclipsing binary—a double star system in which the plane of the stars' orbits is edgewise to observers on earth—it is easy to measure the period of rotation of the star system but usually very difficult to make any accurate determination of the separation a. However, the velocity of one star relative to its companion can easily be deduced from Doppler shifts in the spectra of the two stars. (It is more difficult to determine the absolute rotational velocity by separating it from any recessional velocity the c.m. of the star system may possess.) For a certain eclipsing binary the period is 3.96 days, and the relative velocity is calculated to be 226 km sec^{-1}. What is the total mass of the two stars?

PROBLEM 224

A missile is fired radially outward from the earth's surface with an initial speed v_0.

a. Determine the speed of the missile at some r, $r > R$, where R is the radius of the earth. *Suggestion:* Combine the formula for the radial acceleration $\ddot{r} = -GM/r^2$ with the fact that $\ddot{r}$ can be expressed as $\dot{r}\, d\dot{r}/dr$.

b. How far out did the missile get when its speed relative to the earth dropped to zero, i.e., when $\dot{r} = 0$?

c. What would v_0 have to be for $r = \infty$ when $\dot{r} = 0$? (The answer to this question is known as the "escape velocity.") *Note:* Later we will find a simpler and quicker way of determining the answers to this problem.

Electrical and Magnetic Forces

As indicated at the beginning of this book, the two fundamental properties of matter in the universe are mass and electrical charge. Both affect us through force, mass because of its reluctance to change motion unless we provide a force, charge because the interaction between charges is a source of force.

Forces between charges fall into two categories: one, the force between charges that exists because of the charges themselves—charges push or pull automatically at the sight of another charge; two, the force that arises when charges move relative to other charges. The first kind of force always exists in the presence of charges, whether they are at rest or moving uniformly; in spite of the latter phrase, historically this force has been called the electrostatic force. The second force arises only by virtue of relative motion of different charges, is usually of the order of $1/c^2$ times the value of the corresponding electrostatic force, and is called the electromagnetic or electrodynamic force.

Electrostatic Forces and Fields. If two point charges q_1 and q_2 are a distance r apart in free space,* a force exists between them that is proportional to q_1 and q_2 and inversely proportional to r^2; $F = k(q_1 q_2/r^2)$, where k is a constant of proportionality. This is Coulomb's Law. If we do what would seem natural, and make k a dimensionless constant equal to 1, the units of q are then defined. If F and r are in cgs units of dynes and centimeters, we have the basis for the electrostatic system of units referred to on page 10.

It turns out that what seems natural now will cause us more work, and create more complexity, later. To avoid that, we devise instead the MKSC system of units: we independently define the coulomb as the fundamental unit of electric charge, and we express Coulomb's Law in terms of newtons (N), meters (m), and coulombs (C). Since everything else in the equation is now defined, this results in a new definition of k, which no longer can be dimensionless or equal to 1. In the

* Usually interpreted to mean "in vacuum." It would probably be more accurate to say "in free space" means the charges are placed in an isotropic, continuous medium whose dielectric constant = 1. For normal accuracy, standard air can be considered dielectrically equivalent to free space.

MKSC system $k = 8.99 \times 10^9$ N m^2 C^{-2}. Numerically this $= 10^{-7} c^2$, where c is the speed of light in free space. (This is not just a mysterious coincidence; as we shall see later, both k and the measurement of the coulomb are related to c.) We can reduce future complexity even more by replacing k by $1/(4\pi\epsilon_0)$, in which ϵ_0 is called the permittivity of free space; its value is 8.85×10^{-12} N^{-1} m^{-2} C^2.

Putting this all together, the final and most useful form of Coulomb's Law, in MKSC units, is

$$\vec{F}_1 = \frac{q_1 q_2 \vec{r}_{2-1}}{4\pi\epsilon_0 (r_{2-1})^3}, \tag{3-14}$$

where $\vec{F}_1$ is the electrostatic force on q_1 from q_2, and $\vec{r}_{2-1}$ is the displacement vector from q_2 to q_1. An equivalent form of the above equation, one that will be found in many textbooks, is

$$\vec{F}_1 = \frac{q_1 q_2}{4\pi\epsilon_0 r^2}\vec{u}, \tag{3-15}$$

where $\vec{u}$ is a unit vector in the direction $q_2 - q_1$.

Electrical charges can be either positive or negative. When the charges are alike, the product $q_1 q_2$ is positive, and the Coulomb force is one of repulsion between the charges. When the charges are unlike the product is negative, and the Coulomb force is one of attraction. The signs of the charges automatically take care of this in the above vector equation.

Equation 3-14 looks exactly like the gravitational force equation on page 112; the only difference is in the constant of proportionality. This similarity suggests that everything we learned from the inverse square relation of gravitational phenomena should apply equally well to electrostatic phenomena. This suggestion is valid; we can, by substituting q for m and $1/(4\pi\epsilon_0)$ for G, deduce electrostatic knowledge from gravitational knowledge we have already acquired. In other words, Newton's Law of Gravitation and Coulomb's Law predict exactly the same kind of behavior in their respective areas. Thus we can make the following statements about electrostatics without any further independent investigation:

1. The contribution of q_n toward the electrostatic force on q_1 is independent of the presence of other charges $q_2, q_3 \ldots q_{n-1}$ in the neighborhood, so that

$$\vec{F}_1 = \frac{q_1}{4\pi\epsilon_0}\left(\frac{q_2\vec{r}_{2-1}}{(r_{2-1})^3} + \cdots + \frac{q_n\vec{r}_{n-1}}{(r_{n-1})^3}\right).$$

2. The electrostatic field $\vec{E}$, defined by $\vec{F} = q\vec{E}$, is force per unit charge; its direction at any point is the direction of the force felt by a unit positive charge placed at that point. Field vectors add vectorially as do force vectors; in parallel with the above equation we can write

$$\vec{E}_1 = \frac{1}{4\pi\epsilon_0}\left(\frac{q_2\vec{r}_{2-1}}{(r_{2-1})^3} + \frac{q_3\vec{r}_{3-1}}{(r_{3-1})^3} + \cdots\right).$$

The obvious units for E, from its definition $E = F/q$, are newtons per coulomb: $N\ C^{-1}$; in general we will find that a more useful unit for E is volts per meter. Volts will be defined later.

3. The electrostatic field of a uniformly charged sphere outside the sphere is computed as if the total charge on the sphere was concentrated as a point charge at the center of the sphere (see page 113);

$$E = \frac{Q}{4\pi\epsilon_0 r^2} = \frac{\rho R^3}{3\epsilon_0 r^2},\ r > R,\ Q = \frac{4}{3}\pi R^3 \rho,$$

where ρ is the charge density per unit volume.

4. The electric field inside a uniformly charged spherical shell due to the shell's charge is zero (see problem 202).

5. The electric field of a long straight wire carrying a uniform distribution of charge is perpendicular to the wire and at a distance d from the wire has the value $\lambda/(2\pi\epsilon_0 d)$, where λ is the value of the charge per unit length. (On page 113, substitute λL for M and let L *become* $>> d$.)

6. The electric field of a sheet of charges uniformly distributed is perpendicular to the sheet and has the value $\sigma/2\epsilon_0$, where σ is the charge per unit area. (See page 114 and as before, note that the field is constant and not a function of any reasonable distance from the sheet.)

7. The electric field inside a sphere of radius R that is uniformly filled with charge of density ρ, at a distance r from the center $r < R$, is radially outward and has the value $E_r = (r/R)\ E_R = \rho r/3\epsilon_0$. (See problem 211.)

This information could also have been derived independently from Gauss' Law. For free space the most useful form of this law is $\int \vec{E} \cdot d\vec{S}$* $= \Sigma q/\epsilon_0$—the summation over a chosen closed surface of the components of E perpendicular to that surface is proportional to the total charge inside the surface, the constant of proportionality being $1/\epsilon_0$ in the MKSC system. The term $\vec{E} \cdot d\vec{S}$ is called the flux of E, and the integral $\int \vec{E} \cdot d\vec{S}$ evaluates the total flux through the surface.

We can illustrate the use of Gauss' Law in an example we have already worked —the determination of the electric field of a very large plane sheet holding a uniform positive charge of σ per unit area. For our Gaussian surface we will choose a closed cylinder of cross-sectional area A and of length d, the axis of the cylinder perpendicular to the sheet (Figure 109). From symmetry in this case we would not expect to find E varying in a direction parallel to the sheet, nevertheless we will choose A small enough so that we can feel confident E has the same value through-

*An element of surface $d\vec{S}$ is represented vectorially by a vector perpendicular to the element and of a length that represents the scalar value of the area dS.

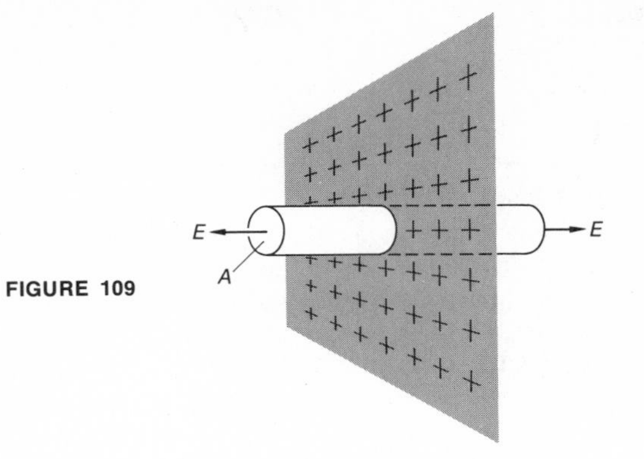

FIGURE 109

out A. Also from symmetry we expect E to be perpendicular to the sheet and hence parallel to the side of the cylinder. Then the total flux of E passes through the two ends, so that $\int \vec{E} \cdot d\vec{S} = 2AE$, which equals the charge inside the cylinder $A\ \sigma/\epsilon_0$. Hence $E = \sigma/2\epsilon_0$ as before.

Gauss' Law and Coulomb's Law produce the same end-product, an evaluation of E; in free space either can be used with equal validity. Which to use in any particular case will normally depend on the availability of symmetry that makes Gauss' Law so easy to apply. Coulomb's Law has the theoretical advantage that it explicitly tells us that the electric field of a stationary charge is an inverse square field and that in free space it is radial from a point charge. Gauss' Law has the practical advantage of providing very easy calculations in many situations, and the theoretical advantage that it is applicable in dielectrics as well as in free space; Coulomb's Law is not in general applicable in dielectrics.

As mentioned earlier, a man-made decision established the coulomb as the unit of electric charge in the MKSC system. Nature seems to prefer a different unit; all of electricity comes in multiples of the charge on an electron, either positive or negative—an electron and a proton each have one such unit, an alpha particle has two positive electron charges, a uranium nucleus has 92 positive electron charges, etc. In fact much of the use of Coulomb's Law will entail charge distributions quoted in terms of numbers of electron charges. For such use it is convenient to put $1/(4\pi\epsilon_0) = 8.99 \times 10^9$ N m^2 C^{-2} together with the charge on an electron $= 1.60 \times 10^{-19}$ C and define a new unit e such that

$$e^2 = 8.99 \times 10^9 \times (1.6 \times 10^{-19})^2 = 2.30 \times 10^{-28} \text{ N m}^2.$$

(Although e is rarely used by itself, it can be seen that e $= 1.52 \times 10^{-14}$ N$^{1/2}$ m; surprisingly, the unit of e contains no dimension of charge. If q_1 consists of m electron charges and q_2 has n electron charges, the Coulomb Law calculation of the force between q_1 and q_2 now reduces to

$$F = \frac{2.30 \times 10^{-28}\ mn}{r^2} \text{ N}.$$

PROBLEM 225

The hydrogen molecule contains two protons 0.74×10^{-10} m apart. How does the electrostatic force of repulsion compare with the gravitational force of attraction between the two protons?

PROBLEM 226

The unexcited hydrogen atom consists of a proton carrying one positive electron charge, and a negative electron outside the proton. Since the electron does not fall into the proton, but seems to stay on the average 0.53 Å from it, it is reasonable to believe that the electron-proton pair must be in some sort of dynamic equilibrium. The most obvious model for this equilibrium is an electron orbiting the proton, with the field of the stationary proton supplying the necessary centripetal acceleration. (This is the starting point for the Bohr model of the atom.) In such a Bohr atom, what must be the frequency of the electron's revolution about the proton? (Frequency $\nu = 1/T = \omega/2\pi =$ number of cycles sec^{-1}.)

PROBLEM 227

In solving problem 226 you probably used nonrelativistic formulae. Now calculate the orbital speed of the electron, to check whether the nonrelativistic approach was valid.

PROBLEM 228

In the Bohr atom discussed in problems 226 and 227, the proton nucleus was assumed to be at rest, and hence Coulomb's Law could be used. Since $m_p = 1836\ m_{\text{el}}$, in an isolated free hydrogen atom is the proton really at rest? *Hint:* It may be helpful to review the binary star discussion on pages 122–123.

PROBLEM 229

If the protons in a uranium nucleus are evenly distributed throughout the volume of the nucleus, they must be about 2.5×10^{-15} m apart. What electrostatic force must then exist between two adjacent protons due to those protons alone?

PROBLEM 230

If the uranium nucleus has a radius of about 7×10^{-15} m, what is the electric field at the (presumed) spherical surface of the nucleus due to the positive charges of the 92 protons in the nucleus? *Hint:* Apply Gauss' Law to a spherical surface just outside the nucleus.

PROBLEM 231

A conducting wire bent in the form of a circle of radius R is uniformly charged with a total charge Q.

a. What is the electric field of the loop on its perpendicular axis at a distance x from the plane of the loop?
b. Where is this field a maximum?

PROBLEM 232

a. By integrating a series of contiguous circular uniformly charged loops of width dr and radii ranging from 0 to R, determine the electric field on its perpendicular axis of a uniformly charged disc of radius R carrying a total charge Q, at a distance x from the plane of the disc.
b. Where is this field a maximum?

PROBLEM 233

An electron is released from rest on the axis R away from a disc of radius R uniformly charged with a total charge $+Q$.

a. With what speed does the electron pass through a very small hole in the center of the disc? (Solve this by integrating $\ddot{x}$ as a function of x.)
b. What is the maximum value the ratio Q/R (in MKSC units) may have for you to avoid having to treat this problem relativistically?

PROBLEM 234

In a mass spectrometer a beam of particles, each of mass m and charge q and with speed v, is to be turned through 90° without any ultimate change in speed. Space limitations dictate that this change must be accomplished within a distance d (Figure 110). The change is to be made electrostatically. What electric field is required if

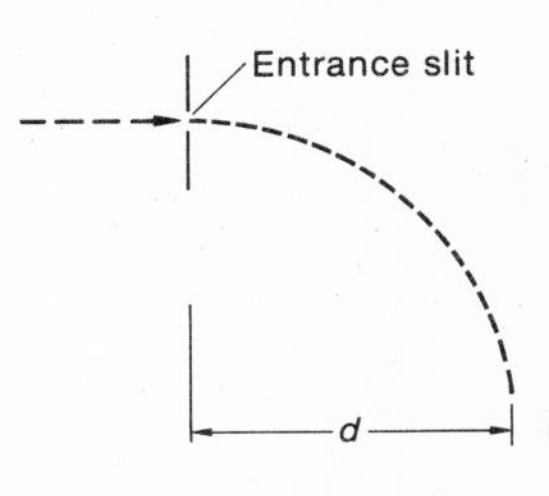

FIGURE 110

a. the field is constant in magnitude but radial in direction, as might be supplied by a pair of appropriately charged concentric curved plates;
b. the field is constant in both magnitude and direction?
(You very probably wouldn't do it as described in (b), but this method is interesting as a comparison. Why isn't method (b) a practical way of solving the problem?)

PROBLEM 235

If the beam of problem 234 represents a current of 1.0×10^{-6} amperes (a current of 1 ampere represents the passage of 1 coulomb of charge per second), what is the electric field of the beam at a distance r perpendicular to the beam?

PROBLEM 236

The average fair-weather electric field at the surface of the earth is about 130 volts m^{-1} vertically downward.

a. What must be the average surface charge density at the earth's surface? (The earth is a conductor, so that all of its net charge resides on the surface.)
b. What average total net charge is carried on the earth's surface, expressed in number of free electrons?

PROBLEM 237

If the average fair-weather electric field at an altitude of 1.5 km above the earth's surface is about 30 volts m^{-1} vertically downward, what must be the average density of the free charges in the earth's atmosphere below an altitude of 1500 m? (*Note:* Part of the necessary data for this problem is contained in the statement of problem 236.)

PROBLEM 238

Two parallel plates are charged with the same uniform charge density σ. Neglecting edge effects, what is the electric field in the region

a. between the plates, and
b. outside the plates, when the plates are i. charged alike; ii. oppositely charged?

PROBLEM 239

A central hole of radius r is cut out of the disc of problem 232 without disturbing the charge density on the rest of the disc.

a. At a point x on the perpendicular axis of the disc, what is now the electric field due to the disc?
b. For $x >> R$, what does the value of this field reduce to?
Hint: Just as we earlier took care of holes in masses by adding a volume of negative mass to a positive mass to obtain a volume of negative mass to a positive mass to obtain a volume of zero mass, we can obtain a region of zero charge by adding a region of one kind of charge to a charge assembly of the opposite sign. Thus the field of a disc with a hole in it and positively charged with a charge density σ is the same as the summation of a field of a complete disc charged with $+\sigma$ and a field of a disc of the size of the hole located at the hole and charged with $-\sigma$.

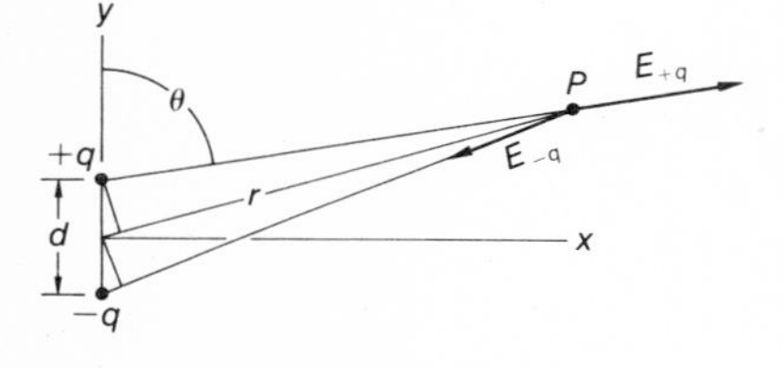

FIGURE 111

Field Derived from a Potential. We have seen in the last group of problems that the field of a single point charge (or of a uniform spherical distribution that can be treated as a single point charge) is easy to evaluate, both in scalar value and in direction. However, when evaluation must be made of two or more point charges, or of an unsymmetrical distribution, it is not that easy.

As an example, consider the dipole: two equal and opposite charges relatively close together. Nature is full of dipoles. To evaluate the field at P of the dipole shown in Figure 111, we note that it must consist of the vector addition of $\vec{E}_{+q}$, pointing away from the dipole, and $\vec{E}_{-q}$, of slightly less scalar value and pointing in a different direction toward the dipole. By computing the x and y components* of $\vec{E}_{+q}$ and $\vec{E}_{-q}$ and adding and squaring and talking the square roots etc., we could ultimately, by main strength and a capacity for grind-through, obtain both the scalar value and direction of $\vec{E}_p$. After we had done such a calculation a couple of times, we would be strongly tempted to look for a simpler way.

*For any x-y plane, no matter how it is revolved about the y-axis shown, the scalar values of the components would be the same. In other words, the third dimension z is not needed to evaluate E. We can take advantage of this fact in this problem to ignore z, and thus simplify the algebra.

In hunting for this simpler way, it should occur to us that adding scalars is always easier than adding vectors. And so we would like to find a scalar that is related to our field vector. And by this time, having studied a little vector calculus, we know that even if we don't find such a scalar, we can invent one.

A vector $\vec{E}$ has direction in space, and therefore has components E_x, E_y, and E_z in Cartesian coordinates (or E_r, E_θ, and E_ϕ in spherical coordinates). $\vec{E}$ can therefore be described by $\vec{E} = E_x\vec{i} + E_y\vec{j} + E_z\vec{k}$, where $\vec{i}$, $\vec{j}$, and $\vec{k}$ are unit vectors along the x-, y-, and z-axes respectively, and E_x, E_y, and E_z are scalar values. We then postulate that there exists some scalar function of coordinates V such that $E_x = -\delta V/\delta x$, $E_y = -\delta V/\delta y$, and $E_z = -\delta V/\delta z$.* Then

$$\vec{E} = -\left[\frac{\delta V}{\delta x}\vec{i} + \frac{\delta V}{\delta y}\vec{j} + \frac{\delta V}{\delta z}\vec{k}\right].$$

This can be summarized by the vector notation $\vec{E} = -\nabla V$. ∇V is called the gradient of V, which seems logical, since $\delta V/\delta x$ gives the change in V in the x-direction, etc.

Since $E_x = -\delta V/\delta x$, etc., we can integrate; we then have

$$\int_x dV = -\int_x E_x\, dx,$$

or, more generally,

$$\int_r dV = \Delta V_r = -\int_r E_r\, dr.$$

This puts us in a position to relate the V that we postulated to the $\vec{E}$ that we already have. Since, for a point charge,

$$E_r = \frac{q}{4\pi\epsilon_0 r^2},$$

$$\Delta V_r = -\int_{r_0}^{r} \frac{q\, dr}{4\pi\epsilon_0\, r^2},$$

or

$$\Delta V_r = V_r - V_{r_0} = \frac{q}{4\pi\epsilon_0}\left(\frac{1}{r} - \frac{1}{r_0}\right).$$

At this point we must stop and think: what shall we choose for our reference limit r_0? Obviously not zero, but we are free to choose any other value. From long experience physicists have learned that in the majority of situations it is most con-

*The $-$ sign is introduced here arbitrarily, without justification. We will see later that this minus sign enables us to give a physical meaning to V that is not essential now; at this time all we need to know about V is that it satisfies the requirement $-\delta V/\delta x = E_x$, etc.

venient to set $r_0 = \infty$ and $V_{r_0} = 0$, i.e., $V = 0$ at infinity. Taking this into consideration, for a point charge we finally have

$$V_r = \frac{q}{4\pi\epsilon_0 r}. \tag{3-16}$$

V_r is called the electrostatic potential of q at a distance r from q. In view of its definition above, an obvious unit for potential would be newton meter per coulomb, or N m C^{-1}. This combination is better known as "volt"; 1 newton meter per coulomb = 1 volt. (Now you can trace where the "volt per meter" unit for E came from.) The dimensions of electrical potential are $ML^2T^{-2}C^{-1}$.

Returning now to our dipole and its diagram, Figure 111, for the electric potential at P,

$$V_P = \frac{+q}{4\pi\epsilon_0(r - \frac{1}{2}d\cos\theta)} + \frac{-q}{4\pi\epsilon_0(r + \frac{1}{2}d\cos\theta)} = \frac{qd\cos\theta}{4\pi\epsilon_0(r^2 - \frac{1}{4}d^2\cos^2\theta)}.$$

For any r that would normally be of interest $\frac{1}{2}d << r$; thus for the normal electrostatic potential of a dipole,

$$V_r = \frac{qd\cos\theta}{4\pi\epsilon_0 r^2} = \frac{p\cos\theta}{4\pi\epsilon_0 r^2} = \frac{\vec{p}\cdot\vec{r}}{4\pi\epsilon_0 r^3}, \tag{3-17}$$

where $p = qd$ is called the dipole moment. The direction of $\vec{p}$ is from $-$ to $+$. Note that the electrostatic potential of a single point charge falls off as $1/r$, but that when you bring two point charges together to form a dipole the electrostatic potential falls off as $1/r^2$.

We can now return to the problem of the field of a dipole at point P. If we wish to work in Cartesian coordinates, we note from our diagram that $\cos\theta = y/(x^2+y^2)^{1/2}$ and $r^2 = x^2 + y^2$. Then

$$V_P = \frac{q\,d\,y}{4\pi\epsilon_0(x^2+y^2)^{3/2}},$$

$$E_x = -\frac{\delta V}{\delta x} = \frac{3xyqd}{4\pi\epsilon_0(x^2+y^2)^{5/2}},$$

and

$$E_y = -\frac{\delta V}{\delta y} = \frac{(2y^2 - x^2)\,qd}{4\pi\epsilon_0(x^2+y^2)^{5/2}}.$$

Then

$$\vec{E} = \frac{qd\sqrt{4y^2+x^2}}{4\pi\epsilon_0(x^2+y^2)^2}$$

at an angle α with the vertical given by

$$\tan\alpha = \frac{3xy}{2y^2 - x^2} = \frac{3\tan\theta}{2 - \tan^2\theta}.$$

Note that $E_x = 0$ for either $x = 0$ or $y = 0$, which is what you should expect.*

In many situations polar coordinates will be more useful than Cartesian coordinates. Evaluating the field of a dipole by means of polar coordinates, we have

$$E_r = -\frac{\delta V}{\delta r} = \frac{2\,qd\,\cos\theta}{4\pi\epsilon_0\,r^3}$$

and

$$E_\theta = -\frac{\delta V}{r\,\delta\theta} = \frac{qd\,\sin\theta}{4\pi\epsilon_0\,r^3},$$

where $\vec{E}_r$ lies in the direction of $\vec{r}$ and $\vec{E}_\theta$ is perpendicular to $\vec{r}$.† Then

$$\vec{E} = \frac{qd\,\sqrt{3\,\cos^2\theta + 1}}{4\pi\epsilon_0\,r^3}$$

at an angle β with the vertical given by $\beta = \theta + \tan^{-1}(\frac{1}{2}\tan\theta)$. (It might be interesting for you to check to see whether you get the same unique $\vec{E}$, regardless of which coordinate system is used to describe it.)

This calculation of the field of a dipole illustrates the statement made on page 127—that the usefulness of Gauss' Law is determined by the symmetry available. If we surround the dipole by a spherical surface, we know that the total flux through that surface must equal zero because the sum of the charges inside the surface equals zero. All that tells us is that some E points out of the surface and some E points inward, but this is of no help in evaluating E.

**Warning:* Remember that these relations were developed for $r >> d/2$; they are not valid close to the dipole. To illustrate this warning, consider the field at the origin: From Coulomb's Law, $E_0 = 2q/\pi\epsilon_0 d^2$ in the $-y$ direction. If you attempt to determine E_0 by evaluating

$$E = \frac{qd\,\sqrt{4y^2 + x^2}}{4\pi\epsilon_0(x^2 + y^2)^2}$$

at $x = 0$ and $y = 0$, it is clear you do not get the correct answer. (What you do get, in fact, is that $E_0 = \infty$!)

†For the use of polar coordinates and their derivatives, see Appendix E.

PROBLEM 240

Referring to the diagram of an electric dipole, Figure 111, what is the potential of the plane $y = 0$, due to the dipole alone?

PROBLEM 241

In view of the warning in the footnote below, there is obviously a lower limit to the value of r for which the dipole field formulae derived from the potential are valid. To determine E_y without incurring a percentage error greater than $\pm$ 0.1 %, how close to the center of the dipole can you apply the above formula

a. on the x-axis;
b. on the y-axis?

PROBLEM 242

Would it have been easier to solve the problem of the field along the axis of a charged disc (problem 232) by first determining the potential along the axis?

PROBLEM 243

A Geiger-Muller tube for detecting ionizing particles consists of a conducting cylinder of radius R, a conducting wire of radius r mounted axially in the cylinder, and a glass envelope surrounding both cylinder and wire. When the wire is maintained at a potential difference $+V$ relative to the cylinder, what is the electric field

a. at the surface of the wire;
b. at the inner surface of the cylinder?

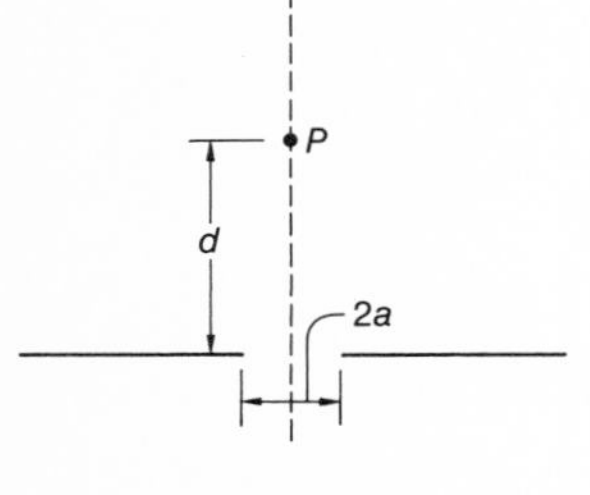

FIGURE 112

PROBLEM 244

A very long, straight rod, uniformly charged with $-\lambda$ coulombs per unit length, is cut into two equal parts and the cut ends separated by a distance $2a$ (Figure 112).

a. What is the field at point P, which is at a distance d from the rod and on a line perpendicular to the rod through the middle of the open section?
b. What is the difference in electric potential between P and a point Y; Y being on the same line through P but at a distance y from the rod? Consider a small enough so the a^3/d^3 is negligible compared to 1.

PROBLEM 245

A solid sphere of radius R is uniformly charged with a total charge $+Q$. (This could be a model for the uranium nucleus, for example.) Compute and graph the electric potential, due to sphere's charge alone, from $r = 0$ to $r > R$.

START OF SOLUTION

This problem can be solved easily by purely mathematical manipulation of $\Delta V = -\int E_r\,dr$. However, just to get a fresh look, let us approach the same computations from a different viewpoint.

If we plot E_r as determined by Gauss' Law, we obtain the graph shown in Figure 113. For $r \leq R$,

$$E_r = \frac{\frac{4}{3}\pi r^3\,\rho}{4\pi\epsilon_0 r^2} = \frac{r\rho}{3\epsilon_0}$$

$$= \frac{r}{R}\frac{Q}{4\pi\epsilon_0 R^2}\ \text{(from } Q = \tfrac{4}{3}\pi R^3\rho\text{)}.$$

For $r > R$,

$$E_r = \frac{Q}{4\pi\epsilon_0 r^2}.$$

From our recognition that $\int E_r\,dr$ is just the area under the curve of E plotted against r, we see that when $r < R$ the relationship is

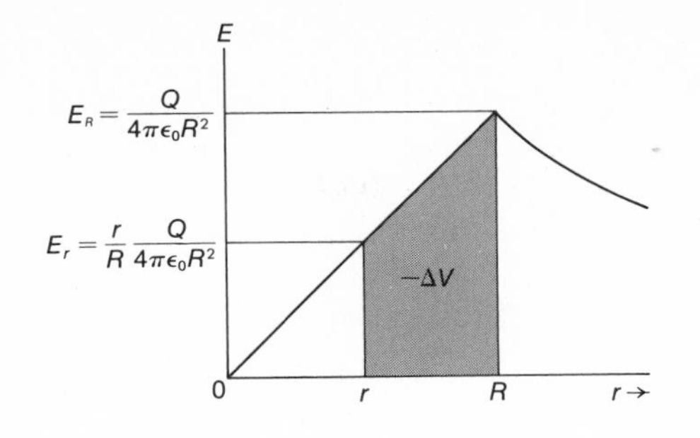

FIGURE 113

linear and the area is very easy to calculate. But by the definition of V, this area is also $-\Delta V$. Hence, from the diagram,

$$\Delta V = V_R - V_r = -\frac{Q}{4\pi\epsilon_0 R^2}\frac{(1+r/R)}{2}(R-r) = -\frac{Q}{8\pi\epsilon_0 R}\left(1-\frac{r^2}{R^2}\right).$$

Relative to $V_\infty = 0$,

$$V_R = \frac{Q}{4\pi\epsilon_0 R}$$

(from page 132). Hence for $r<R$,

$$V_r = V_R + \frac{Q}{8\pi\epsilon_0 R}\left(1-\frac{r^2}{R^2}\right) = \frac{Q}{8\pi\epsilon_0 R}\left(3-\frac{r^2}{R^2}\right).$$

PROBLEM 246

A suggested model for the unexcited hydrogen atom (compare the Bohr model discussed in problem 226) consists of a point charge proton at the center of a spherical electron "cloud" of radius R. The charge density of the cloud is uniform throughout the cloud, and the total charge of the "cloud" equals the charge of the electron.

a. What is E i. inside the cloud, ii. outside the atom;
b. what is V i. inside the cloud, ii. outside the atom?

PROBLEM 247

A large, plane slab of dipoles contains N dipoles per unit area, all alike with dipole moment p and all oriented in the same direction perpendicular to the face of the slab. At a distance z from the slab, z small compared to the lateral dimensions of the slab, due to the dipoles alone, what is the

a. electric potential;
b. electric field?

PROBLEM 248

Imagine a conical surface of cone angle θ erected on a single dipole of moment p, as shown in Figure 114. If r is the distance from the dipole to a point on the conical surface, how does the direction of the field of the dipole at the conical surface vary with r?

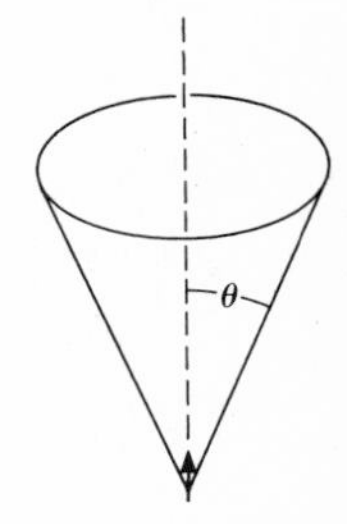

FIGURE 114

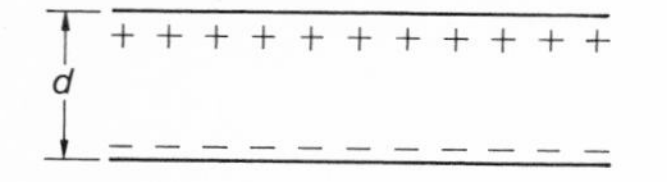

FIGURE 115

PROBLEM 249

Two parallel plates carrying the same charge density but with opposite sign, as shown in Figure 115, are a distance d apart. What potential difference must exist between the two plates?

PROBLEM 250

If the deflection plates of problem 72 were spaced 6.0 mm apart, what potential difference had to be maintained across them to provide the stated acceleration?

FIGURE 116

PROBLEM 251

An electron gun shoots a uniform beam of electrons at a speed u m $\sec^{-1}$ on a line midway between two parallel conducting plates between which a potential difference of V volts is maintained (Figure 116). The width of the effective field between the plates is L m, and the plates are d m apart. It is observed that the beam is deflected through an angle θ.

a. What was the value of V?
b. How far above the axis is the beam when it leaves the region between the plates?

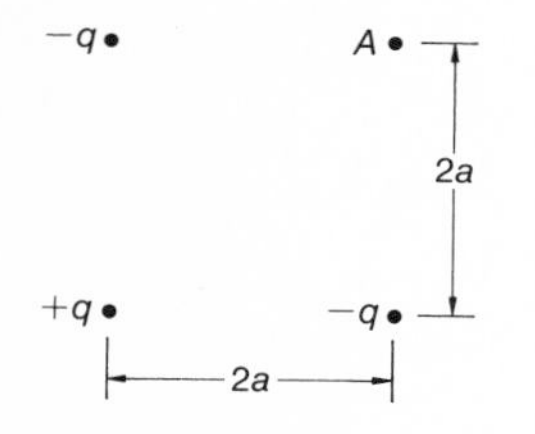

FIGURE 117

PROBLEM 252

An oildrop of known mass m is brought to vertical equilibrium between two parallel horizontal conducting plates d distance apart by adjusting the potential difference between the plates to V volts. How many free electrons were trapped on the drop?

PROBLEM 253

Point charges are placed at three corners of a square $2a$ on a side as shown in Figure 117.

a. What is the electric potential at the fourth corner A?
b. If a point charge $+q$ is brought to A, what force will be required to hold it there?

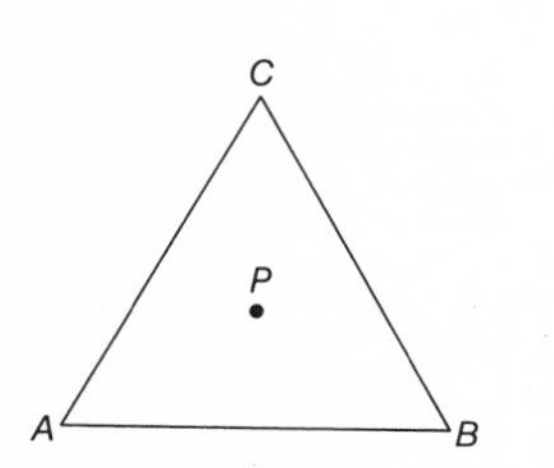

FIGURE 118

PROBLEM 254

In a region originally field-free, a point charge $+q$ is placed at A, an apex of the equilateral triangle ABC whose sides are s long (Figure 118).

a. What is the resulting electric potential at C, assuming $V_\infty = 0$?
b. After an additional identical charge is brought to B, what is the resulting potential at C?
c. When a third identical charge is brought to C, what is the potential at P, the center of the triangle?
d. What charge should now be brought to P to reduce the potential at the corners of the triangle to zero?

PROBLEM 255

What is the gravitational potential of the earth, relative to $V_\infty = 0$, as a function of the distance from the center of the earth r,

a. when $r > R$, the radius of the earth;
b. when $r = R$;
c. when $r = 0$?

By direct analogy with the electric case, $V_{\text{grav},r} = -GM/r$ for $r \geq R$ (if we follow the convention that $V_{\text{grav},\infty} = 0$). The dimensions of V_{grav} are L^2T^{-2}; the usual units are N m kg^{-1}.

PROBLEM 256

At a point on the surface of the earth what is the gravitational potential due to the earth, the sun, and the moon? How much of this is contributed by the sun and how much by the moon? (The answer to this problem is a starting point for computing trajectories for space shots.)

Conductors and Equipotential Surfaces. In view of the Coulomb force of repulsion between like charges, when a uniform distribution of charge is specified—the uniformly charged plane sheet of page 127, for example—the charges are not free to move. (If they were movable, they would be more densely packed near the edges of the sheet rather than uniformly distributed.) Materials in which charges stay where they are put, or at least do not move about easily, are called insulators or dielectrics; we were discussing charges embedded in dielectrics in examples 3, 6, and 7 on page 126 and in several of the problems. However, for most purposes charges moving do more for us than charges sitting, so it is fortunate we have materials, particularly metals, on which negative charges move very easily. Such materials, called conductors, possess two important characteristics resulting from this freedom of movement of electrons:

1. The electrostatic field at the surface of the conductor is perpendicular to the surface.

2. Inside the surface of a conductor the electrostatic field is zero.

It is easy to prove that charges on or in a conductor, because of their freedom of movement, would automatically adjust themselves in such a way that both characteristics would be true.

Since the electrostatic field (which is the gradient of the electrostatic potential V) is perpendicular to the surface of a conductor, V at the surface can only change in a direction perpendicular to the surface; it must remain constant along the surface. (This is electrostatic equilibrium we are discussing; it should not be confused with the dynamical situation in which an impressed field parallel to the surface is used to move charges along the surface, as in the flow of current on a conductor.) Thus a conducting surface must everywhere be at the same electrostatic potential. This includes not only the obvious case of a single metal body but also an array of metallic bodies connected by a conducting metal wire; all the elements of such an array must be at the same electrostatic potential. This applies to the interior of conductors as well. Since inside a conductor the gradient of the potential is zero, the potential must remain constant; it has the same value everywhere inside a conductor as it has at the surface of the conductor.

The interesting converse is that a theoretical equipotential surface in space can be replaced by an actual conducting surface at the appropriate potential without creating any disturbance in the environment. The field of a point charge is radial, hence an equipotential for a point charge is a spherical surface centered on the charge; therefore a metal sphere can be placed around the charge without affecting the external field. Similarly, a line charge can be surrounded by a metal cylinder with the same nondisturbing results.

PROBLEM 257

In view of the statement that inside the surface of a conductor the electric field is zero, what must be E at the surface of a conductor where σ is charge density? *Hint:* Apply Gauss' Law to a small cylindrical box that penetrates the surface.

PROBLEM 258

In problem 238, did it make any difference in your answer whether the plates were insulator or conductors?

PROBLEM 259

Eight small charged raindrops, each of the same size and each by itself at an electrical potential V_0, come together in a storm and form a larger drop. What is the electric potential of the combined drop?

PROBLEM 260

Two metal spheres, one of radius r and one of radius $2r$, hang from long insulating threads over a demonstration lecture table. The spheres are respectively charged with total charges q_1 and q_2, neither q being known as to value. From the known masses of the spheres and the measured angles the threads make with the vertical, the force F and the separation d between the spheres can be determined. Then a shorting wire is touched to both spheres; to everyone's surprise, the spheres do not move or change position in any way. In terms of F and d, what were q_1 and q_2?

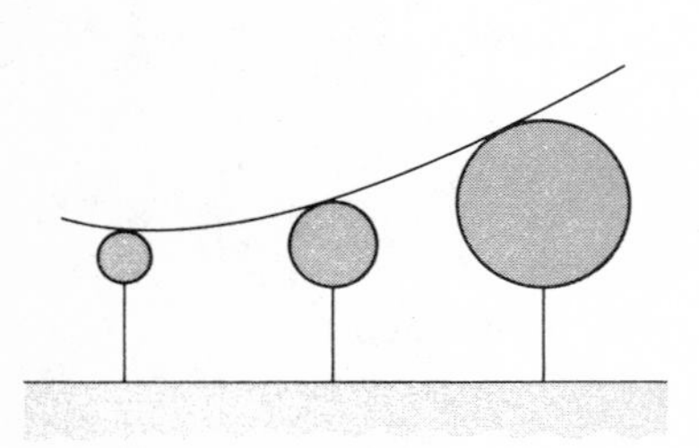

FIGURE 119

PROBLEM 261

For a demonstration you have uncharged metallic spheres of radii R, $\frac{1}{2}R$, and $\frac{1}{3}R$ mounted on insulated stands (Figure 119). Separately you charge the R and $\frac{1}{3}R$ spheres to a potential V above ground. Then with the spheres as widely spaced as possible you lower a fine copper wire so that it touches all of them briefly.

a. What is the subsequent potential above ground of the $\frac{1}{2}R$ sphere?
b. How much of the original charge on the two spheres is now on the $\frac{1}{2}R$ sphere?

PROBLEM 262

If the electric field at the surface of a conductor becomes strong enough to ionize the neighboring dielectric, charge migration takes place between dielectric and conductor; this limits the surface charge density, which in turn places an upper limit on the potential of the conductor.

a. If standard air starts to ionize in the presence of a field of 3.0×10^6 volts m^{-1} (called the breakdown voltage for air), what is the highest operating potential for a Van de Graaff generator whose dome has a diameter of 1.2 m?

b. In working part (a), did you discover any useful relationship between the maximum possible potential of a conductor, the critical radius of the conductor, and the breakdown voltage of the neighboring dielectric?

PROBLEM 263

Static eliminators trailing from the wing of an airplane have a tip radius of about 0.01 mm. What is the maximum electric potential that can develop on a wing equipped with such eliminators?

PROBLEM 264

A long straight wire of radius r and carrying a uniform charge of $+\lambda$ coulombs per unit length is surrounded by a thin-walled metal cylinder of radius R centered on the wire. The cylinder is uncharged and insulated from ground. The wire is maintained at a potential V_0 above ground.

a. What potential relative to ground will the cylinder attain?

b. What will be the electric field at the outer surface of the cylinder?

c. Will an observer at a distance from the wire greater than R have any way of detecting electrically whether or not the cylinder has actually been installed around the wire?

PROBLEM 265

Two charges, $+q$ and $-q$, are held fixed a distance $2d$ apart.

a. What is the value of the electric field in a plane containing the line between the two charges? *Suggestion:* Establish an x-y coordinate system with the x-axis passing through the charges and the center of the coordinate system midway between the charges. Note that such an x-y plane can be revolved about the x-axis without affecting the values of E_x and E_y; thus these values apply at all points on a circle about the x-axis whose plane is perpendicular to that axis.

b. Where could you put a large, thin plane conducting sheet near the charges without creating any disturbance?

c. Relative to $V_\infty = 0$, what would be the electric potential of this sheet?

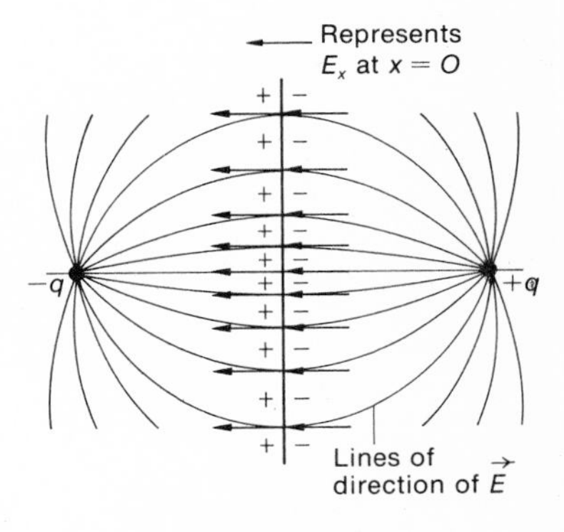

FIGURE 120

Let us take a closer look at the conducting sheet inserted in problem 265. In order to maintain the original field at the two surfaces of the sheet, charges must so arrange themselves on the sheet (Figure 120) that at every point the charge density at that point is just exactly sufficient to provide the necessary E_x at that point (cf. problem 257). Now having discovered that our conducting sheet is actually two

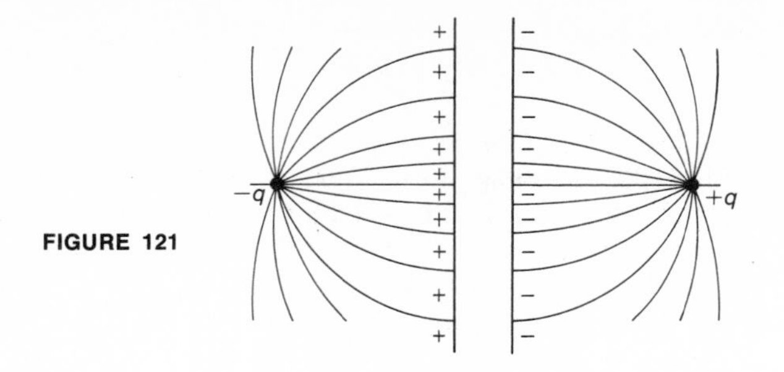

FIGURE 121

thin conducting sheets held in contact, let us separate the two halves of our assembly, as shown in Figure 121. We find to our surprise that nothing happens. It took no work to separate the halves (there was no field between the two sheets),* the electric field configuration did not change, and the total potential energy of the two separate systems remained the same as the original potential energy of the two charge system.

What we have arrived at with either of the halves is a solution to a more difficult problem—the determination of the electric field between a point charge and a plane conductor. The example illustrates—in reverse order—what is called the method of images. To determine the field between a given point charge and a given conductor, fictitious image charges are so placed in relation to the real charge that the resulting field due to both real and image charges creates an equipotential surface coincident with the real surface of the given conductor. Then this resulting field gives exactly the real field between the real charge and the real conductor, and the charge density at any point on the surface of the conductor will be appropriate to the real field at that point.

PROBLEM 266

A point charge $+q$ is d away from a large plane, grounded conductor. If the foot of a perpendicular dropped from the charge to the conductor is made the center of a polar coordinate system on the conductor,

a. what is the charge density on the conductor as a function of r,
b. how large should R be so that a total surface charge of $-0.9q$ is contained within a central area of radius R,
c. how does the field at the conductor at $r = R$ compare with the field at $r = 0$,
d. at the point where $+q$ is located what is the potential due to the charges on the sheet?

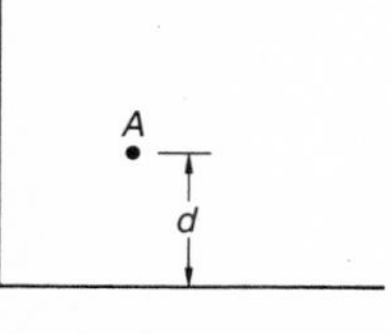

FIGURE 122

PROBLEM 267

A point charge of $+q$ is placed at A, at distance d from both sides of a grounded conducting plate bent at right angles (Figure 122); as seen from A, both sides of the plate seem essentially infinite in extent.

---

*This is strictly true only if the sheets are large enough so that the total surface charge on a sheet is, within any desired degree of accuracy, equal in amount to the point charge in front of the sheet.

a. What is the potential at A due to the surface charges induced on the plate?
b. What force must be supplied at A to hold the charge $+q$ in place there?
Suggestion: If you have any trouble getting started on this, you might review problem 253.

Digression—Force on a Surface Charge. We have already discovered that a point charge q just outside the surface of a conductor feels a force on it of $(\sigma/\epsilon_0)\,q$, where σ is the density of surface charge on the conductor. This suggests a question: is there any force on the surface charge itself?

Consider a Gaussian pillbox constructed on the surface of the conductor, with the layer of surface charge included:

$$E_\sigma = \frac{\sigma}{2\,\epsilon_0}$$

$$E_\sigma = \frac{\sigma}{2\,\epsilon_0}$$

From this charge alone we have the field E_σ as shown; the layer of surface charge is just the same as the sheet of uniform charge discussed on page 127. However, there is something wrong with this picture; it shows a field $\sigma/2\,\epsilon_0$ inside the conductor, whereas we know the field there $= 0$. Hence there must be an external field $\sigma/2\,\epsilon_0$ impressed on the pillbox from charges elsewhere on the conductor, thus:

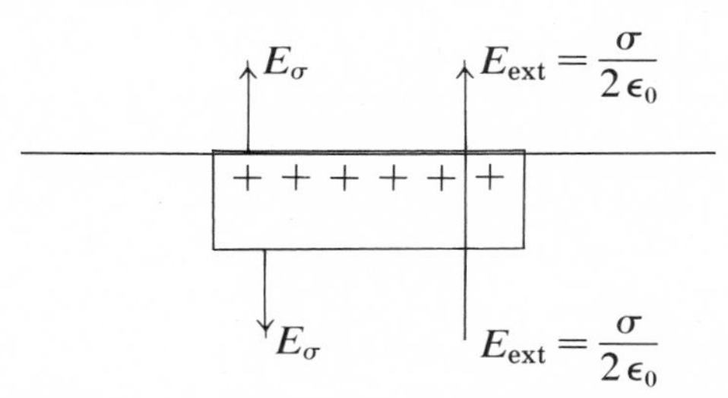

With this additional field the total field outside the conductor is σ/ϵ_0 and the field inside the conductor is zero, as they both must be. However, this external field acts on the surface charge itself, with a resulting force per unit area of $\sigma^2/2\,\epsilon_0$. This force is outward always, regardless of whether σ is positive or negative.

PROBLEM 268
A metal ball of radius 10 cm is charged to a potential of -5×10^3 volts above ground. On the average, what intra-atomic force must be available to hold each conduction electron on the ball?

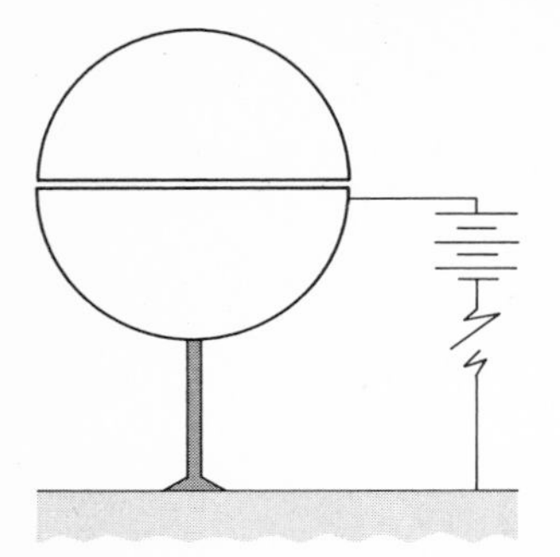
FIGURE 123

PROBLEM 269

A conducting hemisphere of radius R resting on an insulated stand with its diametral plane horizontal is capped by a similar hemisphere (Figure 123). The assembled conducting sphere is then gradually charged; when the sphere reaches a potential V above ground the top hemisphere starts to lift off of the lower one. If this top hemisphere weighed W newtons, what was V?

Comment: Probably you have an intuitive feeling that a uniform force per unit area pushing normally against a hemispherical surface ought to produce the same result as the same force per unit area pushing normally against a flat plate of the same diameter (Figure 124). (If the results weren't the same, you could make a perpetual motion machine by building a barge with hemispherical bosses on one end, as in Figure 125.)

However, until your intuition is backed by a large amount of experience, it is better practice to validate your intuition mathematically, particularly when the mathematics is easy. If f is the normal force per unit area, then from the diagram in Figure 126,

$$F_y = \int_0^{\pi/2} f\, 2\pi R \cos\theta\, R\, d\theta \sin\theta.$$

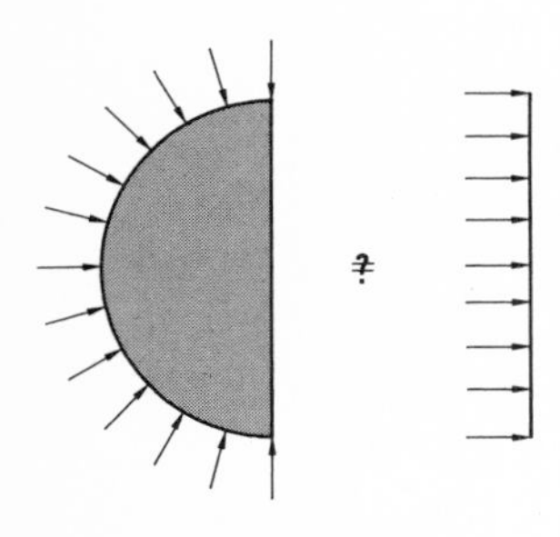

FIGURE 124

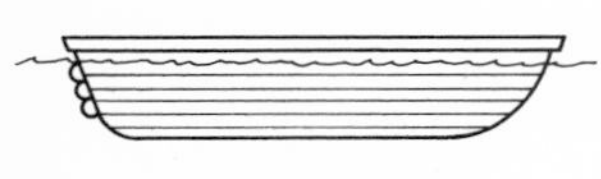
FIGURE 125

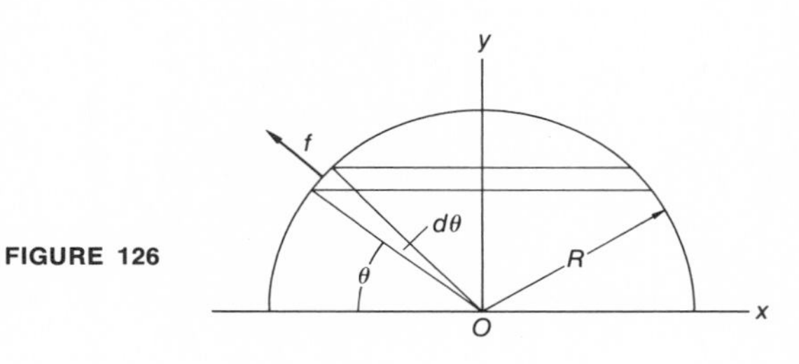

FIGURE 126

PROBLEM 270

Two parallel conducting planes of area A, each uniformly charged with a total charge $+Q$ or $-Q$, are separated a distance d in a vacuum (Figure 127).

a. What force must be used to keep the plates separated?

b. If a thin uncharged conducting sheet is inserted between the plates and parallel to them as shown, neglecting gravity what force is felt by the inserted sheet?

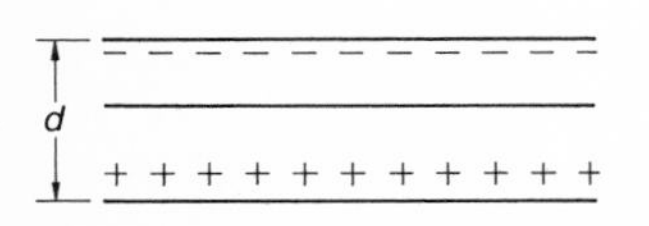

FIGURE 127

PROBLEM 271

The charges $\pm Q$ on the two conducting plates of problem 270 were put there by applying to the plates a potential difference ΔV. What is the ratio of the charge Q to ΔV? *Hint:* Remember that $\Delta V = -\int E_y\, dy$ (y perpendicular to the plates), and assume that E_y is uniform between the plates and 0 outside that region. This assumption leads to acceptable accuracy when $A >> d$.

You will note from the answer to problem 271 that the ratio $Q/\Delta V$ is a function of the geometry only, and for any particular geometry is a constant. This constant is called the capacitance C, and is defined by $C = Q/\Delta V$. The dimensions of C are $M^{-1}L^{-2}T^2Q^2$, and its MKSC unit, coulombs volt^{-1}, is called a farad. A farad is a very large quantity; most capacitors are evaluated in terms of microfarads: $1\ \mu f = 10^{-6}$ farads.

For the most used geometrical forms for a capacitor, C has the following values:

- Two parallel plates of area A separated by a distance $d - C = \epsilon_0 A/d$.
- Two concentric conducting spheres of radii R_1 and R_2, $R_2 > R_1$:

$$C = \frac{4\pi\epsilon_0 R_1 R_2}{R_2 - R_1}.$$

- Two long concentric conducting cylinders of radii R_1 and R_2, $R_2 > R_1$:

$$C = \frac{2\pi\epsilon_0}{\ln R_2/R_1}.$$

PROBLEM 272

a. By making $R_2 \rightarrow \infty$ in the appropriate equation above, determine the capacity relative to infinity of a conducting sphere carrying a charge Q.

b. Disregarding the charges in the atmosphere, what is the capacitance of the earth?

PROBLEM 273

A parallel plate capacitor has a plate spacing in air of 1.0 mm.

a. What is the maximum voltage that can be placed across this capacitor without risking discharge through air breakdown?

b. How much charge per unit area can be put on the capacitor plates?

Capacitors are often assembled in various combinations; the effective capacity of such a combination is given by

$$C_{\text{eff}} = \frac{\text{total charge stored in the combination}}{\text{voltage across the combination}}$$

$$= C_1 + C_2 + C_3 + \cdots \text{ for capacitors combined in parallel;}$$

$$\frac{1}{C_{\text{eff}}} = \frac{1}{C_1} + \frac{1}{C_2} + \frac{1}{C_3} + \cdots \text{ for capacitors combined in series.}$$

PROBLEM 274

It is obvious that the capacity of a parallel combination of capacitors will always be greater than the capacity of any individual

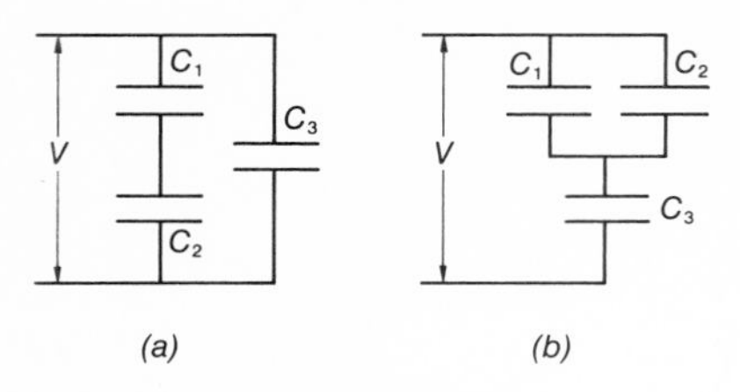

FIGURE 128

capacitor in the combination. Is the converse true for a series combination, i.e., is the capacity of a series combination always less than the capacity of any individual capacitor in the combination?

PROBLEM 275

In Figure 128, what value of C_3 would make the capacity of combination (a), or of combination (b), equal to C_1?

The Effect of a Material Dielectric. The equations we have developed to this point for the calculation of electrostatic forces and fields apply accurately to free space, and are valid within a percentage error of less than 0.1% for normal air and light gases at normal temperatures and pressures. These equations must be modified, however, for charges and fields in the presence of nonconducting liquids and solids.

Nonconductors, called insulators or dielectrics, are made up of molecules that are not only electrically neutral themselves but are so arranged in normal equilibrium that in any small volume the sum of any molecular dipole moments adds to zero. (From this discussion we exclude a small group of materials with permanent polarization.) Thus a dielectric under normal circumstances possesses no charge imbalance and produces no electric field, either externally or internally.

For dielectrics in the presence of an external field this symmetrical order is disturbed, with positively charged nuclei being strained in the direction of the field and the planetary electrons being strained in the opposite direction. The net effect of this strain is to "polarize" the material, with an internal bound positive surface charge appearing at one surface of the dielectric and a similar internal bound negative charge appearing at an opposite surface. The field due to these induced surface charges acts internally to reduce the effect of the external field, so that the net field available to move charges in the dielectric is less than the external field. In relatively simple homogeneous, isotropic dielectrics—the so-called Class A dielectrics that include most of the materials used in electrical applications—this reducing effect is linearly proportional to the external field that caused it.

In the presence of dielectrics we have few opportunities to use Coulomb's Law; only for a point charge buried deep in a dielectric and far from any boundary does the law accurately describe forces on charges, and even then it must be adjusted with a reducing factor $1/K$. However, Gauss' Law is still applicable.

$$\int \vec{E} \cdot d\vec{S} = \frac{\Sigma q}{\epsilon_0},$$

where Σq is the total charge found inside the surface of integration—the bound surface charges as well as the charges that are the source of the external field. For the situation most often encountered, in which an external field is perpendicular to the parallel boundaries of a dielectric, we can indicate the reducing effect of the polarized surface charges by a factor $1/K$, and write Gauss' Law in operational form:

$$\int K\vec{E} \cdot d\vec{S} = \frac{\Sigma q}{\epsilon_0},$$

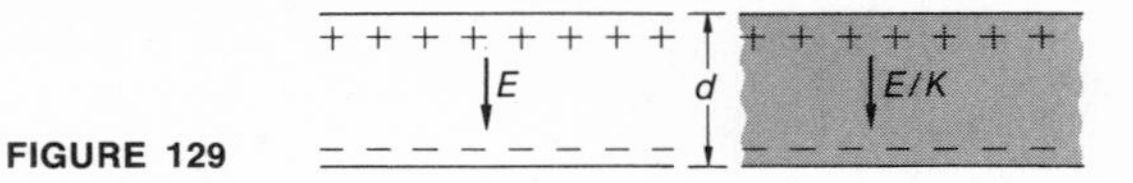

FIGURE 129

where now Σq is an inventory *only* of the charges creating the external field. The coefficient K is experimentally determined for each dielectric material, and is usually called the "dielectric constant." (The word "constant" is misleading, since even for Class A dielectrics K varies with temperature, with strength of external field E, with pressure, and with frequency if E is oscillating.) The coefficient K is always $\geqslant 1$, and is dimensionless.

With the external field perpendicular to the parallel dielectric boundaries, what will concern us most about the reduction of field in dielectrics is its effect on the capacity of capacitors. If we compare two identical parallel plate capacitors each carrying the same charge Q, one filled with free space and one filled between plates with a dielectric of constant K (Figure 129), we see $V_0 = Ed$ for the free space capacitor, whereas for the dielectric capacitor $V_d = Ed/K$. In the dielectric case it took less potential to put a charge Q on the plates than was required for the free space capacitor. Then $C_d = KQ/V = KC_0$; the capacity of a capacitor filled with dielectric is greater by a factor K than that of a capacitor filled with free space.

PROBLEM 276

The parallel plate capacitor of problem 270, still with a charge $\pm Q$ on the plates, is filled with an insulating liquid of dielectric constant K. What is now the force necessary to hold the plates apart?

START OF SOLUTION

Following the discussion on page 141, you probably solved problem 270 the easy way by multiplying $\sigma^2/2\epsilon_0$, the force on unit surface charge per unit area, by the area A. You can do the same in this case, provided you are careful about what the net force on σ is. With a Gaussian pillbox as before, you now must include the bound charge at the surface of the dielectric that results from polarization. Solving $\epsilon_0 \int \vec{E} \cdot d\vec{S} = \Sigma q$, we obtain the field vectors $(\sigma_+ - \sigma_-)/2\epsilon_0$ as shown in Figure 130. Then, since the field inside the conductor must be zero, the pillbox must also be in the presence of an external field E_{ext} from all the rest of the charges of $(\sigma_+ - \sigma_-)/2\epsilon_0$ as shown in Fig. 130. Then the total field in the dielectric must equal $(\sigma_+ - \sigma_-)/\epsilon_0$. But we also know that this field equals $E_0/K = \sigma/K\epsilon_0$. Then the force on the charge on the conductor must be $\sigma^2/2K\epsilon_0$ per unit area, just $1/K$ of the force for free space between the plates. (Later you will have a chance to solve this problem by virtual work methods, and you can decide which method has meaning for you.)

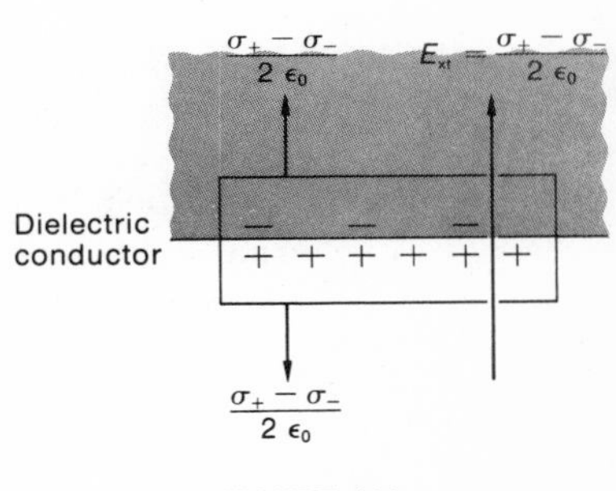

FIGURE 130

PROBLEM 277

A parallel plate capacitor of known capacity C_0 when operated with air spacing, is filled with a dielectric that breaks down when exposed to any electrical field strength greater than E_{cr}. What is the maximum operating potential that can be applied to this capacitor?

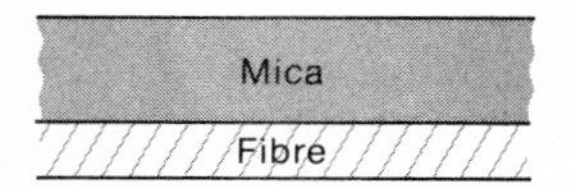

FIGURE 131

PROBLEM 278

An experimental parallel plate capacitor is made up as shown in Figure 131; between the plates there is a 1.0 mm thick sheet of clear mica and a 0.5 mm thick sheet of impregnated fibre. $K_{\text{mica}} = 8.0$, $K_{\text{fibre}} = 2.5$. Tested by itself, the fibre breaks down when subjected to a potential gradient greater than 6.4×10^4 volts cm^{-1}. What is the maximum safe voltage that can be put across the capacitor?

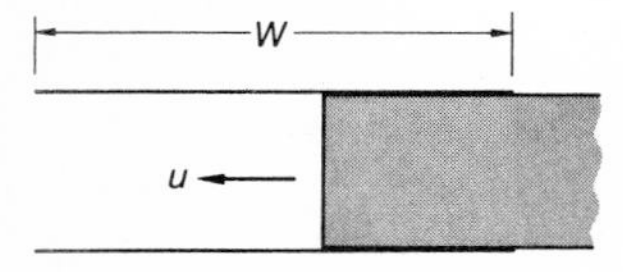

FIGURE 132

PROBLEM 279

A parallel plate capacitor (Figure 132) of width W and capacity C_0 when used with air spacing, is charged to a potential difference V_0 across its plates, and then the charging source is disconnected. A slab of dielectric, of constant K and with the same thickness as the plate spacing, is pushed in between the plates at a constant speed u.

a. At what rate is the capacity changing?

b. At what rate is the potential difference across the capacitor changing?

c. When the slab has been pushed all the way in, what is the potential difference?

PROBLEM 280

A straight conducting wire of radius r carrying a charge λ per unit length is surrounded by a solid cylindrical sheath of radius R made of insulating material of dielectric constant K. What is the electric potential difference between the wire and

a. the outer surface of the sheath,

b. a point outside the sheath $2R$ from the center of the wire?

After the case we have just discussed—the field inside a slab of dielectric when the external field is perpendicular to the slab—the next simplest case is the refraction of the internal field at a dielectric boundary when the external field is not perpendicular to the boundary. In the diagram in Figure 133, the external field $\vec{E}$ is at an angle ϕ with the normal to the boundary between dielectrics ① and ②. Since the

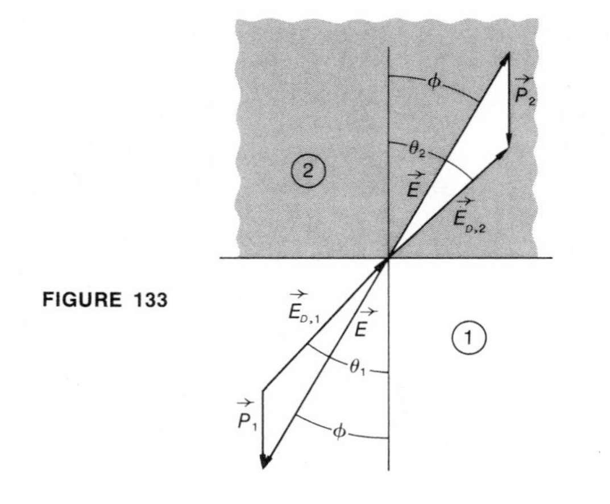

FIGURE 133

vectors $\vec{P}_1$ and $\vec{P}_2$ are perpendicular to the surface at the surface (due to the induced bound surface charges), the components parallel to the surface of both the external field and the net internal field must be equal; hence

$$E_{D,1} \sin \theta_1 = E \sin \phi = E_{D,2} \sin \theta_2. \tag{3-18}$$

Perpendicular to the boundary we have $E_{D,1} \cos \theta_1 = E \cos \phi - P_1$. But from our definition of K on page 144 the net field perpendicular to the boundary $E \cos \phi - P_1$ must equal $(1/K_1)E \cos \phi$. Then $E_{D,1} \cos \theta_1 = (1/K_1)E \cos \phi$. A similar treatment for the other side of the boundary gives $E_{D,2} \cos \theta_2 = (1/K_2)E \cos \phi$. Equating values for $E \cos \phi$, we have

$$K_1 E_{D,1} \cos \theta_1 = K_2 E_{D,2} \cos \theta_2. \tag{3-19}$$

Dividing equation 3-18 by equation 3-19, we finally get

$$K_2 \tan \theta_1 = K_1 \tan \theta_2. \tag{3-20}$$

This tells us that in crossing a boundary between dielectrics the internal field is bent away from the normal when going in a direction of increasing K. (The same result could have been obtained from a Gaussian pillbox enclosing the bound charges on either side of the boundary.)

Force on a Moving Charge. A charge $+q$ moving relative to a magnetic field $\vec{B}$ with a velocity $\vec{v}$ will feel a force equal in magnitude to $qvB \sin \theta$, where θ is the least angle between $\vec{v}$ and $\vec{B}$ in the plane determined by $\vec{v}$ and $\vec{B}$; the force is perpendicular to this plane. The simple way to say this is $\vec{F}_q = q\, \vec{v} \times \vec{B}$. If at the same time the charge is under the influence of an electrical field $\vec{E}$, then the total force acting on the charge is given by

$$\vec{F}_q = q(\vec{E} + \vec{v} \times \vec{B}). \tag{3-21}$$

This is the basic equation for evaluating electrodynamic forces.

Since F, q, and v are all known concepts, equation 3-21 could be used to define B. If we proceeded in this way we would find that B had dimensions $MT^{-1}Q^{-1}$; the standard MKSC unit for B is webers m^{-2}. However, there is available a better operational definition of B which we will come to later. For the moment, therefore, let us accept the experimental fact that there can exist a field that does not affect charges at rest but that provides a force on moving charges. We will call such a field the magnetic field $\vec{B}$.

The most interesting aspect of equation 3-21 is the direction of the force resulting from $q\, \vec{v} \times \vec{B}$. Since this force is perpendicular to the plane containing $\vec{v}$ and $\vec{B}$, it cannot affect the magnitude of v at any time. What the force does provide at all times is an acceleration vector that is perpendicular to $\vec{v}$; as $\vec{v}$ changes direction from this acceleration the acceleration also changes direction so as to stay always perpendicular to the instantaneous $\vec{v}$.

PROBLEM 281

If the change in the direction of the beam in problem 234 is to be made magnetically (the preferred way), what magnetic field $\vec{B}$ is required in the region of the direction change?

PROBLEM 282

Replace the parallel plates providing an electric field in problem 251 by an appropriate magnetic field B that will give exactly the same angular deflection of the beam over the same horizontal distance L. What is B?

PROBLEM 283

A cyclotron consists of two "dees" separated by a narrow accelerating gap across which an electrical potential difference $V \cos \omega t$ is maintained (Figure 134). (A dee is a semicircular half of a very shallow pillbox of very large diameter.) Perpendicular to the plane of the dees is a uniform magnetic field B. A charged particle is injected at low speed into the gap, at right angles to it, as the accelerating voltage across the gap goes through a maximum. If B is adjusted properly the particle travels around a half circle and arrives at the gap π/ω sec later, just in time to receive another maximum acceleration. This process is continued until the particle has attained the desired energy, when it is withdrawn from the dee by passing through a deflection system which provides the field necessary to cancel the particle's centripetal acceleration. Assume a particle of charge q and mass m.

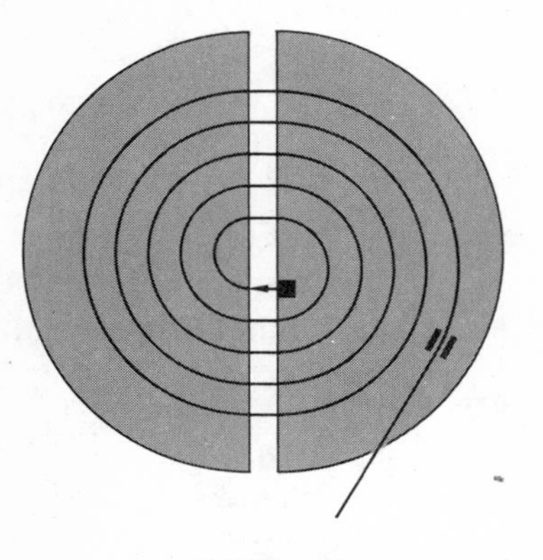

FIGURE 134

a. What is the necessary relation between the angular frequency ω of the accelerating voltage and the magnetic field B? (This is called the cyclotron frequency.)

b. At each crossing of the gap the speed of the particle, and hence the radius of its next half-circle, increases. What adjustment of the angular frequency is necessary to keep up with these increases?

c. The path shown in Figure 134 of a charged particle in a cyclotron looks like a spiral. Is it a spiral, or is there a better description?

Note: For this simple cyclotron, assume nonrelativistic speeds.

PROBLEM 284

As the demand for higher particle energies rises, speeds become relativistic, and the cyclotron equation developed in problem 283 is no longer adequate. You can maintain resonance by accommodating to the relativistically changing momentum in two ways. For a relativistic particle in a cyclotron-type machine to stay in resonance with its accelerating voltage, what is the necessary relationship

a. between B and r if you keep ω constant (this is the principle of the synchrotron) or

b. between ω and r if you keep B constant over the entire area of the dees? (This is the principle of the synchrocyclotron.)

PROBLEM 285

A track in a bubble chamber is a helix with a radius of r and a pitch of p turns per unit length. The direction of the imposed magnetic field is as shown in Figure 135; when viewed in the direction of this field the motion of the particle making the track is clockwise.

a. What was the sign of the charge on the particle?
b. At what angle with the field did the particle move?
c. Is it possible to determine independently both the speed v of the particle and the particle's charge to mass ratio q/m?
d. If the probable identity of the particle, and hence its q/m ratio, can be deduced from the character of the track, what was v?

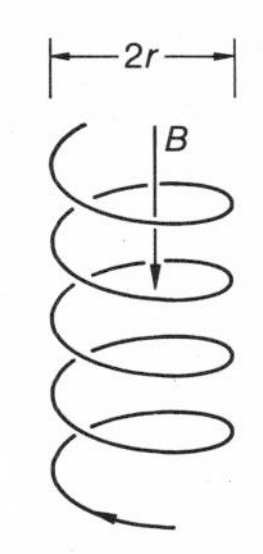

FIGURE 135

PROBLEM 286

Because of the finite size of the entrance and exit openings of a collimating system, a beam of charged particles emerging from such a system can have an angular spread of 2α (Figure 136). You want to form this diverging beam into a cylindrical beam of radius $R = (1 + \sqrt{2})r$, where r is the radius of the exit, by supplying the appropriate magnetic field B in the region beyond the exit. The beam particles each have a speed v.

a. What value of B would you supply?
b. In what direction would you supply B, if the particles are positively charged?

Note: Assume the internal forces between particles in the beam are negligible compared to the force from B that is necessary to accomplish the desired confinement of the beam.

Caution: The geometry of this problem requires as much attention as the physics. For question (a) better start by drawing a circle of radius r representing the exit, then concentric with this, a circle of radius $(1 + \sqrt{2})r$ representing the desired cylinder. Then consider the possible paths of the extreme particles emerging at r with maximum divergence α into the region of the magnetic field; when you recognize that among all these possible paths there is only one, unique path for each extreme particle that satisfies the requirement that the particle shall stay just within a cylindrical surface of radius R, you are then about ready to solve the physics equation for B.

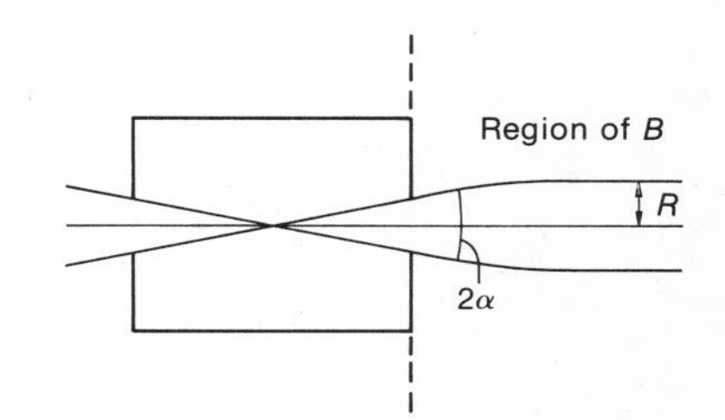

FIGURE 136

PROBLEM 287

An ion mass spectrograph (Figure 137) consists of an evacuated chamber containing a linear ion source L maintained at a positive potential V relative to a base plate, a linear slit S in the base plate, a photographic plate P laid on the base plate, and a uniform magnetic field $\vec{B}$ perpendicular to the plane of the figure in the region above the base plate. Ions emerging from S exactly perpendicular to the base plate arrive at the photographic plate at a distance d from S.

a. What is q/m for these ions as a function of d, V, and B?
b. Because of the finite width of S, many ions emerge at small angles from either side of the perpendicular to the base plate. Do these ions broaden the image on P so as to make precision measurements difficult, or is there a particular line on P that permits precision measurements of q/m?

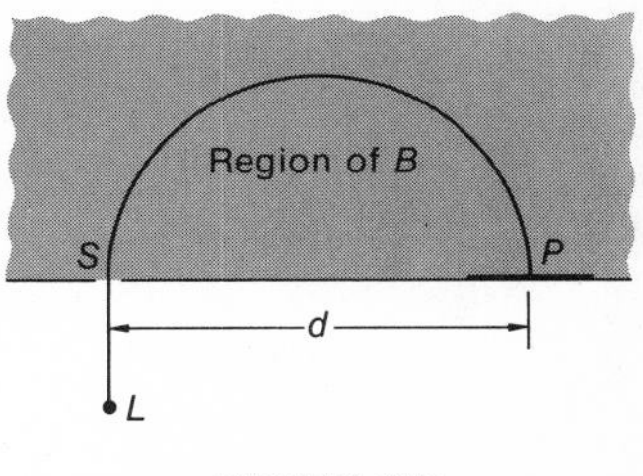

FIGURE 137

PROBLEM 288

A narrow beam of unknown particles moves along the $+z$-axis. When a uniform electric field $\vec{E}$ is turned on in the $+x$ direction, the beam traces out a narrow parabolic path that can be described by $x = Az^2$, A measured and known. Then a uniform magnetic field is established in the $+y$ direction; by careful adjustment of this field the beam is restored to its original narrow path in the $+z$ direction, being held there when the magnetic field has the value B_0.

a. What was the speed of the particles?
b. What was the ratio q/m for the particles?
c. In computing your answers you probably made assumptions that the particles all carry the same charge, that they all have the same speed, and that they are all alike. Are these assumptions valid, and if so, why?

For a point charge, or for a beam where the charges are all alike, both terms of equation 3-21 are applicable. For a conductor, however, the net charge is zero; a conductor is electrically neutral under almost all conditions, so that it feels no net force from any external field $\vec{E}$. Hence for a conductor on which conduction electrons are moving, only the second term in equation 3-21 is applicable. The moving electrons do not normally leave the conductor, so any force felt by them by the action of $q\vec{v} \times \vec{B}$ is transferred to the conductor itself.

A linear flow of charge has a linear density of λ coulombs per unit length; in a small distance dl the total charge is λdl. If dl is small enough so that λ and $\vec{B}$ are constant over dl, then λdl can be substituted for q in the second term of equation 3-21, and we have

$$\vec{F}_{dl} = \lambda dl\, \vec{v} \times \vec{B}. \tag{3-22}$$

The product (λv) has the dimensions of charge $\sec^{-1}$, and is given the special name of "current," symbolized by I. Its dimensions are QT^{-1}, and its MKSC unit is the coulomb $\sec^{-1}$, called the ampere, abbreviated A. Since by original choice $\vec{dl}$ is in the same direction as $\vec{v}$, we can transfer the directional vector indicator to $\vec{dl}$; we finally have, for the force on a length of conductor dl,

$$\vec{F}_{dl} = I\, \vec{dl} \times \vec{B}. \tag{3-23}$$

The total force on a macroscopic conductor would then be found by integrating the above equation over the length of the conductor.

PROBLEM 289

A conducting wire is stretched between two pole pieces, and between the poles there is a strong and uniform magnetic field $\vec{B}$ (Figure 138). How can you make the wire "sing"?

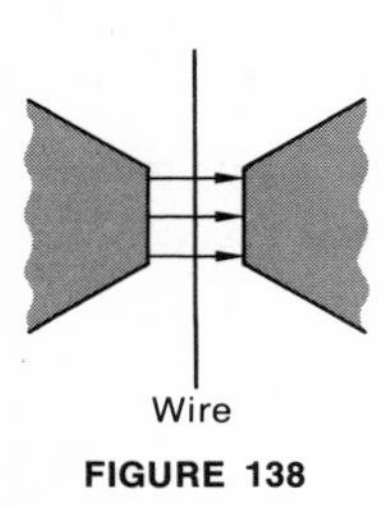

FIGURE 138

PROBLEM 290

A rectangular coil of N turns of wire is suspended from one arm of a beam balance (Figure 139). The lower portion of the coil is in a region of uniform horizontal magnetic field $\vec{B}$; the plane of the coil

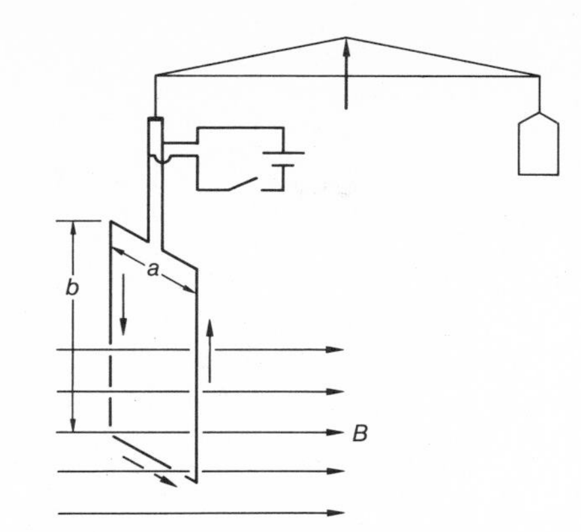

FIGURE 139

hangs perpendicular to this field. When a current I is flowing in the wires of the coil, what does I "weigh"? The direction of movement of the conduction electrons is indicated by the arrows. Do you think this method is better suited to determining I or B?

PROBLEM 291

A loop of wire of length L lies in an irregular shape on the horizontal surface of a lecture demonstration table (Figure 140). When a steady current I is sent through the loop in the direction shown, the loop moves to form a perfect circle.

a. What explanation would you propose?
b. If you could measure the tension in the circled wire, what unknown quantity could you evaluate, and what is its value?

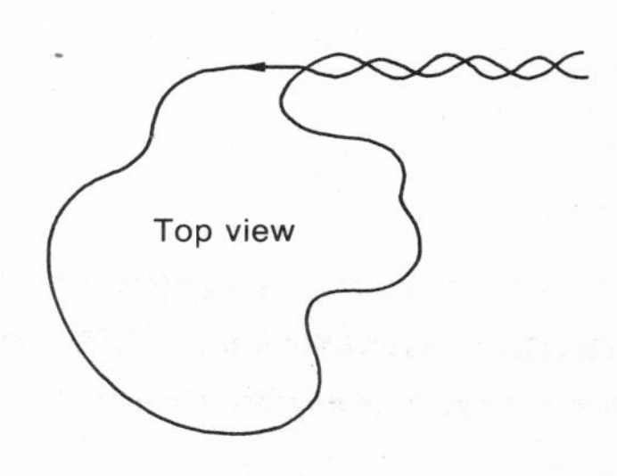

FIGURE 140

The Source of the Magnetic Field. When a moving charge or a flow of charge is subjected to the force described by equation 3-22 or by the second term on the right of equation 3-21, other charges in the vicinity must also be moving. What we have been calling the magnetic field is a convenient shorthand term for describing the effect one set of moving charges has on another set of moving charges.

Before considering mutually moving charges, we start with the following facts:

1. A charge at rest in an inertial coordinate system feels only the normal Coulomb electrostatic force from charges moving with uniform velocity relative to the coordinate system—a point charge q at rest r away from a linear beam of charged particles moving in the system with a velocity $\vec{v}$ feels only the electrostatic force $q\lambda/2\pi\epsilon_0 r$ perpendicular to the beam, where λ is the linear charge density as measured in the coordinate system (see statement 5, page 126).

2. A charge moving with uniform velocity relative to an inertial coordinate system feels only the normal Coulomb electrostatic force from charges at rest in the coordinate system—a point charge q moving with velocity $\vec{v}$ relative to the coordinate system on a line parallel to, and r away from, a line of charge at rest in the system also feels only the electrostatic force $q\lambda/2\pi\epsilon_0 r$ perpendicular to the line of charge.

Of course these are not independent observations; if one is true, the other must also be true.

Let us now consider a combination of aspects from (1) and (2) just given; both the point charge q and the linear charged beam are moving relative to an inertial coordinate system. For convenience in computation we will assume both q and the beam have equal velocities $\vec{v}$, separated by r. (This does not change the physics of the situation, but it does reduce considerably the complexity of the calculation.) If we now look at the charge and the beam in a coordinate system moving with a velocity $\vec{v}$ relative to our original system, both charge and beam are at rest, and the charge must feel an electrostatic force $q\lambda'/2\pi\epsilon_0 r$, where λ' is the "proper" linear density of the beam, its "at rest" density. Since charge does not change value relativistically, any change between λ and λ' must arise solely from changes in length. In determining λ, the length we measure in the fixed coordinate system is $(1-\beta^2)^{1/2}$ times the proper length in the beam; the proper length is thus γ times the fixed system length, and therefore $\lambda' = (1/\gamma)\lambda$. Then the force felt by q in the moving system is $(q\lambda)/(2\pi\epsilon_0 r\gamma)$. To transform this force F' back into the fixed system to obtain F, we must compare dp'/dt' and dp/dt, in this case particularly $dp'_\perp/dt'$ and $dp_\perp/dt$. When we carry out this comparison, we find that $F_\perp = (1/\gamma)F'_\perp$. Applying this, as measured in the fixed coordinate system the force on moving q from the moving beam is

$$F_q = \frac{q\lambda}{2\pi\epsilon_0 r\gamma^2} = \frac{q\lambda}{2\pi\epsilon_0 r}\left(1-\frac{v^2}{c^2}\right)$$

$$= qE - \frac{v^2}{c^2}qE,$$

where E is the normal electrostatic field of a line charge of linear density λ. Both components of this force are perpendicular to $\vec{v}$, and are in opposite directions; if q and λ are both positive or both negative, the principal component qE is repulsion and the modifying term $(v^2/c^2)qE$ is attraction; if q and λ are opposite in sign, the principal force is one of attraction and the modifying term is one of repulsion.

Let us now extend this development by considering the force felt by a moving point charge q in the presence of a wire conductor carrying a steady current. When q and the conductor are at rest in our inertial coordinate system q feels no force; a normal, uncharged conductor has exactly as many electrons as it has protons, and at any time the net charge is zero. Now let us give q the same velocity $\vec{v}$ as the drift velocity $\vec{v}$ of the conduction electrons in the conductor, then look at the situation as seen in a coordinate system moving with $\vec{v}$ relative to our at-rest system. As far as q and the conduction electron beam are concerned, this is exactly the situation discussed above. We could, if we enjoyed doing extra work, determine the force on q in the moving system from the positive linear charge at rest on the conductor. However, we are going to transform forces in the moving system back into the fixed system, and we already know what the force on q from the positive charges will be in the fixed system. Making the transformation and adding forces vectorially (all the forces we have in both systems are perpendicular to $\vec{v}$), as measured in the fixed coordinate system the force on q is given by

$$F_q = \frac{q\lambda_+}{2\pi\epsilon_0 r} + \frac{q\lambda_-}{2\pi\epsilon_0 r} - \frac{q\lambda_-}{2\pi\epsilon_0 r}\frac{v^2}{c^2}$$

$$= -\frac{v^2}{c^2}\frac{q\lambda_-}{2\pi\epsilon_0 r}. \tag{3-24}$$

Thus, for a moving charge in the vicinity of a conductor carrying a steady current, the electrostatic forces cancel out and we are left with a relatively small force resulting from the motion of the charge. Note, as before, that if q is negative the force is one of attraction between q and the conductor; if q is positive the force is one of repulsion.

In equation 3-24, if we substitute I for $(v\lambda_-)$, rearrange slightly, and disregard the sign, we have $F_q = (qvI)/(2\pi\epsilon_0 rc^2)$. When we compare this with the scalar value of the second right-hand term of equation 3-21, we see that for a straight wire conductor carrying a current I what we earlier called B is actually $I/(2\pi\epsilon_0 rc^2)$—an electrostatic correction term to be used when charges move relatively. We can also write this as $B = (2I)/(4\pi\epsilon_0 rc^2)$, and when we put in the value of $4\pi\epsilon_0$ and c^2 we end up with $B = 2I/r \times 10^{-7}$ in MKSC units for the electromagnetic field of a straight wire carrying a current I in free space.

Similar treatment for other forms of moving charge will lead to similar results; what we have so far been calling the magnetic field B is simply a modification of the electrostatic field arising out of the movement of charges, treated relativistically. However, the above evaluation of B does not give us either a satisfactory or complete description of B; for such a description we need three independent equations, each of which presents some aspect of B.

(1)
$$\vec{F}_q = q\,(\vec{v} \times \vec{B}). \tag{3-25}$$

(2)
$$\oint \vec{B} \cdot d\vec{l} = \frac{I}{\epsilon_0 c^2}. \tag{3-26}$$

The integral of the component of $\vec{B}$ around a closed loop is proportional to the current passing through the loop. In free space or in an environment of nonmagnetic materials the constant of proportionality $= 1/\epsilon_0 c^2$.

(3)
$$\int_s \vec{B} \cdot d\vec{S} = 0. \tag{3-27}$$

The total flux of $\vec{B}$ through a closed surface must add to zero, i.e., as much flux must come out of a closed volume as goes in, which means that an isolated magnetic pole as a source of field cannot exist.

Equation 3-27 can be contrasted with Gauss' Law, which states that the flux of $\vec{E}$ through a closed surface depends on charges enclosed by the surface. Equation 3-25 is the operational equation by which we can use $\vec{B}$; equation 3-26 is the equation by which we usually evaluate $\vec{B}$.

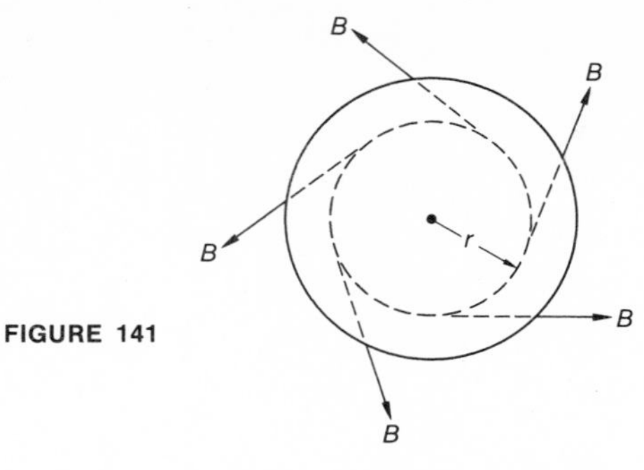

FIGURE 141

The simplest situation for checking the use of equation 3-26 is the evaluation of the magnetic field of a wire carrying a steady current I in free space. We will take as our integration loop a circle of radius r centered on the wire and in a plane perpendicular to the wire. Obviously $\vec{B}$ can be neither radial nor uniformly constant in one direction; either of these possibilities would produce $\oint \vec{B} \cdot \vec{dl} = 0$. $\vec{B}$ must therefore have a net component tangent to the circle; since there is no preferred direction, symmetry would require that the tangent components be everywhere the same around the circle. However, as suggested by the dotted lines in Figure 141, these vectors are tangent to a smaller circle, which suggests making B tangent to whatever circle we draw. When we put this together with the observable fact that a charge q moving parallel to a current-carrying wire is accelerated radially, we are forced to conclude that the only $\vec{B}$ that would satisfy both equations 3-25 and 3-26 is a vector that is perpendicular both to the wire and to the radius vector from the wire; the magnetic field of a current in a long, straight wire is circular about the wire. Then $\oint \vec{B}_r \cdot \vec{dl} = B_r\, 2\pi r = I/\epsilon_0 c^2$, which reduces to $B_r = 2I/r \times 10^{-7}$ in MKSC units as before. All that is left to do is to put in formal instructions about direction—we can't stop to draw a diagram every time we need to evaluate a $\vec{B}$.

In the general case, to allow for other forms of conductor, we will consider a small contribution $d\vec{B}$ from a small element of length $\vec{dl}$ of a line conductor, or of a narrow strip on a sheet conductor. (Note that this dl is different from the one we used for the loop integration.) If $\vec{dl}$ is in the direction of current flow, and $\vec{r}$ is the radius vector from dl to the point in space where we are evaluating $\vec{B}$, then we can get our directional instructions in by saying

$$d\vec{B} = \frac{I\,\vec{dl} \times \vec{r}}{4\pi\epsilon_0 c^2 r^3}. \tag{3-28}$$

B would be determined by integrating this equation over the entire conductor.

(Just to gain confidence that this will really produce the proper answer, and for practice, why don't you try integrating the above equation for a long, straight wire. Let the wire coincide with the x-axis, set $dl = dx$, and remember that the scalar value of $\vec{dx} \times \vec{r} = r\,dx\,\sin(dx, r)$; then evaluate the field at a distance from the wire $d = r\sin(dx, r)$ by integrating from 0 to ∞ and multiplying by 2. Compare your result with the value we obtained earlier by the use of a loop integral.)

When we want to know the force on one current-carrying conductor from the field of another current-carrying conductor, we could combine equations 3-23 and 3-28 and obtain the double integral

$$\vec{F}_{2\text{-}1} = \frac{I_1 I_2}{4\pi\epsilon_0 c^2} \int\int d\vec{l}_2 \times \left(\frac{d\vec{l}_1 \times \vec{r}_1}{(r_1)^3}\right).$$

However, this integral is reasonably easy for very simple geometries only, and even for those geometries it is usually easier to determine fields at different points of conductor 2, and then add forces vectorially.

It should be repeated that these equations apply only to free space or, to use a term that will appear later under Inductance, where the permeability = 1. You should also note that equations 3-23 and 3-28 require that you keep clearly in mind the sign of both current and charge. A real current is actually a movement of the negative conduction electrons; however, following an earlier tradition almost all textbooks treat I as positive and its direction as that of an apparent movement of positive charge in the opposite direction to that of the conduction electrons.

PROBLEM 292

A surveyor doing plane-table mapping of an area where there are overhead electric transmission lines uses a compass that will just barely detect and respond to a 1% change in the horizontal magnetic field. If the normal horizontal component of the earth's magnetic field in the area is 2.4×10^{-5} webers m^{-2}, how close can the surveyor work beneath a line that is 6.0 m above plane-table level and that is carrying a current of 100 A before the compass accuracy of his survey is affected?

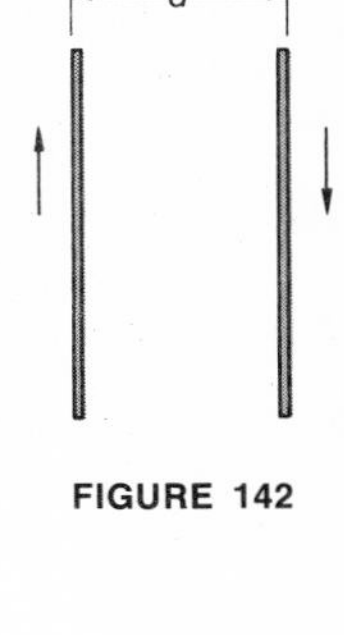

FIGURE 142

PROBLEM 293

Two long parallel metal rods of radius r are supported a distance d apart, center to center; $d >> r$. The rods carry the same current I, but in opposite directions (Figure 142).

a. What is the magnetic field between the wires in their plane?
b. What is the force per unit length on each wire due to the current?

PROBLEM 294

A coaxial conductor circuit consists of a metal rod of radius a concentrically surrounded by a metallic cylindrical sheath of inner radius b and outer radius d. Equal steady currents I flow in opposite directions in the rod and the sheath. Assuming that the current density in each of the two conductors is uniform throughout the cross section, and assuming that the magnetic field is subject to the same description in conducting metals as in free space, evaluate and graph the magnetic field due to the current from $r = 0$ to $r > d$, r being measured from the center of the rod.

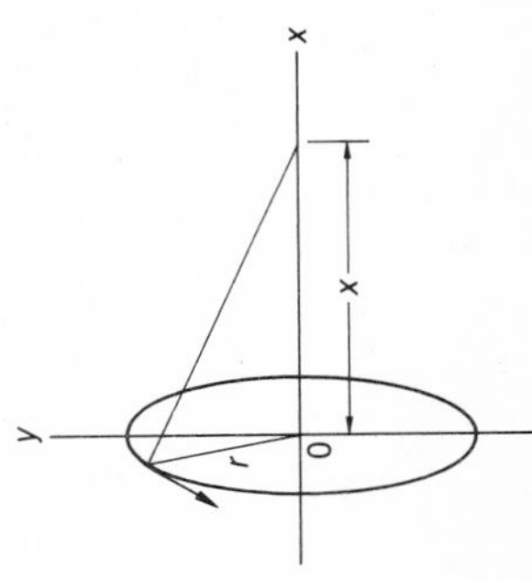

FIGURE 143

When a steady current moves in a circular coil of N turns of wire with an effective radius r (Figure 143), by integration and use of obvious symmetry you can determine that the magnetic field due to the current on the axis of the coil at a distance x from the plane of the coil is axial and has the value $B_x = (NI\,r^2)/[2\epsilon_0 c^2(x^2 + r^2)^{3/2}]$, where I is the current in the wire of the coil.

PROBLEM 295

Two equal plane coils, each of N turns of wire and effective radius R, are mounted coaxially a distance d apart. The coils are so connected that the same current I moves in the same direction in both coils. With coordinates as shown in Figure 144.

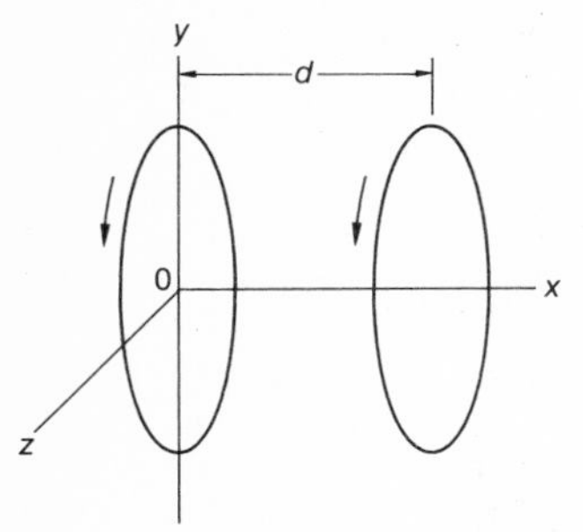

FIGURE 144

a. What is the magnetic field along the x-axis, for $x < d$?
b. For what value of x does $dB/dx = 0$? *Comment:* After you have worked out the formula for dB/dx, don't try to bulldoze an algebraic solution that would lead to $dB/dx = 0$; such a solution is difficult, and what is more relevant, it isn't necessary. Just look at your equation and ponder the symmetry of $(d - x)$ and x. Of course a little intuition about what the answer ought to be helps you ponder.
c. With d^2B/dx^2 evaluated for the x found in question (b), for what value of d does $d^2B/dx^2 = 0$?
d. Using the value of d found in question (c), graph the value of B along the axis in the central region between the two coils.
If you have done this problem correctly you have discovered why this arrangement, called a Helmholtz coil arrangement, is very useful in the laboratory for providing a region of uniform magnetic field.

PROBLEM 296

Two identical copper rings, of radius R, are mounted on the same vertical diameter at right angles to each other; they are insulated from each other at the bottom, where one of the rings is open and each end connected to a source of supply of steady current. At the top there is the switching arrangement shown in Figure 145. A sensitive compass is mounted horizontally at the center of the rings. The rings are oriented so that when a current I is sent through them, and then switched, the compass needle swings from an angle θ with one side of the plane of one of the rings to an equal angle θ with the other side of the same plane. What was the horizontal component of the earth's magnetic field at the center of the rings?

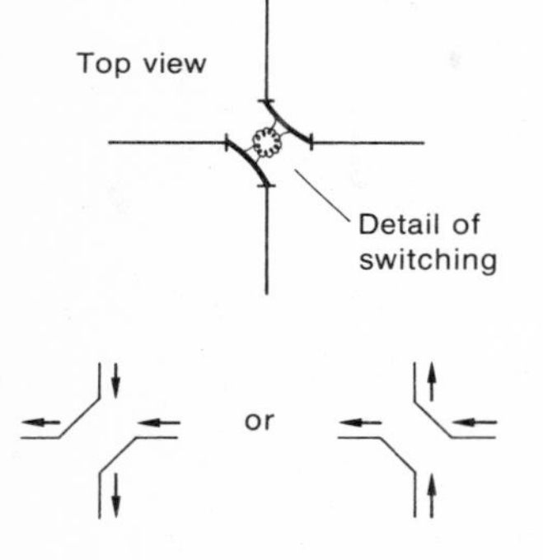

FIGURE 145

PROBLEM 297

The uniformly charged disc of problem 232 is set rotating at a constant angular speed ω about its axis of symmetry.

a. What is the resulting magnetic field B along the axis as a function of the distance x from the plane of the disc?
b. What is it at the center of the disc?
Does your answer to question (a) allow B to $\rightarrow 0$ as $x \rightarrow \infty$, as it should?
Caution: The field is symmetrical about $x = 0$, and your answer to question (a) should indicate this symmetry. If you reduce the algebra in your answer too far you will end up with a term in x, which is not symmetrical for $+x$ and $-x$.

Induced Potential Difference. Equation 3-22 was first applied to the case of charge moving with a speed v along a conductor; the resulting force was applied sideways to the conductor. (This is the operating principle of the electric motor.) What happens when we move an uncharged conductor, on which equal positive and negative charges are just sitting, with a speed v ($v \ll c$) relative to a magnetic field?

The answer is that equation 3-22 still applies: every charge q feels a "sideways" force $q\,\vec{v} \times \vec{B}$, but in this case "sideways" is along the conductor. The charge responds to this force exactly as if subjected to a field $\vec{E} = \vec{v} \times \vec{B}$, or as if there were a potential difference $E\,dl$ between the ends of an element dl of the conductor. Calling this potential difference $d\mathscr{E}$, we have

$$\begin{aligned} d\mathscr{E} &= \vec{E} \cdot \vec{dl} = (\vec{v} \times \vec{B}) \cdot \vec{dl} \\ &= -vB\,dl \end{aligned} \tag{3-29}$$

if v, B, and dl are all perpendicular to each other and the usual right-handed coordinates are used.

Equation 3-29 can be used as it stands, to calculate a potential difference developed across a piece of conductor moving in a magnetic field. However, it can be expanded into much greater usefulness. $v\,dl$ is an element of area da that is exposed per unit time to the field $\vec{B}$, or $v\,dl = da/dt$. But $B\,da$ is the flux of the magnetic field through this element of area, hence $-vB\,dl = -B\,da/dt = -d\phi/dt$, where ϕ is the local magnetic flux. In most cases of interest dl will be only an element of a complete circuit; integrating around this circuit

$$\oint \vec{E} \cdot \vec{dl} = \oint d\mathscr{E} = \mathscr{E} = -\frac{d\Phi}{dt}, \tag{3-30}$$

where $\mathscr{E}$ is the total electromotive force (emf) developed in the circuit and Φ is the total magnetic flux through the circuit. (The term "electromotive force" has only historical validity; it is not a force, but a potential difference.)

In general, the plane of the area A enclosed by the circuit will not be perpendicular to $\vec{B}$. Since $\Phi = \int \vec{B} \cdot \vec{dA}$ always, the general form of equation 3-30 would be

$$\oint \vec{E} \cdot \vec{dl} = \mathscr{E} = -\frac{d}{dt}\int_s \vec{B} \cdot \vec{dA}. \tag{3-31}$$

The negative sign in these two equations came out of the right-handed integration of $(\vec{v} \times \vec{B}) \times \vec{dl}$. However, $\mathscr{E}$ is a scalar and the change $d\Phi$ in the flux can be either positive or negative; thus for nonvector use the sign is not helpful. Then Lenz' Law comes to our aid: $\mathscr{E}$ is in such a direction in the circuit as to induce a current that will of itself establish a magnetic flux in opposition to $d\Phi$. For example, if a circular conductor is expanding in a plane perpendicular to a uniform field $\vec{B}$ pointing out of the paper (Figure 146), the direction of the induced current is indicated by the heavy arrow. The area enclosed by the circuit is increasing, hence $d\Phi/dt$ is positive; to oppose this increase in flux inside the circuit, the current must

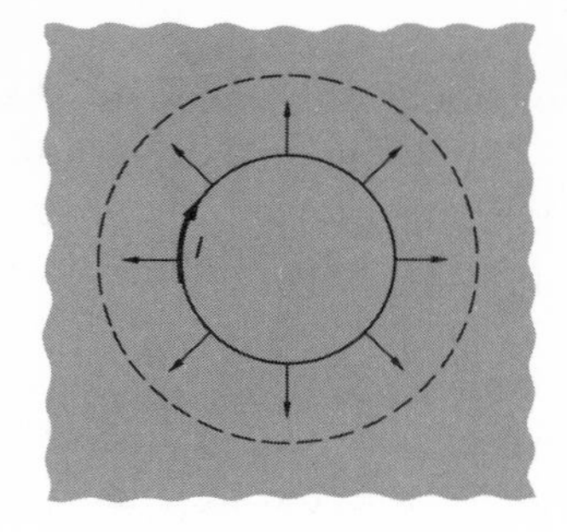

FIGURE 146

be as shown. Note that if we reverse the magnetic field or reduce the area enclosed by the circuit, the induced current is also reversed.

Neither of the above equations places any restriction on how $d\Phi/dt$ is accomplished. We can, as we have already seen, bring about a $d\Phi/dt$ by changing a circuit area in a region where B exists; this is the basic operating principle of the electric generator. We can also, by manipulating the source of B, create a time rate of change of B at the site of a fixed circuit; a $d\Phi/dt$ produced in this way is equally successful in making an induced current flow in a circuit. This latter method is the basic operating principle of the transformer.

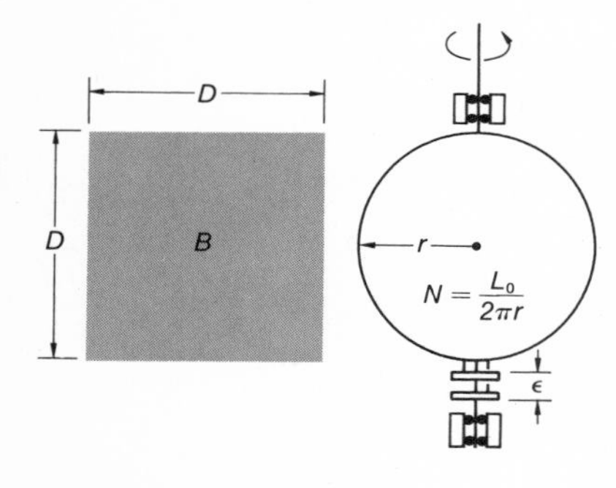

FIGURE 147

PROBLEM 298

The metal blades of a helicopter rotor are 6.0 m long and are attached to a vertical axle that rotates at 420 rpm in normal level flight. When such a helicopter is flying horizontally in a region in which the vertical component of the earth's magnetic field is 4.5×10^{-5} webers m^{-2}, what potential difference develops between the tip and the root of a rotor blade?

PROBLEM 299

You have L_0 m of insulated copper wire, a region of uniform magnetic field that is D m square in cross section, and a motive source capable of driving an axle at R rev sec^{-1}. If you wind the wire into a coil, mount the coil on the axle bringing the ends of the wire out to slip rings on the axle, and place the coil in the appropriate position in the magnetic field (Figure 147),

a. what is the maximum emf you can expect to generate between the slip rings?

b. What difference does it make in the answer for question (a) whether you wind your coil in a circular shape or a square shape?

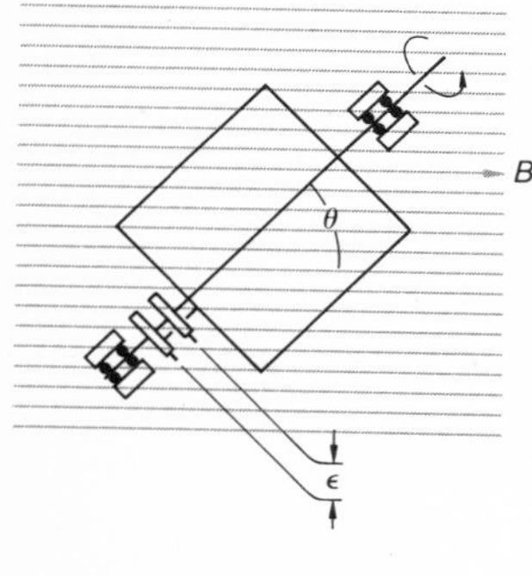

FIGURE 148

PROBLEM 300

A rectangular coil of area A, made of N turns of continuous copper wire, is mounted on a central axis and the two ends of the wire are brought out to slip rings. The coil is made to rotate about the axis with an angular speed ω rad sec^{-1} (Figure 148).

a. What is the potential difference developed between the slip rings when the axis is perpendicular to a uniform magnetic field $\vec{B}$;

b. what is it when the axis is inclined at θ with the direction of the field?
c. Could this arrangement be used to determine the direction of a uniform magnetic field?

PROBLEM 301
The coil of question (a) of problem 300 is held motionless with its plane perpendicular to the direction of the magnetic field, and the field itself is varied according to $B = B_0 \cos \omega t$. What potential difference develops between the slip rings?

PROBLEM 302
The changes with time considered in problems 300 and 301 are now put together; the coil rotates with an angular speed ω about an axis perpendicular to the direction of the magnetic field, and the value of the field itself is varied. If $t = 0$ when the plane of the coil is perpendicular to the direction of the magnetic field, what potential difference develops between the slip rings when

a. $B = B_0 \sin \omega t$,
b. $B = B_0 \cos \omega t$?

PROBLEM 303
In all the preceding problems only the open-circuit emf at the slip rings was asked for. If the slip rings had been connected to an external circuit, this emf would have pushed charges around this circuit and around the generating coil as well. In view of Lenz' Law, what effect would this have had on the calculated emf?

Inductance. When a conductor carrying a current I_1 is in the neighborhood of another conductor, this second conductor is exposed to the magnetic field due to the current I_1. If anything in the situation changes with time—the current I_1, the relative spatial arrangement of the two conductors, the permeability of the space between the conductors—the flux at the second conductor will also change with time, and an emf will be induced in this conductor.

In order to evaluate this emf we need $d\Phi_2/dt$. To get $d\Phi_2/dt$ we need Φ_2, the flux at the second conductor arising from I_1 in the first conductor. The simplest and most direct way to get Φ_2 is to say

$$\Phi_2 = M_{2-1} I_1,$$

where M_{2-1} is a function of the permeability and the relative geometries of the two conductors. In simple cases it can be calculated; in all cases it can be evaluated experimentally. The function M_{2-1} is called the mutual inductance between conductors 2 and 1; the MKSC unit of inductance is the weber A^{-1}, and is called the henry (abbreviated H). The dimensions of inductance are ML^2Q^{-2}.

In most applications of interest the relative geometry remains fixed, so that M_{2-1} is constant. For this case

$$\mathscr{E}_2 = -\frac{d\Phi_2}{dt} = -M_{2-1}\frac{dI_1}{dt}. \tag{3-32}$$

In some special applications a variable mutual inductance is useful; I_1 is kept constant and M is varied. For this case

$$\mathscr{E}_2 = -\frac{d\Phi_2}{dt} = -I_1\frac{dM_{2-1}}{dt}. \qquad (3\text{-}33)$$

If conductor 2 carries a current I_2, a flux Φ_1 will be provided at the site of conductor 1 from the field of I_2. However, for this case we do not have to repeat all of the above work. We can say immediately that Φ_1 is proportional to I_2, just as Φ_2 was proportional to I_1—and what is both remarkable and unexpectedly helpful, the constant of proportionality is *the same for both cases*. Even though the geometry of the two conductors is not at all symmetrical, it is a fact that $M_{1-2} = M_{2-1}$; the inductance really is "mutual." If we call it just M, equation 3-32 then becomes

$$\mathscr{E}_a = -\frac{d}{dt}(MI_b), \qquad (3\text{-}34)$$

where a and b are interchangeable.

Suppose there is no circuit 2, but the current I_1 in circuit 1 changes with time; is circuit 1 immune to this change? The answer is "no"; any current I_1 in circuit 1 creates a field B_1 which provides a flux Φ_1 enclosed by circuit 1, and a $(d\Phi_1/dt$ follows linearly from any dI_1/dt. In parallel to the development for mutual inductance we can say that $\Phi_1 = LI_1$ and

$$\mathscr{E}_{\text{back}} = -\frac{d}{dt}(LI) = -L\frac{dI}{dt}. \qquad (3\text{-}35)$$

L is called the self-inductance of the circuit in which I is flowing. By Lenz' Law the induced emf opposes the potential difference that is driving I through the circuit, hence it is usually called the "back emf." The self-inductance L has the same units and dimensions as the mutual inductance M.

By the application of $\int \vec{B} \cdot d\vec{l} = I/(\epsilon_0 c^2)$ you can determine that the magnetic field near the center of a solenoid of length L and internal radius r is approximately uniform over the cross-sectional area if $L >> r$; its value is $B = nI/(\epsilon_0 c^2)$, where I is the current in the wire winding of the solenoid and n is the number of turns of wire per unit length of the solenoid.

PROBLEM 304
What is the self-inductance per unit length of the above solenoid?

PROBLEM 305
If a small search coil of cross-sectional area a, wound with p turns of wire, is placed at the center of the above solenoid with its plane perpendicular to the axis of the solenoid, what is the mutual inductance of the solenoid and search coil combination?

PROBLEM 306
What is the self-inductance of the coaxial conductor of problem 294? For simplicity of calculation in this problem, assume that the currents are entirely on the outer surface of the wire and the inner surface of the sheath.

PROBLEM 307
The major portion of a circuit is a simple transmission line which consists of two parallel linear conducting rods, each of radius r and a distance d apart—like the arrangement in problem 293. The circuit current moves in opposite directions in the two conductors. What is the self-inductance per unit length of such a transmission system? *Suggestion:* With a current I through the rods, evaluate B between the rods at the plane of the rods, then integrate this value of B over a unit length of the plane between the rods to determine the flux Φ through a unit length of this plane; this will be the flux enclosed by the circuit of unit length. Then L may be calculated from the defining equation $\Phi = LI$. For simplicity, ignore any field in the rods.

The Permeability Constant. In consistency with the identification of the magnetic field as a relativity correction for the electric field (page 153), we have used $1/(\epsilon_0 c^2)$ in our statement of Ampere's Law: $\int \vec{B} \cdot d\vec{l} = I/(\epsilon_0 c^2)$. Since the evaluation of B by this law underlies much that followed, all of the appropriate equations developed have contained the factor $\epsilon_0 c^2$. However, in many texts you will find Ampere's Law stated as $\int \vec{B} \cdot d\vec{l} = \mu_0 I$, and in the subsequent development μ_0 appears where we have $1/(\epsilon_0 c^2)$: μ_0 is called the permeability constant, or the permeability, of free space, and its value is established at exactly $4\pi \times 10^{-7}$, and its unit is called the henry m^{-1}. Since by definition $\mu_0 = 1/(\epsilon_0 c^2)$, the numerical value of $1/(\epsilon_0 c^2)$ is also $4\pi \times 10^{-7}$.

Friction

If solid bodies move relatively to each other in contact, equal and opposite forces are generated in the plane of the surfaces in contact (or, if the surfaces are curved, in the plane tangent to the surfaces at the point of contact). The direction of these forces is such as to oppose the relative motion.

A is moving on the surface B with a relative velocity to the right (Figure 149). As a result of this motion, A feels a real force F_k directed to the left; the surface B

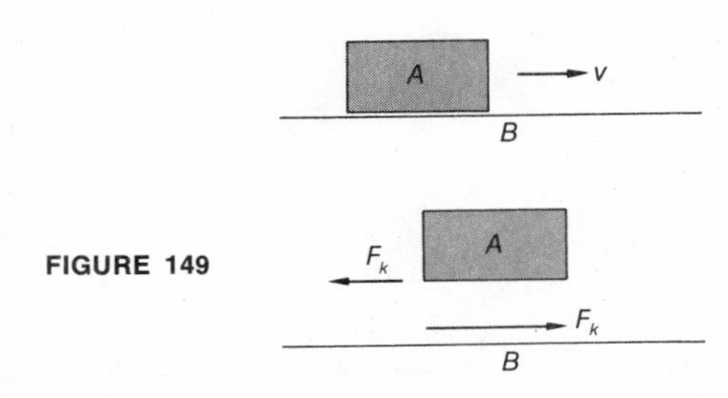

FIGURE 149

feels an equal and oppositely directed force F_k to the right. Because there is actual relative motion, F_k is called a force of sliding friction.

Even if A does not actually move relative to B – if it is at rest on B – there will be opposition to attempted relative motion. According to Newton's Second Law, any small force applied to A parallel to the surface should produce a small acceleration. In practice this does not happen; normally a substantial force must be applied before any acceleration occurs. This means that up to the time acceleration began, ΣF equalled 0, which in turn means that as an applied force was increased, an opposing force was generated up to a certain maximum. This force opposing incipient motion is called static friction, designated F_{st}.

Friction is a function of the two surfaces in contact, and of the condition of these surfaces – whether smooth or rough, lubricated or dry. Experimentally, it is found that the force of friction is proportional to the normal force between the two surfaces, is relatively independent of the area in contact, and, for the relative speeds usually encountered, is independent of the relative speed. All this information is contained in the equation

$$F_k = \mu_k \times N \qquad \text{or } F_{st} = \mu_{st} \times N, \tag{3-36}$$

where N is the normal force between the surfaces, and μ, the dimensionless constant of proportionality, is called the coefficient of friction – either the coefficient of kinetic or sliding friction or the coefficient of static friction. Since μ is dimensionless, F and N must be in the same force units.

For any particular pair of surfaces and any particular N, F_k is constant.* However, F_{st} is not constant; according to the circumstances it can range all the way from 0 up to a maximum value of $\mu_{st} N$. Consider a body resting on an inclined plane (Figure 150). At some plane angle θ_1 the normal force is $mg \cos \theta_1$ and the force resisting motion must equal $mg \sin \theta_1$. If the plane angle is increased to θ_2 and the body still does not move, the resisting force must have increased to $mg \sin \theta_2$. At some θ_{max} the resisting force is insufficient to prevent motion; at that moment $mg \sin \theta_{max} = \mu_{st}\, mg \cos \theta_{max}$ so that $\mu_{st} = \tan \theta_{max}$.

Certain conventions have developed for dealing with friction problems. There is no such thing as a "perfect" surface for which, to any high degree of accuracy, $\mu = 0$. However, surfaces can be prepared for which μ is very small, certainly

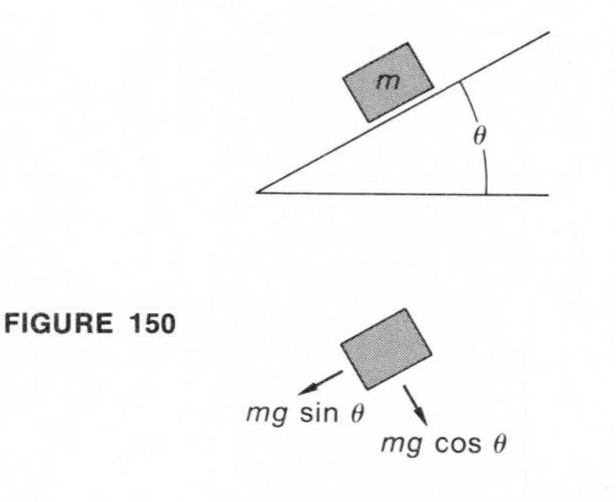

FIGURE 150

*Providing the surfaces stay constant. Obviously, if a surface becomes abraded or loses its lubrication during relative motion, it is then a different surface and μ_k may change.

smaller than the uncertainties in the rest of the data. In such cases, it is customary to speak of such surfaces as "smooth." Whenever you see this word describing a surface, the frictional force is to be considered as negligible. There is no such unanimity about the use of the word "rough"; in some texts it simply means "not smooth," in others it implies that $\mu_{st} = \mu_k = \infty$ (as would be the case, for instance, between meshed cogged gears).

In actual practice, μ_{st} is almost always greater than μ_k, and it is necessary to distinguish between them. However, in many textbook problems no distinction is made; just friction, or the coefficient of friction, is mentioned. In such cases it will be necessary to use the relationship $F = \mu N$, with the μ being considered either static or kinetic as appropriate to that particular problem.

PROBLEM 308

Required to determine a coefficient of sliding friction, and limited to the use of an angle protractor and a stopwatch, students devise the following experiment (Figure 151): block B, originally at rest, is released to slide down a plane at angle θ_1 with the horizontal, and the time of sliding a marked length of the plane is noted. Then the block is mounted on small, frictionless rollers and released from rest to slide down the same length of plane; this time the plane angle θ_2 is adjusted until the time of sliding is the same as before. In terms of the angles θ_1 and θ_2, what was μ_k?

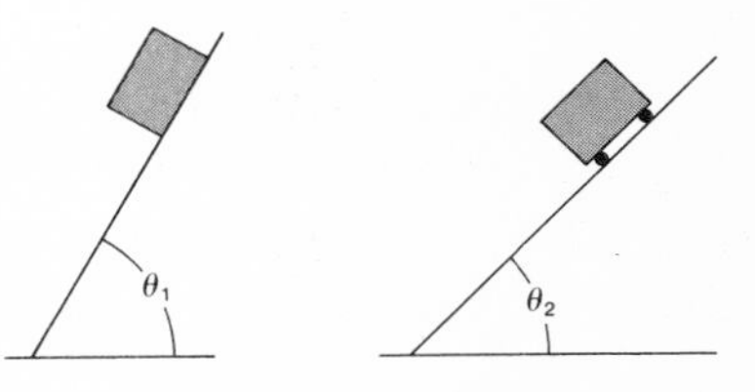

FIGURE 151

PROBLEM 309

Another group of students, faced with the same problem as in problem 308, decided to get an independent check by varying the method. Instead of having the blocks slide the same length on the plane, the blocks were allowed to slide from the same height, as shown in the diagrams in Figure 152. As before, θ_4 was adjusted so that the time of sliding to the foot of the plane was the same as for θ_3. In terms of θ_3 and θ_4, what was μ_k?

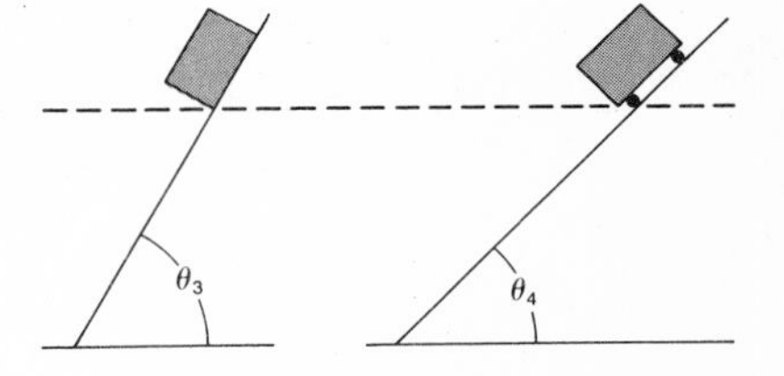

FIGURE 152

PROBLEM 310

a. What is the least time, theoretically, in which a subway train can be driven from rest at one station to rest at another station 1.0 mi away in a level, straight line? The coefficient of friction between the tracks and the tread of the wheels is 0.20; assume that 50% of the weight of the train is carried by the driving wheels but that all of the wheels are used in braking.

b. What was the maximum speed attained?

PROBLEM 311

Two masses are connected by a rigid rod of negligible mass, and lie one above the other on a 30° incline as shown in Figure 153. Between the plane and the lower block there is a coefficient of sliding friction $\mu_{1,k} = 0.25$, and for the upper block $\mu_{2,k} = 0.375$; $m_1 = 200$ gm, and $m_2 = 800$ gm. When the system is released from rest and allowed to slide down the plane, what is the force in the tie bar, and is it tension or compression?

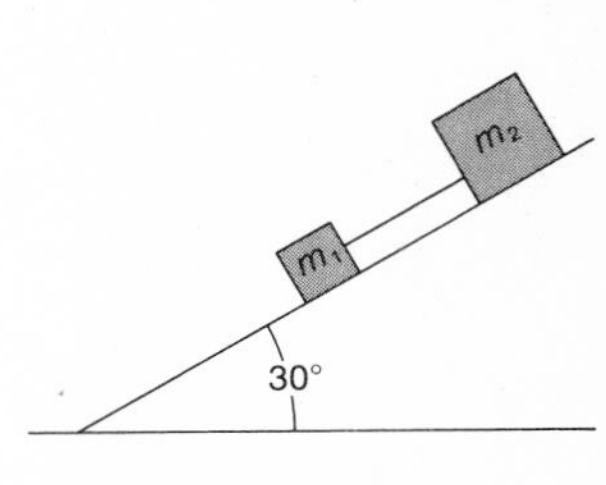

FIGURE 153

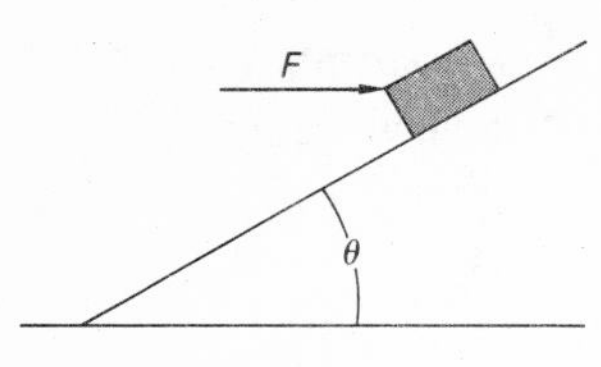

FIGURE 154

PROBLEM 312

A box of weight W is to be moved up an incline at θ with the horizontal (Figure 154); there is a coefficient of sliding friction μ_k between box and incline. If F is applied horizontally, and is just sufficient to move the box at constant speed, what is F?

PROBLEM 313

The same arrangement exists as in problem 312; the force F is applied as shown in Figure 155 by a rope with a breaking strength of B lbwt.

a. At what angle α with the horizontal should the rope be pulled to move the box up the incline with maximum acceleration?
b. What is this maximum acceleration?

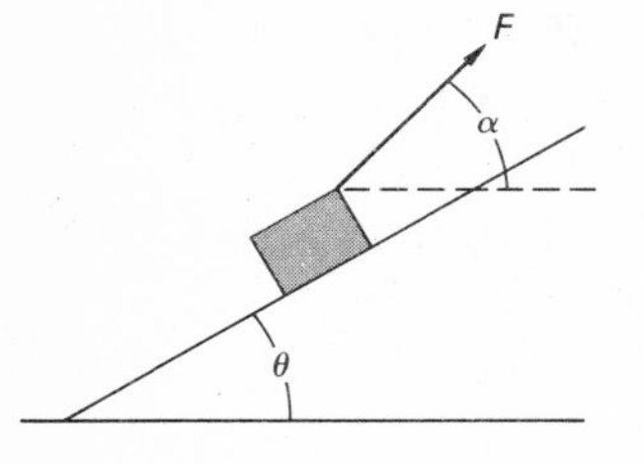

FIGURE 155

PROBLEM 314

A mechanism mounted at the rear end of the flat, horizontal floor of a test car is capable of projecting a mass m along the lengthwise axis of the car with a fixed initial speed relative to the car. When the car is stationary and the mass is so projected, it is observed to hit the forward end of the car in 3.0 sec. When the car is accelerating forward with an acceleration of 200 cm sec^{-2} and the mass is so projected, it just barely reaches the forward end of the car before coming to rest relative to the car. If the floor of the car is 9.0 m long, what was the coefficient of sliding friction between the mass and the floor of the car?

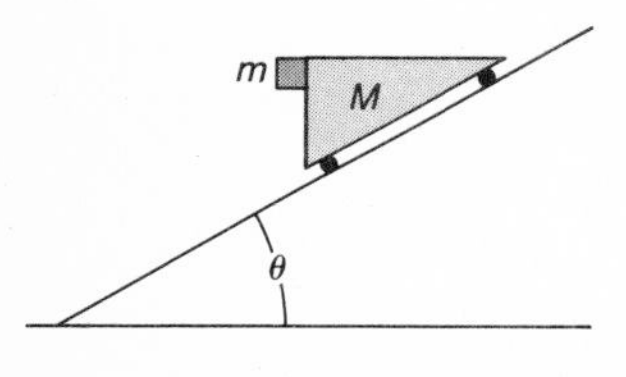

FIGURE 156

PROBLEM 315

For purposes of design analysis, a delayed triggering device can be represented by the simplified elements shown in Figure 156; a wedge of mass M is free to slide, with negligible friction, down an incline at θ with the horizontal, and a mass m is free to move down the forward vertical face of M. There is a coefficient of sliding friction μ_k between m and M, and the system is released from rest when m is at the top of the face of M.

a. Write all the equations of motion and the constraint equations necessary to determine the motions of m and M.
b. Calculate the value of μ_k such that the total motion of m is at θ with the vertical.

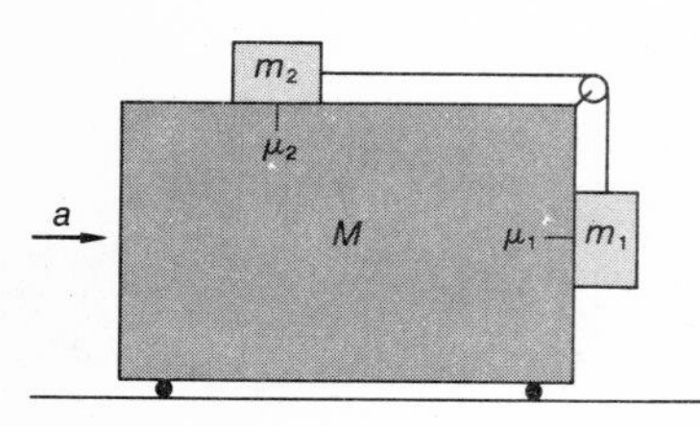

FIGURE 157

PROBLEM 316

A variation of problem 170 is as shown (Figure 157); μ_1 and μ_2 are coefficients of static friction. What is the range of accelerations that may be given to M such that m_1 and m_2 will not move relative to M?

PROBLEM 317

When a flatbed truck is braked to a stop, crates in the rear of the truck may slide forward unless the truck is braked gently. When the brakes are jammed on so that the tires skid on dry pavement, it is determined that a particular truck stops with uniform acceleration in a distance $S = v^2/20$, where S is in ft and v, the speed of the truck just at the start of braking, is in mi hr^{-1}. It is observed that during such a skidding stop, loose crates slid forward in the truck.

a. What can be said about the coefficient of friction between crate and truck bed, compared to the coefficient of friction between tires and pavement?
b. Construct a v-t diagram showing motion of both truck and crate.
c. Where on this diagram is the distance the crate slide in the truck?
d. At what point did the sliding crate have the greatest speed relative to the truck?
e. What was the coefficient of friction between tires and pavement?

PROBLEM 318

In a certain firing mechanism a mass m, originally at rest, is projected by a released spring across a horizontal surface and hits a firing button (Figure 158). There is a coefficient of sliding friction μ between mass and surface. If it is desired that the mass have at least mv of horizontal momentum when it hits the button, what must be the minimum speed of projection v_0 when the horizontal surface is moving upward with an acceleration of 4g?

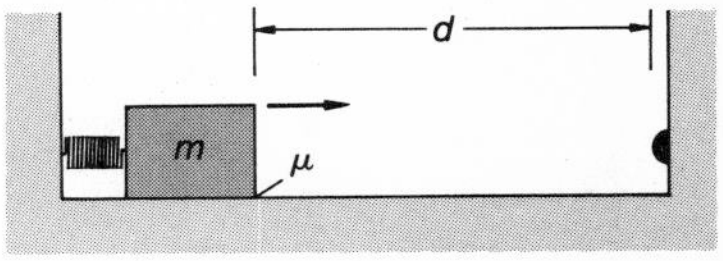

FIGURE 158

Viscosity

When adjacent layers, or lamina, of liquids or gases move relatively to each other, an internal force opposes this relative motion.

The diagram in Figure 159 shows the cross section of a gas or liquid flowing to the right without turbulence, i.e., the velocity vectors of adjacent layers are continuous and either parallel or nearly parallel. (The usual description of this condition is "streamlined.") With the velocity profile shown, layer B has a speed greater than that of layer A; because of this relative motion, layer B feels a retarding force at the interface with A. Conversely, at the interface a force acts on, and attempts to accelerate, A. These forces are equal in value but opposite in direction, and are parallel to the velocity vector at the interface.

If no other forces act on these layers, these forces resisting relative motion of the layers would soon reduce the relative speed to zero. However, if outside forces are available to maintain a velocity profile such as the one shown in Figure 159, these forces must just exactly counterbalance, layer by layer, the internal forces on those layers.

What are these outside forces? The driving force that layer B needs to offset the drag from A can come from gravity—streams flowing downhill, liquids mov-

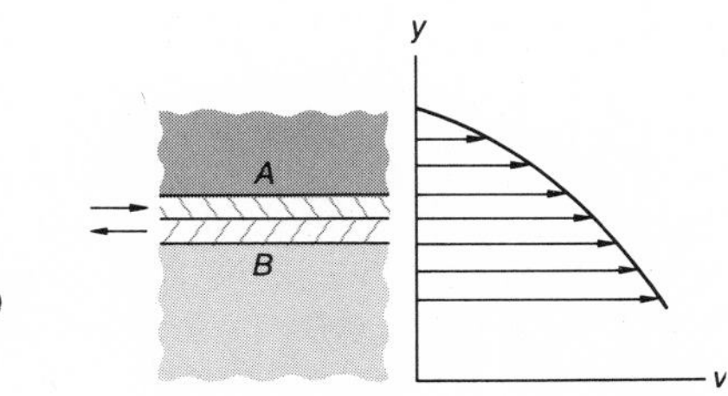

FIGURE 159

ing in vertical pipes—or from pressure differences in the direction of motion. The retarding force needed by layer A to offset the accelerating force from layer B comes from the layer above A. And *that* layer is retarded by the layer above it, etc., until finally we come to the solid boundary that contains the gas or liquid. The solid exerts a force on the infinitesimal layer adjacent to it that is sufficient to keep that layer at rest relative to the solid; at the solid-containing boundary for gases and liquids, the velocity vector is zero.* Thus the outside forces that offset the internal forces and serve to maintain a velocity gradient are the driving force from gravity or a pressure difference, and a retarding force from the walls containing the moving material.

Returning to the diagram (Figure 159), the retarding force on B that must be offset to maintain B's motion is directly proportional to the interface area between A and B and to the velocity gradient at the interface. Thus

$$F = \eta A \frac{dv}{dy}, \tag{3-37}$$

where η, the constant of proportionality, is called the coefficient of viscosity. The coordinate y is perpendicular to the interface, and dv/dy is just the slope of the velocity profile at the interface.

There exists a characteristic η for every gas or liquid. η is independent of the internal pressure (over normal ranges), but is very sensitive to temperature. The dimensions of η are $ML^{-1}T^{-1}$. The standard unit for η was developed in the cgs system; it is called the poise. 1 poise $= 1$ gm cm^{-1} sec$^{-1} = 1$ dyne sec cm^{-2}.

Equation 3-37 is applicable only when the surfaces in relative motion are plane, as in problem 319, or have cylindrical symmetry, as would be the case in streamlined flow in a round pipe. For many irregular surfaces the only practical way to determine the force opposing relative motion is by experiment.†

Gases and liquids may flow between solid walls at rest, in which case a velocity profile exists in the flowing material and a driving force is needed to maintain the flow. Or a solid may move through gases or liquids at rest, creating a velocity profile in the region of its passage and thereby inducing a force opposing its motion. A case of particular interest is that of a sphere moving through a medium: raindrops falling, balloons rising, beads sinking in a cooling solution, etc (Figure 160). For a sphere moving slowly enough to avoid creating turbulence, Stokes developed the following formula, known as Stoke's Law, for the force resisting the sphere's motion:

$$F = 6\pi r \eta v,$$

where r is the radius of the sphere, η is the coefficient of viscosity of the medium, and v is the speed of the sphere relative to the medium. (As an example, the factor

*This appears inconsistent with the data in problem 124, in which the speed at the bank was given as v_b, $v_b \neq 0$. Even in that problem the water speed exactly at the bank would $= 0$; however, for any distance from the bank that was significant in comparison with the dimensions of the boat, the minimum speed *was* v_b.

†That is one of the reasons we build wind tunnels.

$6\pi r\eta$ divided by the mass of the balloon appears as the 0.01 coefficient of v in problem 56.)

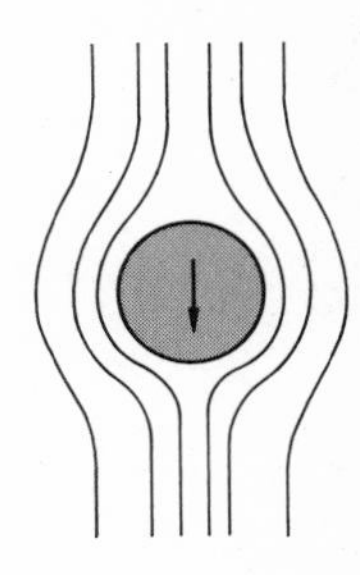

FIGURE 160

PROBLEM 319

To measure the viscosity of CO_2 vapor, a rectangular block of solid CO_2, 3.0 cm wide and 10 cm long, with a flat lower face, is placed on a horizontal belt (Figure 161). The CO_2 block rests on a supporting cushion of subliming CO_2 vapor 0.20 mm above the belt. A sensitive hairspring balance is connected as shown. When the belt is run at a linear speed of 1.6 m sec^{-1}, the balance reads 31 dynes. What is η for CO_2 vapor at the temperature of the vapor in the experiment?

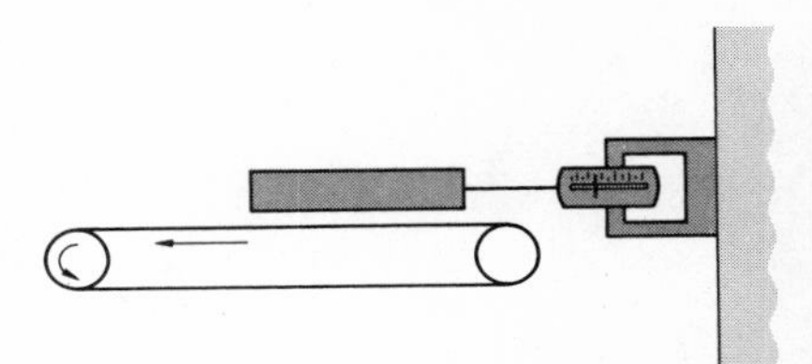

FIGURE 161

SOLUTION

There is enough data to allow for an immediate substitution in equation 3-37. We do not know anything about the velocity profile in the gas layer between block and belt, but we do not need to know. (If we did, it would be reasonable to assume that it was linear.) What we do know is that between the two surfaces that are in relative motion there is a speed difference of 1.6 m sec^{-1}, and this difference is communicated to the top and bottom surfaces of the gas layer. So for the complete gas layer $dv/dy = (1.6 \times 10^2)/(0.2 \times 10^{-1})$. Then $\eta = 31/(3 \times 10 \times 8 \times 10^3) = 1.3 \times 10^{-4}$ poise. (This value is about right for CO_2 vapor at $-20°C$.)

PROBLEM 320

A thin, parallel-sided sheet of solid CO_2, 1.3 mm thick and of 1.55 gm cm^{-3} density, falls edgewise down a slot 1.4 mm wide between two parallel vertical walls. The pressure of the subliming vapor holds the sheet centrally in the slot. If η for the CO_2 vapor is that found in problem 319, how far has the sheet fallen 1.0 sec after release from rest?

PROBLEM 321

A fluid, with viscosity coefficient η, flows at a constant rate in streamlined flow in a cylindrical tube of internal radius R as the result of a driving pressure ΔP per unit length of the tube.

a. What is the speed of flow at a distance r ($r < R$) from the center of the tube?
b. What does the velocity profile in diametral cross section look like?

PROBLEM 322

In problem 321

a. what is the rate of flow of liquid in the tube;
b. how does the average speed of the liquid compare with the maximum speed?

PROBLEM 323

Ethyl alcohol, for which $\eta = 1.2 \times 10^{-2}$ poise, is flowing through a horizontal glass tube 20 cm long and of internal diameter 2.0 mm

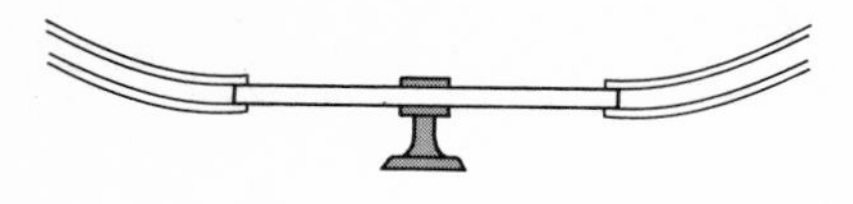
FIGURE 162

(Figure 162). The glass tube is connected at either end to flexible plastic tubing. When 4.0 cm^3 sec^{-1} of alcohol are flowing through the tube, what horizontal force must be supplied by the clip holding the glass tube?

PROBLEM 324

A $33\frac{1}{3}$ rpm phonograph is turned on, and the record changer operated to drop a 12-inch record on the turntable. Unknown to the owner, however, as a joke, a short collar has been mounted with a setscrew on to the turntable spindle, so that the collar holds a dropped record at rest 1.0 mm above the turntable surface. When the turntable is rotating, what is the torque on a record resting on the collar? ($\eta_{air} = 1.8 \times 10^{-4}$ cgs units for standard temperature and pressure.)

Stoke's Law is inapplicable for a sphere moving through standard air when the product rv is > 0.14; r the radius of the sphere and v the speed of the sphere relative to the air, both in cgs units. For $0.14 < rv < 70$, the drag force in air is variable as the physics of the relative motion changes from pure slip to head-on momentum transfer. For $70 < rv < 21 \times 10^3$, the drag force on a sphere moving in air is essentially constant at $F_D = 0.2\,\rho A v^2$, where ρ is the density of the air and A is the cross-sectional area of the sphere—all in cgs units.

PROBLEM 325

A small raindrop condenses out of a cloud 2×10^3 m above ground and starts falling. ("Small" means small enough that Stoke's Law applies.) Neglect the buoyant effect of the air but not viscous drag from the air, and assume ρ_{air} to be constant.

a. What is the terminal speed of the drop?
b. How soon after the drop started falling did it reach 99.9% of terminal speed?
c. How far had it fallen when it reached 99.9% of terminal speed?
d. In falling, drops can decrease in size by evaporation or increase by accretion. Which of these possibilities would result in a higher terminal speed?

PROBLEM 326

What is the upper limit to the size of the raindrop in problem 325 that would make Stoke's Law applicable even at terminal speed?

PROBLEM 327

A raindrop is formed near the lower surface of a cloud 2.0×10^3 m above ground, and by bumping into other drops grows very quickly to a 0.1-cm radius. Assume a drag force of $0.2\rho A v^2$ from the start of motion. Also assume, again, that the buoyancy of the air is negligible, and that ρ_{air} is constant.

a. How long did the drop take to reach the ground?
b. With what speed did it arrive there?

c. How soon after the start of motion did the assumption about the drag force actually become valid?
d. If you had assumed Stoke's Law applied at the start, how long would that assumption have remained valid?

Hydrostatic Pressure and Buoyancy

In a gravitational field, a weight force in the direction of the field is applied to every point mass in the field. If the point mass does not accelerate, then there must exist a force on the mass that is equal to the weight but opposite in direction.

When the mass in question is an obvious, ponderable object, this supporting force is clearly visible (cf. problem 164). But what about the water in a tall glass graduate that is standing vertically on a laboratory table? The upper 3 cm of that water isn't accelerating, so it must be supported by some upward force equal to its weight. The upper 6 cm of the water weighs twice as much as the upper 3 cm,* so it must be supported by an upward force twice as large as the supporting force for the upper 3 cm. Clearly this supporting force must increase linearly with increasing depth of the supported column of water. We can state this analytically by saying

$$F_{\text{sup}} = \rho A h g,$$

where A is the horizontal cross-sectional area of water column, h is the height of the column being supported, and ρ is the volume density of the water. (Since $\rho A h$ is the mass of the column, the supporting force F_{sup} is just equal to the weight of the column.)

If we divide both sides of the equation by A, we get

$$\frac{F}{A} = \rho g h,$$

where F/A is the force required to support a column of unit cross section. This is so useful and flexible a concept that it is given a special name—pressure, denoted P, with dimensions of $ML^{-1}T^{-2}$. As we shall see later, force per unit area, or pressure, has wider usefulness than just evaluating the support of a column of liquid. When it is used in this liquid support sense, it is customary to call it hydrostatic pressure. Thus,

$$P_{\text{hyd}} = \rho g h. \tag{3-38}$$

Solids have an internal structure that enables them to transmit forces in one direction. Liquids and gases have no such structure, hence any pressure applied to a liquid or gas is transmitted equally well in all directions. This omnidirectional aspect is what makes the concept of pressure so useful in dealing with liquids and

---

*We ignore here the infinitesimal change in the gravitational field and an infinitesimal change in density of the water.

gases; it also ensures that the force felt by a surface subject to hydrostatic pressure is always perpendicular to the surface.

The hydrostatic pressure given by equation 3-38 equals the weight of the unit cross section of fluid above the level at which the pressure is measured; it does not, in most cases, equal the total pressure at that level. Usually the upper surface of the unit column feels a downward force from the ambient pressure above it. If we call this ambient pressure B, then the total pressure at a depth h in the fluid must offset this external force on the column as well as the weight of the column. Hence the total pressure in a liquid at a depth h is given by

$$P_h = B + \rho_\ell gh. \tag{3-39}$$

Let us return to our column of liquid in a container and look at an irregular volume V of the liquid (Figure 163). That volume has weight, and it is also subject to a multitude of forces from the surrounding hydrostatic pressure on each element of surface. It should be easy to see, without having to resort to surface integration, that in a summation the horizontal forces will cancel each other and the vertical forces will just add up to the weight of the volume, $\rho_\ell Vg$.

Now, without disturbing the rest of the liquid, let us replace the liquid in volume V by a solid body, of density ρ_s (Figure 164). The liquid external to V continues to supply the same hydrostatic pressure as before, so that the solid body experiences a supporting force $\rho_\ell Vg$. Since the weight of the solid body is $\rho_s Vg$, the net force on the solid body is $Vg(\rho_\ell - \rho_s)$. If $\rho_\ell > \rho_s$, the body accelerates upward unless restrained; if $\rho_s > \rho_\ell$, the body moves downward unless restrained.

The force provided by the hydrostatic pressure from the surrounding medium is called the buoyant force. This buoyant force on an immersed body A is just exactly the force, in direction and magnitude, that would have been available to the fluid displaced by A. When the entire fluid is at rest in a gravitational field the fluid displaced by A would have been at rest also, and would only have needed a force to support its weight; for this the buoyant force is just $\rho_\ell Vg$. When the fluid is accelerating, the displaced volume of fluid would have accelerated also, so that the buoyant force

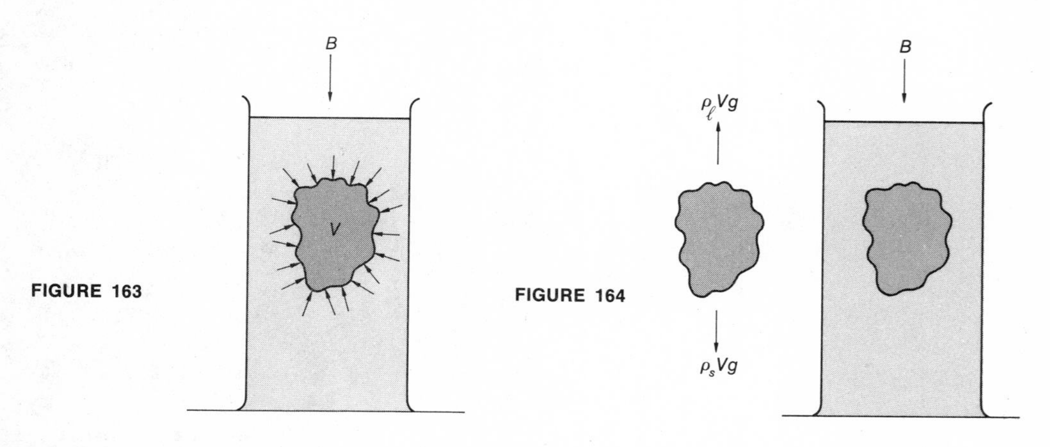

FIGURE 163

FIGURE 164

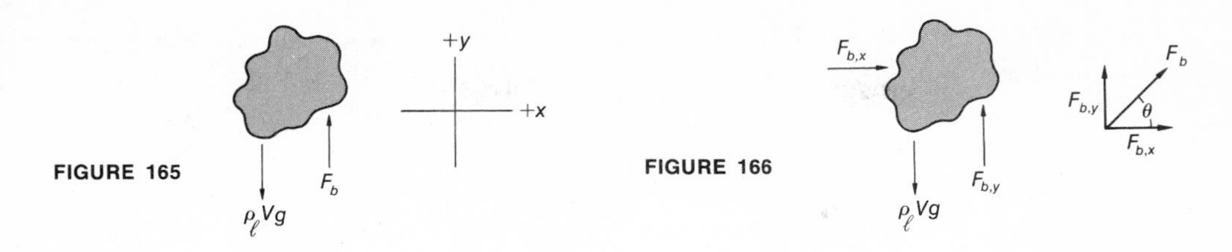

FIGURE 165

FIGURE 166

would have been required to provide for this as well. In such circumstances, F_b must equal $\rho_\ell \ V\vec{g}$ plus $\rho_\ell \ V\vec{a}$. (This equation must be expressed in vector form, since $\vec{g}$ and $\vec{a}$ might not be in the same direction.)

For example, let us consider a volume V of liquid somewhere in the tall graduate previously discussed. The graduate is being accelerated upward with an acceleration a. From the free-body diagram in Figure 165, $\Sigma F_y = F_b - \rho_\ell \ Vg = \rho_\ell \ Va$, so that $F_b = \rho \ V(g + a)$. Since this buoyant force is supplied by the hydrostatic pressure, in the accelerated case the hydrostatic pressure must everywhere be increased by the ratio $(g + a)/g$ over the pressure in the unaccelerated case. In both cases (at rest or vertically accelerated) the hydrostatic pressure increases with depth; planes of equal pressure—the isobars—are horizontal (the upper liquid surface is an isobaric plane of zero hydrostatic pressure); and the buoyant force is perpendicular to the isobaric planes and in the direction of decreasing pressure.

Now let us discontinue the vertical acceleration and accelerate the graduate in the x-direction with a. Now from the free-body diagram (Figure 166), $\Sigma F_x = F_{b,x} = \rho_\ell \ Va$ and $\Sigma F_y = F_{b,y} - \rho_\ell \ Vg = 0$. From this,

$$F_b = \rho_\ell V \ \sqrt{a^2 + g^2}, \text{ at } \theta = \tan^{-1} \frac{g}{a},$$

as shown. In this case the isobaric planes and the free liquid surface are at θ with the vertical.

We can include all of these cases, at rest or accelerated, by again turning to vectors for a more complete and general statement and saying that

$$\vec{F_b} = \rho_\ell \ V(\vec{g} + \vec{a}). \tag{3-40}$$

In many practical cases the $\vec{a}$ in this equation is induced by rotary motion; the centrifuge is an obvious example.

The Center of Buoyancy. An obvious question arises—where does the force of buoyancy appear to act? As can be seen from the previous discussion, it actually acts in little ΔF's all over the submerged surface of the body, the ΔF's being perpendicular to the isobars. However, by a process very similar to that employed in defining the center of gravity (or by a careful surface integration if you prefer rigor to intuitive reasoning), you can determine that all the little ΔF's acting on their own little ΔA's produce the same effect on the submerged body as would a single F_b, defined by equation 3-40, that was applied at the center of mass of the submerged portion of the body. For obvious reasons, such a point is called the center of buoyancy.

Note carefully that the center of buoyancy is at the center of mass of only that portion of the total body that is submerged. If an object is entirely under the surface of a liquid, or is entirely surrounded by a gas, the center of buoyancy coincides with the center of mass of the entire body. However, if a body is partly submerged in a liquid and partly projects into the atmosphere above the liquid, it is subjected to two buoyant forces—one from the liquid and acting through the center of mass in the liquid, and one from the fluid of the atmosphere and acting at the center of mass there. Since the density of a normal atmosphere is 10^{-3} or less of the density of most liquids, unless three-figure accuracy or better is called for it is customary to ignore $F_{b,\text{atmo}}$ in comparison with $F_{b,\text{liq}}$.

PROBLEM 328

In problem 325 you ignored the buoyancy of the air. How much error did this introduce into your calculation of the terminal speed of a small raindrop?

PROBLEM 329

There is a pleasant story for children about a Dutch boy who stopped a leak in the dike facing the North Sea by holding his finger in the hole, and thus saved The Netherlands from inundation. If his finger had a cross-sectional area of 1 cm² and the leak was 4 m below the outside sea level, what force did the boy have to exert to keep his finger in the hole? The density of sea water = 1.03 gm cm^{-3}.

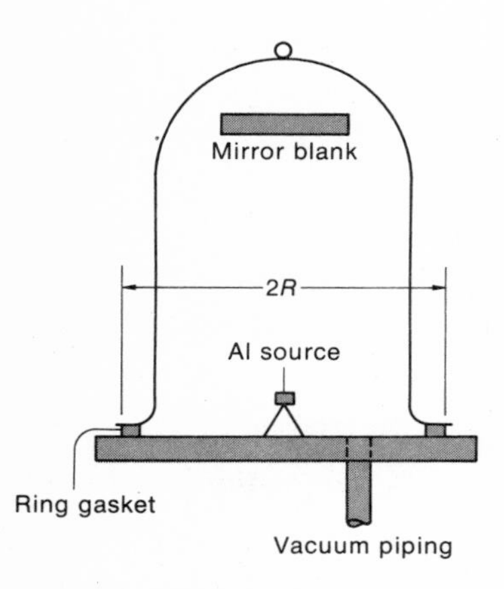

FIGURE 167

PROBLEM 330

Excellent first-surface mirrors for optical instruments can be made by molecular deposition of aluminum in a vacuum. For small mirrors the mirror blank and the aluminum source can be placed inside a bell jar which is then placed on a gasket ring on a metal table as in Figure 167. The gasket has an external radius of R and a width w. When the atmospheric pressure is B_0 and the bell jar is evacuated, what is the pressure on the gasket? $R >> w$.

PROBLEM 331

After a storm in the North Pacific, round hollow glass balls that have broken loose from Japanese fishing nets can often be found washed ashore on Oregon beaches. Assume such a ball weighed 520 gmwt and had a diameter of 12 cm.

a. When floating freely in ocean water of 1.025 specific gravity what percentage of the ball's volume was above water?
b. How many such balls per 100 m would be required to support a fishing net that weighed, in the water, 560 gmwt per m?

PROBLEM 332

If you need a quick, two-figure accuracy determination of the density of an unknown liquid, and you don't have a hydrometer or a balance handy—and you're not afraid of the possibility that the liquid may have a corrosive vapor—you can do the following. Place two straws in your mouth side-by-side, lean over so that the straws can lead vertically down into two adjoining beakers, one containing distilled water and the other containing the unknown liquid (Figure 168). Gently suck on the straws until the higher column of liquid nearly reaches your mouth. Have someone mark on the straws the simultaneous heights of both columns and the position of the surface levels in the beakers. If the water column was h_w long and the column of the unknown liquid was h_ℓ long, what was the density of the unknown liquid?

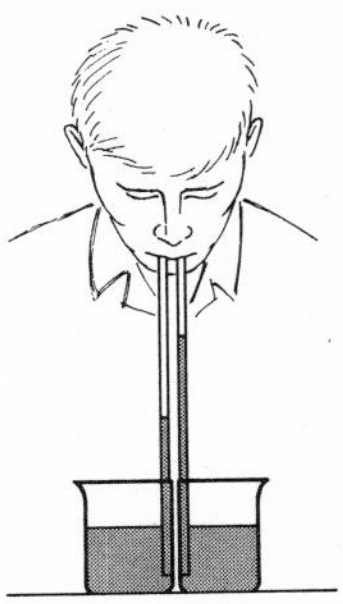

FIGURE 168

PROBLEM 333

Even a normal, healthy person with well-developed lungs finds it almost impossible to expand his lungs for breathing against a pressure differential of more than about one-twentieth of a standard atmosphere. How far below the surface of the water can such a person swim, using a snorkel for breathing?

PROBLEM 334

A standard mercury barometer shows a mercury height of 76.0 cm when the barometer is at rest on the surface of the earth where the atmospheric pressure is 1.013×10^6 dynes cm^{-2}. How high will the mercury rise in the barometer if it is placed in a space capsule that is in a stable orbit around the earth?

PROBLEM 335

A student practicing with scuba diving equipment in the school's freshwater swimming pool finds he must wear one lead weight of mass 0.50 kg and volume 50 cm^3 to achieve neutral buoyancy. When he goes into ocean water, of specific gravity 1.025, he finds he must wear five such weights to reach neutral buoyancy. Equipped for scuba diving, but not including the weights, how much does the student weigh?

PROBLEM 336

Mr. and Mrs. J. Harold Gottet celebrate their fortieth wedding anniversary by giving a formal party on the terrazo terraces surrounding their 30 ft $\times$ 55 ft swimming pool. That evening the air temperature rises to 97°F.; to provide some relief Mr. Gottett, who owns stock in a nearby ice and cold-storage plant, has the plant dump thirty 200-lb cakes of ice into the water in the pool.

a. Before the ice was dumped in, the water level in the pool was 3.0 inches below the drain channels; after all the ice was floating in the pool (ice has a density of 0.92 gm cm^{-3}), where was the water level?
b. Next morning, after all the ice had melted and 1.0 inch of water had evaporated during the hot night, where was the water level?

PROBLEM 337

For extremely accurate weighing in air on equal-arm balances, the buoyancy of the air must be taken into account. (This would not be necessary in the very unlikely case of the volume of the object being weighed equalling the volume of the weights used.) If the volume of an object being weighed is V and its apparent weight is W, and if ρ_a is the density of the air and ρ_w the density of the material used in the weights, what is the error in W?

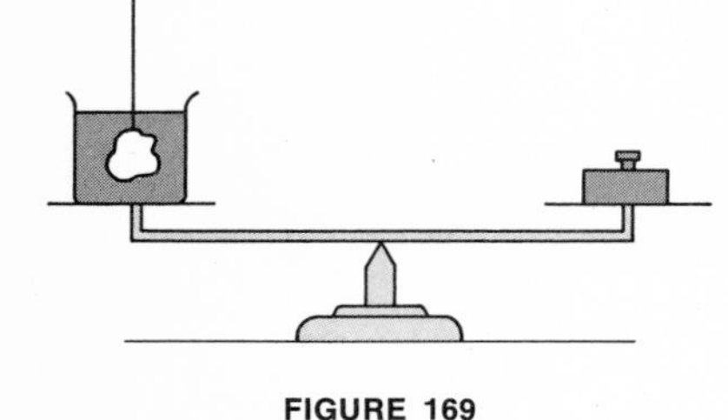

FIGURE 169

PROBLEM 338

A beaker partially filled with liquid of density ρ is placed on one platform of an equal-arm balance; a weight W on the other arm is required for balance. Then an object suspended from a wire is lowered into the liquid until completely immersed but does not touch the beaker (Figure 169). An additional weight w must be added to W to maintain balance. Then the object is lowered to the bottom of the beaker and the wire is removed. An additional weight of $3w$ must now be added to the previous weights to maintain balance.

a. What was the weight of the object?
b. What was its density?
c. What was the tension in the wire when the object was suspended in the liquid?

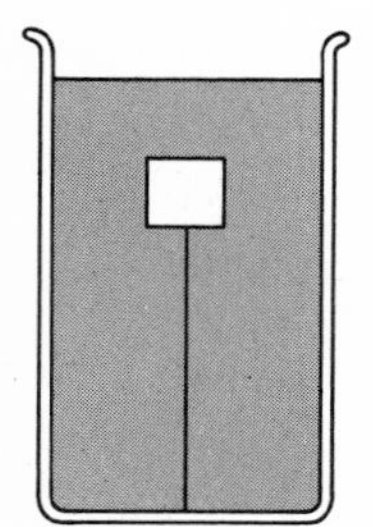

FIGURE 170

PROBLEM 339

In a space rocket a vertical tank of horizontal cross section 3.0 m^2, contains 18×10^3 kg of liquid oxygen. Immediately after lift-off the rocket accelerates upward at 3.3 m sec^{-2} relative to the earth. What is the pressure on the tank bottom at that time? (The data given are reasonably appropriate for a Titan rocket.)

PROBLEM 340

In problem 176 the ascenseur boxes contain barges of significantly different sizes and weights. Is it then still correct to say that the boxes are in balance when their water levels are equal?

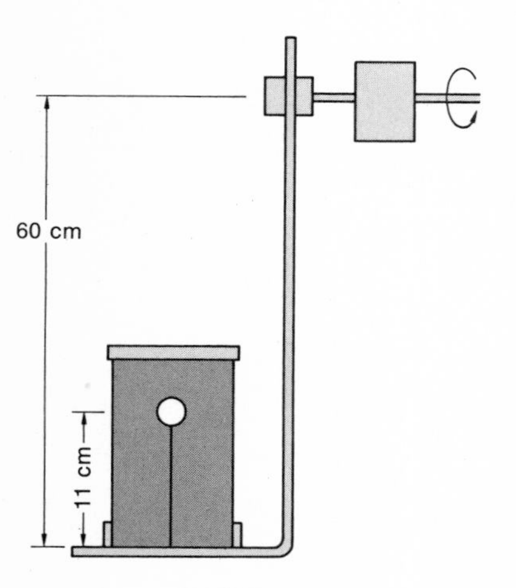

FIGURE 171

PROBLEM 341

A solid is held below the surface of an incompressible liquid of density greater than the density of the solid by a slender cord cemented to the bottom of the liquid container (Figure 170). When the container is at rest, the tension in the cord is T_0. Derive the equation which relates the tension in the cord T to the vertical acceleration of the container.

PROBLEM 342

A ping-pong ball weighs 2.0 gmwt and has an external volume of 24 cm^3. Such a ball is cemented to a fine nylon filament, the other end of which is cemented to the bottom of a glass jar (Figure 171). When the filament is fully extended, the center of the ball is 11 cm from the bottom of the jar. The jar is filled with distilled water and securely capped.

a. When the jar is resting upright on a laboratory bench, what is the tension in the filament?
b. The jar is then clamped to the ledge of an arm that can be rotated in a vertical plane. When the speed of such rotation is 45 rpm, what are the maximum and minimum tensions in the filament?

PROBLEM 343
In getting the apparatus of problem 342 set up and adjusted for demonstration, it was discovered under stroboscopic illumination that at the top of the circle, when the jar was upside down, at a certain critical rotational speed the filament went slack. What was that critical speed?

PROBLEM 344
The apparatus of problem 342 is now turned through 90°, so that the arm carrying the jar rotates in a horizontal plane (Figure 172). When it is rotating at the same speed as before—45 rpm—what is the tension in the filament and its direction?

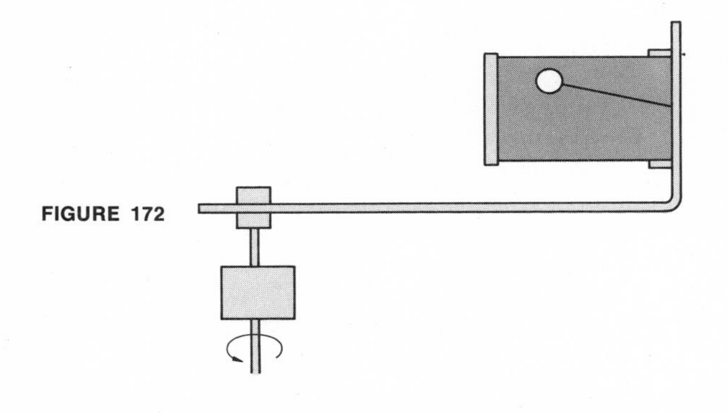
FIGURE 172

PROBLEM 345
Liquid viscosities in the higher ranges can be determined by timing the fall of a metal ball through the liquid in a tall vertical container. If a suitable ball is chosen, after release from rest at the top of the liquid the ball reaches a constant terminal speed before arriving at an upper mark on the cylinder; this speed can be measured by timing the fall of the ball between the upper mark and a mark y below it (Figure 173). Develop a formula for η in terms of this time t and the constants of the experiment.

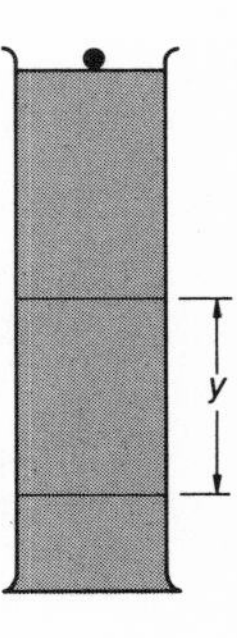

FIGURE 173

PROBLEM 346
An object of specific gravity 0.900 is dropped into a deep, fresh-water lake, entering the water with a velocity of 7.00 m sec^{-1} downward. If the object is perfectly streamlined, so that you feel you can ignore drag or viscosity effects.

a. how soon after entering the water should the object resurface,
b. how deep did the object go?

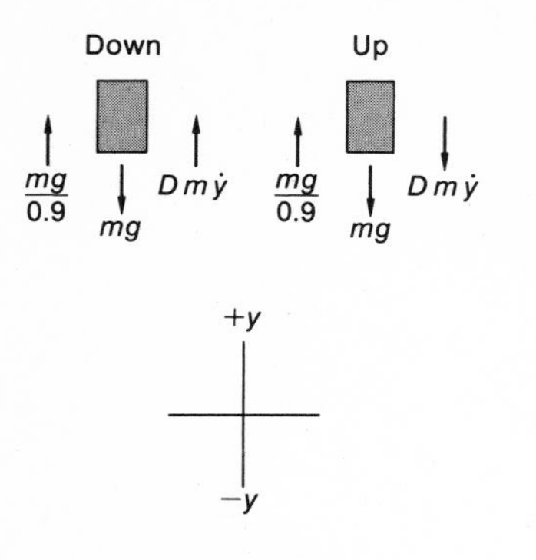

FIGURE 174

PROBLEM 347

Actually, the object in problem 346 resurfaced in exactly 11.0 sec, which indicated the existence of a drag force. If you assume that the drag force was linear with speed, and can be expressed in the form $Dm\dot{y}$, what was the value of D?

SOLUTION

The free-body diagrams are as shown in Figure 174, and the equation of motion is

$$m\ddot{y} = -mg + \frac{mg}{0.9} - Dm\dot{y}.$$

This equation applies to both diagrams; on the down trip, $\dot{y}$ is negative in the coordinate system chosen, so that $-Dm\dot{y}$ is upward as shown. On the up trip, $\dot{y}$ is positive, so $-Dm\dot{y}$ is downward as it should be. After the cancelling and rearranging, the equation becomes

$$\ddot{y} = \frac{d\dot{y}}{dt} = \frac{g}{9} - D\dot{y},$$

$$\int_{V_0}^{\dot{y}} \frac{d\dot{y}}{(g/9) - D\dot{y}} = \int_0^t dt,$$

$$-\frac{1}{D} \ln \frac{(g/9) - D\dot{y}}{(g/9) - DV_0} = t;$$

or

$$\dot{y} = \frac{g}{9D} - \frac{1}{D}\left(\frac{g}{9} - DV_0\right)e^{-Dt},$$

$$\int_0^y dy = \frac{1}{D}\int_0^t \left[\frac{g}{9} - \left(\frac{g}{9} - DV_0\right)e^{-Dt}\right] dt,$$

$$y = \frac{1}{D}\left[\frac{g}{9}t + \frac{1}{D}\left(\frac{g}{9} - DV_0\right)\left(e^{-Dt} - 1\right)\right].$$

When $y = 0$, $t = 0$, or $t = 11$.

If we choose the latter value and put in values for g and V_0,

$$0 = \frac{1}{D}\left[11.98 + \frac{1}{D}(1.089 + 7D)\,(e^{-11D} - 1)\right].$$

Here the only unknown is D, which you are to solve for. The equation looks confusing, so the first thing to do is simplify and rearrange it so that you end up with

$$11.98 - 7\,(1 - e^{-11D}) = \frac{1.089}{D}\,(1 - e^{-11D}).$$

If you call the left-hand side f_1 and the right-hand side f_2, then of course $f_1 = f_2$. You do this so you can plot f_1 and f_2 against assumed values of D, separately (Figure 175). If the curves for f_1 as a function of D and f_2 as a function of D intersect, then $f_1 = f_2$ at the intersec-

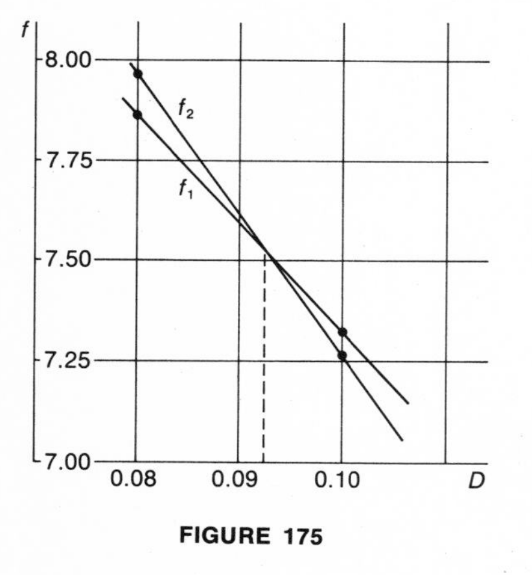

FIGURE 175

tion and the value of D for that intersection is a root of the equation. If there is only one intersection (which is what you hope for), that root is the unique solution of the equation.

Set up a table for f_1 and f_2 against D, and for a starter, assume $D = 0.1$.

D	f_1	f_2
0.1	7.31	7.27
0.08	7.88	7.97
0.092	7.53	7.53

(In this kind of solution it helps to have a little judgment about the probable range of values.) If the curves for f_1 and f_2 are to intersect, we have to find a range of paired values for f_1 and f_2 so that at one end of the range $f_1 > f_2$, and at the other end, $f_2 > f_1$.

For $D = 0.1$ $f_1 > f_2$, so we want to find a pair of values for which $f_2 > f_1$; let's guess that decreasing D will do that, and for a trial shot, assume $D = 0.08$. For this value f_2 is $> f_1$ so we have established limits. Since the spread of values is not great, as a first approximation we can assume straight lines in this limited region for f_1 and f_2. Plotting the values and drawing the lines, the intersection occurs at $D = 0.092$. Inserting this value in the table as a check, we see that $f_1 = f_2$. Hence $D = 0.092$ is a solution of our equation; from the nature of the physical situation (not from the graph, since that represents too small a sample of the f vs. D relationship) we are probably safe in assuming this is a unique solution. Putting in this value, the equations we have developed for $\dot{y}$ and y become

$$\dot{y} = 11.84 - 18.84e^{-0.092t},$$

$$y = 11.84t - 204.8\,(1 - e^{-0.092t}).$$

PROBLEM 348

a. How far down in the water did the object of problem 347 go?
b. Why did the down trip take less time than the up trip?
c. At what value of $\dot{y}$ would the up trip become a motion at constant speed?

NEWTON'S SECOND LAW FOR ROTATIONAL MOTION

The Moment of Inertia

When masses are made to turn or rotate, we find by experience that the same turning effort produces greater motion when applied to a uniform, solid sphere than when applied, for example, to the same amount of mass arranged in dumbbell form. We are thus forced to conclude that for rotational application, the inertial property appearing in Newton's Second Law must depend, not only on mass, but on how that mass is distributed in space.

This requirement can be met quite simply. Let us consider a collection of point masses m_i, rigidly connected so that they do not move relative to each other. If ℓ_i, the perpendicular distance of the mass m_i from the axis of rotation, is known, then we can express the inertial property of that collection, relative to that axis of rotation, by

$$I_{A\text{-}A} = \Sigma m_i \ell_i^2,$$

where A-A is axis of rotation, and $I_{A\text{-}A}$ is called the moment of inertia relative to that axis. Its dimensions are ML^2. If the body is a continuous distribution of mass, rather than a collection of mass points, then the summation must be accomplished by integration, over the volume of m, and we have

$$I_{A\text{-}A} = \int \ell_m^2 \, dm. \tag{3-41}$$

Since $I_{A\text{-}A}$ varies for every one of an infinite number of possible A-A axes, we face the possibility of having to reintegrate every time we consider a new body or a new axis. There ought to be a better way—and there is. We can take advantage of the fact that the c.m. (center of mass) of a rigid body stays fixed relative to the body. If we consider an axis through the c.m., then the summation, or integration of equation 3-41 will be a constant for that axis. If we choose useful axes, we can tabulate for any body a list of constant moments of inertia for that body.

For a c.m. axis of rotation, we rewrite our equation defining moment of inertia as

$$I = \Sigma m_i r_i^2$$

or

$$I_{\text{c.m.}} = \int r^2 \, dm, \tag{3-42}$$

where r is the perpendicular distance from an element of mass to an axis of rotation through the c.m.

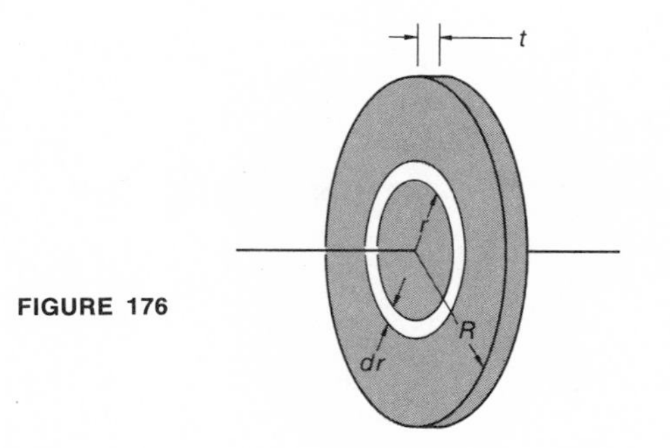

FIGURE 176

For example, let us determine the moment of inertia of a uniform, solid disc of radius R about a transverse axis through its center (Figure 176). At a distance r from the center we will take as our dm the mass of an annular ring of width dr, circumference $2\pi r$, and thickness t. Our dm is then $2\pi r dr \, t\rho$, and

$$I_{\text{c.m.}} = \int_0^R 2\pi r\, dr\, t\rho r^2$$

$$= 2\pi t\rho \int_0^R r^3\, dr = \frac{\pi R^4 t\rho}{2}$$

$$= \frac{mR^2}{2}.$$

In similar fashion, we can arrive at a tabulation of the following moments of inertia for axes through the c.m.:

- For a solid, uniform disc about a transverse axis: $MR^2/2$; about an axis parallel to the face of the disc: $MR^2/4$ if the disc is so thin that $R >> t$.
- For a thin-walled hoop about a transverse axis: MR^2; about an axis in the plane of the hoop: $MR^2/2$.
- For a uniform solid sphere: $\frac{2}{5}MR^2$.
- For a thin, uniform spherical shell: $\frac{2}{3}MR^2$.
- For a uniform thin rod about a transverse axis: $\frac{1}{12}ML^2$; about a longitudinal axis: O.
- For a thick-walled, uniform hollow cylinder of external radius R, internal radius r, and mass M, about a longitudinal axis: $M(R^2 + r^2)/2$. *Note*: This result was obtained by adding to the I of a solid rod of radius R and density ρ the I of a solid rod of radius r and density $(-\rho)$, the latter to allow for the negative contribution of the axial hole.

Even though we have now tabulated moments of inertia only for a c.m. axis, we are not restricted to rotations about that axis. The extreme usefulness of the $I_{\text{c.m.}}$ is evident when we become aware of the transfer theorem, which states that

$$I_{A\text{-}A} = I_{\text{c.m.}} + Md^2. \tag{3-43}$$

In this equation, A-A is any rotational axis you choose, $I_{\text{c.m.}}$ is the moment of inertia about an axis through the c.m. *that is parallel to the axis A-A*, and d is the perpendicular distance between the two axes. This equation should be memorized.

With the help of equation 3-43 we can now expand our tabulation of moments of inertia. For example, the moment of inertia for a solid uniform disc about an axis A-A, at the edge and parallel to the transverse c.m. axis (Figure 177), would be, by equation 3-43,

$$I_{A\text{-}A} = \frac{MR^2}{2} + MR^2 = \frac{3}{2}MR^2.$$

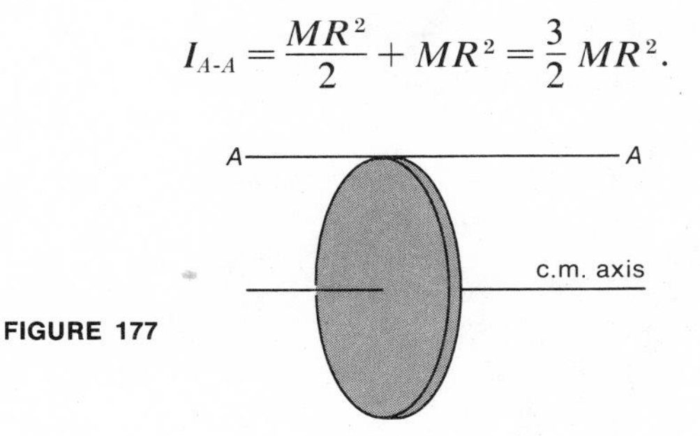

FIGURE 177

PROBLEM 349

What is the moment of inertia of a uniform, solid round rod, of mass M and radius R, about an axis through its center of mass and perpendicular to the axis of the rod?

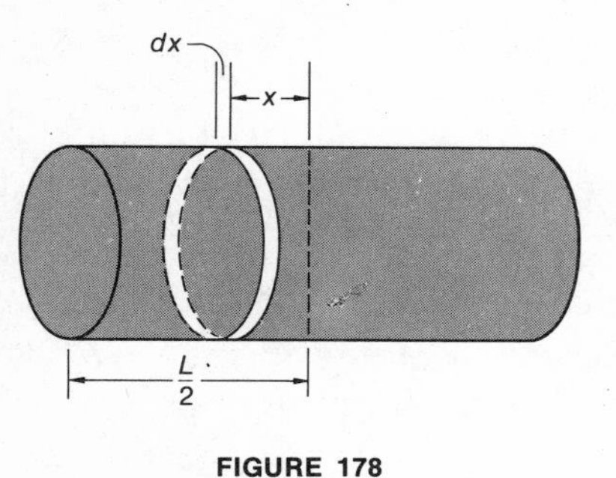

FIGURE 178

SOLUTION

We can consider the bar to be made up of a very large number of thin discs dx wide (Figure 178). The first item of page 179 applies; so the $I_{\text{c.m.}}$ of one of these discs is $MR^2/4 = \rho\pi R^2 dx R^2/4$. Thus, by the transfer theorem, the $I_{A\text{-}A}$ of such a disc located x from the center of the whole rod would be

$$\frac{\rho\pi R^4\,dx}{4} + \rho\pi R^2\,dx \times x^2.$$

Then for the rod as a whole,

$$I_{A\text{-}A} = 2\int_0^{L/2}\left(\frac{\rho\pi R^4}{4} + \rho\pi R^2 x^2\right)dx = \frac{M}{4}\left(R^2 + \frac{L^2}{3}\right).$$

PROBLEM 350

Compute the moments of inertia for the following:

a. A uniform solid sphere of mass M and radius R, about an axis tangent to the surface of the sphere.

b. A long, thin, uniform rod about an axis perpendicular to the rod through its end.

Tabulations of the moments of inertia for various bodies, in reference to various possible axes, are often simplified by the use of a radius of gyration. Instead of a c.m.-axis definition of I as in equation 3-42, the moment of inertia can be defined by

$$I = Mk^2, \tag{3-44}$$

where M is the total mass of the object and k is the radius of gyration for that object in reference to some particular axis. The value of k must be such that, for a particular axis,

$$Mk^2 = I_{\text{c.m.}} + Md^2.$$

Thus for the rod of problem 349, $k^2 = R^2/4 + L^2/12$. For the sphere of problem 350, $k = R(7/5)^{1/2}$; and for the rod, $k = L/3^{1/2}$.

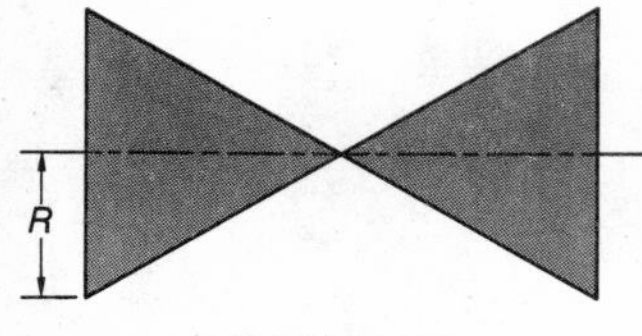

FIGURE 179

PROBLEM 351

Two equal, uniform, and solid right-circular cones, each of mass m and base radius R, are mounted apex to apex on a common axis, as shown in Figure 179. What is the radius of gyration of the assembly relative to this common axis?

PROBLEM 352

a. What is the moment of inertia about axis A-A of the spool shown in Figure 180? The outer discs are each of mass M, and the solid cylindrical axle has a mass $\frac{1}{4}M$ and a radius of $R/6$.
b. In designing an operating mechanism in which two-figure accuracy was sufficient, could you afford to ignore the moment of inertia of the axle: i. if the spool was to rotate about axis A-A; ii. if the spool was to roll without slipping on surface G-G?

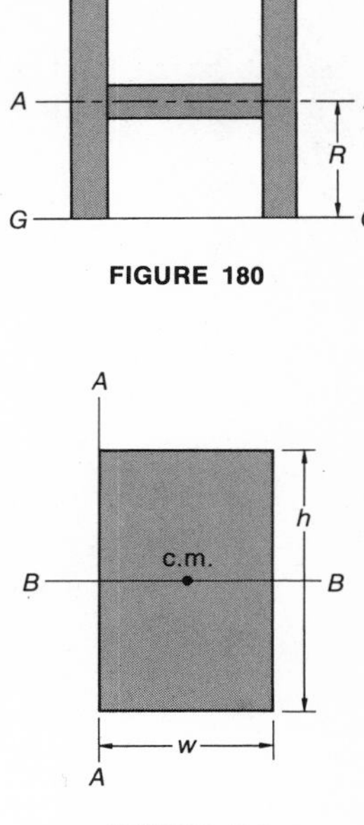

FIGURE 180

PROBLEM 353

What is the moment of inertia of a solid, uniform rectangular panel, h high, w wide, and of mass m (Figure 181),

a. about an axis A-A through a long edge;

b. about an axis B-B through its c.m. and perpendicular to a long edge;

c. about an axis through the c.m. and perpendicular to the panel?

FIGURE 181

Torque

Just as we had to allow for the spatial distribution of mass in arriving at an understanding of rotational inertia, we find we must consider distance from the axis of rotation in computing our turning effort. For a given body turning about a given axis, we discover by experiment that our turning effort seems to increase linearly with the applied force, and also linearly with the perpendicular distance between the axis of rotation and the line of application of the force. We are therefore tempted to say that, if we symbolize our turning effort, or torque, by τ, then

$$\tau = Fd. \tag{3-45}$$

Equation 3-45 is limited to those physical situations in which both F and d are in a plane perpendicular to the axis of rotation, and d, the distance between the axis of rotation and the line of F,* is measured perpendicular to F (Figure 182).

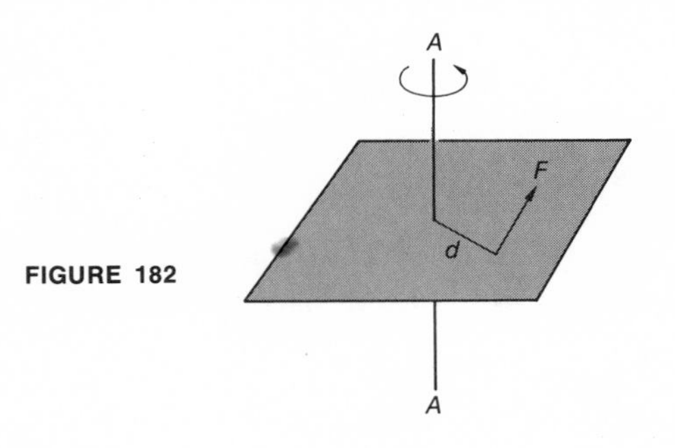

FIGURE 182

*For this usage, d is often called the "lever arm" of the force; the product Fd is often called the "moment of the force."

Although many simple and common problems in rotation correspond to this arrangement, we need some statement that is more general, and that also carries with it information about directions. Such a statement, in three dimensions, is

$$\vec{\tau} = \vec{d} \times \vec{F}. \tag{3-46}$$

$\vec{F}$ is the force applied, $\vec{d}$ is the displacement from some chosen reference point to the point of application of $\vec{F}$, and $\vec{\tau}$ is a vector perpendicular to the plane of $\vec{d}$ and $\vec{F}$. Mathematically, equation 3-46 gives us all we need to describe any turning effort we may encounter; physically, equation 3-46 implies turning effort, and hence rotation, about a point in space. This should make us feel a little queasy, since almost any turning we can visualize is related to an axis and not to a point; even the spin of an elementary particle has an oriented axis. However, in its application in an x-y-z coordinate system equation 3-46 breaks down to the following physically comfortable statements:

$$\begin{aligned} \tau_x &= d_y F_z - d_z F_y, \\ \tau_y &= d_z F_x - d_x F_z, \\ \tau_z &= d_x F_y - d_y F_x. \end{aligned} \tag{3-47}$$

In this tabulation, d_y is the y-component of $\vec{d}$, F_z the z-component of $\vec{F}$, etc.; $d_y F_z - d_z F_y$ describes, in a right-handed sense, the turning effort about the x-axis. τ_x is the component of $\vec{\tau}$ in the positive x-direction, etc. (Describing turning effort by a vector along the axis about which turning takes place is consistent with describing rotation by a vector along the axis of rotation, which we learned to do earlier.) τ_x is perpendicular to the y-z plane which contains d_y, F_z, d_z, and F_y; each statement of equation 3-47 leads us back almost to the simplification discussed earlier under equation 3-45.

The dimensions of the torque τ are ML^2T^{-2}; in MKS units torque is expressed in meter newtons (m N).

Rotational Dynamics

In consistency with the nonrelativistic definition of linear momentum as $\vec{p} = m\vec{v}$, we define angular momentum L as

$$\vec{L} = I\vec{\dot{\theta}} = I\vec{\omega}. \tag{3-48}$$

Then, for rotational motion, Newton's Second Law takes the form

$$\vec{\tau} = \frac{d\vec{L}}{dt} = \frac{d}{dt}(I\vec{\omega}). \tag{3-49}$$

Since we will usually deal with rigid bodies and an axis of rotation fixed relative to this rigid body, I will usually be constant. Then equation 3-49 reduces to

$$\vec{\tau} = I\frac{d\vec{\omega}}{dt} = I\vec{\ddot{\theta}} = I\vec{\alpha}. \tag{3-50}$$

For equation 3-50, τ and I are both computed relative to some axis of rotation, and α is the angular acceleration about that axis.

The use of equation 3-50 involves the same procedures already developed for linear dynamics. Appropriate free-body diagrams are drawn with all forces shown, linear and rotational coordinates are chosen, dynamical equations are written, constraint relationships are established, and then the algebra is worked through to solve for the desired unknowns. At this point it ought to be pointed out that the rotational coordinate that goes with the normal x-y-z linear coordinate system in a right-handed sense is $\circlearrowleft^{+}$, i.e., counterclockwise is positive. It should also be helpful at this time to review problem 132.

PROBLEM 354

In problem 317 it was assumed for simplicity that the truck bed stayed level during the braking.

a. What actually happens?
b. Does this affect any of the answers to problem 317?

PROBLEM 355

A cylinder is observed to be rolling freely at constant speed on a horizontal surface.

a. Is it rolling without slipping (note two possible cases)?
b. What is the frictional force between cylinder and surface if the surface is not "smooth"?

PROBLEM 356

A solid cylinder whose axis is horizontal rests transversely on a movable surface of a plane inclined 30° with the horizontal (Figure 183). The contact is such that the cylinder always rolls without slipping on the surface. What acceleration must be given to the movable surface so that the axis of the cylinder does not move?

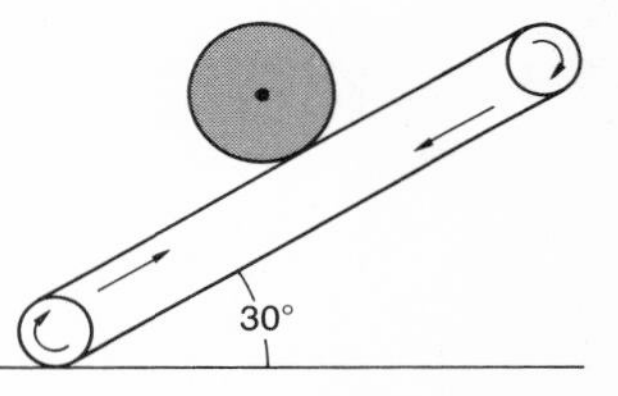

FIGURE 183

PROBLEM 357

A solid spool, which can be considered a solid, uniform cylinder of effective radius r and mass m, is supported on an incline as shown (in Figure 184) by a string which supplies a force F upward and parallel to the incline. The surface of the incline is smooth, but when the force F is a certain critical value, the center of mass of the spool does not move.

a. What is the critical value of F?
b. What is the angular acceleration of the spool?
c. How fast is the string accelerating?

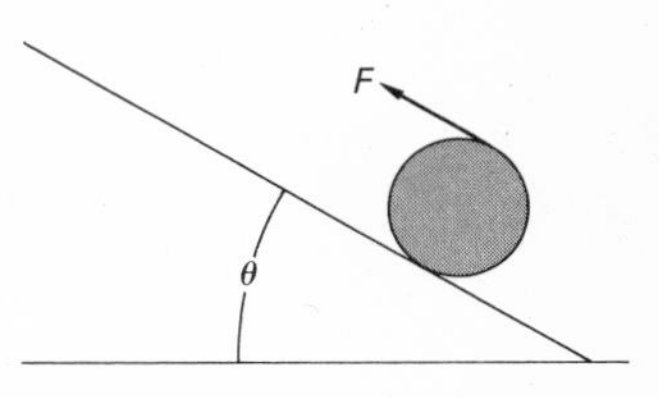

FIGURE 184

PROBLEM 358

The surface of the incline in problem 357 is now altered to provide just barely enough static friction so that the spool does not move at all, either linearly or rotationally, when pulled on by the string with a force F'.

a. What is the value of the force F'?

b. What is the value of the friction force between spool and incline surface?
c. What is the direction of this friction force?

PROBLEM 359
The string in problem 358 breaks.

a. With what linear acceleration does the center of the spool now move down the incline?
b. What effective frictional force exists between spool and incline during the descent?

PROBLEM 360
The 160-lb attendant on the carrousel of problem 133 sees someone attempting to jump on the carrousel while it is moving, and runs out along a radius at a constant speed of 10 ft sec^{-1} to prevent this. During his run the carrousel's speed is maintained at 6.0 rpm.

a. At what rate does the torque demand on the carrousel's driving mechanism change?
b. What was the maximum additional torque required?

PROBLEM 361
Instead of the idealized simplification presented in problem 182, consider the same arrangement, with the same values, more realistically. This time there is a coefficient of sliding friction = 0.20 between the surface and the masses and pulley resting on it; the pulley, which can be considered as a thin rim, has a mass of 0.30 kg. Only the light frame holding the axle of the pulley is in contact with the surface, so that friction affects the linear motion of the pulley but not its rotational motion. Now what is the acceleration of the pulley?

PROBLEM 362
A uniform, solid sphere, rolls, without slipping, down a plane that is inclined at θ above the horizontal.

a. How does the linear acceleration of its center of mass compare with the linear acceleration of a block mounted on frictionless rollers of negligible mass and allowed to slide down the same plane?
b. What is the value of the frictional force between sphere and plane that ensures that the sphere will roll without slipping?
c. What is the direction of this force?

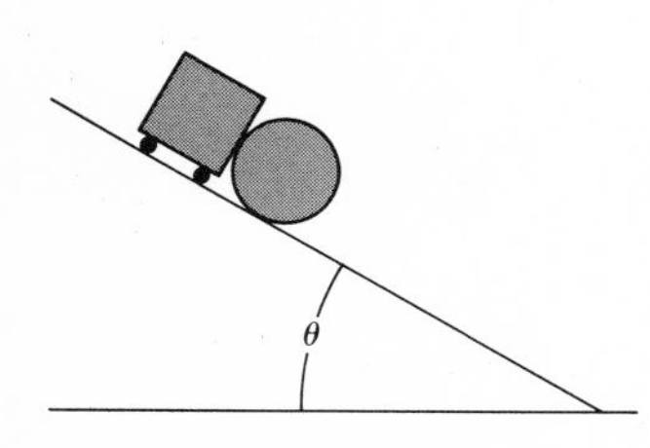

FIGURE 185

PROBLEM 363
The block and sphere of problem 362 are now placed together on the same plane (the block directly uphill from the sphere and touching it; Figure 185) and released from rest. If the sphere again rolls without slipping on the plane, but there is no friction between the sphere and the block, with what acceleration will the combination move down the plane,

a. if the masses of block and sphere are equal, and
b. if they are unequal?

PROBLEM 364

The situation in problem 363 is repeated, but this time the block is reversed, so that the face touching the sphere presents frictional resistance to the motion of the sphere against it. When the combination is released from rest, the sphere again rolls without slipping, but the acceleration of the combination is exactly the same as for the sphere alone. What is the coefficient of sliding friction between block and sphere?

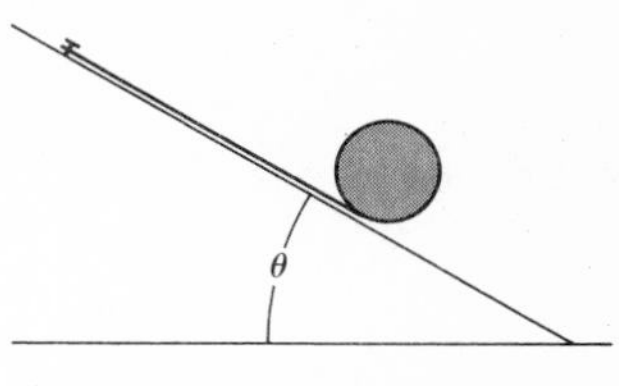

FIGURE 186

PROBLEM 365

With the arrangement shown in problem 363, the mass of the sphere was 0.50 kg, that of the block was 0.40 kg, and the acceleration of the combination was measured as 3.7 m sec^{-2}. What was the angle θ?

PROBLEM 366

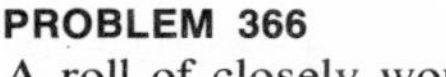

A roll of closely wound flexible tape is held by a tack at the upper end and allowed to unroll down an incline of angle θ with the horizontal (Figure 186). If the unwound tape was L long, how much time did it take for the complete unrolling process?

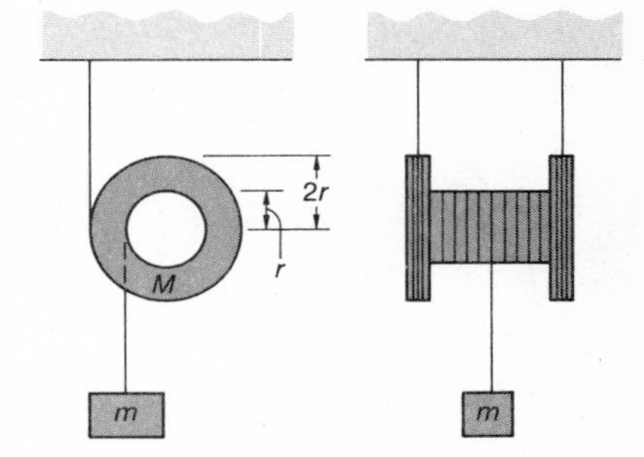

FIGURE 187

PROBLEM 367

A mass m is hanging from a spool of mass M as shown in Figure 187. The radius of gyration of the spool about its central axis is $\sqrt{2}\, r$.

a. When the system is released, what is the acceleration of the center of mass of the spool?
b. What is the limiting value of this acceleration as the ratio $m/M \rightarrow 0$?

PROBLEM 368

A demonstration experiment in dynamics is set up as shown. A stiff, uniform slender beam, of mass M and length L, is placed at its midpoint on a fulcrum that is mounted on the platform of a weighing scale (Figure 188). At the ends of the beam are placed masses m_1 and m_2, $m_2 > m_1$; to keep the beam from moving, a brace is placed under the beam at the m_2 end. (The dimensions of m_1 and m_2 are $<< L$.)

a. What is the scale reading with the brace in place?
b. What is it immediately after the brace is removed?

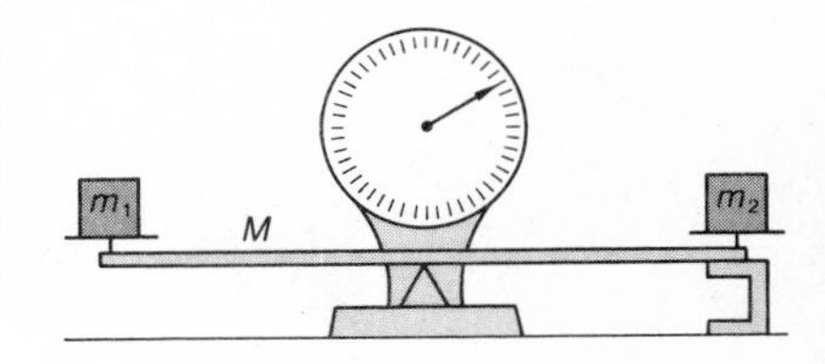

FIGURE 188

PROBLEM 369

A phonograph record is dropped from rest in a horizontal plane centrally onto a horizontal turntable rotating at a constant angular speed ω. When dropped, the disc makes and maintains uniform contact with the table over its entire area.

a. If the disc is not to "slip" more than 9° on the turntable before reaching speed ω, what must be the minimum coefficient of sliding friction between disc and turntable?
b. Through what angle did the turntable turn during this process?
c. Use $R = 15$ cm and $\omega = 33\frac{1}{3}$ rpm to check whether the answer you got for question (a) sounds reasonable.

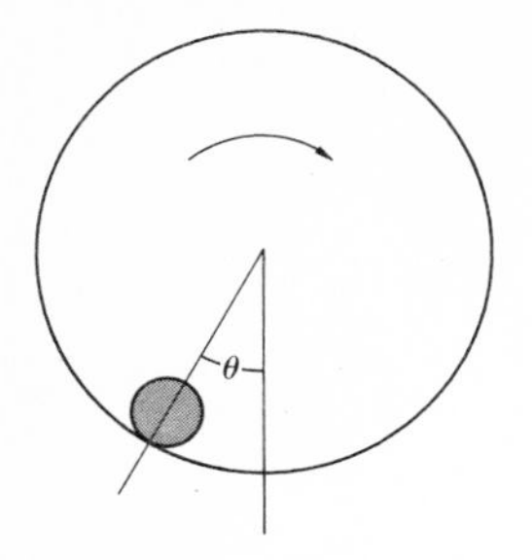

FIGURE 189

PROBLEM 370

A large circular rim of radius R rotates in a vertical plane about its central axis, and a small steel ball rolls freely, without slipping, on the inside surface of the rim (Figure 189).

a. When the rim is rotating with constant ω, where does the c.m. of the ball come to rest?
b. When the rim is rotating with a constant angular acceleration α, what angle will the radius of the rim that passes through the c.m. of the ball make with the vertical?

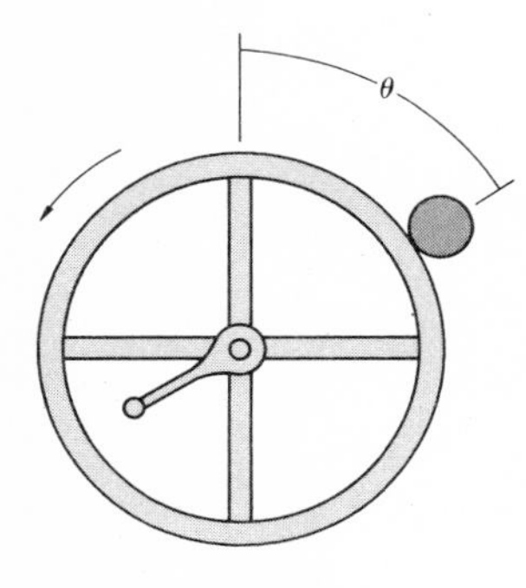

FIGURE 190

PROBLEM 371

While touring the Los Angeles Museum of Science and Industry, a sophomore physics major comes upon a group of liberal arts majors who are examining a rigid, rimmed wheel that is mounted on a horizontal axis and can be turned by means of a crank. He notices that one of the group is carrying a rubber handball, and offers to bet the members of the liberal arts group he can make the handball stay in equilibrium on the rim at any angle less than 90° below the vertical that they specify (see Figure 190). Naturally they make the bets, and he then proceeds to win the money. If m and r were the mass and radius of the handball, and M and R the same for the rim, and assuming the mass of the spokes was negligible and that the ball always rolled without slipping on the rim, what is the relation between the position angle θ and the torque the physics major had to provide to win the money?

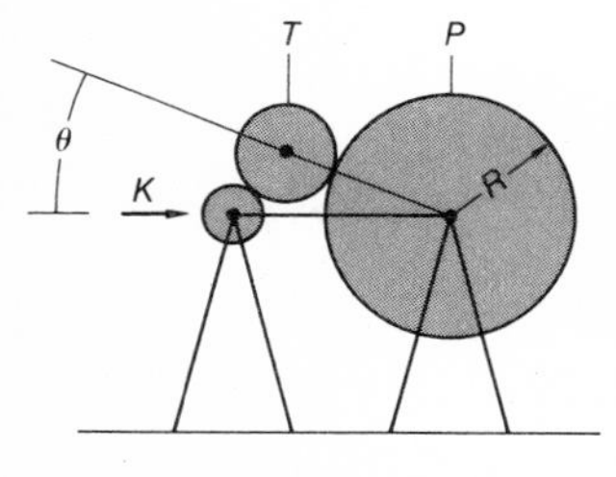

FIGURE 191

PROBLEM 372

An adaptation of an inking arrangement for a printing press is as shown in Figure 191. K is a firmly supported, but idling, inking roller of negligible moment of inertia; P is a driven press roller firmly supported; and T is a transfer roller freely floating between K and P. T is a solid cylinder of radius r and mass M; it always rolls without slipping on both K and P, and the geometry is such that the line of centers TP is θ above the horizontal. What is the maximum angular acceleration A that can be given to P without T losing contact with K?

PROBLEM 373

A uniform cylinder of radius R is spinning about its horizontal longitudinal axis with an angular speed ω_0. It is then gently placed on a horizontal plane. The coefficient of sliding friction between the material of the cylinder and the material of the plane is μ. How far will the cylinder move along the plane before it reaches a condition of rolling without slipping?

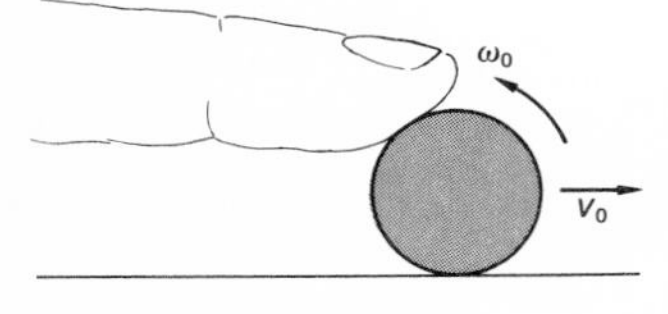

FIGURE 192

PROBLEM 374

An amusing table-top trick is to press a fingertip down on a marble lying on the table in such a way that the marble is squeezed out from under the finger with a linear speed v_0 along the table, and with simultaneously a rotational speed ω_0, as shown in Figure 192. If the coefficient of sliding friction between marble and table top is μ,

a. what should be the relationship between v_0 and ω_0 for the marble to slide and then come to a dead stop, with linear and rotational motion simultaneously reduced to zero?
b. When the marble stops, how far is it from where it started?
c. What should be the relationship between v_0 and ω_0 for the marble to slide to a stop linearly and start a return motion with a rotational speed of $\omega_0/2$?

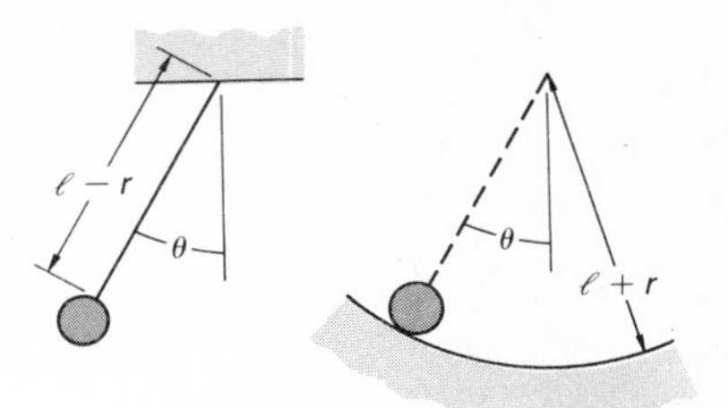

FIGURE 193

PROBLEM 375
A solid ball, of radius r, is cemented to a string of length $(\ell - r)$ and allowed to swing as a pendulum from a fixed point (Figure 193). An identical ball is allowed to roll without slipping in a cylindrical trough of radius $\ell + r$. If both balls are released from rest when their radius vectors are at the same angle θ with the vertical,

a. how do their linear speeds at the bottom of their motions compare;
b. how do their times for a cyclic operation, i.e., to return to their starting position, compare? *Warning:* Their motion is not simple harmonic.

Solve this by Newtonian dynamics; $\Sigma\tau = I\ddot{\theta}$, $v = \ell\dot{\theta}$, etc. Later, in problem 552, you will be able to compare this solution with an energy solution.

PROBLEM 376
A standard method for determining the viscosity of liquids such as oils is illustrated in Figure 194. Between two concentric cylinders, inner one rotatable, a liquid to be tested is poured to a depth h. The inner cylinder rotates when the cord is pulled; to pull the cord at a constant speed v a pull F is required. What is η for the liquid, in terms of F, v, and the constants of the equipment?

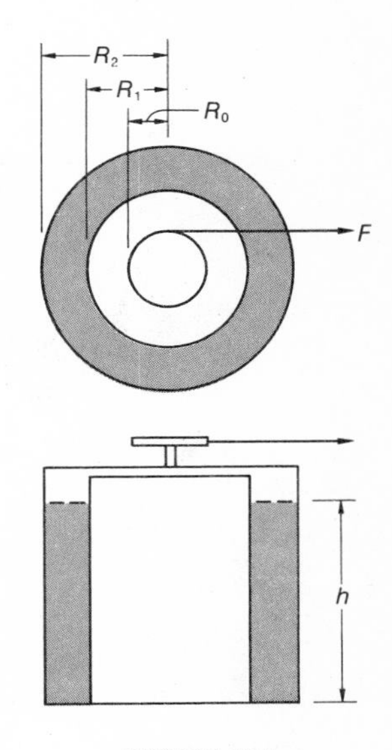

FIGURE 194

EQUILIBRIUM

In every-day life the word equilibrium implies a static balance: things stay put, and don't move or fall down. To the scientist or engineer, equilibrium implies no change of momentum; it means that $d\vec{p}/dt = 0$ or $d\vec{L}/dt = 0$. This of course includes the every-day definition; it also says that a constant mass moving at a constant velocity is also in equilibrium. Note that this is a vector statement; a constant mass moving in a circular path at constant speed is not in equilibrium.

Since $d\vec{p}/dt$ or $d\vec{L}/dt$ equals 0, equilibrium is defined analytically by a reduced form of Newton's Second Law:

$$\Sigma\vec{F} = 0 \text{ or } \Sigma\vec{\tau} = 0. \tag{3-51}$$

Statics

In applying equation 3-51 to a particular static situation, the choice of a coordinate system has a great effect on the clarity and simplicity of the calculations. The

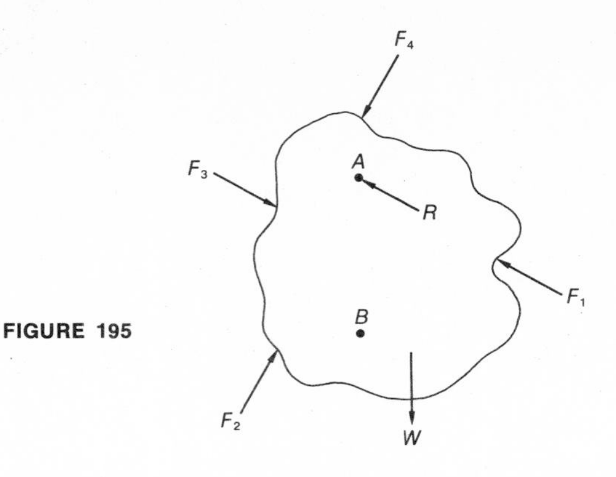

FIGURE 195

development of judgment about appropriate coordinate systems will pay large dividends. With a little thought, the choice of a good Cartesian coordinate system will be fairly obvious. What is usually not so obvious is the best choice of a rotational coordinate system. A body known to be in rotational equilibrium about one fixed axis is also automatically in rotational equilibrium about all other axes fixed in space that are parallel to the first axis. The body shown in Figure 195 is in equilibrium about an axis through A; i.e., $\Sigma\tau_A = 0$. Because our local space is isotropic, homogeneous, and linear, we can prove that for an axis through B that is parallel to the axis through A, $\Sigma\tau_B = 0$ also. (In computing $\Sigma\tau_B$, note that we must include the torque of the reaction force at A.) This parallel-axis possibility presents us with a confusing number of choices; making the best choice becomes a profitable game. For an illustration of this, see the second solution to problem 377.

Although equations 3-51 were derived from Newton's Second Law, these new equilibrium equations, and their application to statics, are not restricted to use with absolute units, as were the original equations in their dynamical applications. Since the right hand side of the equations are without units, the only units are those of ΣF or $\Sigma\tau$. These can be any units you please, provided the same units are used throughout the summation. Thus you could tabulate your forces in leash-pulls-by-a-Great-Dane-dog-chasing-a-cat (lpbaGDdcac) if such a unit seemed useful to you and understandable to others. Because of this freedom about units, problems in statical equilibrium are as likely to be expressed in practical units as they are in absolute units.

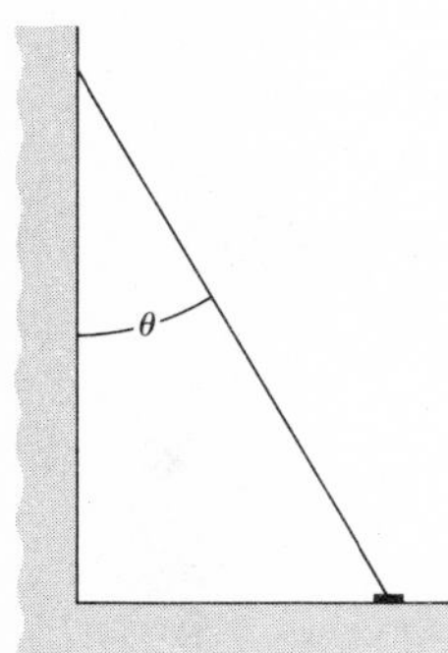

FIGURE 196

PROBLEM 377

A standard straight ladder can be treated as a uniform beam, of length L and weight W. Pads on its feet provide a coefficient of friction of 0.8 relative to the surface of the floor. If the ladder is placed against a smooth vertical wall at an angle θ to the wall (Figure 196), what is the maximum value of θ for the ladder to stay in place?

SOLUTION

Obviously the problem requires that, up to the critical angle θ, the ladder be in static equilibrium: it must not move linearly, and it must not rotate about any axis. (See Figure 197.) We take care of the first requirement by $\Sigma F_y = F_N - W = 0$ and $\Sigma F_x = F_1 - \mu F_N = 0$. Since neither of these equations involves θ, we are not very far along toward a solution to the problem. Turning to the second requirement by investigating torques about an axis through P, we have $\Sigma \tau_P = F_N\, L \sin\theta - \mu F_N\, L \cos\theta - W\,(L/2) \sin\theta = 0$; from this, $\tan\theta = \mu F_N/(F_N - W/2)$. From ΣF_y we have $F_N = W$, so that, finally, we get $\theta = \tan^{-1} 2\mu$.

Would it have made any difference if we had investigated torques about some other axis, say an axis through 0? Then we would have had $\Sigma \tau_0 = W\,(L/2) \sin\theta - F_1\, L \cos\theta = 0$; from this, $\tan\theta = 2F_1/W$. From ΣF_x we have $F_1 = \mu F_N$ and from ΣF_y we have $F_N = W$, so that, finally, we again get $\tan\theta = 2\mu W/W = 2\,\mu$.

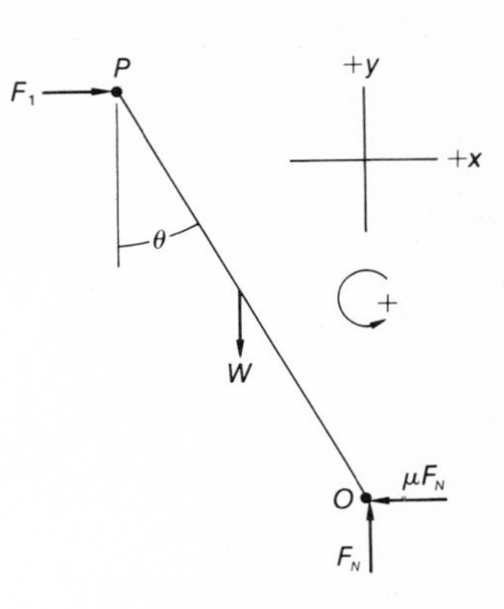

FIGURE 197

Now let us look at problem 377 again, this time bearing in mind the statement about equilibrium about parallel axes—*any* parallel axes. (See Figure 198.) If the ladder is in equilibrium about an axis through P it is also in equilibrium about an axis through A. Hence $\Sigma \tau_A = 0$. As the lines of action of both F_1 and W pass through A, these two forces offer no torque about A. (In fact, this is why we chose A.) So we are left with $\Sigma \tau_A = F_N(L/2) \sin\theta - \mu F_N\, L \cos\theta = 0$, which, without any further calculation, immediately gives us $\theta = \tan^{-1} 2\mu$.

A further refinement of the second form of solution leads us to consider R, the floor reaction whose components are μF_N and F_N. Its line of action must pass through A, so that

$$\tan\phi = \frac{\mu F_N}{F_N} = \mu = \frac{(L/2)\sin\theta}{L\cos\theta} = \tfrac{1}{2}\tan\theta;$$

hence $\tan\theta = 2\mu$ as before. This may not be a shorter solution than the other but it does offer the interesting information that the line of action of R does not lie along the axis of the ladder.*

Finally, for the answer, $\theta = \tan^{-1} 2\mu = \tan^{-1} 1.6 = 58°$.

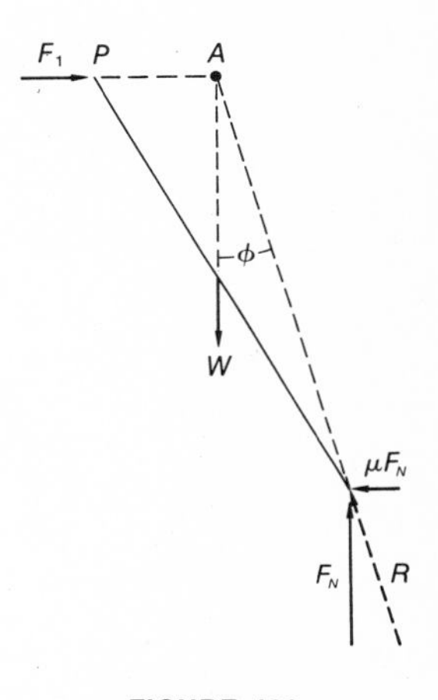

FIGURE 198

*How much time to spend on distilling auxiliary information from a problem is very often moot. In a straightforward engineering or scientific calculation, an efficient, direct approach to a specific answer is usually all that is justified or worth paying for. However, in research and development work you never know in advance where the hint of a new idea may come from; in these areas a comprehensive approach, with side explorations, may be very productive. This applies to teaching and learning as well. Many of an instructor's most effective teaching devices, many of a student's clearest insights, come from such excursions outside of a direct line to an answer.

PROBLEM 378

An engine block can be lifted out of an auto body by two A-frames supporting a uniform, heavy cross beam with a chain hoist at its midpoint (Figure 199). If the beam, hoist, and engine block together weigh W, what is the stress at the joint J in one of the A-frames? Assume that all joints are pin-connected, and that the weight of the A-frame members is negligible compared to W.

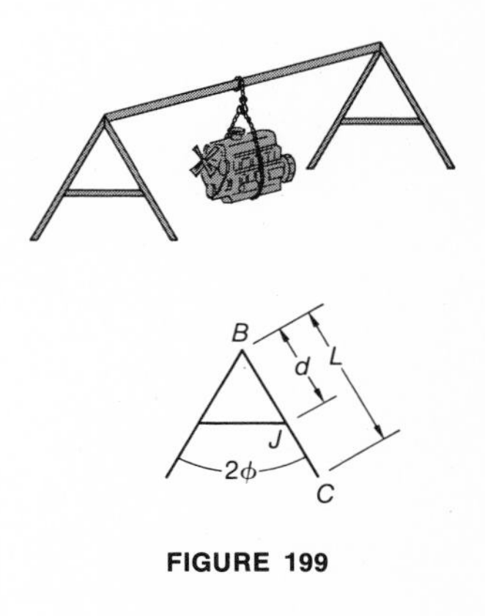

FIGURE 199

START OF SOLUTION

To get at J we must look at the member BC by itself. In assigning forces to BC, we can reduce our work by paying attention to symmetry. Obviously each A frame supports $W/2$; then from symmetry

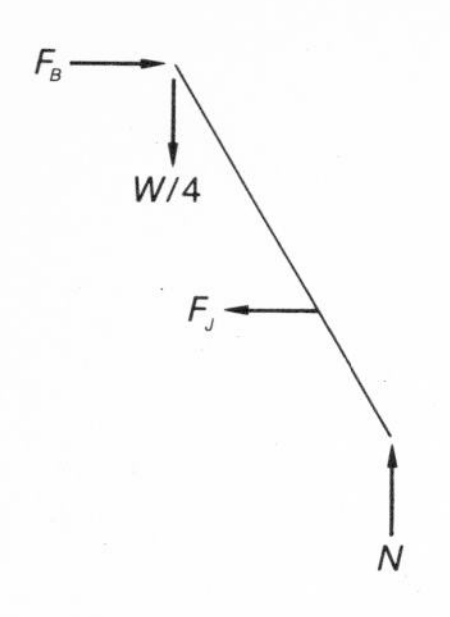

FIGURE 200

you would expect that each leg of an A frame must support $W/4$. As far as F_B, the force on BC from the other leg, is concerned, there is nothing in the information you have that says F_B can't be like this,

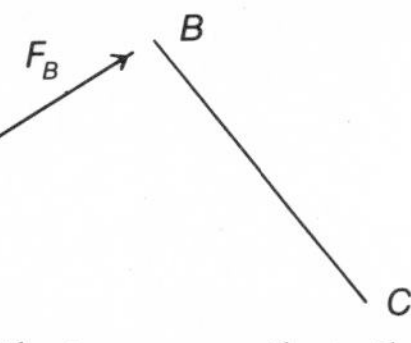

at an angle. However, that means that the other member feels a force $-F_B$ like this,

F_B

which obviously violates symmetry. The only way to preserve symmetry is for F_B to be horizontal. Thus the force diagram for BC must be as shown in Figure 200. (All these conclusions can be developed analytically, but that takes more time and entails more work. Recognizing the requirements of symmetry will many times permit useful shortcuts in your work.)

PROBLEM 379

A stiff beam of length L, whose c.m. is aL from the foot of the beam (a being a fraction < 1), rests on a horizontal floor and leans against a vertical wall. The coefficient of static friction between beam and floor is μ_1, and between beam and wall is μ_2. What is the maximum angle θ between beam and wall for which the beam will stay put? *Comment:* This problem can be ground through by the straightforward summation methods applicable to equilibrium. However, for one moment-axis the force diagram does all the physics for you, and you need only three lines to solve the geometry.

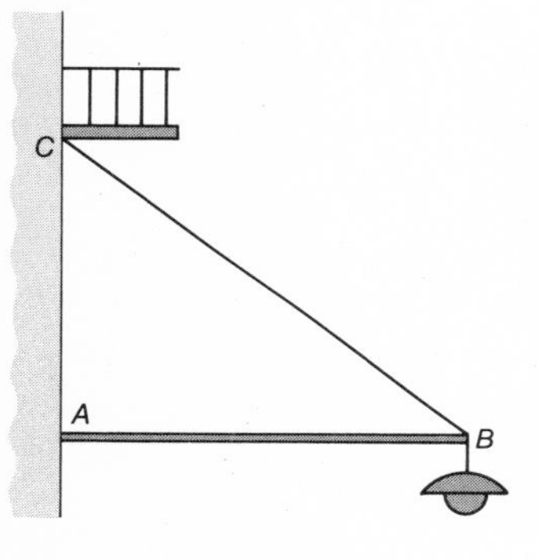

FIGURE 201

PROBLEM 380

A street lamp of weight W is to be supported from the side of a building by a boom AB and guy wire BC, as shown in Figure 201. A balcony limits how high C can be placed, so that the maximum ratio AC/AB that can be obtained is 3/4. Assume weight of boom is negligible relative to W.

a. The guy wire BC must be designed for what minimum tensile strength?

b. The boom AB must be designed for what minimum compression?

SOLUTION

All of the forces we are interested in—weight W, tension T in the guy wire, force F in the boom—come together at joint B, and serve to keep the fitting at B in equilibrium. We can consider this fitting to be a point, of no appreciable weight; hence we are not required to worry about its rotation. From the free-body diagram of the point at B (see Figure 202), we have $\Sigma F_y = T \sin\theta - W = 0$ and $\Sigma F_x = F - T\cos\theta = 0$. From the first equation $T = W/\sin\theta = W/\sin\tan^{-1} 3/4$

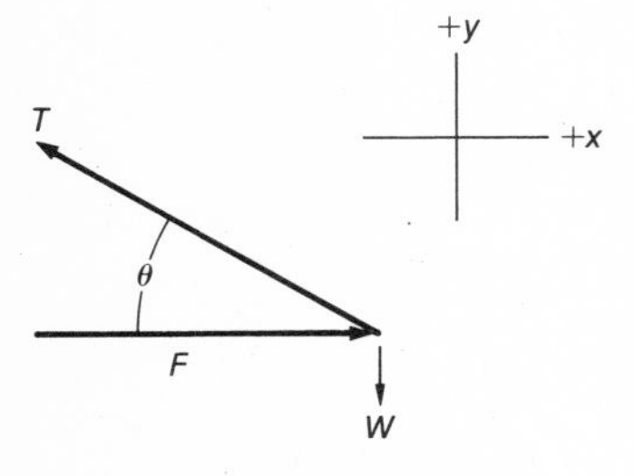

FIGURE 202

$= 5W/3$. Then, we rearrange both of the summation equations and set $T \sin \theta = W$ and $T \cos \theta = F$. Dividing the second equation by the first, we get $\cot \theta = F/W$ or $F = W \cot \theta = 4W/3$.

FIGURE 203

PROBLEM 381

To start his small son to swinging, a father pulls horizontally back on the swing seat until the L-long taut rope of the swing is at θ with the vertical, and then releases his hold (Figure 203). The son weighs 60 lbwt, and relative to this weight the seat and rope are negligible; $L >> $ dimensions of boy.

a. What maximum force did the father exert?
b. What was the tension in the rope just before release?
c. What was the tension in the rope just after release?
d. Where did the maximum tension in the rope occur?

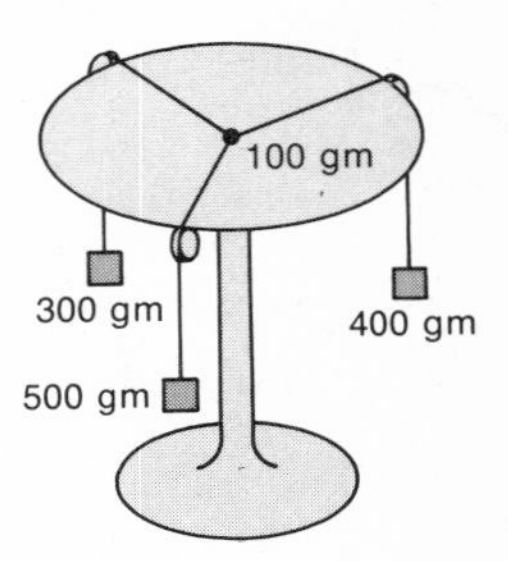

FIGURE 204

PROBLEM 382

A small 100-gm mass rests at the center of a standard demonstration force table, held there in equilibrium under the action of horizontal forces from three weights (300 gm, 400 gm, and 500 gm) hanging from threads running over frictionless pulleys (Figure 204).

a. To produce the greatest initial acceleration of the 100-gm mass, which of the three threads would you burn?
b. What initial acceleration would result?

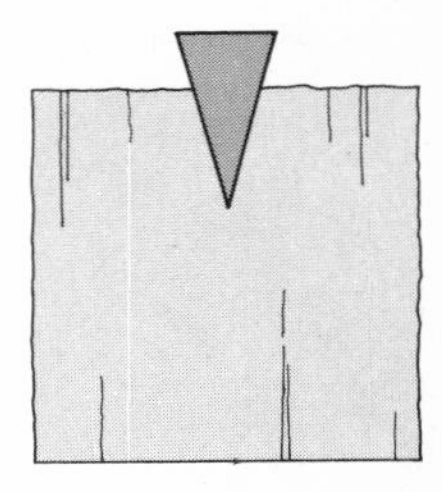
FIGURE 205

PROBLEM 383

A steel wedge, of wedge angle θ, is driven into a log to split it (Figure 205). What must be the minimum coefficient of static friction between wedge and wood so that the wedge does not pop out of the log after it is driven in? Neglect weight of wedge.

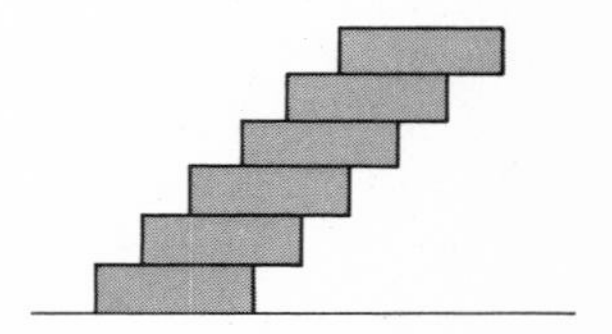
FIGURE 206

PROBLEM 384

A uniform brick of length L is laid on a smooth horizontal surface. Other equal bricks are now piled on the first brick, as shown in Figure 206; the sides of the bricks form a continuous plane, but the ends are offset at each brick from the previous brick by a distance L/a, where a is an integer. How many bricks altogether, including the bottom brick, can be used before the pile topples over?

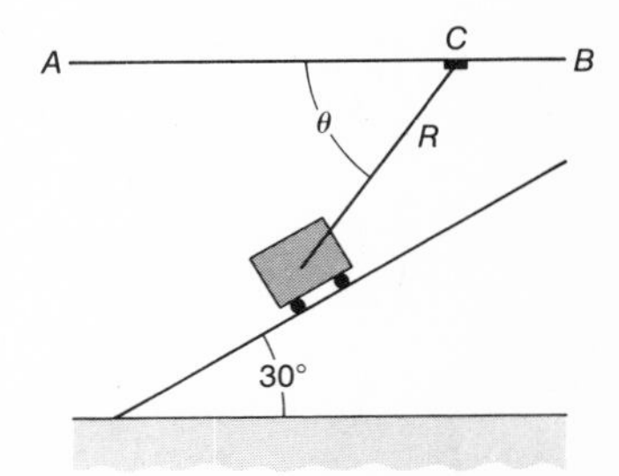

FIGURE 207

PROBLEM 385

A cart of weight W is restrained from sliding down a 30° incline by a rope R (Figure 207). For values of θ between 30° and 60° the rope holds, but when the clamp C is moved toward A so as to increase θ, the rope breaks when θ reaches 60°. What is the breaking strength of the rope?

PROBLEM 386

A thin-walled, smooth open-ended tube, of internal diameter $3R$ and weight W, is held upright on a horizontal surface while two identical smooth balls, of radius R, are dropped into the tube, as shown in Figure 208. What is the maximum weight each of the balls can have, if the tube is to stay upright without being held?

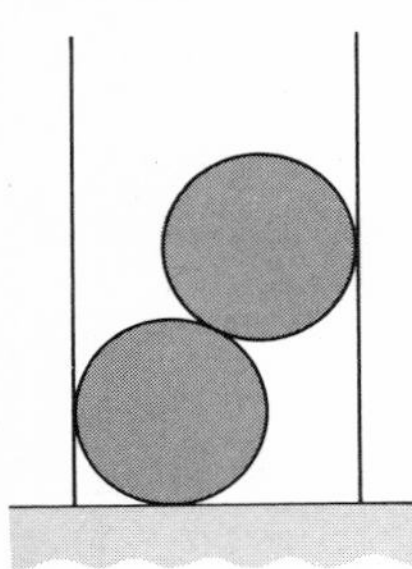

FIGURE 208

PROBLEM 387

In an arrangement like that of problem 386, the top ball has the maximum permissible weight determined in that problem, but the bottom ball is replaced by one of the same radius but weighing twice as much as the top ball.

a. What now happens to the equilibrium of the tube?
b. The balls in (a) are now reversed in position. What now happens to the equilibrium of the tube?

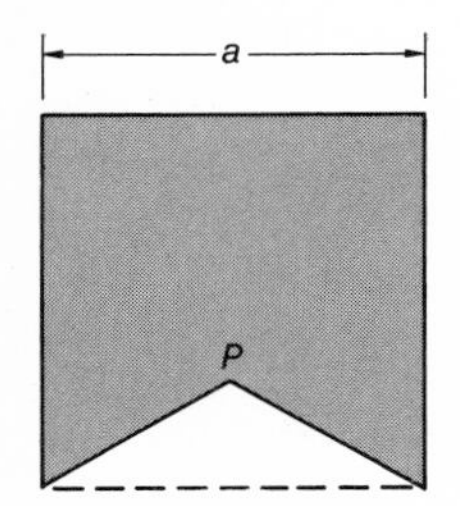

FIGURE 209

PROBLEM 388

An isosceles triangle is cut from one edge of a square of uniform sheet metal, as shown in Figure 209, so that the remaining piece, when suspended from the apex P of the cut, will remain in equilibrium in any orientation. What should be the altitude of the cut-out triangle?

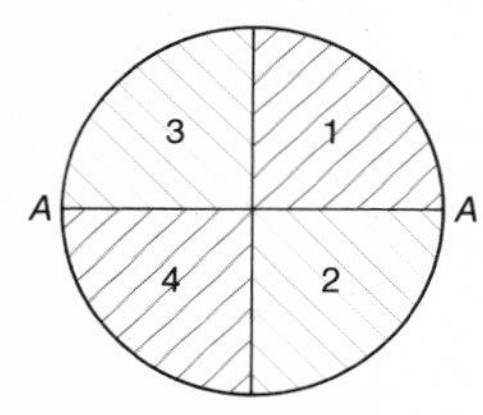

FIGURE 210

PROBLEM 389

A solid cylinder has a density which varies by uniform quadrants, with the numbers shown in Figure 210 indicating relative densities. If the cylinder is placed on a horizontal surface and allowed to come to equilibrium, what angle will the line A-A make with the surface?

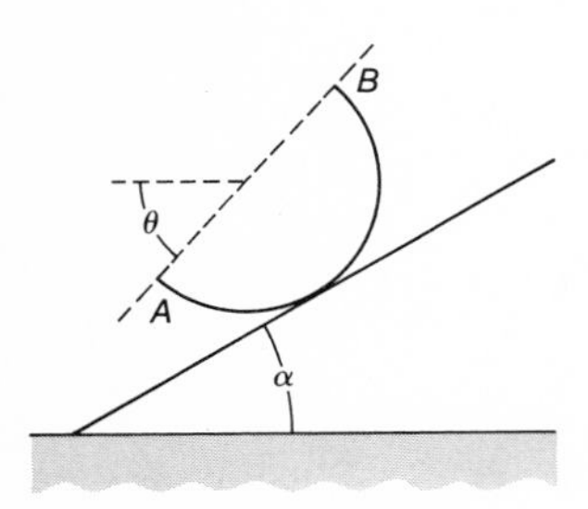

FIGURE 211

PROBLEM 390

An open-ended semicylindrical trough, of weight W and radius R, is placed on a rough inclined plane, as shown in Figure 211. The plane is at an angle α with the horizontal.

a. What is the angle θ which the diameter AB makes with the horizontal when equilibrium is reached?
b. What is the upper limit for α for this to be a possible problem? *Two hints:* Don't forget the theorems of Pappas; and try considering torques about the line of contact between the trough and the plane.

PROBLEM 391

Logs dumped into a holding pond float flat in the water, even though many of them entered the water upright. Why don't the logs that entered the water vertically remain vertical?

PROBLEM 392

Long straight logs fitted with an iron casting at one end are stacked on the deck of a supply ship that is taking them to an off-shore oil-drilling platform. A lashing fails, and one of the logs rolls overboard. It floats upright with one-sixth of its length projecting above the water surface. If the volume of the casting can be considered negligible compared to the volume of the log, what is the lower limit for the weight of the casting?

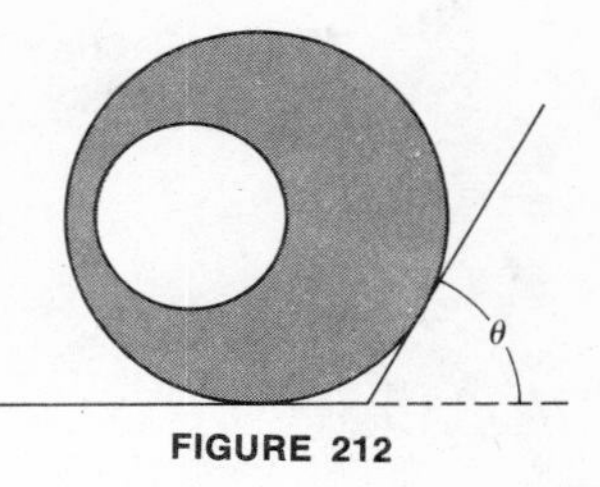

FIGURE 212

PROBLEM 393

A uniform cylinder of radius R has a longitudinal hole of radius $\frac{1}{2}R$ bored in it, the axis of the hole parallel to the axis of the cylinder and $R/3$ from it. The cylinder is placed in an angle between a smooth inclined wall and a horizontal floor, as shown in Figure 212. The axes of the hole and the cylinder lie in the same horizontal plane. What is the minimum coefficient of static friction between cylinder and floor necessary to prevent the cylinder from turning?

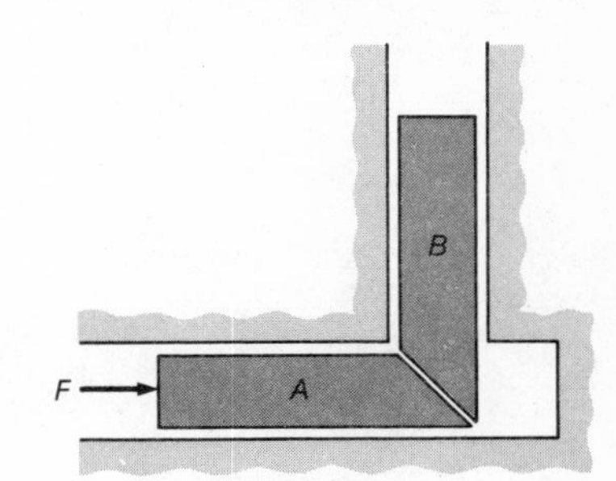

FIGURE 213

PROBLEM 394

A side view of a simplified form of vertical latch is as shown in Figure 213. The lower member A can be pushed forward in its horizontal channel, thus raising member B, of mass m, upward in its channel. The sides of the channels are smooth, but at the interface of A and B, which is at 45° with the horizontal, there exists a static coefficient of friction μ. What is the least force F, applied horizontally to A, that will start motion of the latch?

PROBLEM 395

A summer research party on Malespina glacier has at its disposal a jeep which is always loaded so that 2/3 of its total weight of 2250 lbwt is distributed on the rear wheels and 1/3 on the front wheels. One day while being driven by a student, it crashes through a snow cover into a crevasse and comes to rest in a horizontal position between ice walls, as shown in Figure 214—luckily with the student still in the driver's seat and unharmed. A rope is lowered to rescue him, but when he starts to transfer his weight from the seat to the rope the jeep starts to move; when he releases the rope the jeep returns to its delicately balanced horizontal position. How much does the student weigh?

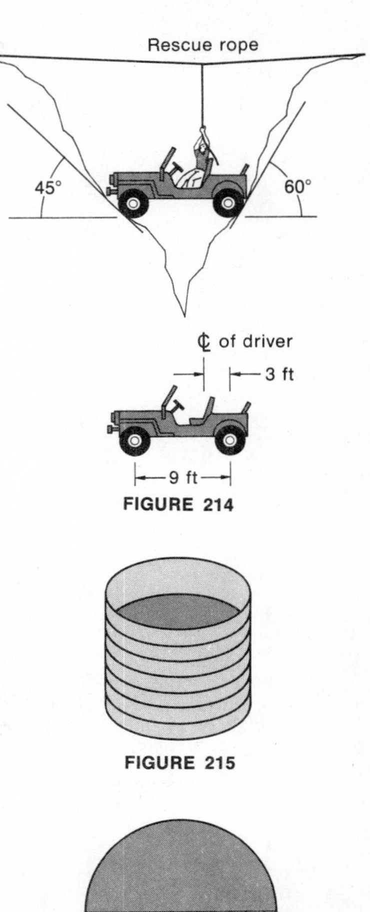

FIGURE 214

FIGURE 215

FIGURE 216

PROBLEM 396

A vertical cylindrical tank, to hold a volume V of liquid of density ρ, is to be made of bands of sheet steel, each w wide, welded or riveted together horizontally. The steel in each band needs to be strong enough to withstand the bursting force at the bottom of that band (Figure 215). Assume T is the tensile strength per unit cross-sectional area of the steel used, and r is the radius of the tank.

a. What thickness of steel sheet should be used in the bottom band?
b. What total volume of steel would be required for the bottom band?

Hint: Consider a semicircular layer of liquid dy thick and y deep below the liquid surface. On that layer will exist two horizontal forces; one from the pressure at that depth times $2r\ dy$ (the cross-sectional area of the edge of the layer) and the other, in the opposite direction, supplied by the confining steel band segment at the ends of the diameter of the layer (Figure 216). Since the steel band segment and its enclosed semicircular layer of liquid are in equilibrium, these forces must be equal.

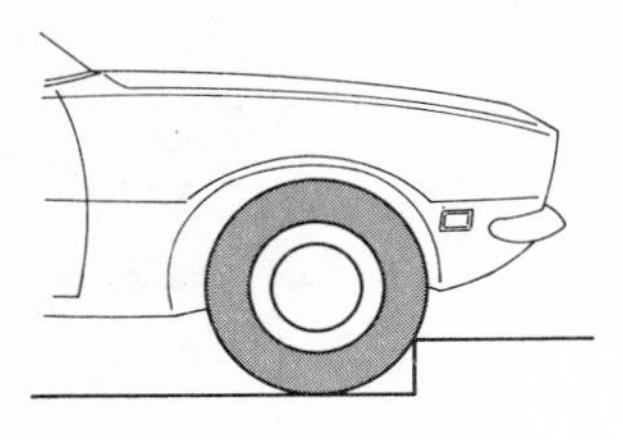

FIGURE 217

PROBLEM 397

An automobile of weight W and wheelbase L is parked head-on against a curb which is h high (Figure 217). The tires of the automobile are $2h$ in radius. If the c.m. of the automobile is midway between front and rear axles, what initial torque must be supplied by the drive shaft to start the automobile moving up over the curb? Assume no loss of torque between drive shaft and wheels. *Hint:* The torque must be such as to induce at the ground contact point of the rear wheels (assuming rear-wheel drive) a forward force sufficient to move the car up over the curb. This necessary force must be determined before you can compute the torque producing it.

PROBLEM 398

A circular loop of wire carrying a steady current I is placed in a uniform magnetic field B; the diametral axis of the loop is perpendicular to the field. What is the torque about this axis when the plane of the loop is

a. perpendicular to the field;
b. parallel to the field?

PROBLEM 399

Would it have made any difference in your answer to problem 398 if the loop had been rectangular instead of circular?

PROBLEM 400

A galvanometer coil consists of N turns of fine wire assembled in a rectangular frame of area A. The coil is suspended by a torsion wire of torsion constant τ_0; a pointer is mounted on the coil perpendicular to its plane. When no current is passing through the coil it hangs in equilibrium with its plane parallel to a uniform magnetic field B.

a. When a steady current I is sent through the coil, what is the relation between θ, the angular departure from its equilibrium position, and I?
b. If a scale under the pointer is in units of I, what is the sensitivity of the scale, i.e., what is the ratio of a small change dI in I to a small change $d\theta$ in θ?
c. The relationship between I and θ that you found in question (a) is not linear, and this affects the sensitivity in question (b). Do you have any good suggestions for making the I vs. θ relationship linear?

PROBLEM 401

Two parallel wires a distance $2a$ apart carry a steady current I in opposite directions (Figure 218). Midway between the wires is the axis of a rectangular loop of wire with dimensions $2b \times d$ as shown. The same current I moves in the loop.

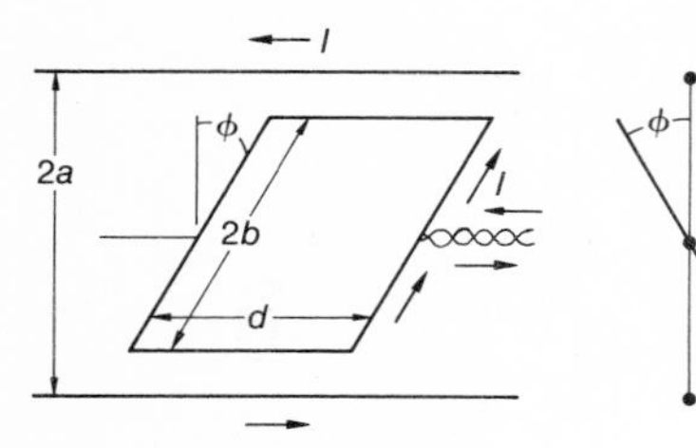

FIGURE 218

a. What is the torque tending to rotate the rectangular loop about its axis as a function of ϕ, the angle the plane of the loop makes with the plane of the wires?
b. For what ϕ is the torque a maximum?
c. What is the maximum torque?

PROBLEM 402

A long, uniform stick of wood weighing 60 lbwt is lowered into still water, as shown in Figure 219. When the system has come to equilibrium, i.e., when the rope is vertical and the stick is motionless, the stick is observed to be exactly half submerged.

a. What is the tension in the rope?
b. What is the density of the wood?

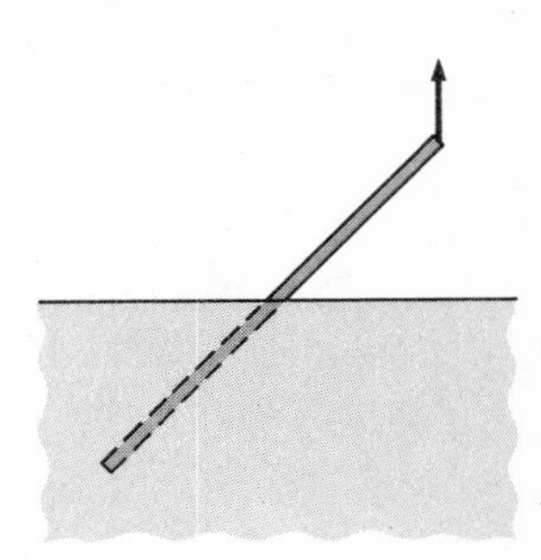

FIGURE 219

PROBLEM 403

The cross section of a power boat at rest in the water is shown in Figure 220. The draft $d = \frac{1}{3}$ of the beam b, and the distance a from the water line to the chine $C = \frac{1}{9}b$. For simplicity in calculation assume the center of gravity (c.g.) of the section is in the center of the boat at the water line. When the boat is tipped so that one chine is at the water line, as shown in Figure 221, what is the righting moment, per unit length of the boat, tending to restore the boat to level trim?

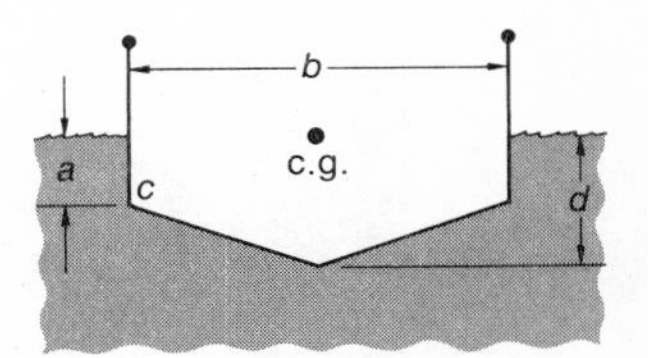

FIGURE 220

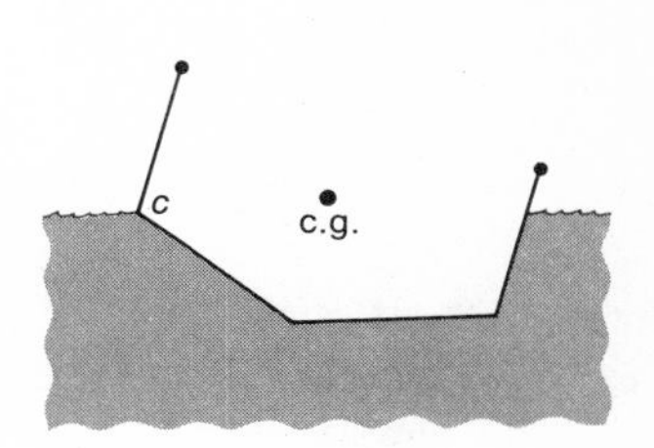

FIGURE 221

PROBLEM 404

A semicircular block of weight W and radius R is cut out of a thick, nonuniform slab of wood. When placed on its circular face on a horizontal surface, the block comes to rest with its diametral edge AOB at an angle θ with the horizontal, as shown in Figure 222. When a uniform bar of weight P and length R is placed on the diametral portion AO, the system comes to equilibrium with the edge AOB horizontal, as shown in Figure 223. When the bar is moved to the OB portion of the block edge, the system comes to equilibrium with AOB at θ with the vertical, as shown in Figure 224. When the block alone is freely suspended from a horizontal axis congruent with the edge B, it hangs with its diametral edge AOB at θ with the vertical as shown in Figure 225.

a. What is the value of P?
b. What is the angle θ?
c. Where is the c.m. of the block located?

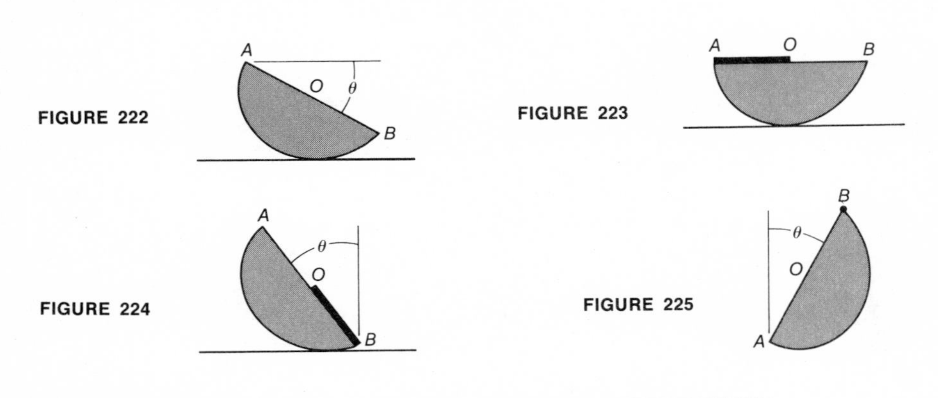

FIGURE 222

FIGURE 223

FIGURE 224

FIGURE 225

PROBLEM 405

A uniform, stiff bar with slightly rounded ends and of length L is hung from a fixed point on a vertical wall by a string that is also of length L. If the free end of the bar is placed against the wall (Figure

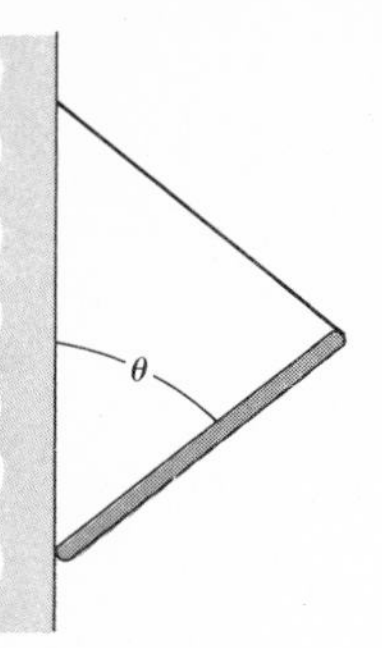

FIGURE 226

226), in repeated experiments the bar always falls down unless θ, the angle between the bar and the wall, is $\geqslant 75°$. What is the coefficient of static friction between the bar and the wall? *Suggestion:* This can be solved either algebraically or geometrically by vectors. The geometric solution is less work.

PROBLEM 406

How much frictional torque must be built into the mechanism of the upper double sheave of the differential pulley of problem 136, so that a maximum-rated load of W will not sink of its own weight when the chain is left untended?

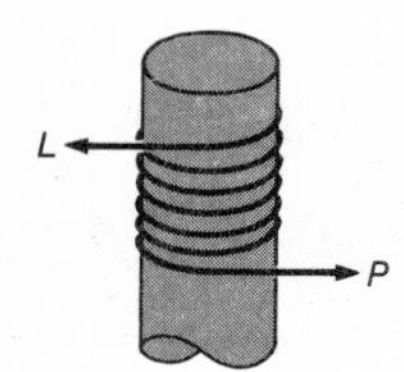

FIGURE 227

PROBLEM 407

A very large load L can be controlled by a very small pull P, by a rope snubbed around a fixed post, as shown in Figure 227. This practice has many applications in construction work and in marine situations, and is the basis for belaying techniques in mountain climbing. If the coefficient of friction between rope and post is 0.40 and the rope is wound three times around the post, what pull P is required to hold a load L of 3.0 tonwt?

START OF SOLUTION

Consider a very small arc of the rope that intersects a very small angle $d\theta$ (Figure 228). When the rope is at equilibrium under a tension T, the normal force between the rope and the surface of the post is given by $N = 2\ T \sin\ (d\theta/2) \simeq T\ d\theta$ (for $d\theta$ small, $\sin d\theta \simeq d\theta$). With an additional pull dT at one end of the arc ($dT << T$), the rope will still stay in equilibrium as long as $\mu T\ d\theta = dT$. When this equation is integrated (T between the limits of pull P and load L, and θ between the limits of 0 and 6π) the solution of the problem is at hand.

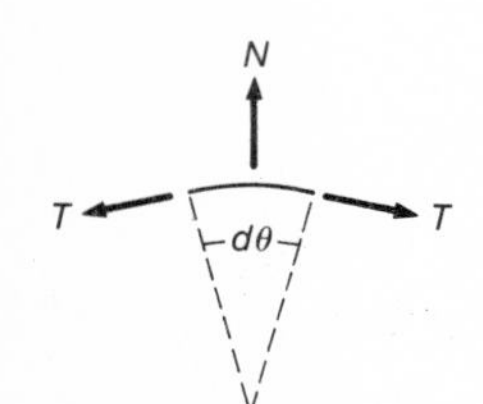

FIGURE 228

PROBLEM 408

A heavy crate is moved across a floor at constant speed (Figure 229).

a. Allowing for friction between crate and floor, what is the least force you can apply to accomplish the desired movement?
b. At what angle with the horizontal do you apply this least force? *Note:* Be careful of signs.

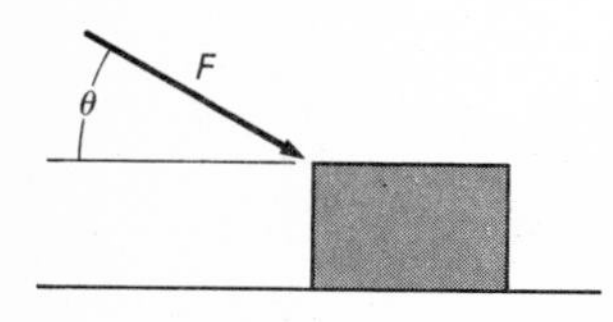

FIGURE 229

PROBLEM 409

A heavy crate of weight W is to be dragged at constant speed across a shop floor by a line from a chain hoist mounted as shown in Figure 230. Assuming the weight of this line is negligible (so as not to worry about sag of the line), what is the force in the line as a function of x, the horizontal distance from the crate to the hoist? There is a coefficient of friction μ between crate and floor.

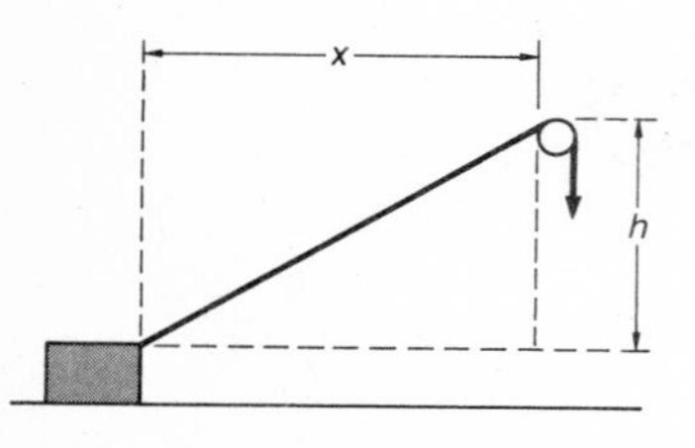

FIGURE 230

PROBLEM 410

Smooth, identical logs are piled in a stake truck. The truck is forced off the highway and comes to rest on an even keel lengthwise but with the truck bed at angle θ with the horizontal. As the truck is unloaded, the removal of one log (shown dotted in Figure 231) leaves the remaining three logs in an incipiently unstable condition, i.e., if θ were infinitesimally smaller the logs would fall down? What was θ?

PROBLEM 411

An ornament for a courtyard at a World's Fair is made up of four identical smooth metal spheres, each weighing $2\sqrt{6}$ tonwt. The spheres are arranged as shown in Figure 232; three are resting on a horizontal surface and touching each other, the fourth resting freely on the other three. The bottom three are kept from separating by spot welds at the points of contact. Allowing for a factor of safety of 3 (this means you make the welds three times as strong as you calculate is necessary), how much tension should the spot welds be capable of providing?

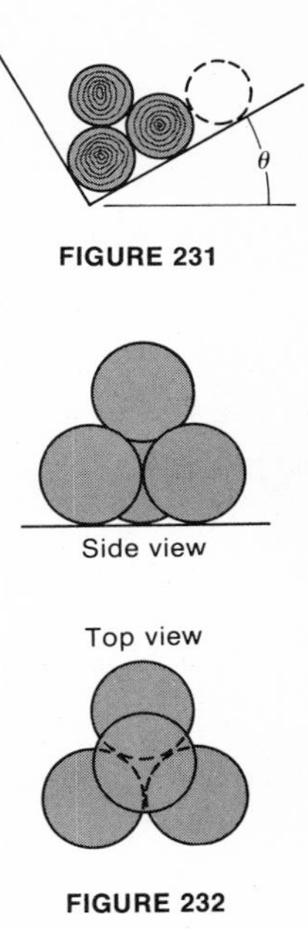

FIGURE 231

FIGURE 232

The Method of Cuts

In analyzing stresses in simple, multimember frameworks such as bridge trusses, by repeated applications of $\Sigma F = 0$ and $\Sigma\tau = 0$, you can work your way across the framework joint by joint. This usually entails laborious calculation; it also has the disadvantage that an error made in the early stages of the calculation is repeated in the subsequent work. There is a simpler way—the method of cuts.

To illustrate, let us look at a simple bridge truss at the moment a truck of weight W is passing over the joint D (Figure 233). The truss is to be considered a rigid framework of stiff members connected at their ends by transverse steel pins (Figure 234). For a first design calculation, the members are regarded as weightless.

The exterior forces on the structure are the truck's weight W and the supporting reaction forces at A and E (Figure 235). (Unless the load is accelerating, there is

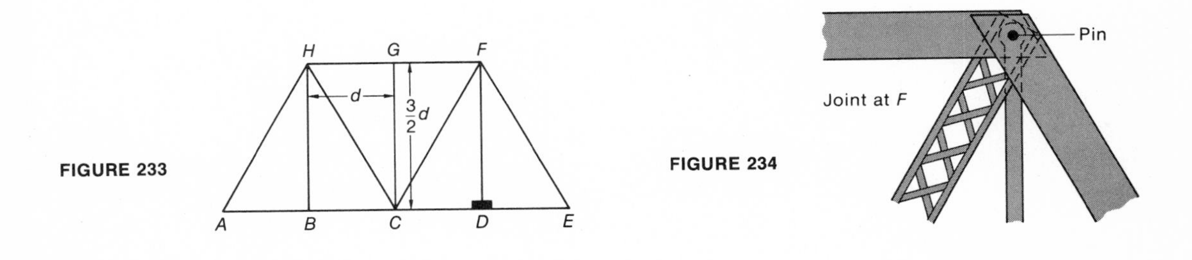

FIGURE 233

FIGURE 234

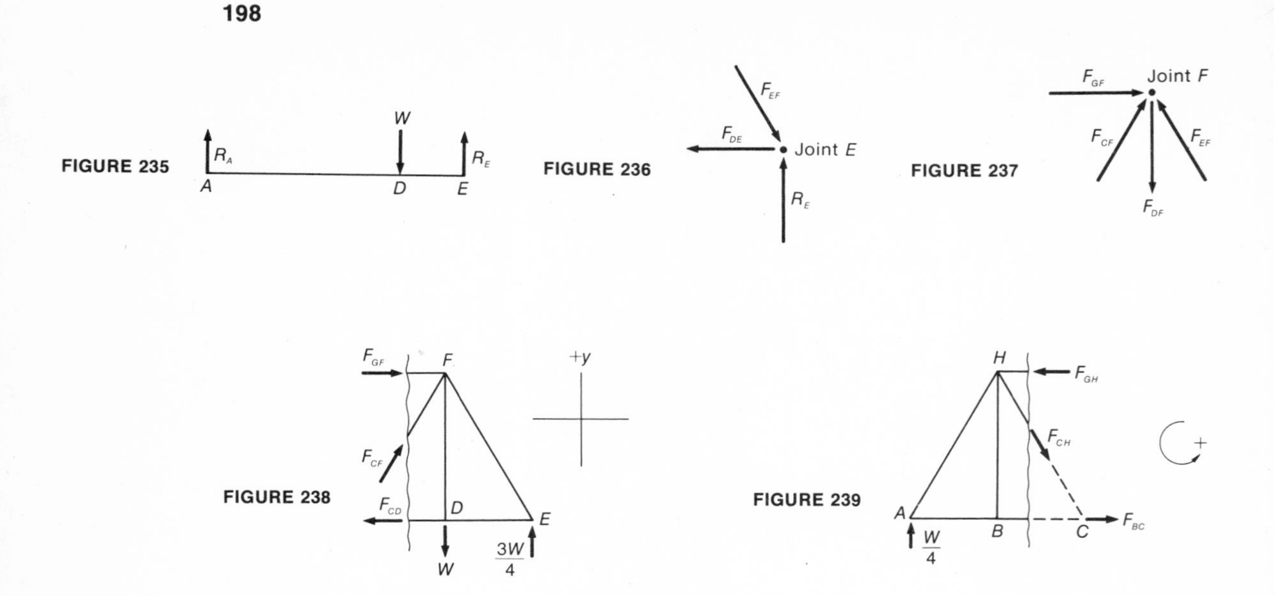

FIGURE 235

FIGURE 236

FIGURE 237

FIGURE 238

FIGURE 239

no horizontal force on the load from the bridge, and hence none acting on the bridge.) The internal forces must lie axially along the members, because the connecting pins cannot provide any torque about their axes. By simple moments, $R_A = W/4$ and $R_E = 3W/4$.

Suppose that we want to determine the forces in members GH and CF. We could start at joint E, and by application of $\Sigma F_x = 0$ and $\Sigma F_y = 0$, find F_{EF} (Figure 236). We could then proceed to joint F (Figure 237). Since we now know F_{EF}, there are only three unknowns at joint F. However, we have only two independent conditions for equilibrium: $\Sigma F_x = 0$ and $\Sigma F_y = 0$. Before we can go any further we find we must go down to joint D, where we are able to write enough independent equations to solve for F_{FD}. Then going back to joint F, where we now have only two unknowns, we can solve for F_{GF}. Knowing F_{GF}, we can proceed to joint G, where we are able to solve for F_{GH}. Obviously, joint-by-joint analysis is tedious business.

Now let us start all over, and try cuts instead of joints. Since we want to know the force in CF, we will make a vertical cut through the C-D panel and consider the structure to the right of that cut (Figure 238). We must not disturb the equilibrium of the cut portion, however, so we restore to the cut members forces that were in these members before the cut. Applying $\Sigma F_y = 0$, $3W/4 + F_{CF} \cos \tan^{-1} 2/3 - W = 0$, or $F_{CF} = \sqrt{13}/12W$ compression. Now let us make a vertical cut in the B-C panel, this time considering the structure to the left of the cut (Figure 239). We see that if we take moments about a transverse axis through C, we will have $\Sigma\tau_C = F_{GH}\,(3d/2) - (W/4)\,2d = 0$, or $F_{GH} = W/3$ compression. Thus by the method of cuts we have found the two forces we were interested in with only two moves.

In setting up the force diagrams for either method, we guessed the directions of the unknown forces. If we had guessed incorrectly, we would have gotten a minus sign in our answer, which would automatically have told us, "turn it around."

PROBLEM 412
In the four-panel truss just described,

a. by inspection, what are the forces in the members BH and CG;
b. where was the load when the force in the top members HG and GF was a maximum?

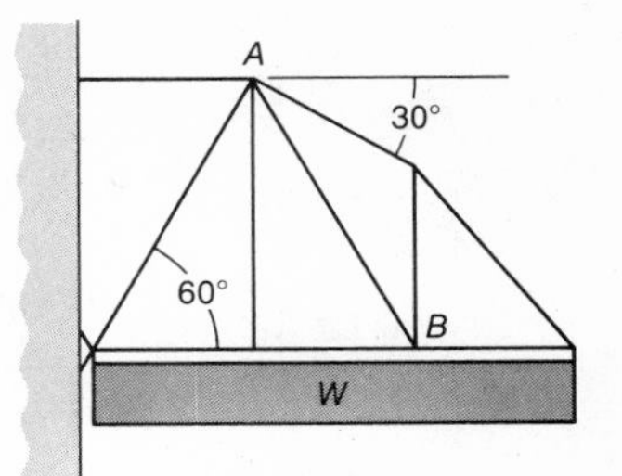

FIGURE 240

PROBLEM 413
A cantilever framework supporting a sign of weight W is mounted on the side of a building as shown in Figure 240. All the horizontal members are of equal length. What is the force in the member AB, in amount and whether tension or compression? The joints are pin-connected, and the weight of the members is negligible relative to W.

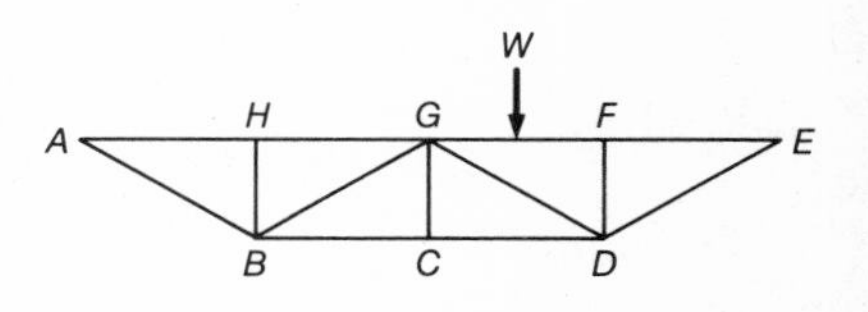

FIGURE 241

PROBLEM 414
A transverse floor beam for a highway bridge is as shown in Figure 241. The beam is vertically supported at A and E. Each panel is twice as wide as it is high. The joints are pin-connected, and for a first approximation, the structural members are assumed to be weightless. If a load W is applied midway between F and G, what are the forces in members FD, CD, and GD?

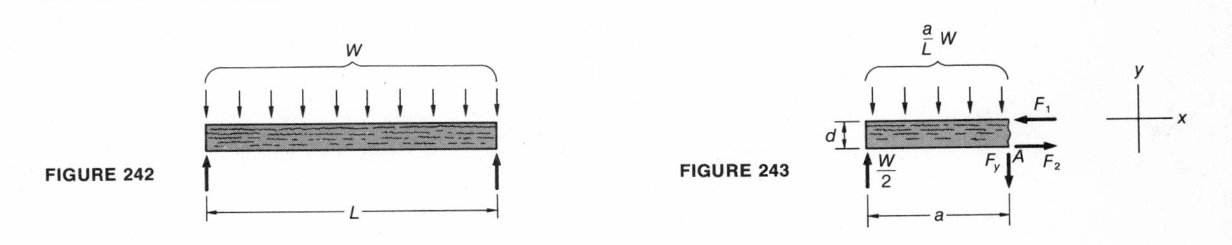

FIGURE 242

FIGURE 243

The structures in problems 413 and 414 illustrate the two standard classes of beams: the simply supported beam (problem 414), in which the load is applied between the supports resulting in the upper web of the beam being under compression and the lower web in tension; and the cantilever beam (problem 413), supported at one end with the load outboard of the support, resulting in the upper web being in tension and the lower web in compression. A vertical cut should enable us to examine the torque created by these opposing forces.

To do this, let us look at the primordial beam, the first bridge in human history —a log across a stream (Figure 242). From its own weight W it has a distributed load between supports. Obviously each support reaction must equal $W/2$, vertically. A cut at A made a from the left end, $a < L/2$, is diagrammed in Figure 243. Forces F_1, F_2, and F_y were exerted on the section by the part that has been removed. To achieve y equilibrium, F_y must $= -W(1/2 - a/L)$. By taking moments about an axis through A, we find that to achieve rotational equilibrium F_1 must $= (aW/2d)(1 - a/L)$, directed as shown in Figure 243. Finally, for x equilibrium, F_2 must $= -F_1$.

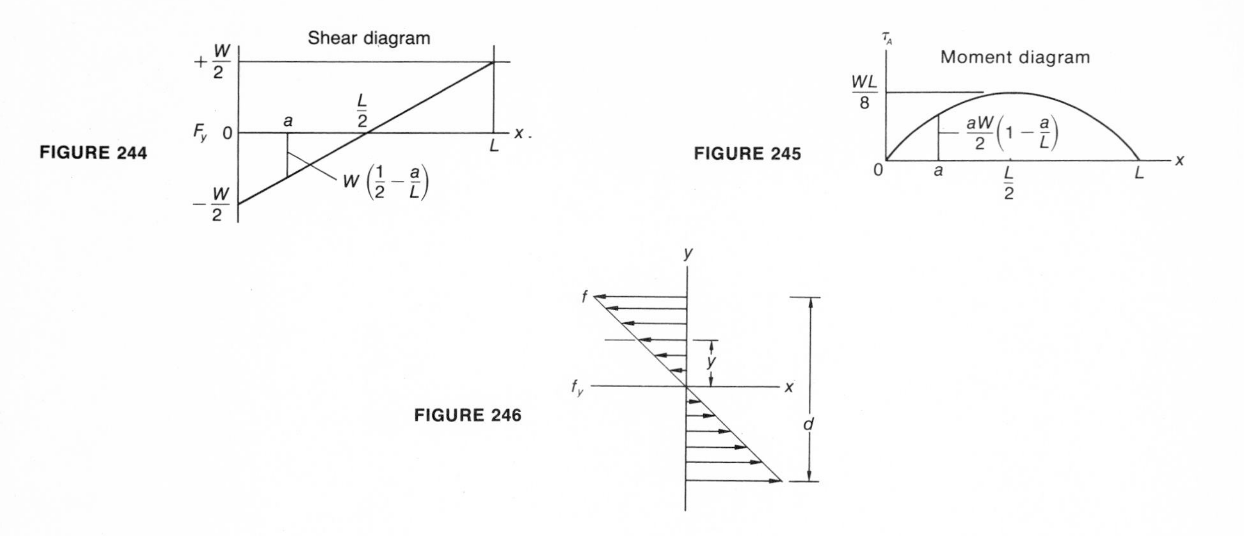

FIGURE 244

FIGURE 245

FIGURE 246

The shear force at section A, F_y, can be seen from its equation to vary along the log, as shown in Figure 244. Obviously a beam needs to be strongest in its shear stress capability near its points of vertical support.

The horizontal torque or moment at section A is seen to be $\tau_A = F_1 d = \frac{1}{2} a W (1 - a/L)$. The moment is zero at the support section, hence this section is not required to supply any torque. The moment is a maximum at $L/2$ (Figure 245), which shouldn't surprise anyone who has crossed a ditch on a thin board.

For the usual beam, including the log we started out with, F_1 and F_2 are not discrete forces applied at the edge of the beam. Instead, the necessary moment is supplied by a summation of horizontal stress contributions across the section. For a homogeneous, uniform beam of width w horizontal stress would be distributed as shown in Figure 246; f_y, the stress, or force per unit area, at y from the neutral axis $= (2y/d)f$, and f is the maximum stress. The necessary moment for section A per unit width is then obtained from

$$2\int_0^{d/2} \frac{2y}{d} f\, y\, dy = \frac{f\, d^2}{6}.$$

Since this moment must equal $[\frac{1}{2} aW (1 - a/L)]/w$,

$$f = \frac{3aW}{wd^2}\left(1 - \frac{a}{L}\right).$$

PROBLEM 415

When a load of $2W$ is placed on the beam of Figure 242 at $2/3\ L$ from the left support, what do the shear and moment diagrams look like?

PROBLEM 416

A beam of length L and uniformly distributed weight W is supported from one end as a cantilever (Figure 247). What do its shear and moment diagrams look like?

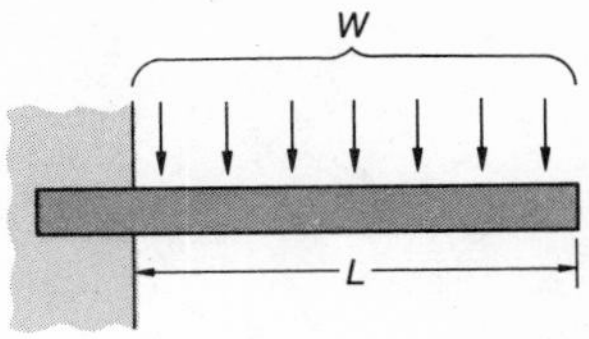

FIGURE 247

PROBLEM 417

In California beach areas where the land often rises steeply back from the beach, many houses are built with sun decks supported by a cantilever portion of the main floor beams of the house (Figure 248). If such beams extend 6.0 ft beyond the foundation wall on which they rest, and each beam must support a constant floor load of 50 lbwt ft^{-1} and a possible load of two 200 lbwt persons leaning on the railing looking at the view, what are the minimum sheer and moment requirements for the beam section at the foundation wall?

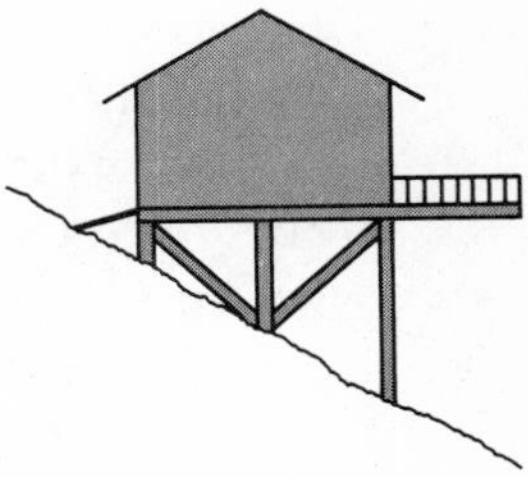

FIGURE 248

PROBLEM 418

The essential structure of an erecting crane is as shown in Figure 249. The upper portion of the boom is 30 ft long and weighs 1.8 tonwt, and its c.m. is 20 ft down from the pulley at the top. The weight of the ball and hook at the end of the hoisting cable, plus the average weight of the extended cable itself, is 0.4 tonwt. If the boom is designed for a maximum safe bending moment of 90 ft tonwt at the section where the guy wire is attached, what graph would you post in the crane operator's cabin controlling the relation between the angle θ and the maximum permissible load to be hoisted at that angle?

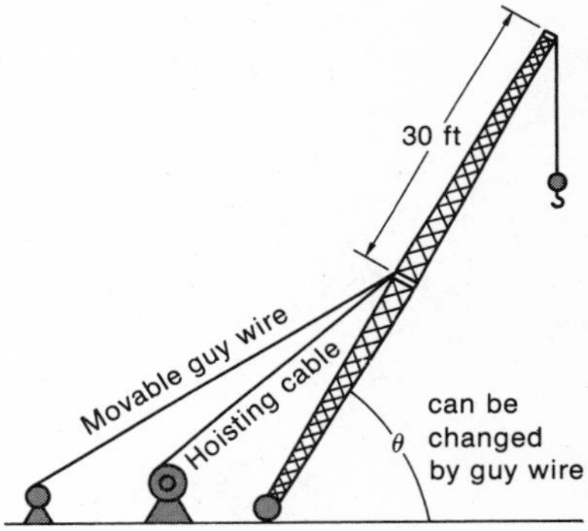

FIGURE 249

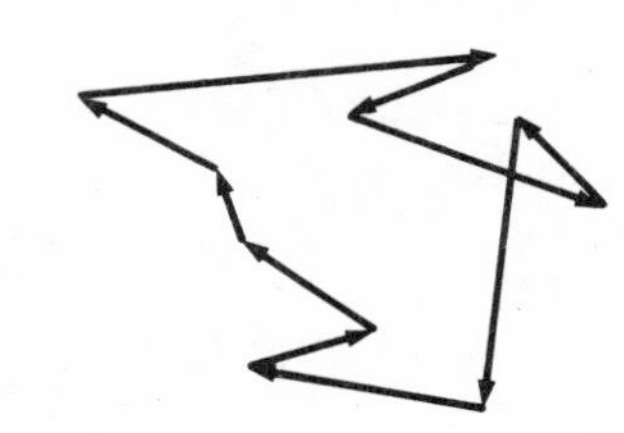

FIGURE 250

The Vector Polygon

For most analyses of equilibrium at a point, the force component summations, such as $\Sigma F_x = 0$, offer the easiest and most informative approach. However, these summations are derivative from the basic statement about equilibrium at a point: $\Sigma \vec{F} = 0$. Summation of the actual vectors, instead of their components, is sometimes an economical approach, particularly when the physical situation is, or can be reduced to, a two-dimensional problem. Summation of the actual vectors is a graphical method (which is why it usually is limited to two dimensions); it entails adding force vectors tail to head *in any order* until the head of the final vector reaches the tail of the first vector (Figure 250), making $\Sigma \vec{F} = 0$. This process has been called "closing the vector polygon."

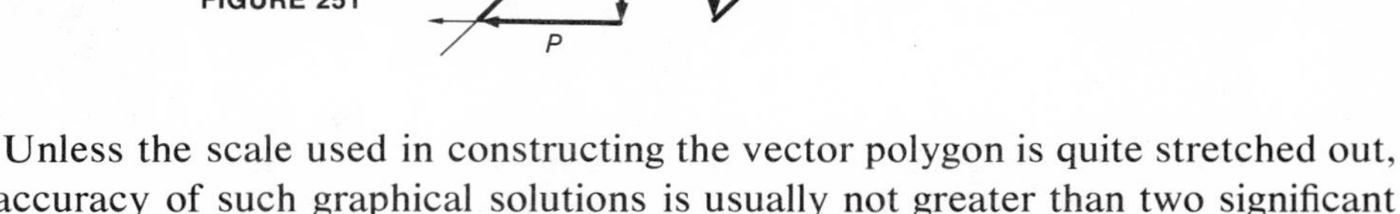

FIGURE 251

Unless the scale used in constructing the vector polygon is quite stretched out, the accuracy of such graphical solutions is usually not greater than two significant figures. In professional applications of this graphical method—in structural analysis, for instance—this is sufficiently accurate for most purposes.

For example, consider problem 381—the boy in a swing. Here the point is the boy; the force vectors (Figure 251) are the father's unknown pull $\vec{P}$, the boy's known weight $\vec{W}$, and the unknown tension in the rope $\vec{T}$. Starting with known vector $\vec{W}$, you can draw in the known directions of $\vec{T}$ and then $\vec{P}$ or of $\vec{P}$ and then $\vec{T}$. In either case you get exactly the same triangle. Where these two directions intersect is what closes the vector polygon and determines the unique vectors for $\vec{T}$ and $\vec{P}$.

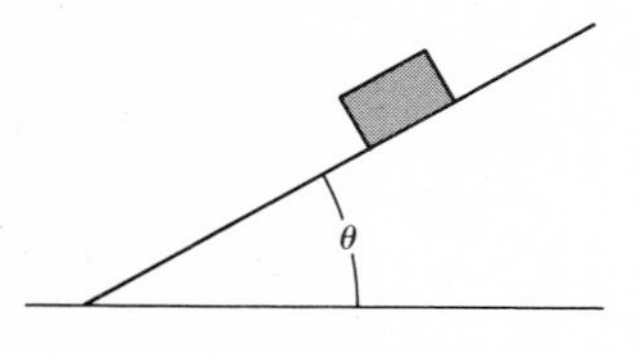

FIGURE 252

PROBLEM 419

A block of mass m is at rest on a plane inclined at θ with the horizontal (Figure 252). What is the minimum coefficient of static friction between block and plane for this to be possible? *Comment:* The correct graphical solution to this problem should remind you that μN and N form the sides of a right triangle whose apex angle is $\tan^{-1}\mu$.

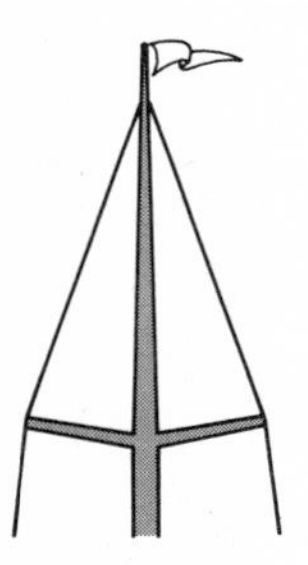

FIGURE 253

PROBLEM 420

The shrouds of a sailboat mast are attached to tangs near the masthead and run down at 20° with the mast to the spreaders, and from there at a lesser angle down to the deck (Figure 253). If the two shrouds are each under 200 lbwt tension, what is the compression load due to the shrouds alone in the mast just above the spreaders? *Comment:* This is a simple, reasonable physics problem, but it would be of little use to a designer of sailboat masts. There is a heavy compression load on masts, but most of it comes from the halyards, some of it from the stays, and relatively little from the shrouds.

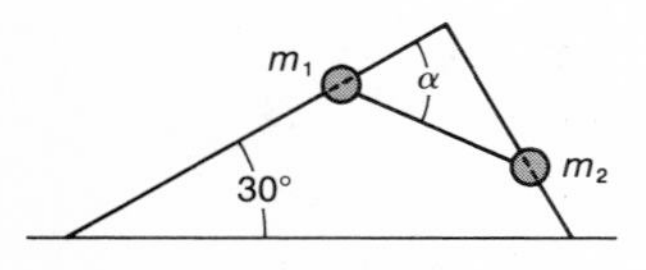

FIGURE 254

PROBLEM 421

A smooth rod is bent at a right angle and mounted in a vertical plane on a horizontal base, as shown in Figure 254. Beads, of mass $m_1 =$ 100 gm and $m_2 = 300$ gm, slide without friction on the rod, and are connected by a cord of negligible weight. When the system is in static equilibrium,

a. what is the tension in the cord;
b. what is the value of α?

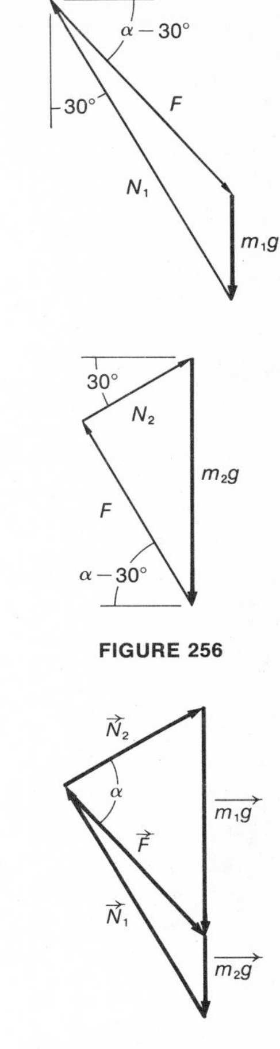

FIGURE 255

FIGURE 256

FIGURE 257

SOLUTION

Each of the particles is independently in equilibrium, but there are too many unknowns for us to be able to solve for either particle separately. However, the force in the cord is common to both particles, and we can use that as a constraint relationship to cancel the independence of the particles. Starting with m_1, the forces holding it in equilibrium are its known weight m_1g acting vertically downward, the unknown force N_1 from the rod it rests on acting at 30° with the vertical (in the absence of friction the force from the rod must be perpendicular to the rod), and the unknown force F from the cord acting at $(\alpha - 30°)$ below the horizontal (Figure 255). There are three unknowns in this diagram—N_1, F, and α—hence no unique solution is possible.

A similar diagram can be constructed for m_2 (Figure 256): known weight m_2g acting vertically downward, unknown force N_2 from the rod acting at 30° with the horizontal, and unknown force F acting at $(\alpha - 30°)$ above the horizontal. This diagram also contains three unknowns, hence no unique solution is possible.

However, we can put the two diagrams together by the F common to both, with Figure 257 as the result. Now all of the unknown values can be scaled from this diagram.

In Figure 257 you should note that the outside polygon, consisting of the vertical vector $\overrightarrow{(m_1 + m_2)g}$ and $\vec{N}_1$ and $\vec{N}_2$ at right angles to each other, is an equilibrium polygon for the complete two-bead-plus-cord system.

Following up the suggestion that vector polygon can be added to vector polygon to obtain a composite polygon for an entire system, you should be able to construct the composite polygon for the sign-supporting cantilever framework of problem 413 (whose external forces are shown in Figure 258) just about as fast as you can manipulate ruler and protractor. It would be useful to try it, just to find out how easy graph-

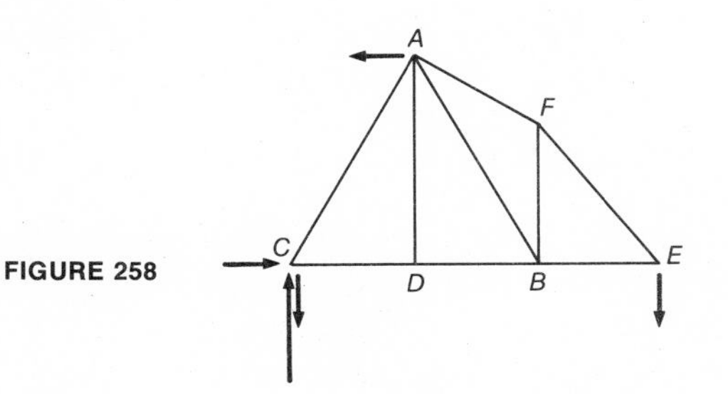

FIGURE 258

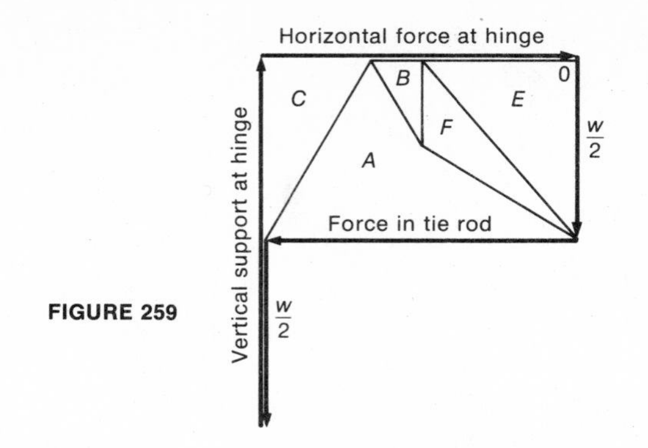

FIGURE 259

ical solutions can be for equilibrium at a point. Your result, on a much larger scale for accuracy, should look like Figure 259. The polygon begins and closes at O, and you will note the comforting fact that the external forces add to zero as they should. The individual joint polygons were all developed in the clockwise direction. The letter in a polygon indicates the joint for which that polygon demonstrates equilibrium.

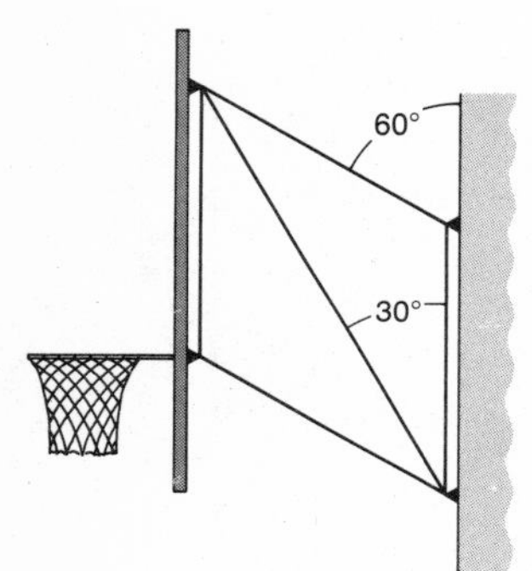

FIGURE 260

PROBLEM 422
A basketball backstop of weight W is mounted out from a gymnasium wall by the parallelogram framework shown in Figure 260. Assuming the weight of the framework itself is negligible compared to W, sketch the force polygon for the forces in the framework. *Caution:* A little thought ahead of time will make this much simpler than it may seem at first. Can you suggest a way to improve the design?

PROBLEM 423
A simple bridge on a country road is made up of three equal equilateral triangles as shown in Figure 261. The bridge carries a uniform floor load W, which for design purposes can be treated as load $W/2$ at B and $W/4$ at A and C. Sketch the force polygon for the forces in the structural members from the floor load alone.

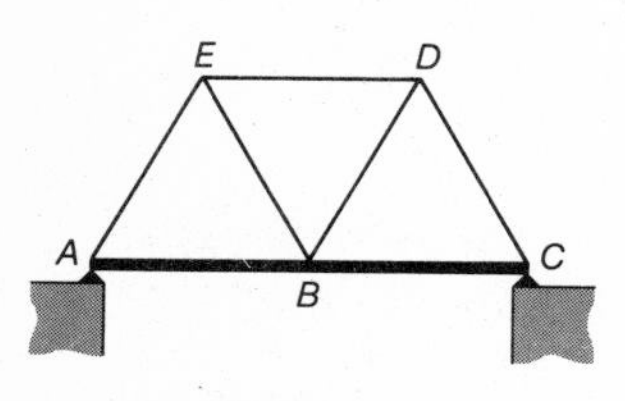

FIGURE 261

PROBLEM 424
If a farm tractor of weight W came to a stop in the middle of span BC of the bridge in problem 423, what effect would that have on the force polygon for the bridge? Don't guess; better construct the polygon and find out accurately.

THE HARMONIC OSCILLATOR

The harmonic oscillator represents a simple and clear-cut operation in the cyclic storage and release of energy; it would therefore seem appropriate to postpone study of simple harmonic motion until we come to the chapters on energy. However, the oscillator's motion is a direct and obvious consequence of Newton's Second Law, and the force or torque causing this motion has a unique characteristic that makes the identification of a harmonic oscillator very easy. It is therefore equally appropriate to introduce the subject of simple harmonic motion in this chapter.

Simple Harmonic Motion (SHM)

There are many cyclic motions—motions that return to some reference position at regular time intervals—but only one such motion has been given the name "simple harmonic." This is a cyclic motion whose displacement from an equilibrium position is described in general terms by $x = A \cos \omega t + B \sin \omega t$. We saw earlier that this general form can be made simpler by certain conditions prevailing at the start of the motion. For example, when $x = x_0$ and $\dot{x} = 0$ at $t = 0$ (the usual starting condition for a simple oscillator), then $A = x_0$, $B = 0$, and our general equation reduces to $x = x_0 \cos \omega t$. On the other hand, if $x = 0$ and $\dot{x} = v_0$ at $t = 0$ (a possible but less frequent starting condition), then $A = 0$ and $B = v_0/\omega$ and our general equation reduces to $x = (v_0/\omega) \sin \omega t$. If, at $t = 0$, $x = x_0$ and $\dot{x} = v_0$ (an unusual but possible set of initial conditions) then $x = x_0 \cos \omega t + (v_0/\omega) \sin \omega t$.

Even this equation can be simplified. Let $x_0 = A_0 \cos \phi$ and $v_0/\omega = A_0 \sin \phi$. We then have $x = A_0 \cos \omega t \cos \phi + A_0 \sin \omega t \sin \phi$, which leads to $x = A_0 \cos (\omega t - \phi)$. In this form

$$A_0 = \sqrt{x_0^2 + \frac{v_0^2}{\omega^2}},$$

and

$$\phi = \tan^{-1} \frac{v_0}{\omega x_0}.$$

In some instances this is a very useful maneuver, for it converts a sum of two harmonic motions into one motion, with the same angular frequency but with a phase shift.

Harmonic Motions Calculated

Undamped Motions

If we differentiate twice with respect to time any of the SHM equations for x, we get $\ddot{x} = -\omega^2 x$. This result has two interesting aspects: (1) the displacement x and the acceleration $\ddot{x}$ are always oppositely directed, with the acceleration always attempting to move the oscillator back to its equilibrium position; and (2) the maximum displacement A does not appear, which tells us that the frequency of oscillation is not related to how far from equilibrium the oscillator moves. Your common sense, or even your intuition, might suggest that the further an oscillator moves from equilibrium the longer it will take to return; if so, the equation $\ddot{x} = -\omega^2 x$

tells you that your common sense can't be trusted when dealing with SHM. (What this equation doesn't tell you, but the first derivative equation $\dot{x} = \omega A \cos \omega t$ does tell you, is that the bigger the maximum excursion A the faster the oscillator must move to complete its oscillation in the required time.)

From $\ddot{x} = -\omega^2 x$ we can go to $m\ddot{x} = -m\omega^2 x$. If only a single driving force is acting on the oscillator, and if the oscillator's motion is nonrelativistic (which is true most of the time), this single force is $= m\ddot{x}$, so that we have

$$m\ddot{x} = F = -m\omega^2 x = -kx, \tag{3-52}$$

where $k/m = \omega^2$. We studied this oscillator in problem 52; the c of those equations is the k/m or the ω^2 in these equations.

Aside from the energy exchange that occurs during an oscillation (which we will study later), we are usually interested in two things about an oscillator—its position as a function of time (the $x = A \cos \omega t + B \sin \omega t$ statement), and the period or frequency. The angular frequency ω is related to the period by $\omega T = 2\pi$, or $T = 2\pi/\omega$. When we put in these values for ω, the result for a simple undamped linear harmonic oscillator is

$$T = 2\pi \sqrt{m/k}. \tag{3-53}$$

Thus, when we know the mass of an oscillator and the k that evaluates its driving force, we know immediately what its period is.

Oscillators can depart from an equilibrium position by rotation as well as by linear movement. By the usual transformation from linear to rotational parameters, for a simple undamped rotational harmonic oscillator we have

$$I\ddot{\theta} = \tau = -I\omega^2 \theta = -\tau_0 \theta, \tag{3-54}$$

where $\tau_0/I = \omega^2$. τ_0, the torsional constant, has the dimensions ML^2T^{-2}. And for the period of such an oscillator, we have

$$T = 2\pi \sqrt{I/\tau_0}. \tag{3-55}$$

The identifying mark of a simple, undamped harmonic oscillator is the $-(k/m)x$ or the $-\omega^2 x$ character of its linear acceleration or the $-(\tau_0/I)\theta$ or the $-\omega^2 \theta$ character of its angular acceleration, or the $-kx$ or the $-\tau_0 \theta$ character of its driving mechanism. As soon as you can identify a force or a torque or an acceleration as being proportional to, and opposed to, a displacement, you know you are dealing with SHM. Once you know that, all other wanted information flows automatically out of equations 3-52–3-55.

PROBLEM 425

A straight, frictionless tube is bored through the earth between New York City and Honolulu. Assuming a uniform earth, how long will it take a crate of papayas, released from rest at the mouth of the tube in Honolulu, to reach New York City?

PROBLEM 426

After the First Planetary War, King Terra I assumed command of what was left of the earth's civilization and established an impregnable headquarters at the center of the earth. In order to control all communication, he destroyed all surface transportation and all of the tubes along earth chords such as the one mentioned in problem 425. As replacements he established radial tubes which led through his headquarters. Thus a crate of papayas bound for New York had to go from Honolulu to the center of the earth and be rerouted from there to New York. (Since papayas were still available, obviously not all aspects of civilization were destroyed.)

a. With what speed did a crate of papayas pass through the transfer point?

b. If the transfer was made instantaneously, how much longer did the Honolulu-center-New York delivery take compared to the direct trip from Honolulu to New York?

PROBLEM 427

When the power boat of problem 403 is pushed vertically down in the water below its normal trim and then released, with what frequency does it bob up and down? Assume all cross sections are as shown in that problem; also assume no drag from the water.

PROBLEM 428

A vibration-testing machine consists of a motor, a movable testing tray mounted on a platform, and a link between motor and tray (Figure 262). The link fits over a stud on a driving disc which is mounted on the shaft of the motor. Driving discs of various effective radii are available. To test an item mounted on the tray to an acceleration of 12 g, what relation must exist between r, the effective radius of the disc used, and ω, the angular speed of the disc?

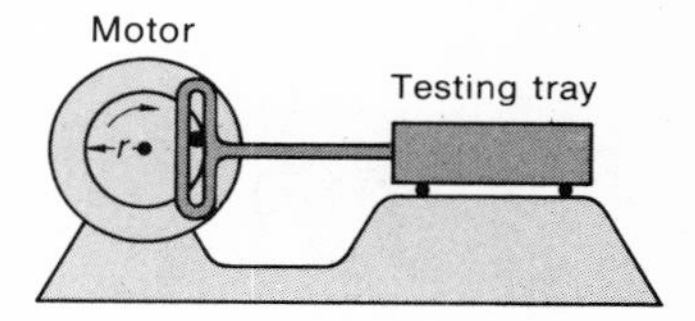

FIGURE 262

When forces and torques can be controlled quite exactly—as is possible when they are derived from gravitational or electric fields—and masses and moments of inertia can be accurately determined, then the other variable in the SHM equation, the period or the frequency, acquires accuracy also. Thus harmonic oscillators—from the swinging cathedral chandelier by which Galileo timed his pulse to the modern piezoelectric crystal—have become standard instruments for marking off time intervals.

The Pendulum. If a mass is moved from an equilibrium position in a gravitational field, or a charge in an electric field, a force develops that tends to restore equilibrium. Suppose we have a mass suspended from a frictionless pivot P as shown in Figure 263. Unless the c.g. of the mass (which is d from P) is directly below P, a torque exists about P from the weight Mg, and this torque is $\tau_P = -Mgd \sin \theta$. The $-$ sign must be inserted; if θ is measured positive clockwise, the torque is

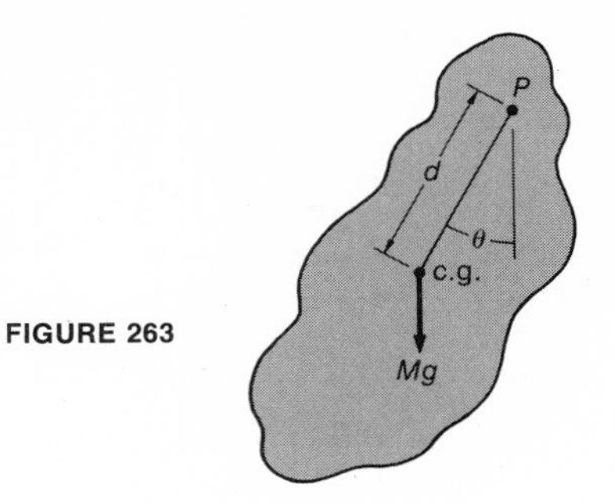

FIGURE 263

counterclockwise, and hence negative. This value for τ_P does not agree with equation 3-54 and hence does not indicate SHM. However, if we expand sin θ as a series in θ (found in almost any physics handbook or calculus text) we can write

$$\tau_P = -Mgd\left(\theta - \frac{\theta^3}{6} + \frac{\theta^5}{120} - \frac{\theta^7}{5040} \cdots\right).$$

If we restrict our amplitude of motion to a θ small enough so that $(-\theta^3/6 + \theta^5/120 - \theta^7/5040 \cdots)$ is less than the permissible error in our timing,* then we can ignore these terms and consider only

$$\tau_P = -Mgd\,\theta \text{ for } \theta \text{ small.} \tag{3-56}$$

This does satisfy the requirement for SHM, so that we can move immediately to

$$T = 2\pi\sqrt{\frac{I_P}{M\,g\,d}}\dagger \quad \text{for } \theta \text{ small.} \tag{3-57}$$

The oscillator described by equations 3-56 and 3-57 is usually called a physical pendulum, and most pendula are of this type.

For demonstration purposes (not often for accurate measuring) we can reduce the physical pendulum to a simpler form. If we hang a small mass on the end of an inextensible string or fine wire L long (the mass of the string or wire negligible compared to the pendulous mass, and the pendulous mass itself so relatively small in size that its moment of inertia about its own central axis is negligible) then I_P becomes simply ML^2. For such an oscillator we have

$$T = 2\pi\sqrt{\frac{ML^2}{MgL}} = 2\pi\sqrt{\frac{L}{g}}. \tag{3-58}$$

To no one's surprise, this is called the simple pendulum formula.‡

---

*In many practical applications, restricting θ to $6° = 0.1$ radian satisfies this condition.

†From transfer equation 3-40, $I_P = I_{c.m.} + M\,d^2$.

‡You may occasionally be confused in different texts by finding "ideal" pendulum and "simple" pendulum apparently interchangeable terms. An "ideal" pendulum is just that—a point mass suspended from a massless string. A "simple" pendulum actually exists, and is a practical approximation of an ideal pendulum.

PROBLEM 429
A pendulum is to be made up of a solid, uniform sphere of radius R and mass M, supported by a cord of negligible mass and of length $L-R$. If the percentage error resulting from the use of the simple pendulum formula is not to exceed 0.10%, what is the minimum length of cord that must be used?

PROBLEM 430
A straight, uniform rod of length L is hung from a frictionless pivot at one end. For small amplitude oscillations in a vertical plane, what is its period?

In solving problem 430 you had to use the physical pendulum equation

$$T = 2\pi \sqrt{\frac{I_{\text{c.m.}} + Md^2}{Mgd}},$$

with $d = L/2$. As $d \rightarrow 0$, i.e., as you move the suspension point in along the rod, the period can be seen to $\rightarrow \infty$. The question then might occur to you, what happens between $d = L/2$ and $d = 0$; it becomes a little difficult to imagine a linear transition, say, from

$$T = 2\pi \sqrt{\frac{2L}{3g}} \text{ to } T = \infty.$$

PROBLEM 431
To put the question under discussion in another way, we know the maximum limit for the period is ∞; is there any minimum period for a rod of length L?

a. If there is, where must we put the suspension point to obtain it?
b. What is the minimum period?
c. What is the longest rod we can use as a physical pendulum and still obtain as short a period as 1.0 sec?

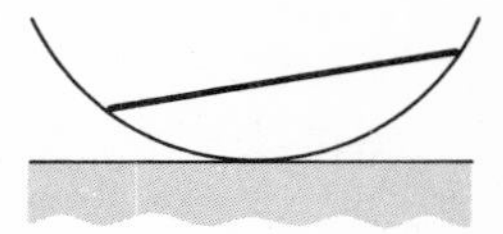
FIGURE 264

PROBLEM 432
A stiff, thin rod of length L is placed in a smooth spherical bowl of radius R. As the rod was originally placed in the bowl slightly off-center (Figure 264), it began an oscillatory sliding motion. What was the period of this motion?

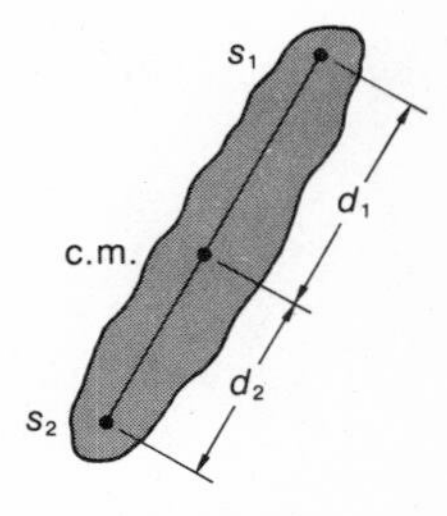

FIGURE 265

PROBLEM 433
In the Kater reversible pendulum, two suspension points s_1 and s_2 can be found along a straight line through the c.m., for which the periods for small oscillations of the pendulum are exactly equal (Figure 265). In terms of the distance between these two points, what is g? (This is a standard method for evaluating g, by timing T and measuring $d_1 + d_2$.)

FIGURE 266

PROBLEM 434

A uniform, solid sphere of radius r rolls without slipping in a spherical bowl of radius R, $R >> r$ (Figure 266). The axis of symmetry of the bowl is vertical. If the ball is given a small displacement from the bottom of the bowl and then released, what is the period of the ball's rolling motion?

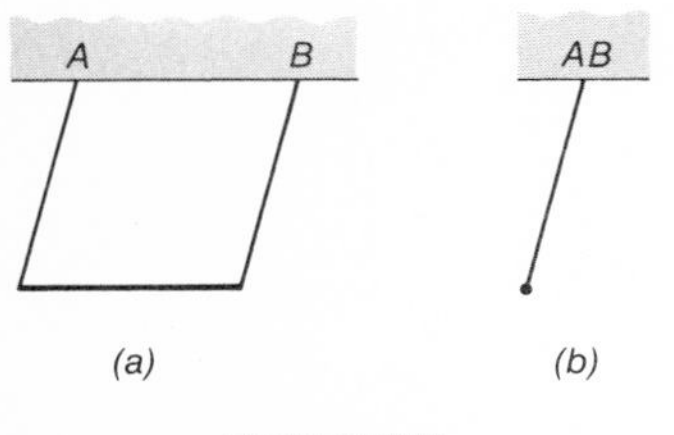

FIGURE 267

PROBLEM 435

Three identical stiff, uniform rods, each of length L, are frictionlessly hooked together and suspended from a horizontal axis. For small amplitudes, what is the period of oscillation of the assembly

a. when it swings in its own plane (Figure 267, *a*),
b. when its plane swings about the horizontal axis AB (Figure 267, *b*)?

PROBLEM 436

A mechanic balancing a flywheel mounted on a frictionless horizontal axis noted that the wheel always sought the same "at rest" orientation, and when displaced slightly from this equilibrium position it oscillated with a period T. By trial and error he found he could balance the wheel perfectly by fastening a small mass m to a particular point on the wheel a distance r from the axis. What was the final moment of inertia of the wheel about its axis?

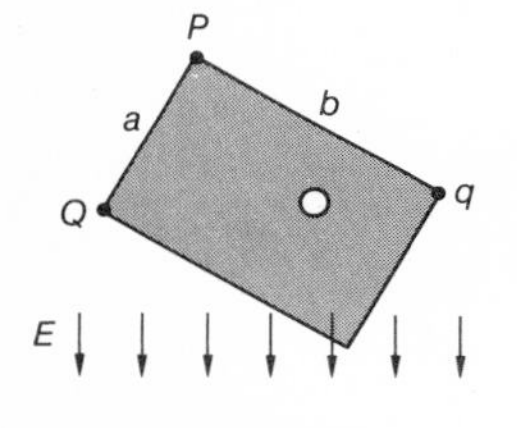

FIGURE 268

PROBLEM 437

A uniform circular disc of radius R is to be mounted on a horizontal axis to swing as a pendulum in its own plane, in small amplitude.

a. How far from the center of the disc should the axis be mounted for the frequency of oscillation to be a maximum?
b. What is this maximum frequency?

PROBLEM 438

To make an automatic beam interrupter, a rectangular, perfectly insulating plastic sheet, size $a \times b$ and of mass m, with a small hole in it, is hung by one corner from a frictionless pivot P. A very small area at two opposite corners is gold-plated, and these areas are given charges q and Q as shown in Figure 268. The assembly is placed in a region of uniform electric field E parallel to the plane of the sheet and set swinging in its own plane. To operate the interrupter at a particular frequency ν, what should E be? Consider gravitational forces negligible in comparison with electrical forces; for an axis perpendicular to the sheet, review problem 353.

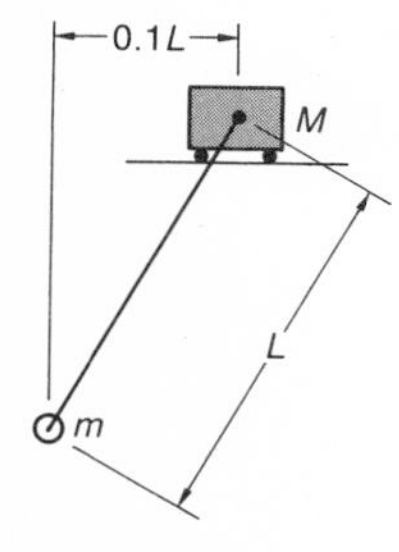

FIGURE 269

PROBLEM 439

The basic mechanism at the back of an animated advertising display is as shown in Figure 269. A mass m hangs on a wire stirrup of negligible mass, length L from the c.m. of a car of mass $M = 9m$; the car can move relatively frictionlessly on a horizontal track. To start the animation, the car and hanging mass are separated horizontally a distance $0.10\ L$ as shown, and then released.

a. What is the frequency of oscillation of the system?
b. What is the horizontal amplitude of the motion of the hanging mass? *Suggestion*: It will prove useful to review page 118.

PROBLEM 440

A carefully adjusted pendulum clock is sent aloft as part of the payload carried by a rocket that accelerates vertically with an acceleration of 3 g.

a. During the acceleration does the clock change its time-keeping rate?
b. If it does, what is the change?

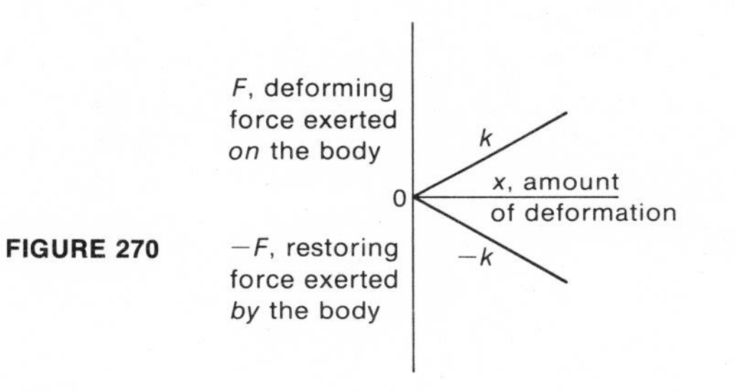

FIGURE 270

Hooke's Law Oscillators. So far our oscillators have been driven by $-kx$ forces (or by $-\tau_0\theta$ torques), which developed out of the geometrical constraints of a situation rather than directly from the forces. Let us now look at some oscillators driven by forces whose $-kx$ aspect is a natural part of the forces themselves.

All solid bodies resist deformation, but only elastic bodies "fight back" by attempting to return to their undeformed state and exerting in the process a force on their environment. Most elastic bodies obey Hooke's Law, $F = kx$; a deforming force F is proportional to the amount of deformation x, and k, the constant of proportionality, is the slope of the F-x graph. The dimensions of k are MT^{-2}; the MKSC unit for k is N m^{-1}.

When elastic bodies "fight back" they exert a restoring force $-F = -kx$ on whatever they are in contact with; this restoring force is a mirror image of the deforming force (Figure 270). This $-kx$ aspect makes the restoring force an obvious candidate for driving an SHM oscillator. The deformations giving rise to restoring forces useful for SHM fall principally into two categories: linear extension along an axis of symmetry, and twists or rotations about an axis of symmetry.

The best known elastic body that undergoes linear extension under force is the common coil spring. Actual coil springs have mass, and extend by the twisting and bending of the spring wire; ideal springs are massless and appear to undergo simple

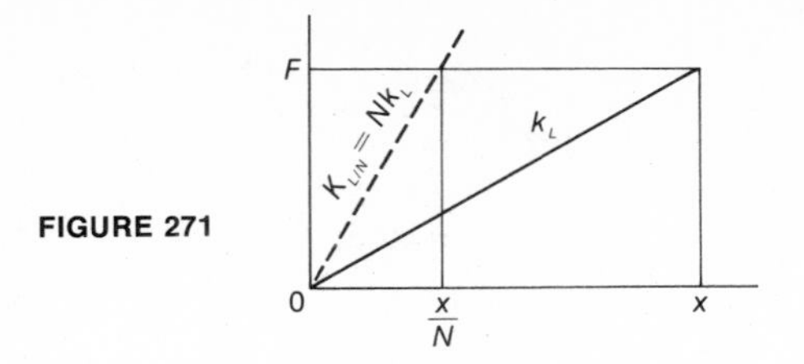

FIGURE 271

linear changes in length. For most oscillatory purposes the differences between actual springs and ideal springs are so slight that it is possible to predict accurately the behaviour of spring-driven oscillators by assuming the springs used are ideal.

A linear coiled spring obeys Hooke's Law through a wide range of extension. With springs, the k of this law is called the spring constant. For any particular spring k is a constant,* but it is also a function of the length of a spring. For example, under a linear force F a spring of normal, unstretched length L extends a distance x beyond length L. A segment of the spring L/N long contributes $1/N$ of this extension, or x/N, although subject to the same total linear force F. Hence for that segment the F-x graph is shown dashed in Figure 271, and it can be seen that k for a length $L/N = N\,k_L$.

*When the spring is extended so far that the F-x graph loses its straight-line character and k does not stay constant, we are beyond the range of the present discussion.

PROBLEM 441

An ideal spring of unstretched length 25 cm is hung vertically, and a weight hanger of mass 70 gm is firmly attached at the lower end (Figure 272). When a 130-gm mass is placed on the weight hanger, the spring's length becomes 32 cm. Then the 130-gm mass falls off the weight hanger.

a. What is the frequency of the ensuing SHM of the weight hanger?
b. What is the amplitude of this motion?

FIGURE 272

SOLUTION

When a force of 70g + 130g = 200g pulls on the spring it extends 32–25 = 7 cm. Hence $k = 200\text{g}/7$ dynes cm^{-1}.

a. Then, for the 70-gm mass,

$$\nu = \frac{1}{T} = \frac{1}{2\pi}\sqrt{\frac{k}{m}} = \frac{1}{2\pi}\sqrt{\frac{200\times 980}{7\times 70}} = \frac{10}{\pi}\ \text{oscillations sec}^{-1}.$$

b. When the 70-gm mass is in equilibrium, the spring is extended $x = 70\text{g} \times 7/200\text{g} = 2.45$ cm. When the 130-gm mass fell off the weight hanger the spring was extended $7 - 2.45 = 4.55$ cm beyond this equilibrium point for the 70-gm mass. Hence the amplitude of the SHM of the weight hanger = 4.55 cm. The weight hanger thus oscillates ± 4.55 cm from equilibrium, falling to 32 cm below the top of the spring and rising to 22.9 cm from the top of the spring.

PROBLEM 442

A mass M is hung from the lower end of an ideal coil spring of unstretched length L and allowed to come to rest (Figure 273). Then it is pulled down to a point $\frac{3}{2}L$ below the top of the spring and released; the subsequent SHM is timed to have a period of T.

a. What was the force constant of the spring?
b. What was the amplitude of the SHM?
c. At $t = T/4$, what was M's velocity and acceleration?

FIGURE 273

PROBLEM 443

A solid cylindrical roller of mass M rolls without slipping on a horizontal surface. A horizontal ideal spring, spring constant k, is attached to the axis of the roller by a suitable bail of negligible mass (Figure 274). When the roller is displaced from its equilibrium position and then released, what is the period of the resulting motion?

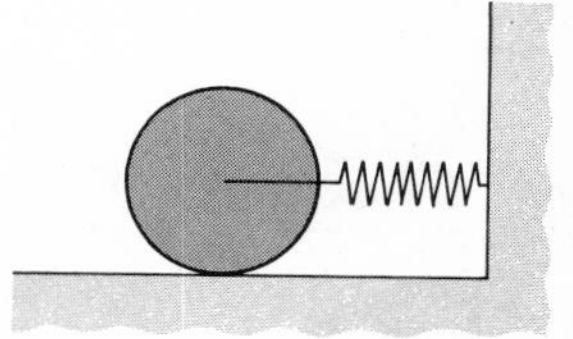

FIGURE 274

PROBLEM 444

An ideal spring of unstretched length L and spring constant k is fastened between supports that are L apart on a horizontal, frictionless surface. A mass M is then clamped to the middle of the spring.

a. When the mass is drawn aside perpendicular to the original line of the spring [(a), Figure 275] and then released, is the resulting motion SHM?
b. When the mass is displaced in the direction of the original line of the spring [(b), Figure 275] and then released, is the resulting motion SHM?
c. If either motion was SHM, what was its period?

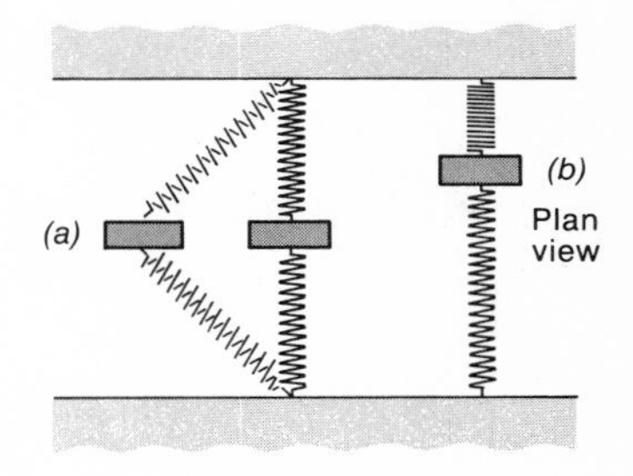

FIGURE 275

PROBLEM 445

A mass M is attached to two equal ideal springs, each of spring constant k, in three different ways as shown in Figure 276; (a) hung from springs in series; (b) hung from springs in parallel; and (c), set between springs on a frictionless horizontal surface. For each case, what is the period for SHM?

PROBLEM 446

In problem 445 the two springs were equal. Actually, it takes a lot of time and trouble to select two equal springs from a bin of similar springs. So to be more realistic, reconsider the arrangement in (a), Figure 276, this time with unequal springs of spring constants k_1 and k_2.

a. Now, what is the period of SHM?
b. Would your answer be different for a horizontal arrangement in series?

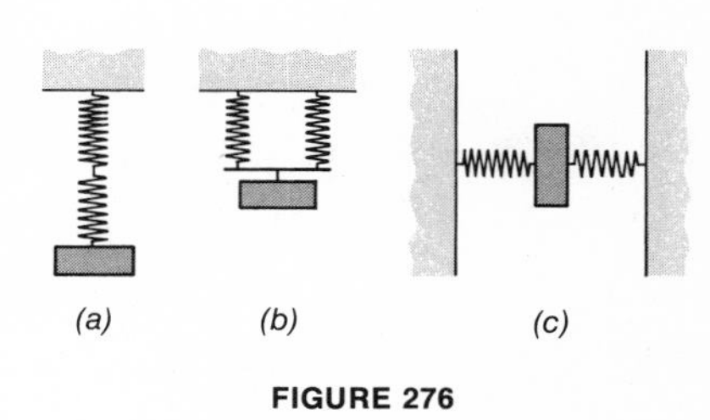

FIGURE 276

PROBLEM 447

A mass M on small frictionless wheels is allowed to oscillate transversely in a cylindrical trough of radius R, with an amplitude $A << R$, as shown at the top of Figure 277. Then two identical masses m,

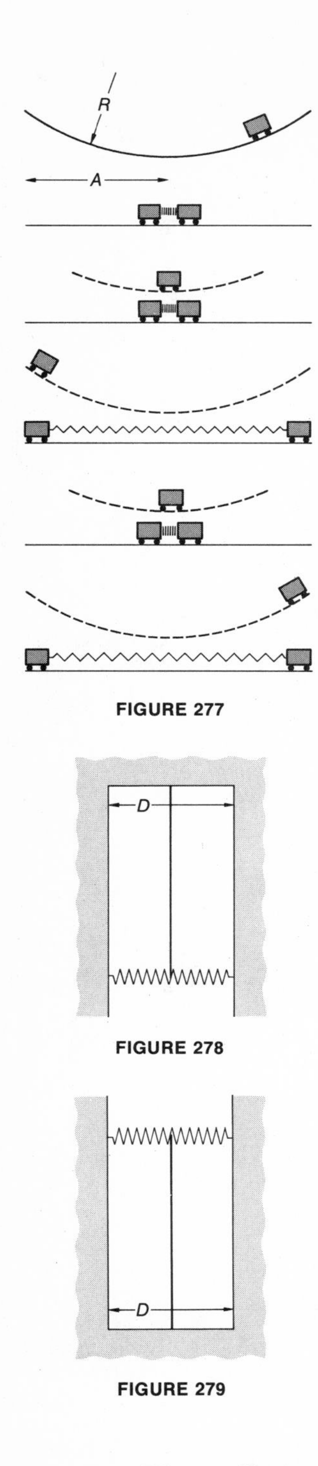

FIGURE 277

FIGURE 278

FIGURE 279

also mounted on small frictionless wheels, are connected by a compressed ideal spring of spring constant k, and placed on a horizontal surface adjacent and parallel to the path of M. At the instant that M passes the bottom of the trough the cocked spring between the two masses m is released, and the subsequent motion for quarter cycles of M is shown in the remaining three panels of Figure 277.

a. What was the value of m?
b. If the masses m with cocked spring had been placed inside the cylindrical trough alongside the path of M and released as before, with the subsequent motion showing exactly the same pattern as before, what then was m?

PROBLEM 448

A stiff, straight thin rod of mass M and length L is suspended midway between two vertical walls D apart (Figure 278). Its lower end is attached to the middle of a spring mounted perpendicular to the walls. The spring has a spring constant k and an unstretched length D.

a. What is the period of oscillation of the rod after its lower end is displaced a small distance sideways, perpendicular to the walls?
b. What is the period of oscillation if the displacement is parallel to the walls. *Hint:* Unless you have a strong intuition about this question, you will probably need to expand both $\cos\theta$ and $\sin\theta$ as a series in θ.

PROBLEM 449

The arrangement in problem 448 is now turned upside down, so that the rod may act as an inverted pendulum (Figure 279).

a. When the upper end of the rod is given a small displacement perpendicular to the walls, what is its period of oscillation?
b. In order for the inverted pendulum in part (a) to oscillate, how stiff must the spring be, i.e., what must be the minimum value for k?

The oscillations of an elastic body in twisting about an axis of symmetry are exactly analogous to the linear oscillations we have been studying. With τ_0 substituted for k, I for m, and θ for the variable coordinate instead of x, the methods of calculation for torsional oscillations are exactly the same as those we have learned for linear oscillations.

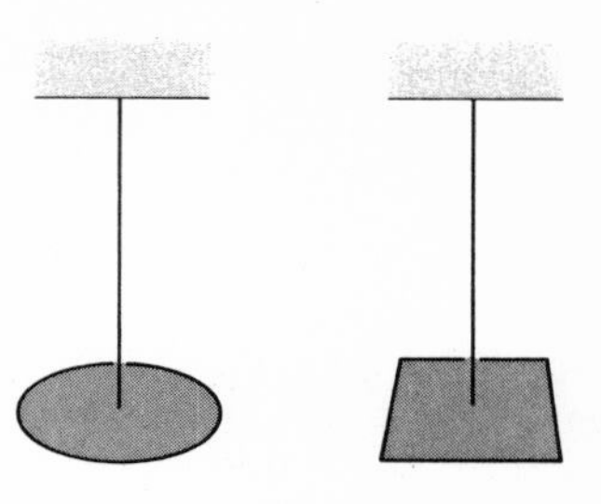

FIGURE 280

PROBLEM 450

A torsion pendulum, a simple pendulum, and a mass oscillating at the end of a spring are all adjusted to have the same frequency of oscillation on earth. When all three are taken to the moon, how do their periods compare?

PROBLEM 451

A circle and a square of equal masses are cut from the same sheet of uniform material. They are then mounted in turn on the lower end of a vertical torsion pendulum, at their centers of mass (Figure 280). How do their periods compare for oscillation in a horizontal plane?

PROBLEM 452

A uniform, solid cylindrical plate of mass M and radius R is suspended from its center in a horizontal plane from a vertical elastic rod that is rigidly clamped at its upper end. When the plate is given a small twist about the axis of the rod and then released, it oscillates in SHM with a period T. Then the cylinder is removed from the rod and an unknown mass of irregular shape is suspended at its center from the rod. When this mass was given a small twist and released, it oscillated in a horizontal plane in SHM with a period $2T$. What was $I_{\text{c.m.}}$ for the unknown mass?

PROBLEM 453

To calibrate a torsion suspension, a stiff, straight uniform rod of mass M and length L is hung at its midpoint from the suspension. Then the rod is set oscillating in a horizontal plane and its period accurately timed. If you want the accuracy of your determination of τ_0 to be within $\pm 0.1\%$, and your measurements of M, L, and T are better than that, how large must the radius of the suspended rod be, in terms of L, before you need to take it into account in your computation of τ_0? *Hint*: It may prove useful to compare the item on the uniform thin rod in the tabulation on page 179 with problem 349.

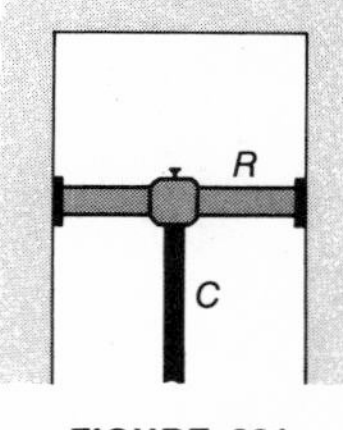

FIGURE 281

PROBLEM 454

You are shown an assembly at rest consisting of a solid horizontal metal bar R firmly fastened at both ends, and a heavy bar C which is firmly attached to R at its midpoint by a collar with a setscrew (Figure 281). You are told the oscillation of C in a vertical plane perpendicular to R has just been timed to have a period T. Confident in your knowledge of physics, you boast that you can say what the value of T was, without repeating the experiment. Your challenge is accepted, with the restriction that you may not remove R or weigh or measure either R or C.

You first loosen the setscrew and determine that C swings freely on R; for small oscillations you measure its period as P. You then

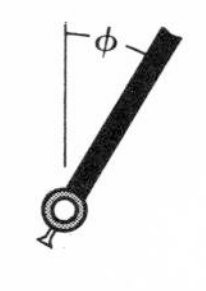

FIGURE 282

bring C to a delicate balance vertically above R at its midpoint, and tighten the setscrew. You then allow C to come to stable equilibrium at an angle ϕ with the vertical, $\phi<\pi/2$ (Figure 282). After four lines of calculation, you announce the correct value for T. What was it? (Neglect any consideration of the moment of inertia of R about its own axis of symmetry.)

Possibly the parenthetical statement at the end of problem 454 about neglecting the moment of inertia of the rod R made you curious about how much this neglect amounted to. In the torsion pendulum formula, I does not normally contain any contribution from the torsion rod itself. Nor did the earlier formula, for the period of a mass on a spring, indicate any inclusion of the mass of the spring in the value for m. Of course these neglected masses share in the SHM, but in most practical cases this share can be neglected in a simple, first-order approximation solution. If you are interested in a more accurate solution, one that takes these neglected masses into account see problem 958 in the Supplementary Problems at the end of this book.

Coupled Oscillators and Normal Modes. So far the oscillators we have studied have had only one degree of freedom. However, from the example of a car spring with an auto body vibrating at one end of it and the unsprung wheel and axle assembly vibrating at the other end, to the example of atoms in crystals interacting with each other during vibrations, the real world is full of coupled oscillators.

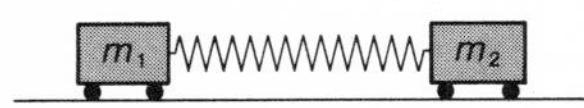

FIGURE 283

PROBLEM 455

A spring of unstretched length L and spring constant k is attached at either end to m_1 and m_2, unequal masses at rest on a horizontal frictionless surface (Figure 283). The masses are pushed toward each other, compressing the spring a distance D, and then released. What motion ensues?

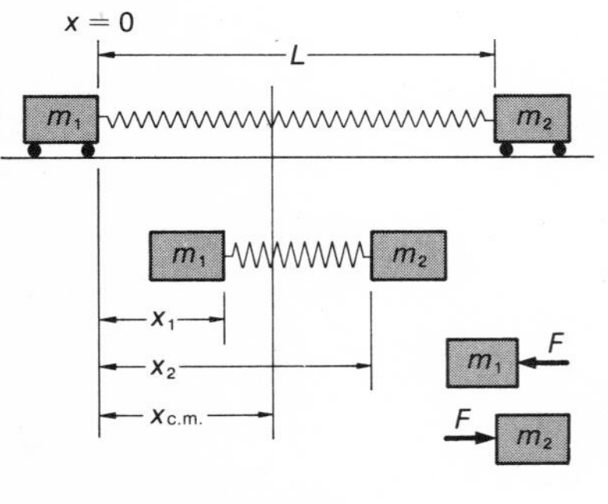

FIGURE 284

SOLUTION BY THE COMPREHENSIVE METHOD

From the free body diagrams (Figure 284) we can write $-F = m_1\ddot{x}_1$ and $F = m_2\ddot{x}_2$, where F is the restoring force in the spring and the 0 of the x-coordinate system is at the equilibrium position of m_1. (There are at least two other equally suitable coordinate centers.) It is clear that our essential problem is to evaluate F.

To do that, we look at the assembly when the masses have returned to their equilibrium positions, i.e., when m_1 is at $x = 0$ and m_2 at $x = L$. Since at this instant we know where the masses are, we can determine where the c.m. is: $x_{c.m.} = m_2L/(m_1 + m_2)$. Since there are no outside horizontal forces on the system, and since the system was at rest at $t = 0$, the system will stay at rest and the c.m. remains fixed at $[m_2/(m_1 + m_2)]L$. Since the c.m. stays fixed, m_1 and m_2 must vibrate relative to this point in space with equal periods but 180° out of phase. Then the point at $x_{c.m.}$ on the spring must also be at rest; then m_1 is vibrating at the end of an anchored spring of un-

stretched length $[m_2/(m_1 + m_2)]L$, and m_2 at the end of a spring of unstretched length $m_1L/(m_1 + m_2)$. Then for m_1 the applicable spring constant $k_1 = [(m_1 + m_2)/m_2]k$, and for m_2, $k_2 = [(m_1 + m_2)/m_1]k$.
With these values for k_1 and k_2 we are now ready to evaluate F:

$$-F = -\frac{m_1 + m_2}{m_2} kx_1 = m_1\ddot{x}_1,$$

$$F = \frac{m_1 + m_2}{m_1} k(L - x_2) = m_2\ddot{x}_2.$$

When we solve and put in limits,

$$x_1 = \frac{m_2D}{m_1 + m_2} \cos \sqrt{\frac{k(m_1 + m_2)}{m_1m_2}}\, t,$$

$$x_2 = L - \frac{m_1D}{m_1 + m_2} \cos \sqrt{\frac{k(m_1 + m_2)}{m_1m_2}}\, t,$$

and

$$T = 2\pi \sqrt{\frac{m_1m_2}{k(m_1 + m_2)}}.$$

SOLUTION BY THE SHORT METHOD

If we are interested only in the period of the coupled oscillation, there is a short cut. We start again with $-F = m_1\ddot{x}_1$ and $F = m_2\ddot{x}_2$. Also $F = kx$, where x is the change in length of the entire spring. From the diagram in Figure 284, $x = x_1 + L - x_2$. When we differentiate this twice with respect to time we have $\ddot{x} = \ddot{x}_1 - \ddot{x}_2$. From $-F = m_1\ddot{x}_1$, we have $-kx = m_1\ddot{x}_1$ and $-m_2kx = m_1m_2\ddot{x}_1$; similarly, $m_1kx = m_1m_2\ddot{x}_2$. By subtracting we get $-(m_1 + m_2)kx = m_1m_2(\ddot{x}_1 - \ddot{x}_2) = m_1m_2\ddot{x}$.

Rearranging,

$$-kx = \frac{m_1m_2}{m_1 + m_2}\ddot{x}.$$

This equation has the solution

$$x = D \cos \sqrt{\frac{k(m_1 + m_2)}{m_1m_2}}\, t;$$

as before, the period

$$T = 2\pi \sqrt{\frac{m_1m_2}{(m_1 + m_2)k}}.$$

The equation $-kx = [m_1m_2/(m_1 + m_2)]\ddot{x}$ describes the motion of an apparent single body of mass $m_1m_2/(m_1 + m_2)$; we have thus reduced our two-body problem to a one-body problem. It therefore seems appropriate to call $m_1m_2/(m_1 + m_2)$ the "reduced" mass; it is frequently denoted by the symbol μ. In any problem involving two bodies and a central force (the force in the spring was a central force in problem 455) the use of "reduced" mass may simplify the calculation.

PROBLEM 456

We place the two masses and spring of problem 455 in a horizontal air trough (hence frictionless) and set them in linear motion by giving m_1 a strong shove toward m_2. At the instant of maximum compression of the spring the masses are $L/2$ apart and both are moving with a linear speed v. What is the subsequent motion of each mass?

PROBLEM 457

By spectroscopic analysis it can be determined that the two protons in the H_2 molecule vibrate along the line joining them with a frequency of 1.3×10^{14} vib sec^{-1}.

a. What must be the force constant of the hydrogen molecule?

b. The diagram in Figure 285 describes, for the H_2 molecule, the relation between the interproton force F and the interproton distance x. What is the slope of the curve in the vicinity of x_0, the equilibrium distance between the protons?

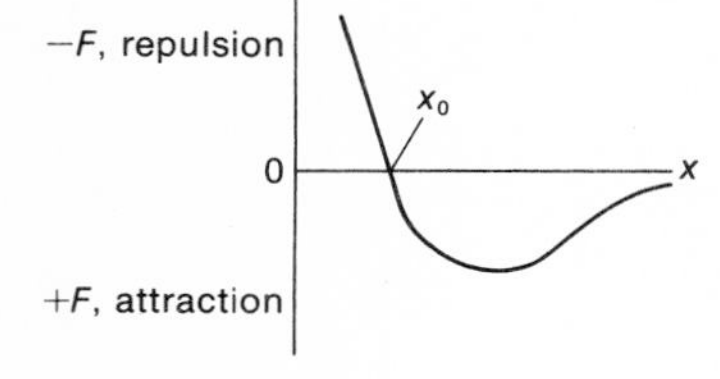

FIGURE 285

PROBLEM 458

From energy considerations the force constant of a CO molecule is calculated to be 1.8×10^3 N m^{-1}. What vibrational frequency would you expect for carbon monoxide?

Let us now advance a step and consider a CO_2 molecule. This is a symmetric linear molecule, thus: O—C—O. We can make a mechanical model of CO_2 by putting three masses in a frictionless horizontal air trough and connecting them with equal springs (Figure 286). What free vibrations can such an assembly have?

Before starting any calculation, let's see what we can find out just by looking at the problem (Figure 287). Since no outside forces act horizontally on the assembly, the c.m. must not move horizontally. This means that the three masses may not move in unison in the same direction; (*a*) is forbidden. We can, however, imagine the following possibilities that would keep the c.m. fixed: m_2 remains stationary and the outside m_1 masses oscillate equally relative to m_2 but with opposite phases (*b*); and the motion of m_2 in one direction is just sufficient to balance the motion of both m_1 masses in the other direction (*c*). For the general case there is a third possibility: two adjacent masses move in one direction with the other mass moving in the opposite direction. However, such a motion could not be maintained under the conditions of symmetry prevailing in the system under study.

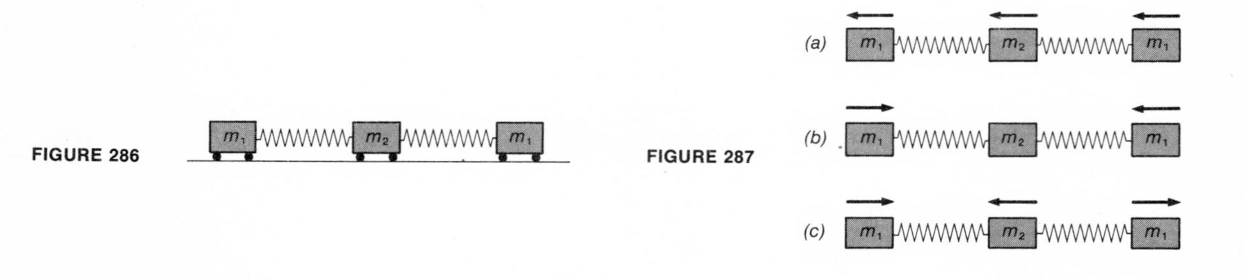

FIGURE 286

FIGURE 287

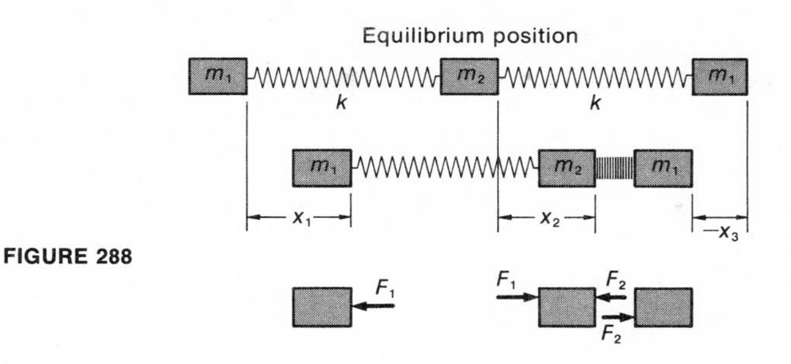

FIGURE 288

For case (b), without doing any calculation we can say that if $x_1 = A \cos \omega t$ (x_1 the horizontal coordinate for the left-hand m_1) then $x_2 = 0$ and $x_3 = -A \cos \omega t$. For case (c) we need a little algebra; to ensure that the c.m. does not move we must have $m_1x_1 + m_2x_2 + m_1x_3 = 0$. If $x_1 = x_3$ (a reasonable first-order guess at this preliminary stage), then $x_2 = -(2m_1x_1)/m_2 = -(2m_1x_3)/m_2$. Then for case ($c$) motion, if $x_1 = A \cos \omega t$, $x_2 = -(2m_1/m_2) A \cos \omega t$ and $x_3 = A \cos \omega t$.

A word must be said here about ω, which is seen to govern the motion of all three bodies; why can't ω be different for the different bodies? What we are studying are the free vibrations of the system, motions that, once started, can be maintained by the internal forces of the system. It can be proved (by invoking the fixity of the c.m.) that only repetitive motions with the same frequency satisfy this criterion of natural maintainability. Such motions are called normal modes, and they are characterized by the fact that, for any particular pattern of motion, ω is the same for all the moving bodies.

Let us now see what we can calculate. With three bodies the equations of motion will be much simpler if we measure the x-coordinate for each body from that body's equilibrium position. As seen in Figure 288, the change in length of the left-hand spring is $x_1 - x_2$, and the change for the right-hand spring is $x_2 - x_3$. Then

$$m_1\ddot{x}_1 = -k(x_1 - x_2);$$

$$m_2\ddot{x}_2 = k(x_1 - x_2) - k(x_2 - x_3);$$

$$m_1\ddot{x}_3 = k(x_2 - x_3).$$

Rearranging,

$$\ddot{x}_1 + \frac{k}{m_1}x_1 = \frac{k}{m_1}x_2;$$

$$\ddot{x}_2 + \frac{2k}{m_2}x_2 = \frac{k}{m_2}(x_1 + x_3);$$

$$\ddot{x}_3 + \frac{k}{m_1}x_3 = \frac{k}{m_1}x_2.$$

Let us assume $x_1 = A \cos \omega t$, $x_2 = B \cos \omega t$, and $x_3 = C \cos \omega t$. Then $\ddot{x}_1 = -\omega^2 A \cos \omega t$, etc.

If we substitute these terms in our equations, and cancel out the common $\cos \omega t$ term, we have

$$\left(-\omega^2 + \frac{k}{m_1}\right) A = \frac{k}{m_1} B; \tag{3-59}$$

$$\left(-\omega^2 + \frac{2k}{m_2}\right) B = \frac{k}{m_2}(A + C); \tag{3-60}$$

$$\left(-\omega^2 + \frac{k}{m_1}\right) C = \frac{k}{m_1} B. \tag{3-61}$$

For equation 3-60 if $B = 0$, $A = -C$, which confirms our earlier guess about case (b). For $B \neq 0$, from equations 3-59 and 3-61 we see that $A = C$. We have left only to evaluate B when $B \neq 0$. From equations 3-59 and 3-60,

$$-\omega^2 + \frac{k}{m_1} = \frac{k}{m_1}\frac{B}{A};$$

$$-\omega^2 + \frac{2k}{m_2} = \frac{k}{m_2}\frac{A + C}{B} = \frac{k}{m_2}\frac{2A}{B}.$$

We subtract to eliminate ω^2 and solve for B:

$$B = -\frac{2m_1}{m_2} A.$$

Happily, this confirms what we had gotten earlier from c.m. considerations. After all these calculations, we actually haven't learned anything new. However, we don't yet know about ω.

Looking at either equation 3-59 or 3-61, when $B = 0$, $-\omega^2 + k/m_1 = 0$, and $\omega = \sqrt{k/m_1}$. So for normal mode No. 1,

$$B = 0,\ A = -C, \text{ and } \omega_1 = \sqrt{\frac{k}{m_1}}.$$

Then $x_1 = A \cos \sqrt{k/m_1}\, t$, $x_2 = 0$, and $x_3 = -A \cos \sqrt{k/m_1}\, t$, and the frequency of normal mode No. 1 is given by

$$\nu_1 = \frac{1}{2\pi}\sqrt{\frac{k}{m_1}}.$$

When $B = -(2m_1/m_2)A$ and $C = A$, from equation 3-60 we have $\omega^2 = k/m_1 + 2k/m_2$. Hence for normal mode No. 2,

$$B = -\frac{2m_1}{m_2} A,\ A = C, \text{ and } \omega_2 = \sqrt{\frac{k}{m_1} + \frac{2k}{m_2}}.$$

Then for normal mode No. 2,

$$x_1 = A \cos \sqrt{\frac{k}{m_1} + \frac{2k}{m_2}}\, t,$$

$$x_2 = -\frac{2m_1}{m_2} A \cos \sqrt{\frac{k}{m_1} + \frac{2k}{m_2}}\, t,$$

and

$$x_3 = A \cos \sqrt{\frac{k}{m_1} + \frac{2k}{m_2}}\, t,$$

and

$$\nu_2 = \frac{1}{2\pi}\sqrt{\frac{k}{m_1} + \frac{2k}{m_2}} = \frac{1}{2\pi}\sqrt{\frac{k}{m_1}\left(1 + \frac{2m_1}{m_2}\right)}.$$

PROBLEM 459

What would you expect for the ratio of the frequencies of the two linear modes of vibration of

a. the CO_2 molecule;
b. the XeF_2 molecule, also a linear molecule?*

*The answers you get using the mass and spring model differ by about 10% for real CO_2 molecules, and about 5% for real XeF_2 molecules, from the ratios determined from spectroscopic data. The fact that you can come within 5% illustrates the usefulness of such models; the fact that you can miss by 10% illustrates their limitations. However, the size of the "miss" often gives a clue to the reason for the deficiency of the model. For the two molecules in the problem there is an additional bond between the outside atoms that could have been represented by a weaker spring connecting the two outer masses, which explains the difference between the actual frequency ratios and the theoretical ratios derived from the mass and spring model. The difference between the actual ratios can be explained by difference in valence force from the central atom.

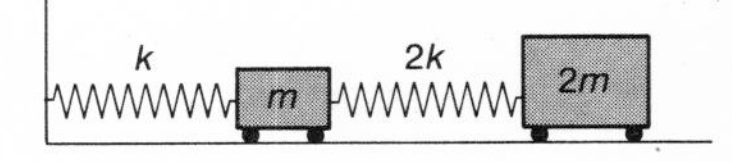

FIGURE 289

PROBLEM 460

For the coupled oscillators shown in Figure 289 there are two normal modes of oscillation. What are the displacement equations and frequencies for these modes?

PROBLEM 461

The masses and springs of problem 460 are now hung as shown in Figure 290, and set in vertical oscillation. What adjustments do you have to make to adapt the answers to problem 460 to this situation?

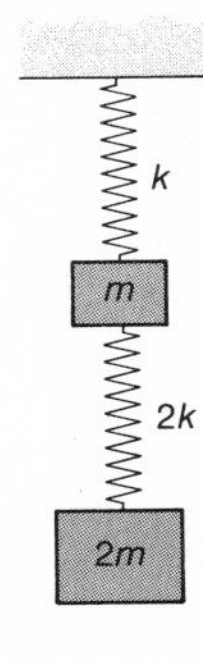

FIGURE 290

PROBLEM 462

Masses M_1 and M_2 are placed in a horizontal, frictionless air trough and connected with each other and with an anchor point by unequal

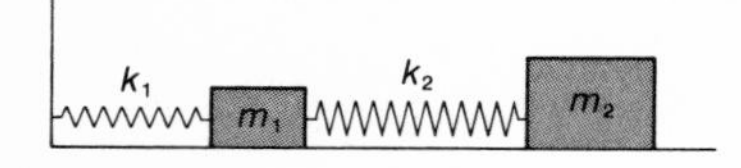

FIGURE 291

springs, as shown in Figure 291. When starting positions are carefully adjusted so that $x_{1,0} = -x_{2,0}$, it is observed that during oscillations $x_1 = -x_2$ at all times. (x_1 and x_2 are departures from equilibrium positions.)

a. What must be the ratio M_1/M_2?
b. What was the frequency of the motion?

PROBLEM 463

An instructor wishing to demonstrate normal modes places masses of 540 gm and 270 gm in a frictionless horizontal air trough (Figure 292). The masses are 10 cm wide between spring attachment points and are connected with springs of unstretched length 20 cm and spring constants k and $3k$ as shown; $k = 6600$ dynes cm^{-1}.

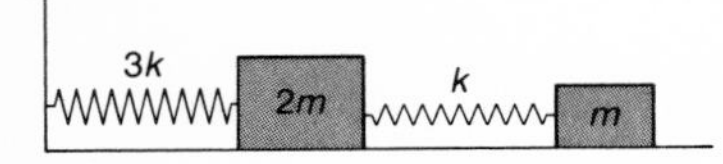

FIGURE 292

a. The instructor holds the 540-gm mass with its left end 16 cm from the anchor end; where should he hold the left end of the 270-gm mass so that on release the masses will immediately move in normal modes?
b. What are the periods of these normal modes?

Damped Motions

We have limited our discussion of harmonic oscillators so far to frictionless situations. It is time to enlarge our study to include retarding forces of friction and viscosity, forces which are present in most real situations.

Constant Friction. It was stated earlier that for most normal speeds and surfaces, mechanical friction is independent of the relative speed of the surfaces in contact. With that in mind, let us see what friction does to a simple, one-body oscillator.

PROBLEM 464

A mass m rests on a horizontal surface and is connected to a fixed point by a horizontal spring of spring constant k (Figure 293). There is a coefficient of friction μ between mass and surface. When the mass is displaced X_0 from its equilibrium position and released, what is its subsequent motion?

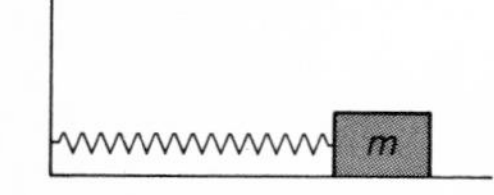

FIGURE 293

SOLUTION

The mass is released at $x = X_0$, with the arrangement shown in Figure 294; it first moves from right to left, and the frictional force is directed from left to right. This is true for the odd-numbered half-cycles, from $\omega t = 0$ to $\omega t = \pi$, from $\omega t = 2\pi$ to $\omega t = 3\pi$, etc. For the alternate half-cycles the frictional force is directed from right to left. We can carry both of these possibilities in our work by the use of both + and − signs. Thus for the equation of motion we have

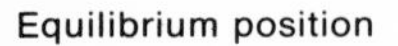

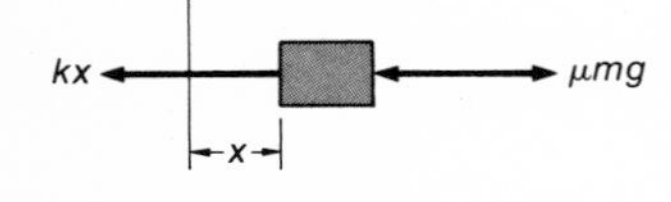

FIGURE 294

$$m\ddot{x} = -kx \pm \mu\, mg \text{ or } \ddot{x} + \frac{k}{m}\left(x \mp \frac{\mu\, mg}{k}\right) = 0.$$

We can put this into more manageable form by changing variables, thus: let $y = x \mp \mu\, mg/k$. Then $\ddot{y} = \ddot{x}$, and we have $\ddot{y} + (k/m)y = 0$.

As before, let us assume $y = A \cos \omega t$.* Then $\ddot{y} = -\omega^2 y$, so that $\omega = \sqrt{k/m}$. Then $y = x \mp \mu\, mg/k = A \cos \sqrt{k/m}\, t$. When $t = 0$, $x = X_0$; then $X_0 \mp \mu\, mg/k = A$.

Finally, we have for the solution of our equation of motion,

$$x = \pm \frac{\mu\, mg}{k} + \left(X_0 \mp \frac{\mu\, mg}{k}\right) \cos \sqrt{\frac{k}{m}}\, t.$$

It is not a general or final solution, however; because of the reversal of the friction force, the solution must be applied a half-cycle at a time.

For the first half-cycle, from $\omega t = 0$ *to* $\omega t = \pi$,

$$x = \frac{\mu\, mg}{k} + \left(X_0 - \frac{\mu\, mg}{k}\right) \cos \sqrt{\frac{k}{m}}\, t,$$

and at $\omega t = \pi$,

$$x = -\left(X_0 - \frac{2\mu\, mg}{k}\right).$$

Thus, in one-half cycle the amplitude has diminished by $2\mu\, mg/k$, and so the starting amplitude for the second half-cycle is $X_0 - 2\mu\, mg/k$.

By a development similar to that for the first half-cycle you can determine, for $\omega t = \pi$ to $\omega t = 2\pi$, that

$$x = -\frac{\mu\, mg}{k} + \left(X_0 - \frac{3\mu\, mg}{k}\right) \cos \sqrt{\frac{k}{m}}\, t$$

and at $\omega t = 2\pi$,

$$x = X_0 - \frac{4\mu\, mg}{k}.$$

Each half-cycle starts with an amplitude $2\mu\, mg/k$ less than the starting amplitude of the preceding half-cycle.

Finally, for motion during the $N + 1$ half-cycle,

$$x = (-1)^N \frac{\mu\, mg}{k} + \left(X_0 - \frac{(2N+1)\,\mu\, mg}{k}\right) \cos \sqrt{\frac{k}{m}}\, t,$$

where N is the number of half cycles already completed.

It is important to note that friction affects only the amplitude of motion; the angular frequency $\sqrt{k/m}$ remains the same as for the frictionless case. Later we will relate the decrease in amplitude to the work being done against friction. The diagram in Figure 295

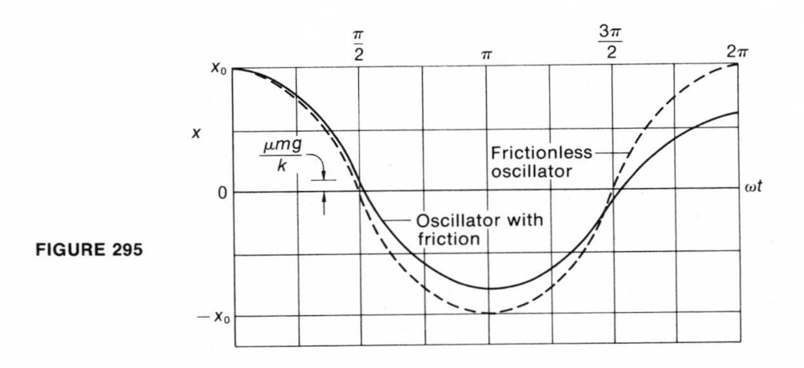

FIGURE 295

*We simplify our work here by remembering that the $B \sin \omega t$ term $= 0$ when $\dot{x} = 0$ at $t = 0$.

depicts the motion of the above oscillator and, for comparison, the motion of a similar but frictionless oscillator. It is interesting to note that the oscillator with friction, like a good pro football team on defense, has a roving center.

PROBLEM 465

What must be the relation among k, X_0, and μ so that the oscillator in Figure 295 can complete at least 10 full cycles before it is brought to rest by friction? (Note that when $kx \leqslant \mu\, mg$, the mass is no longer accelerated.)

Drag Proportional to Speed. We have just seen that mechanical friction, because of its constant aspect, attentuates oscillations rather quickly. Retarding forces that decrease as the speed of an oscillating mass decreases provide less restraint to an oscillator's motion. Such forces are present in many practical oscillators.

PROBLEM 466

The frictional force that affected the motion of m in problem 464 is removed by floating m on a film of oil. This has the effect of replacing a constant frictional force $\mu\, mg$ by a viscous force that is proportional to speed. Now, when m is displaced horizontally from equilibrium and released, what motion ensues?

SOLUTION

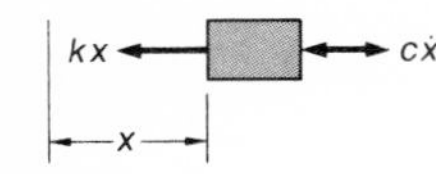

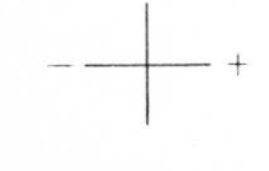

FIGURE 296

The statement of the forces in the free-body diagram (Figure 296) are simpler in this case than in problem 464. When $\dot{x}$ is $-$, $c\dot{x}$ is in the $+$ direction; when $\dot{x}$ is $+$, $c\dot{x}$ is in the $-$ direction. Hence for the entire motion, the equation of motion is $m\ddot{x} = -kx - c\dot{x}$, or

$$\ddot{x} + \frac{k}{m}\, x + \frac{c}{m}\, \dot{x} = 0.$$

In many texts it is customary to denote the term c/m by γ, i.e., $c = m\gamma$. Sooner or later we will need to indicate k/m by ω_0^2. (By now you should recognize that the natural angular frequency of a simple, undamped oscillator is $\sqrt{k/m}$, which we will henceforth designate by ω_0.) Making these changes to place our statement in line with standard notation, we have

$$\ddot{x} + \omega_0^2\, x + \gamma\dot{x} = 0.* \tag{3-62}$$

Since the assumption that $x = A \cos \omega t$ has worked so well in the past, our first instinct is to try it in this instance. And we run into trouble immediately. For if $x = A \cos \omega t$, then we also have $-\,\omega^2 A \cos \omega t + \omega_0^2\, A \cos \omega t - \omega\gamma A \sin \omega t = 0$. With that result we can get as far as $\tan \omega t = (\omega_0^2 - \omega^2)/\gamma\omega$, which doesn't seem a very promising way to find out what ω is. So then we remember that, although we have been getting by with $x = A \cos \omega t$ for the simpler

*As used here, $\gamma\dot{x}$ is the specific viscous force, i.e., the viscous force per unit mass. Thus one could call γ the coefficient of specific viscous resistance. The dimension of γ is T^{-1}.

cases, the general expression is $x = A \cos \omega t + B \sin \omega t$. When we use that expression, $\dot{x} = -\omega A \sin \omega t + \omega B \cos \omega t$, $\ddot{x} = -\omega^2 A \cos \omega t - \omega^2 B \sin \omega t$. Substituting these values in our equation of motion, and collecting separately coefficients of $\cos \omega t$ and $\sin \omega t$, we have $(-\omega^2 A + \gamma\omega B + \omega_0^2 A) \cos \omega t + (-\omega^2 B - \gamma\omega A + \omega_0^2 B) \sin \omega t = 0$. The only way this equation can $= 0$ for any and all values of t is for the coefficients of $\cos \omega t$ and $\sin \omega t$ to $= 0$ separately. Thus,

$$(\omega_0^2 - \omega^2)A = -\gamma\omega B,$$

and

$$(\omega_0^2 - \omega^2)B = \gamma\omega A.$$

These look so much like the equations we worked with for coupled oscillators that we decide to try the same method here. Solving for A and B, we get $B^2 = -A^2$, or $B = \pm A\sqrt{-1}$. Setting $\sqrt{-1} = i$, and using $B = \pm iA$ in the equations above to solve for ω, we get

$$\omega = \pm \frac{i\gamma}{2} \pm \sqrt{\omega_0^2 - \frac{\gamma^2}{4}}.$$

We have now come to a point where we need to be reminded of what we learned in mathematics about i* and the imaginary world it inhabits. There are many places in physics where we will meet $\sqrt{-1}$, and the complex number system has been developed to help us at such times. Some basic statements about that system are useful now.

- A complex number $\hat{A} = A + iB$, where A is the real part of $\hat{A}$—the part we can measure and work with in the real world—and iB is the imaginary part that exists in the mathematical world.
- If $\hat{A} = A + iB$, then $\hat{A}^* = A - iB$, and $\hat{A} \times \hat{A}^* = (A + iB)(A - iB) = A^2 + B^2$. $\hat{A}^*$ is called the complex conjugate of $\hat{A}$, and we see that a complex number multiplied by its complex conjugate produces a real number.
- $e^{\pm i\alpha} = \cos \alpha \pm i \sin \alpha$; $\cos \alpha$ is the real part of $e^{\pm i\alpha}$.

Returning to our calculation for

$$\omega = \pm \frac{i\gamma}{2} \pm \sqrt{\omega_0^2 - \frac{\gamma^2}{4}},$$

we are immediately faced with the question: is ω complex, as it appears, with

$$\sqrt{\omega_0^2 - \frac{\gamma^2}{4}}$$

being real? There is nothing we know so far that says $\gamma/2$ must always be less than ω_0. In fact, we can see that there are three possibilities to consider: $\omega_0 > \gamma/2$, $\omega_0 = \gamma/2$, and $\omega_0 < \gamma/2$.

Case I: $\omega_0 > \gamma/2$, *the normally damped oscillator.* When $\omega_0 > \gamma/2$, $\omega_0^2 - \gamma^2/4$ is intrinsically positive. For simplicity in writing, let us replace $\omega_0^2 - \gamma^2/4$ by ω_γ^2; then $\omega = \pm i\gamma/2 + \omega_\gamma$. If we put this equa-

---

*The imaginary i is not to be confused within $\vec{i}$, the unit vector in the x direction.

tion, plus $B = \pm iA$, back into our original postulated equation, we get

$$x = A \cos \omega t \pm iA \sin \omega t = A\ e^{\pm i\omega t} = A\ e^{\pm i(\pm i\gamma/2 + \omega_\gamma)t}$$
$$= e^{-(\gamma/2)t}\ (A_1\ e^{i\omega_\gamma t} + A_2\ e^{-i\omega_\gamma t}). \tag{3-63}$$

At this point it is obvious that x is a complex number, and probably A is, too. To allow for this condition we transform to complex notation, thus:

$$\hat{x} = e^{-(\gamma/2)t}\ [(A_1 + iB_1)(\cos \omega_\gamma t + i \sin \omega_\gamma t) + (A_2 + iB_2)(\cos \omega_\gamma t - i \sin \omega_\gamma t)].$$

If we expand this equation and take only the real part,

$$x = e^{-(\gamma/2)t}\ (C_1 \cos \omega_\gamma t + C_2 \sin \omega_\gamma t),$$

where $C_1 = A_1 + A_2$, etc. When we put in the boundary conditions that $x = X_0$ and $\dot{x} = 0$ at $t = 0$, we find that $C_1 = X_0$ and $C_2 = \gamma X_0/2\omega_\gamma$. (It would be more general to let $\dot{x} = v_0$ at $t = 0$, but to keep terms as simple as possible we have taken advantage of the fact that usually $\dot{x} = 0$ at the start of a motion.)

Finally we reach a usable solution:

$$x = e^{-(\gamma/2)t} \left(X_0 \cos \omega_\gamma t + \frac{\gamma X_0}{2\omega_\gamma} \sin \omega_\gamma t \right). \tag{3-64}$$

This equation is valid as it stands, and appears in this form in several textbooks. However, it can be converted into a form that provides more obvious information; let $X_0 = a \cos \phi$ and $\gamma X_0/2\omega_\gamma = a \sin \phi$. This makes $\phi = \cos^{-1} \omega_\gamma/\omega_0$, and $a = \omega_0 X_0/\omega_\gamma$. Then

$$x = e^{-(\gamma/2)t} \frac{\omega_0}{\omega_\gamma} X_0 (\cos \omega_\gamma t \cos \phi + \sin \omega_\gamma t \sin \phi)$$
$$= e^{-(\gamma/2)t} \frac{\omega_0}{\omega_\gamma} X_0 \cos (\omega_\gamma t - \phi). \tag{3-65}$$

Equations 3-64 and 3-65 describe what is called normally damped harmonic motion. (See curve I, Figure 297.) This motion has an angular frequency ω_γ that is less than the frequency ω_0 available without damping, and has a phase lag ϕ compared to the motion of the undamped oscillator. Although the oscillator was released at

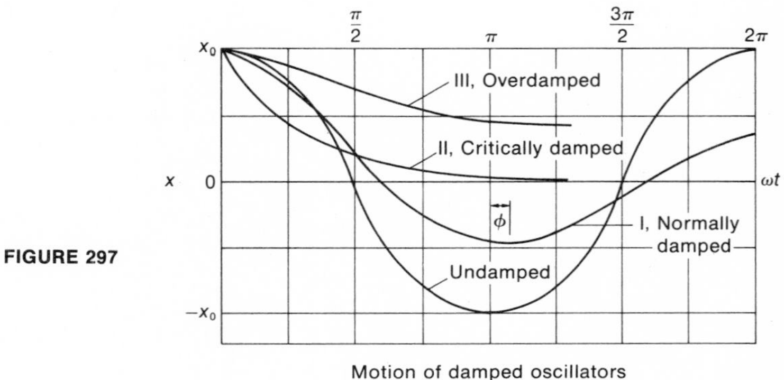

FIGURE 297

Motion of damped oscillators

$x = X_0$ it has an apparent amplitude $(\omega_0/\omega_\gamma)X_0$ greater than X_0, and its actual amplitude during motion is confined within the decreasing envelope $e^{-(\gamma/2)t}$.

Case II: $\omega_0 = \gamma/2$, *the critically damped oscillator.* When $\omega_0 = \gamma/2$, $\omega_\gamma = 0$. In such a case, equation 3-64 reduces to

$$x = X_0\, e^{-(\gamma/2)t}. \qquad (3\text{-}66)$$

Equation 3-66 describes critically damped harmonic motion. (See curve II, Figure 297.) There is no oscillatory term; the motion simply degrades, with a slope on the x-t graph equalling $-(\gamma/2)x$.

Case III: $\omega_0 < \gamma/2$, *the overdamped oscillator.* When $\omega_0 < \gamma/2$, $\omega_0^2 - (\gamma^2/4)$ is negative, ω_γ^2 as defined above is negative, and $(\omega_\gamma^2)^{1/2}$ is imaginary. We therefore must replace the ω_γ by $i\omega_\gamma$. (Note that this redefines ω_γ as $(\gamma^2/4 - \omega_0^2)^{1/2}$, and also implies that $\gamma/2 > \omega_\gamma$.) Going back to equation 3-56 to make this replacement, we have

$$x = e^{-(\gamma/2)t}\,(A_1\, e^{-\omega_\gamma t} + A_2\, e^{\omega_\gamma t}).$$

When we put in our boundary conditions, we find that

$$A_1 = -X_0\, \frac{(\gamma/2) - \omega_\gamma}{2\omega_\gamma}$$

and

$$A_2 = X_0\, \frac{(\gamma/2) + \omega_\gamma}{2\omega_\gamma}.$$

Then

$$x = e^{-(\gamma/2)t}\left(X_0\, \frac{(\gamma/2) + \omega_\gamma}{2\omega_\gamma}\, e^{\omega_\gamma t} - X_0\, \frac{(\gamma/2) - \omega_\gamma}{2\omega_\gamma}\, e^{-\omega_\gamma t}\right)$$

$$= X_0 \left[\frac{(\gamma/2) + \omega_\gamma}{2\omega_\gamma}\, e^{-(\gamma/2 - \omega_\gamma)t} - \frac{(\gamma/2) - \omega_\gamma}{2\omega_\gamma}\, e^{-(\gamma/2 + \omega_\gamma)t}\right].$$

This is the equation describing overdamped motion. (See curve III, Figure 297.) Again there is no oscillatory term. The mass simply subsides toward equilibrium position at a slower rate than when critically damped; its motion is the sum of two motions, one with a larger amplitude and a slower degradation, and one with a smaller amplitude but a faster degradation.

The method by which we got from our original equation of motion with damping to the working equations 3-64 and 3-65 is not the most elegant method available. However, the method does show that even when you don't know, or have forgotten, the more sophisticated approaches, you can bulldoze your way through unknown territory and come out with a useful answer.

Now that we have done the bulldozing, we should not have to travel that route again. Hereafter, when we are faced with differential equations of motion (or with differential equations describing physical activity other than motion) that are of the general linear form of the equation with which we started this section, we ought to be able to proceed immediately to a statement similar to equation 3-63: $\hat{x} = \hat{A}_1 e^{i\omega t} + \hat{A}_2 e^{-i\omega t}$. From that point on your work will entail relating the A's and the ω's to the data of your particular problem.

PROBLEM 467
How long does it take a critically damped oscillator to reduce its amplitude to $1/e$ of its original amplitude? (You will occasionally see this time referred to as the oscillator's "relaxation" time.)

PROBLEM 468
When the log of problem 392 fell overboard it entered the water vertically, casting end down, and subsequently bobbed up and down vertically in SHM in a calm sea. It was observed that in 28 sec of such motion the minimum height above water reached by the top of the log increased from $L/30$ to $2L/15$, L being the length of the log.

a. What was the average value of γ?
b. The change in height took place in exactly 5 cycles. From that can you make a good guess about the length of the log?

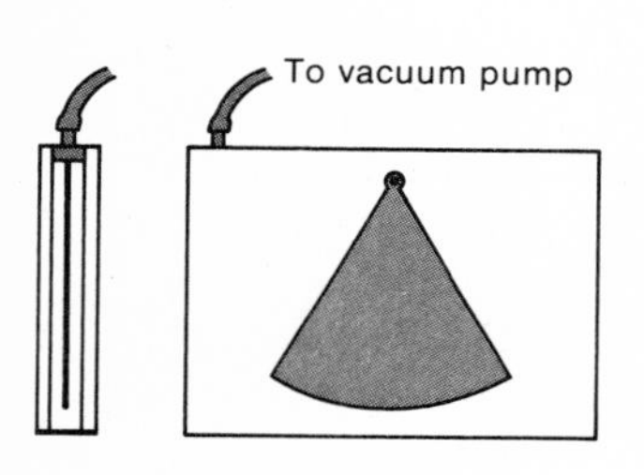

FIGURE 298

PROBLEM 469
A sector of a circle is cut from a stiff, plane sheet of metal and is mounted at the circle's center to swing as a pendulum in its own plane. It is placed centrally in a case whose sides parallel to the pendulum are plane plate-glass sheets (Figure 298); the separation between a pendulum surface and the adjacent glass is d. When the case is evacuated and the pendulum set swinging from θ_0, it is observed that bearing friction reduces the maximum amplitude by 1.0% of θ_0 every 41 sec. Then standard air is admitted to the case and the pendulum again set swinging from θ_0; it is now noted that at the end of 41 sec the maximum amplitude has been reduced to 0.90 θ_0. What is the viscosity of the air in the case, in terms of the separation d and the surface density σ of the metal used in the pendulum?

Forced Oscillations. We have seen that a damped oscillator left to itself gradually loses motion; if we want to maintain the motion of a damped oscillator, we must drive it. It is obvious we cannot do this driving with a constant force; for maintenance of uniform motion the applied force and the oscillation must be in some sort of synchronism.

Let us consider the damped oscillator of the previous section, now being driven by an oscillating force. For the equation of motion we will have

$$m\ddot{x} = -kx - c\dot{x} + F_0 e^{i\omega t},$$

which we can transform to

$$\ddot{x} + \omega_0^2 x + \gamma\dot{x} = \frac{F_0}{m} e^{i\omega t}.$$

Recognizing our need for synchronism, let us assume $\hat{x} = (A + iB)\, e^{i\omega t}$. (We use this form, rather than the general form discussed previously, which contained an $e^{-i\omega t}$, to avoid unnecessary work. Because of the $e^{i\omega t}$ character of our force, any term in our description of $\hat{x}$ that had an $e^{-i\omega t}$ in it would have to have a zero for a coefficient. By recognizing that ahead of time we save ourselves some work.) Performing our usual differentations and cancelling out the $e^{i\omega t}$ we have $-\omega^2 (A + iB) + \omega_0^2 (A + iB) + i\gamma\omega (A + iB) = F_0/m$. This leads to

$$A + iB = \frac{F_0}{m\,[\omega_0^2 - \omega^2 + i\gamma\omega]} = \frac{F_0\,[\omega_0^2 - \omega^2 - i\gamma\omega]}{m\,[(\omega_0^2 - \omega^2)^2 + \gamma^2\omega^2]}.$$

If we substitute this value for $A + iB$ back into our earlier assumption about x, and at the same time write $e^{i\omega t} = \cos \omega t + i \sin \omega t$, we have

$$\hat{x} = \frac{F_0\,\{[(\omega_0^2 - \omega^2) \cos \omega t + \gamma\omega \sin \omega t] + i\,[(\omega_0^2 - \omega^2) \sin \omega t - \gamma\omega \cos \omega t]\}}{m\,[(\omega_0^2 - \omega^2)^2 + \gamma^2\omega^2]}.$$

Taking only the real part, we finally have

$$x = \frac{F_0\,[(\omega_0^2 - \omega^2) \cos \omega t + \gamma\omega \sin \omega t]}{m\,[(\omega_0^2 - \omega^2)^2 + \gamma^2\omega^2]}, \tag{3-68}$$

a useful description as it stands. However, if as before we let

$$\cos \delta = \frac{(\omega_0^2 - \omega^2)}{\sqrt{(\omega_0^2 - \omega^2)^2 + \gamma^2\omega^2}}$$

and

$$\sin \delta = \frac{\gamma\omega}{\sqrt{(\omega_0^2 - \omega^2)^2 + \gamma^2\omega^2}},$$

we can after rearrangement arrive at

$$x = \frac{F_0 \cos (\omega t - \delta)}{m\,\sqrt{(\omega_0^2 - \omega^2)^2 + \gamma^2\omega^2}}, \tag{3-69}$$

where $\delta = \tan^{-1} \gamma\omega/(\omega_0^2 - \omega^2)$ and x is the departure from equilibrium position when the damped oscillator is driven by a force $F_0 \cos \omega t$.

The substitution of $\cos \delta$ and $\sin \delta$ for the coefficients of the terms in ωt is a valid manipulation, but there is a way of identifying δ that is more informative. Going back to our complex value for x, we can separate terms so that

$$\hat{x} = \frac{F_0\, e^{i\omega t}}{m\,\sqrt{(\omega_0^2 - \omega^2)^2 + \gamma^2\omega^2}} \left[\frac{\omega_0^2 - \omega^2}{\sqrt{(\omega_0^2 - \omega^2)^2 + \gamma^2\omega^2}} - i\,\frac{\gamma\omega}{\sqrt{(\omega_0^2 - \omega^2)^2 + \gamma^2\omega^2}}\right]$$

The terms within the brackets look like a $(\cos + i \sin)$ operation on an angle; calling

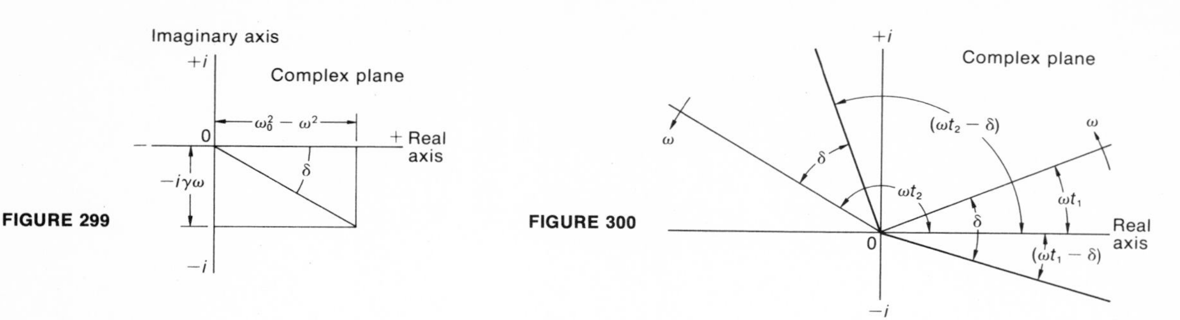

FIGURE 299

FIGURE 300

that angle δ, the terms in brackets then become $\cos \delta + i \sin \delta$ which can be summarized as $e^{-i\delta t}$. Hence

$$x = \frac{F_0\, e^{i\omega t}\, e^{-i\delta}}{m\, \sqrt{(\omega_0^2 - \omega^2)^2 + \gamma^2\,\omega^2}} = \frac{F_0\, e^{i(\omega t\, -\, \delta)}}{m\, \sqrt{(\omega_0^2 - \omega^2)^2 + \gamma^2\,\omega^2}},$$

which leads to the same real result as in equation 3-69.

Complex scalar quantities like δ can be understood more clearly by considering them in a complex plane coordinate system which has an imaginary i axis perpendicular to a real axis (Figure 299). In this plane the coordinates for δ are $(\omega_0^2 - \omega^2)$ and $-i\gamma\omega$, and it is easy to see the relationships for $\sin \delta$, $\cos \delta$, and $\tan \delta$. For any particular situation these coordinates do not change with time, so that δ is a constant.

However, in our evaluation of x we are not interested in δ alone, but in $(\omega t - \delta)$, and obviously this does change with time. Since ωt can be identified from $e^{i\omega t} = \cos \omega t + i \sin \omega t$, ωt can also be plotted on a complex plane, and hence $(\omega t - \delta)$ can be so plotted. The change in ωt, and hence in $(\omega t - \delta)$, with time can be depicted by having the radius vector for ωt rotate counterclockwise about the origin, dragging the $(\omega t - \delta)$ vector behind it (Figure 300).

We can now combine the angular relationship shown in Figure 300 with an appropriate scale for the length of the rotating radius vector, and turn our complex plane into a device for visualizing how complex physical relationships change with time. For example, the amplitude of the motion of m discussed above is

$$\frac{F_0}{m\, \sqrt{(\omega_0^2 - \omega^2)^2 + \gamma^2\,\omega^2}}.$$

Making our radius vector for $(\omega t - \delta)$ equal in scale length to that amplitude (Figure 301), the projection of the radius vector on the real axis gives the value of x as determined by equation 3-69.

When we plot the amplitude of x against the ratio ω/ω_0, we get the interesting graph shown in Figure 302. From either equation 3-69 or the graph we can obtain the following information about forced oscillators:

1. The response of the oscillator lags the application of the force by a phase angle δ. For $0 \leqslant \omega < \omega_0$, $\delta < \pi/2$; when $\omega = \omega_0$, $\delta = \pi/2$; for $\omega > \omega_0$, $\delta \rightarrow \pi$.

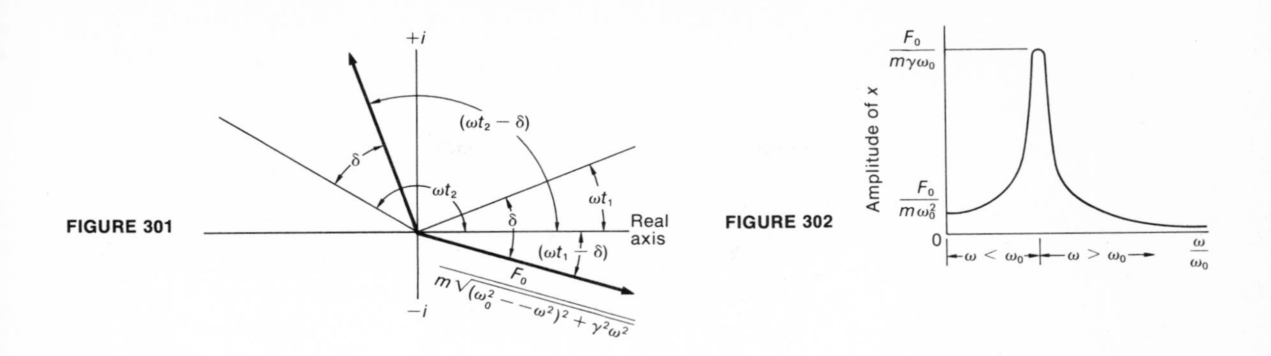

FIGURE 301

FIGURE 302

2. For values of ω very near to ω_0, the response increases dramatically; the steepness of the increase, and its ultimate value, are governed by the value of γ. The condition for maximum response—that $\omega = \omega_0$—is called resonance.

3. When $\omega << \omega_0$, the amplitude of response is very nearly that of the unforced, undamped oscillator; for $\omega > \omega_0$, the amplitude of response approaches zero. In other words, for $\omega >> \omega_0$, the oscillator is practically motionless.

Equation 3-69 is the steady-state solution of the equation of motion of a forced oscillator, and is usually the only solution we are interested in when dealing with forced oscillators. It is called the steady-state solution because its maximum amplitude of response does not change with time. However, a mathematician would recognize that it was an incomplete solution, because it contains no constants of integration. (There should have been two, since we got our zero-order solution from a second order differential equation.) We can repair this deficiency by adding to our steady-state solution the transient solution of the normally damped oscillator (equation 3-65). (Since the equation of motion for the normally damped oscillator equalled zero, we are thus just adding zero to our forced oscillator equation of motion, which doesn't change the description of our forced oscillator.) Thus the complete solution for the response of a forced oscillator is

$$x = e^{-(\gamma/2)t}\frac{\omega_0}{\omega_\gamma}X_0 \cos(\omega_\gamma t - \phi) + \frac{F_0 \cos(\omega t - \delta)}{m\sqrt{(\omega_0^2 - \omega^2)^2 + \gamma^2\omega^2}}. \qquad (3\text{-}70)$$

Because of the $e^{-(\gamma/2)t}$, the first term degrades with time, leaving only the second term to describe the steady-state motion.

PROBLEM 470

We alter the vibration-testing machine of problem 428 by inserting a spring of spring constant k between the driving linkage and the testing tray, as shown in Figure 303. The tray has a mass m. When the motor has an angular speed ω,

a. what is the motion of the testing tray;

b. how does the maximum acceleration felt by an object mounted in the tray compare with the maximum acceleration felt in the directly driven case described in problem 428?

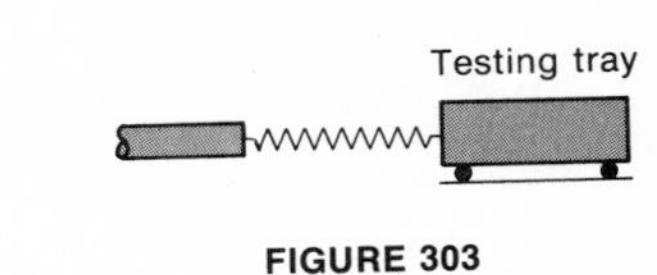

FIGURE 303

PROBLEM 471

The log of problem 468 is floating at rest in a calm sea when wave action from a distant storm arrives. The waves passing the log give to the water surface at the log a vertical motion described by $y = y_0 \cos(2\pi/T)t$. When T is timed at 10.5 sec, what is the total vertical movement of the top of the log? *Hint:* The driving force comes from the change in the buoyant force on the log due to rise or fall of the water surface relative to its equilibrium position on the log. (You will need data that can be found in the answers to problem 468.)

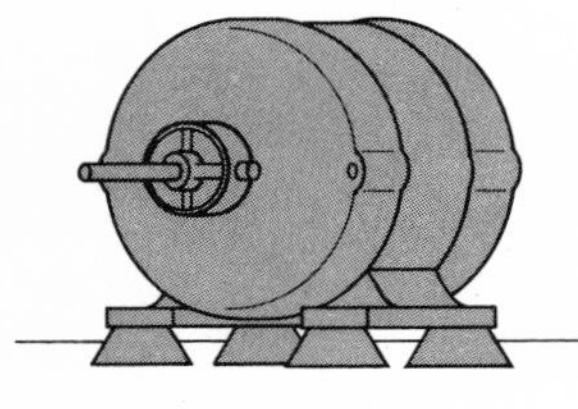

FIGURE 304

PROBLEM 472

Almost any commercial electric motor has some rotational imbalance that, when the motor is running, results in a vibration with the frequency of the motor. If you wished to isolate this vibration from the environment by using flexible motor mounts, as shown in Figure 304, you would look for motor mounts of what value for k? The mass of the motor is M, and its speed, ω.

PROBLEM 473

As a result of complaints about building vibration from office workers in an office above a machine shop, you are assigned to determine which machines are causing the trouble. Starting with the hope that only one machine may be the offender, and assuming that machine can be identified by the vibrational frequency, you hang a simple spring with weight hanger from a floor joist and start slowly piling small weights on the hanger. As the spring stretches a distance L it starts violent oscillations and the weights all fall off. What was the frequency of vibrations of the floor joist?

4

THE INTEGRATION OF FORCES

WORK

In our discussion of equilibrium, the term "equilibrium" was applied equally aptly to bodies at rest and to bodies whose vector momenta remained unchanged. For both cases, it sufficed to say $\Sigma F = 0$ and $\Sigma \tau = 0$. It is now time, however, to discuss an important difference between these two cases.

For a body at rest in equilibrium nothing happens—the body goes nowhere, the forces on it do nothing except exist. However, a body moving with unchanging momentum requires that any forces acting on it must move also; we can also observe that inevitably other things happen. If a body is pushed across the floor at constant speed, the push moves with the body and both the body and the floor become warmer (and perhaps slightly smoother). If a body is pushed up a smooth ramp at constant speed, the push must move with the body and the body is moved farther from the center of the earth. If a wheel on a threaded shaft is turned at constant angular speed, the turning effort must rotate with the wheel and the turning shaft compresses a clamp. In all of these cases, and in many more you can think of, there is a direct relation between a movement of a force and the result of that movement.

This relationship is even more obvious in dynamical situations. We have already discussed a number of problems in which a force accelerated an object; here again the force had to move with the object and the result was a change in the object's momentum.

In all of these illustrations the results varied, but the cause was the same—a force moved through a distance, or the lever arm of a force moved through an angle. This common cause for all of the various "results" was the product of force times distance moved, or torque times angle turned. This product is called work. This is a limited, technical use of the term "work," and bears no relation to the common use of the term. In this limited sense work is essentially described by $W = FS$ or $W = \tau\theta$.

The equation $W = FS$ has very limited usefulness, because it does not convey the necessary restriction that, to do work, a force and its displacement must be colinear. A body moving on a horizontal surface is supported by a vertical force, but this vertical force does not move (a force translated parallel to itself does not "move"), does no work, is not involved in any results. In problem 312 a force pushed horizontally on a box that moved up an incline at constant speed. If we resolve F into its components parallel to and perpendicular to the incline, we see that only the component parallel to the incline moves (Figure 305). Thus only the $F \cos \theta$ part of F does work.

FIGURE 305

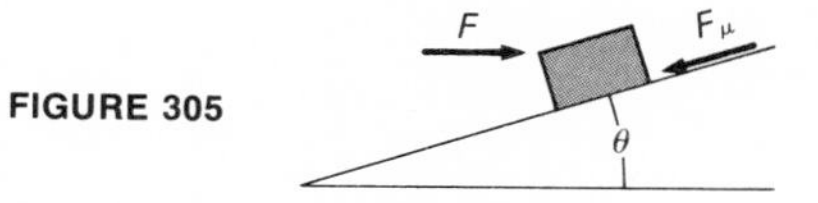

Is that the only work done? No, there is the force of friction parallel to the plane also; that force retreats with the box as it moves up the incline, so that it does work $F_\mu \times (-S)$. Thus you can have negative work—work done against a force—as well as positive work—work done by a force.

All of these aspects of the relations among work, force, and distance moved can be brought together in the statement

$$W = \int_{S_1}^{S_2} \vec{F} \cdot \vec{ds}. \tag{4-1}$$

This is the vector dot product, which is a scalar; work has only a numerical value, but no direction.

The formal dot product reduces to

$$W = \int F \cos (F, ds)\, ds, \tag{4-2}$$

where (F, ds) is the least angle between $\vec{F}$ and $\vec{ds}$, measured in the plane that contains both $\vec{F}$ and $\vec{ds}$. When F is not a function of the displacement and $\int_0^S ds$ is a linear displacement, then the above equation reduces to

$$W = F \cos (F, S)\, S.$$

Similarly, for rotational motion

$$W = \tau\theta,$$

where the vectors for $\vec{\tau}$ and $\vec{\theta}$ lie along the same axis.

In most cases the work we will be interested in is the positive work being done by the mechanical or electrical forces being applied in the direction of the displacement. However, when we are attempting to make an inventory of the changes in potential energy, to be discussed later, we will be interested in the negative work done by the conservative* forces against the motion of a body. An example of such negative work can be found in the body being pushed up a plane (Figure 305): $-mgd \sin \theta$ (where d is the distance the body moves along the plane) is being done by the conservative gravitational force.

When F is expressed in newtons and S in meters, or when τ is given in meter newtons, then the standard MKSC unit of work is the newton meter (N m). One N m is also called 1 joule. In some cases you will find N m restricted to use as a work unit, and joule restricted to use as an energy unit. However, from now on we shall become increasingly aware that work and energy are freely interchangeable; it therefore seems proper to regard N m and joule as also freely interchangeable, using whichever is most appropriate for the given data.

In the cgs system the unit of work is 1 dyne cm = 1 erg. Obviously 1 joule = 10^7 ergs. In the practical English system the unit of work is 1 lbwt ft. (You will frequently see this term reversed, as ft lbwt.) The dimensions of work are ML^2T^{-2}. (You will find another useful unit of work in problem 477.)

*Conservative forces are natural forces that are a function of position only. So far we have discussed three forces that fit this category: gravitational, electric, and perfectly elastic. (Actually, elastic forces are a mechanical manifestation of the electric forces that are developed in solids when the solids are deformed from a state of internal equilibrium.)

PROBLEM 474

A steamer trunk of weight W rests on the floor of the pier shed of the French Line in New York City. There is a coefficient of sliding friction μ between trunk and floor. If the baggage porter moves the trunk a distance S to the correct collecting position, how much work does he do

a. if he leans down and pushes horizontally;
b. if he leans *on* the trunk, at 30° with the horizontal, when he pushes?

PROBLEM 475

Later, a stevedore put the trunk of problem 474 on a wheeled cart and pushed the cart up a ramp and through a loading port in the side of the S.S. *France*. If he pushed horizontally, how much work did he do in moving the trunk and the cart (which weighed one-fourth as much as the trunk) a distance S along the ramp, which sloped up at an angle θ with the horizontal?

PROBLEM 476

A boy who owns a bicycle that weighs one-fifth as much as he does lives at the foot of a street that slopes upward 6° above the horizontal. One day he visits a friend who lives at the foot of a street with a 6.°4 slope upward. When he tries out his friend's new bike he decides that he gets just as tired (i.e., he feels he has done just as much work in the technical sense) riding to the top of the block on his friend's street as he does riding to the top of an equal block at home on his own bike. What was the weight of his friend's bike?

PROBLEM 477

A nonconducting belt runs vertically upward at a constant speed in a region in which there exists a constant electric field E vertically upward. How much work is required to carry a free electron upward on the belt to a height h?

SOLUTION

The physics of the problem is simple. Because of the upward field E, there is on the electron a downward force $q_{el}E$; then, to move the electron upward at constant speed an upward force of $q_{el}E$ must be supplied by the belt (ignoring the weight of the electron as negligible compared to the electric force). Then, to carry the electron to a height h the belt must supply $W = q_{el}Eh$.

The interesting part of this problem comes when we assign units to the work done. You will recall that one of the MKSC units in which E could be expressed was volts per meter; hence if h is in meters, the product Eh is simply volts. Then $W = q_{el}Eh$ is electron charge times volts, which we shorten to electron volts. Thus work, and its equivalent energy, can be expressed in the MKSC system as electron volts. In much of modern physics the electron volt, abbreviated ev, is the most useful unit of energy. For larger units of energy we have 1 Mev $= 10^6$ ev, and 1 Gev (or in older usage 1 Bev) $= 10^9$ ev.

We can relate the ev unit of work to our standard MKSC unit of work—the joule—as follows. 1 ev $= 1\ q_{el} \times 1$ unit of $E \times 1$ m. But the fundamental unit for E is newton per coulomb. Putting that, and the value of q_{el} in coulombs into the equation, we have 1 ev $= 1.60 \times 10^{-19}$ C $\times$ 1 N C^{-1} $\times$ 1 m $= 1.60 \times 10^{-19}$ N m $= 1.60 \times 10^{-19}$ joules $= 1.60 \times 10^{-12}$ ergs.

PROBLEM 478

A cylindrical redwood log of length L and weight W, to be used in constructing a pier, is lowered vertically between guides into the water until it floats in equilibrium with half of its length projecting above water. How much work is required later to sink it vertically until only one-tenth of its length is above water?

PROBLEM 479

In the problem of the chain hoist and the crate, problem 409, how much work was done in dragging the crate from the point where the line made an angle of 30° with the horizontal to the point where the line was at 60° with the horizontal? The crate weighed 300 kgwt, the coefficient of friction between crate and floor was $1/\sqrt{3}$, and $h = 2.7$ m.

PROBLEM 480

The usual starting condition for an harmonic oscillator is for $x = X_0$ and $\dot{x} = 0$ at $t = 0$. Since the restoring force on the oscillator equals zero only at $x = 0$, work had to be done to get the oscillator out to $x = X_0$ and ready to start. How much work was required to do this

a. if the oscillator was driven by a spring of spring constant k;
b. if the oscillator was a simple pendulum of length L and was started from θ_0?

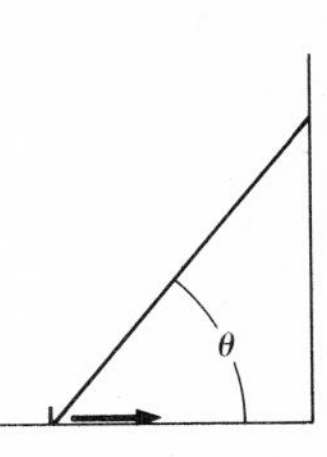

FIGURE 306

PROBLEM 481
A uniform ladder rests at an angle θ with the horizontal against a smooth vertical wall and a peg on a smooth horizontal floor. If a horizontal force is applied to the foot of the ladder as shown in Figure 306, how much work is required to push the ladder to rest in an upright position against the wall?

PROBLEM 482
The foot of a ladder is an awkward place to push. To be more practical, the ladder at rest at an angle θ in problem 481 is now grasped at a point d along the ladder from its foot and pushed horizontally (Figure 307). Now how much work was done to move the ladder to rest in an upright position against the wall?

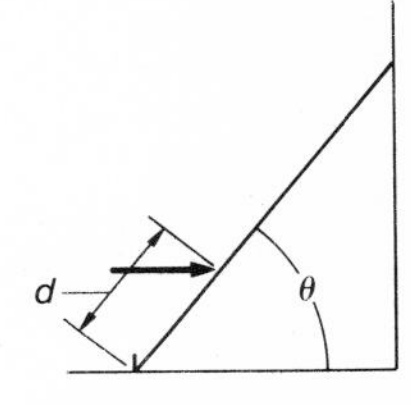

FIGURE 307

PROBLEM 483
To be even more realistic, we now admit that there is a coefficient of sliding friction μ between the ladder and the floor; the wall is still considered smooth. Again the ladder at rest at an angle θ is grasped at a point d along the ladder from its foot and pushed horizontally. Now how much work was done to move the ladder to rest in an upright position against the wall?

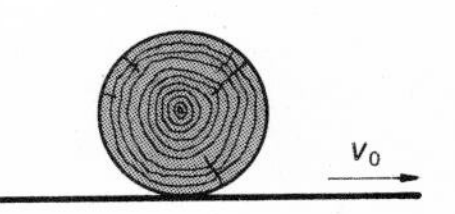

FIGURE 308

PROBLEM 484
A conveyor belt is running horizontally at a constantly maintained linear speed v_0. Uniform, solid cylindrical logs are gently lowered at rest transversely onto the belt and there released (Figure 308). There is some friction between belt and log. When a log reaches the condition of rolling without slipping on the belt,

a. how fast is the axis of the log moving?
b. How much work against friction has been done per log?

Equilibrium Solutions by the Method of Virtual Work

We have already discussed the calculation for equilibrium by making the vector sum of the applied forces, or the scalar sum of the force or torque components, equal to zero. There is an analogous method for equilibrium determination based on making equal to zero the scalar sum of the work done during an infinitesimal deformation. This method is called virtual work—"virtual" because no work actually is done. Instead we devise a thought operation in which we imagine a deformation, or dislocation, which induces both positive and negative work. We of course require that this imagined operation shall make no real change in the physical situation, which means that the scalar sum of the induced work shall equal zero.

For example, if you solved problem 481 by $W = \int_{L\cos\theta_0}^{0} F\,dx$, you first had to obtain F as a function of x. In other words, you had to solve problem 481 as an equilibrium problem first. You obtained this value of F—which had to be

$$F = \frac{W \cot\theta}{2} = \frac{Wx}{2\sqrt{L^2 - x^2}}$$

either by $\Sigma\tau = 0$ (the easiest way) or by a combination of $\Sigma\tau = 0$ and $\Sigma F = 0$. Now let us evaluate F by the virtual work method.

Let us imagine pushing the foot of the ladder in horizontally a distance $-dx$ (Figure 309); this would raise the c.m. of the ladder a distance dy. Looking at the

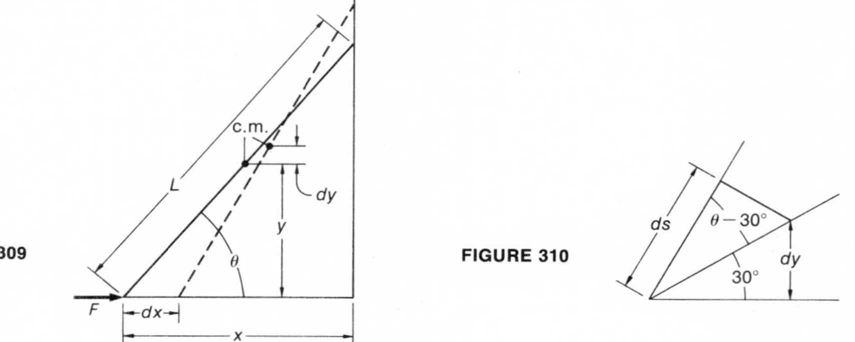

FIGURE 309

FIGURE 310

geometry we have $x = L\cos\theta$ and $-dx = L\sin\theta\,d\theta$, and $y = (L/2)\sin\theta$ and $dy = (L/2)\cos\theta\,d\theta$. The virtual work $F(-dx)$* and virtual work $-W\,dy$ were done when we pushed; no work was done by the perpendicular forces at the floor and the wall, since these forces did not move in their own direction. Virtual work must sum to zero, so we have $F(-dx) - W\,dy = 0$. When we substitute in these values, we get $F\,L\sin\theta\,d\theta - W\,(L/2)\cos\theta\,d\theta = 0$; from which we get

$$F = \frac{W}{2}\cot\theta = \frac{Wx}{2\sqrt{L^2 - x^2}}$$

as before.

For another example, let us look again at problem 385. If in imagination we shorten the rope a distance ds (Figure 310), we do virtual work $F\,ds$. (We assume that the rope is in tension, so that the force in the rope is in the direction of the shortened distance ds, and hence the virtual work is positive.) By shortening the rope an amount ds we raise the cart a distance dy, where

$$dy = \frac{ds}{\cos(\theta - 30)}\sin 30.$$

*The change in x, dx, is − because the change is opposite to the direction of increasing x. It may be clearer to you if you just recognize that F moves in its own direction, and hence the work it does must be positive. In other words, the virtual work done by F is $F|dx|$.

Then

$$F\,ds - \frac{W\,ds \sin 30}{\cos(\theta - 30)} = 0,$$

or

$$F = \frac{W \sin 30}{\cos(\theta - 30)}.$$

For $\theta = 60°$, this reduces to

$$F = \frac{W}{\sqrt{3}}.$$

As you can guess from these examples, the method of virtual work entails reducing a physics problem to a problem in what might be called simple linear differential geometry, in which a relationship among infinitesimal displacements provides a key to the solution. Virtual work is always used to evaluate a force or a torque. The analytical statement for virtual work—that $\Sigma\, dW = 0$—can be rephrased by saying that the work performed by a force or torque undergoing an infinitesimal displacement must equal the work resulting from that displacement.

PROBLEM 485

The automobile of problem 397 has a dead battery, so in order to get it up over the curb you and several friends lean down and push horizontally on the rear bumper, which is just $2h$ above the pavement.

a. What initial force must the group supply to just start the front wheel up over the curb? *Suggestion:* Use the method of virtual work. As the front wheels start to roll about the point of contact with the curb edge, there is an easy determination of the relation between the distance the car is raised and the distance the horizontal push at the rear bumper moves.

b. Why does the answer to this problem seem inconsistent with the answer to problem 397?

PROBLEM 486

The poor man's stump puller (poor in that he has more timber than he has money to buy dynamite) is as shown in Figure 311. The framework is made of heavy timbers; TS and AB are taut chains with A and C at midpoints. When a force F is applied at C directly downward, what is the pull P on the stump?

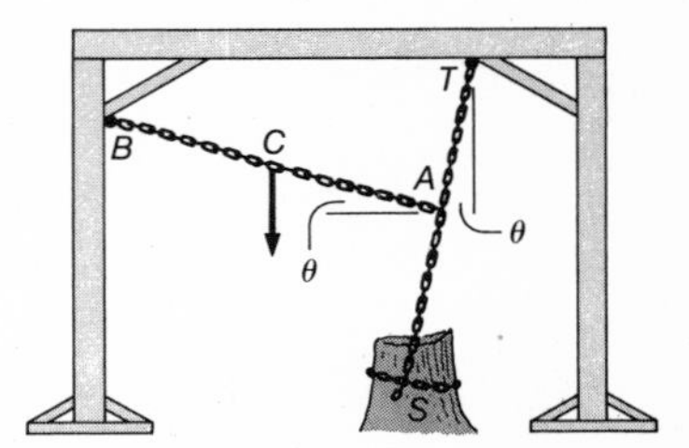

FIGURE 311

PROBLEM 487

A simple roof truss is made up of three equal equilateral triangles as shown in Figure 312. The truss members are pin-connected and of negligible weight. If a load W is suspended from the middle of member BC, evaluate the force in member DE by the method of virtual work.

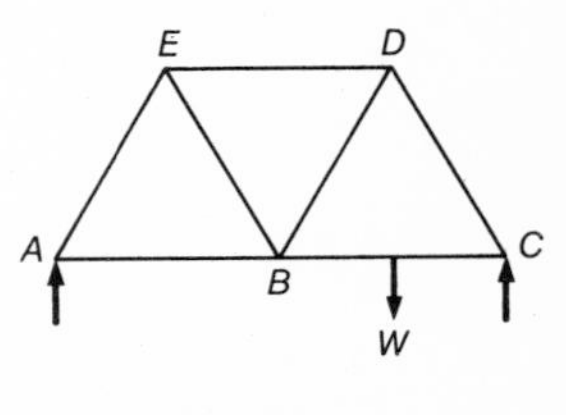

FIGURE 312

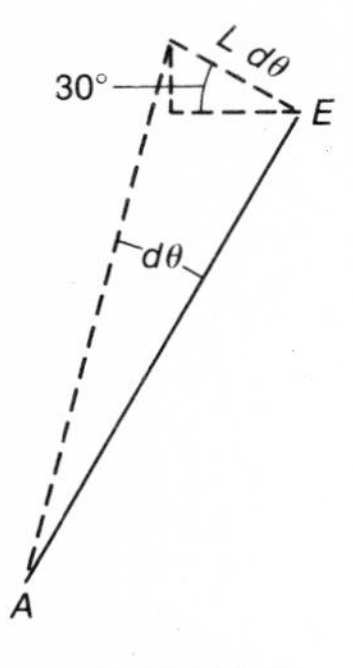

FIGURE 313

START OF SOLUTION

Since we are interested in the force in DE, only DE can be allowed to change length. This means that triangles ABE and BCD must remain as unchanged, rigid structures. Thus to change the length of DE, these rigid triangles must rotate about A and C respectively. This rotation will cause B to move up or down, which in turn will raise or lower W. If we rotate AB counterclockwise an infinitesimal angle $d\theta$ (which of course also means that BC rotates clockwise by $d\theta$), B will rise a distance $L\,d\theta$, where L is the length of one of the sides of a triangle (Figure 313). At the same time, E will move a distance $L\,d\theta$ perpendicular to AE, which brings about a change in length for DE of $L\,d\theta \cos 30$. The same change of length simultaneously occurs at D. Thus the force in DE times $2L\,d\theta \cos 30$ must equal the work done in raising W. At this point the only question still remaining is whether F_{DE} is tension or compression. If it is tension, work must be done on DE to lengthen it; if it is compression, DE will do work when lengthening. If you get a $-$ sign in your answer it means you guessed incorrectly.

PROBLEM 488

Solve problem 411, the World's Fair ornament problem, by the method of virtual work. If you had already solved this problem by the customary $\Sigma F = 0$ methods, you will probably find on comparison that the virtual work method is simpler.

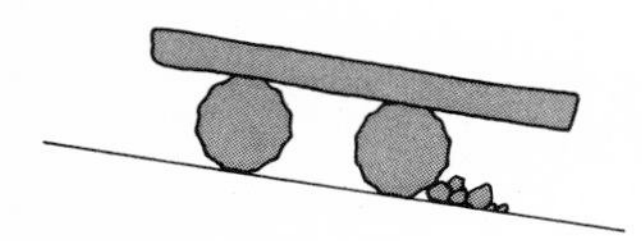

FIGURE 314

PROBLEM 489

On the Ionian coast of Italy, near Crotone where Pythagoras was born, are many remnants of the rich Greek civilization that flourished there in the centuries BC. A modern peasant comes upon the ruins of an old temple, and finds a fallen marble lintel resting on two pieces of fallen marble pillar, as shown in Figure 314. The hard ground has a slope of about 6°, but rubble behind a pillar fragment keeps the group from moving. The peasant decides he can use the lintel, so he hitches a team of oxen to the uphill end of the lintel, with the pull from the oxen parallel to the ground.* How much pull did the oxen have to exert to move the lintel uphill at constant speed, using the pillar fragments as rollers? The lintel weighed $2W$, and the pillar fragments each W.

*By so doing the peasant ran a heavy risk. The Italian government has a strict law forbidding just such pillaging of Greek and Roman antiquities.

Work is of fundamental interest in science and technology because it is a prime mechanism by which desirable physical changes can be made in a physical situation. Work is usually classified by its results: (1) conservative work,* in which

- the speed of a body is increased, or
- a body is moved in a field to a position of different potential, or
- the shape of a perfectly elastic body is changed;

(2) nonconservative work, which is an expense instead of an asset, and is irreversible.

*Conservative work is a capital asset; it is work that is reversible, that is available for recapture at full value. Doing conservative work may entail changing the form of some of the energy in a system, but it does not result in any change in the total amount of usable energy in the system.

Work against friction, and against viscosity, are obvious examples of nonconservative work. It is possible in theory to compute the probability of a sandpapered surface reassembling the sawdust and restoring the original surface and offering back to the outside world the energy that went into the sandpapering, but there is only about an n^{-n} ($n \to \infty$) probability that you will ever be interested in such a computation.

It would be pleasant to be able to make the statement about conservative work increasing the speed of the body symmetrical, by including under conservative work the negative work done in decreasing the speed of a body. However, such negative work is not always recapturable. If you stop a body by letting it push against an ideal spring, the negative work done by the spring (or if you would rather look at it this way—the work done on the spring) is stored in the spring and is available for reuse. However, if you brace yourself to catch and stop your small son who is running down the walk to greet you, the physical work you do is not stored, is not recapturable, and is not conserved. (Note that this work isn't "lost"; it can be accounted for, but it is not reusable.) This type of nonconservative work is covered more extensively as energy lost in inelastic collisions; see page 275.

THE CONSEQUENCES OF WORK

Kinetic Energy

When we do work to change the linear speed of a body, for the $\vec{F}$ in $W = \int \vec{F} \cdot \vec{ds}$ we choose the force that we have learned governs the changes of motion — $\vec{F} = \vec{dp}/dt$. Then

$$W = \int \frac{\vec{dp}}{dt} \cdot \vec{ds} = \int dp_s \frac{ds}{dt} = \int v_s \, dp_s.$$

To be as general as possible, let us use the complete relativistic form for p_s;

$$p_s = \frac{mv_s}{\sqrt{1 - v_s^2/c^2}},$$

and

$$dp_s = \frac{m \, dv_s}{(1 - v_s^2/c^2)^{3/2}}.$$

When we substitute this in the above and make the appropriate change of variables, we have

$$W = \int_{v_0}^{v_1} \frac{m \, v_s \, dv_s}{(1 - v_s^2/c^2)^{3/2}} = \frac{mc^2}{\sqrt{1 - v_1^2/c^2}} - \frac{mc^2}{\sqrt{1 - v_0^2/c^2}}. \tag{4-3}$$

The term

$$\frac{mc^2}{\sqrt{1 - v^2/c^2}} = \frac{mc^2}{\sqrt{1 - \beta^2}} = \gamma \, mc^2$$

is the total linear relativistic energy E of the body of mass m. You will note that when the mass is at rest and $v = 0$, the mass still has $E_0 = mc^2$. This is often called the rest energy, or the rest mass energy.

Another useful way of interpreting $E_0 = mc^2$ is as an equation defining a conversion factor between energy E_0 and the inertial property of mass; E_0/c^2 is a measure of inertia. We will find this interpretation helpful later on when we try to understand why light can exert pressure, and we have to accept the experimental fact that photons possess momentum, and hence have the property of inertia, even though we know their mass is zero.

Coming back to equation 4-3, we have $W = E_1 - E_0 = \Delta E$. If we expand the terms in equation 4-3 by the usual method we have

$$\begin{aligned}\Delta E &= mc^2 (1 + \tfrac{1}{2}v_1^2/c^2 + \tfrac{3}{8}v_1^4/c^4 + \ldots) \\ &\quad - mc^2 (1 + \tfrac{1}{2}v_0^2/c^2 + \tfrac{3}{8}v_0^4/c^4 + \ldots) \\ &= \tfrac{1}{2}mv_1^2 (1 + \tfrac{3}{4}v_1^2/c^2 + \ldots) - \tfrac{1}{2}mv_0^2 (1 + \tfrac{3}{4}v_0^2/c^2 + \ldots).\end{aligned}$$

The collection of terms $\frac{1}{2}mv^2 (1 + \frac{3}{4}v^2/c^2 \ldots)$ is called the linear kinetic energy of m, symbolized by T. This definition applies over the complete range of speeds from $v = 0$ to $v \rightarrow c$. However, it is obvious that for mundane speeds the ratio v^2/c^2 is negligible compared to 1; hence for all reasonable requirements for accuracy and for all except relativistic speeds, the definition of linear kinetic energy reduces to the familiar

$$T_{\text{lin}} = \tfrac{1}{2}mv^2. \tag{4-4}$$

Putting the above all together, for any particular mass m we have $W = \Delta E = \Delta T$. This statement applies to both positive and negative changes in linear speed; if ΔT is negative, then W is negative. This can be expressed graphically:

$$\text{work} \rightleftarrows \text{change in kinetic energy.}$$

The rotational analogue of $\int \vec{F} \cdot d\vec{s}$ is $\int \vec{\tau} \cdot d\vec{\theta}$, and the rotational analogue of $T_{\text{lin}} = \frac{1}{2}mv^2$ is $T_{\text{rot}} = \frac{1}{2}I\omega^2$. (This can be obtained independently or by a transformation from linear to rotational coordinates.) In general a mass can have both linear and rotational kinetic energies at the same time—a wheel on your moving automobile has a linear kinetic energy of the c.m. of the wheel plus a rotational kinetic energy about a transverse axis throught that c.m. Thus we should write, for a complete and general statement of the relation between work and change in kinetic energy, that

$$W = \Delta T_{\text{kin}} + \Delta T_{\text{rot}} = \tfrac{1}{2}mv_1^2 + \tfrac{1}{2}I_1\omega_1^2 - \tfrac{1}{2}mv_0^2 - \tfrac{1}{2}I_0\omega_0^2. \tag{4-5}$$

The dimensions of energy, ML^2T^{-2}, and the MKSC units for energy, newton meter or joule or electron volt, are the same as for work.

PROBLEM 490

Traffic police investigating an automobile collision determine that car A left skid marks 140 ft long on dry pavement before hitting car B.

The tires on car A were considered good enough to produce a coefficient of sliding friction against dry pavement of 0.70. What was the minimum speed the police calculated for car A before its brakes were applied?

PROBLEM 491
If you solved part (a) of problem 369, the phonograph record problem, by starting with $\tau = I\ddot{\theta}$, now try it by starting with $\int \tau \, d\theta = \frac{1}{2}I\omega^2$. Compare the amount of work and the clarity of solution in the two methods.

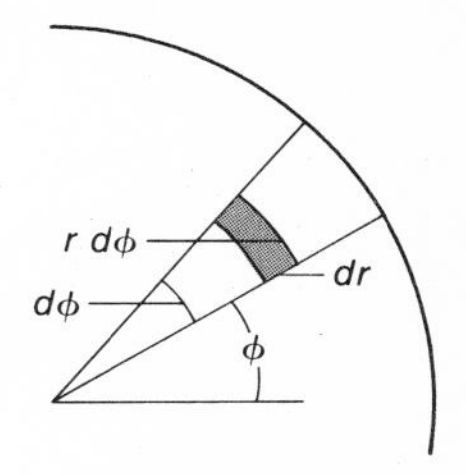

FIGURE 315

START OF SOLUTION
As can be seen from the diagram in Figure 315, an element of mass of the disc is $\sigma r\, dr\, d\phi$, the weight of such an element is $\sigma r\, dr\, d\phi\, g$; the horizontal force acting on this element is $\mu\sigma r\, dr\, d\phi\, g$; and the torque of this force about the vertical axis of the disc is $r\mu\sigma r\, dr\, d\phi\, g$. Thus

$$\int \tau d\theta = \int_0^{\pi/20} \left(\int_0^R \int_0^{2\pi} \mu\sigma g r^2 \, dr\, d\phi \right) d\theta,$$

which integrates to $\mu mgR/30$ after setting $\pi R^2\sigma = m$.

The alternative solutions to the phonograph problem (369 and 491) illustrate the interesting fact that a large proportion of dynamical problems can be solved either by the application of Newton's Second Law or by energy considerations. However, there is no standard test to tell you ahead of time which method will provide the simplest or the most informative solution. Most scientists and engineers who do much problem solving tend to look for an energy solution first, possibly because their training has given them an automatic first reliance on the principle of the conservation of energy, possibly because their experience with the kind of problems they normally face has taught them that an energy inventory usually provides an easier solution.* Probably you will develop a bias for one of these methods on the basis of how you "see" a problem: some people have an instinctive feeling for forces and torques and "see" a problem in dynamical terms, others have an instinctive sympathy for the concept of energy balance.

*In situations in which all of the energy cannot be accurately catalogued, it is of course not possible to use an energy inventory method of solution.

PROBLEM 492
Problems 165, 166, and 178 were originally given to you as Newton's Second Law problems. Now solve them from the point of view of

work = change in kinetic energy,

and compare your current and former solutions.

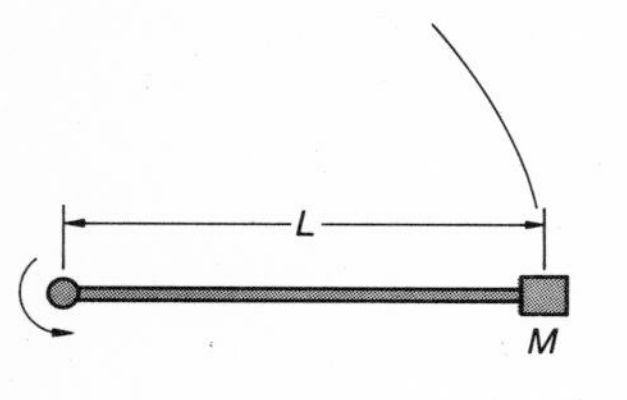

FIGURE 316

PROBLEM 493

A small block of mass M resting on a smooth horizontal surface is fastened to the end of a rigid bar also of mass M and of length L (Figure 316). The other end of the bar is connected to a vertical axle. The fastener which holds the block to the bar breaks under any force greater than F_0. When the axle is revolved, causing the block to move in a horizontal circle,

a. what is the axle's angular speed when the block breaks loose from the rod?

b. How much energy had been supplied to the system when the block broke loose? For simplicity, assume L remains constant throughout.

PROBLEM 494

The block-bar-axle assembly of problem 493 is now placed on a different horizontal surface, where the coefficient of sliding friction between block and surface is μ.

a. What now is the angular speed of the axle when the block breaks loose?

b. How far from the axis of rotation does the block come to rest? Again assume that L remains constant throughout.

PROBLEM 495

a. If you want to hold the percentage error in your calculation of T below 1.0%, for what maximum value of v can you still use the simple, nonrelativistic formula for T?

b. What is the linear kinetic energy, in *ev*, of an electron moving with the maximum speed determined in question (a)?

Potential Energy of Fields

A mass released in a gravitational field, or a charge in an electric field, "falls" toward a lower potential in accordance with the field forces on it, and acquires kinetic energy in the process. We could include this behavior under our "work done by forces = change in kinetic energy" considered in the previous section. However, it has been found economical in the long run to invent an intermediate concept—the concept of potential energy. We create potential energy by doing work or investing kinetic energy; we harvest potential energy by letting it do work or create kinetic energy.

Early in this book we discussed free fall in the earth's gravitational field; we learned how to compute the speed, and hence the kinetic energy, of a ball dropped from rest at the top of a cliff of height h. Where did this gradually increasing kinetic energy come from? Of course the immediate answer is that it is produced by the work being done on the ball by the force of the gravitational field. But before we can fully accept that answer we need to know the answer to an antecedent question: since the natural place for the ball is on the ground, at the lower potential, how did it get to a higher potential, at the top of the cliff, in the first place?

If it was carried up from the ground, a vertical force mg had to be supplied, and mgh of work had to be done. If it was thrown up, it had to be given a vertical speed of $v = \sqrt{2gh}$, which means that it had to be given a kinetic energy of vertical motion $= \frac{1}{2}mv^2 = \frac{1}{2}m\,(2gh) = mgh$. So in either case there had to be provided mgh of energy to offset the negative work (−mgh) of the force of gravitation.

When the ball is at rest at the top of the cliff our investment in energy is mgh. We know that from the top of the cliff the ball can fall and acquire kinetic energy. But as long as it stays at rest on the cliff it is only potentially capable of acquiring kinetic energy. This makes it logical to call our investment potential energy—an energy capability not yet called on to produce.

The dimensions of potential energy, and the usual MKSC units, joules or electron volts, are the same as for work or kinetic energy. Potential energy is of course a scalar.

Let us now put all this in quantitative form by defining the potential energy U by the equation

$$\int dU = -\int \vec{F}_F \cdot \vec{ds}, \tag{4-6}$$

where $\vec{F}_F$ is the natural force exerted by a field. The negative sign appears in front of the integral because we wish to adopt the convention that potential energy increases positively when we do work against a field.

Thus in the example about getting the ball to the top of the cliff,

$$\int dU = U_{\text{cliff}} - U_{\text{ground}} = -\int_0^h -mg\,dy = mgh.$$

(The natural force is $-mg$ because it is opposite in direction to the direction of dy.)

If h is so large that we have doubts about the constancy of the gravitational field between the surface of the earth and the top of the cliff, then we can go to the general form of the gravitational force:

$$\int dU = U_{\text{cliff}} - U_{\text{ground}} = -\int -\frac{GmM}{r^2}\,dr = GmM\left[-\frac{1}{r}\right]_R^{(R+h)}$$

$$= \frac{GmMh}{R(R+h)}$$

$$= \frac{GmM}{R^2}\,\frac{R}{(R+h)}\,h = mgh\,\frac{R}{R+h}.$$

(Again the natural force F_F is $-\,GmM/r^2$ because the force is opposite in direction to the direction of dr.)

Or consider the following question: how much work was done in problem 254, part (b), in bringing the charge $+q$ in from infinity to point B? Since work against natural forces involves a change in potential energy, we have

$$\Delta U = \Delta W = -\int_{\infty}^{s} \frac{q^2}{4\pi\epsilon_0 r^2}\,dr = \frac{q^2}{4\pi\epsilon_0 s}.$$

(It is worth noting that our answer is unaffected by our choice of path from infinity to B; an integration along all possible paths produces the one unique answer. Forces that permit this kind of integration are often called conservative forces, another name for what we have been calling forces derivable from a field.)

A more interesting question to ask about problem 254 is the following: how much work was required to bring in to P the charge whose value was computed in part (d) of that problem? You ought to work this out by straightforward integration, for the sake of your self-respect as a physics problem solver as well as to appreciate what follows. Choose a direction from infinity to P, determine an E_r along this direction resulting from the charges at A, B, and C, multiply this E_r by the charge $-(2/\sqrt{3})\,q$ to get the force F_r, and integrate $F_r\,dr$ from infinity to P. If you do everything correctly you should obtain for an answer $\Delta U = -3q^2/2\pi\epsilon_0 s$.

The examples given illustrate a general and very useful relationship: $\Delta V_{\text{elec}} = (1/q)\,\Delta U$ and $\Delta V_{\text{grav}} = (1/m)\,\Delta U$; the scalar potential is simply the potential energy per unit charge or per unit mass. Thus the scalar potential is a general value, whereas potential energy is a particular value tied to a particular charge or to a particular mass. The usefulness of this arises as follows: we can easily compute a difference in scalar potential between two points in space, and can then compute the difference in potential energy developed in moving along any path between these two points, simply by multiplying the difference in potential by the appropriate amount of charge or mass. Thus in the question posed about problem 254, the change in potential energy that occurred when the charge $-(2/\sqrt{3})q$ was brought in from ∞ to P is simply $-(2/\sqrt{3})q \times 3\sqrt{3}\,q/4\pi\epsilon_0 s$ (from part (c) of problem 254) $= -(3q^2/2\pi\epsilon_0 s)$, a much simpler way of getting the answer than the integration suggested earlier.

Since both potential and potential energy are defined by integrals, we always compute a change ΔV or ΔU; there is no absolute scale for V or U. However, in most cases it is convenient to establish by fiat a point at which V or $U = 0$; then in reference to such a point ΔV and V, or ΔU and U, are equivalent. (Whatever reference zero is chosen for V applies to the related U as well, since the only difference between V and U is a factor q or m.) Earlier it was indicated that ∞ is a convenient choice for $V_{\text{elec}} = 0$; another frequent choice is "ground," a conducting surface in circuits to which all potentials are referred. In extraterrestrial dynamics ∞ is also a standard choice for zero for V or U gravitational, but in the laboratory the convenient reference level will change from problem to problem. In the earlier discussion about the ball and the cliff it was most convenient to set $U_{\text{grav}} = 0$ at the ground; in satellite computations $U_{\text{grav}} = 0$ would probably be at the center of the earth; in a free-fall experiment in the laboratory you would probably make your zero reference point the release point for the falling weight.

PROBLEM 496

A small steel ball originally at rest is allowed to fall in a vacuum through a height h near the surface of the earth. What is the percentage increase in the ball's inertial mass at the end of the fall? (This is a coarse mechanical analogy of a test of General Relativity made possible by the Mössbauer technique.) *Suggestion:* if $m = E/c^2$, $\Delta m = \Delta E/c^2$.

PROBLEM 497

If the difference in water levels between the two canals connected by the ascenseur in problem 176 is 8.0 m,

a. what change in potential energy had occurred when the overfilled barge box reached the lower level?
b. Normally, when the machinery worked correctly, how would this ΔU have been taken care of?

PROBLEM 498

A cylindrical cork of height h and cross-sectional area A floats with its axis vertical in a vertical, cylindrical container of water. The total volume of water is equal to the volume of the cork. The cork is a loose fit in the cylinder, i.e., $A_{\text{cork}} \simeq A_{\text{water}}$, but water can flow past the cork. If the cork normally floats half-submerged, how much work is required to push it to the bottom of the container?

PROBLEM 499

What is the electric potential energy of an isolated conducting sphere of radius R carrying a total charge Q? *Hint:* Since the potential of the sphere with a charge q on it is $q/4\pi\epsilon_0 R$ the work to bring a charge dq from ∞ to R is $V\,dq = (q\,dq)/(4\pi\epsilon_0 R)$. For a total charge Q on the sphere, this should be integrated from 0 to Q.

PROBLEM 500

What is the electric potential energy of a solid, uniformly charged sphere of radius R and total charge Q? *Hint:* Follow the hint for problem 499, but this time develop both V_q and dq as functions of r, and integrate from 0 to R.

PROBLEM 501

Using insulating material that carries a charge of ρ coulombs per unit volume you mould a sphere of radius R. The material turns out not to be perfectly insulating, so that in time all of the charge migrates to the surface of the sphere.

a. How much energy was lost by this migration?
b. Where do you think this energy went?

PROBLEM 502

a. Graph the potential energy of a mass m in the gravitational field of the earth, outside the earth. (The simplest way to do this would be to copy the appropriate portion of the graph of problem 255, and multiply the ordinates of that graph by m.)

What can you say about a mass m whose E vs. r relationship
b. falls on the curve you have just drawn;
c. lies between this curve and the axis $E = 0$;
d. lies above the axis $E = 0$?
Suggestion: To answer these questions, particularly (c) and (d), think about what it must mean for a body to possess energy in excess of its potential energy of position in a field.

PROBLEM 503
For the combined charged raindrop of problem 259,

a. what was its increase in electrical potential energy over the total electrical potential energy of the eight original drops;
b. do you have any suggestion about where this additional energy could have come from?

PROBLEM 504
Return to the three conducting spheres demonstration discussed in problem 261.

a. How much energy was originally expended in charging the R and the $R/3$ spheres?
b. What fraction of this energy was lost in the process of equalizing the potential among the three spheres?
c. What happened to this lost energy?

PROBLEM 505
Considering the very strong intranuclear attractive force at very short distances (the force that holds the nucleus together against the Coulomb forces from the proton charges), it might seem reasonable to postulate a distribution of mass, and also therefore of charge, somewhat like the diagram in Figure 317, for which $\rho_r = \rho_0(1 - r/r_0)$. Assume that this distribution applies to the Na nucleus, which has 11 protons and a radius of about 3×10^{-15} m.

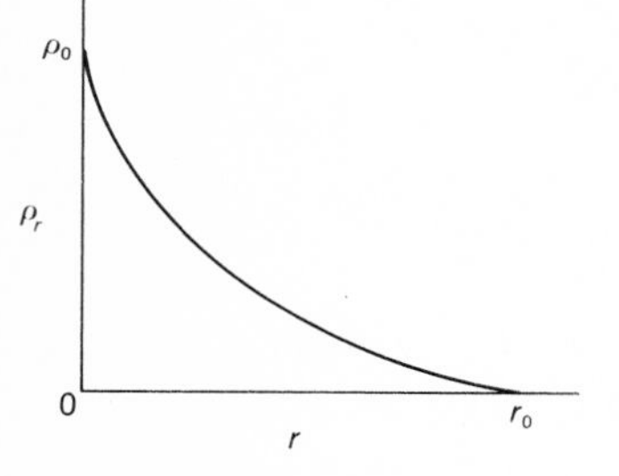

FIGURE 317

a. Integrate this charge distribution over the entire nucleus and equate the result to the total charge on the nucleus to obtain ρ_0.
b. With this value inserted into the above formula, determine the electrical potential energy for the Na nucleus having the assumed distribution.
c. Compare this with the potential energy computed under the assumption that the Na nucleus is a uniformly charged sphere.

Potential Energy of Elastic Bodies

Elastic bodies were referred to in the discussion of Hooke's Law oscillators; they are now worth a deeper study as a medium for an interchange among work, kinetic energy, and potential energy.

All bodies deform under stress. The term "elastic body" is usually limited to those situations in which the deformation is linear with stress. Thus the defining equation for an "elastic" body is stress $\propto$ strain, or

$$\text{stress} = \text{constant of proportionality} \times \text{strain}. \tag{4-7}$$

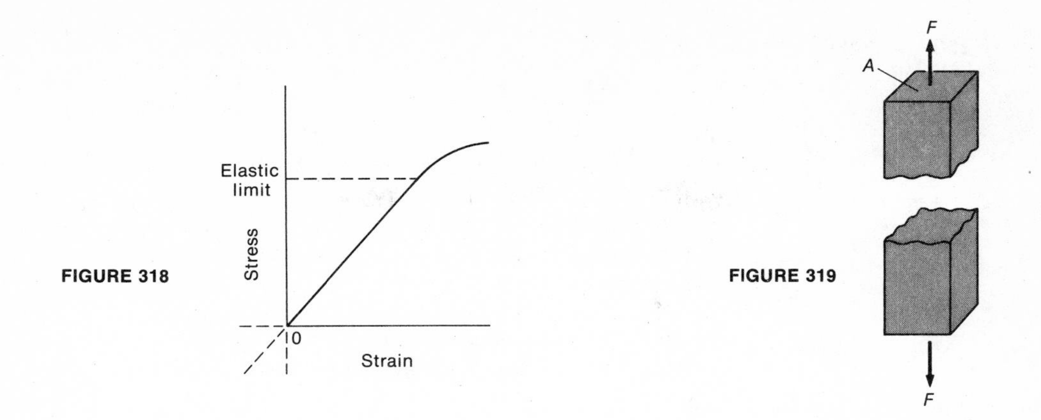

FIGURE 318

FIGURE 319

(In conformity with standard usage, the word "strain" replaces the word "deformation," which has been used up to this point for its descriptive value.) Equation 4-7 is the generic form of Hooke's law.

The behavior of bodies under stress is shown in the graph in Figure 318. The region in which the body behaves elastically under the above definition lies between zero stress and the elastic limit—the region in which the graph is a straight line. For clarity only one quadrant, in which both stress and resulting strain are positive, is shown; for physical accuracy the third quadrant indicated by the dashed lines should also be shown; negative stress produces a corresponding negative strain. A steel bar under tension will elongate in accordance with the graph in the first quadrant; a column made of the same steel under compression will shorten by following the same graph extended into the third quadrant.

Linear Stress

Linear stress occurs when two equal but opposite forces collinear with an axis of symmetry are applied at opposite ends of a body causing it to lengthen or shorten along the axis of symmetry.

The simplest form of Hooke's Law for evaluating linear stress-strain is $F = kx$, where F is a linear force, x is the resulting change in length, and k is the proportionality constant applicable to the one particular body for which the equation is being used. A more general statement applicable to all bodies under linear stress is the equation

$$\frac{F}{A} = Y\frac{\Delta\ell}{\ell_0}. \tag{4-8}$$

In this equation the stress is F/A, the force per unit area (Figure 319). The strain $\Delta\ell/\ell_0$ is the actual extension divided by the original unstressed length, or the extension per unit length. The constant of proportionality Y is a characteristic of the material in a body, and is called Young's modulus, the stretch modulus, or sometimes the elastic modulus.* We can arrange equation 4-8 so that it looks like $F = kx$: $F = (AY/\ell_0)\Delta\ell$, which indicates that for any particular body k could be computed from AY/ℓ_0. (This relationship does not apply to springs, which lengthen under stress by twisting and not by linear strain.)

* The dimensions of Young's modulus are $ML^{-1}T^{-2}$; the usual units are newtons m^{-2} or lbwt in^{-2}, the latter a practical engineering unit.

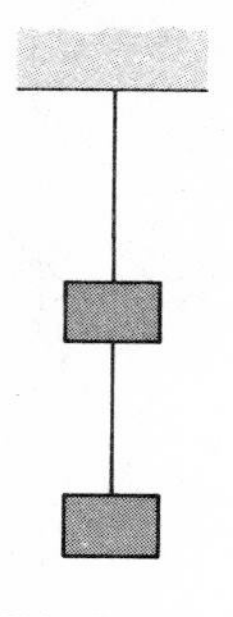

FIGURE 320

PROBLEM 506

A steel cable 20.00 ft long with effective cross section of 0.37 in^2, is suspended from a fixed point.

a. When a weight of 4.0 tonwt is hung from the lower end of the cable, how long does the cable become? Ignore the weight of the cable itself.

b. Then an identical cable is hung from the first weight, and another 4.0 tonwt is suspended from this second cable (Figure 320). What is now the combined length of the two cables? ($Y_{steel} = 2.7 \times 10^7$ lbwt in^{-2})

PROBLEM 507

A uniform spring has an unstretched length ℓ_0, a spring constant k, and a total mass m. How long is the spring when it hangs vertically from one end?

SOLUTION BY THE NONCALCULUS METHOD

Imagine the spring massless, with a mass m suspended from its lower end; its length would be $\ell_0 + mg/k$. Then imagine the spring cut into two equal parts, with a mass $m/2$ suspended from each part (similar to the arrangement in Figure 320); its length would then be

$$\frac{\ell_0}{2} + \frac{(m/2)g}{2k} + \frac{\ell_0}{2} + \frac{mg}{2k} = \ell_0 + \frac{mg\,(1+2)}{4k}.*$$

Cut into four equal parts, with a mass $m/4$ hung at the end of each part, the spring's length would be $\ell_0 + [mg(1+2+3+4)]/4^2k$. Cut into n equal parts with m/n hung at the end of each part, its length would be $\ell_0 + [mg(1 + 2 + \ldots + n)]/n^2k$, which reduces to $\ell_0 + mg/k\ [n(n/2 + 1)]/n^2$. As $n \to \infty$ the physical situation approaches that of a spring whose mass m is uniformly distributed along its length, and the length approaches $\ell_0 + mg/2k$.

SOLUTION BY THE CALCULUS METHOD

Consider a section dy long, from which hangs a mass $[(\ell - y)/\ell]m$. On that section, the linear force is then $[(\ell - y)/\ell]mg$ and the force constant is $(\ell/dy)k$; hence the change in length for that section is $d(dy) = [(\ell - y)\,mg/k\ell^2]\ dy$. When we integrate dy from 0 to Δy and y from 0 to ℓ, we arrive at $\Delta y = mg/2k$ and hence $\ell = \ell_0 + mg/2k$.

*Review page 212. If a spring of length ℓ has a spring constant k, a half length of that spring has a spring constant $2k$, etc.

PROBLEM 508

In problem 493 the assumption was made that the length of the rod with attached mass did not change during rotation. What was the actual length of the rod just before the mass broke loose, if the rod had a cross section A and a stretch modulus Y? *Suggestion:* The calculus method illustrated in problem 507 is the straightforward way of approaching this problem.

Shear Stress

If forces are applied in the plane of opposite faces of a body, instead of perpendicular to these faces as in linear stress, a body twists out of its original shape, as shown in Figure 321. Each thin plane parallel to the stressed faces moves parallel to itself and relative to its neighbors. In this situation the stress is F/A as before, and the strain is defined as $x/d = \tan \beta$. For most materials the strain, even for that up to $F/A =$ the elastic limit, is so small that $\tan \beta$ can usually be replaced by β. Thus for shear,

$$\frac{F}{A} = n \tan \beta \simeq n\beta, \tag{4-9}$$

where n is the shear modulus (dimensions again $ML^{-1}T^{-2}$).

Shear is a characteristic of solids only; there is no ordered structure among the molecules or atoms of liquids or gases that could provide forces resisting changes of shape. This fact is usually used as the crucial discrimination between solids and fluids.

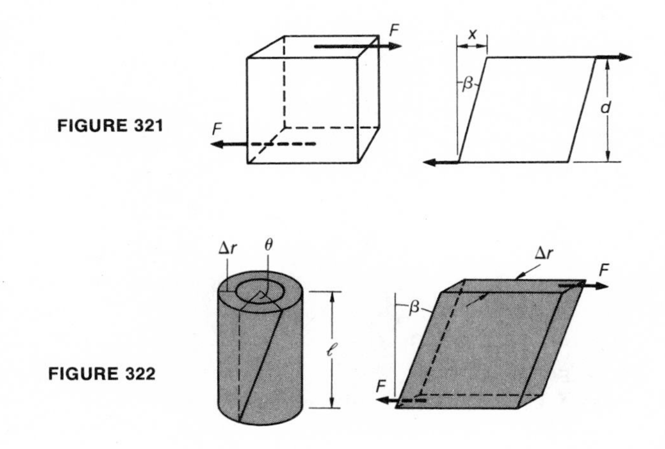

FIGURE 321

FIGURE 322

Torsion

Shear appears less often as the straightforward shape deformation illustrated in Figure 321 than it does as the underlying explanation for the elastic twisting about the axis of symmetry of tubes and rods. Consider a tube anchored at one end and with a tangential torque applied at the other end. If we unroll the tube we get the shape shown in Figure 322, for which $F/2\pi r \Delta r = n \tan \beta$. (From equation 4-9.) We are usually more interested in θ, the amount of twist at one end relative to the other end, than we are in β; from the diagram in Figure 322 we can see that $x = r\theta = \ell \tan \beta$. When we make the substitution,

$$\frac{F}{2\pi r \Delta r} = n \frac{r\theta}{\ell}.$$

Usually we are more interested in the torque causing the twist than we are in F specifically; in terms of this torque

$$\tau_r = Fr = n\,\frac{2\pi r^3\theta\Delta r}{\ell}. \tag{4-10}$$

If we wish to retain consistency with our earlier dimension of stress as $ML^{-1}T^{-2}$, we can rearrange equation 4-10 and identify the torsional stress on a tube as $\tau_r/2\pi r^2\,\Delta r$ and the strain as $(r/\ell)\theta$; then

$$\frac{\tau_r}{2\pi r^2\Delta r} = n\,\frac{r}{\ell}\theta.$$

Thus, for a tube undergoing torsional stress, the torsion modulus is the same as the shear modulus.

A solid rod is a series of adjacent tubes each of thickness dr. We can thus move from a tube to a rod in torsion by integrating the tube formula from $r = 0$ to $r = R$, the radius of the rod. The total torque on the rod is then

$$\tau = \int_0^R n\,\frac{2\pi r^3\theta\,dr}{\ell} = \frac{n\pi R^4\theta}{2\ell}. \tag{4-11}$$

We will not rearrange this equation as we did for the tube formula, since it is customarily expressed as $\tau = \tau_0\theta$, where

$$\tau_0 = \frac{n\pi R^4}{2\ell}. \tag{4-12}$$

τ_0 is called the torsion constant for a particular rod of radius R and length ℓ.

PROBLEM 509

In designing a metal torsion bar suspension for an automobile wheel you are required, within a working length of bar ℓ, to provide a torque τ with a corresponding twist in the bar of $\theta°$. As a preliminary check, you compare a solid bar with a tube which has a wall thickness of 0.10 r; both rod and tube are made of the same material and meet the stated requirements.

a. What is the ratio of the metal used in the tube to that in the rod?
b. What is the ratio of the radius of the tube to that of the rod?

PROBLEM 510

A drive shaft consists of two segments (Figure 323): a solid rod 4.0 ft long of 1.0-in radius and made of steel for which $n = 10 \times 10^6$ lbwt in^{-2}, and a solid rod 1.5 ft long of 0.75-in radius and made of steel for which $n = 12 \times 10^6$ lbwt in^{-2}. The two segments are connected axially through a universal joint which transmits torque without itself developing any measurable twist. When a torque of 360 ft lbwt is being transmitted by the drive shaft,

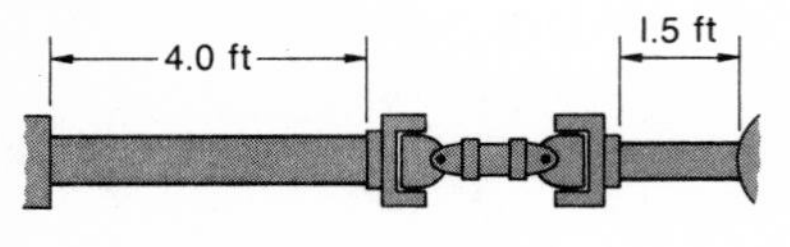

FIGURE 323

a. what is the total twist between the driving and the driven ends of the shaft?
b. What fraction of this twist appears across the shorter segment?

PROBLEM 511
When a steel rod is heated and then quenched, the elastic constants for the material in the rod are found to vary over the cross section, being larger at the surface than near the center. If a certain rod of radius R cm is given this heat treatment its shear modulus varies with r according to the formula $n_r = 7.8 \times 10^{11}\,(1 + 0.48r/R)$ dynes cm^{-2}. How much torque is required to twist a length L m of this rod through an angle θ?

Elastic Volume Change

In shear a body changes shape but not volume; it is now time to consider how bodies can change volume but not shape. This is accomplished by subjecting a body—solid, liquid, or gas—to a change in ambient pressure that is the same at all points on the body. Such a pressure provides a force on the body that is perpendicular to the surface of the body at every point; a change in this force produces an elastic change in the volume of the body. The stress is a change in pressure ΔP, and the strain is the relative change in volume $\Delta V/V$. Then volume elasticity is described by

$$\Delta P = -B\,\frac{\Delta V}{V}. \tag{4-13}$$

The minus sign is necessary to indicate that a positive ΔP induces a negative ΔV. B is the volume or bulk modulus; as with the other elastic moduli, its dimensions are $ML^{-1}T^{-2}$.

When the volume of a mass changes, its density changes also; this change in density is often of more interest than the volume change. From $m = \rho V$ we get $\Delta V/V = -\Delta\rho/\rho$, so equation 4-13 can also be written

$$\Delta P = B\,\frac{\Delta\rho}{\rho}. \tag{4-14}$$

The differential form of this relationship is

$$dP = -B\,\frac{dV}{V} = B\,\frac{d\rho}{\rho}. \tag{4-15}$$

PROBLEM 512
The average depth of the world's oceans is 3800 m, and the average hydrostatic pressure at that depth is 380 atmospheres. (One atmosphere = 1.013×10^5 N m^{-2}.) What is the density of sea water at this depth if sea water at the surface and at the same temperature has a density of 1025 kg m^{-3}? (For sea water $B = 2.2 \times 10^9$ N m^{-2}.)

PROBLEM 513

A buoyant life-saving ring, with sealed surface and a normal density of 0.70 gm cm^{-3}, is attached to a yacht that sinks after an encounter with a hurricane in the Caribbean. The life ring finally becomes detached and free as the yacht reaches a depth of 3800 m; however, instead of rising the life ring floats in equilibrium at that depth.

a. What must have been the bulk modulus for the material of the life ring?

b. Is the life ring in stable equilibrium at that depth, i.e., if it is slightly displaced vertically by local currents will it return to the 3800 m level?

c. Does the answer to question (b) validate the suggestion that objects that have sunk below the ocean surface can be found permanently distributed at various depths below the surface?

PROBLEM 514

Geophysicists currently believe that the earth's core has a radius of about 3500 km and is composed of molten iron that ranges in density from about 9 gm cm^{-3} at the surface of the core to at least 12 gm cm^{-3} at the center of the earth. (For comparison, the average density of the earth as a whole is 5.5 gm cm^{-3}.) Assuming that iron follows Hooke's Law over this entire range, what is the increase in the earth's internal pressure between $r = 3500$ km and $r = 0$? (For the high pressures involved, $B_{\text{iron}} \simeq 12 \times 10^{12}$ dynes cm^{-2}.)

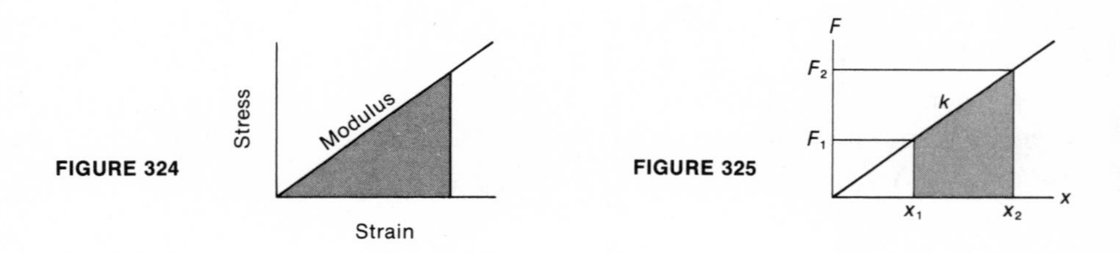

FIGURE 324

FIGURE 325

Strain-produced Changes in Potential Energy

The change in potential energy of an elastic body is a direct consequence of the work done on or by the body or the kinetic energy absorbed or delivered by the body. Elastic bodies provide many examples of the interchanges possible among these forms of energy.

The change in potential energy of an elastic body undergoing strain is the area under the stress-strain graph (Figure 324). This area can be expressed in several ways: by stress × strain, by modulus × strain², or by stress²/modulus.

Two kinds of stress produce work: One, a simple linear spring for which $F = kx$ (Figure 325);

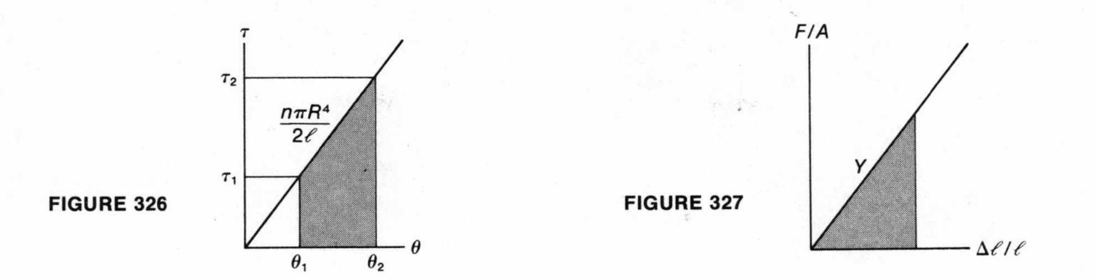

FIGURE 326

FIGURE 327

$$\Delta U = \int_{x_1}^{x^2} F\,dx = \int_{x_1}^{x^2} kx\,dx = \tfrac{1}{2}k(x_2^2 - x_1^2)$$

$$= \frac{F_2^2 - F_1^2}{2k}$$

$$= \frac{F_2 + F_1}{2}(x_2 - x_1)$$

$$= F_{\text{av}}\,(x_2 - x_1)$$

$$= \text{work.}$$

And two, a rod under torsional stress, for which $\tau = (n\pi R^4\theta)/(2\ell)$ (Figure 326);

$$\Delta U = \int_{\theta_1}^{\theta_2} \tau\,d\theta = \frac{n\pi R^4}{2\ell}\int_{\theta_1}^{\theta_2} \theta\,d\theta = \frac{n\pi R^4}{4\ell}(\theta_2^2 - \theta_1^2)$$

$$= \tfrac{1}{2}(\tau_2\theta_2 - \tau_1\theta_1)$$

$$= \frac{\ell}{n\pi R^4}(\tau_2^2 - \tau_1^2)$$

$$= \frac{\tau_2 + \tau_1}{2}(\theta_2 - \theta_1)$$

$$= \tau_{\text{av}}\,(\theta_2 - \theta_1)$$

$$= \text{work.}$$

Two kinds of stress produce work per unit volume: one, a linear stress for which $F/A = Y(\Delta\ell/\ell)$ (Figure 327);

$$\Delta u = \int_0^{\Delta\ell/\ell} \frac{F}{A}\,d\left(\frac{\Delta\ell}{\ell}\right) = \int_0^{\Delta\ell/\ell} Y\,\frac{\Delta\ell}{\ell}\,d\left(\frac{\Delta\ell}{\ell}\right) = \tfrac{1}{2}Y\left(\frac{\Delta\ell}{\ell}\right)^2$$

$$= \tfrac{1}{2}\frac{1}{Y}\left(\frac{F}{A}\right)^2$$

$$= \tfrac{1}{2}\frac{F\,\Delta\ell}{A\ell}$$

$$= \text{work per unit volume.}$$

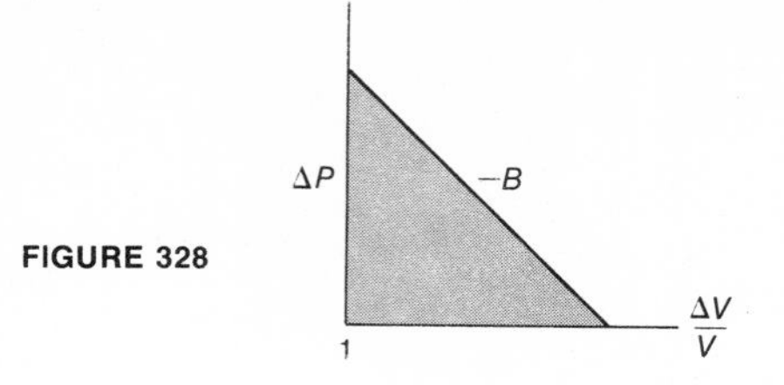

FIGURE 328

And two, a volume stress for which $\Delta P = -B\ (\Delta V/V)$ (Figure 328);

$$\Delta u = \tfrac{1}{2}\Delta P\,\frac{\Delta V}{V} = \tfrac{1}{2}B\left(\frac{\Delta V}{V}\right)^2$$

$$= \tfrac{1}{2}\frac{1}{B}\,(\Delta P)^2$$

$$= \text{work per unit volume.}$$

(With a bulk modulus that remains constant over a range of pressure or volume changes, the relation between finite changes ΔP and ΔV follows Hooke's Law. However, if we express this relation in differential form $[dP = -B\ (dV/V)]$, the integration gives us the relation $V = V_0\ e^{-(P-P_0)/B}$; the relation between P and V is not linear. We will explore further a nonlinear relationship between P and V when we come to the gas laws and the thermodynamic storage of energy.)

PROBLEM 515

A simple harmonic oscillator is displaced from its equilibrium position $x = 0$ to a point $x = A$ and then released.

a. Construct a graph showing the relation between the kinetic and potential energies of the oscillating mass at any x. Show on this graph the average potential energy and the average kinetic energy. *Suggestion:* Review problem 64 as a check.

b. At what x has the speed of the oscillating mass risen from 0 to half its maximum value?

c. What proportion of the maximum potential energy was expended to enable the mass to acquire this half speed?

PROBLEM 516

The two cable lengths and the two weights of problem 506 are connected in sequence just as in that problem, and lie on level ground. How much energy has been stored in the cables when the free end has been hoisted high enough so that the bottom weight is just clear of the ground? Ignore weight of cables. *Caution:* If you are tempted to use answers from problem 506, check significant figures first.

PROBLEM 517

A slingshot consisting of a Y-shaped handle and two rubber bands obeying Hooke's Law, each of natural length d and force constant k, is used to shoot a pebble of mass m vertically into the air. If the pebble reached a height h, what was the stretched length ℓ of the rubber bands? Neglect mass of the rubber bands and of the pad joining them.

PROBLEM 518

A rope to be used for hoisting is tested for load vs. stretch, with the results shown in the graph in Figure 329. An exactly similar rope is hung from a firm support, and a weight of 1.0×10^3 lbwt is attached to the free end. The weight is then lifted and dropped. From what maximum height can the weight be dropped without breaking the rope? *Note:* The test results were obtained from static loading. For this problem assume that the rope behaves in the same way under impact loading.

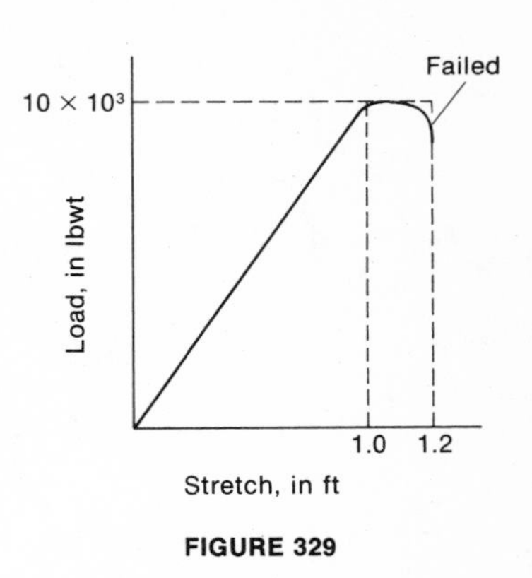

FIGURE 329

PROBLEM 519

A ball of mass 1.0 kg is supported 0.80 m above a retaining cup that is firmly mounted on top of a vertical coiled spring 0.80 m long whose spring constant is 147 N m^{-1}; the cup and spring are relatively massless (Figure 330). When the ball is released it drops into the cup and stays there.

a. How far below its original position does the ball sink?
b. How close to its original position does the ball rise?
c. What is its amplitude of oscillation? Its period?

Suggestion: A coordinate system based on the original position of the ball will offer the easiest solution to questions (a) and (b).

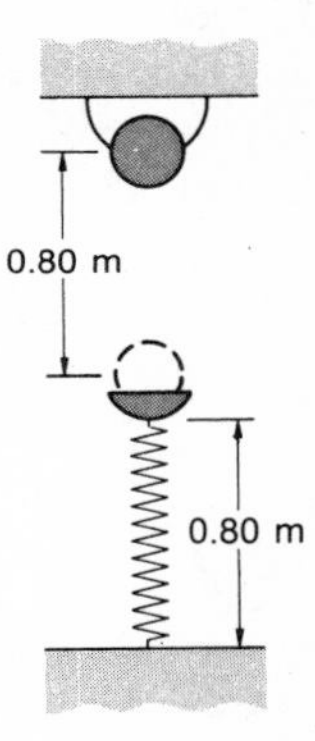

FIGURE 330

PROBLEM 520

A small mass of 1.0 kg is held in place by a clamp; suspended from this mass is an ideal spring of negligible mass, of unstretched length 0.30 m. and with a spring constant 49 N m^{-1}; suspended from the spring is a small mass of 0.50 kg (Figure 331). When the clamp is released and the system falls, what is the kinetic energy of the system at the first instant the spring returns to a length of 0.30 m? *Hint:* The two masses at the ends of the spring oscillate relative to their c.m., which is falling freely. If you need to compute these oscillations, a review of page 212 may be helpful.

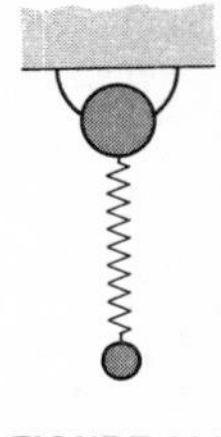

FIGURE 331

PROBLEM 521

A brass wire and a steel wire, each of unstretched length ℓ and radius r, are soldered together, end to end. The composite wire, 2ℓ long, is suspended from one end; to its other end a mass m is attached. After release m is found in equilibrium $(2\ell + h)$ below the suspension point for the wire.

a. How much energy is stored in each of the original wires? Consider the masses of the wires as negligible compared to m.
b. If the sum of the energies stored in the wires does not equal the gravitational potential energy lost by m in sinking from 2ℓ to $(2\ell + h)$ below the suspension point, how do you account for the difference?
c. To have an equal amount of energy stored in each wire, what should have been their relative lengths to start with?

d. If the wires are given these relative lengths, keeping the total unstretched composite length still 2ℓ, how much did each wire lengthen when the mass m was suspended from the composite wire? ($Y_{\text{brass}} = 9.0 \times 10^{10}$ N m^{-2}; $Y_{\text{steel}} = 20 \times 10^{10}$ N m^{-2}.)

SUGGESTED START OF SOLUTION

For $\ell_{\text{br}} = \ell_{\text{st}} = \ell$,

$$W_{\text{br}} = \tfrac{1}{2}\frac{1}{Y_{\text{br}}}\left(\frac{F}{A}\right)^2 A\ell$$

and

$$W_{\text{st}} = \tfrac{1}{2}\frac{1}{Y_{\text{st}}}\left(\frac{F}{A}\right)^2 A\ell.$$

Then

$$\frac{W_{\text{br}}}{W_{\text{st}}} = \frac{Y_{\text{st}}}{Y_{\text{br}}} = \frac{\Delta\ell_{\text{br}}}{\Delta\ell_{\text{st}}} \left(\text{from } \frac{F}{A} = Y_{\text{br}}\frac{\Delta\ell_{\text{br}}}{\ell} = Y_{\text{st}}\frac{\Delta\ell_{\text{st}}}{\ell}\right),$$

and

$$W_{\text{br}} + W_{\text{st}} = W_{\text{br}}\left(1 + \frac{\Delta\ell_{\text{st}}}{\Delta\ell_{\text{br}}}\right),\ \Delta\ell_{\text{st}} + \Delta\ell_{\text{br}} = h, \text{ etc.}$$

PROBLEM 522

An elevator weighing 2700 lbwt is suspended from a steel cable that weighs 1.5 lbwt ft^{-1} and has a cross-sectional area of 0.40 in^2. When the elevator is being accelerated upward at 8 ft sec^{-2}, how much energy is stored in the cable at the instant its working length is 200 ft? *Double caution:* First, the weight of the cable is not negligible compared to that of the elevator; second, since the mass of the cable is also being accelerated, the strain along the cable is not uniform.

START OF SOLUTION

Since the strain is not uniform along the cable, the energy stored in each infinitesimal element of length dy must be computed separately and then added (integrated) to determine the total energy storage.

Consider an element dy that is y below the suspension point (Figure 332); this element must both support and accelerate all of the mass below it. Thus $F_y = [M + \lambda(\ell - y)]\,(g + a)$, where M is the mass of the elevator and λ is the mass per unit length of the cable. When we substitute in equation 4-8, we get

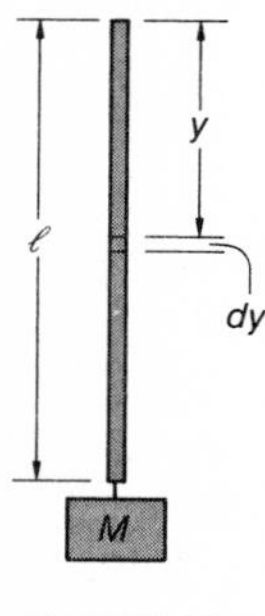

FIGURE 332

$$\frac{d(dy)}{dy} = \frac{F}{AY} = \frac{[M + \lambda(\ell - y)]\,(g + a)}{AY}.$$

The energy stored in the element is then

$$d(\Delta U) = \tfrac{1}{2}Y\left[\frac{d(dy)}{dy}\right]^2 A\,dy,$$

where $A\,dy$ is the volume of the element. For the entire cable we then have

$$\Delta U = \frac{AY}{2}\int_0^{\ell}\left[\frac{d(dy)}{dy}\right]^2 dy.$$

PROBLEM 523

A cylindrical thin-walled tube is placed on a solid metal surface, and a loose-fitting metal mandrel is placed at the upper end, as shown in Figure 333. There is a static coefficient of friction μ between the tube and the metal surfaces at either end; n and Y are the shear and the Young's moduli for the material of the tube. An axial force is applied to the mandrel; then a torque about the axis is applied to the mandrel until slippage impends at the ends of the tube. What was the ratio of the work done in torsion to the work done in compression? Ignore the weight of the mandrel in comparison to the axial force.

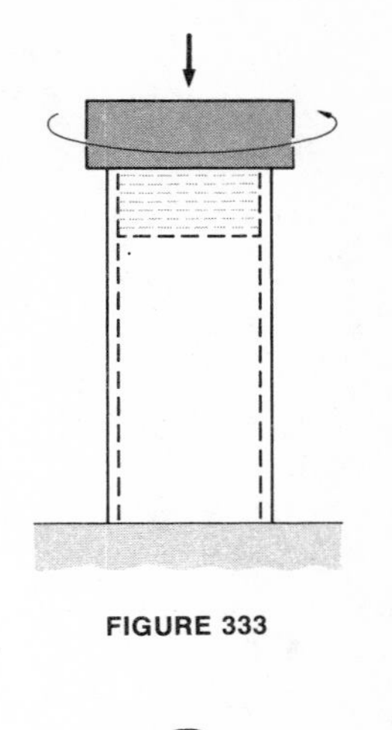

FIGURE 333

PROBLEM 524

A semicircular disc is soldered to the midpoint of a horizontal rod, the plane of the disc being normal to the rod; R and M are the radius and mass of the disc. The ends of the rod are rigidly clamped. Figure 334 shows the rod in four positions. In (*a*) the rod is unstressed, with the disc in unstable equilibrium above it. The disc is then rotated to position (*b*); on release it snaps up to (*c*), where it comes momentarily to rest. It then oscillates in the quadrant between (*b*) and (*c*); when these oscillations are damped out the disc finally comes to rest at (*d*). At no time is the rod stressed beyond its elastic limit. What is the torsional constant for the entire rod? *Comment:* In spite of the apparent lack of hard data, there is one change that is illustrated that you know about completely. You might find the answer to problem 214 useful, and don't forget the difference between the torsional constant for a whole rod and for a half rod.

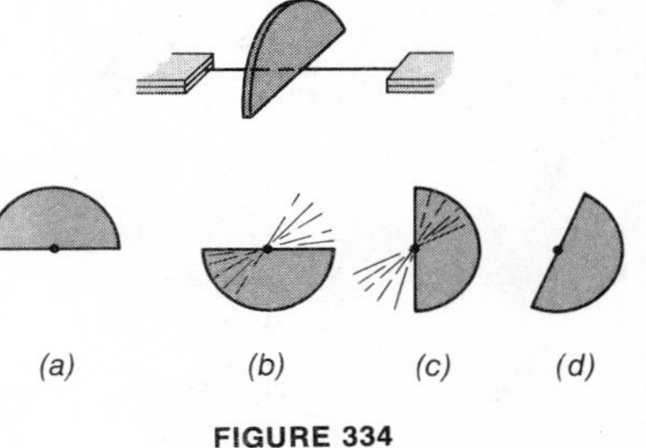

FIGURE 334

PROBLEM 525

Water stands 5 m deep in a vertical well whose cross-sectional area is 1 m^2. How much more elastic potential energy does the water in the well possess compared to the same amount of water at the same temperature spread 1 mm deep over the surface of a lake? (For water $B = 2.0 \times 10^9$ N m^{-2}.)

Power and Efficiency

In most practical situations the rate at which work is done may be as important as the amount of work done; certainly you wouldn't hire a gardener who would take half a day to mow your lawn if you could hire an energetic college student who would do as good a job in an hour and a half.

To introduce this aspect of time into work done, we define power by $P = dE/dt$. Here E stands for any of its possible forms W or T or U; a shorthand for $P = dW/dt$ or dT/dt or dU/dt, or for any combination $d(W + T + U)/dt$. The dimensions of P are $M\ L^2T^{-3}$; the usual units are joule sec^{-1} = watt, or lbwt ft sec^{-1}. A frequently used practical unit is 1 horsepower (HP) = 550 lbwt ft sec^{-1}. Occasionally it is convenient to work backward from power units to energy units; for example, you will sometimes find data for energy given in watt sec (1 watt sec = 1 joule) or in kilowatt hours (1 kwh = 3.6×10^6 joules.)

The fundamental definition of P can be put into more direct operational form for some particular situations:

1. Rate of work done, force constant;

$$P = \frac{d(FS)}{dt} = F\frac{dS}{dt} = Fv$$
$$= \frac{d(\tau\theta)}{dt} = \tau\frac{d\theta}{dt} = \tau\omega. \qquad (4\text{-}16)$$

2. Rate of change of kinetic energy, mass constant;

$$P = \frac{dT}{dt} = \frac{d(\frac{1}{2}mv^2)}{dt} = mv\frac{dv}{dt} = mva$$
$$= \frac{d(\frac{1}{2}I\omega^2)}{dt} = I\omega\frac{d\omega}{dt} = I\omega\alpha. \qquad (4\text{-}17)$$

3. Rate of change of kinetic energy, constant velocity;

$$P = \frac{dT}{dt} = \frac{d(\frac{1}{2}mv^2)}{dt} = \tfrac{1}{2}v^2\frac{dm}{dt}. \qquad (4\text{-}18)$$

4. Rate of change of potential energy, constant field;

$$P = \frac{dU}{dt} = \frac{d}{dt}\left[-\int \vec{F}_F \cdot d\vec{s}\right] = -\vec{F}_F \cdot \vec{v}. \qquad (4\text{-}19)$$

5. Rate of change of potential energy, constant potential;

$$P = \frac{d}{dt}(mV_{\text{grav}}) = V_{\text{grav}}\frac{dm}{dt}$$
$$= \frac{d}{dt}(qV_{\text{elec}}) = V_{\text{elec}}\frac{dq}{dt} = V_{\text{elec}}I. \qquad (4\text{-}20)$$

(The confusion arising from different meanings for the same symbol has been commented on before. In examples 1 through 3, v represents speed; in 4, v is velocity; and in 5, V is potential.)

PROBLEM 526
A tugboat moving by itself through the water requires 100 HP at the propeller to move at 10 knots. When it is towing a large ship, 900 HP must be delivered to the propeller to keep the combination moving at 10 knots. What is the force in the towline?

PROBLEM 527
When a Van de Graaff accelerator is operating with a dome potential of 5×10^6 volts, the charging current to the dome averages 100

μamps. Ignoring the loss in belt friction (this is too large an item to ignore in an actual design computation), what power is being used to maintain the dome potential?

PROBLEM 528

A shaft 4.00 cm in diameter is turning at 1.80×10^3 rpm in a sleeve bearing 10.0 cm long. The shaft is centered in the bearing by oil pressure, the oil has a viscosity of 1.05 poise, and the clearance between the shaft and the bearing is 0.100 mm. How much power is being dissipated in the bearing?

PROBLEM 529

The average flow of a river at a particular canyon site is 4×10^3 ft^3 sec^{-1}. If it is believed that the river can be dammed in the canyon to provide a difference in water levels of 66 ft without impairing the flow of the river, what is the theoretical hydroelectric power possibility of the site?

PROBLEM 530

A ship's steel propeller shaft has a diameter of 10.0 cm and a length of 5.00 m. If the shaft's twist is not to exceed 5° when the shaft is turning at 400 rpm, what is the maximum power the shaft can transmit?

PROBLEM 531

The potential energy of the charged sphere with imperfect insulation (problem 501) is found to change with time according to the graph shown in Figure 335; this graph can be described by $U - U_f = \Delta U e^{-\alpha t}$, where $\Delta U = U_o - U_f$ is the total potential energy change computed in problem 501. If half of this change occurred in the first 3.5 sec,

a. what does the graph of P vs. t look like, and what is its equation?
b. Is this equation dimensionally correct, i.e., does it have the dimensions of dU/dt?
c. What was the power dissipation at $t = 2.1$ sec?

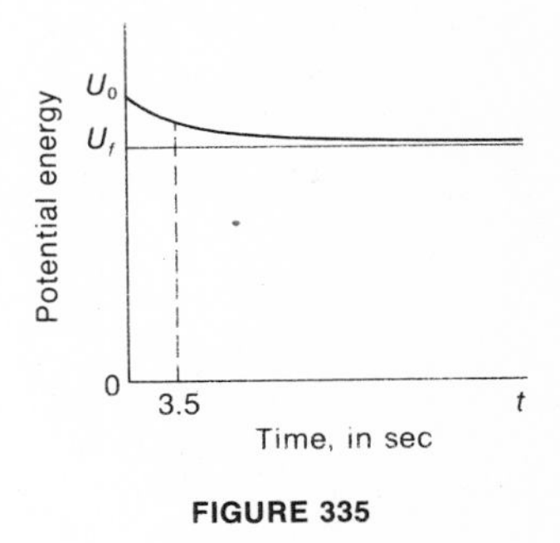

FIGURE 335

PROBLEM 532

At any time t how much power is being dissipated in damping friction by the normally damped oscillator of equation 3-64? *Suggestion:* You can do this the long way, by solving for $T_t = \frac{1}{2}m\dot{x}^2$ and $U_t = \frac{1}{2}m\omega_0^2 x^2$, adding these to get E_t, and then differentiating to get $P = dE_t/dt$. Or you can do it the short way by remembering from the definition of γ that the frictional force is $\gamma m\dot{x}$, and hence the power dissipated in friction $= \gamma m\dot{x}^2$.

PROBLEM 533

A mass m slides down an incline set at an angle θ with the horizontal (Figure 336); there is a coefficient of sliding friction μ between m and

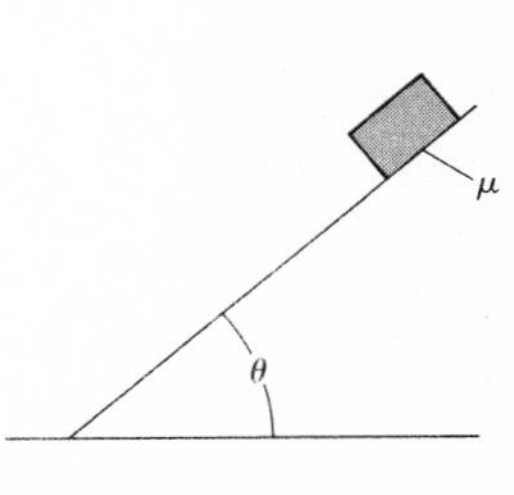

FIGURE 336

the plane. As a function of the time after release, what is the rate at which m is

a. gaining kinetic energy;
b. losing potential energy?
c. Do the two answers balance each other? If not, why not? (Assume $\tan\theta > \mu$.)

Problem 533 represents the normal course of any real mechanical operation; more energy is paid in to the operation than is paid out. You never get back in return as much as you have invested. This ratio of out/in is useful data about any action involving energy or power; it has the name "efficiency" and is defined by

$$\text{Efficiency} = \frac{\text{output of useful energy}}{\text{input of energy}} = \frac{E_{\text{out}}}{E_{\text{in}}}$$

$$= \frac{\text{power delivered}}{\text{power supplied}} = \frac{P_{\text{out}}}{P_{\text{in}}} \qquad (4\text{-}21)$$

(In the unusual case in which E_{out} and E_{in} are not the same functions of t, then energy efficiency $\neq$ power efficiency.) Efficiency is usually expressed in percent, and of course can never be greater than 100 percent.

As an example, consider problem 533. At any time t the kinetic energy gained by the mass m is $\Delta T = \frac{1}{2}mg^2t^2 (\sin\theta - \mu\cos\theta)^2$ and the potential energy given up is $\Delta U = \frac{1}{2}mg^2t^2 \sin\theta(\sin\theta - \mu\cos\theta)$. Thus the efficiency of the operation in problem 533 $= \Delta T/\Delta U = 1 - \mu\cot\theta$.

Equation 4-21 can be rearranged as $E_{\text{out}} = E_{\text{in}} \times \text{efficiency}$. It is also useful sometimes to know that the percentage loss of energy $\Delta E/E = (100 - \text{efficiency})$ percent if efficiency is expressed in percent.

PROBLEM 534
A hydroelectric plant was ultimately built at the site investigated in problem 529. When the effective head of water and the flow were the same as those given in that problem, the maximum output of the generators was 1.8×10^4 kw. What was the overall efficiency of the plant?

PROBLEM 535
Considering the definition of efficiency, what is the efficiency of the operation of slowing an automobile by braking? This question does not refer to the efficiency of the braking mechanism itself, but to the overall ratio of the gain in mechanical energy to the input in mechanical energy involved in the act of braking.

PROBLEM 536
In view of the necessary frictional torque that must be available in a differential pulley (see problem 406), what is the maximum possible efficiency of such a pulley?

PROBLEM 537
In problem 312,

a. what was the efficiency of the operation;
b. for what angle θ would the efficiency be a maximum;
c. what is this maximum efficiency?

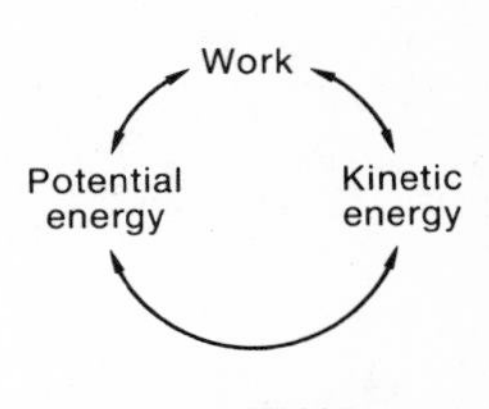

FIGURE 337

EXAMPLES IN THE CONSERVATION OF ENERGY

It is now about time to emphasize that work, kinetic energy, and potential energy (and rest mass energy if you wish to include that separately) are all just different forms of the same thing—energy—and are easily convertible among themselves (Figure 337). Thus for an isolated system we can always write

$$E_o + U_o + \text{work performed} = E_f + U_f + \text{work accomplished.}$$

This is one form of the statement of the Conservation of Energy. In this form E includes both the kinetic energy T and the rest mass energy mc^2. Since most situations are nonrelativistic, the statement usually reduces for practical operation to

$$T_0 + U_0 + \text{Work}_{\text{in}} = T_f + U_f + \text{Work}_{\text{out}}. \quad (4\text{-}22)$$

The following problems are examples of this interchange among the forms of energy.

PROBLEM 538
One of the simplest, and at the same time one of the most penetrating, demonstrations ever developed was Galileo's use of a simple pendulum and a peg to show what we would now call an example of the conservation of mechanical energy.

A small solid ball at the end of a thread of length L is mounted in front of a vertical board on which has been painted a horizontal line a distance D below the pendulum support. When the ball is drawn up to the line and released (Figure 338), it swings up just to the line on the other side. If a small peg is placed on the line at P before the ball is released, the ball will still swing up just to the line. With the peg in place, if the ball is released above the line it will swing around the peg and rise above the line. From how far above the line should the ball be released so that it just barely completely circles around the peg without the thread becoming slack at any time?

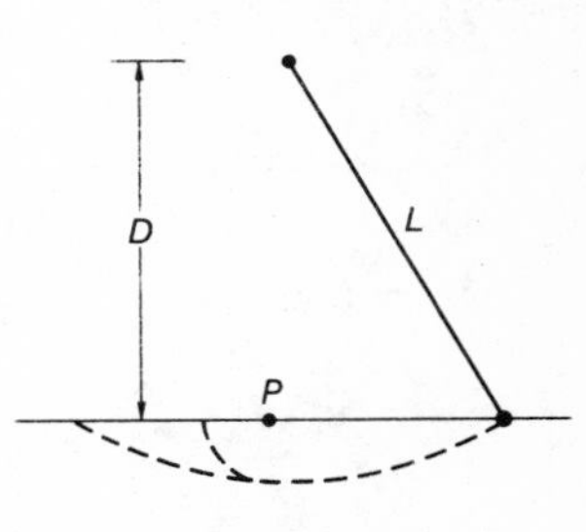

FIGURE 338

PROBLEM 539

A champion high jumper can clear a bar about 7 ft above his ground take-off point.

a. Estimate what upward speed he must impart to his body to make such a jump. *Suggestion:* Visualize the probable location of the c.m. of his body at take-off and at the time he clears the bar.
b. If the spring with which he imparts this upward speed to his body takes place in $\frac{1}{3}$ sec, what power must a 160-lb high jumper supply in clearing 7 ft?

PROBLEM 540

a. Where on the graph for problem 502 will you find a point appropriate to a mass that at the surface of the earth had been given the minimum amount of energy necessary to escape from the field of the earth?
b. What is the "velocity of escape" for a space probe; for a hydrogen molecule?

PROBLEM 541

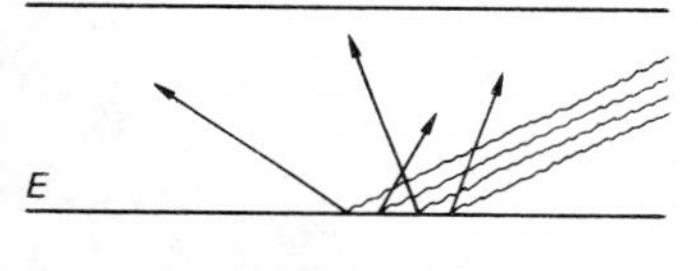

FIGURE 339

Essential features of Millikan's apparatus for measuring energies in photoelectric emission are parallel metal plates C and E; the space between these plates is evacuated. When monochromatic light shines on E, electrons are ejected in random directions but with a unique kinetic energy that is characteristic of the frequency of the light and of the surface of E (Figure 339). The ejected electrons that arrive at C constitute an electric current that can be measured; by adjusting the voltage between C and E to a particular value V, this current can be reduced to zero. In terms of V, what is the characteristic kinetic energy of the ejected electrons?

PROBLEM 542

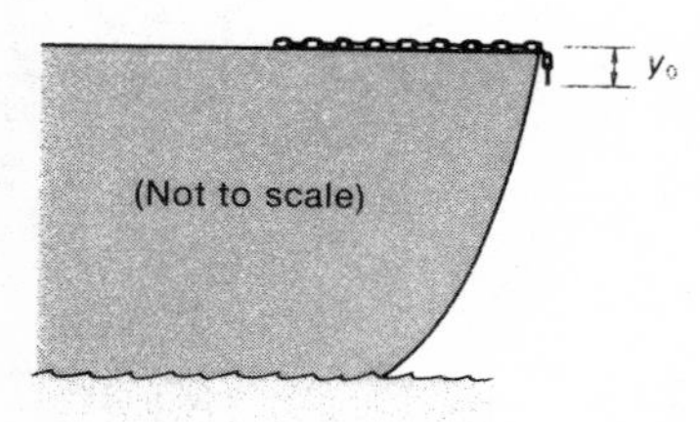

FIGURE 340

A helper around a boatyard lays out a length L of chain on the smooth, horizontal deck of a large boat, preparatory to bending on an anchor (Figure 340). The height of the deck above the water is greater than L. The helper does not notice that he allowed a short length y_0 to overhang the bow of the boat. When he walks away to get the anchor the chain starts to slip off of the deck. How fast was the chain moving at the instant the end of it disappeared over the side? (For two-figure accuracy assume that the chain links turn straight downward as they leave the deck.)

PROBLEM 543

An electron is released from rest at a distance a above a grounded conducting plane. How fast is it moving when it is 0.1 a above the plane?

PROBLEM 544

At a premixed concrete plant an endless bucket hoist (Figure 341) delivers sand at the rate of 1600 lb min^{-1} to a dump point 20 ft above the bottom of the hoist where the buckets are scooping up sand from rest. Neglecting friction, when the buckets are moving at a speed of 3 ft sec^{-1} what power is required to operate the hoist?

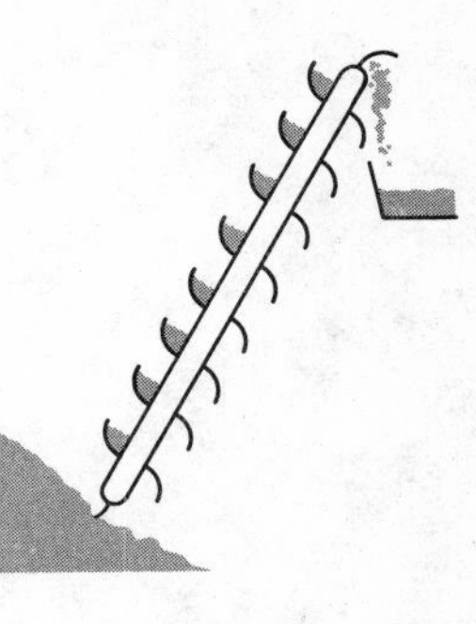

FIGURE 341

PROBLEM 545

A drop of water of mass 18 mg falls away from the bottom of a charged conducting sphere of radius 20 cm, carrying with it a charge of 1.0×10^{-9} C and leaving on the sphere a uniformly distributed charge of 2.5×10^{-6} C. What is the speed of the drop after it has fallen 30 cm?

PROBLEM 546

A thin-walled open-ended concrete cylinder when tamped full of earth is found to weight 242 lbwt. When the filled cylinder rolls without slipping down a delivery ramp, its speed at the bottom is 10% faster than the speed attained by the empty cylinder rolling down the same ramp without slipping. How much did the earth in the cylinder weigh?

PROBLEM 547

A proton moves with a speed of 5×10^6 m sec^{-1} directly toward a free proton originally at rest.

a. What is the distance of closest approach for the two protons?
b. How fast is the incident proton moving when it gets this close to the other proton?
c. What would be the answers to the above questions if the target particle had been a free alpha particle at rest, instead of a proton?

PROBLEM 548

Across the gap between the dees of a cyclotron a voltage difference $V \sin \omega t$ is maintained. If the gap is narrow enough so that the passage time across it is negligible compared to $\omega/2\pi$, and if a charged particle of charge q and mass m arrives at the gap in resonance with maximum voltage, by how much is its path radius changed every revolution?

PROBLEM 549

An elevator of 4.0 tons total mass when loaded is designed to be uniformly accelerated upward from rest to 18 ft sec^{-1} in 3.0 sec, and then to continue upward at that constant running speed.

a. What is the maximum horsepower required?
b. What is the running horsepower required?

c. How much energy in HPh had been expended when the elevator passed a level 99 ft above its starting point?
d. If the driving mechanism was 63% efficient, how many kwh had been delivered to the driving mechanism between the time of the start and the time the elevator passed the 99 ft level?

PROBLEM 550

A circular loop of wire carrying a steady current I is mounted on a frictionless axis coincident with a diameter of the loop. The loop is set rotating about this axis with an angular speed ω_o, and then a uniform magnetic field B is established perpendicular to the loop's axis; the region of the uniform field is large enough to encompass the entire loop.

a. If the field was established at the instant the plane of the loop was parallel to the field, how fast was the loop rotating after $(2n + 1)/2$ rotations?
b. If energy changes occurred during rotation of the loop, where did the energy come from or go to?

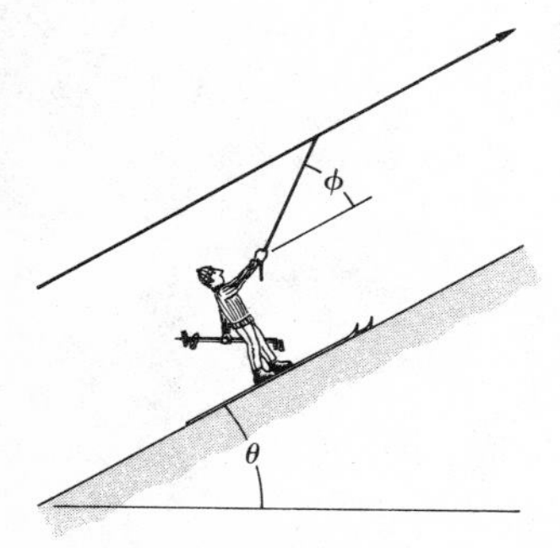

FIGURE 342

PROBLEM 551

You are asked to design a ski tow, which consists of a long endless cable to which are attached short handhold cables, as shown in Figure 342. It is estimated that the average skier will weigh W lbwt and that there will be no more than N skiers along the tow at any one time. The tow is to operate at a steady v ft sec^{-1}. The coefficient of sliding friction between skis and surface is estimated at μ. If the cable drive is expected to be E% efficient, what power capacity should you specify?

PROBLEM 552

You were asked to solve problem 375 by the methods of Newtonian dynamics. Now solve the same problem by conservation of energy methods; for your own information, compare the amount of work involved in the two methods.

PROBLEM 553

In a linear accelerator the potential difference between proton source and target is 5.0×10^6 volts. Protons are emitted at the source with negligible kinetic energy.

a. With what speed do the protons arrive at the target?
b. With what momentum?

Suggestion: With energies of this magnitude relativistic speeds are a possibility, so it might save time to use relativistic formulae from the start of your solution.

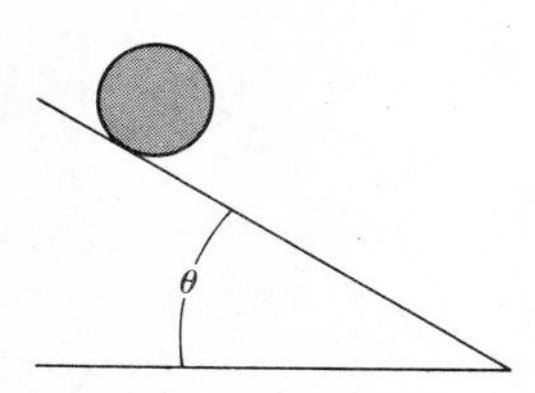

FIGURE 343

PROBLEM 554

A uniform solid disc is released from rest and rolls down an incline at θ with the horizontal (Figure 343). The angle θ is adjusted until the disc barely rolls without slipping, i.e., for any greater angle the disc partly rolls and partly slips. The disc is then replaced by a thin-walled cylinder made of the same material as the disc; the friction coefficients remain the same, whether sliding or static. When the cylinder is released from rest it moves down the incline partly rolling and partly slipping. What fraction of the cylinder's available energy is being dissipated during slipping? *Suggestion:* The problem can be solved either by computing the work done against friction, or by computing a difference in energies. If you are careful not to make an error in calculating the "slip" between cylinder and plane, the former method is probably simpler.

IMPULSE

When we apply a force F to a body, by Newton's Second Law we have, after rearrangement, $\vec{F}\,dt = d\vec{p}$. When we integrate,

$$\int_{t_0}^{t_1} \vec{F}\,dt = \Delta\vec{p}, \qquad (4\text{-}23)$$

where $\Delta\vec{p}$ is a finite change in linear momentum during the time interval $(t_1 - t_0)$.

The integral $\int \vec{F}\,dt$ is called linear impulse, and is often indicated by the symbol $\vec{J}$. If $\vec{F}$ is constant, or a known function of time, the integration can easily be performed. However, in many practical cases an instant-by-instant value of F is not known; the integral as a whole, or the value of $\vec{J}$, can be inferred only from an observed $\Delta\vec{p}$. (One has only to look at high-speed photographs of the impact of a golf-club head with a golf ball to appreciate the difficulty of giving an instantaneous statement about what is happening inside the $\int F\,dt$ integral; yet the result, the changes in momentum of the ball and club-head, are relatively easy to measure.) In fact, equation 4-23 is more often used in this direction: an observed change in momentum permits a calculation of either the total impulse or of an average force assumed to be acting over a known time interval.

In rotational coordinates we define angular impulse similarly:

$$\int \vec{\tau}dt = \Delta\vec{L}. \qquad (4\text{-}24)$$

As $\Delta\vec{p}$ or $\Delta\vec{L}$ will be in absolute units, F or τ must also be expressed in absolute units.

Equations 4-23 and 4-24 can be used not only when F or τ are continuous; they are also applicable in situations involving rapid, discrete changes in momentum such as occur in machine-gun fire or in the operation of an air-driven jack hammer. In such applications, the summation of a series of brief individual force actions is considered equivalent to an action of a steady average force.

PROBLEM 555
Just before the ball of problem 45 was hit it was moving horizontally at 98 ft sec^{-1}. What was the impulse supplied by the bat?

PROBLEM 556

A ball of mass m is dropped from a height h_0 onto a concrete floor, and rebounds to a height h. What was the impulse supplied to the ball by the floor?

PROBLEM 557

A steel worker closing rivets pushes against his air-driven riveting hammer with a steady force F. The plunger of the hammer makes 10 strokes per second, and on each stroke is in contact with the rivet tail for only 0.01 sec. What is the average force being applied to the rivet?

PROBLEM 558

A 2.0 lb hammer head hits a nail while moving at a speed of 24 ft sec^{-1}. The wood into which the nail was driven offered an average resistance to the passage of the nail of 50 lbwt. How long was the hammer in effective contact with the nail head?

PROBLEM 559

Rain has fallen on the Atlantic coast of Panama at the rate of 2.1 cm min^{-1}.* If the rain fell with a vertical speed of 9.0 m sec^{-1} (a reasonable value for the terminal speed of a hard rain), and rebounded from a flat, horizontal roof with a vertical speed of 1.0 m sec^{-1}, what was the vertical force per square meter felt by the roof?

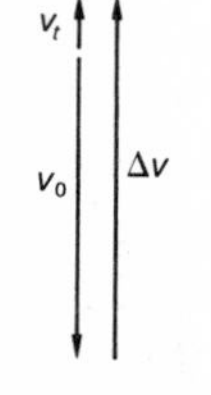

FIGURE 344

SOLUTION

Let F = the force supplied by 1 m^2 of roof that is necessary to produce the observed change in momentum of the rain falling on that square meter (Figure 344). (For a steady rain, it is reasonable to assume F is constant.) For simplicity, take $\int dt = 1.0$ sec. Then

$$F \int dt = F \times 1 = \Delta(mv) = \frac{2.1 \times 10^{-2} \times 1 \times 10^3}{60} \times (9+1),$$

and

$$F = 3.5 \text{ N m}^{-2} \text{ upward}$$

(10^3 kg m^{-3} is the density of the falling water). F is the force felt by the rain. F_R, the force felt by the roof, is equal and opposite to this; $F_R = 3.5$ N m^{-2} downward.

If you think we cut corners by specifically taking one square meter of roof and 1.0 sec of time, let's do it more generally. Let A be the horizontal area of the roof, and F the force per unit area provided by the roof. Then

$$\int FA\, dt = FA \int dt = \Delta(mv) = \int \frac{dm}{dt}\, dt \times \Delta v$$

$$= \frac{2.1 \times 10^{-2} \times A \times \int dt \times 10^3}{60} \times (9+1)$$

and

$$F = 3.5 \text{ N m}^{-2} \text{ upward, as before.}$$

---

*The rate of rainfall is measured by the depth of water accumulated in a vertical-sided reservoir in unit time.

PROBLEM 560

A device for giving individuals free vertical motion in the air is in the development stage. It consists essentially of a back-pack motor and a pair of counter-rotating fan blades to blow air downward (Figure 345). If the effective cross-sectional area of the fan is 12 ft^2, what should be the speed of the downward-moving column of air to support a man who, with back-pack, weighs 270 lbwt?

FIGURE 345

PROBLEM 561

In riot control, fire hoses are sometimes used to make people move. The nozzle of a fire hose has a cross-sectional area of 20 cm^2, and ejects a solid stream of water at a speed of 40 m sec^{-1}. What force is felt by a person who receives a direct hit from the stream,

a. if he is stationary when hit;
b. if he was fleeing directly away from the stream at a speed of 7.0 m sec^{-1}?

PROBLEM 562

A butcher holding a long string of small link sausages upright just above the scale pan offers to charge the customer for just one-half of the maximum reading of the scale after he releases the string (Figure 346). The customer, not knowing much physics, eagerly agrees. How much more did the customer pay than the proper charge? (Assume no overrun of the scale reading.) *Hint:* At any moment the scale pan must support the weight of the sausage that has already arrived there, and must also absorb the momentum per second that is still arriving.

FIGURE 346

PROBLEM 563

The South American bolas, used to catch large animals, consists of two round stones connected by a thong of length L and of negligible mass compared to the mass m of each of the stones. In use, one of the stones is grasped and the bolas whirled in a horizontal circle until a suitable angular speed has been attained (Figure 347); it is then released to fly toward its target. Assume the throwing hand describes a circle of radius r every second.

FIGURE 347

a. After release, what is the angular speed of the bolas?
b. With what linear speed does its c.m. move toward the target?
c. What is the angular momentum of the thrown bolas?
d. If the advancing stone of the bolas makes inelastic impact with a relatively stationary object of large mass just as the thong of the bolas is perpendicular to the velocity of the c.m. of the bolas (Figure 348), what is the impact impulse the large object feels?
e. Following the impact described in (d), with what angular speed does the thong start to wrap around the stationary object?

FIGURE 348

PROBLEM 564

In part (b) of problem 129 we assumed for simplicity that the flapper mechanism provided a constant negative angular acceleration. We now decide to question the realism of such an assumption, and we frame our question in two different ways.

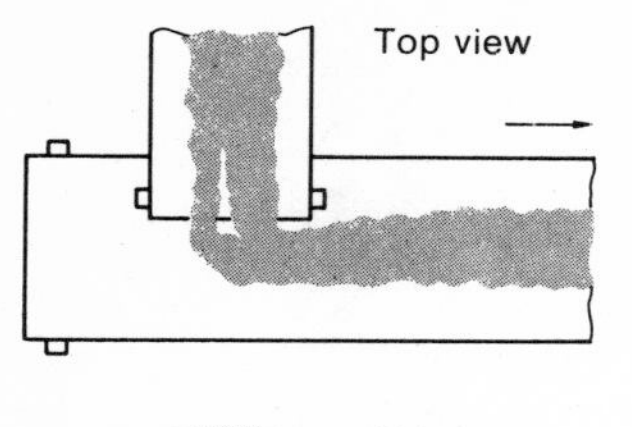

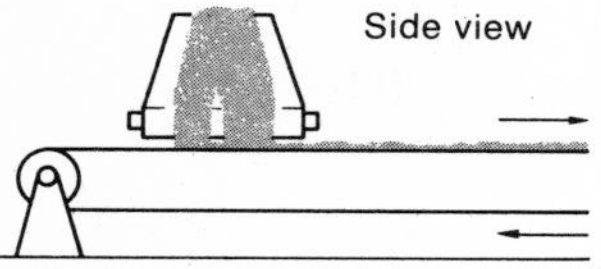

FIGURE 349

a. If the assumption of uniform acceleration actually was correct, what does this say about how the force of the individual impulses from the flapper mechanism must vary with the speed of the wheel?
b. If, on the other hand, the force from each individual impulse from the flapper mechanism remained at a constant value, what must have been the relation between the acceleration of the wheel and its speed?

PROBLEM 565
In the process area of the gypsum mill at Plaster City, California, a belt conveyor running southward drops 20 lb of material sec^{-1} onto a belt conveyor moving horizontally eastward at a speed of 5.0 ft sec^{-1} (Figure 349). If half the power supplied to the eastward conveyor is used to overcome frictional losses, what is the total power necessary to keep the eastward belt running?

PROBLEM 566
A flexible, uniform rope of total length L and total mass M lies in a loose coil on the ground. When one end of the rope is lifted with a constant vertical speed v (Figure 350),

a. what power is being supplied (note that the power is a function of y, the distance of the end above the ground)?
b. How much of this power is being dissipated, i.e., is not being conserved?
c. How do you account for this loss of power?

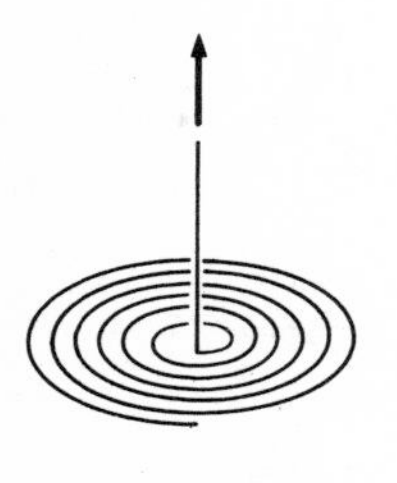

FIGURE 350

PROBLEM 567
On page 8 we arrived by dimensional analysis at a formula for the force exerted by a wind on a flat surface perpendicular to the direction of the wind; this formula included a dimensionless constant k. Can you now make an estimate of the probable upper and lower limits for the numerical value of this constant?

PROBLEM 568
A long, club-like object lying at rest on a smooth horizontal surface is struck a horizontal blow perpendicular to the longitudinal centerline of the object at a distance d from its c.m. (Figure 351). What is the resulting *initial* motion of each point along the centerline of the club? *Suggestion:* Center your x-y-z coordinate system at the c.m., with the y-axis along the original position of the centerline and the z-axis perpendicular to the surface. Then each point on the centerline can be identified by its original y-coordinate.

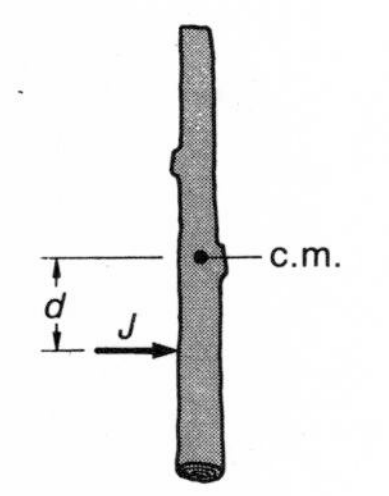

FIGURE 351

SOLUTION
We learned earlier that a useful aspect of the concept of the c.m. was that the c.m. of a body moved as if all of the external forces on the body were applied at the c.m. This statement can be expanded to include rotational motion: a body rotates about an axis through the c.m. in accordance with the sum of the torques of the external forces about

this axis. In applying these statements to any particular situation *all* external forces, including any induced restraint forces, must appear in the summation of forces or torques. For example, an analysis of the motion of the door of the president's office being kicked open by a raiding group of nonstudents must include the induced force at the hinges.

These statements about forces can easily be extended to impulses. The c.m. of a body acquires a change in linear momentum as if all the external impulses were applied at the c.m., and the body acquires a change in angular momentum about an axis through the c.m. in accordance with the sum of the angular impulses about that axis [(*a*), Figure 352].

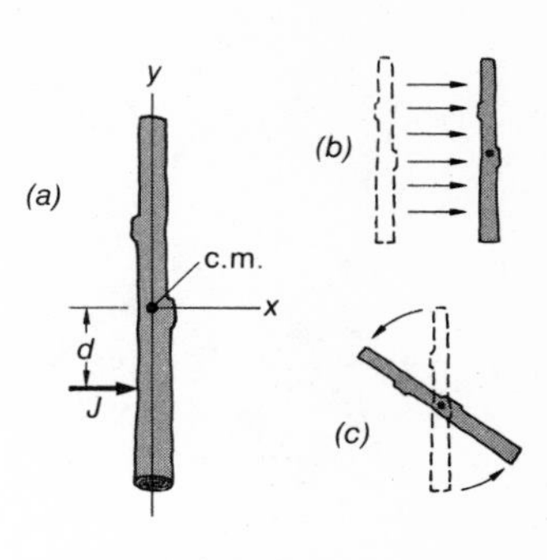

FIGURE 352

Returning now to the body in our problem,

$$\overrightarrow{\Delta p} = mv_{x,\text{c.m.}}\vec{i} = \vec{J}.$$

so

$$v_{x,\text{c.m.}} = \frac{J}{m},$$

and

$$\Delta\vec{L} = I_0\vec{\omega}_0 = \vec{d} \times \vec{J},$$

so

$$\omega_{0,z} = \frac{Jd}{I_0},$$

From this we can see that the total initial motion of the club is compounded of a motion of the entire club in the *x*-direction [(*b*), Figure 352] and at the same time a rotation of the club about the *z*-axis [(*c*), Figure 352]. Then the total initial motion of a point on the centerline whose coordinate is *y* is given by

$$\dot{x}_y = V_{x,\text{c.m.}} - \frac{yJd}{I_0} = J\left(\frac{1}{m} - \frac{yd}{I_0}\right).$$

For $y = 0$, $\dot{x} = \dfrac{J}{m}$;

for $y = \dfrac{L}{2}$, $\dot{x} = J\left(\dfrac{1}{m} - \dfrac{Ld}{2I_0}\right)$;

for $y = -\dfrac{L}{2}$, $\dot{x} = J\left(\dfrac{1}{m} + \dfrac{Ld}{2I_0}\right)$.

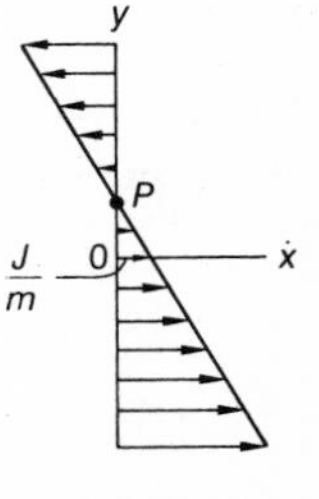

FIGURE 353

The graph of this motion is shown in Figure 353.

The most interesting point in the graph is at $y_P = I_0/md$, where $\dot{x} = 0$. In its initial motion the club moves apparently in pure rotation about an axis through *P*. For such an axis, the impulse point at $y = -d$ is often called the center of percussion. For any particular body the axis point *P* and the center of percussion are reciprocal points: if a perpendicular blow to the object at rest is delivered at y_P, the point at $y = -d$ will be found to be the location of an instantaneous axis of pure rotation. (As a useful exercise, you might prove this statement.)

These reciprocal points at y_P and d, where $y_P d = I_0/m$*, are not new to you; you have already seen them at d_1 and d_2 in problem on the Kater pendulum.

*As used here, y_P and d are both intrinsically positive, being measured from the c.m. in opposite directions. It is not possible for the two reciprocal points to be on the same side of the body relative to the c.m.

PROBLEM 569
Where should the straight rod hung from one end as a pendulum in problem 430 be tapped to start it in SHM without any initial horizontal reaction at the suspension?

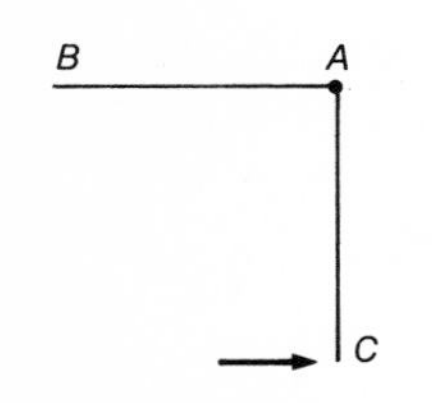

FIGURE 354

PROBLEM 570
A long, thin, stiff bar of mass m and length L lies on a smooth horizontal surface. It is struck at one end with a horizontal impulse J that is perpendicular to the bar. When the bar has turned end for end for the first time, how far has it moved from its original position?

PROBLEM 571
Two equal, uniform, stiff rods AB and AC are joined with a frictionless hinge at A and are placed on a smooth horizontal table with AC perpendicular to AB (Figure 354). AC is hit by a horizontal blow at C, perpendicular to AC. What is the ratio of the linear velocities of AB and AC immediately following impact?

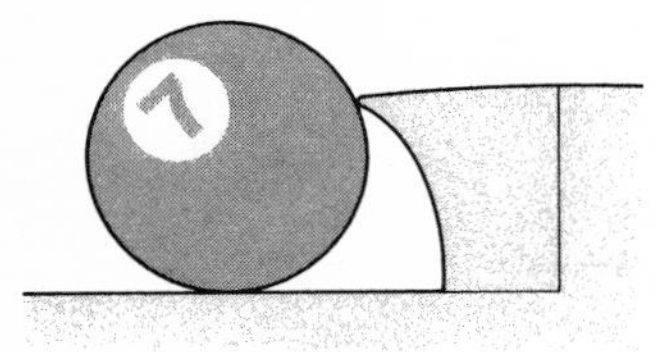

FIGURE 355

PROBLEM 572
A billiard ball rolls without slipping on the surface of a billiard table (Figure 355). How high above this surface should the point of the cushion be, in terms of the diameter of the ball, so that the ball rebounds from the cushion without slipping?

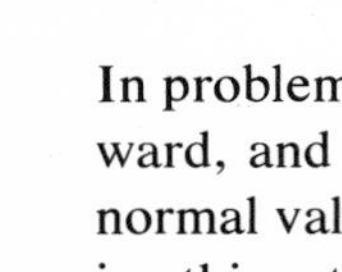

---

u

v = eu

FIGURE 356

The Coefficient of Restitution

In problem 556, the ball arrived at the floor with a speed $u = \sqrt{2gh_0}$ vertically downward, and rebounded with a speed $v = \sqrt{2gh}$ vertically upward (Figure 356). For normal values of u, and for most colliding masses, the ratio v/u will be constant. Calling this ratio the dimensionless constant e, we have $v = -eu$ ($-$ because v and u are in opposite directions).

In this simple example one of our colliding objects, the floor, was at rest in our fixed coordinate system. However, anyone who has seen an automobile accident involving two cars or has watched the tackling in a pro football game knows that

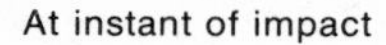

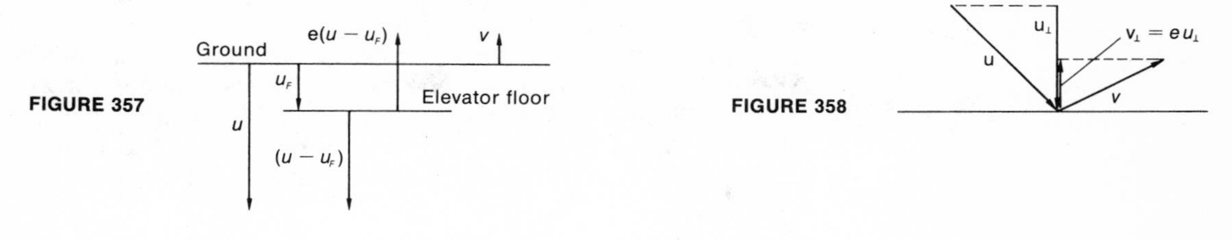

FIGURE 357

FIGURE 358

neither object in a two-body collision is usually at rest, relative to a fixed observer. To broaden the scope of our definition of e, let the floor in our example be a massive elevator floor moving vertically downward with a constant speed u_F relative to the ground (Figure 357). If the ball is moving with a speed u at the instant it reaches the floor, it will collide with the floor at a speed $(u - u_F)$ relative to the floor. It will rebound from the floor with a speed $(v - u_F)$ relative to the floor; v is the rebounding speed measured relative to the ground. Then,

$$(v - u_F) = -e(u - u_F). \qquad (4\text{-}25)$$

When two objects collide, in the line of collision the relative speed after collision is e times the relative speed before the collision, and in the opposite direction.

Equation 4-25 is not a vector equation. The coefficient of restitution e is a ratio only of the relative speeds that are directly in the line of collision. For most collisions this direction is perpendicular to the mutual surface at contact; for round objects this means the collision direction is along the line of centers at contact. Thus for a ball bouncing on a pavement at an angle (Figure 358), equation 4-25 can be written only for $u_\perp$ and $v_\perp$, the components of the ball's velocity perpendicular to the pavement.

The value of e is always ≤ 1. (Why should you reject a value of $e > 1$?) The statement that a collision is "perfectly elastic" is another way of saying that $e = 1$. (As we shall see a little later, it is also a way of saying that no energy is lost in the collision.) When a collision is reported as "inelastic," $e = 0$; kinetic energy has been lost, and the relative speed after impact is zero. Inelastic impact occurs when you fire a test bullet into the wooden plug of a ballistic pendulum, or when you drop a soft fried egg on the floor.

PROBLEM 573

A ball is dropped onto a solid, firmly anchored, smooth inclined plane, with which it makes a perfectly elastic collision (Figure 359). The angle θ of the plane with the horizontal is adjusted so that the range R of the rebound trajectory is a maximum. What is the value of θ? *Comment:* You can think your way to the correct answer without putting pencil to paper. Try it.

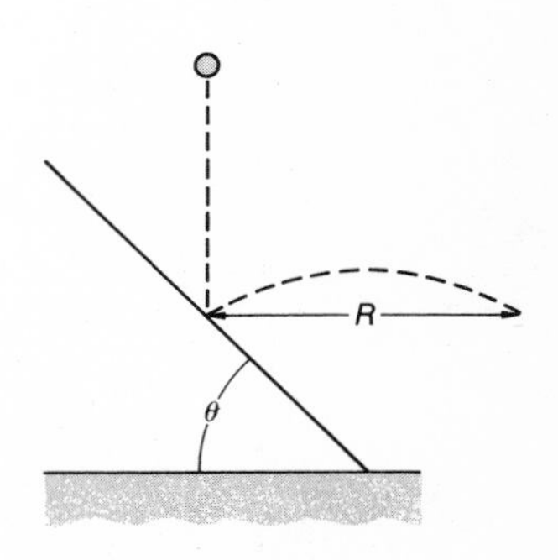

FIGURE 359

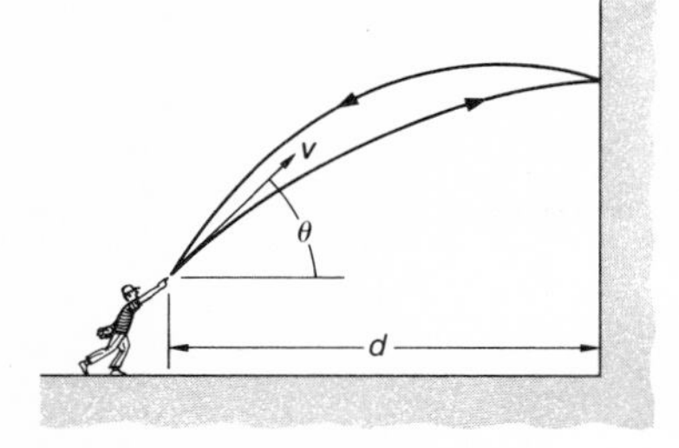

FIGURE 360

PROBLEM 574

A boy playing "catch" by throwing a ball against a vertical wall wants the ball always to return to the same point from which he threw it, as shown in Figure 360. For this to happen, what is the necessary relationship between the speed v with which he throws the ball and the angle θ of the throw? The coefficient of restitution between ball and wall is 0.75.

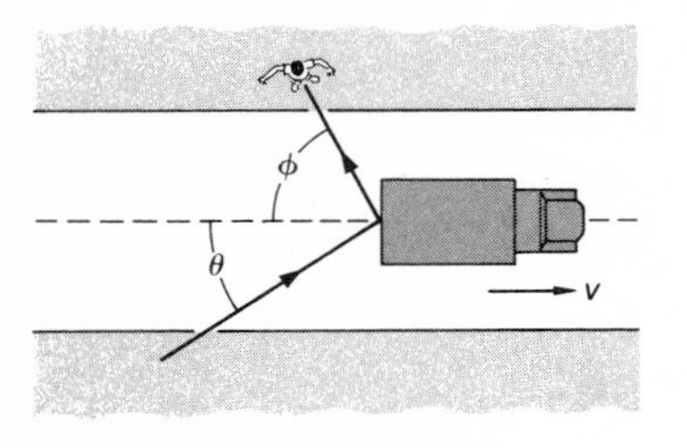

FIGURE 361

PROBLEM 575

During a civil disturbance a police paddy wagon is moving down the middle of a street with a speed v. A bottle is thrown at the wagon with a horizontal speed u at an angle θ with the centerline of the street. The bottle bounces off the vertical rear door of the wagon at an angle ϕ with the centerline and hits an innocent bystander on the opposite sidewalk (Figure 361). If the coefficient of restitution between bottle and wagon was e, and there was essentially no friction, what was v? *Suggestion:* Consider the collision between bottle and wagon in a coordinate system moving with the wagon.

PROBLEM 576

For high impact speeds a steel ball may exhibit a variable coefficient of restitution against a steel plate; this variation may be expressed by the relation $e = e_o h_o g/(v^2 + h_o g)$, where h_o is an experimentally determined constant that has the dimension of length, and v is the speed of the ball at impact with the plate. If such a ball is dropped on a steel plate that is immovable,

a. to what maximum height will the ball bounce, regardless of the height from which it is dropped?

b. From what height was the ball dropped to obtain this maximum bounce?

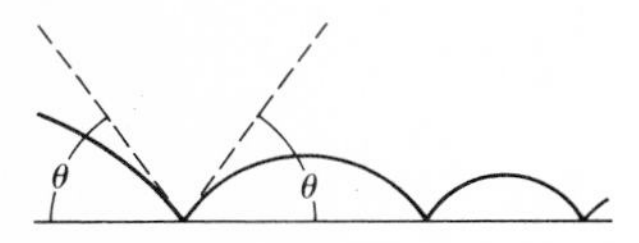

FIGURE 362

PROBLEM 577

A mass is bouncing forward on a horizontal surface (Figure 362). The coefficient of restitution between the mass and the surface is e. What must be the coefficient of sliding friction μ, such that all the bounces are symmetrical with each other, i.e., such that the angles with the horizontal of the velocity vectors at the surface, both arriving and departing, are always equal? *Hint:* If there is a $\vec{J}$ perpendicular to the surface of contact, there can be a $\mu\vec{J}$ parallel to that surface.

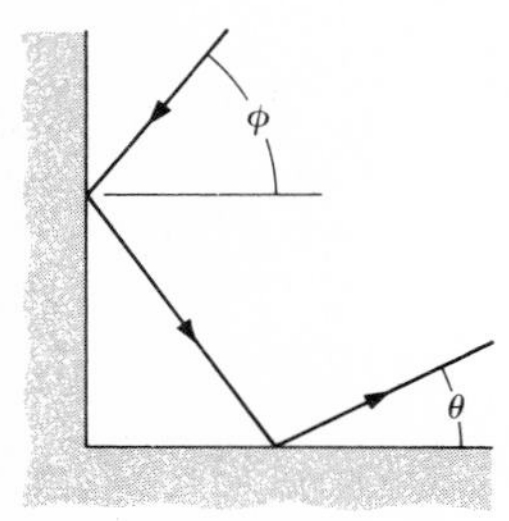

FIGURE 363

PROBLEM 578

A billiard ball is rolled into the corner of a billiard table, as shown in Figure 363, with an initial speed u at an angle ϕ with the normal to rail first hit. For simplicity (and fairly realistically), assume no tangential friction and that e for a ball-rail impact is the same everywhere on the table.

a. What is the angle θ of the rebound out of the corner? (Would the

answer to this explain the presence of the small diamond markers set in the rail of a billiard table?)
b. If $e = 0.90$, how many complete corner rebounds like the one shown are possible before the speed of the ball drops to $0.6\,u$?

Since the kinetic energy of an object is related to its speed, and the coefficient of restitution measures changes in relative speed, you might guess that there is some way to use e as an indicator of changes in kinetic energy. If you did, you guessed correctly.

In problem 556, the ball of mass m arrived at the floor with a relative speed u, and a kinetic energy of $\frac{1}{2}mu^2$. It rebounded with a relative speed of $v = eu$ and a kinetic energy of $\frac{1}{2}mv^2 = \frac{1}{2}me^2u^2$. Thus, $T_f = e^2T_0$. As a result of the collision, the ball lost kinetic energy $\Delta T = (1 - e^2)\,T_0$; the percentage loss, $\Delta T/T_0 = (1 - e^2)$.

In the revision of that problem, in which the floor moved with a constant speed u_F, the final kinetic energy measured *relative to the floor* was $T_f = \frac{1}{2}me^2(u - u_F)^2$, the original kinetic energy *relative to the floor* was $T_0 = \frac{1}{2}m(u - u_F)^2$, the energy lost was $\Delta T = (1 - e^2)T_0$, and again the percentage loss was $\Delta T/T_0 = (1 - e^2)$. This is a simple introduction to the fact that whenever we investigate energy loss in collisions we will find $(1 - e^2)$ as a factor in the calculations.

Note the italic phrases in the paragraph above: the collision took place in the floor coordinate system, and any calculation based upon e must relate to the system in which the collision took place. (This particular example—the moving massive floor—is a special case of a collision in a center-of-mass coordinate system to be discussed later.)

PROBLEM 579
A golf ball driven 200 yd in the air over level ground lands on a horizontal rock surface and bounces into the air for another 50 yd. What was the coefficient of restitution between ball and rock?

PROBLEM 580
a. When a ball dropped onto a solid, horizontal floor has bounced n times, how much of the kinetic energy available at the first impact has disappeared? Express your answer in terms of e.
b. A ball dropped from a height h above a solid, horizontal floor is observed after 6 bounces to rise only to $0.3\,h$. What was e for the collisions between ball and floor?

PROBLEM 581
A metal surface that is to be lightly peened for a decorative effect is mounted on a heavy slab that is moved slowly in a horizontal direction under a line of falling steel balls (Figure 364). The surface is placed at an angle θ with the horizontal so that the balls will bounce out of the way of the balls still coming down. When the horizontal speed of the surface is negligible compared to the speed of the falling

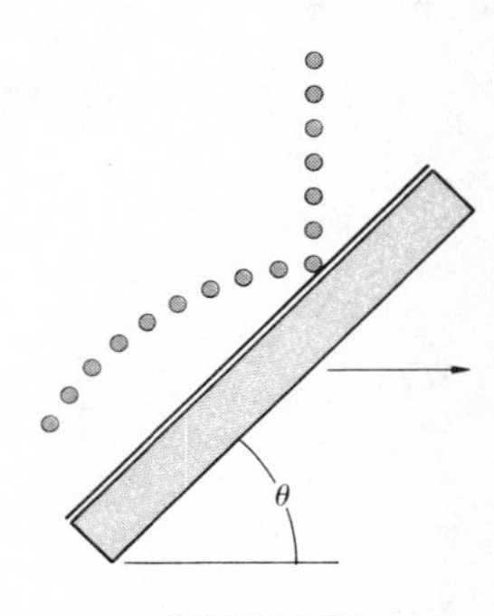

FIGURE 364

balls as they reach the surface, it is noted that the balls rebound from the surface horizontally.

a. What was the coefficient of restitution between balls and metal surface?
b. What percentage of the kinetic energy of the balls at impact is lost in impact?

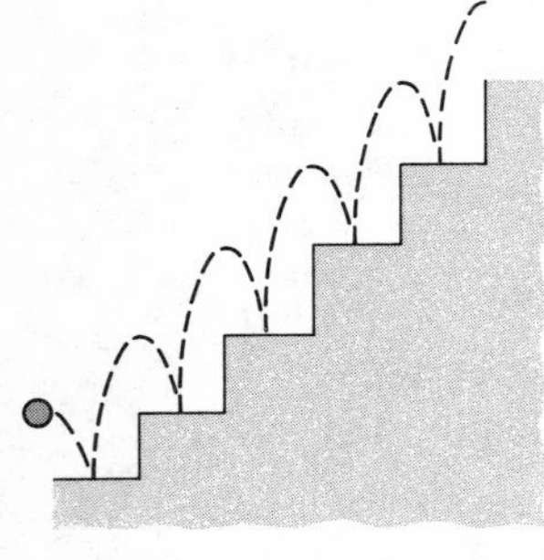

FIGURE 365

PROBLEM 582
A ball bounces down some smooth steps, in a plane perpendicular to the front edges of the steps (Figure 365). It is observed that the ball always bounces exactly in the middle of each step, and that after each bounce it rises to the height of the previous step. If the steps are 1.0 ft high and 1.0 ft deep,

a. what must be the value of the coefficient of restitution between the ball and the steps?
b. What is the velocity of the ball at the top of its trajectory after each bounce?

CONSERVATION OF MOMENTUM

Linear Momentum, Nonrelativistic—Including Collisions

If an isolated system is not acted upon by any external forces, or if the sum of the external forces $= 0$, then for the system $\int \vec{F}\,dt = 0$, and hence $\Delta\vec{p}$ must also $= 0$; linear momentum must remain constant, it must be conserved. If components of such a system possess at any time individual linear momenta $\vec{p}_{1,o}, \vec{p}_{2,o}, \ldots \vec{p}_{n,o}$ and at some later time possess $\vec{p}_{1,t}, \vec{p}_{2,t}, \ldots \vec{p}_{n,t}$, then

$$\vec{p}_0 = \vec{p}_{1,0} + \vec{p}_{2,0} + \ldots + \vec{p}_{n,0} = \vec{p}_t = \vec{p}_{1,t} + \vec{p}_{2,t} + \ldots + \vec{p}_{n,t}.$$

The linear momentum of the system must remain unchanged; the sum of the momenta of the individual parts of the system must remain unchanged. If components of the system interact with each other, by Newton's Third Law the interaction increases the momentum of one component by exactly the same amount it decreases the momentum of another component; internal interactions in a system do not change the sum of component momenta.

PROBLEM 583
Two gliders, one of mass 300 gm and one of mass 500 gm, are coupled together and sent down a horizontal air trough at a speed of 40 cm $\sec^{-1}$. As the gliders pass a certain point on the trough an electrical connection is made that fires an explosive cap between the gliders. If the explosion separates the gliders and causes the 300-gm glider to come to a stop, what is the post-explosion motion of the 500-gm glider?

PROBLEM 584

In a fixed, linear air trough a glider of mass m_1, moving with a speed u_1 relative to the trough, collides with another glider of mass m_2 moving in the opposite direction with a speed u_2. Assume a coefficient of restitution e between gliders.

a. What were the speeds of the gliders immediately after impact?
b. How much of the original kinetic energy disappeared in the collision?

PROBLEM 585

The simple, everyday way of comparing masses is to place them on a beam balance. This method is accurate to the extent that you can assume that all masses on the balance are subject to exactly the same gravitational field. However, the fundamental method for comparing masses, one that requires no assumptions, employs an inverse collision between the masses to be compared.

A standard 1.000-kg mass and an unknown mass M are placed in a horizontal air trough, with a short, compressed spring between them. When the string holding the spring in compression is burned, the masses move in opposite directions in consequence of the central impulse applied equally to each mass. For a certain measured time interval the standard mass traversed a distance of 0.811 m, and M, a distance of 0.520 m. What was the mass of M in kg?

PROBLEM 586

Before the days of air troughs and other approximately frictionless devices, in comparing masses by the method discussed in problem 585 allowance had to be made for friction. A standard mass of 1.000 kg and an unknown mass M are placed on a horizontal surface with a compressed spring between them. Previous to this, separate experiments have determined that the ratio of the friction coefficients for each mass relative to the surface is $\mu_{std}/\mu_M = 0.79$.

When the string holding the compressed spring is burned, it is observed that the standard mass slides 0.81 m before coming to a stop, and that M slides 0.52 m before stopping. What was the mass of M in kg?

PROBLEM 587

A football player weighing 190 lbwt comes through a hole in the scrimmage line with the ball at a speed of 26 ft sec^{-1}. Ten yards beyond the line he is tackled by a 208-lb player moving at a speed of 21 ft sec^{-1} at an angle of 120° with the path of the ball carrier; at the impact both players lost their footing. In what direction from the point of impact did they fall?

PROBLEM 588

Two equal masses moving with equal speeds collide inelastically and move on together with half their original speed. What must have been the angle between the two original velocities? *Comment:* You can work this out in a straightforward analytical manner in five or six steps. Or you can stop and do a little thinking and see the answer without putting pencil to paper.

PROBLEM 589

Two coupled, loaded freight cars have been shunted on to a team track branching off from a main railroad line, and are coasting along with a speed u_1. A switch engine sends an exactly similar freight car down the team track toward the other cars with a speed u_2, $u_2 > u_1$. A brakeman on this third car is supposed to operate the handbrake and slow the car down, but before he can do this he has a heart attack and falls off the car. When the third car crashes into the other two cars the automatic coupler operates and the three cars move on coupled together. (If you think railroad cars are too old-fashioned for a modern problem, then consider a glider in an air trough overtaking two similar, connected gliders moving in an air trough.)

a. How much energy had to be absorbed by the coupling mechanisms?
b. Would more, or less, energy have been lost if there had been originally only one car on the team track, instead of two?
c. If the original situation had been reversed, and two coupled cars had crashed into one car, how would the kinetic energy lost have compared with the correct answer to question (a)?
d. In general, if m_2 is a multiple of m_1, what is the ratio m_2/m_1 for which the reduced mass $m_1m_2/(m_1 + m_2)$ is the least?
e. How do the energy losses compare for the case of $u_1 = 5$ mi hr^{-1} and $u_2 = 10$ mi hr^{-1}, and for the case of $u_1 = 9$ mi hr^{-1} and $u_2 = 14$ mi hr^{-1}?

PROBLEM 590

As a traffic officer you are investigating a collision between a Chevrolet and a Cadillac, at a right-angled intersection in a 35 mi hr^{-1} zone, and you have assembled the following information: (1) the Chevrolet was travelling eastward and made skid marks 50 ft long from all four tires before the collision; (2) after the collision the Cadillac, wheels locked, slid 28 ft before coming to a stop 30° N of E of the impact point; (3) the Chevrolet bounced off the Cadillac and, bumper and broken wheel scraping, slid about 20 ft north of the impact point before coming to a stop; (4) the Chevrolet driver testified he was moving at 30 mi hr^{-1} when he saw the Cadillac and jammed on his brakes: (5) the Cadillac driver said he was driving north at 35 mi hr^{-1} and only saw the Chevrolet in time to jam on his brakes just as the cars hit; (6) the Cadillac weighed 1.5 times as much as the Chevrolet; and (7) the condition of tires and pavement indicated a probable coefficient of sliding friction of 0.70 for the Cadillac and 0.58 for the Chevrolet. To which driver did you issue a citation, and for what infractions? (Isn't it surprising, and maybe a little discouraging, to see how much more complicated a tabulation of information is required to describe a real situation compared to what is required for an ideal laboratory situation?)

PROBLEM 591

A lightweight, portable platform with stairs, for working on an airplane, weighs 600 lbwt. With the platform touching the fuselage of an airplane as shown in Figure 366, a workman weighing 160 lbwt stands on the bottom step listening to instructions from his foreman. Then, thinking the brakes on the platform are set, he runs up the steps. In fact, the brakes had not been set and the platform rolls freely. When the workman comes up against the guard rail at the back of the platform, how far away from the fuselage does he find himself? *Hint:* How far has the c.m. of the platform-man system moved?

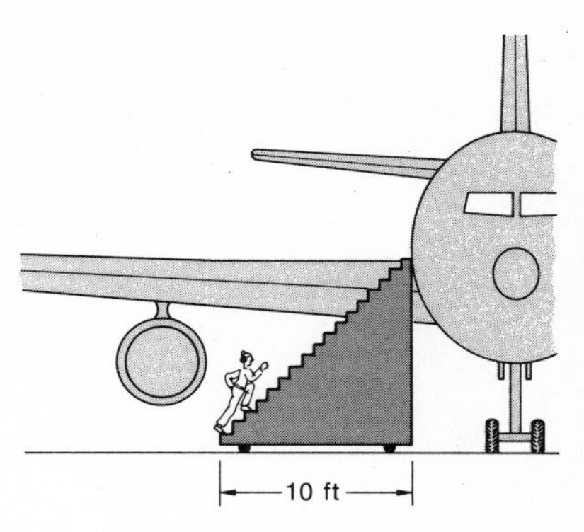

FIGURE 366

PROBLEM 592

A 500-gm glider is at rest on a horizontal linear air trough. To demonstrate a collision between two equal masses, an instructor mounts a 200-gm mass on top of a 300-gm glider and sends this assembly moving toward the 500-gm glider with a speed u. Both gliders are equipped with spring bumpers whose coefficient of restitution is essentially $= 1$. At the collision, the 200-gm mass jarred loose from its glider and slid onto the 500-gm glider, where it stayed. The empty 300-gm glider rebounded from the collision with a speed $-u/6$.

a. What was the post-collision motion of the loaded glider?
b. What percentage of the original kinetic energy disappeared in the collision?
c. What do you think happened to this lost kinetic energy?

PROBLEM 593

A body of mass m collides head-on with a body of mass M at rest. Even though the collision may be perfectly elastic, so that no energy is lost from the m-M system, m itself will lose kinetic energy in an amount that depends on the ratio m/M.

a. Calculate the percentage loss of the original kinetic energy of m as a function of the ratio m/M.
b. For what ratio m/M would this percentage loss be greatest?
c. For very low percentage loss, would it be better to have $m \ll M$ or $m \gg M$?
d. You want to slow down fast neutrons by surrounding the neutron source with a moderator whose function is to provide nuclei for the neutrons to collide with. With the above answers in mind, for the moderator would you choose beryllium (9 atomic mass units) or graphite (carbon, 12 atomic mass units)? A neutron has a mass of 1 atomic mass unit.

PROBLEM 594

A smooth rod bent into a semicircle is mounted in a vertical plane, with its ends on a horizontal line. Small beads are placed on the two ends and released simultaneously. After their first collision they are observed to rise only to a level $R/2$ above the bottom of the semicircle (Figure 367). What was the coefficient of restitution between the beads?

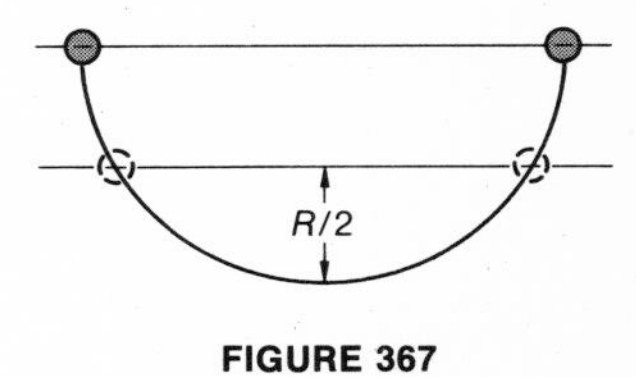

FIGURE 367

PROBLEM 595

The same experiment as in problem 594 was performed with the same equipment at a later date. By then the rod had become tarnished, so that there was a coefficient of friction μ between rod and beads; the beads rose only to $R/3$ above the bottom of the semicircle after their first collision. Assuming e had not changed, what was μ?

PROBLEM 596

An atom of radium C′ (atomic weight, 214) decays at rest into an atom of radium D (atomic weight, 210) and a helium atom (atomic weight, 4); during the decay 7.83 Mev of energy is released and appears as kinetic energy of the decay products. What is the speed of these products? (1 atomic mass unit $= 1.66 \times 10^{-27}$ kg.)

PROBLEM 597

At any instant a rocket in flight can be viewed as an isolated system whose linear momentum must be conserved. This conservation can be expressed by $dm/m = -(1/v_0)\, dv$, where m is the instantaneous mass of the rocket, v_0 is the speed relative to the rocket nozzle with which the burned fuel is being ejected, and dm and dv are changes in the rocket's mass and speed in an infinitesimal time dt.

a. If the rocket started from rest with a total mass M_0, what percentage of this mass must be available as fuel for the rocket to reach a speed v_o? Consider this as happening in free space, so that you can neglect gravitational forces.

b. At what speed of the rocket is the rocket motor most efficient? *Suggestion:* Don't work (b) out—think it out!

So far our collisions and momentum exchanges have been confined to very simple linear or rectangular situations. In general, however, for a system, $\vec{p}_o$ and $\vec{p}_t$ may have components at various angles. For these more complicated systems there exists a standard analytical procedure.

To illustrate, let us consider two air pucks colliding on a frictionless horizontal surface. (There should be no loss in generality in confining ourselves to two dimensions; the extension to a third dimension follows the same pattern and can be made quite easily if necessary.) From conservation of linear momentum we can write

$$m_1u_{1,x} + m_2u_{2,x} = m_1v_{1,x} + m_2v_{2,x},$$

$$m_1u_{1,y} + m_2u_{2,y} = m_1v_{1,y} + m_2v_{2,y}, \tag{4-26}$$

where the u and v speeds are respectively before and after the collision, measured in an inertial coordinate system—our own laboratory system normally.

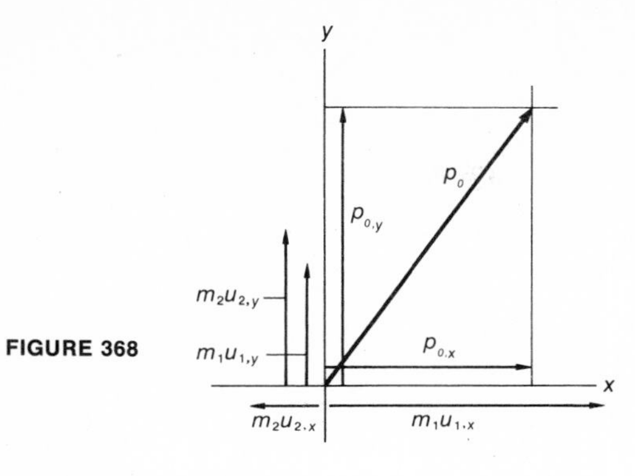

FIGURE 368

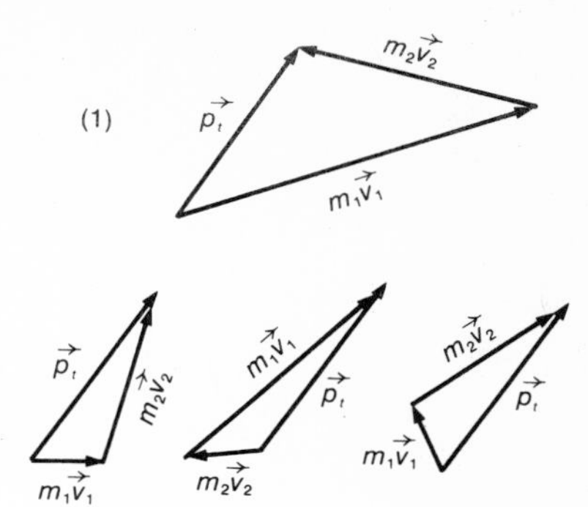

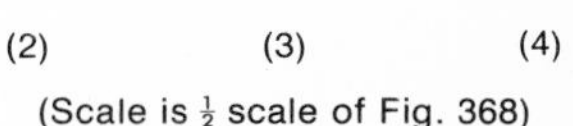

FIGURE 369

(Scale is ½ scale of Fig. 368)

(Note that when inserting actual scalar values for u and v in equations 4-26, with the value itself you also insert a + or − sign to indicate direction in the chosen coordinate system; this makes equations 4-26 linear vector equations. Thus for the example shown in Figure 368, $p_{o,x} = m_1u_{1,x} - m_2u_{2,x}$, the minus sign indicating that $u_{2,x}$ is in the negative direction.)

Usually we know m_1 and m_2, and in most cases we know $\vec{u_1}$ and $\vec{u_2}$. Thus we have two independent equations with the four unknowns $v_{1,x}$, $v_{1,y}$, $v_{2,x}$, and $v_{2,y}$; to solve for unique values we need two more independent equations or two more items of information. Almost always, one of these two items has to do with energy. We can always write a third independent equation:

$$\tfrac{1}{2}m_1u_1^2 + \tfrac{1}{2}m_2u_2^2 = \tfrac{1}{2}m_1v_1^2 + \tfrac{1}{2}m_2v_2^2 + \Delta T. \tag{4-27}$$

Sometimes we can measure v_1 and v_2, sometimes we know e so that we can deduce something about ΔT or about a ratio u/v; usually we can make equation 4-27 work for us in some way.

The final necessary item of information usually comes from a knowledge of some direction—a direction at impact, or a direction of m_1 or of m_2 after impact (which will give us a value for $v_{1,x}/v_{1,y}$ or for $v_{2,x}/v_{2,y}$ in the coordinate system we are using).

The necessity for knowing a direction is illustrated in Figure 369. Taking the same precollision data, $\vec{p_t}$ must $= \vec{p_o}$. However, $\vec{p_t}$ by itself can be the result of any of the four possibilities shown, or of an infinite number of other possibilities not shown; we must know a direction of $m_1\vec{v_1}$ or of $m_2\vec{v_2}$ to discriminate among them. *Question:* In view of the original data which determined $\vec{p_o}$, is there anything about diagram (*1*), Figure 369, that makes you distrust it? *Answer:* By comparison with the earlier diagram for $\vec{p_o}$, even without accurate measurement it is easy to see that $m_1v_1 > m_1u_1$ and $m_2v_2 > m_2u_2$. Thus $\tfrac{1}{2}m_1v_1^2 + \tfrac{1}{2}m_2v_2^2$ is greater than $\tfrac{1}{2}m_1u_1^2 + \tfrac{1}{2}m_2u_2^2$—the kinetic energy after collision is greater than before the collision. In the absence of any special explanation justifying this unusual state of affairs, diagram (*1*) should be rejected as a possibility.

PROBLEM 598

An air puck of mass m_1, sliding eastward on a frictionless, horizontal surface with a speed u_1, collides with a motionless puck of mass m_2. At the instant of collision the center of m_2 was on a line $\theta°$ S of E from the center of m_1. What was the post-collision motion of both pucks? Assume a coefficient of restitution e between pucks, and no friction.

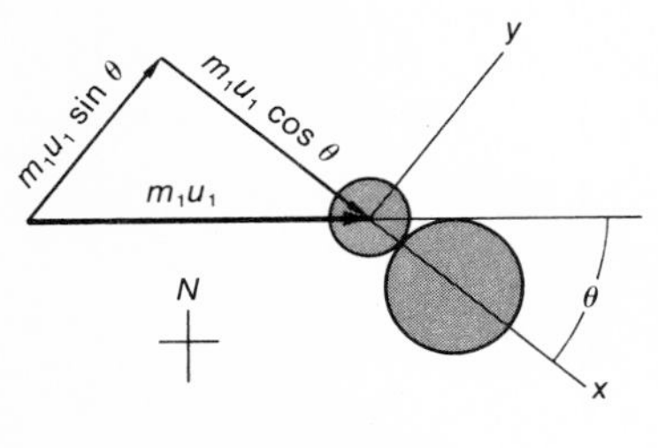

FIGURE 370

SOLUTION

For ease of description let us establish a coordinate system with the x-axis along the line of centers at impact, as shown in Figure 370. At impact the original momentum of the system m_1u_1 has components $p_{0,x} = m_1u_1 \cos\theta$ and $p_{0,y} = m_1u_1 \sin\theta$. $p_{0,y}$ is a grazing momentum that is not involved in contact with m_2; hence m_1 retains after impact its original momentum in the y-direction. However, m_1, moving in the x-direction with a speed of $u_1 \cos\theta$, collides head-on with the motionless m_2. This linear collision is exactly the same as the linear collision of the gliders in problem 584. By making appropriate use of the answers to that problem, we have

$$v_{2,x} = \frac{m_1u_1(1+e)\cos\theta}{m_1+m_2}.$$

$$v_{1,x} = \frac{u_1(m_1 - em_2)\cos\theta}{m_1+m_2}.$$

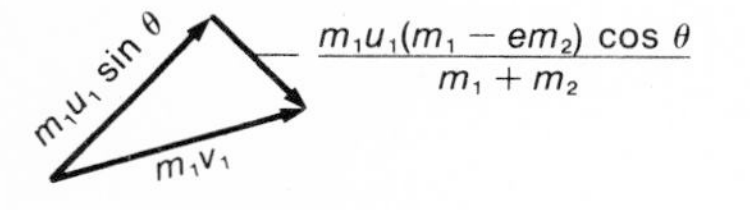

FIGURE 371

This is final speed of m_2, along a line $\theta°$ S of E. Assuming that $m_1 > em_2$, the post-collision momenta vectors for m_1 are as shown in Figure 371. Thus, after the collision m_1 moves with a speed

$$v_1 = \frac{u_1}{m_1+m_2}\sqrt{m_1^2 + m_2^2(\sin^2\theta + e^2\cos^2\theta) + 2m_1m_2(\sin^2\theta - e\cos^2\theta)}$$

at an angle

$$\phi = \theta + \tan^{-1}\frac{(m_1 - em_2)\cot\theta}{m_1+m_2} \text{ E of N.}$$

In collisions of the kind discussed in problem 598, two conclusions particularly should be noted:

1. If initially at rest, the struck object always leaves the collision along the direction of the line of centers at impact, providing there is no appreciable friction at the interface during impact.

2. Any loss in kinetic energy is subtracted only from the kinetic energy directly involved in the collision.

Thus for the present example, and utilizing the results of problem 584,

$$\Delta T = \frac{m_1m_2}{2(m_1+m_2)}(1-e^2)u_1^2\cos^2\theta.$$

PROBLEM 599

An air puck of mass m_1 sliding on a horizontal surface collides with another puck of mass m_2 at rest. It is observed that after the collision the pucks move at right angles to each other.

a. If it is assumed that the collision is perfectly elastic, what must be the ratio m_1/m_2?
b. If e for the pucks is known, and is < 1, what must be the ratio m_1/m_2?

PROBLEM 600

For the pucks of question (b) of problem 599, what percentage of the original kinetic energy disappeared? *Suggestion:* You can do this the long way, or you can take advantage of conclusion 2 following the solution to problem 598.

PROBLEM 601

An air puck of mass m_1 sliding on a horizontal surface collides with a mass m_2 at rest. After the collision, m_2 is observed to move along a line at θ with the original direction of m_1, and m_1 to move at right angles to its former path (Figure 372).

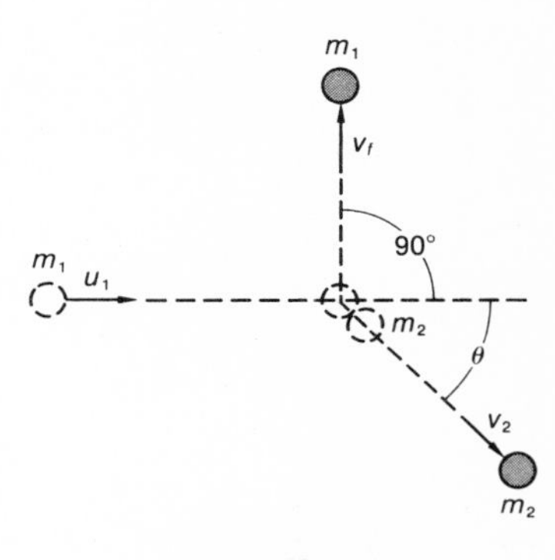

FIGURE 372

a. If you assume that the collision was perfectly elastic, what must have been the ratio m_1/m_2?
b. If, however, you knew that $m_1/m_2 = 1/2$, what must have been the coefficient of restitution e for the collision?
c. And, what percentage of the original kinetic energy was lost in the collision?

PROBLEM 602

In the Italian game of boccie, played on a flat field of level sand or ground, the object is to place your balls nearer a smaller target ball, called the jack, than are your opponent's balls. In the process, it is legal to make your ball collide with an opponent's ball and knock it away. When propelling your ball, you must stand behind a restraining line at the opposite end of the field.

It is your turn to roll, and your opponent's balls marked with an X are situated as shown (Figure 373) relative to the jack. The jack is so located on the field that the total angle of approach available to you at the restraining line is as shown. For maximum favorable effect, in what direction should you try to roll your next ball? Assume e between balls $= 1$. *Comment*: As in most games, there is probably no unique answer. However, there is one clear choice that would enhance your scoring position significantly.

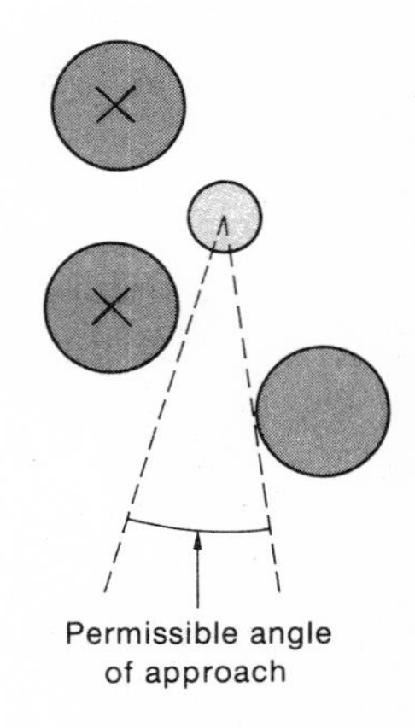

FIGURE 373

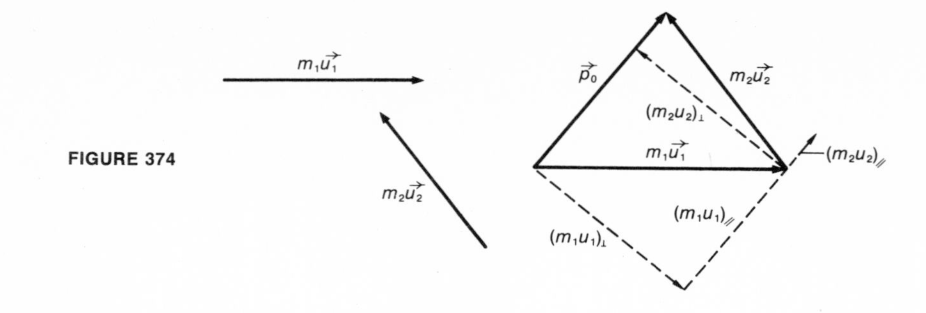

FIGURE 374

The Center-of-mass Coordinate System

When the size of colliding particles is so small that the distance between centers at impact is negligible, a coordinate system based on the center of mass (c.m.) of the colliding particles can be a very useful labor-saving device.

Let us consider the motion of two such particles about to collide. Looking at them in the plane determined by their momenta, we see them as shown in Figure 374. The vector $\vec{p}_0$, the sum of these momenta, can be evaluated by the addition of $(m_1u_1)_\parallel$ and $(m_2u_2)_\parallel$, components in the direction of $\vec{p}_0$. It can be seen from the diagram that components perpendicular to $\vec{p}_0$, $(m_1u_1)_\perp$ and $(m_2u_2)_\perp$, add to zero.

As seen from our inertial viewpoint (usually called the laboratory system in collision studies), the system's c.m. moves with linear momentum $\vec{p}_0 = (m_1 + m_2)\vec{v}_{\text{c.m.}}$, where $\vec{v}_{\text{c.m.}}$ is the velocity of the c.m. in our laboratory system. In a coordinate system moving with $v_{\text{c.m.}}$ – say a coordinate system attached to the c.m. – the only momentum visible is the sum of $(m_1u_1)_\perp + (m_2u_2)_\perp$ which, as you can see, adds to zero. We can make our notation slightly less cumbersome by using capital letters for speeds measured in the c.m. system; thus $m_1\vec{U}_1 + m_2\vec{U}_2 = 0$. Thus in a c.m. system *the total momentum is always zero.* From the viewpoint of the c.m., the particles are either moving toward a head-on collision or receding in opposite directions from such a collision. After the collision, $m_1\vec{V}_1 + m_2\vec{V}_2 = 0$, where $V_1 = eU_1$ and $V_2 = eU_2$.

Note that it is not required that $\vec{V}_1 = -e\vec{U}_1$ or that $\vec{V}_2 = -e\vec{U}_2$; the requirement that the momentum in the c.m. system is zero does not imply any requirement that the direction of $\vec{V}$ be related to the direction of $\vec{U}$. Hence the rebound from the collision may be in any direction, as shown in Figure 375. All of these, and an infinite number not shown, are possible inventories of post-collision momenta in the c.m. system. (In the last two possibilities shown, e is obviously < 1.)

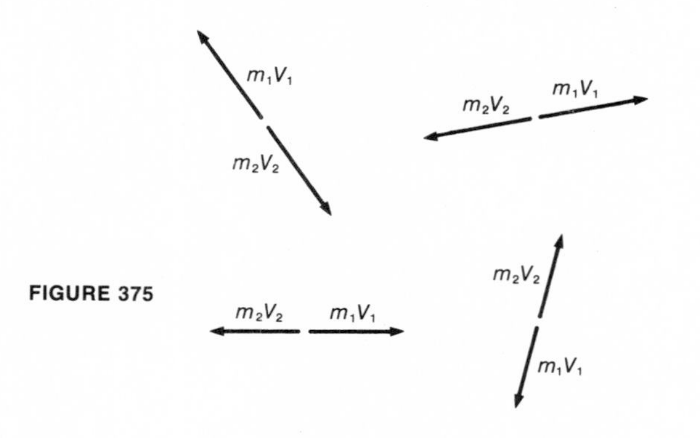

FIGURE 375

We don't see what happens in the c.m. system; what we see in the laboratory system are $\vec{v}_1$ and $\vec{v}_2$, described by $\vec{v}_1 = \vec{V}_1 + \vec{V}_{\text{c.m.}}$ and $\vec{v}_2 = \vec{V}_2 + \vec{V}_{\text{c.m.}}$ To get to these final equations, we need to go in order through the following steps:

$$\begin{aligned}
\vec{V}_{\text{c.m.}} &= \frac{m_1\vec{u}_1 + m_2\vec{u}_2}{m_1 + m_2} \\
\vec{U}_1 &= \vec{u}_1 - \vec{V}_{\text{c.m.}} = \frac{m_2(\vec{u}_1 - \vec{u}_2)}{m_1 + m_2} \\
\vec{U}_2 &= \vec{u}_2 - \vec{V}_{\text{c.m.}} = \frac{-m_1(\vec{u}_1 - \vec{u}_2)}{m_1 + m_2} \\
V_1 &= eU_1 = \frac{em_2|(\vec{u}_1 - \vec{u}_2)|}{m_1 + m_2} \\
V_2 &= eU_2 = \frac{em_1|(\vec{u}_1 - \vec{u}_2)|}{m_1 + m_2} \\
\vec{v}_1 &= \vec{V}_1 + \vec{V}_{\text{c.m.}} \\
\vec{v}_2 &= \vec{V}_2 + \vec{V}_{\text{c.m.}}.
\end{aligned} \tag{4-28}$$

Note that although the fourth and fifth equations are scalar, in the last two equations we return to vector statements. Between these two groups we take advantage of our flexibility about direction in the c.m. system to insert some independent statement about direction. This direction insertion, and the reduction of the vector equations 4-28 to very simple nonvector operations, is easily accomplished by the use of vector circle diagrams.

These circle diagrams are based on the fact that although $\vec{V}_1$ and $\vec{V}_2$ must lie along the same line but in opposite directions, the direction of this line is not determined within the c.m. system. To illustrate this, consider two concentric circles of radii V_1 and V_2 (Figure 376). *Any* collinear pair of radii will satisfy the requirement that $m_1\vec{V}_1 + m_2\vec{V}_2 = 0$. If to this circle diagram we then add the vector $\vec{V}_{\text{c.m.}}$, with its head at the center of the circles, we have all we need to solve the equation $\vec{v}_1 = \vec{V}_{\text{c.m.}} + \vec{V}_1$ and $\vec{v}_2 = \vec{V}_{\text{c.m.}} + \vec{V}_2$, providing we possess one additional item of information

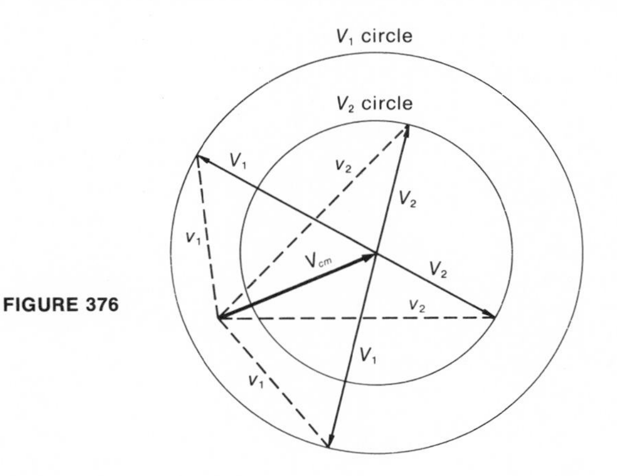

FIGURE 376

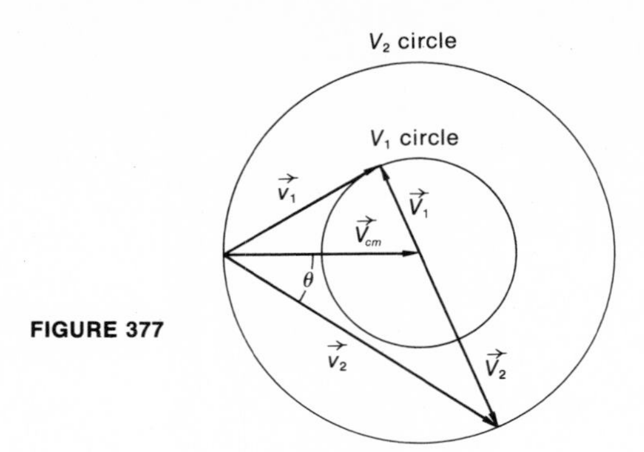

FIGURE 377

that enables us to select one particular vector triangle from all the possible vector triangles. (In practice, we would draw $\vec{V}_{\text{c.m.}}$ first and then erect the V_1 and V_2 circles on the head of this vector.)

As an illustration, let us redo problem 598 as a particle collision, with $m_1 > m_2$, with $e = 1$ for simplicity, and with the observation that m_2 leaves the collision at an angle θ with the direction of the original motion of m_1. Then

$$\vec{V}_{\text{c.m.}} = \frac{m_1\vec{u}_1}{m_1 + m_2};$$

$$\vec{U}_1 = \vec{u}_1 - \vec{V}_{\text{c.m.}} = \frac{m_2\vec{u}_1}{m_1 + m_2};$$

$$\vec{U}_2 = 0 - \vec{V}_{\text{c.m.}} = \frac{-m_1\vec{u}_1}{m_1 + m_2}.$$

For $e = 1$, $V_1 = U_1$ and $V_2 = U_2 = V_{\text{c.m.}}$ Drawing $\vec{V}_{\text{c.m.}}$, and erecting on it circles of radii V_1 and V_2 (all to the same scale, of course), we draw $\vec{v}_2$ from the tail of $\vec{V}_{\text{c.m.}}$ at the required angle θ (Figure 377). Its intersection with the V_2 circle determines $\vec{V}_2$, and this in turn determines $\vec{V}_1$. The vector $\vec{v}_1$ is then found as shown. By this process we have arrived graphically at the closed vector triangles $\vec{v}_1 = \vec{V}_{\text{c.m.}} + \vec{V}_1$ and $\vec{v}_2 = \vec{V}_{\text{c.m.}} + \vec{V}_2$.

Since the interior angle between $\vec{V}_{\text{c.m.}}$ and $\vec{V}_1$ is 2θ, the value of v_1 can be written down immediately from the geometry of the appropriate triangle. It might be interesting to compare this value with the value obtained in problem 598, after making allowance for $e = 1$.

PROBLEM 603

A slow neutron (mass = 1 amu*) moving through "heavy" water strikes a deuteron (mass = 2 amu) and is deflected through 90°. What percentage of the original kinetic energy of the neutron was acquired by the deuteron?

* 1 Atomic mass unit (amu) equals approximately the mass of 1 neutron or 1 proton; 1 amu = 1.66×10^{-27} kg.

PROBLEM 604

A particle of unknown mass M moving through a gas passes an atom and knocks loose one of the planetary electrons of the atom. The binding energy of the electron in the atom is so small compared to the energy of M that the collision can be considered perfectly elastic. Figure 378 shows a cloud chamber picture of the collision. What was the mass of M, in terms of m_{el}, θ, and ϕ?

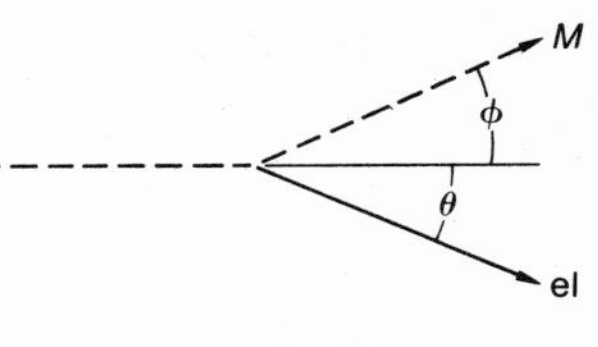

FIGURE 378

PROBLEM 605

A stream of neutrons of speed u (u nonrelativistic) is sent eastward through a cloud chamber. A proton formed from a prior collision moves northward at what is estimated to be a speed of $u/\sqrt{3}$, and is struck elastically by one of the neutrons. After the collision the proton makes a cloud-chamber track due eastward. Where did the neutron go after the collision, and with what speed? *Hint:* Remember, $\vec{V}_{c.m.}$ can be broken down into $V_{c.m.,x}$ and $V_{c.m.,y}$, and the vector equations $\vec{U}_1 = \vec{u}_1 - \vec{V}_{c.m.}$ etc. can be reduced to component equations $U_{1,x} = u_{1,x} - V_{c.m.,x}$, etc.

PROBLEM 606

A particle of mass m_1 moving with a laboratory speed u_1 collides with a particle of mass m_2 moving in the same direction with a laboratory speed u_2.

a. How much of the kinetic energy of the system is visible before collision to an observer in the c.m. coordinate system? Call this energy $T_{c.m.}$; this is the kinetic energy that will be involved in the collision.

b. Where is the rest of the original kinetic energy, and what is its value?

c. If there is a coefficient of restitution e between m_1 and m_2, how much kinetic energy is visible after the collision to an observer in the c.m. system? Call this energy $T'_{c.m.}$.

d. What percentage of the original energy seen in the c.m. system was lost in the collision?

e. What percentage of the original kinetic energy in the laboratory system was lost in the collision?

Comment: You can work these out in a straightforward manner, or you can take jumps by applying what you learned on page 282 and from problem 600.

PROBLEM 607

A particle of mass $2m$ moving on a smooth horizontal surface collides with a particle of mass m at rest. One-sixth of the original kinetic energy is lost in the collision. What is the maximum possible angle of divergence between the paths of the two particles after the collision?

PROBLEM 608

A moving mass m_1 collides with a stationary mass m_2 nonrelativistically.

a. Draw post-collision vector circle diagrams for the following possibilities: (1) masses equal, collision elastic; (2) $m_1 > m_2$, collision elastic; (3) $m_1 < m_2$, collision elastic; (4) masses equal, energy lost in the collision; and (5) $m_1 > m_2$, energy lost in the collision.
b. If the post-collision paths form an acute angle between them, which of the above possibilities might be true?

PROBLEM 609

a. In the diagrams called for in (a,2) and (a,3) of problem 608, what is the value of the ratio $V_1/V_{c.m.}$ in terms of m_1 and m_2?
b. If a stream of particles moving nonrelativistically is sent through a gas, and it is observed that the maximum deflection of any of the moving particles is about 15°, what can you conclude about the ratio of the mass of a moving particle to the mass of a gas molecule? Assume that the thermal motion of the gas molecules is negligible compared to the speed of the particles.

Relativistic Collisions

We have already learned that for three-figure accuracy we need to go to relativistic formulation when speeds get as high as about 10^7 m sec^{-1}. Such speeds are easily reached in particle collisions—both those occurring naturally (cosmic rays) and those induced in modern particle accelerators. Practically all modern physics textbooks contain cloud-chamber or bubble-chamber or photographic-emulsion or spark-chamber pictures of such collisions.

High-speed particle collisions are governed by three requirements:

1. conservation of linear momentum;

2. conservation of total energy (the collisions are considered to be perfectly elastic, so that the total energy content—kinetic plus mass energy—of any colliding system remains constant throughout the interaction); and

3. the momentum and energy inventories must be expressed relativistically.

You will remember that the relativistic expressions for linear momentum and energy are $p = \gamma mv$ and $E = \gamma mc^2 = mc^2 + T$. Putting these together we get

$$p^2c^2 = E^2 - m^2c^4. \qquad (4\text{-}29)$$

This is the controlling equation for all relativistic particle collisions. It can be applied separately to components of a colliding system or to the entire system; the only requirement is that the right-hand side of the equation, $E^2 - m^2c^4$, and the left-hand side, p^2c^2, must apply to the same masses at the same time.

PROBLEM 610

A particle of mass M at rest disintegrates to form two particles of rest masses m_1 and m_2. What are the resulting kinetic energies T_1 and T_2 of the disintegration products?

START OF SOLUTION

Almost always it is best to start by looking at the momentum. In this case $p_o = 0$, hence $p_f = 0$. But $p_f = p_1 + p_2$, so $p_1 = -p_2$. Then $p_1^2c^2 = p_2^2c^2$, which leads to $p_1^2c^2 = E_1^2 - m_1^2c^4 = p_2^2c^2 = E_2^2 - m_2^2c^4$. Next we look at the energy; we started with M at rest, so the only energy we have in the system is Mc^2. Thus $E_1 + E_2 = Mc^2$, or $E_2 = Mc^2 - E_1$. When we substitute this in equation 4-29, we get $E_1^2 - m_1^2c^4 = (Mc^2 - E_1)^2 - m_2^2c^4$. Solving this for E_1, and then recognizing that $E_1 = m_1c^2 + T_1$, we finally get

$$T_1 = \frac{[(M - m_1)^2 - m_2^2]c^2}{2M}$$

PROBLEM 611

In a high-energy proton accelerator neutral pions π° can be created in a proton-proton collision: $p_i + p_r \rightarrow p + p + \pi^\circ$; p_i is the incident high-energy proton and p_r is the target proton at rest before the collision. What minimum kinetic energy must be given to the incident proton to bring about this production of a neutral pion? (Proton rest mass = 938.3 Mev; π° rest mass = 134.9 Mev.)

START OF SOLUTION

To supply the minimum amount of energy before the collision, we must ask for the least amount of energy after the collision. This occurs when the collision products move along together with the same velocity (if their velocities diverge while at the same time maintaining the required linear momentum, their speeds must be greater). Hence for minimum energy we can treat the post-collision $(p + p + \pi^\circ)$ as one composite particle. Let us consider first the incident proton, which provides all the momentum. For it we can write $p_o^2c^2 = (m_pc^2 + T)^2 - m_p^2c^4$. Then for the system after collision we can write $p_f^2c^2 = E_f^2 - (2m_p + m_\pi)^2c^4$. But $p_f = p_o$; E_f, which is all the energy of the system after collision, must equal the energy before the collision, which was the energy of the incident proton $(m_pc^2 + T)$ plus the energy of the target proton (m_pc^2). Putting all this together, we should get $p_o^2c^2 = (m_pc^2 + T)^2 - m_p^2c^4 = p_f^2c^2 = (2m_pc^2 + T)^2 - (2m_p + m_\pi)^2c^4$. When you solve this for T you should get $T = 2m_\pi c^2 + m_\pi^2c^2/2m_p$. You should be interested to note that it took more than $2m_\pi c^2$ of energy to create a particle of rest mass $m_\pi c^2$.

PROBLEM 612

Photons, or gamma rays, are particles of radiation possessing no mass. In the light of equation 4-29, what must be the linear momentum p_ν of a photon?

PROBLEM 613

A gamma ray striking a free electron at rest can produce an electron-positron pair: $\nu + e^- \longrightarrow e^- + e^+ + e^-$. What is the minimum gamma ray energy needed for this reaction? (Electron rest mass $m_e = 0.511$ Mev.)

PROBLEM 614

A particle of rest mass m moving at relativistic speed v is captured by a free particle at rest also of rest mass m.

a. What is the rest mass of the composite particle?
b. What is the speed of the composite particle?

Suggestion: Your work and your answers will be less messy if you make your computations in terms of γ.

PROBLEM 615

In a hydrogen atom at rest in a gas a planetary electron drops from an excited level to one that is ΔE lower in energy, and a photon is emitted in the process. What is the energy of the photon? *Warning:* Conservation of linear momentum requires that the hydrogen atom gain as much linear momentum as the photon does, which means that some of the available energy must go to the recoiling hydrogen atom.

PROBLEM 616

A proton essentially at rest captures a slow neutron, forming a deuteron and emitting a photon of 2.26 Mev energy in the process: $p + n \longrightarrow {}_1H^2 + \mu$.

a. What is the resulting speed of the deuteron?
b. What is the rest mass of the deuteron? (Rest mass $m_p = 1.00759$ amu; rest mass $m_n = 1.00898$ amu.)

Angular Momentum

The techniques we have learned for dealing with linear momentum apply exactly and equally as well to angular momentum: $\int \vec{\tau} dt = \Delta \vec{L}$, an angular impulse produces a change in angular momentum; for an isolated system $\Delta \vec{L} = 0$, angular momentum must be conserved. This conservation of angular momentum must take place even when energy is not being conserved.

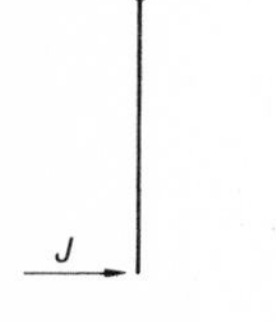

FIGURE 379

PROBLEM 617

A uniform, stiff bar of mass m and length L hangs at its upper end from a frictionless horizontal axle. If the bar is struck transversely at its lower end, what minimum impulse J must be supplied to get the bar to make a complete rotation about the axle (Figure 379)? Hint:

The angular impulse determines the angular speed with which the bar leaves its rest position. Then energy considerations must be used to determine whether the bar completes the desired rotation.

PROBLEM 618

An electron is moving in a plane that is perpendicular to a uniform magnetic field B; there is no electrostatic field in the area. What is the angular momentum of the electron in terms of its kinetic energy T?

PROBLEM 619

At the moment the bat of problem 555 hit the ball the bat's motion was essentially pure rotation about a vertical axis through the handhold on the bat. If the contact between ball and bat occurred at the center of percussion for that particular handhold, for a right-handed batter the angular momentum of the bat was $Md^2\omega$ vertically upward $-M$ the mass of the bat and ω its angular speed. Assuming the bat was swinging freely, what was its motion after it hit the ball? (From problem 45, the initial speed of the ball after being hit was 80 ft sec^{-1} upward.)

START OF SOLUTION

Let us refer all angular momentum to the point of the handhold on the bat. From problem 555 we find that the initial linear momentum of the ball was 98 m, where m is the mass of the ball; at d away from the handhold of a right-handed batter its initial angular momentum is then 98 md vertically downward. Hence $\vec{L}_0 = Md^2\omega - 98\ md$ upward. The final linear momentum of the ball was 80 m upward, hence its final angular momentum would be 80 md backward, away from the pitcher. Since $\vec{L}_f$ must $= \vec{L}_0$, $\vec{L}_{\text{final, bat}}$ must be as shown in Figure 380.

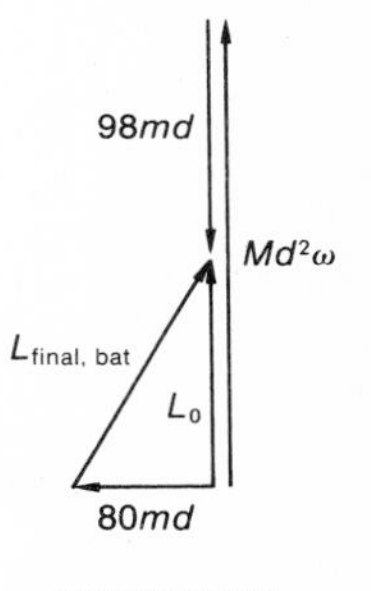

FIGURE 380

PROBLEM 620

The electron in the unexcited Bohr hydrogen atom (problems 226 and 227) is assumed to revolve about the proton nucleus at an average distance of 0.53 Å from its center.

a. What is the angular momentum of such an electron, in MKSC units?

b. In the same Bohr model the proton nucleus is a sphere with a radius of about 1×10^{-5} Å. If you assume that the hydrogen atom at rest should exhibit no angular momentum, what must be the angular speed of the proton?

c. Does the answer to question (b) pose electromagnetic difficulties? The orbiting electron creates a magnetic field in the environment of the proton (see paragraph preceding problem 295); is a revolving spherical charge stable in such an environment?

PROBLEM 621

A form of magnetron for determining the ratio q_{el}/m consists of a cylindrical conductor of inner radius R concentric with a cathode filament of radius a (Figure 381). A potential difference V is main-

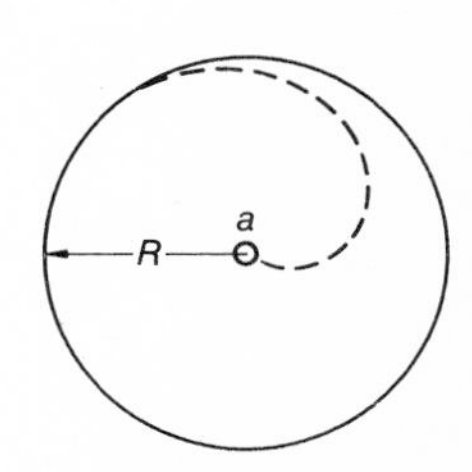

FIGURE 381

tained between cathode and cylinder and a constant and uniform magnetic field B parallel to the filament and cylinder is established inside the cylinder. Electrons boiled off of the filament at negligible speeds are accelerated toward the cylinder, providing a filament-cylinder current which can be measured. This current can be reduced to zero by adjusting V; at the critical value of V the electrons are turned by the magnetic field so that they just miss reaching the wall of the cylinder.

a. What is the ratio q_{el}/m in terms of this critical V and the constants of the apparatus?
b. At its point of tangency with the cylinder what is the radius of the electron path that just misses the cylinder?

START OF SOLUTION
The magnetic field does not affect the kinetic energy of the electron, but it does control the angular momentum; the electrostatic field does not affect the angular momentum of the electron, but it does control its kinetic energy. The crux of the solution lies in expressing these two statements in equation form. Taking the second statement, the easier one, first we find at R from the center that $\frac{1}{2}mv_R^2 = q_{el}V$. For the first statement we note that for any r, $a < r < R$, an electron will have a radial velocity dr/dt and a velocity $r(d\theta/dt)$ perpendicular to the radius; although both of these velocities will create $q\,\vec{v} \times \vec{B}$ forces from the magnetic field, only $q_{el}\,(d\vec{r}/dt \times \vec{B})$ will provide a torque about the center of the filament. Thus if L_r is the angular momentum of the electron about the center,

$$\frac{dL_r}{dt} = rq_{el}\frac{dr}{dt}B.$$

When we cancel the dt and set up the integral we get

$$\int_0^{mv_R R} dL_r = \int_a^R q_{el}\,Br\,dr.$$

Then v_R can be eliminated between the two statements.

PROBLEM 622
The stage area of the Ahmanson Theatre in Los Angeles can be provided with a rotating stage—essentially a uniform disc of mass 2×10^3 kg. In rehearsing for a musical show the director has the drive clutch released so the stage can rotate frictionlessly. On the stage he places a man and a woman sideways at opposite ends of a diameter; each of these weighs 70 kgwt (she's the comedy relief, so she's big). The man faces north and the woman south, with the stage motionless. At a musical cue the man starts running forward at a uniform speed around the perimeter of the stage; when he arrives next to the woman he stops. In which direction are the actors now facing?

PROBLEM 623
A long thin uniform rod of length L lies at rest on a frictionless horizontal surface. It is struck by a horizontal impulse J perpendicular to the rod.

a. Where should J strike so that the energy acquired by the rod is evenly divided between translational energy and rotational energy?
b. What is the total energy acquired?
c. When the rod is struck so as to meet the requirement in question (a), where on the rod is there a point that appeared instantaneously at rest as motion started?

PROBLEM 624
A framework made up of four equal members, each of length d and mass m, is held in a diamond shape by a massless cross member OP, as shown in Figure 382. While in this shape the framework is set rotating in its own plane with an angular speed ω_o about a transverse frictionless axis through P. Then the member OP breaks, and the framework collapses, also shown in Figure 382. How much energy was released in the collapse?

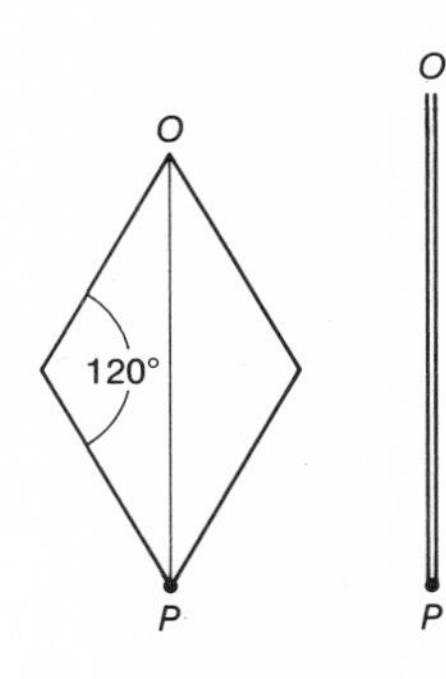

FIGURE 382

PROBLEM 625
Consider the Bohr model of the hydrogen atom discussed in problem 226.

a. What is E_r, the sum of the electron's kinetic and potential energies as a function of its orbital radius r?
b. What is the electron's angular momentum L_r as a function of E_r?
c. If the angular momentum is required to equal $n\hbar$, where $\hbar$ = Planck's constant h divided by 2π and $n = 1, 2, 3 \ldots\ldots$, what is E_n in terms of n?
d. If the electron moves from an $(n+1)$ orbit to an n orbit, how much energy is released?

PROBLEM 626
After the two actors of problem 622 met, the stage was started rotating; when it reached an angular speed of ω_0 the clutch was again disengaged. The man then walked from the rim of the stage to its center R away.

a. How much work did he have to do to reach the center?
b. How fast was the woman moving when the man reached the center?
c. To get a simple answer to these questions, you had to disregard something as negligible—what was it?

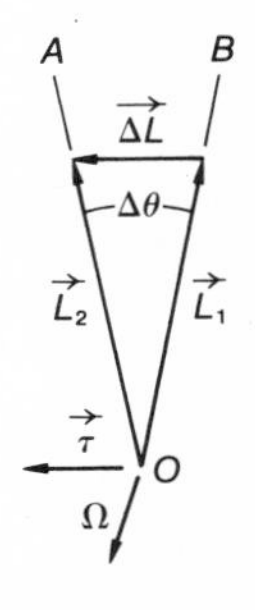

FIGURE 383

Precession

So far we have tacitly assumed that $\vec{\tau}$ and $\vec{L}$ were in the same direction, so that $\vec{\tau} = d\vec{L}/dt$ was used only to evaluate changes in the scalar value of L. However, there are many interesting situations in which L changes direction.

Let us consider a body spinning about its axis with an angular momentum L (Figure 383). At one moment the axis is in the direction OA, and the angular momentum vector is $\vec{L}_1$; Δt later the axis is in the direction OB, and the angular momentum vector is $\vec{L}_2$. In the time Δt the angular momentum has changed by $\vec{\Delta L} = \vec{\Delta\theta} \times \vec{L}$; there must be a torque in direction $\vec{\Delta L}$ to account for this. Then $\vec{\tau} = d\vec{L}/dt = d\vec{\theta}/dt \times L = \vec{\Omega} \times \vec{L}$, where Ω is the angular rate at which direction L is changing and is usually called the rate of precession. If, in the diagram, $\vec{L}$ stays in the plane of the paper, $\vec{\Omega}$ is perpendicular to the plane of the paper and $\vec{\Delta L}$ and $\vec{\tau}$ are in the plane of the paper. Thus, for changes in the direction of angular momentum we must write

$$\vec{\tau} = \vec{\Omega} \times \vec{L}. \tag{4-30}$$

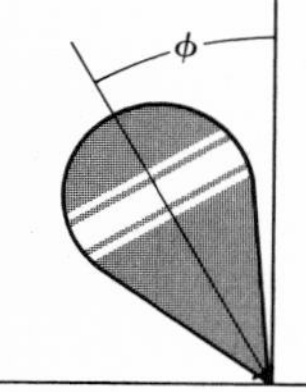

FIGURE 384

PROBLEM 627

A top of mass M and radius of gyration k about its axis of symmetry is spinning at R rev sec^{-1} with its axis at an angle ϕ from the vertical (Figure 384). Its c.m. is d from its point. At what rate is the top precessing about the vertical?

SOLUTION

The weight of the top provides a torque about the point of $Mgd \sin \phi$ (Figure 385); this torque is perpendicular to the plane determined by the axis of the top and the vertical, which is also the plane determined by L and Ω. Thus, $Mgd \sin \phi = \Omega L \sin \phi$, or $\Omega = 2\pi gd/k^2R$.

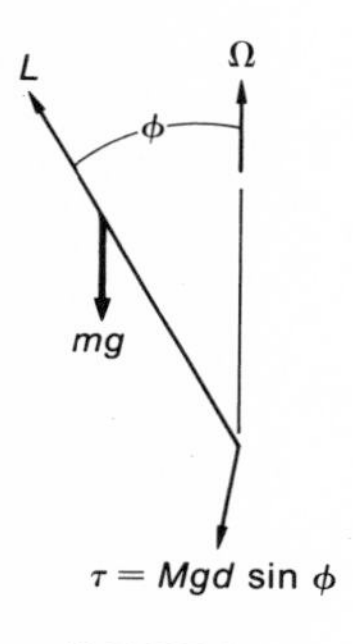

FIGURE 385

PROBLEM 628

A pilot who is used to twin-engine planes with counterrotating propellers is assigned to fly a plane whose twin propellers both rotate in a clockwise direction as viewed from in front of the plane.

a. When the pilot first attempted a left turn in level flight by rudder action alone, what happened?

b. After that experience, which told the pilot what was different about this plane, he did what when he wanted to make a left turn in level flight? Additional information: the pilot was a good student of physics.

In problem 627, about a top, the principal moment of inertia and the angular speed were both referred to the same axis; for axes perpendicular to this principal axis the angular speed, and hence the angular momentum, were zero. However, in cases where the angular speed vector and a principal moment of inertia axis are not collinear, components of L must be considered separately.

PROBLEM 629

A uniform, slender, stiff rod of mass M and length L is mounted at its center at an angle θ on a horizontal axle which rotates about its linear axis with an angular speed ω (Figure 386). If the axle bearings are d apart, what force perpendicular to the axle must each bearing be capable of supplying?

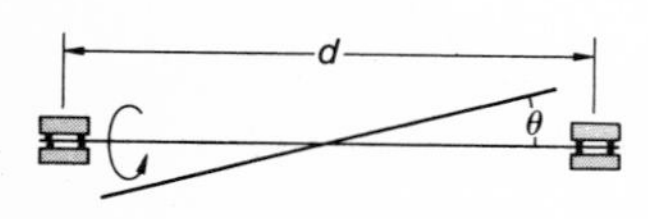

FIGURE 386

SOLUTION

The moment of inertia of the rod about its linear axis is essentially zero, hence its angular momentum about this axis is also zero (Figure 387). Its moment of inertia about a transverse axis through its center is $1/12\ ML^2$; the component of ω about this axis is $\omega \sin \theta$. The component of ω about a third axis perpendicular to those already considered is zero. Hence the total angular momentum of the bar equals $1/12\ ML^2\omega \sin \theta$. We know that this is precessing about the axle with an angular speed $\Omega = \omega$; hence the torque that is being supplied to the axle is $\tau = \omega \times L_\perp$. Then $Fd = 1/12\ ML^2\omega^2 \sin \theta \cos \theta$, or $F = (ML^2\omega^2 \sin \theta \cos \theta)/12d$.

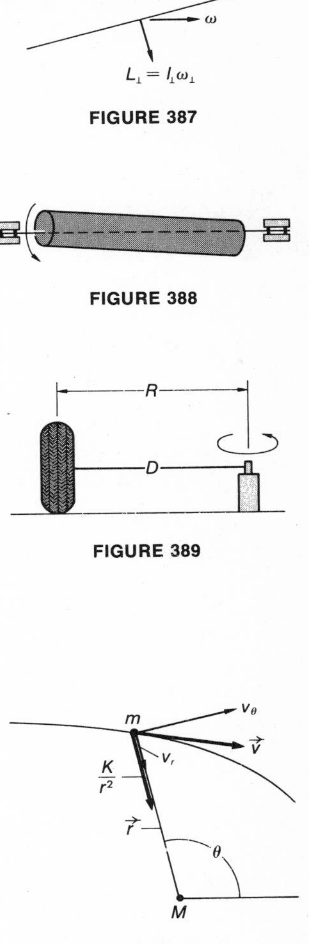

FIGURE 387

FIGURE 388

FIGURE 389

PROBLEM 630

An unbalanced crankshaft can be represented by a uniform cylindrical rod of radius r, length L, and mass M that is mounted at a small angle δ with its axis of rotation (Figure 388). When the shaft is rotating with an angular speed ω, what is the torque developed across the main bearings?

PROBLEM 631

In a tire-testing machine a mounted tire of mass M, radius r, and moment of inertia I about its central axis is caused to roll on a horizontal surface by a driving axle D. The driving axle is driven about a vertical axis with an angular speed ω counterclockwise as viewed from above (Figure 389).

a. With what force does the tire press on the horizontal surface?
b. How much would this force change if the direction of the driving axle motion was reversed?

FIGURE 390

Elastic Collisions Without Contact

In the collisions we have studied so far the reactions were the result of direct contact, and the colliding bodies moved linearly at specific angles relative to each other both before and after contact. Nature provides another class of collisions in which bodies react with each other to change their motions without contact and without loss of energy. However, there is a "collision" in the sense that motions change; these changes are not linear, but are conic in their geometry.

Let us consider the motion of a mass m whose trajectory in the neighborhood of a stationary mass M is as shown in Figure 390. At some instant m is r from M and has a velocity $\vec{v}$; the direction of $\vec{r}$ is measured from an arbitrarily chosen baseline. At that instant m is subject to an attractive force K/r^2; this force must supply both a radial acceleration and a centripetal acceleration, the latter because of the component v_θ. Hence

$$-\frac{K}{r^2} = m\ddot{r} - mr\dot{\theta}^2. \qquad (4\text{-}31)$$

Considering this as an isolated system, both angular momentum and energy must be conserved:

$$L = mr^2\dot{\theta} = \text{constant}; \tag{4-32}$$

and

$$E = \tfrac{1}{2}mv_r^2 - \frac{K}{r} = \text{constant}. \tag{4-33}$$

These three equations contain all the information we need to describe completely the motion of m in the field of M. (M has been assumed to stay stationary during the collision. In most practical cases $M >> m$, so this assumption does not introduce appreciable error. In those cases where the c.m. is not essentially coincident with M, we can regain accuracy in our computations by substituting the reduced mass $mM/(m+M)$ for m in the equations.)

For solving the equation of motion, equation 4-31, we need to evaluate $\ddot{r}$; from equation 4-32 $\dot{\theta} = L/mr^2$. Since it is fairly obvious that r and θ must be related, we can start with $\dot{r} = dr/dt = dr/d\theta\ d\theta/dt = dr/d\theta\ (L/mr^2)$. Then

$$\ddot{r} = \frac{d^2r}{d\theta^2}\dot{\theta}^2 + \frac{dr}{d\theta}\left(-\frac{2L}{mr^3}\right)\dot{r} = \frac{L^2}{m^2r^4}\left[\frac{d^2r}{d\theta^2} - \frac{2}{r}\left(\frac{dr}{d\theta}\right)^2\right].$$

When we substitute these values in equation 4-31 and pull everything together, we get

$$-\frac{Km}{L^2} = -\frac{1}{r} + \frac{1}{r^2}\frac{d^2r}{d\theta^2} - \frac{2}{r^3}\left(\frac{dr}{d\theta}\right)^2. \tag{4-34}$$

The solution to this is not obvious. However, whenever you are faced with an array of variables in the form $1/x^n$, it is useful to try to change the variable to its reciprocal. So in this case let $z = 1/r$. Then $dz/d\theta = -(1/r^2)\ dr/d\theta$, so that $dr/d\theta = -1/z^2\ dz/d\theta$. By the same process $d^2r/d\theta^2 = (2/z^3)\ (dz/d\theta)^2 - (1/z^2)\ d^2z/d\theta^2$. We substitute these values in equation 4-34 and rearrange, and finally get

$$\frac{d^2z}{d\theta^2} + \left(z - \frac{Km}{L^2}\right) = 0. \tag{4-35}$$

We have already learned how to solve equations like this, so we can write that

$$z = \frac{1}{r} = \frac{Km}{L^2} + A\sin\theta + B\cos\theta. \tag{4-36}$$

This is one form of the polar equation for a conic section; it can be made more informative with a little more work.

Although we do not yet know the specific form of m's path about M, there are some things we do know: there will be one point on the path that is closer to M than all other points on the path—let us call this distance of closest approach d (Figure 391); because of the conservation requirements for both angular momentum and en-

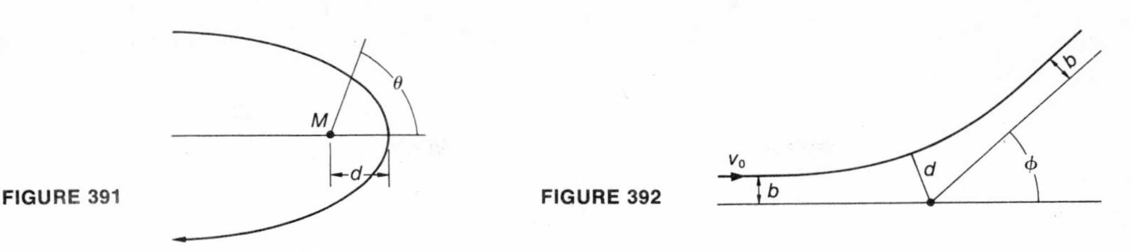

FIGURE 391

FIGURE 392

ergy the path must be symmetrical about some line; this axis of symmetry must be the same line along which we measure d. The $\theta = 0$ baseline was arbitrarily established without any physical significance; let us now decide that our axis of symmetry shall serve as our $\theta = 0$ baseline. By so doing, for $\theta = 0$ we have $r = d$, $\dot{r} = 0$, and $v_d = \dot{\theta} d$. Using these values, by judicious use of some of the work just done we find that $A = 0$ and $B = 1/d - Km/L^2$. Hence finally we have

$$\frac{1}{r} = \frac{Km}{L^2} + \left(\frac{1}{d} - \frac{Km}{L^2}\right) \cos\theta, \tag{4-37}$$

and

$$r = \frac{L^2/Km}{1 + [(L^2 - Kmd)/Kmd] \cos\theta}. \tag{4-38}$$

This is the equation of a conic section whose eccentricity $e = (L^2 - Kmd)/Kmd$. When $e = 0$, i.e., when $L^2 = Kmd$, m moves about M in a circle of radius L^2/Km. When $0 < (L^2 - Kmd)/Kmd < 1$, m moves in an ellipse with M at one focus and with a semimajor axis $a = Kmd^2/(2Kmd - L^2)$ and a semiminor axis $b = Ld/\sqrt{2Kmd - L^2}$. For $e > 1$, the motion is along a hyperbola with M at the focus; this is the motion that is similar to a collision reaction.

Often it is desirable to express m's motion in terms of E rather than L. Evaluating E at the distance of closest approach, $E = L^2/2md^2 - K/d$. Using this relationship to eliminate L^2 we get

$$r = \frac{2d(1 + Ed/K)}{1 + [(2Ed + K)/K] \cos\theta}. \tag{4-39}$$

Now that the algebra is done, it is time to give K a physical identification. For motion in gravitational fields, $K = GmM$; in electrostatic fields when q_1 and q_2 are of opposite sign, $K = q_1 q_2/4\pi\epsilon_0$.

For q_1 and q_2 of the same sign, $K = -q_1 q_2/4\pi\epsilon_0$ and the force between m and M is one of repulsion. This results in scattering: when positive charged particles approach a fixed positively charged particle they "scatter" in accordance with equation 4-38 with K changed to $-K$. In scattering, the parameters of interest are the incident energy T, the impact parameter b, and the scattering angle ϕ. The latter two are illustrated on the diagram in Figure 392.

To get these parameters into our equation of motion, return to equation 4-36, which now reads $z = 1/r = -(Km/L^2) + A \sin\theta + B \cos\theta$, and reevaluate the constants. At $r=\infty$ we know the following: $\theta=\pi$ or $\theta=\phi$, $\dot{r}=v_0$, $L=mv_0b$. Then $z=0=-(Km/m^2v_0^2b^2)-B$, or $B=-K/2Tb^2$. Also,

$$\frac{dz}{d\theta}=-\frac{1}{r^2}\frac{mr^2}{mv_0b}\dot{r}=-\frac{1}{b}=-A,\text{ or } A=\frac{1}{b}.$$

Then

$$\frac{1}{r}=\frac{1}{b}\sin\theta-\frac{K}{2Tb^2}(1+\cos\theta). \qquad (4\text{-}40)$$

We also have

$$z=0=\frac{1}{b}\sin\phi-\frac{K}{2Tb^2}(1+\cos\phi),$$

so that

$$\cot\frac{\phi}{2}=\frac{2Tb}{K}. \qquad (4\text{-}41)$$

Remember that in equations 4-40 and 4-41, $K=|q_1|\ |q_2|/4\pi\epsilon_0$.

PROBLEM 632

An electron is accelerated to a linear speed v; the line of $\vec{v}$ passes within a perpendicular distance r from a fixed positive charge Q. If both the electron and Q are treated as point charges, can Q be given sufficient charge so as to pull the electron into a head-on collision?

PROBLEM 633

A celestial object is observed at a very great distance from the sun to be moving with speed v on a course that, projected, would miss the sun by a distance D. Later it is observed that the object just grazed the surface of the sun at a distance R from its center. What must have been the value of v?

PROBLEM 634

A body is moving about the earth in a circular orbit of radius R. How much additional energy ΔE must be given to it to send it out into space in an open orbit?

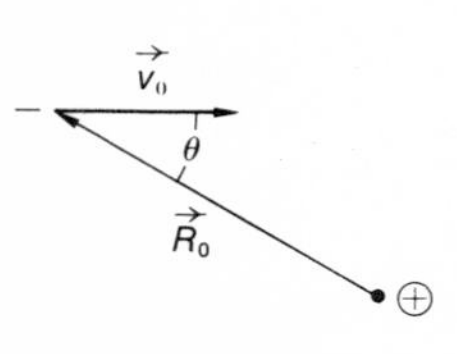

FIGURE 393

PROBLEM 635

A new comet is measured by astronomers on earth to be moving at a speed v_0 when it is R_0 from the center of the earth (Figure 393); the angle between $\vec{R}_0$ and $\vec{v}_0$ is $(\pi-\theta)$. How far from the center of the earth will the comet be at its closest distance?

PROBLEM 636

The astronomers announce that the comet of problem 635 is making a one-shot appearance; that it is on an open trajectory and will never return. For them to be able to make such a statement, what relation must they have observed between v_0 and the other parameters of the motion? *Hint:* Review the restriction on E for an open path.

PROBLEM 637

A celestial object is tracked in a hyperbolic orbit about the sun as shown in Figure 394. The impact parameter is b and the distance of closest approach is d. What can you say about v_0, the speed with which the object entered the trajectory at a very great distance from the sun?

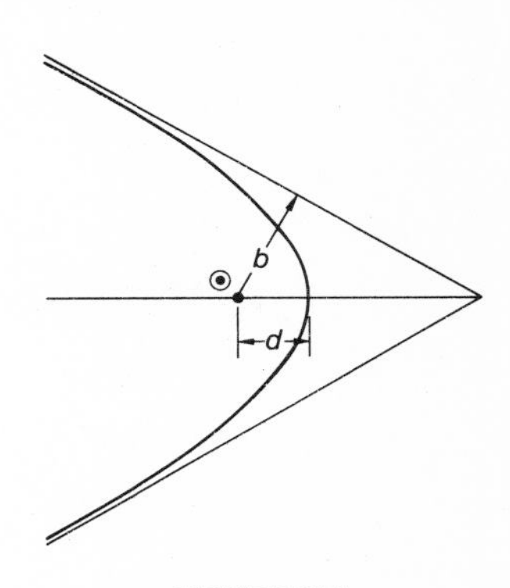

FIGURE 394

PROBLEM 638

In scattering, what is the value of the distance of closest approach d, in terms of T and the scattering angle ϕ? If m is a proton and M a nucleus, would d be the radius of the nucleus?

PROBLEM 639

In an effort to determine the upper limit to the value of the radius of a nucleus, an experiment is proposed in which positively charged particles are shot at a nucleus, and their impact parameter is adjusted until the deflection angle $\phi = 0$.

a. Is the proposal worth considering from a practical standpoint?
b. Is the proposal worth considering from a theoretical standpoint?

PROBLEM 640

A beam of identical positively charged particles is directed toward a nucleus. The particles in the beam are assumed to be uniformly distributed over the cross section of the beam; the density of this distribution and the cross-sectional area A of the beam are known. If $P\%$ of the incident particles are deflected into collectors surrounding the nucleus, what is the probable collision cross-sectional area of the nucleus?

5

APPLICATIONS OF THE CONSERVATION OF ENERGY

THE STORAGE OF ENERGY IN ORBITS

The orbits for m discussed in pages 295 to 297 all represent, from a coordinate system based on M, a closed system of $(M+m)$ in which energy is conserved. The most interesting of these orbits, particularly in the days of man-made satellites and space travel, are those that are closed and $e < 1$. For these closed orbits we have already learned that the semi-major axis $a = Kmd^2/(2Kmd - L^2)$. We substitute $L^2 = 2md^2E + 2Kmd$, and get the very informative statement:

$$a = \frac{K}{-2E} \text{ or } E = -\frac{K}{2a}.^* \tag{5-1}$$

This equation tells us several things:

1. in a closed orbit the total energy is negative. Since $E = T + U = -K/2a$, and T is always positive, U must be negative and its absolute value $|U|$ must be $> T$.

2. An obvious corollary to statement (1) is that as the major axis increases the $T - |U|$ difference decreases. At the limit, as $a \rightarrow \infty$, $T \rightarrow |U|$, and we have the necessary condition for determining the "velocity of escape" from a closed orbit. (A review of problem 492 and the graph you drew for question (c) of that problem may be helpful at this point.)

---

* Remember that $K = GmM$ or $|q_1q_2|/4\pi\epsilon_o$, the latter providing q_1 and q_2 are of opposite sign. Gravitational orbits and electrostatic orbits follow exactly the same description.

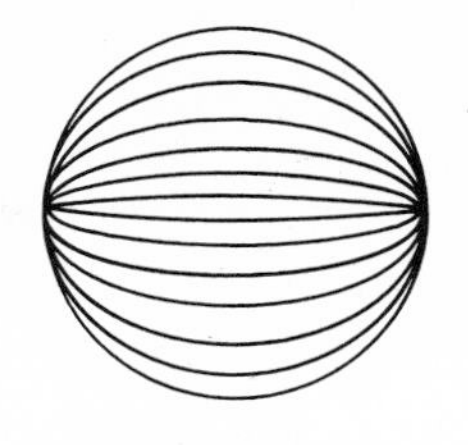

FIGURE 395

All of these orbits are supplied with the same amount of energy

3. All closed orbits of the same major axis possess exactly the same amount of total energy (Figure 395).

Kepler's first two laws are included in what we have already discussed beginning on page 295. Let us now see Kepler's Second Law (that the areal speed of a body moving in an orbit about another body is constant) to develop Kepler's Third Law, which relates the period of the motion to the size of the orbit.

From Kepler's Second Law,

$$\frac{dA}{dt} = \frac{d}{dt}\left(\frac{r^2 d\theta}{2}\right) = \tfrac{1}{2} d v_p \text{ at perigee}$$

(perigee is the point of closest approach, apogee is at the opposite end of the major axis at the point of maximum excursion). But

$$\frac{dA}{dt} \text{ also} = \frac{\pi a^2 \sqrt{1-e^2}}{T},$$

where T is the period of the motion and $\pi\alpha^2\sqrt{1-e^2}$ is the total area enclosed in the orbit. Also at perigee we can write $\frac{1}{2}mv_p^2 - K/a(1-e) = -K/2a$ with $d = a(1-e)$. Solving for v_p,

$$v_p = \sqrt{\frac{K(1+e)}{ma(1-e)}}.$$

We pull this together and get

$$\frac{dA}{dt} = \frac{1}{2}a(1-e)\sqrt{\frac{K(1+e)}{ma(1-e)}} = \frac{\pi a^2\sqrt{1-e^2}}{T}.$$

Solving this for T we get

$$T = \frac{2\pi a^{3/2}}{\sqrt{K/m}} = \frac{2\pi a^{3/2}}{\sqrt{GM}} \tag{5-2}$$

for gravitational motion. Equation 5-2 is frequently seen in the form

$$T^2 = \frac{4\pi^2}{GM}a^3$$

and in that form is known as Kepler's Third Law (cf. page 122).

PROBLEM 641

If a closed orbit is circular, how does the kinetic energy T compare with

a. the total energy E;
b. the potential energy U?

PROBLEM 642

How much energy is stored in the earth's orbit for each kg of the earth's mass?

PROBLEM 643

A satellite is in a three-hour orbit about the earth. In terms of its energy in the three-hour orbit, how much additional energy is required to boost it into a twenty-four hour orbit?

PROBLEM 644

a. What is the length of the semi-major axis of the orbit of a three-hour satellite of the earth?
b. Then what is the actual value of the energy required in problem 643 to boost the satellite into a twenty-four hour orbit?

PROBLEM 645

An earth satellite has a period of 90.0 min and at perigee it is 200 km above the earth's surface.

a. How far above the earth's surface is it at apogee?
b. How fast is it moving: i. at perigee, and ii. at apogee?

PROBLEM 646

A space vehicle sits on the moon's surface. What is the minimum amount of energy, per kg of mass of the vehicle, necessary to get the vehicle off the moon and back under control of the earth's gravitational field? Does it make any difference in your answer whether the vehicle was sitting on the back side of the moon or on the side facing the earth? *Suggestion:* It may help to refer to problem 205.

PROBLEM 647

In question (c) of problem 625 the total energy of the planetary electron in the hydrogen atom was determined as a function of the orbital number n.

a. When the electron is in its fundamental orbit, i.e., the atom is unexcited and $n = 1$, how much energy in electron volts is required to ionize the atom by removing the electron completely from it?
b. How does your answer compare with published values for the

binding energy (or the ionization potential) of the unexcited hydrogen atom?

PROBLEM 648

By 1984 the near extraterrestrial space had become so cluttered with debris, mainly burnt-out final stage boosters, that the U.N. obtained an agreement among all nations that all such stages should be separated from the payload vehicle in such a way that at separation, the rejected stage was momentarily at rest relative to the earth. Thus the stage would fall directly toward the earth with the hope that it would be vaporized on reentering the atmosphere. A payload plus final stage, weighing together M kgwt, are orbiting the earth in a circular parking orbit 200 km above the earth's surface. If the final stage, weighing m kgwt, is rejected in accordance with the above agreement,

a. what will be the semi-major axis of the new orbit for the payload?
b. What will be the new period for the payload?

PROBLEM 649

A newly discovered celestial object is tracked around the sun; at its closest approach it is observed to be 1.1×10^6 km from the center of the sun and is moving at a speed of 500 km sec^{-1}. What can you say about its orbit?

PROBLEM 650

The data on the celestial object of the above problem is reevaluated, and it is discovered that an error was made in reporting the speed of the object; the corrected speed at perigee was 488 km sec^{-1}.

a. Is the object actually a comet on a closed orbit about the sun?
b. If so, what is the semi-major axis of its path?
c. What is the eccentricity of its path?
d. What is the semi-minor axis of its path?
e. How long a time will elapse before it reaches perigee again, provided its motion is not perturbed by the planets?

PROBLEM 651

A satellite is in a closed circular orbit about the earth. Draw an energy vs. orbit radius graph showing the potential energy U_r, the kinetic energy T_r, and the total energy E_r (compare with problem 502). Then assume that the total energy decreases by an amount ΔE, but that the satellite's orbit remains circular. Making use only of the fact that $T = -E$ for circular orbits (see problem 641), show by the diagram that T must increase by an amount ΔE, and U must decrease by an amount $2\Delta E$. In other words, use the graph to demonstrate the proof of the well-known satellite paradox—that when you take energy away from a satellite it speeds up.

THE STORAGE OF ENERGY IN FIELDS

Nature provides us with gravitational, electrostatic, and magnetic fields, all of which can act as a source of energy. Gravitational fields are relatively static and not under our control; electric and magnetic fields outside of matter are under our control.

Electrostatic Fields

Let us first ask how the parallel plate capacitor of Problems 270 and 276 got its charge of $+Q$ on one plate and $-Q$ on the other. We will start with two neutral, uncharged, parallel conducting plates each of area A and a distance d apart (Figure 396); for generality consider the space between the plates filled with a dielectric of

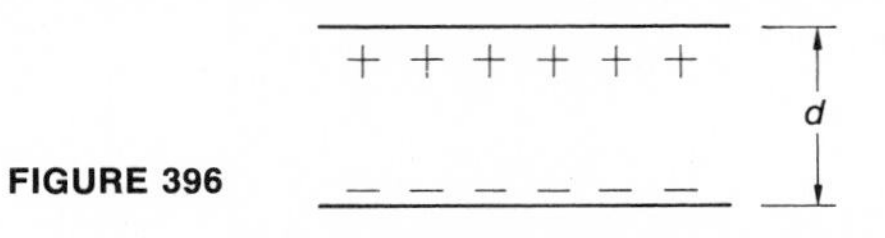

FIGURE 396

constant K. From the bottom plate we will extract a small amount of positive charge dq and carry it to the upper plate. The $-dq$ left behind will spread over the lower plate creating a small downward field $dq/KA\epsilon_0$; if we keep dq small the work to carry a $+dq$ upward against this field can be ignored. Repeating the process a number of times we will get a charge of $+q$ spread over the upper plate and $-q$ spread over the lower plate, with a downward field $q/KA\epsilon_0$ between the plates; to carry the next $+dq$ to the upper plate will require work to the amount of $dq(q/KA\epsilon_0)d$. Thus to get $+Q$ on the upper plate with $-Q$ left on the lower plate, we must supply

$$W = \int_0^Q dq \frac{q}{KA\epsilon_0} d = \frac{1}{2} \frac{Q^2 d}{KA\epsilon_0}. \tag{5-3}$$

By this work we have created potential energy U_{el}. There are a number of ways we can evaluate this energy:

1. Since $Qd/KA\epsilon_0$ equals the potential difference V between the plates, we can say

$$U_{\text{el}} = \tfrac{1}{2} QV. \tag{5-4}$$

2. Since the capacitance of the two plates is $C = Q/V$,

$$U_{\text{el}} = \tfrac{1}{2} CV^2 = \tfrac{1}{2} \frac{Q^2}{C} = \tfrac{1}{2} CE^2 d^2. \tag{5-5}$$

3. When we rearrange equation 5-5 and divide through by Ad, we get

$$\frac{U_{\text{el}}}{Ad} = \frac{U_{\text{el}}}{\text{vol}} = u_{\text{el}} = \tfrac{1}{2} K\epsilon_0 E^2, \tag{5-6}$$

where vol is the volume occupied by the field E and u_{el} is the energy density in the field.

These formulae were developed from consideration of a parallel plate capacitor, but they are accurately applicable to any electrostatic situation. Equations 5-4 and 5-5 are helpful in calculations involving capacitors; equation 5-6 is of more general usefulness.

PROBLEM 652

A parallel plate capacitor filled with dielectric of constant K is charged with $\pm Q$ on its plates, and then the charging source is disconnected. Using the method of virtual work, find the force required to keep separated a pair of parallel plates carrying equal but opposite charges. *Hint*: The work $F\,dy$ done by moving one plate perpendicular to its plane an infinitesimal distance dy must account for the resulting change dU_{el} in the energy stored in the field of the capacitor. After you have finished, compare your solution with that of problem 276; does one seem simpler and more direct to you than the other?

PROBLEM 653

A rotary variable capacitor has a capacity range from C_{max} to C_{min}, with $C_{min} = 0.1\ C_{max}$. The capacitor is set at C_{max}; its plates are charged with $\pm Q$; it is then disconnected from the charging source.

a. What was the charging voltage V_0?

b. What was the voltage V_f across the capacitor when the plates were rotated to the C_{min} position?

c. How much work was required to rotate the plates?

PROBLEM 654

A conducting sphere of radius R and carrying a total charge $+Q$ is immersed in an infinite, homogeneous, isotropic dielectric of constant K.

a. What is the radius R of a surrounding, concentric spherical surface in the dielectric which encloses one-half of the field energy of the charged sphere? *Hint:* Compare the result of integrating u_{el} over the enclosed area to the result of integrating u_{el} over all space.

b. How does your answer for question (a) compare with the location in space at which the electrostatic potential is one-half that at the surface of the charged sphere?

PROBLEM 655

The rotary variable capacitor of problem 653 has the same initial conditions as before—charged to $\pm Q$ when set at C_{max}. While still connected to the charging source it is rotated to C_{min}.

a. What was Q_f, the final charge on the plates?

b. How much energy was transferred between source and capacitor during the rotation?

c. How much work was required to rotate the plates?

The solution to question (c) problem 655 can be generalized for capacitors by an adaptation of equation 4-22:

$$U_{\text{el},o} + W_{\text{in}} = U_{\text{el},f} + W_{\text{out}}.$$

For capacitors not connected to an external circuit there is no opportunity for the capacitor to do work, hence $W_{\text{out}} = 0$. Then for that situation W_{in} must equal

$$\Delta U_{\text{el}} = \Delta\left(\tfrac{1}{2}\frac{Q_o^2}{C}\right) = \tfrac{1}{2}Q_o^2\,\Delta\left(\frac{1}{C}\right).$$

Note that when ΔC is negative, the work necessary to change C is positive, and vice versa.

For capacitors connected to an external circuit, the capacitor in changing can do work, either by returning charge to a charging source (as in problem 655), or by rearranging charge and hence changing U_{el} somewhere in the circuit (as in problem 656).

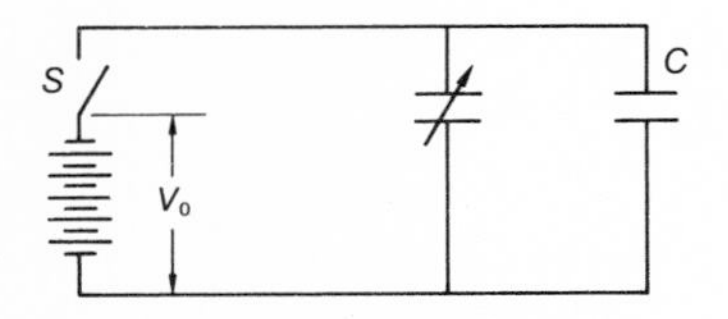

FIGURE 397

PROBLEM 656
With the switch S closed, the variable capacitor (Figure 397) is adjusted to have exactly the same capacity C as the fixed capacitor. After the switch is opened the variable capacitor is changed to a capacity fC, $f < 1$.

a. What was the resulting voltage across either capacitor?
b. How much work was expended in moving the variable capacitor?

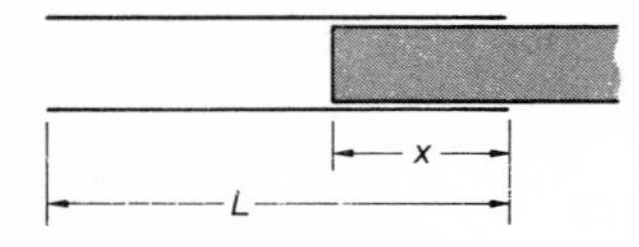

FIGURE 398

PROBLEM 657
A parallel plate capacitor filled with empty space has a capacitance C (Figure 398); it is charged to a potential difference V_o and then disconnected from the charging source. Then a slab of dielectric of constant K, essentially as thick as the plate spacing, is inserted part way between the plates. As a function of x, the distance of insertion, what is

a. V_x, the voltage across the plates,
b. the energy stored in the field of the capacitor?

START OF SOLUTION
In this problem we have the equivalent of two capacitors in parallel, one empty and one filled with dielectric; the crux of the problem is to determine how they share the total charge. Since they are in parallel, the voltages across them must be equal: $V_e = E_e d = \sigma_e d/\epsilon_o = E_d d = \sigma_d d/K\epsilon_o$, hence $\sigma_d = K\sigma_e$. (The subscript e refers to the empty capacitor, and the subscript d to the dielectric capacitor.) Also, from the requirement that the total charge must remain constant we can

write that $[(L-x)/L]A\sigma_e + (x/L)A\sigma_d = Q_0 = CV_0$. When we combine these two equations we arrive at

$$Q_e = \frac{L-x}{L}A\sigma_e = \frac{(L-x)CV_0}{L + x(K-1)}$$

and

$$Q_d = \frac{xKCV_0}{L + x(K-1)}.$$

From the geometry we have $C_e = [(L-x)/L]C$ and $C_d = (xK/L)C$. We put these together and get, for question (a),

$$V_x = \frac{Q_e}{C_e} = \frac{Q_d}{C_d} = \frac{L}{L + x(K-1)}V_0.$$

PROBLEM 658

An instructor demonstrating problem 657 lost his hold on the dielectric when it was halfway in, i.e., when $x = L/2$. If the slab had a mass m and the plates were smooth, how fast was the slab moving when it completely filled the capacitor?

Magnetic Fields

Energy is stored in a magnetic field by the same mechanism—work—that stores energy in an electric field. It takes work to build up a current against the back emf, and this work is stored in the magnetic field resulting from the current. This work is recaptured when the movement of charges slows down; in this case dI/dt changes sign, and what was formerly back emf now becomes a "forward" emf attempting to keep the charges moving. Thus the magnetic field serves as a circuit flywheel.

The energy stored in a self-inductance is given by

$$U = \tfrac{1}{2}LI^2. \tag{5-7}$$

By the time a current through an inductance L reaches a value I, $\frac{1}{2}LI^2$ of energy has been stored there. If a current I_1 moves in an inductance L_1, and a current I_2 in an inductance L_2, and there is a linkage between these elements such as to provide a mutual inductance M between them, then the total energy stored is

$$U = \tfrac{1}{2}L_1I_1^2 + \tfrac{1}{2}L_2I_2^2 + MI_1I_2. \tag{5-8}$$

Equations 5-7 and 5-8 are accurate for static fields, and can be used for approximations even for slowly changing fields. A more general statement about energy storage, one that applies under all conditions, is

$$u = \frac{\epsilon_0 c^2}{2}B^2 \text{ or } U = \frac{\epsilon_0 c^2}{2}\int B^2 dV, \tag{5-9}$$

which is exactly analogous to equation 5-6 for storage in electric fields.

PROBLEM 659

A circuit contains two inductances L_1 and L_2 in series, so arranged that there is no flux linkage between them. What is the total self-inductance of the circuit?

PROBLEM 660

Calculate the self-inductance per unit length of the coaxial conductor of problem 306 by first determining the energy density in the space between the conductors, and then combining equations 5-7 and 5-9. Does your answer agree with the answer to problem 306? Which method seems to you less complicated or less work?

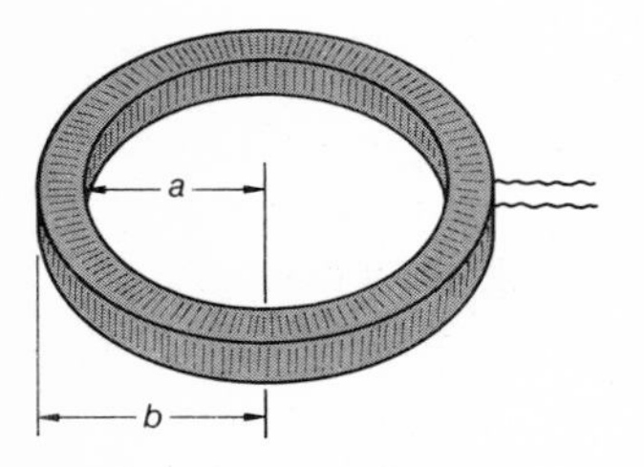

FIGURE 399

PROBLEM 661

A torus of rectangular cross section and made of nonmagnetic material is wound with N turns of wire (Figure 399). What is the self-inductance of such a coil?

PROBLEM 662

In problem 307 the flux inside the conducting rods themselves was ignored in computing the inductance. If the current I in a rod of radius r is assumed to be distributed uniformly over the cross section there will be a circular magnetic field at every point of the cross section, and hence by equation 5-9 an energy density will exist inside the rod. By computing U per unit length from this energy density determine the amount of inductance per unit length per rod that was neglected in your calculation for problem 307. *Note:* We are still ignoring the flux in one conductor arising from the current in the other conductor, but for $d >> r$ this really *is* negligible.

THERMAL STORAGE OF ENERGY

When we discussed kinetic energy and momentum of masses, even when the mass was that of an electron, we were considering individual masses whose motion could be explicitly and separately identified. However, in large assemblies of atoms or molecules it is neither useful nor possible to follow individual masses; only the statistical average of the aggregate can be determined. Rules for the evaluation of these statistical averages constitute the subject of kinetic theory—an important segment of the broad subject of thermodynamics.

The Gas Laws

Rules for evaluating the effect of thermal motion are most easily developed for confined gases. We start with the following idealized simplifications:

1. All of the confined gas molecules are alike.

2. These molecules are treated as point masses whose dimensions are negligible compared to the dimensions of the gas container.

3. All collisions are elastic, whether between molecules or with the walls of the container.

4. The only forces between molecules are the collision impulses.

5. The molecular motions are completely random, so that one direction is as equally probable as any other.

We also start with a closed rectangular box of dimensions x, y, and z filled with $\mathcal{N}$ molecules all alike. (We would get the same results using any closed container of any shape, but the geometrical development is easier with a rectangular box.) When a molecule with velocity components v_x, v_y, and v_z hits a y-z wall it will experience a change of linear momentum $\Delta(mv_x) = 2mv_x$, and it will rebound and return to repeat this change every $2x/v_x$ sec (the time it takes to go to the opposite wall and back again). Thus the force felt by this molecule from the wall, and hence the force felt by the wall from this one molecule, is $f_x = \Delta p_x/\Delta t = 2mv_x/(2x/v_x) = mv_x^2/x$. But on the average $v_x^2 = v_y^2 = v_z^2 = \frac{1}{3}(v_x^2 + v_y^2 + v_z^2) = \frac{1}{3}v^2$, so that $f_x = \frac{1}{3}mv^2/x = \frac{2}{3}(1/x)(\frac{1}{2}mv^2)$. When we sum the effect from all $\mathcal{N}$ molecules, the total force on the y-z wall is given by

$$F_x = \tfrac{2}{3}\frac{1}{x}\mathcal{N}\,(\tfrac{1}{2}mv^2)_{\text{av}} = \tfrac{2}{3}\frac{1}{x}\mathcal{N}E_{\text{av}},$$

where E_{av} is the average linear kinetic energy per molecule. Dividing both sides of the equation by yz we get

$$\frac{F}{yz} = P = \tfrac{2}{3}\frac{\mathcal{N}}{xyz}E_{\text{av}} = \tfrac{2}{3}\frac{\mathcal{N}}{V}E_{\text{av}} = \tfrac{2}{3}n_0E_{\text{av}}, \tag{5-10}$$

where V is the volume of the container and n_0 is the number of molecules per unit volume. The product n_0E_{av} is then the density of the average kinetic energy in the gas. If we call this energy density u, or the total internal kinetic energy $U = uV$, we see that

$$P = \tfrac{2}{3}u, \text{ or } PV = \tfrac{2}{3}U = \tfrac{2}{3}\mathcal{N}E_{\text{av}}. \tag{5-11}$$

What we feel as pressure, what produces a push outward on the walls of the container, is thus simply a measure of the density of the confined kinetic energy, and the product PV is a measure of the total kinetic energy within the container. This internal kinetic energy of random motion is called thermal energy, or more simply, heat.

Obviously $U = \mathcal{N}E_{\text{av}}$ depends upon $\mathcal{N}$ as well as upon E_{av}, and it is difficult to count molecules directly. We avoid this difficulty by counting units of mass instead, and then converting units of mass into units of moles by the relationship $N = M/MW$; N is the number of moles (or gram-molecules), M is the total mass of the confined gas in grams, and MW is its molecular weight. In doing this we count molecules indirectly by virtue of a remarkable fact deduced by Avogadro—that one mole of *any* substance contains exactly the same number of molecules: $N_0 = 6.023 \times 10^{23}$ molecules. Thus if we know M we know N, and knowing N we get $\mathcal{N} = NN_0$. Thus equation 5-11 can also be written

$$PV = \tfrac{2}{3}NN_0E_{av}. \tag{5-12}$$

This equation is physically correct but limited in its usefulness because E_{av} and N_0E_{av} are not easy to evaluate directly. However, you will note from equation 5-12 that if P and N are held constant, V and E_{av} are linearly related. Thus by establishing a scale for changes in V we can obtain a related scale for changes in E_{av}.* The device by which we do this is called a thermometer, and the scale we read on a thermometer is called a temperature scale. We are free to establish any temperature scale we please; over the years the one that has been found most useful is the linear scale defined by

$$PV = NRT, \tag{5-13}$$

where P and V are in MKS units, N is the number of moles, R is called the universal gas constant and is a standard unit of energy equalling 8.31 joules mole^{-1} (°K)$^{-1}$, and T is a scale number called the absolute thermodynamic temperature expressed in number of degrees Kelvin (°K). The term T is obviously the number of 8.31 joule units of thermal energy per mole contained in any gas whose state is described by the product PV. The only dimensions on the right-hand side of the equation are supplied by R, which has the dimensions of energy ML^2T^{-2}. (The symbol N stands for a ratio of masses, and T is a pure scale number—both dimensionless.)†

The odd value of 8.31 joules for the unit of molar thermal energy is the result of a choice for the size of the scale unit for T. This unit was chosen to provide a difference of 100 units, or 100°K, between the thermal energy of one mole of gas in equilibrium with melting pure ice and the thermal energy of the same gas in equilibrium with boiling water—both measured at standard barometric pressure. These are the same calibration points for a spread of 100 units on the popular Centigrade temperature scale; thus a °C and a °K have the same "size." The zeros of the scales are displaced, however, as can be seen on the graph (Figure 400).

The equation $PV = NRT$ and those preceding it were developed with the aid of the idealizations listed on pages 308–309; hence they describe the state of an "ideal" gas. Real molecules are not "ideal"; in general they are not point masses, they do occupy space, and they possess intramolecular force fields. However, the corrections for these realities are so small in relative value that even for a gas of polyatomic molecules $PV = NRT$ is still an adequate evaluation of the linear kinetic energy content in most practical situations.

(As we make the transition from "ideal" to real gases, we need not change our concepts of P, V, R, and T; U, however, must be redefined. For an ideal gas, U was

---

* If it proved more convenient we could hold V and N constant and obtain a linear relationship between ΔP and ΔE_{av}. If we did this, we would find that the scale relationship was exactly the same as between ΔV and ΔE_{av}.

†As mentioned in Chapter 1, for convenience in dimensional analysis T is often assigned an artificial dimension θ. Then R has the dimensions $ML^2T^{-2}\theta^{-1}$.

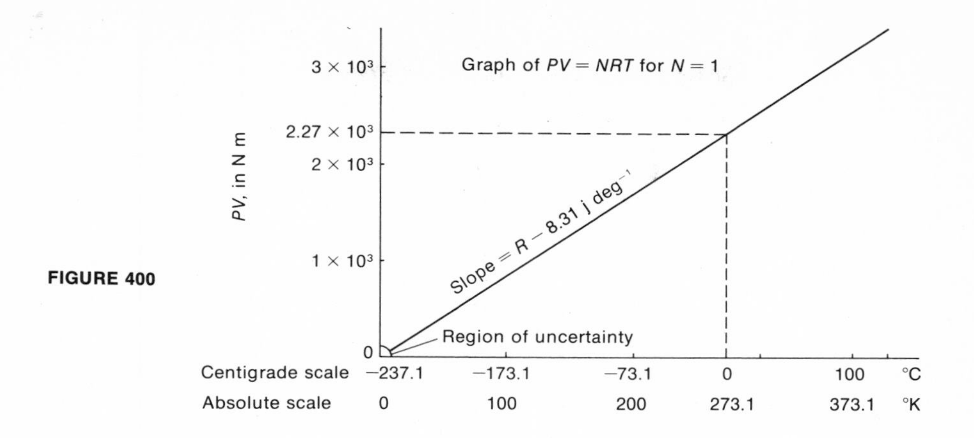

FIGURE 400

simply the total linear kinetic energy of the confined molecules—linear kinetic energy is the only form of kinetic energy available to a point mass. However, complex molecules can possess both rotational kinetic energy and vibrational energy. These additional energies are not measured by T, but they must be included in the thermal energy U.)

Equation 5-13 is often used in the form

$$PV = \mathcal{N}kT, \tag{5-14}$$

where as before $\mathcal{N}$ is the total number of molecules in the confined gas, and $k = R/N_0 = 1.38 \times 10^{-23}$ joules molecule^{-1} (°K)$^{-1}$ is called the Boltzmann constant. The term k appears most usefully as the proportionality constant in the linear relationship between the average linear kinetic energy of a molecule and the temperature scale, thus:

$$E_{av} = \tfrac{3}{2}kT. \tag{5-15}$$

Note that E_{av}, which we measure with a temperature scale, equals $\frac{1}{2}m(v^2)_{av}$, and not $\frac{1}{2}m(v_{av})^2$; there is a difference in value and in physical meaning between the root mean square speed $\sqrt{(v^2)_{av}}$ and v_{av}, usually denoted v_{rms} and $\bar{v}$. Can you tell by mathematical intuition which is greater?*

Note also that the units used in any application of equation 5-15 must be consistent; if the MKS value for k is used, the m of E_{av} must be expressed in kg, and v_{rms} in m sec^{-1}. Since in many applications it is more convenient to express m in grams and v_{rms} in cm sec^{-1} you will often find k in cgs units: 1.38×10^{-16} ergs molecule^{-1} (°K)$^{-1}$.

PROBLEM 663

By inspection of the diagram in Figure 400,

a. what is the volume occupied by one mole of an ideal gas at STP (standard temperature and pressure; 0°C and 1 atmosphere = 1.013×10^5 N m^{-2} or 14.7 lbwt in^{-2}.)

---

*$\bar{v} = 0.92\, v_{rms}$; see page 344.

b. For one mole of a gas at STP, how many molecules are there per cm^3? (This is known as Loschmidt's number.)
c. If a molecular diameter is about 2.5×10^{-8} cm, what percent of the volume of one mole at STP is empty space?

PROBLEM 664
a. Develop a formula for v_{rms} for the molecules of any gas in terms of T and the molecular weight of the gas.
b. What temperature would the "escape velocity" from the earth (See problem 540) represent for: i. hydrogen, and ii. helium?

PROBLEM 665
Equal masses of two different ideal gases A and B are confined in equal fixed volumes that are thermally insulated from each other: $MW_A = 4\ MW_B$. By a method not pertinent here it is determined that $\bar{v}_A = \bar{v}_B$.
a. What must be the ratio of T_A/T_B?
b. What must be the ratio of P_A/P_B?

PROBLEM 666
The two volumes in problem 665 are now placed in thermal contact with each other but are kept insulated from their environment. When the two gases come to thermal equilibrium with each other,
a. what is the final equilibrium temperature T for both gases, in terms of the original temperature T_A of gas A?
b. What is the final ratio P'_A/P'_B?

START OF SOLUTION
Since the volumes are fixed and insulated from the environment the total internal energy content cannot change. Assuming this internal energy is all linear kinetic energy,

$$U_o = \tfrac{3}{2}P_AV + \tfrac{3}{2}P_BV = 3\ P_AV \text{ (from the answer to (b) in problem 665)}$$

$$= 3\ N_ART_A = 3\frac{M}{MW_A}RT_A$$

$$= U_f = \tfrac{3}{2}N_ART + \tfrac{3}{2}N_BRT = \tfrac{3}{2}RT\frac{M}{MW_A}\left(1 + \frac{MW_A}{MW_B}\right)$$

$$= \frac{3 \times 5}{2}RT\frac{M}{MW_A}.$$

Then for question (a), $T = 2/5\ T_A$. You will find that a large proportion of the work in gas law and kinetic theory problems consists of algebraic juggling.

PROBLEM 667
An automobile tire containing 1.5×10^{-2} m^3 of dry air at a tire gauge pressure of 34 lbwt in^{-2} blows out when the temperature of the con-

fined air reaches 82°C. When the escaping air came to equilibrium with the outside atmosphere, which was at 20°C and 14.7 lbwt in^{-2} absolute pressure, what volume did it occupy? (Gauge pressure is pressure in excess of the ambient pressure.)

PROBLEM 668

The essentials of a roughing pump to reduce pressure in a large bell jar are a cylinder and reciprocating piston, as shown in Figure 401. The working volume of the cyclinder is 320 cm^3 for a piston stroke in either direction; the volume of the bell jar is 1.0×10^4 cm^3. If the piston is originally at rest at one end of the cylinder and all enclosed volumes are in equilibrium with the standard atmospheric pressure outside,

a. what will be the pressure in the bell jar: i. after one one-way stroke of the piston; ii. after one round-trip stroke of the piston?

How many complete round-trip strokes are required to reduce the pressure in the bell jar from
b. atmospheric pressure to half-atmospheric pressure;
c. half-atmospheric to one-fourth atmospheric?
Assume the piston operation is slow enough so that all changes take place isothermally, that the flap valves operate on very small differences of pressure, and that the flap valve to the bell jar is more sensitive than are the cylinder valves to the atmosphere.

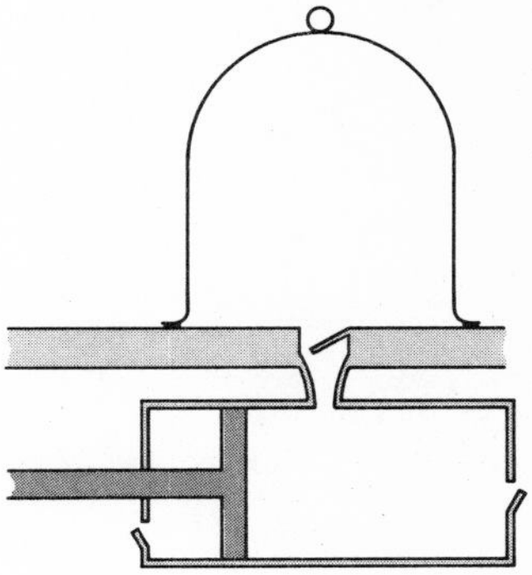

FIGURE 401

PROBLEM 669

A 50-liter tank is connected to a 15-liter tank through a short tube that contains a pressure release valve that allows gas to pass from the larger tank to the smaller tank only when the pressure in the larger tank exceeds that in the smaller tank by 88 cm Hg*. If at 17°C the larger tank contains gas at atmospheric pressure and the smaller tank is evacuated, what will be the pressure in the smaller tank when a fire in the shop where the tanks are stored heats both tanks to 162°C?

* Pressure is often measured by a mercury barometer, with a linear relationship between the height of the mercury column in the barometer and the outside pressure. For standard atmospheric pressure the barometer reads 76 cm Hg.

PROBLEM 670

Two bulbs, one twice the volume of the other, are connected by a short tube containing an insulating porous plug that permits equalization of pressure but not of temperature between the bulbs (Figure 402). Originally both bulbs contain oxygen under a pressure of 760 mm Hg and at a temperature of 27°C. Then the small bulb is immersed in an ice and water bath at 0°C and the large bulb is placed in a steam bath at 100°C. What is the final pressure inside the system? Assume relative volume changes are negligible. *Suggestion:* There are several ways to start problems of this kind. In this case probably the most direct way is to note that the ratio of the final and original pressures in one of the bulbs is equal to the ratio of the final and orig-

FIGURE 402

inal number of moles in that bulb times the ratio of the final and original absolute temperatures in that bulb. Then the only work required is the determination of the mole ratio; since the sum of the moles in the two bulbs does not change, this ratio determination is not difficult.

PROBLEM 671

A closed cylinder 90 cm long contains ideal gases on either side of a freely moving but gas-tight, heat-insulating piston. When the gases on both sides of the piston are at 24°C, the piston is one-third of the length of the cylinder from one end. When the gas in the smaller enclosure is heated to 105°C, the other gas remaining at 24°C, how far did the piston move?

PROBLEM 672

At ordinary temperatures nitrogen tetroxide *partially* dissociates into nitrogen dioxide as follows: $N_2O_4 \rightarrow 2\ NO_2$: one mole of dissociated nitrogen tetroxide becomes two moles of nitrogen dioxide. Into an evacuated flask of volume 250 cm³, 0.90 gm of liquid N_2O_4 at 0°C is placed. When the temperature in the bulb has risen to 27°C, all of the liquid has vaporized and the total pressure in the bulb is 96 cm Hg. What percentage of the nitrogen tetroxide has dissociated?

SOLUTION

When all of the liquid has vaporized we have in the bulb gaseous nitrogen tetroxide and gaseous nitrogen dioxide, both occupying the full volume at the equilibrium temperature T. Thus if n_1 is the final number of moles of N_2O_4 and n_2 the final number of moles of NO_2, we can write $P_1V = n_1RT$ and $P_2V = n_2RT$, where P_1 and P_2 are respectively the independent partial pressures of the gas N_2O_4 and of the gas NO_2. If there is no chemical reaction among gases in a mixture of gases in an enclosure, the Law of Partial Pressures instructs us to add the independent partial pressures to obtain the actual total gas pressure in the enclosure. Thus if we add the two equations we just arrived at, we obtain $(P_1 + P_2)V = PV = (n_1 + n_2)RT$. If n is the original number of moles of N_2O_4 before any dissociation, and p is the percentage of n that dissociates, then $n_1 = n(1 - p)$ and $n_2 = 2\,pn$. Putting this all together we end up with $p = (PV/nRT) - 1$.

The numerical value for p, namely $p = 0.31$ or 31 percent, can be obtained by expressing the given data for P and V in the usual MKS units. However, the liter-atmosphere system of units used by chemists works very well for problems of this kind: if P is expressed in terms of standard atmospheres and V in liters, R has the numerical value 0.082. If we solve for p in these units we have

$$p = \frac{96/76 \times 0.25}{(0.9/92) \times 0.082 \times 300} - 1 = 1.31 - 1 = 0.31.$$

In summing the partial pressures to produce $PV = (n_1 + n_2)RT$, we had to keep n_1 and n_2 separate to meet the need in problem 672. Generally, however, for a mixture of nonreacting gases it is simpler to write a comprehensive equation of state $PV = NRT$ for the entire mixture, with P representing the total pressure in the enclosure and N the sum of the number of moles of all of the gases in the mixture.

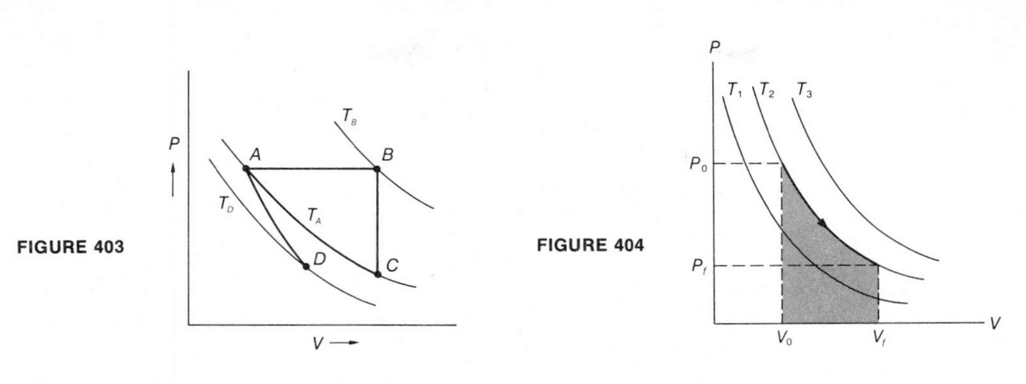

FIGURE 403

FIGURE 404

Work Done during a Change of State

For any particular mass of gas if T is kept constant, equation 5-13 (or 5-14) reduces to $PV=$ constant, and a plot of P vs. V will be a hyperbola. A different temperature for the same gas would produce a different hyperbola, so that $PV=NRT$ (or $\mathcal{N}kT$); can be represented by an infinite family of T hyperbolas on a P-V plot. Thus all the parameters of the state of a gas, P, V, and T, can be represented by one point on a P-V graph.

By providing for the requisite exchange of energy we can change the state of a gas, can move at will from one point on a P-V graph to another.* Some examples (Figure 403): (1) from A to B, at constant P with increases in V and T; (2) from A to C along an isothermal at constant T with decrease in P and an increase in V; (3) from C to B, V constant, with increases in P and T; (4) from A to D along an adiabatic (see page 316) with an increase in V and decreases in P and T. Although the changes of state that occur in actual thermodynamic engines are not as ideally simple as these illustrated, such engines are almost always designed to approximate the ideal.

Isothermal Changes. A mass of gas at T_2 is confined at a pressure P_o in a volume V_o which is fitted with a smoothly moving piston. If we let the pressure push the piston out slowly, at the same time making sure that the temperature stays constant at T_2, the state of the gas will move along the T_2 isotherm to a new volume V_f and a new pressure P_f. During this process the gas will have done work, by pushing the piston through the distance necessary to change the volume from V_o to V_f. Since any area on the P-V diagram (Figure 404) has the dimensions of energy or work, the work done during this expansion must equal the gray area under the change of state curve, or $W=\int_{V_o}^{V_f} P\,dV$. Since at any point on this change of state curve $P=NRT_2/V$,

$$W=\int_{V_0}^{V_f}\frac{NRT_2}{V}\,dV=NRT_2\ln\frac{V_f}{V_0}=P_0V_0\ln\frac{V_f}{V_0}=P_fV_f\ln\frac{V_f}{V_0}. \tag{5-16}$$

We are now faced with an interesting question: since the temperature of the gas remained constant, its internal energy remained constant; then where did the work come from? The answer is that it had to come from some outside source of

*This statement is true if we operate above the critical temperature, which will be discussed later.

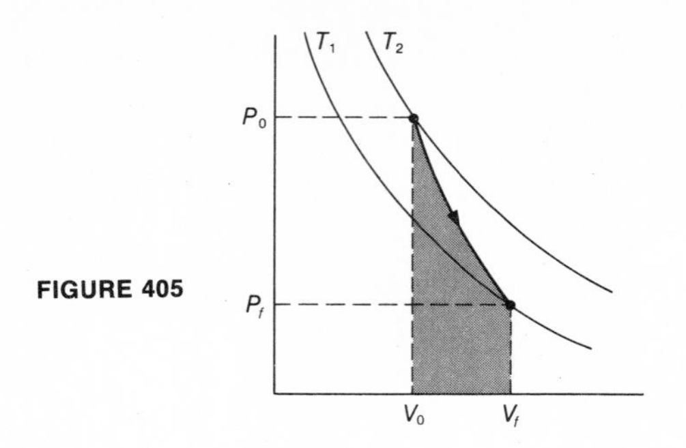

FIGURE 405

energy. It is not possible to carry a gas through an isothermal expansion unless it is in contact with a reservoir of thermal energy, and the work of isothermal expansion is done with energy drawn from this reservoir.

The reverse operation—an isothermal compression—also requires an external reservoir, which in this case is usually called a heat sink. The work done on a gas during compression must be allowed to move into this sink if the compression is to be carried out isothermally. Thus, in isothermal changes of state an ideal gas acts simply as a catalyst, or as a sort of middleman, in transferring energy without in any way suffering any change in its own internal energy.

Adiabatic Changes. In view of the discussion on isothermal changes, an obvious question is, what happens if this outside agency, the reservoir or the sink, isn't there? If the gas is confined in a completely insulated volume, or allowed to expand so suddenly that it reaches V_f before any energy can move into it from some external source—what happens then?

In expanding from V_0 to V_f the gas again does work $W = \int_{V_0}^{V_f} P\,dV$. This work must come from the only source of energy available to the gas—its own internal energy—which means that it must drop from T_2 to some lower temperature T_1, and its state must move along some path as indicated in the diagram in Figure 405. The equation for this path is

$$PV^\gamma = \text{constant} \tag{5-17}$$

so that $PV^\gamma = P_0V_0^\gamma = P_fV_f^\gamma$. Then the work done when the state of the gas goes from P_0, V_0, T_2 to P_f, V_f, T_1, using only the internal energy of the gas in the process, is indicated by the gray area in Figure 405. This is called an adiabatic process, and the work done is given by

$$W = \int_{V_0}^{V_f} P\,dV = \int_{V_0}^{V_f} \frac{P_0V_0^\gamma}{V^\gamma}\,dV = \frac{P_0V_0^\gamma}{\gamma - 1}\left(V_0^{(1-\gamma)} - V_f^{(1-\gamma)}\right) = \frac{1}{\gamma - 1}(P_0V_0 - P_fV_f). \tag{5-18}$$

In any adiabatic process the term γ will appear. As we shall see later, γ is a ratio of specific heats, and is a function of the complexity of a molecule. $PV = NRT$ (or $\mathcal{N}kT$) applies with equal validity to all gases, whether monatomic, diatomic, or polyatomic, since it defines a temperature T which is solely a measure of the linear kinetic energy of the c.m. of a gas molecule. However, only monatomic gases approximate the ideal point mass and possess only linear kinetic energy; more complex molecules can have other forms of energy than that measured by T, and these other forms of energy must be included in the evaluation of the total available internal energy of a gas. The γ is a kind of inverse index of this availability—inverse because as a molecule increases in complexity and in number of ways of storing energy, the value of γ decreases. For monatomic molecules—the noble gases Na, K, and Hg, $\gamma = 5/3$; for diatomic molecules—air, H_2, CO, N_2, and O_2, $\gamma = 7/5$; for polyatomic molecules, $7/5 > \gamma > 1$. These are the classical values for γ that are valid in the temperature ranges normally encountered. (For diatomic molecules at very low temperatures $\gamma \rightarrow 5/3$; for heavy complex molecules at high temperatures γ is significantly less than 1.4.) For the usual problems involving γ, these classical values are sufficiently accurate.

Since gases other than monatomic may possess internal energy U greater than their linear kinetic energy, equation 5-11 must be rewritten for the general case. We can make the necessary change by replacing $\frac{2}{3}$ in that equation by $(\gamma - 1)$; then for an evaluation of U that applies to all gases we can say

$$U = \frac{1}{\gamma - 1} PV = \frac{1}{\gamma - 1} NRT. \qquad (5\text{-}19)$$

(Note that this reduces to $U = \frac{3}{2}PV$ for monatomic gases, in agreement with equation 5-11).

With U properly defined we are now in a position to adapt equation 4-22, the conservation of energy statement, to thermodynamic situations. If during a thermodynamic process the total internal energy of a system changes by ΔU, an amount of energy Q enters the system from outside, and the system does an amount of work W, then

$$\Delta U = Q - W. \qquad (5\text{-}20)$$

In the use of this equation care must be taken with signs: ΔU is positive if the internal energy increases, Q is positive if it is an addition to the system's energy, W is positive if work is done *by* the system. For example, in an isothermal compression, W would be negative, and from what was said earlier, Q would also be negative.

From equation 5-19, $\Delta U = [1/(\gamma - 1)]\ NR\Delta T$, so we also have the relationship

$$Q - W = \frac{1}{\gamma - 1} NR\Delta T. \qquad (5\text{-}21)$$

PROBLEM 673

By combining $PV = NRT$ and $PV^{\gamma} = \text{constant}$,

a. determine the relationship in an adiabatic change between i. pressure and temperature, and ii. volume and temperature.

b. Develop a formula for the work done in an adiabatic change of state, in terms of the initial and final temperatures.

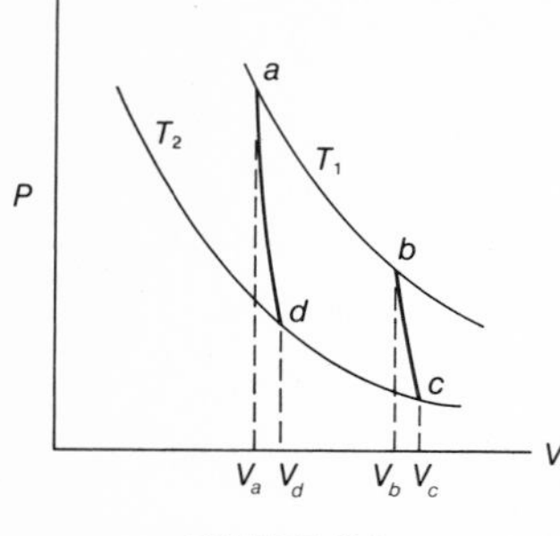

FIGURE 406

PROBLEM 674

Two different adiabatic paths for the same gas intersect two isotherm-als T_1 and T_2, as shown in Figure 406. How does the ratio V_a/V_d compare with the ratio V_b/V_c?

PROBLEM 675

Return to the tire of problem 667, which blew out. (Blowouts usually happen so rapidly that the change is adiabatic.)

a. What work was done by the released air as a result of the blowout?
b. How much energy did the blown-out air have to absorb from the surroundings to bring it into equilibrium with the environment?

SOLUTION

A necessary start in problems of this kind is a *P-V* diagram of the changes that take place (Figure 407).

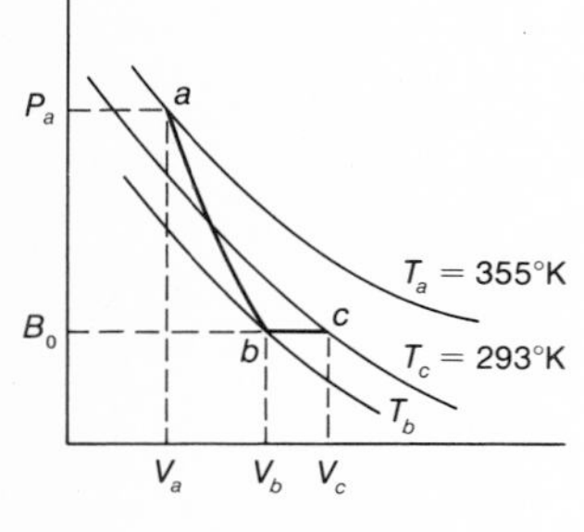

FIGURE 407

From an initial known P_a, V_a, and T_a from the data of problem 667, the air ($\gamma = 7/5$) follows the adiabatic path *ab* until its pressure is re-duced to atmospheric pressure B_0. In expanding from V_a to V_b the air does work, which is represented by the area under the path from *a* to *b*.

a. From equation 5-18,

$$W = \frac{1}{\gamma - 1}(P_a V_a - B_0 V_b)$$

$$= \frac{P_a V_a}{\gamma - 1}\left(1 - \frac{B_0}{P_a}\frac{V_b}{V_a}\right)$$

$$= \frac{P_a V_a}{\gamma - 1}\left[1 - \left(\frac{B_0}{P_a}\right)^{1-1/\gamma}\right] \qquad \left[\text{from equation 5-17, } \frac{V_b}{V_a} = \left(\frac{P_a}{B_0}\right)^{1/\gamma}\right]$$

$$= \frac{5}{2} \times \frac{48.7}{14.7} \times 1.01 \times 10^5 \times 1.5 \times 10^{-2}\left[1 - \left(\frac{14.7}{48.7}\right)^{2/7}\right]$$

$$= 3.6 \times 10^3 \text{ j.}$$

b. After the adiabatic expansion the air comes to equilibrium with its environment along the path *bc*, expanding from V_b to V_c and increas-ing in temperature from T_b to T_c; both T_c and V_c are known from prob-lem 667.

$$Q = B_0(V_c - V_b) + \frac{NR}{\gamma - 1}(T_c - T_b)$$

$$= B_0V_c\left(1 - \frac{V_b}{V_c}\right) + \frac{B_0V_c}{\gamma - 1}\left(1 - \frac{T_b}{T_c}\right)$$

$$= B_0V_c\left(1 - \frac{V_b}{V_c}\right)\left(\frac{\gamma}{\gamma - 1}\right) \qquad \left(\text{since } \frac{V_b}{T_b} = \frac{V_c}{T_c}\right)$$

$$= 1.01 \times 10^5 \times 4.1 \times 10^{-2}\left(1 - \frac{3.5}{4.1}\right) \times \frac{7}{2}$$

$$= 2.1 \times 10^3 \text{ j.}$$

PROBLEM 676

In a Diesel engine oil is sprayed into a cylinder in which air has been heated by compression so that its temperature is at or above the oil's flash point. If for the oil being used this flash point is 600°C, and if the air originally was at standard atmospheric pressure and 18°C,

a. what minimum compression ratio is required to heat the air?
b. What minimum pressure must the oil injection pump be capable of providing?

PROBLEM 677

A thermally conducting, rigid partition divides an insulated container into volumes of 6.0 liters and 4.0 liters. 5.0 gm of neon at 27°C are placed in the larger volume and 1.0 gm of hydrogen at 177°C in the smaller volume.

a. What equilibrium temperature was reached in the container?
b. What was the final pressure of the neon, of the hydrogen?

PROBLEM 678

A mole of ideal monatomic gas in a cylinder with a movable piston is originally at P_1, V_1, and $T_1 = 27$°C. Then the gas is slowly heated until 8.31 watthours of energy have been added to it; at the same time it is allowed to expand at constant pressure to a new state P_1, V_2, and T_2.

a. What is the value of T_2?
b. What is the value of the ratio V_2/V_1?

PROBLEM 679

The amount J joules of work are required to compress adiabatically to half volume 0.1 mole of mercury vapor, the compression starting from known values of P_0 and V_0.

a. How much work will be required to compress nitrogen gas adiabatically to half volume, the compression starting from the same initial values P_0 and V_0?
b. If the amount of the nitrogen gas was 0.2 moles, what was the ratio of the original temperatures $T_{0,\mathrm{Hg}}/T_{0,\mathrm{N_2}}$?
c. Could both compressions be depicted on the same P-V diagram?

PROBLEM 680

N moles of an ideal diatomic gas are undergoing compression in a cylinder fitted with a piston of area A, the piston moving with a constant speed v.

a. If the cylinder is perfectly insulated except for a heat leak which is adjusted so as to keep the gas temperature constant, at what rate must heat be flowing through the leak at the instant the gas pressure reaches P?
b. If the leak is suddenly closed just as the gas pressure reaches P, what will be the rate of temperature increase immediately thereafter?

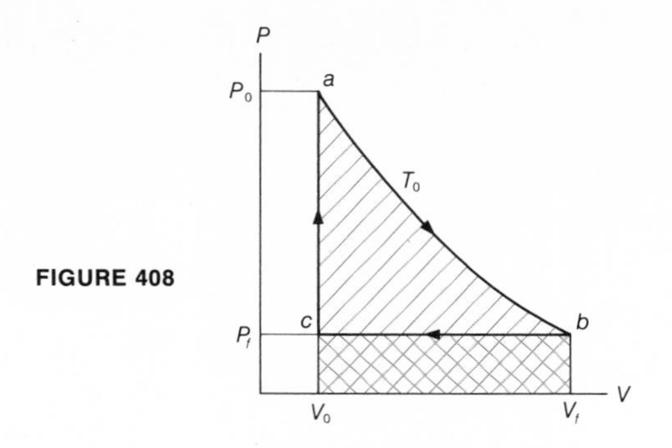

FIGURE 408

Cycles. In normal use, successive changes of state are linked together to form a closed operation—a cycle that brings the gas, or whatever the thermodynamic working substance is, back to its starting condition, ready to repeat. In such a cycle, work will be done, heat will be absorbed from reservoirs and rejected to heat sinks, changes in internal energy will occur. However, since a cycle returns the working substance to its initial condition, no net change of internal energy can take place; a necessary defining condition for a cycle is that $\Delta U = 0$.

Let us consider a simple cycle, as shown in Figure 408.

1. A confined gas expands isothermally along the path ab from a state P_0, V_0, T_0 to a state P_f, V_f, T_0. During this expansion the gas does work and absorbs heat from a reservoir at T_0. From equation 5-16,

$$W_{ab} = P_0V_0 \ln \frac{V_f}{V_0} = P_0V_0 \ln \frac{P_f}{P_0}.*$$

This work is represented by the hatched area. Since $\Delta U = 0$ along an isothermal, from equation 5-20,

$$Q_{ab} = W_{ab} = P_0V_0 \ln \frac{P_f}{P_0}.$$

2. The gas is compressed at constant pressure along the path bc to its original volume V_0. During this compression work is done on the gas:

$$W_{bc} = P_f(V_0 - V_f) = -V_0(P_0 - P_f).$$

This work is represented by the crosshatched area. On this path the gas is also cooled to a temperature $T_c = (P_f/P_0)T_0$; from equation 5-19,

$$\Delta U_{bc} = \frac{NR}{\gamma - 1}(T_c - T_0) = -\frac{V_0}{\gamma - 1}(P_0 - P_f).$$

---

* Equations 5-13 and 5-17, and the answers to problem 673 will be used as needed to establish desired relationships.

During this compression heat is also rejected to a heat sink; from equation 5-20,

$$Q_{bc} = W_{bc} + \Delta U_{bc} = -V_0(P_0 - P_f) - \frac{V_0}{\gamma - 1}(P_0 - P_f) = -\frac{\gamma V_0}{\gamma - 1}(P_0 - P_f).$$

3. The gas is heated at constant volume to its original temperature T_0, its state moving along the path ca. $W_{ca} = 0$.

$$\Delta U_{ca} = \frac{NR}{\gamma - 1}(T_0 - T_c) = \frac{V_0}{\gamma - 1}(P_0 - P_f).$$

$$Q_{ca} = W_{ca} + \Delta U_{ca} = \frac{V_0}{\gamma - 1}(P_0 - P_f).$$

During this cycle the net work accomplished, represented by the area that is hatched only once, is

$$W_{\text{net}} = W_{ab} + W_{bc} = P_0 V_0 \ln \frac{P_0}{P_f} - V_0(P_0 - P_f).$$

During the cycle the heat invested in the process is

$$Q_{\text{absorbed}} = Q_{ab} + Q_{ca} = P_0 V_0 \ln \frac{P_0}{P_f} + \frac{V_0}{\gamma - 1}(P_0 - P_f).$$

(Note that the heat Q_{bc} that was rejected to the heat sink cannot be counted as an asset to this cycle. The only way Q_{bc} can be used is by cycles in cascade, each cycle operating at a lower initial temperature than the one ahead of it in the cascade. Thus the heat rejected by one cycle can be stored in the reservoir of the following cycle.)

When we write equation 4-21 in thermodynamic terms we have

$$\text{Efficiency} = \frac{W_{\text{net}}}{Q_{\text{abs}}}. \tag{5-22}$$

For this cycle under discussion,

$$\text{Efficiency} = \frac{P_0 V_0 \ln P_0/P_f - V_0(P_0 - P_f)}{P_0 V_0 \ln P_0/P_f + [V_0/(\gamma - 1)](P_0 - P_f)}$$

$$= \frac{T_0 \ln T_0/T_c - (T_0 - T_c)}{T_0 \ln T_0/T_c + [1/(\gamma - 1)](T_0 - T_c)},$$

where T_0 and T_c are the highest and lowest temperatures of the cycle.

An energy balance sheet can be constructed for any cycle; if the inventory doesn't balance, we know an error in calculation has been made somewhere. Since for a complete cycle $\Delta U = 0$, then ΣQ should equal ΣW. For the cycle just discussed, the balance sheet would look like this:

Q	W
$Q_{ab} = P_0 V_0 \ln P_0/P_f$	$W_{ab} = P_0 V_0 \ln P_0/P_f$
$Q_{bc} = -\dfrac{\gamma V_0}{\gamma - 1}(P_0 - P_f)$	$W_{bc} = -V_0(P_0 - P_f)$
$Q_{ca} = \dfrac{V_0}{\gamma - 1}(P_0 - P_f)$	
$\Sigma Q = P_0 V_0 \ln P_0/P_f - V_0(P_0 - P_f)$	$\Sigma W = P_0 V_0 \ln P_0/P_f - V_0(P_0 - P_f)$

PROBLEM 681

A cycle similar to the one just discussed starts from the same state values P_0, V_0, and T_0 and expands to the same P_f as before (Figure 409); however, for this cycle the expansion is adiabatic instead of isothermal.

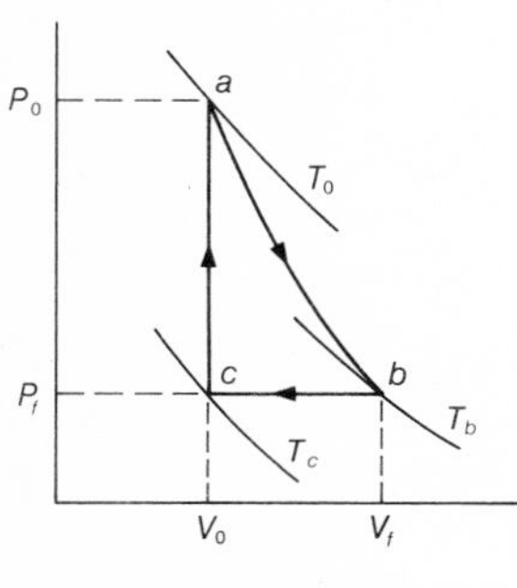

FIGURE 409

a. Can you tell, before you make any calculations, whether the net work done will be greater, equal to, or less than the work done by the isothermal cycle?
b. What is the net work done by this cycle?
c. What is the efficiency of this cycle?
d. Assume a reasonable numerical value for the ratio P_0/P_f and compute and compare the efficiencies for the adiabatic and the isothermal cycle for a diatomic gas. Can you predict ahead of time how the comparison will come out?

PROBLEM 682

An amount of ideal gas is carried through the following cycle, starting at P_0, V_a, T_0: First, expanding at constant pressure along a path ab to P_0, V_b, T_1; second, expanding along an isothermal path bc to P_f, V_c, T_1; third, compressed at constant pressure along the path cd to P_f, V_d, T_0; and fourth, compressed along the isothermal path da back to P_0, V_a, and T_0.

a. Sketch this cycle on i. a P-V diagram, ii. a P-T diagram, and iii. a V-T diagram. Label the points on these diagrams to correspond to the given data.
b. Indicate by crosshatching any area on any of the diagrams that represents the net work done during the cycle.
c. Where on the P-V diagram is heat i. being absorbed from a reservoir, and ii. being discarded into a heat sink?

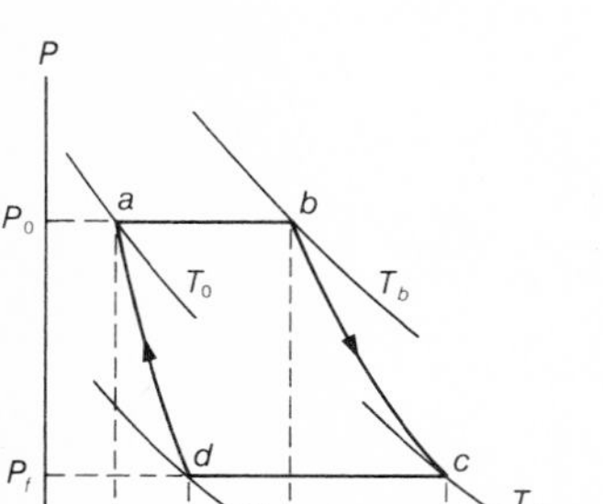

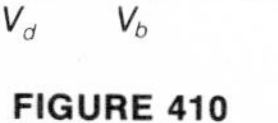

FIGURE 410

PROBLEM 683

N moles of a gas are used as a working substance in a cycle consisting of two isobars (paths of constant pressure) and two adiabatics, as shown in Figure 410. (This is a rough approximation of a steam engine cycle.)

a. What is the efficiency of the cycle?
b. How is this efficiency affected by the complexity of the gas molecules?
c. For a given engine, how could you increase this efficiency?

PROBLEM 684
A cycle consisting of two adiabatics and two paths of constant volume is shown in Figure 411. (This is a fair approximation to the two power strokes of a gas engine.)

a. What is the efficiency of this cycle?
b. How could you improve this efficiency?

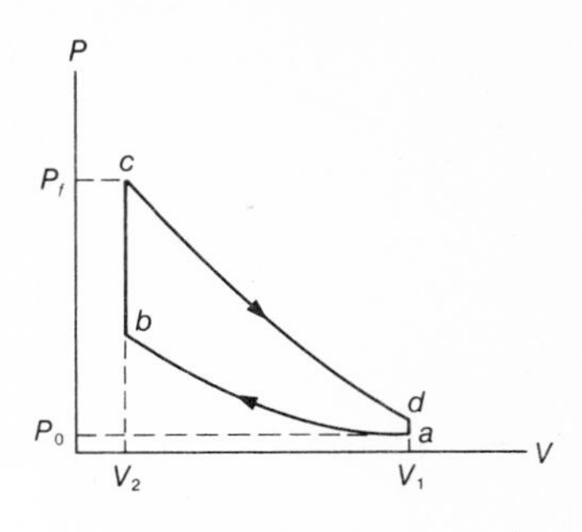

FIGURE 411

PROBLEM 685
A cycle starting at P_0, V_a, T_0 consists of two isothermals and two adiabatics, as shown in Figure 412.

a. What is the ratio of the heat Q_0 absorbed from the reservoir at T_0 to the heat Q_1 rejected to the heat sink at T_1?
b. What is the value of the net work done during the cycle?
c. What is the efficiency of the cycle?
Suggestion: The answer to problem 674 will be useful.

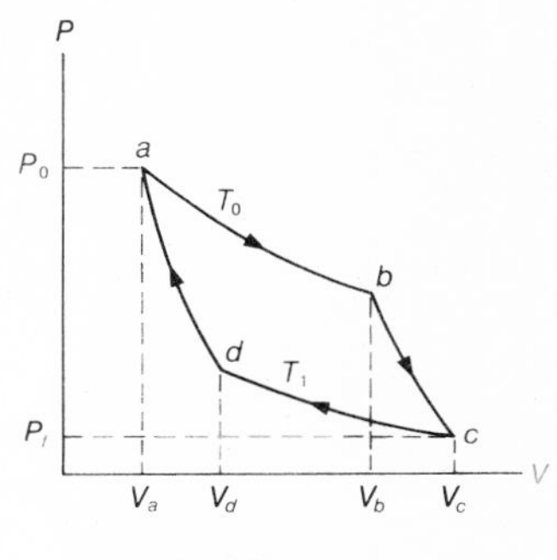

FIGURE 412

The cycles discussed so far have been traversed clockwise on a *P-V* diagram; positive work was done by the cycle at the expense of heat absorbed from a reservoir. We could just as easily traverse the cycle in the counterclockwise direction. When we do this we extract heat from a reservoir at T_1, and then do work on the cycle in order to discharge heat at a higher temperature T_0. The work done by the cycle is negative, and the heat extracted from the lower reservoir is positive. This is in principle the operation of a refrigerator. (Most actual refrigerators operate with both gas and liquid phases and are concerned with heats of vaporization, which will be discussed later.)

PROBLEM 686
The cycle of problem 685 is operated in the reverse direction—adiabatic expansion *ad*, isothermal expansion *dc*, adiabatic compression *cb*, isothermal compression *ba*.

a. How much heat was extracted from the reservoir at T_1?
b. How much heat was delivered to the reservoir at T_0?
c. How much work was required to operate the cycle?
d. What was the "effectiveness ratio"—the ratio of the heat extracted to the work required, per cycle?
e. Does the effectiveness of the operation increase or decrease with increasing difference between T_0 and T_1?

PROBLEM 687
What is the value of the effectiveness ratio Q_{ext}/W_{req} for the cycle of problem 686 for

a. a refrigerator working between an inside temperature of 41°F and a room temperature of 77°F;
b. a freezer in the same room with an inside temperature of −13°F?

The Carnot Cycle. The cycle of problems 685 and 686 is called a Carnot cycle, and is the basic cycle in theoretical thermodynamics. On the isothermal legs of the cycle there is no change in the internal energy of the working substance, on the adiabatic legs there is no intake of energy. Since the changes of internal energy and the works done on the two adiabatic legs are equal in value but opposite in sign, the net work done is solely a function of the difference in temperature of the two isothermals. The efficiency of this cycle—see answer to question (c) of problem 685—is the maximum efficiency possible for any thermodynamic cycle; proof of this can be found in most textbooks.

Any path of change of state that can be depicted on a P-V diagram can be duplicated to within any desired accuracy by a series of isothermal and adiabatic differential changes (Figure 413); any cycle can be reproduced by a series of Carnot cycles. The Carnot cycle is the archetype of all reversible cycles.

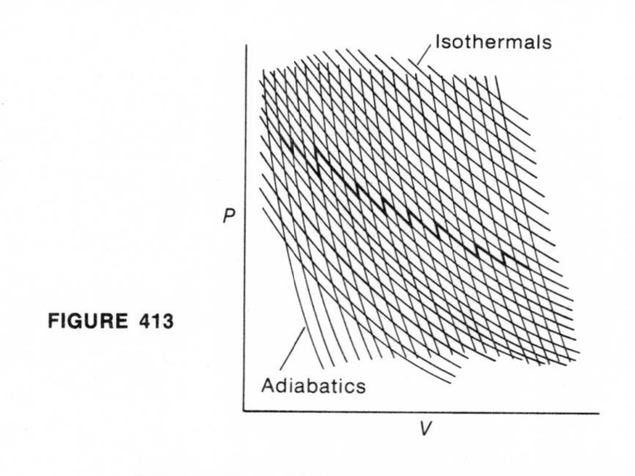

FIGURE 413

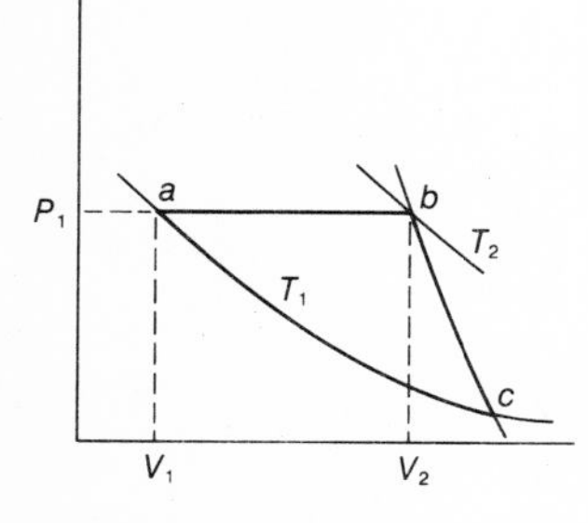

FIGURE 414

PROBLEM 688
A confined gas at P_1, V_1, T_1 is allowed to expand at constant pressure to a volume V_2.

a. Compute, in terms of the ratio V_2/V_1, the amount of heat Q absorbed by the gas during expansion by considering the work done and the change of internal energy along *ab*, the path of constant pressure (Figure 414).
b. By considering the heat absorbed, make the same computation for the isothermal plus adiabatic path *acb*.
c. How do the two values of Q compare? Could you have predicted this result?

PROBLEM 689
Two identical Carnot cycle engines operate between maximum and minimum volume limits V_c and V_a and between maximum and minimum temperatures T_1 and T_2 (Figure 415). $V_c/V_a = e^3$ and $T_1/T_2 = e$ (e the natural logarithm base). Engine 1 operates on monatomic gas, engine 2 on an equal number of moles of diatomic gas.

a. What is the ratio of the cutoff volumes $V_{b,1}/V_{b,2}$?
b. What is the ratio of the work per cycle W_1/W_2?
c. How do the efficiencies of the two engines compare?

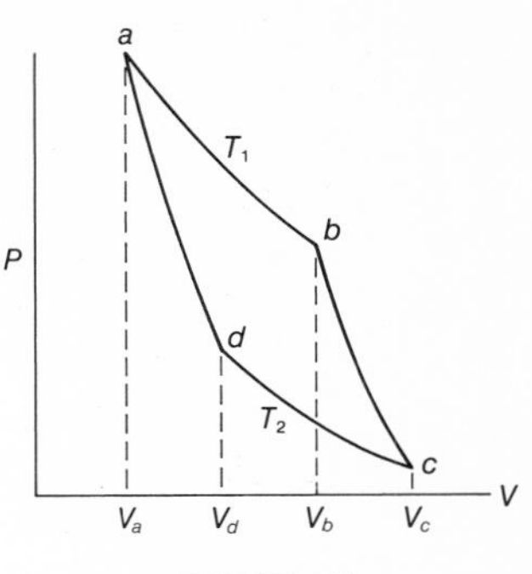

FIGURE 415

PROBLEM 690
Three identical Carnot engines, each having the same amount of the same working substance, operate in cascade between an intake temperature for engine 1 of 1200°K and a heat-sink temperature for engine 3 of 300°K. The heat sink of one engine is the intake reservoir for the following engine.

a. How should the total temperature difference of 900°K be divided among the three engines if the work performed by each shall be the same?
b. What should the temperature division be for the efficiency of each engine to be the same?
c. What is the total work done per completed cycle per set of 3 engines for either arrangement?
d. Which arrangement—equal work or equal efficiency—gives the greater overall efficiency for the set of three engines?

Reversible and Irreversible Cycles, and Entropy. The cycles discussed so far are ideal cycles. They can be accurately plotted on a P-V diagram because of these assumptions: first, that the working substance is completely homogeneous, with all volume elements at the same condition at the same instant; second, that the temperature difference between the working substance and either the reservoir or the heat sink (there must be some temperature difference for a heat flow across a boundary) is of differential order only, so that both working substance and its environment can be considered to be at the same temperature; and third, that changes take place so slowly that the rate at which heat energy moves across a boundary is exactly equal to the rate at which work is being done. It is customary to call such cycles reversible; the assumptions listed make it possible to move in either direction along a cycle path with exact knowledge.

Ignoring for the moment the work done by a reversible cycle, the heat energy Q_1 dumped into the heat sink is less than the heat energy Q_0 taken from the reservoir; even for the most efficient cycle the inevitable result is that the total heat energy of the environment is reduced—which means that the capability for doing work in the future is reduced. In the long run this would not bother us if all processes were 100 percent efficient and reversible; we could carry the work produced by our first cycle through an indefinite cascade of operations, and as a final operation put it to running a refrigerator which would take Q_1 out of the heat sink and put Q_0 back into the reservoir. (Use the heat engine of problem 685 to drive the refrigerator of problem 686 and prove to yourself that this is possible.)

However, real processes in the real world are not 100 percent efficient nor are they reversible; the work of burning rubber off of a tire in a skid, the work of a bulldozer clearing a land area for a freeway, the work of a train whistle dissipating into the atmosphere, the work of putting pencil marks on a paper in solving a problem, are not reversible, they could not be recovered to drive a refrigerator. All real processes are irreversible, hence any use of a heat engine to operate a real process reduces the overall heat capability of the universe.

This inevitable flow of heat energy downhill is the substance of the Second Law of Thermodynamics. For an analytical statement of the Second Law a new condition of state S, called the entropy, is needed. Entropy is defined by

$$dS = \frac{dQ}{T}. \tag{5-23}$$

The units of S are joules $(°K)^{-1}$. We will not be concerned with the absolute value of S; as with potential energy, we can adjust the $S = 0$ level to our convenience. ΔS is a measure of the decrease in heat energy capability in any real process. The difference in entropy $\Delta S_{A\text{-}B}$ for the working substance between two conditions of state, A and B, is obtained from

$$\Delta S_{A-B} = \int_A^B \frac{dQ}{T} \tag{5-24}$$

This can be evaluated only for reversible paths, since only for such paths is dQ/T determinable at every point. By carrying out this integration we discover another similarity between S and the potential energy U:$\Delta S_{A\text{-}B}$ has the same value for every one of the infinite number of reversible paths between A and B. By an extension of equation 5-24, for the working substance for a complete cycle

$$\Delta S_{\text{cycle}} = \oint \frac{dQ}{T} = 0. \tag{5-25}$$

This is true for the working substance even for irreversible paths; dQ/T cannot be carried out point by point along an irreversible path, but to bring the working substance back to its original condition $\oint dQ/T$ must $= 0$.

Although our integration of $\int dQ/T$ must be carried out for the working substance, since we know the conditions of state only for the working substance, we usually are more interested in ΔS of the environment. Following the convention that heat flowing into a region is + for that region, and − if it flows out, the following relationships hold for the environment:

For reversible paths	For irreversible paths
for the output (high temperature) part of a cycle	
$\Delta S_{\text{envir, rev}} = -\Delta S_{\text{work sub}}$	$\Delta S_{\text{envir, irr}} > \Delta S_{\text{envir, rev}}$

For reversible paths	For irreversible paths
for the return (low temperature) part of a cycle	
$\Delta S_{\text{envir, rev}} = -\Delta S_{\text{work sub}}$	$\Delta S_{\text{envir, irr}} < \Delta S_{\text{envir, rev}}$

For a reversible cycle, $\Delta S_{\text{envir}} = 0$; for an irreversible cycle, $\Delta S_{\text{envir}} > 0$.

The working substance and the environment together constitute what some textbooks call the "universe"; $\Delta S_{\text{universe}} = \Delta S_{\text{work sub}} + \Delta S_{\text{envir}}$.

PROBLEM 691

For the expansion from V_1 to V_2 described in Problem 688, what is the change in entropy for the working substance between a and b

a. along path ab;
b. along path acb?

SOLUTION

a. In order to integrate dQ/T we need to express dQ as a function of T. For the path ab we select an intermediate point P_1, V, T between a and b. In going from a to this point the working substance has absorbed heat

$$Q_T = (\Delta U + W)_T = \frac{NR}{\gamma - 1}(T - T_1) + P_1(V - V_1) = \frac{\gamma NR}{\gamma - 1}(T - T_1).$$

Differentiating, $dQ = [\gamma NR/(\gamma - 1)]\ dT$. Hence,

$$\Delta S_{a\text{-}b} = \frac{\gamma NR}{\gamma - 1} \int_{T_1}^{T_2} \frac{dT}{T} = \frac{\gamma NR}{\gamma - 1} \ln \frac{T_2}{T_1} = \frac{\gamma P_1 V_1}{(\gamma - 1)T_1} \ln \frac{T_2}{T_1}.$$

b. For the path acb we need to consider only the ac portion, since dQ equals 0 along the adiabatic portion cb. Along the isothermal path, T_1 is constant, so it can be taken out in front of the integral. Since U is also constant along the isothermal, we are left with

$$\Delta S_{a-c-b} = \frac{1}{T_1} \int_a^c dQ = \frac{1}{T_1} W_{ac} = \frac{1}{T_1} NRT_1 \ln \frac{V_c}{V_1} \text{ (from equation 5-16)}$$

$$= NR \ln \left(\frac{T_2}{T_1}\right)^{\gamma/(\gamma-1)} = \frac{\gamma P_1 V_1}{(\gamma - 1)T_1} \ln \frac{T_2}{T_1}.$$

You will note that the answers for questions (a) and (b) agree, as they should.

PROBLEM 692

The confined gas of problem 688 at P_1, V_1, T_1 is heated at constant volume until it reaches a point d on the same T_2 isotherm that passed through point b in problem 688 (Figure 416).

a. What is ΔS_{a-d}?
b. How does this compare with ΔS_{a-b} from problem 691?

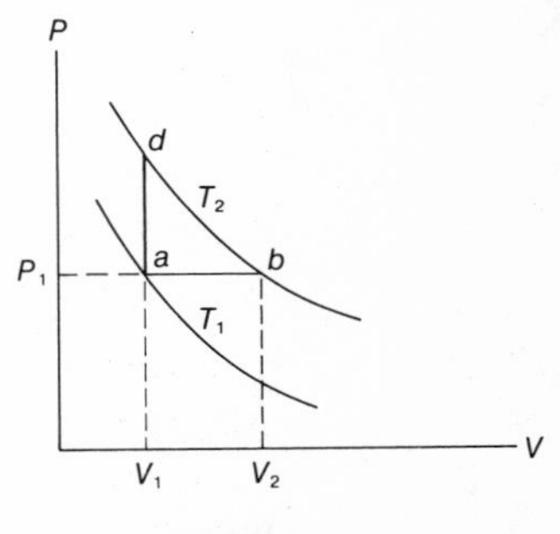

FIGURE 416

PROBLEM 693

For the isobar-adiabatic cycle of problem 683, what is the change in entropy

a. of the working substance along each of the four separate paths of the cycle;
b. of the environment for the complete cycle?

PROBLEM 694

A thermodynamic system is carried around a closed cycle $ABCA$; AB is an isothermal path, BC is reversibly adiabatic, and CA is an intermediate path.

a. What is the change in entropy of the universe if the path CA is reversible?
b. What is the change if the path CA is irreversible?
c. How is the entropy of the system affected by (a) or (b)?

PROBLEM 695

A Carnot engine operates between temperatures of 777°C and 147°C from an initial state of $P_0 = 8.0$ As and $V_0 = 1.0$ L, with a volume at maximum expansion of 25 L. If the working substance is nitrogen, what is its change of entropy

a. during the expansion half-cycle,
b. during the compression half-cycle?

PROBLEM 696

In a truck service station a compressed air tank of 8.0 ft^3 capacity holds dry air at 75 lbwt in^{-2} pressure when the equilibrium temperature is 19°C. Over a holiday the tank developed a slow leak and all of the air leaked out isothermally. What was the resulting change in entropy

a. of the air originally confined in the tank,
b. of the universe?
c. If due to work at the leak or to other causes the air escape was not reversible, what then would be the answer to question (b)?

Degrees of Freedom and the Evaluation of γ

Reversible paths can be constructed to any desired accuracy out of a sequence of isobars and paths of constant volume. This is as useful a construction as the isothermal-adiabatic construction discussed earlier. As we saw in problems 691 and 692, an absorption or rejection of heat along an isobar or a path of constant volume is linearly related to a change in temperature—$dQ = c\ dT$, where c is a constant of proportionality. More specifically we can say for a working substance of N moles,

$$\text{for an isobaric path, } dQ_p = c_p N\, dT;$$
$$\text{for a path of constant volume, } dQ_v = c_v N\, dT. \qquad (5\text{-}26)$$

c_p and c_v are called specific heats per mole, or molar specific heats. Their units are joules mole^{-1} (°K)$^{-1}$.

To review work already done in solving problems 691 and 692, for the same dT

$$\Delta Q_p = c_p N\, dT = \Delta U + W = \frac{NR}{\gamma - 1}\, dT + NR\, dT = \frac{\gamma NR}{\gamma - 1}\, dT,$$

and (5-27)

$$\Delta Q_v = c_v N\, dT = \Delta U = \frac{NR}{\gamma - 1}\, dT.$$

From these we get $c_p/c_v = \gamma$, which we know can vary in value, and $c_p - c_v = R$, which has a fixed value. We must then conclude that variations in γ must result from variations in U. The existence of these variations has already been mentioned on pages 316–317.

A monatomic gas has an internal energy of $U = \frac{3}{2}R$ per mole per °K. This energy must be limited to the linear kinetic energy of the molecules—this is the only kind of energy a point mass can have. (In talking about internal energy we ignore external potential fields.) Since for a point mass there are three independent linear coordinates, on the average this $U = \frac{3}{2}R$ will be divided $\frac{1}{2}R$ per coordinate. Such an independent coordinate is called a degree of freedom, and a fundamental fact of nature is that on the average, energy will divide itself equally among as many available degrees of freedom as it can.

A diatomic molecule has the same three degrees of freedom for motion of its c.m. as has the monatomic molecule. In addition, the diatomic molecule has appreciable moments of inertia about two independent axes of rotation perpendicular to the molecular axis, giving it two degrees of rotational freedom. It also can vibrate in SHM in the direction of the molecular axis, which gives it two degrees of vibrational freedom (one kinetic and one potential). Thus a diatomic molecule possesses seven degrees of freedom, each capable of storing $\frac{1}{2}k$ of energy per molecule per °K; a mole of such molecules can store $\frac{1}{2}$ R of energy per °K.

The storage capacities of the various degrees of freedom are all equal at $\frac{1}{2}k$, or $\frac{1}{2}R$, but the probabilities for energy actually being stored are not equal. In a diatomic molecule the two vibrational degrees possess a high energy threshold—which means a high temperature—which must be reached before they can be activated; the two rotational degrees have a lower threshold, and the three linear kinetic energy degrees the lowest of all. Thus you would expect that c_v, which from equation 5-27 is a direct measurement of the internal energy stored per degree of temperature, would vary

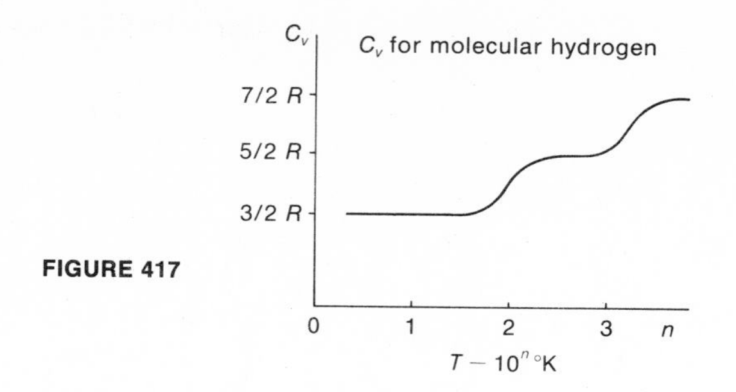

FIGURE 417

from $\frac{3}{2}R$ to $\frac{5}{2}R$ to $\frac{7}{2}R$ with rising temperature. That c_v does vary exactly in this way is shown on the graph of the experimental data for molecular hydrogen (Figure 417). Note that the temperature scale is logarithmic; note also that in the normal range of temperatures, say from −50°C to 800°C, c_v for $H_2 = \frac{5}{2}R$. From $\gamma = c_p/c_v = (c_v + R)/c_v$ we thus calculate the following for a diatomic gas: at very low temperatures $\gamma = (\frac{3}{2}R + R)/\frac{3}{2}R = \frac{5}{3}$; for normal, medium range of temperatures $\gamma = (\frac{5}{2}R + R)/\frac{5}{2}R = \frac{7}{5}$; for high temperatures $\gamma = \frac{9}{7}$. For a more general rule, if D is the number of degrees of freedom activated at a certain temperature, then for that gas at that temperature $\gamma = (D + 2)/D$.

PROBLEM 697

To raise the temperature of 0.1 moles of a certain ideal gas from 20°C to 30°C at constant atmospheric pressure 37 joules of energy are required.

a. What are c_p and γ for the gas in that temperature range?

b. If the volume is maintained while the gas is allowed to cool down to its original temperature of 20°C, how much energy will leave the gas?

PROBLEM 698

When heat equal to 1.00×10^4 Btu is absorbed by a confined gas, it is observed that work equivalent to 2.90×10^3 Btu is done by expansion at constant pressure. For the gas at the operating temperature, what was γ, c_p, and c_v? (The British thermal unit (Btu) is a unit of energy frequently used in engineering calculations; 1 Btu = 1055 joules.)

SOLUTION

From equation 5-27,

$$\gamma = \frac{Q_p}{Q_v} = \frac{Q_p}{Q_p - W} = \frac{10^4}{10^4 - .29 \times 10^4} = 1.41.$$

Assuming the gas is ideal,

$$c_p - c_v = c_p - \frac{c_p}{\gamma} = \frac{(\gamma - 1)}{\gamma} c_p = R.$$

Substituting in the value for γ, $c_p = 28.6$ joules mole^{-1} (°K)$^{-1}$ = 2.7×10^{-2} Btu mole^{-1} (°K)$^{-1}$. $c_v = c_p - R = 20.3$ joules mole^{-1} (°K)$^{-1}$ = 1.93×10^{-2} Btu mole^{-1} (°K)$^{-1}$.

PROBLEM 699

For a certain gas, $\gamma = 1.31$ and over a reasonable temperature range and at normal pressures its internal energy vs. temperature graph is described by the equation $U/N = 29.4T + B$, where U/N is the internal energy per mole and B is a constant for that particular gas.

a. What are the values for c_p and c_v?
b. How complex do you think the gas molecules are?
c. What evidence is there that this is not an ideal gas?

PROBLEM 700

A smoothly lined cylinder of cross-sectional area 200 cm² and internal volume 11.2 L sits upward, top open, in standard dry air at 0°C (Figure 418). A smooth piston of mass 20.0 kg is placed at the top of the cylinder and released, suddenly compressing the air trapped in the cylinder. Assume air can be treated as an ideal gas and that the compression is accurately described by the P-V diagram in Figure 419, where $V_c = 0.936\ V_0$.

a. What was the value for γ?
b. What was c_p?
c. How much heat was rejected by the enclosed air in coming to equilibrium with its environment?

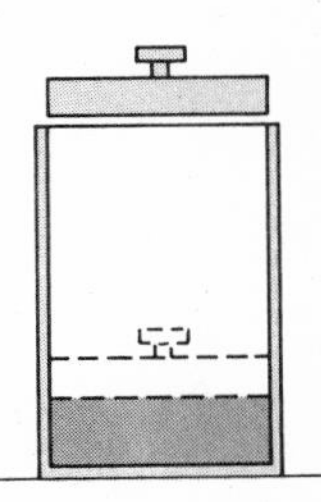

FIGURE 418

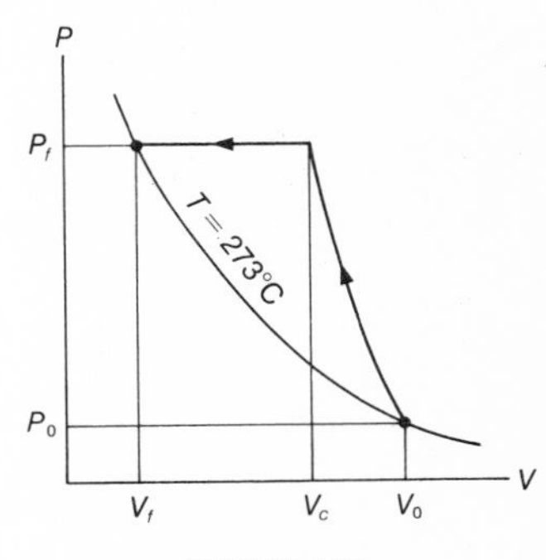

FIGURE 419

PROBLEM 701

In doing problem 700 did you check to see whether the work done by the atmospheric pressure plus the loss in potential energy of the piston exactly equalled the heat energy rejected to the environment? If these energy accounts do not balance, what do you think is the explanation? How do you think the real change of state diagram for the process compares with the diagram in Figure 419?

Speed and Energy Distributions

So far the only molecular speed we have paid attention to is

$$\sqrt{\langle v^2 \rangle} = v_{\text{rms}} = \sqrt{\frac{3kT}{m}},$$

from equation 5-15. However, in any assemblage of a large number $\mathcal{N}$ of molecules of a confined gas at an equilibrium temperature T we should expect to find a broad range of speeds.

By experiment we do indeed find such a range. The experimental results are plotted as a histogram; exact numbers of molecules at exact speeds are not known,* but a count of molecules possessing speeds in the range from some v to $v + dv$ can be

* Since the facts of a large assemblage of molecules are statistical facts only, an exact evaluation of any particular parameter should not be expected.

made. These counts are plotted as rectangular blocks on the histogram, as in Figure 420, with width dv and height $n_{(v)}$, the count in that particular range. When all $\mathcal{N}$ molecules are counted we will have a large number of these blocks lying side by side perpendicular to the v-axis.

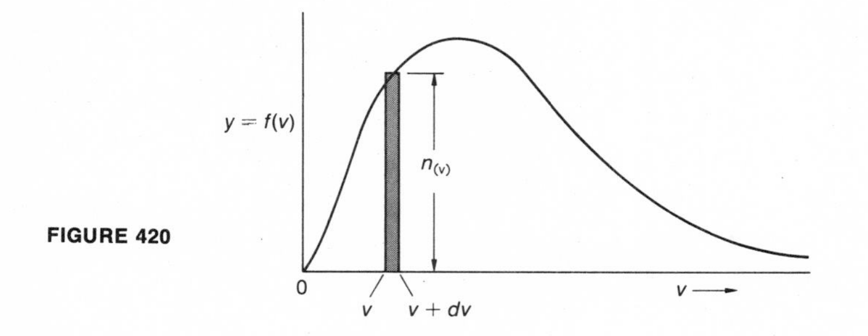

FIGURE 420

If a smooth curve is drawn through the center of the tops of the blocks, the area under the curve will equal the area of all of the blocks. The equation of this curve will be some function of v, $y = f(v)$, and an area of one of the blocks, which originally had been $n_{(v)}\, dv$, can now be expressed as $f(v)\, dv$. Then we can say that the fraction of all $\mathcal{N}$ molecules in the speed range v to $v + dv$ is $n_{(v)}/\mathcal{N}$, and this fraction must also equal the fractional area $f(v)\, dv / \int_0^\infty f(v)\, dv$. Thus the basic equation for describing the experimental results of studies in molecular speeds is

$$\frac{n_{(v)}}{\mathcal{N}} = \frac{f(v)\, dv}{\int_0^\infty f(v)\, dv}. \tag{5-28}$$

(Remember, $n_{(v)}$ is the number of molecules in the speed range dv around v.)

The problem now facing us is to find the form of the function $f(v)$. First we make equation 5-28 as simple as possible by adjusting our scale and any constants of integration so that $\int_0^\infty f(v)\, dv = 1$. Next, noticing the declining curve, we expect some $e^{-\phi(v)}$ effect. Finally (after reading Maxwell's paper No. XX, "Illustrations of the Dynamical Theory of Gases") we try

$$f(v) = \frac{4}{\sqrt{\pi}} \left(\frac{m}{2kT}\right)^{3/2} v^2\, e^{-mv^2/2kT}. \tag{5-29}$$

We find that this fits our experimental curve perfectly, so for our master equation for the distribution of molecular speeds at an equilibrium temperature T we have

$$\frac{n_{(v)}}{\mathcal{N}} = \frac{4}{\sqrt{\pi}} \left(\frac{m}{2kT}\right)^{3/2} v^2\, e^{-mv^2/2kT}\, dv. \tag{5-30}$$

By integrating both sides of equation 5-30, we should be able to get the total number of molecules in the macroscopic speed range v_1 to v_2 by

$$n_{v_2 - v_1} = \int_{v_1}^{v_2} \mathcal{N} \frac{4}{\sqrt{\pi}} \left(\frac{m}{2kT}\right)^{3/2} v^2\, e^{-mv^2/2kT}\, dv.$$

Unfortunately, $\int v^2 e^{-av^2} dv$ is not integrable, and can only be evaluated for the limits 0 to ∞.* Thus a determination of $n_{v_2-v_1}$ can only be made when $v_2 - v_1 = dv$ is small and v can be approximated by $(v_1 + v_2)/2$. For an illustration of the limits to this process, see problem 705.

---

* Some useful integrals:

$$\int_0^\infty v^2 e^{-av^2} dv = \frac{\sqrt{\pi}}{4} a^{-3/2}; \int_0^\infty v^3 e^{-av^2} dv = \tfrac{1}{2} a^{-2}; \int_0^\infty v^4 e^{-av^2} dv = \frac{3\sqrt{\pi}}{8} a^{-5/2}; \int_0^\infty \sqrt{x}\, e^{-ax} dx = \frac{\sqrt{\pi}}{2} a^{-3/2}.$$

PROBLEM 702

a. For a large assemblage of gas molecules at an equilibrium temperature T, what is the average speed $\bar{v}$?
b. What would $\bar{v}$ be for nitrogen in equilibrium at a room temperature of 20°C?

SOLUTION TO THE FIRST PART

If $n_{(v)}$ molecules are associated with the speed v, then the weighted sum of all the speeds is $\Sigma n_{(v)} v = \int_0^\infty \mathcal{N} f(v)\, dv\, v$. Then

$$\bar{v} = \frac{1}{\mathcal{N}} \int_0^\infty \mathcal{N} f(v)\, v\, dv = \frac{4}{\sqrt{\pi}} \left(\frac{m}{2kT}\right)^{3/2} \int_0^\infty v^3\, e^{-mv^2/2kT}\, dv = \sqrt{\frac{8kT}{\pi m}}.$$

PROBLEM 703

For a large assemblage of gas molecules at an equilibrium temperature T,

a. what is the mean square speed $\overline{v^2}$?
b. What is the root mean square speed $\sqrt{\overline{v^2}} = v_{\text{rms}}$? Does your answer check with what you already know about v_{rms}?
c. What is v_{rms} for nitrogen at room temperature?

PROBLEM 704

For a large assemblage of gas molecules at an equilibrium temperature T,

a. what is the most probable speed v_{mp}?
b. What is v_{mp} for nitrogen at room temperature?
Suggestion: The most probable speed would be the speed possessed by more molecules than any other speed, hence it must be the speed for which $y = f(v)$ is a maximum.

PROBLEM 705

For a large assemblage of nitrogen molecules at an equilibrium temperature of 20°C,

a. what fraction of the total number of molecules will be found with a speed range of i. $v_{\text{rms}} \pm 1$ percent, ii. $v_{mp} \pm 1$ percent. Are your answers consistent with the definition of v_{mp}?

b. What fraction of the total number of molecules will be found with a speed range of i. $v_{\text{rms}} \pm 5\text{m sec}^{-1}$, ii. $v_{mp} \pm 5\text{m sec}^{-1}$?

PROBLEM 706

With the aid of the relationship $E = \frac{1}{2}mv^2$ (E is used rather than T to avoid confusion with the temperature T), change variables in equation 5-30 from v to E and determine the fractional distribution $n_{(E)}/\mathcal{N}$ of the kinetic energy in a large assemblage of gas molecules at an equilibrium temperature T.

PROBLEM 707

In view of the distribution function for E developed in problem 706, for a large assemblage of molecules at an equilibrium temperature T,

a. what is $\bar{E}$? Does your answer agree with what you already know about $\bar{E}$?
b. What is E_{mp}?

FIGURE 421

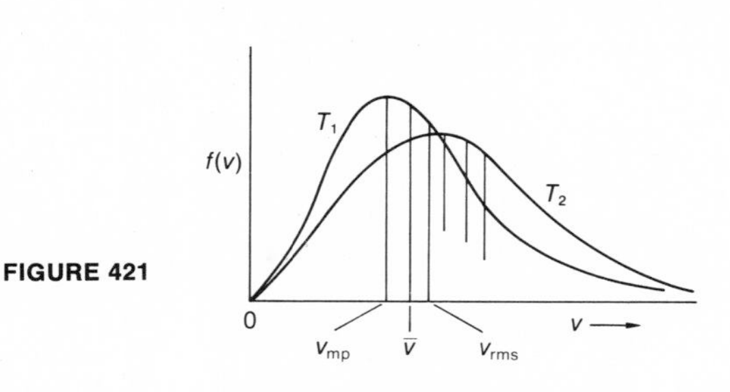

The particular speeds evaluated in problems 702 to 704 are shown on the graph (Figure 421) for a fixed amount of gas at two different equilibrium temperatures T_1 and T_2, $T_2 > T_1$. The area under both curves must be the same. Useful relationships among these speeds are:

$$v_{mp} = 0.82\, v_{\text{rms}}, \quad \text{and} \quad \bar{v} = 0.92\, v_{\text{rms}}.$$

PROBLEM 708

A mass of gas at an equilibrium temperature T_1 is heated to a new equilibrium temperature T_2. It is observed that at the latter temperature v_{mp} has the same numerical value as v_{rms} had at T_1. What was ΔT?

PROBLEM 709

Two different amounts of the same gas are separately confined; one amount is at an equilibrium temperature T_1, the other at T_2. It is observed that $v_{mp,2}$ for the gas at $T_2 = v_{\text{rms},1}$ for the gas at T_1. It is also observed that $n_{(v),1}$ at $v_{\text{rms},1} = n_{(v),2}$ at $v_{mp,2}$. How do the masses of the two separate assemblages compare?

PROBLEM 710

By an experimental method not pertinent here you determine that in an enclosed gas made up of identical molecules at an equilibrium temperature T, a certain number n_a of these molecules possess approximately akT of energy each, and a number n_b of the molecules possess $2akT$ energy each. If $n_a + n_b = f\mathcal{N}$, where f is a fraction of the total number of molecules $\mathcal{N}$, what is $n_a/\mathcal{N}$?

PROBLEM 711

A confined gas at an equilibrium temperature T is made up of identical molecules.

a. If the energy distribution $f(E)$ vs. T is plotted, what is the value of the ordinate $f(E)$ at $\bar{E}$?
b. What fraction of the total number of molecules will possess energy within 1.0 percent of $\bar{E}$?

Probability. The fraction $n_{(v)}/\mathcal{N}$ of equation 5-30 is a measure of probability; if out of $\mathcal{N}$ molecules $n_{(v)}$ of those molecules possess speeds between v and $v + dv$, then the probability that any molecule will be found in that speed range is $n_{(v)}/\mathcal{N}$. Equation 5-30 is a specialized form of a more general probability statement

$$\frac{n}{n_0} = e^{-E/kT}. \tag{5-31}$$

If n_0 molecules contained in unit volume at an equilibrium temperature T possess a range of energies, the fraction of n_0 that will possess any particular energy E (E being included in the energy range), or the probability that any particular molecule will possess energy E, is $e^{-E/kT}$. As thus stated, equation 5-31 permits all possible values of E, and is applicable as a continuous function in those physical situations in which continuity is possible. However, applied to physical situations that are subject to the rules of quantum mechanics, equation 5-31 is discontinuous, and E is permitted to have only discrete levels such as E_0, $E_0 + \hbar\omega$, $E_0 + 2\hbar\omega$, etc.

PROBLEM 712

An isothermal atmosphere in a uniform gravitational field consists of a mixture of several different gases. What is the partial pressure of one of these constituent gases as a function of the height above a reference datum?

SOLUTION

At y above the datum a molecule of mass m has a potential energy of mgy. Then the number of molecules per unit volume at a height y above the datum compared to the number per unit volume at $y = 0$ is $n_y/n_0 = e^{-mgy/kT}$. Also $n_y m/n_0 m = \rho_y/\rho_0$ and from $PV = (m/MW)RT$ at T constant, $P \propto \rho$. Hence, $P_y/P_0 = \rho_y/\rho_0$ and $P_y = P_0 e^{-mgy/kT} =$

$P_0 e^{-MW\, gy/RT}$, where P_y and P_0 are the partial pressures at y above the datum and at the datum for a gas whose molecular weight is MW. If MW is the molecular weight for standard air the above answer becomes $P_y = B_0 e^{-MW\, gy/RT}$. This is the standard barometric equation; B_0 is the barometric pressure of standard air at sea level and $= 1.013 \times 10^5$ N m^{-2}.

PROBLEM 713

Dry sand on a beach on a warm day is statistically subject to thermal motion. What fraction of the grains of sand in the top layer, average mass per grain about 1.0 milligram, will be observed at any moment to be at least 1.0 mm above a normal resting place?

PROBLEM 714

Equal masses of He and N_2 at an equilibrium temperature T are injected into a vertical cylinder of height H and constant cross section; the cylinder sits in a uniform gravitational field. (For simplicity, assume $mgH \gg kT$, so that $e^{-mgH/kT} \simeq 0$.) As a function of the height h above the bottom of the cylinder what is the "mix" of the two gases?

PROBLEM 715

Adjacent vibrational states of energy for a diatomic molecule can be expressed by $E_n = E_0 + n\hbar\omega$, where $n = 0, 1, 2, \ldots$. For an assemblage of such molecules at an equilibrium temperature T,

a. how do the number of molecules in adjacent states compare, i.e., what is the probable value of n_{n+1}/n_n?
b. If for a particular kind of molecule $\hbar\omega \simeq 10^{-19}$ joules, what is the ratio $(n_{n+1})/n_n$ for i. $T = 300$°K, ii. $T = 10^4$°K?
c. Do your answers for (b) seem consistent with the diagram for c_v for H_2 on page 330?

Mean Free Path. The velocity distribution of equation 5-30 can be accounted for only by a continuing process of random collisions between molecules. (Maxwell developed the equation by starting with a statistical study of random collisions.)

If molecules collide, they must have dimensions, and these dimensions are related to the distances molecules travel between collisions. For a target molecule at rest—unrealistic but an easier situation to analyze—this distance between collisions is given by

$$\lambda = \frac{1}{\pi d^2 n_0} = \frac{1}{\sigma n_0}, \qquad (5\text{-}32)$$

where λ is the length of the mean free path between collisions, d is the distance of closest approach between the c.m.'s of the two colliding molecules (cf. page 296 et seq.), n_0 is the number of molecules per unit volume, and σ is the collision cross section. For the more realistic case, when the target molecule moves with any appreciable speed relative to the projectile molecule, the mean free path is reduced by about a factor of $\sqrt{2}$, thus:

$$\lambda = \frac{1}{\sqrt{2}\pi d^2 n_0} = \frac{1}{\sqrt{2}\sigma n_0}. \qquad (5\text{-}33)$$

The time between collisions τ, or the frequency of collisions ν, are often of more interest than λ. By obvious algebra,

$$\tau = \frac{\lambda}{\bar{v}} = \frac{1}{\sqrt{2}\sigma n_0 \bar{v}} \text{ and } \mu = \frac{1}{\tau} = \sqrt{2}\sigma n_0 \bar{v}. \qquad (5\text{-}34)$$

The mean free path is also related to the viscosity of a gas η (see page 166) by the relationship

$$\eta = \tfrac{1}{3}\rho \bar{v} \lambda. \qquad (5\text{-}35)$$

Since the viscosity of a gas is relatively easy to measure, equations 5-33 and 5-35 can be combined to provide a method for determining σ.

PROBLEM 716

For most purposes for which high-vacuum containers are used it is desirable that the mean free path of the molecules in the container be greater than the significant dimensions of the container. If the collision cross section for air molecules is about 2.8×10^9 barns ($1 \text{ barn} = 1 \times 10^{-28} \text{ m}^2$), what should be the maximum allowable air pressure at an equilibrium temperature of 27°C inside

a. the walls of a thermos flask that are 1.0 cm apart,

b. a tank for coating telescope mirrors by the evaporation of aluminum, when the vaporizing elements are 1.0 m away from the surface of the mirror?

PROBLEM 717

By combining equations 5-33 and 5-35,

a. determine what effect a change in pressure has on viscosity and hence on viscous damping.

b. How is the damping factor changed when the air pressure in a glass enclosure housing an astronomical pendulum clock is reduced from B_0 to $0.1\ B_0$?

PROBLEM 718

An electron with an initial kinetic energy of 2.0×10^{-5} Mev is injected into an atmosphere of helium which has a temperature of 27°C and a pressure of 1.0 mm of Hg.

a. Assuming the helium molecules are essentially at rest relative to the moving electron, what is the average collision rate of the electron at the beginning of its excursion into the helium? (For electrons the collision cross section in helium is about 4×10^8 barns.)

b. By comparing average speeds check the validity of the assumption made in question (a).
c. At what average electron speed would you consider the assumption no longer valid?
d. How many collisions must occur before the critical speed you chose in question (c) is reached? (The electron loses energy at approximately an average rate of $2m_{el}/m_{He}$ of its incident energy at every collision—see Appendix I.)

PROBLEM 719
Experiments to measure the coefficient of viscosity are performed separately on He gas and on N_2 gas; it is noted that η for N_2 at 113°C exactly equals the value of η for He at 27°C. If the collision cross section for He–He is known to be about 1.5×10^9 barns, what is the collision cross section for N_2–N_2?

If the diameter of a small hole in the wall of a gas container is small compared to the mean free path of the molecules of the gas, the gas will leak out at a rate

$$F = \frac{n\bar{v}A}{4} \tag{5-36}$$

where F is the molecular flux in molecules sec^{-1} and A is the area of the hole. (See Appendix H.)

PROBLEM 720
A minute hole of area 2×10^{-18} m^2 develops in the glass wall of a vacuum pumping apparatus. At what rate does outside air at standard barometric pressure and 27°C leak through the hole?

PROBLEM 721
A tank of volume V containing He at a high pressure P in equilibrium with the ambient temperature T develops a small leak of area A.

a. How long does it take for the pressure in the tank to fall to 0.9 P? For simplicity, neglect any leak of outside air *into* the tank.
b. Using reasonable values of $V = 20$ L, $T = 27$°C and $A = 1 \times 10^{-18}$ m^2 (about as large as A can be for equation 5-36 to be valid at high pressure) evaluate your answer to question (a).

PROBLEM 722
Equal numbers of moles of He and N_2 at an equilibrium temperature of 27°C are injected into a volume of 10 m^2. One side of the enclosure contains 1.0×10^{12} openings each of cross-sectional area 3.0×10^{-19} m^2; the environment outside these openings is maintained at near vacuum. Twenty-four hours after the injection what is the ratio n_{He}/n_{N_2} of the number of molecules of each gas remaining in the enclosure?

Vapors and Relative Humidity

Above a certain critical temperature a real gas changes conditions of state along a hyperbola on the P-V diagram essentially as predicted by the ideal gas laws. Below this critical temperature a gas can condense to a liquid and a liquid can freeze to a solid. In the clear area of the diagram (Figure 422), a substance would be a gas; in the diagonally hatched area, it would be a liquid or a solid, in the horizontally hatched area it is a mixture of gas and liquid. It is customary to use the word vapor for a gas in equilibrium with its liquid in this area.

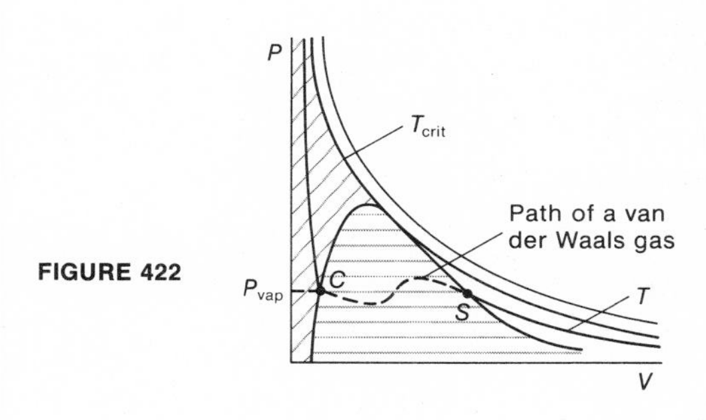

FIGURE 422

As a gas is compressed along an isotherm T it reaches a particular pressure at S called the vapor pressure P_{vap}; the gas is then saturated and on the point of condensing. During further isothermal compression the pressure remains at P_{vap} and along the path S-C the vapor continuously condenses at just exactly the rate necessary to keep ρ_{sat}, the density of the saturated vapor, constant. (With P and T constant, ρ must also remain constant.) At C, all the vapor has condensed and the volume is entirely occupied by liquid. For higher temperatures, but with T still less than T_{crit}, it is clear from the diagram that both P_{vap} and ρ_{sat} will increase with T.

Following an expansion along the same isotherm T, at C the substance is all liquid but just at the point of starting to evaporate. Along the path C-S it evaporates at constant P_{vap} and ρ_{sat}, until at S all the liquid has evaporated and the volume is filled with saturated vapor. On further expansion the pressure starts to drop and the substance becomes a gas.

The critical temperature of water is above 374°K, hence at temperatures and pressures we are normally accustomed to, water exists in a vapor-liquid state, and our atmospheric pressure normally contains a partial pressure component of P_{vap}. Some values for P_{vap} and ρ_{sat} for water at different temperatures are given in Appendix A.

We are usually less aware of the actual value of ρ_{sat} for water vapor (except when it rains or we have fog) than we are of the ratio of the water vapor we actually have to the water vapor we could have at the ambient temperature. This ratio, called the relative humidity r, is defined by

$$r = \frac{\rho_{act}}{\rho_{sat}} \text{ or } r = \frac{P_{act}}{P_{sat}}, \tag{5-37}$$

where ρ_{act} and P_{act} are the actual, existing water vapor density and pressure, and ρ_{sat} and P_{sat} are the values for saturated vapor in equilibrium with liquid water at the ambient temperature. It is customary to use the density definition of r in situations where the density is controlled, as in closed containers, and the pressure definition in situations where the pressure is controlled, as in experiments conducted in large open areas.

PROBLEM 723

A laboratory worker is cleaning a glass tube to be used in a mercury barometer when he is interrupted by a request for help from another worker. When he returns to his own work he assembles the barometer and fills it with Hg, forgetting he had not dried the tube. As a result the completed barometer has a film of water on top of the Hg column. What correction should be made to the barometric reading B when the ambient temperature is 25°C?

PROBLEM 724

When the ambient relative humidity is 80% and the temperature is 25°C, a flask originally in equilibrium with the atmosphere is closed and sealed. As the ambient temperature slowly falls (assume temperature inside and outside the flask the same), will dew form first on the inside or the outside of the flask, and at what temperature?

START OF SOLUTION

From table on page 444 of Appendix A, at 25°C $\rho_{sat} = 23.0 \times 10^{-6}$ gm cm^{-3}, hence $\rho_{act} = 0.8 \times 23.0 \times 10^{-6} = 18.4 \times 10^{-6}$ gm cm^{-3} which $= \rho_{sat}$ at 21.1°C (from the table). Hence dew will form on the inside when the temperature drops to 21.1°C. By carrying out the same kind of computation for the outside, this time comparing pressures rather than densities, the outside dew point can be determined.

PROBLEM 725

The air in an open room is at 30°C and 60% relative humidity. In the room sits an old-fashioned refrigerator with its door open, freshening after being cleaned. The door of the refrigerator is then closed and the cooling mechanism turned on.

a. What is the temperature inside the refrigerator when dew first starts to form?

b. When the inside temperature reaches −5°C, how much frost has formed? (Frost is frozen dew.) The inside volume of the refrigerator is 1.0 m^3.

PROBLEM 726

A flask of volume 10.0 L contains air, water, and water vapor. The flask is closed when the temperature is 27°C, the total pressure in the flask is 1.0 As, and the mass of liquid water is 0.574 gm.

a. What is the partial pressure of the air in the flask?

b. What is the total mass of water in the flask?

c. When the flask is heated until all the water has evaporated, what is the temperature in the flask?
d. What is the total pressure inside the flask at this temperature?
e. If the flask is now heated to an additional 50°, what is the total pressure in the flask?

PROBLEM 727

A mixture of air and water vapor with a temperature of 30°C, a total pressure of 760 mm Hg, and a relative humidity of 80% is compressed isothermally.

a. What is the pressure in the mixture when condensation begins? If the compression is continued until one-half of the water vapor has condensed,
b. What is the resulting total pressure?
c. What is the resulting mixing ratio ρ_{H_2O}/ρ_{air}, where ρ_{H_2O} is the density of the water vapor in the air?

SOLUTION

From the table on page 444 in Appendix A, at 30°C $P_{vap} = 31.9$ mm Hg and $\rho_{sat} = 30.4 \times 10^{-6}$ gm cm^{-3}. Then before compression $P_{0,H_2o} = 0.8 \times 31.9 = 25.5$ mm Hg, so that $P_{0,air} = 760 - 25.5 = 734.5$ mm Hg. To achieve condensation isothermally the water vapor must be compressed so that $P_{f,H_2O} = P_{vap,30°C} = P_{0,H_2O}/0.8$; the necessary compression ratio must then be $P_{f,H_2O}/P_{0,H_2O} = 1/.8 = 1.25$.

a. When condensation begins, $P_{total} = 1.25\ P_{0,air} + 31.9 = 1.25 \times 734.5 + 31.9 = 918 + 31.9 = 950$ mm Hg.
b. Since the water vapor density remains constant during an isothermal compression, when half of the water vapor has condensed then the volume of water vapor must also have been halved, hence the compression ratio must = 2. As a result, $P_{total} = 2 \times 918 + 31.9 = 1868$ mm Hg.
c. At 30°C and 1836 mm Hg pressure, $\rho_{air} = 2.91 \times 10^{-3}$ gm cm^{-3}:

$$\frac{\rho_{H_2O}}{\rho_{air}} = \frac{30.4 \times 10^{-6}}{2.91 \times 10^{-3}} = 1.0\%.$$

PROBLEM 728

Hydrogen at 0°C is admitted to an evacuated container until the pressure reaches 415 mm Hg. Oxygen at 0°C is then added until the total pressure in the container is 760 mm Hg. The mixture is then exploded, forming some water vapor. The container is then cooled until condensation just starts to appear on the inside walls.

a. At that point, what is the temperature in the container?
b. What is the total pressure?

START OF SOLUTION

The first step is to determine whether, in addition to the water vapor formed, there is any H_2 or O_2 left over after the explosion. Originally $N_{H_2} = 415\ V/RT$ and $N_{O_2} = 345\ V/RT$. (Never mind that these do not seem to be compatible units; they can be adjusted later if necessary.)

Since we need only half as many moles of O_2 as of H_2 to form H_2O, there is obviously an excess of O_2 which amounts to

$$N'_{O_2} = \frac{345\ V}{RT} - \frac{1}{2}\frac{415\ V}{RT} = \frac{275\ V}{2RT}.$$

For the water vapor formed we have

$$\rho_{H_2O} = \frac{MW \times N_{H_2O}}{V} = \frac{MW \times 415}{RT}$$

$$= \frac{18 \times 10^{-3} \times 415/760 \times 1.01 \times 10^5}{8.3 \times 273}$$

$$= 439.5 \times 10^{-6}\ \text{gm cm}^{-3}.$$

This is ρ_{sat} at 91.0°C – the answer to question (a).

The Storage of Heat Within Liquids and Solids

Liquids and solids in thermal contact with each other exchange heat energy just as gases do if they are not all at the same temperature. This heat exchange is governed by the relationship

$$\Delta Q = cm\ \Delta T, \tag{5-38}$$

where c is the specific heat of a substance and m is its mass. Equation 5-38 can be written separately for each mass in a heat exchange; in such an array of equations heat must be conserved. If there are only static exchanges of heat, the total sum of the ΔQs, in and out, plus and minus, must equal zero; if mechanical energy losses are involved which appear as heat, such as work done against friction, these must be included in the heat exchange inventory also. Note that temperatures are not conserved – the sum of the ΔTs is not necessarily equal to zero. This can be seen more clearly by rearranging equation 5-38 to $\Delta T = \Delta Q/cm$; the changes in temperature of a substance depend upon its specific heat and its mass as well as upon ΔQ.

The specific heat c for liquids and solids corresponds usually to c_p for gases, since most heat exchanges in liquids and solids take place at constant pressure. However, very little work of expansion is done in adding heat to a liquid or solid (although the expansion itself may be considerable – see page 344), hence for a liquid or a solid there is little difference between c_v and c_p – which is the reason it is usually just designated c. The MKS unit for c would be joule kg^{-1} (°K)$^{-1}$. However calorimetry, the study of heat energy exchanges, developed using an older cgs unit, the calorie, for heat energy; thus in much of the literature and in many textbooks c is given in calories gm^{-1} (°C)$^{-1}$. 1 calorie = 4.19 joules $\simeq R/2$. Average values of c for some substances can be found in Appendix A.

PROBLEM 729

A solid is made up of molecules held in place by intermolecular forces; thus a molecule in a solid is essentially a three-dimensional harmonic oscillator. Assuming that per degree of freedom a solid stores energy at the same rate as does a gas, what do you estimate would be the limiting value for c for a solid, in calories mole^{-1} (°C)$^{-1}$? Your answer contains the essence of the Law of Dulong and Petit.

PROBLEM 730

You have no thermometer available, and you are trying to follow a recipe that calls for one cup of water at 50°C. If, except for a thermometer, your kitchen is well equipped, what is the simplest thing you can do to solve your problem?

PROBLEM 731

Lead BB's, small round lead shot, are made by pouring molten lead through a screen at the top of a shot tower. The individual drops falling through the screen assume a spherical shape from surface tension and freeze in that shape as they cool in falling toward a vat of water at the bottom of the tower. The average temperature of a solid shot as it reaches the water is about 200°C. If the screen is 60 m above the water surface and the vat held 1.0 m^3 of water originally at 15°C, what was the water temperature after 100 kg of shot had fallen? Assume that the mixing effect from the sinking shot ensures thermal equilibrium of both shot and water. How necessary was it to know the tower height?

PROBLEM 732

2.0 L of water at 20°C are placed in an electric tea kettle also at 20°C. The mass of the tea kettle is 1.5 kg and its self-contained heating element is rated at 1.5 kw. After 8.5 min of heating the element is disconnected, and the kettle, element and water all come to thermal equilibrium just as the water reaches the boiling point. (This is known as efficient management.) Ignoring any heat radiation from the kettle as negligible, what is the overall specific heat of the tea kettle? Use $c_{H_2O} = 1.01$ cal gm^{-1} deg^{-1} as the average specific heat for water over the temperature range 20°C to 100°C.

PROBLEM 733

In grinding and shaping a machine tool bit you press the bit with a force of 3.0 lbwt against a grinding wheel 6.0 inches in diameter which is rotating with a speed of 1400 rpm; the coefficient of friction between tool bit and wheel is 0.70, the bit has a mass of 30 gm and a specific heat of 0.12 cal gm^{-1} (°C)$^{-1}$.

a. If you held the bit against the wheel for 5.0 sec, how much heat must be accounted for?
b. if 80% of the heat appeared in the tool bit, what was its increase in temperature?

Thermal Expansion

As stated earlier, although the work of thermal expansion of liquids and solids is negligible compared to the heat input, the expansion itself may not be negligible. Almost all substances expand linearly with heat according to the relationship

$$\ell_t = \ell_0 \, [1 + \lambda(t - t_0)], \tag{5-39}$$

where ℓ_0 is the length at some reference temperature t_0 and λ is the coefficient of linear expansion for the substance in question over the temperature range encountered. For volume expansion there is a similar relationship,

$$V_t = V_0 \, [1 + \alpha(t - t_0)]. \tag{5-40}$$

λ and α are usually given in tables (see Appendix A) in units of $(°C)^{-1}$, and are usually average values over a normal temperature range.

PROBLEM 734

a. What is the first-order relationship between λ and α?
b. What is the probable form of an equation describing how a plane area changes with temperature?

PROBLEM 735

A steel bar 1.0 m long and 10 cm^2 in cross section is heated from 20°C to 40°C while being prevented from expanding by a mechanical constraint. If the constraint breaks, how much energy does the bar deliver? *Hint*: $\Delta\ell = \ell_0\lambda\Delta t$, from equation 5-39.

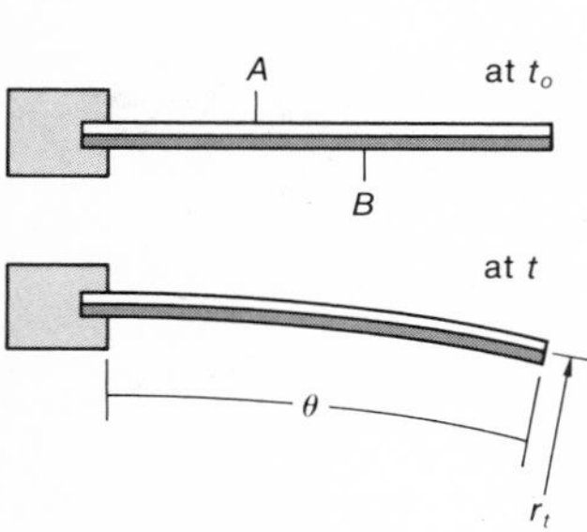
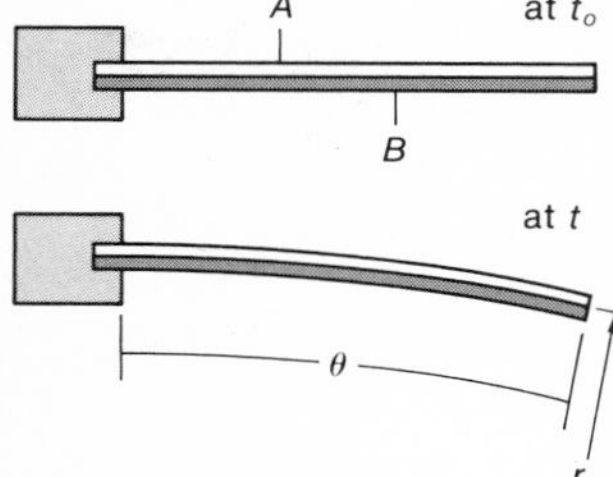

FIGURE 423

PROBLEM 736

A thermostat element is constructed by riveting together two metal strips A and B, each of thickness d. The strips are straight and of equal length at some reference temperature t_0 and $\lambda_A > \lambda_B$, both based on t_0. When the element is heated above t_0 it bends as shown in Figure 423. Assuming the center line of each strip expands as if the strip were free, what is r_t, the radius of bend measured to the center line of the element as a function of temperature t? *Suggestion:* Evaluate ℓ_t for each strip separately in terms of r_t and θ, and then eliminate θ by algebraic manipulation.

FIGURE 424

PROBLEM 737

A cylindrical brass vessel 10.0 cm deep is filled to a height h with mercury (Figure 424); both vessel and Hg are originally at the same temperature. What should h be so that the distance between the Hg surface and the upper edge of the vessel does not change with changes in temperature? Neglect terms containing products of coefficients of expansion—these are second-order effects.

PROBLEM 738

You are assigned to design an indicating instrument which consists essentially of an evacuated glass bulb into which equal volumes of mercury and a nonvolatile oil are introduced. At a specified °C reference temperature the volumes of the Hg and the oil are each V and the volume of the glass bulb is V_0, $V_0 > 2V$. What should V be, in terms of V_0, so that the residual volume not occupied by Hg or oil remains constant while the bulb and its contents are exposed to a wide range of temperatures?

PROBLEM 739

A long uniform bar having a mean linear coefficient of thermal expansion λ has one end maintained at 0°C; from this end the temperature of the bar increases linearly with distance until it reaches T at the far end. When the bar was entirely at 0°C its length was ℓ_0.

a. Sketch a graph of the temperature of the bar t as a function of x, the distance from the cold end.

b. For a small element of length dx at x, by what amount $d(dx)$ does this element change in length when its temperature is increased from 0°C to t°C?

c. By how much did the bar lengthen from ℓ_0 when it was heated as described? *Caution*: When you integrate the function you obtained in question (b) you get the change in length in terms of the heated length, which is unknown. This change in length must then be converted to a change in length referred to ℓ_0.

PROBLEM 740

A horizontal slab of steel has drilled in it a vertical conical hole that tapers to a point at the lower face of the slab. Into this hole is placed a solid brass sphere (Figure 425). At an equilibrium temperature of 0°C the diameter of the hole at the upper face is 5.0 cm, the slab is 10 cm thick, and the radius of the brass sphere is 2.0 cm. When the system is heated uniformly such that its temperature is rising at a rate of 50°C sec^{-1}, what is the upward speed of the center of mass of the brass sphere? *Suggestion*: Determine first how the cone angle of the hole changes with temperature. After that the problem is a simple one of elementary differentiation.

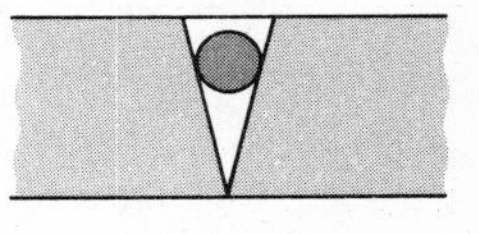

FIGURE 425

THE STORAGE OF ENERGY IN LIQUID SURFACES

Within a distance below a liquid surface equal to a few molecular diameters the summation of intermolecular forces at any point is asymmetrical; as a result a molecule at that point feels a net force inward, away from the surface. There are two interesting consequences of this force field close to the surface: a potential gradient is perpendicular to the surface, so that any attempt to extend the surface entails work in bringing additional molecules up to the surface from within the liquid; and the pressure below the surface is greater than at the surface. (This increase in pressure is not related to the hydrostatic pressure ρgh that was discussed earlier.)

From the first consequence it is clear that a liquid surface represents a storage of potential energy, and an increase in surface area requires an increase in stored potential energy. Thus

$$U_{\text{surf}} = \sigma A \text{ and } \Delta U_{\text{surf}} = \sigma \Delta A, \tag{5-41}$$

where A is the surface area, σ is an experimentally determined coefficient of surface tension, and U_{surf} is called the free surface energy. In MKS units σ is given in joules m^{-2}; however σ usually appears in tables in cgs units of ergs cm^{-2}.* The coefficient σ is a function not only of the liquid but also of what is on the other side of the interface; unless otherwise stated it is usually standard air.

(Although U_{surf} measures the work required to form a surface, it does not represent all of the energy invested in forming an isothermal surface. Molecules that reach the surface after coming up through the force field have less average energy than the molecules down in the liquid, hence the surface tends to be cooler than the liquid. Additional heat energy must then be supplied by the environment to make the surface isothermal with its liquid.)

Since natural processes move in a direction to reduce the potential energy of a system to a minimum, an obvious corollary to the first consequence is that a liquid free to move will assume a shape of minimum surface area. Thus the falling lead drops of problem 731, rain drops before they are pushed out of shape by air pressure, beads of mercury on a glass plate, all assume a spherical shape.

The most interesting aspect of the second consequence is that the pressure under a liquid surface varies with the curvature of the surface. If P_p is the pressure under a plane liquid surface and P_r is the pressure under a curved surface,

$$P_r = P_p + \sigma \left(\frac{1}{R_1} + \frac{1}{R_2} \right),$$

where R_1 and R_2 are the principal radii of curvature of the surface. R is + if the liquid surface is convex (the center of curvature lies within the liquid), and R is − if the surface is concave (the center of curvature lies outside the surface).

*Or as dynes cm^{-1}. This unit is derived from the original experimental method of measuring σ, in which a tensile force parallel to the surface did work in pulling to enlarge the surface. Hence the name surface "tension" is still in use.

PROBLEM 741

Mercury in a barometer tube of radius 0.10 cm has a convex meniscus that meets the walls of the tube at an angle of 40° (Figure 426). What correction Δh in mm should be made to the reading h in mm to give a truer measure of the barometric pressure B_0? When in contact with a vacuum at normal temperatures, $\sigma_{\text{Hg}} = 480$ ergs cm^{-2}.

SOLUTION

From symmetry, $R_1 = R_2 = r/\cos\alpha$. Thus the pressure underneath the meniscus due to its curvature $= P_p + (2\,\sigma\cos\alpha)/r$. Comparing the pressure just below the flat surface of the mercury in the cistern with the equal pressure at a point on the same level inside the tube, $B_0 + P_p = P_p + (2\,\sigma\cos\alpha)/r + \rho gh$ from which

$$B_0 = \rho g\left(h + \frac{2\,\sigma\cos\alpha}{\rho gr}\right).$$

Then,

$$\Delta h = \frac{2 \times 480 \times 0.766}{13.6 \times 980 \times 0.1} = 5.6 \text{ mm}.$$

(Ignored in this problem is a correction for the fact that with a curved meniscus the hydrostatic pressure is not exactly determined by measuring to the top of the meniscus.)

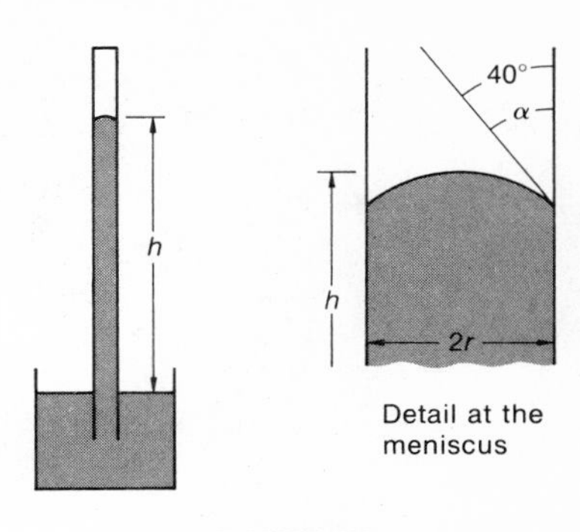

FIGURE 426

PROBLEM 742

What is the air pressure inside a soap bubble of radius R if the ambient pressure is B_0? *Caution:* There are two parallel surfaces to a soap bubble, one concave and one convex.

Suggestion: You should get the same value for the pressure at a point inside the film regardless of whether you approach that point by crossing the outside surface from the outside or by crossing the inside surface from the inside.

PROBLEM 743

Two separate bubbles formed from the same material come together and form a double bubble, as shown (Figure 427). The radii of the component bubbles are R and r, $R > r$.

a. What is the radius of curvature of the internal film common to both bubbles?

b. Toward which bubble is this film concave?

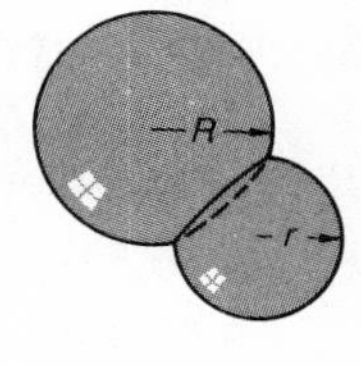

FIGURE 427

PROBLEM 744

A soap bubble is slowly and isothermally blown in STP air to a radius R. How much work was done?

PROBLEM 745

A spherical drop of mercury of radius 2.5 mm is resting on a smooth, clean horizontal surface. How much work is required to cut this drop into eight equal smaller spherical drops? (For Hg in contact with air, $\sigma_{Hg} = 520$ ergs cm^{-2}.)

PROBLEM 746

If a bubble could be blown in a vacuum by filling it with an ideal gas, what would be the ratio of the energy stored in the bubble film to the internal energy of the enclosed gas?

PROBLEM 747

a. What is the maximum electrical potential (relative to $V_\infty = 0$) that you can apply to a water drop without causing it to break up? (At break-up the disruptive electrical forces on the surface just equal the cohesive surface tension forces.)

b. Just prior to break-up, what was the ratio of the electrical potential energy of the drop to its free surface energy?

THE TRANSPORT OF ENERGY IN ELECTRICAL CIRCUITS

In modern technology the most obvious transport of energy is by electrical currents through electrical circuits. The amount of this flow of energy is determined by the electrical potential difference V available to the circuit from some electromotive source, and by the current I being driven through the circuit by V. The basic equation governing this flow states that the total algebraic sum of potential differences around a circuit must equal zero; the sum of potential "sinks" where energy is extracted from the circuit exactly equals the sum of the potential differences supplied to the circuit by electromotive sources.

Energy vs. Power. Energy has the dimensions of Vq, hence to be consistent in talking about the transport of energy we should discuss flow of charge q rather than current I. However, in most practical cases both the measurement and usage of electrical energy are better understood in terms of $I = dq/dt$ than in terms of q. Thus when we discuss the movement of current I driven by a potential difference V we are discussing the product VI which measures a transport of power rather than of energy.

Direct Current Circuits

Steady-state Conditions

The simplest form of the basic energy-flow equation occurs in describing the steady state in any element of a direct current circuit:

$$\mathrm{V}_i - I_i R_i = 0, \tag{5-42}$$

where $-I_iR_i$ is the negative potential difference developed across the i'th circuit which has a resistance R_i carrying a current I_i, and V_i, the potential difference supplied to the element, is a function of the total circuit potential difference V.

If the same current moves through all of the circuit elements they are said to be in series (*above*, Figure 428), and equation 5-42 reduces to

$$V - I(R_1 + R_2 + \cdot\cdot\cdot + R_n) = 0.$$

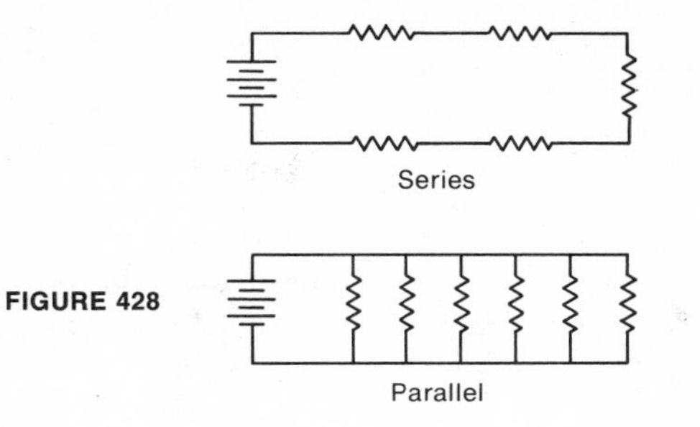

FIGURE 428

For elements in parallel (*below*, Figure 428): $V = I_1R_1 = I_1R_2 = \cdots = I_nR_n, I = I_1 + I_2 + \cdots I_n$, and equation 5-42 reduces to

$$V - IR_{\text{eff}} = 0,$$

where

$$\frac{1}{R_{\text{eff}}} = \frac{1}{R_1} + \frac{1}{R_2} + \frac{1}{R_3} + \cdots + \frac{1}{R_n}. \tag{5-43}$$

The power absorbed by the i'th circuit element is the product V_iI_i, where V_i is the potential difference across the element and I_i is the current through the element. Since V_i is also equal to I_iR_i we can obtain the more familiar equation

$$P_i = I_i^2R_i. \tag{5-44}$$

In a resistor this power appears as heat, and it is usually referred to as the heat loss, or the I^2R loss, of a resistor.

PROBLEM 748

An automotive service shop has available a DC charging source of negligible internal resistance that produces a constant voltage of 24 volts. A 12-volt battery and a 6.0-volt battery, each with an internal resistance of 0.04 Ω per rated volt, are connected in series to be brought up to full charge.

a. How much resistance must be placed in series with the batteries to hold the charging current to 4.0 A?
b. How much power will be wasted in this resistance?

PROBLEM 749

A DC laboratory microammeter with a full-scale reading of 10^3 μA has an internal resistance of 60 Ω.

a. What is the voltage drop across this instrument when it reads 650 μA?
b. What resistance would you have to put in series with this instrument to be able to use it as a DC voltmeter with a full-scale reading of 250 volts?
c. How would the make-shift instrument described in question (b) differ from a stock DC voltmeter advertised in a catalog as a "1000 ohm per volt" voltmeter?

d. How could you use the microammeter described at the beginning of this problem as a DC ammeter reading 0.50 A at full scale?

SOLUTION

a. Across the instrument $V = IR = 650 \times 10^{-6} \times 60 = 0.039$ volts.
b. Since the maximum current that can be allowed through the instrument is 10^3 μA, the resistance of the instrument plus an external series resistance must be sufficient to hold to 10^3 μA any current driven through the combination by an external potential difference. Thus $250 = 10^3 \times 10^{-6}(R + 60)$, from which $R \simeq 2.5 \times 10^5$ Ω, or 10^3 Ω per volt full scale.
c. There is no essential difference; any DC voltmeter is simply a sensitive current-measuring device in series with a high resistance and fitted with an appropriate scale.
d. Since at a full current of 0.50 A only 0.001 A can be allowed through the instrument, a parallel path must be provided for the balance of 0.499 A. The potential difference across both paths must be the same, so $10^3 \times 10^{-6} \times 60 = 0.499 \times R$, or $R = 0.1203$ Ω. Thus a shunt of 0.1203 Ω resistance placed in parallel with the instrument will enable it to evaluate DC currents up to a maximum of 0.50 A (Figure 429).

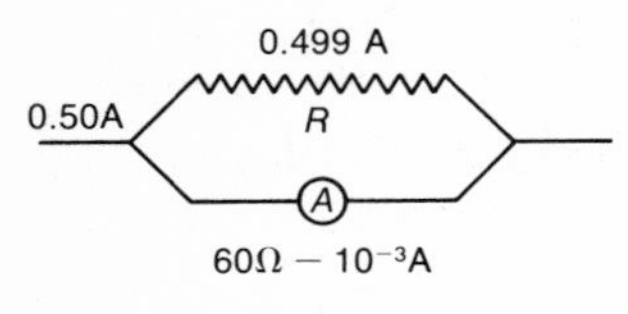

FIGURE 429

PROBLEM 750

When the switch is closed, for what value of R will the ammeter A read zero (Figure 430)? Assume negligible internal resistance in the batteries.

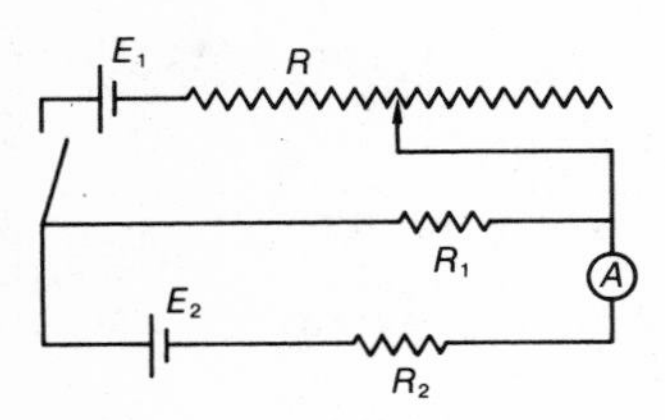

FIGURE 430

PROBLEM 751

The Wheatstone bridge, a standard instrument for measuring resistances, is as shown in Figure 431; R_x is the resistance to be measured, R_2 is usually a standard resistance, and R_1 and R_3 are treated as ratio resistances. G is a galvanometer, a special form of a very sensitive microammeter. When the bridge is "in balance" no current moves through the galvanometer, and then $R_x = (R_3/R_1)\,R_2$. How would you have to change this equation if you exchanged the positions of the battery and the galvanometer?

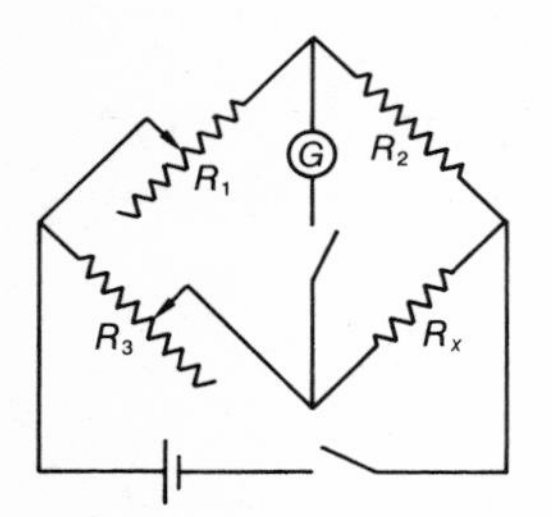

FIGURE 431

PROBLEM 752

The resistance of a conductor changes with temperature according to the relation $R_t = R_{t_0}[1 + \alpha(t - t_0)]$, where t_0 is some chosen reference temperature and α is an experimentally determined constant for that reference temperature and for that particular material of the resistor. A Wheatstone bridge can be made into a remote indicating thermometer by connecting an R_t whose temperature is to be measured into the bridge as shown in Figure 432. The bridge is balanced when both bridge and R_t are at t_o, and the bridge is kept in balance by changing R_2 as R_t changes with temperature, and these changes in R_2 are calibrated in terms of the temperature t of R_t. What is the equation governing this calibration?

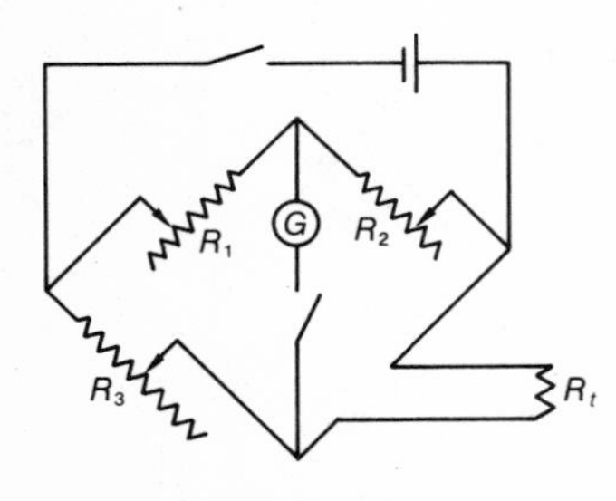

FIGURE 432

PROBLEM 753

A Wheatstone bridge is used to measure an unknown resistance R_x when R_x and the bridge are both at equilibrium with the laboratory temperature of 0°C. When the laboratory warms up to 25°C and both R_x and the bridge have come to equilibrium with this new temperature, the new measured value of R_x is found to be 0.40% greater than before. If the temperature coefficient of resistance α, referred to 0°C, of the material of the bridge resistors is 5.0×10^{-5} $(°C)^{-1}$, what is the temperature coefficient of resistance, referred to 0°C, of the material of which R_x is made?

PROBLEM 754

A pair of telegraph wires each having a resistance of 13.2 Ω mi^{-1} forms a telegraph line 10 mi long. When a "ground" occurs on the line (as might happen during a storm or when an irresponsible hunter shoots at an insulator) the following procedure is used to locate it: at one end the wires are connected together, and a bridge circuit is established at the other end as shown in Figure 433. If the bridge balances for $R_1 = 58$ Ω and $R_2 = 87$ Ω, what is the distance d to the ground?

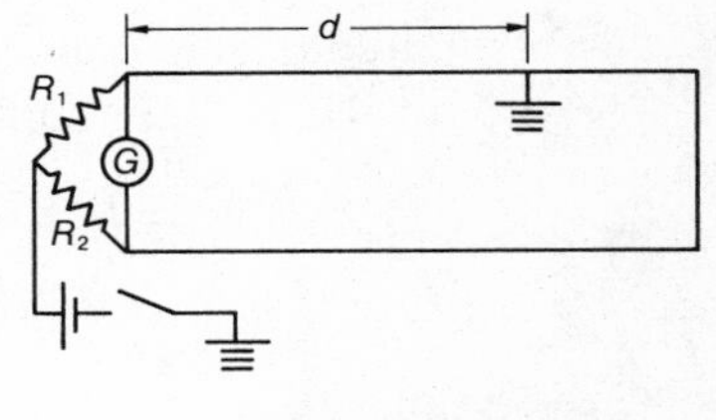

FIGURE 433

PROBLEM 755

A transmission line has a total resistance R and is required to deliver a constant amount of power P_L at the load end. The supply end must thus deliver power to the line at a rate $P_0 = P_L + I^2R$, the latter term being the line loss. P_L and R are fixed in amount but the supply voltage V can be varied.

a. What is the efficiency P_L/P_0 of the transmission, in terms of P_L, R, and V? Or, equivalently, what is the loss ratio I^2R/P_0? *Suggestion:* The easiest way to an answer in simple form is to set up the desired ratio in terms of V, R, and I and then eliminate I by the relation $VI - I^2R = P_L$.

b. What does your answer tell you about the relation between efficiency and high transmission voltages?

PROBLEM 756

In the circuit shown in Figure 434 the battery E_1 has an emf of 5.0 volts and an internal resistance of 0.50 Ω, and the battery E_2 has an emf of 1.0 volts and an internal resistance of 0.10 Ω. $R_1 = 3.5$ Ω, $R_2 = 1.0$ Ω, and $R_3 = 1.0$ Ω. When the switches are closed,

a. what does the voltmeter V_1 read,
b. what does the voltmeter V_2 read,
c. how much power is being dissipated in R_3?

Suggestion: The sum of the voltages plus voltage drops around *any* closed circuit must equal zero; thus you can write this summation for the external circuit or for either of the two internal circuits. Also current must be conserved: the total current into a junction must equal the total current out of that junction. (These two statements

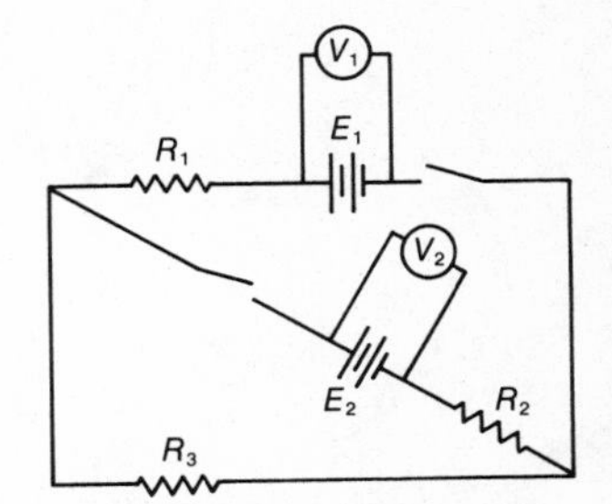

FIGURE 434

are codified in most texts as Kirchhoff's Laws.) Thus you could write for the external circuit $E_1 - I_0(r_1 + R_1) - (I_0 - I_i)R_3 = 0$ and for the upper internal circuit $E_1 - I_0(r_1 + R_1) - I_i(r_2 + R_2) - E_2 = 0$, where I_0 is the current through R_1 and I_i is the current through R_2.

Initial Conditions

In steady-state conditions with direct current the only potential sinks are provided by the resistances in the circuit. When the current is changing, however, (which even direct current does immediately after the opening or closing of a circuit) any inductances or capacitances in the circuit also contribute to the sum of the potential differences around the circuit.

PROBLEM 757

If the switch in the circuit shown in Figure 435 is closed at $t = 0$, as a function of the time (and, for simplicity, assuming negligible resistance in the battery and negligible current through the voltmeters),

a. what is the current in the circuit?
b. What are the voltage readings in V_R and V_L?
c. What is the rate at which energy is being dissipated in the resistor?
d. What is the rate at which energy is being stored in the field of the inductor?
e. What is the total power being supplied by the battery?
f. What was the reading V_L at the instant the switch closed?

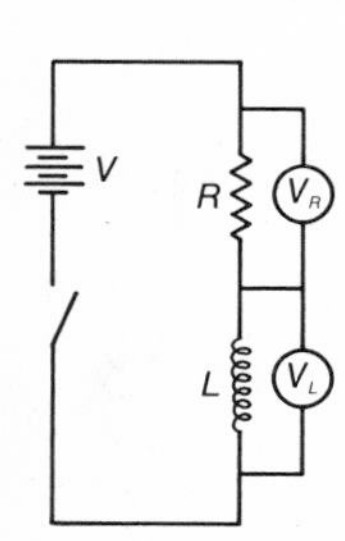

FIGURE 435

SOLUTION

Summing up the potential differences around the circuit (remember that the back emf of an inductance is $L\dot{I}$), $V - IR - L\dot{I} = 0$. Rearranging, $dI/dt + (R/L)(I - V/R) = 0$ so that

$$\frac{dI}{I - V/R} = -\frac{R}{L}\,dt.$$

At $t = t$, $I = I_t$ and at $t = 0$, $I = 0$. By integrating between these limits, we get the following answers.

a. $I_t = (V/R)(1 - e^{-(R/L)t})$.

b. $V_{R,t} = RI_t = V(1 - e^{-(R/L)t})$ and $V_{L,t} = L\dot{I} = V\, e^{-(R/L)t}$.

c. $P_{R,t} = I_t^2 R = (V^2/R)(1 - e^{-(R/L)t})^2$.

d. $P_{L,t} = (d/dt)U_{L,t} = (d/dt)(\frac{1}{2}LI_t^2) = LI_t\dot{I}_t =$
$L(V/R)(1 - e^{-(R/L)t})(V/L)\, e^{-(R/L)t} = (V^2/R)\, e^{-(R/L)t}(1 - e^{-(R/L)t})$.

e. $P_{t,\text{total}} = P_{R,t} + P_{L,t} = (V^2/R)(1 - e^{-(R/L)t})$.

f. At $t = 0$, $V_L = V$. This follows from the formula for $V_{L,t}$ but even without the formula the result could have been predicted: $V_{L,t} = L\dot{I}$; at $t = 0$, $I = 0$ (hence $V_R = IR = 0$) and $\dot{I}$ is at maximum.

PROBLEM 758

If the switch in the circuit shown in Figure 436 is closed at $t = 0$, as a function of time (and assuming negligible resistance in the battery and negligible current through the voltmeters),

a. what is the current in the circuit?
b. What are the voltage readings V_R and V_C?
c. What is the rate at which energy is being dissipated in the resistor?
d. What is the rate at which energy is being stored in the field of the capacitor?
e. What is the total power being supplied by the battery?
f. What was the reading V_R at the instant the switch closed?

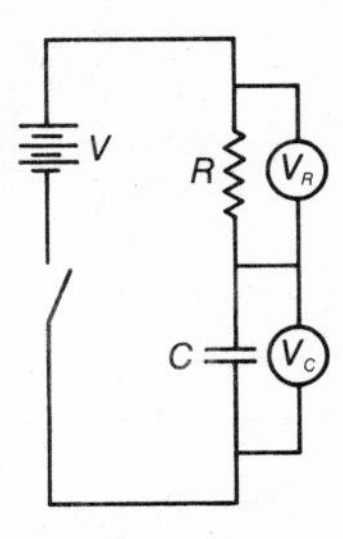

FIGURE 436

PROBLEM 759

If the switch in the circuit shown in Figure 437 was closed at $t = 0$ (and assuming negligible internal resistance in the battery and negligible current through the voltmeter),

a. what was the current through R immediately after $t = 0$?
b. What was the current through R as a function of time after $t = 0$?

When steady-state conditions are reached,

c. what is the reading of the voltmeter V_R,
d. how much energy has been stored in the field of the capacitor?

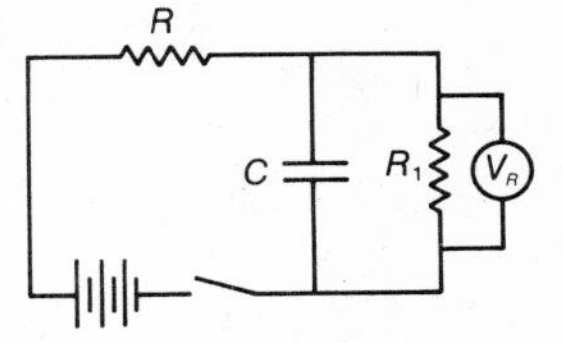

FIGURE 437

PROBLEM 760

Originally the switch was at A, but at $t = 0$ it was moved to B. At t large, when equilibrium has been reached (Figure 438),

a. what is the potential difference across either capacitor?
b. How much energy has been dissipated as heat in the resistor R?

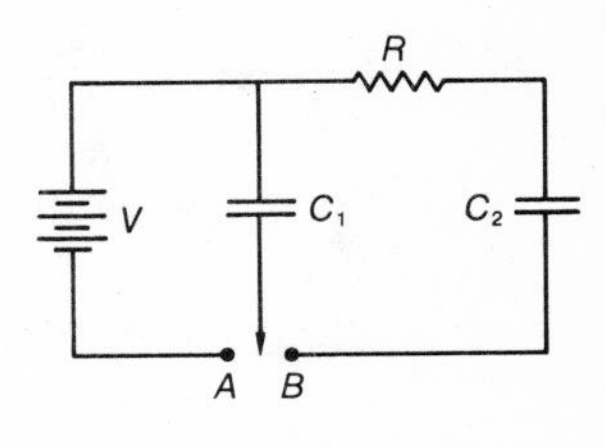

FIGURE 438

PROBLEM 761

The flashing lights seen on barricades at street work and construction sites are driven by a relaxation oscillator circuit as shown in Figure 439, with N being a neon lamp. The lamp is nonconducting until the voltage across it (and across the capacitor also) rises to a value of 45 volts whereupon it "ignites," dropping its resistance almost instantly to zero and discharging the capacitor. Then the lamp again becomes nonconducting, and the process repeats.* If V is 60 volts and $C = 0.10\ \mu F$, what should R be for the lamp to flash every 1.5 sec? *Hint*: The answer to questions (b) of problem 758 may be useful.

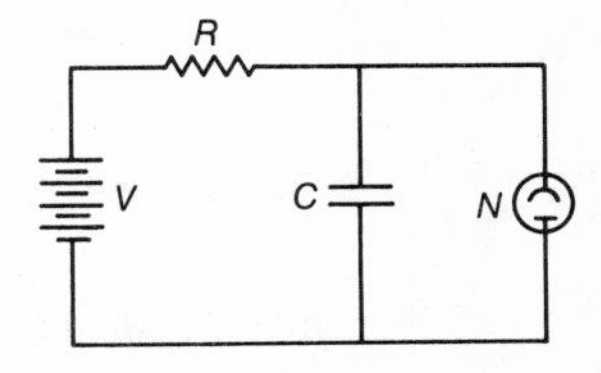

FIGURE 439

*To keep the statement of the problem simple the existence of a necessary minimum voltage across the lamp at all times is ignored. In actual practice, whenever the lamp "fires" and discharges the capacitor, the voltage across both does not drop to zero but only to the value of a voltage sufficient to maintain the lamp in readiness to ignite. Thus the real zero level for voltage is below the zero implied in the problem. However, this should make no difference in your solution.

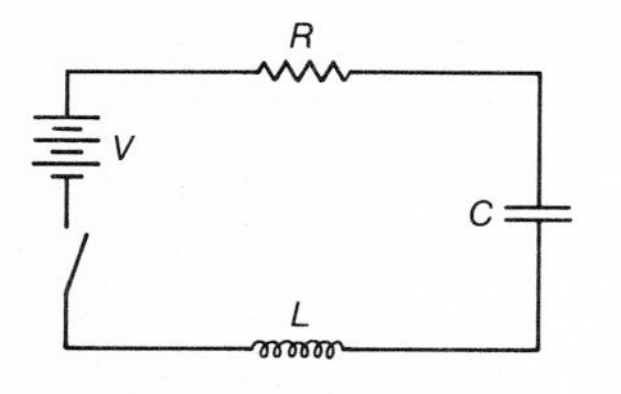

FIGURE 440

PROBLEM 762
Originally the switch was open and the capacitor uncharged (Figure 440). Then at $t = 0$ the switch was closed.

a. For $t > 0$, write the equation summing the potential differences around the circuit.
b. Solve this equation for the special case of $R^2 = 4L/C$.
c. For this special case what are the instantaneous values of the voltages across the capacitor, the resistor, and the inductor? Do your computed values of these voltages add to V as they should?
Comment: This problem does not possess major significance but it does have one very attractive aspect—you don't have to do any work to solve it, because you already know the solution. For question (a) you just look at the diagram and write down $V - \dot{q}R - q/C - L\ddot{q} = 0$; changing variables to $Q = (q - VC)$ transforms this to $\ddot{Q} + (R/L)\dot{Q} + (1/LC)Q = 0$. You have already seen this equation and its solution; it is the same as equation 3-62, with R/L corresponding to γ and $1/LC$ to ω_0^2. The special values given in question (b) correspond to the critically damped oscillator, hence the solution you need is equation 3-66 with the appropriate change of the constants. You might as well review the section on SHM clear through forced oscillations, for you will find from now on that the electrical circuit equations you will develop are exactly alike the SHM equations and have the same form of solution.

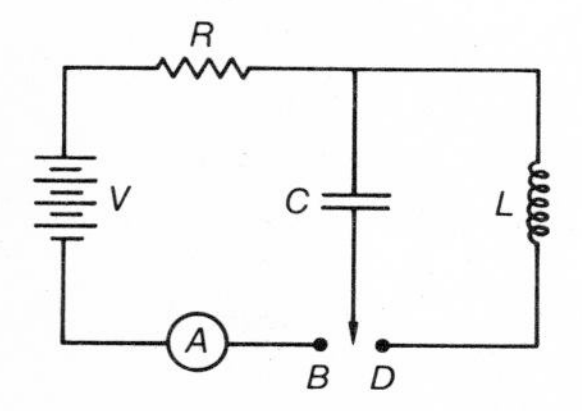

FIGURE 441

PROBLEM 763
With the capacitor uncharged the switch is thrown to B (Figure 441). When the reading of ammeter A drops to zero the switch is then thrown to D at $t = 0$. For $t > 0$,

a. what are the instantaneous values of the potential differences across the capacitor and the inductor, and do these add to zero as they should for the LC loop?
b. What is the frequency of the oscillation of charge in the LC loop?
c. What is the maximum value of the current in the LC loop?

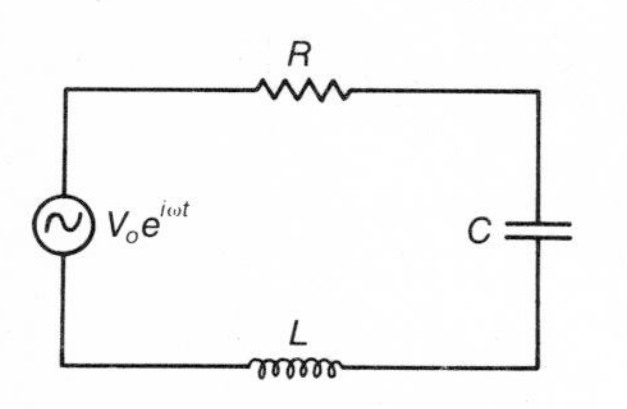

FIGURE 442

Alternating Current Circuits

Let us reconsider problem 762, with the battery replaced by a source of alternating emf of constant angular frequency ω. The amplitude of the emf can be described by either $V_0 \cos \omega t$ or $V_0 \sin \omega t$, depending on when we start our clock. However, remembering our experience in SHM in trying to find solutions to equations containing differentials of varying order, we play it safe by describing the emf by the more general $V_0 e^{i\omega t}$ (Figure 442).

Summing up the potential differences around the circuit we get

$$\dot{q}R + \frac{q}{C} + L\ddot{q} = V_0 e^{i\omega t}.$$

This is nothing new; it is our driven oscillator equation (discussed with solution on pages 228–231 with $\omega_0^2 = 1/LC$ and $\gamma = R/L$. We could make these substitutions in equation 3-62 and obtain a usable evaluation for q. However, we can learn more by going again through the solution.

Assuming $\hat{q} = \hat{A}e^{i\omega t}$ and solving by the usual methods we get

$$\hat{q} = \frac{V_0 e^{i\omega t}}{(i\omega)^2 L + i\omega R + (1/C)}.$$

However, in alternating current circuits we almost always are interested in the current I rather than in the charge q. Differentiating, we get

$$\hat{I} = \frac{d\hat{q}}{dt} = \frac{i\omega V_0 e^{i\omega t}}{(i\omega)^2 L + i\omega R + (1/C)} = \frac{V_0 e^{i\omega t}}{R + i[\omega L - (1/\omega C)]}. \tag{5-45}$$

This can be put in the form

$$\hat{I} = \frac{\hat{V}}{\hat{Z}} \text{ or } \hat{V} = \hat{I}\hat{Z}, \tag{5-46}$$

where $\hat{Z} = R + i(\omega L - 1/\omega C)$ is called the complex impedance of the circuit.

Multiplying both numerator and denominator of equation 5-45 by the complex conjugate of the denominator we get

$$\hat{I} = \frac{V_0 e^{i\omega t}\left[R - i\left(\omega L - \frac{1}{\omega C}\right)\right]}{R^2 + \left(\omega L - \frac{1}{\omega C}\right)^2} = \frac{V_0 e^{i\omega t} e^{-i\delta}}{\sqrt{R^2 + \left(\omega L - \frac{1}{\omega C}\right)^2}}, \tag{5-47}$$

where

$$\cos\delta = \frac{R}{\sqrt{R^2 + \left(\omega L - \frac{1}{\omega C}\right)^2}},$$

or

$$\tan\delta = \frac{\omega L - \frac{1}{\omega C}}{R}.$$

Taking the real part of equation 5-47, we finally arrive at an evaluation for I:

$$I = \frac{V_0 \cos(\omega t - \delta)}{\sqrt{R^2 + \left(\omega L - \frac{1}{\omega C}\right)^2}}. \tag{5-48}$$

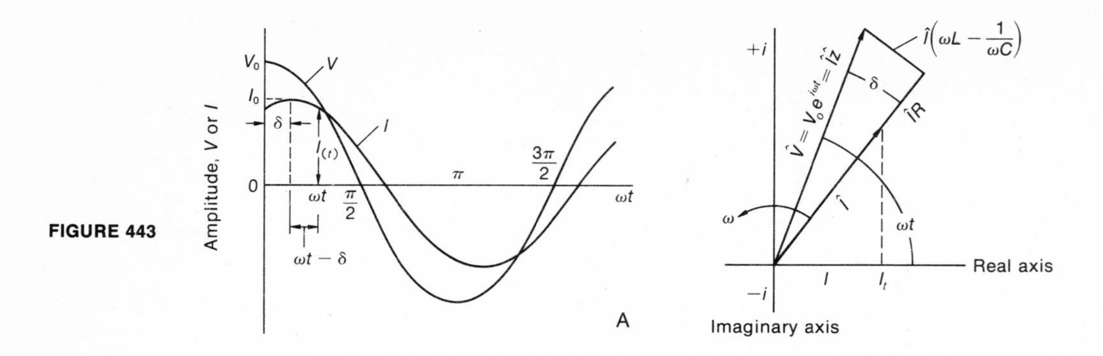

FIGURE 443

This can also be written

$$I = \frac{V_0 \cos(\omega t - \delta)}{Z},$$

where

$$Z = \sqrt{R^2 + \left(\omega L - \frac{1}{\omega C}\right)^2}$$

is the real impedance of the circuit. Note the similarity between equation 5-48 and equation 3-69, which describes the forced oscillator; the equations are not exactly alike because I corresponds to $\dot{x}$, not x.

These relationships are plotted in graphs A and B (Figure 443), A being the more conventional cosine graph, B the complex plane graph. Both graphs depict the same value for I at the same instant in time.

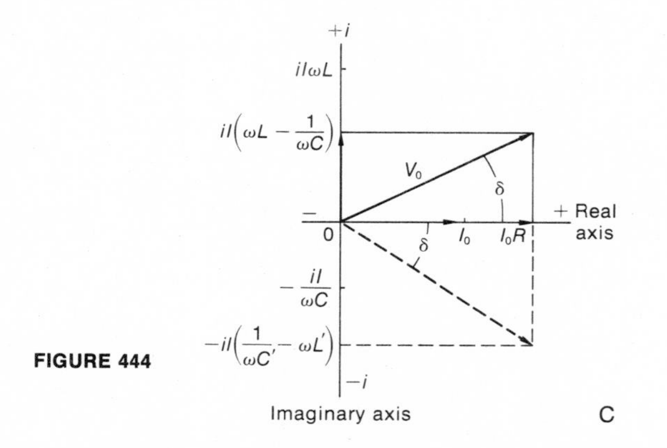

FIGURE 444

The phase angle δ between I_0 and V_0 is best illustrated on the complex plane by plotting the potential differences around the circuit for the instant when I_{real} equals I_0 (graph C, Figure 444). When $\omega L > 1/\omega C$, V_0 leads I_0 by the angle δ, as shown by the solid lines; when $\omega L' < 1/\omega C'$, I_0 leads V_0, as shown by the dashed lines. This should have been expected; an inductance opposes the growth or decay of current, so the potential difference across an inductance opposes and delays the influence of the external emf. Thus in a circuit dominated by an inductance you would expect the current to lag the external emf. A capacitance acts as a current sink and presents an opposite influence from an inductance; thus in a circuit dominated by a capacitance you would expect the current to lead the external emf.

Measurement of Current and Voltage. The current I_0 and the voltage V_0 discussed so far in this section are the amplitudes of the cosine (or sine) functions $I_0e^{i\omega t}$ and $V_0e^{i\omega t}$. In normal usage, however, what we mean when we speak of AC current and voltage, and what we measure, are not I_0 and V_0 but an I_{eff} and a V_{eff}, usually just indicated by I and V.

The AC measuring devices actually respond to I^2, not I; in use AC circuits deliver power which is also a function of I^2. It is therefore natural to measure current in terms of I_{eff} defined by

$$I_{eff} = \left[(I^2)_{av}\right]^{1/2} = \left[\frac{1}{2\pi}\int_0^{2\pi}(I_0 \cos \omega t)^2 \, d\omega t\right]^{1/2} = \left[\frac{I_0^2}{2}\right]^{1/2} = \frac{I_0}{\sqrt{2}}. \tag{5-49}$$

Similarly, $V_{eff} = V_0/\sqrt{2}$. Hereafter when we speak of current I or voltage V in AC circuits we will mean I_{eff} or V_{eff}.

Since I_{eff} and V_{eff} (I and V) are derived as an average their values do not fluctuate over a cycle. Hence for practical purposes we can redraw C (Figure 444) as a static graph D (Figure 445) not dependent on time, with δ remaining the same as before.

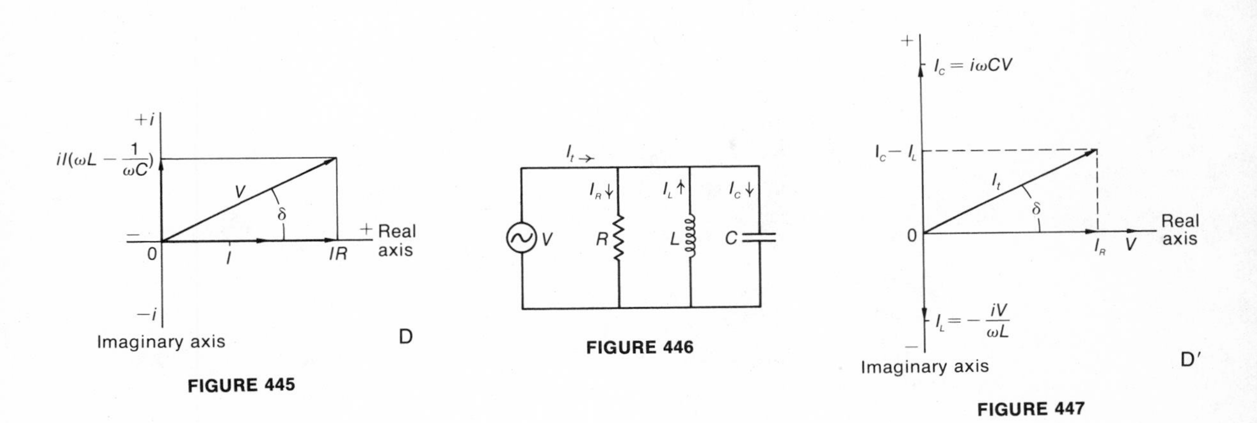

FIGURE 445

FIGURE 446

FIGURE 447

Elements in Parallel. Complex plane diagrams similar to Figures 444 and 445 can also be drawn for circuits containing elements in parallel. In the circuit shown in Figure 446 V is common to all elements and can be plotted on the real axis. $I_RR = V$ and I_R is in phase with V. For the inductor $V = iI_L\omega L$ and $I_L = -iV/\omega L$; for the capacitor $V = I_C/i\omega C$ and $I_C = i\omega CV$. Thus for the parallel circuit the current is complex, with out-of-phase components, in contrast to the series circuit where the voltage was complex. For the parallel circuit these components are plotted in D′ (Figure 447). If $I_C > I_L$, the total current I_t leads the voltage by

$$\delta = \tan^{-1}\frac{\omega C - 1/\omega L}{R}.$$

Note that the current components in the capacitor and the inductor are in opposite directions; this is indicated by the opposite signs for the value of I_C and I_L, and required by the fact the potential differences around the LC loop of the circuit must add to zero.

Power and Power Factor

In either a capacitor or an inductor energy is stored during one-half cycle and released during the following half-cycle; averaged over a full cycle energy is not stored in a capacitor or an inductor. Hence any average power delivered to our original AC series circuit (Figure 442) is absorbed only in the resistor R. As with DC circuits this power is evaluated by $I^2R = I \times IR$; however, note from diagram D (Figure 445) that in the AC circuit $IR = V \cos \delta$ is not in synchronism with V. (IR and V were in synchronism in the DC circuits discussed earlier.) The power capability of the source of emf is measured by the product $V \times I$—the source must be able to move charge at a rate I with an electromotive force V—but it delivers energy only at a rate $VI \cos \delta$. The ratio of the power delivered to the required power capability is thus equal to $\cos \delta$; this ratio is called the power factor. Obviously for efficient use of generating equipment a high power factor at the load end of the circuit is most desirable.

PROBLEM 764

In a series AC circuit similar to that shown in Figure 442, V_0, ω, R, and L are fixed in value but C is adjustable.

a. For what value of C is maximum average power delivered to the resistor?
b. What is the value of this maximum average power?
c. For what value of C will the current lead the voltage by 30°?

PROBLEM 765

A voltage V_1 is applied across a pure resistance R and the resulting current I is measured. Then an inductance L is placed in series with the resistor, and it is found that the voltage must be changed to V_2, $V_2 > V_1$, in order that the current remain the same as before.

a. What must have been the ratio V_2/V_1?
b. What average power was delivered to the resistor? Was this power the same for both voltages?
c. By what phase angle did the current through the resistor lag V_2?

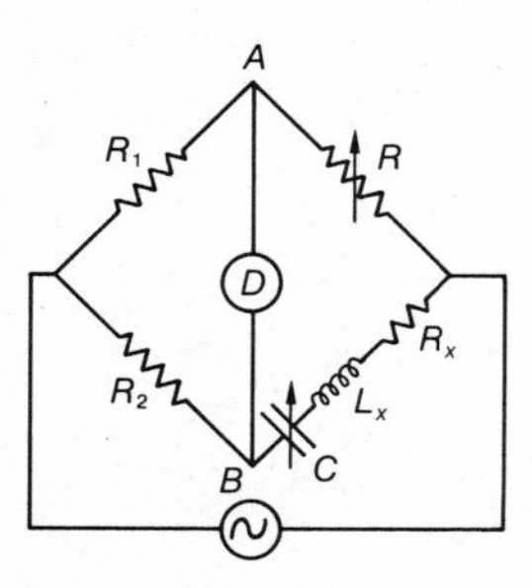

FIGURE 448

PROBLEM 766

A bridge for measuring inductance is as shown in Figure 448. The resistance arms R_1 and R_2 are pure resistance and have equal values. R is a variable standardized pure resistor, C is a variable standardized pure capacitor, L_x is the inductance to be measured, and R_x is the unknown effective resistance of the inductor. D is a detector: a pair of headphones if the frequency used is in the audible range, or some form of tuned analyzer. When a voltage of frequency ν is applied across the bridge, R and C are adjusted by trial until the detector indicates no voltage between A and B. If the standard resistor and capacitor read R_S and C_S when this null balance occurs, what are the values of L_x and R_x? *Suggestion:* Any equation summing complex potential differences can be split into two equations, one equating real values and one equating imaginary values.

PROBLEM 767

In the network shown in Figure 449 the frequency of the voltage source is varied until a null indication is obtained in the detector D. What is this critical frequency in terms of the capacitances and inductances of the circuit?

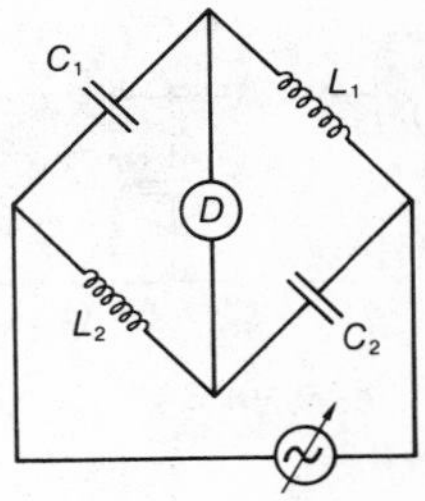

FIGURE 449

PROBLEM 768

A closed circuit consists of a resistance of 10 Ω, an inductance of 2.0 henrys, and a voltage source, all in series. When the source provides a constant voltage of 45 volts DC,

a. what is the steady state value of the current in the circuit?
b. How long after the voltage was turned on did it take the current to reach 75% of this steady-state value?
c. When an AC voltage of 126 volts at 10 cycles sec^{-1} is added in parallel to the constant DC voltage, what is the maximum measured current at any time in the circuit?

PROBLEM 769

A source of constant DC emf, with an internal resistance R_S, is connected to a pure resistance load R_L by a transmission line whose resistance is negligible compared to R_L.

a. What should be the value of the ratio R_L/R_S for maximum power delivery to the load?
b. What is the overall efficiency of the system when the power transfer is maximum?

PROBLEM 770

A source of constant emf, of angular frequency ω, is connected to a load by a transmission line of negligible impedance. The source possesses an internal resistance R_S and an internal inductance L_S, and the load possesses an internal capacitance C_L as well as an internal resistance R_L. What should be the relations among the impedance elements of the circuit for maximum power delivery to the load? *Suggestion:* Start by redrawing the series circuit in Figure 442 as shown in Figure 450.

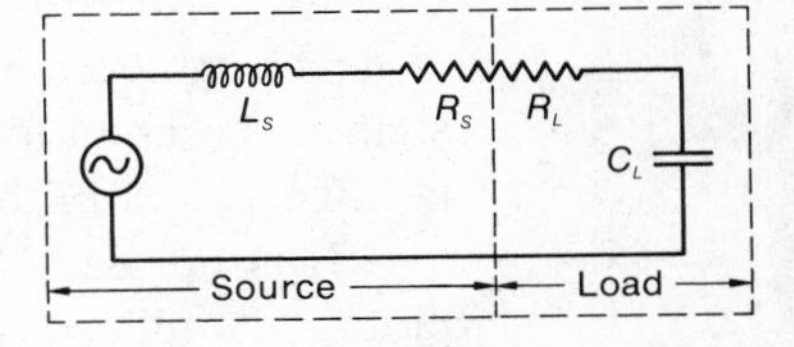

FIGURE 450

PROBLEM 771

In the parallel circuit shown in Figure 446, $V = 60$ volts AC, $R = 120$ Ω, $L = 0.65 \times 10^{-2}$ henrys and $C = 0.039$ μfarad.

a. For what frequency of the AC voltage will the power factor = 1?
b. At this frequency what average power is being delivered to the circuit by the source?
c. How does this supply of average power change with changing frequency of the source?

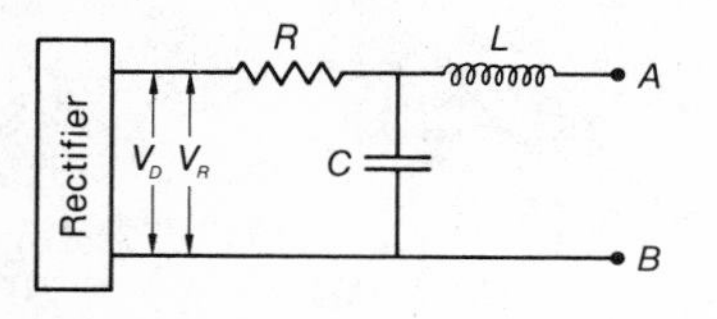

FIGURE 451

PROBLEM 772

The output of a power rectifier contains a DC voltage V_D and a ripple voltage V_R at a frequency of 120 Hz. If a filter circuit is inserted between the rectifier and the load terminals *A-B*, as shown in Figure 451, what is the open-circuit voltage at these terminals when $R = 500\ \Omega$, $C = 50\ \mu$ farad, and $L = 0.01$ henry?

THE TRANSPORT OF ENERGY BY STEADY STATE HEAT FLOW

Temperature T is a measure of a substance's content of energy of random thermal motion. If adjacent volumes of substances not insulated from each other possess different temperatures there exists an unstable distribution of thermal energy, and a flow of heat will be induced in an attempt to achieve thermal equilibrium. It is reasonable to postulate the following: this flow of heat will be "downhill," in the direction of the negative temperature gradient or perpendicular to the isothermal contours; it will be proportional to the gradient; it will probably be different for different substances. All of these postulations are confirmed by experiment. Hence the steady-state flow of heat across an isothermal surface of area A is described by

$$u = -k\,A\,\frac{dT}{dr}, \tag{5-50}$$

where u is the flow of thermal energy per unit time, $\vec{r}$ is perpendicular to the isothermal surface, and k is a characteristic of the substance through which the heat is flowing and is called the thermal conductivity or the coefficient of heat conduction. The dimensions of k are $MLT^{-3}\theta^{-1}$ and its units usually found in tables are cal sec^{-1} cm^{-1}(°C)$^{-1}$ or watts cm^{-1}(°C)$^{-1}$ or Btu ft^{-1}hr^{-1}(°F)$^{-1}$. The latter is a unit much used in heat engineering: 1 Btu ft^{-1}hr^{-1}(°F)$^{-1}$ = 1.73×10^{-2} watts cm^{-1}(°C)$^{-1}$. The minus sign in the equation is necessary because r is usually measured in the direction of heat flow, and in this direction dT is negative.

It is not easy to determine the temperature gradient at a single surface, and in practically all cases we will be more interested in a flow of heat through a finite body, which means a flow across at least two bounding surfaces. In the simplest possible case—a homogeneous slab with parallel plane faces d apart with a heat flow perpendicular to the faces (Figure 452)—after making the reasonable assumption that the temperature gradient in the slab is linear, we can reduce equation 5-50 to

$$u = -kA\,\frac{T_2 - T_1}{d} = kA\,\frac{T_1 - T_2}{d}. \tag{5-51}$$

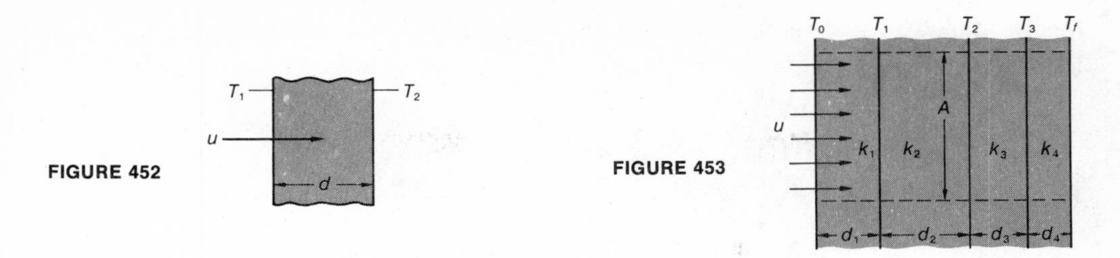

FIGURE 452

FIGURE 453

T_1 and T_2 are the temperatures exactly at the plane faces of the slab, and $T_1 > T_2$.

The next simplest case is that of parallel plane slabs of different materials. Assuming that adjacent faces are in thermal contact and possess the same temperature (Figure 453), we can write

$$u = k_1 A \frac{T_0 - T_1}{d_1} = k_2 A \frac{T_1 - T_2}{d_2} = \ . \ . \ . \ k_4 A \frac{T_4 - T_f}{d_4}.$$

Rearranging,

$$T_0 - T_1 = \frac{d_1}{k_1}\frac{u}{A}$$

$$T_1 - T_2 = \frac{d_2}{k_2}\frac{u}{A}$$

$$T_2 - T_3 = \frac{d_3}{k_3}\frac{u}{A}$$

$$T_3 - T_f = \frac{d_4}{k_4}\frac{u}{A}.$$

Adding and rearranging,

$$T_0 - T_f = \frac{u}{A}\left(\frac{d_1}{k_1} + \frac{d_2}{k_2} + \frac{d_3}{k_3} + \frac{d_4}{k_4}\right). \tag{5-52}$$

When the substances through which heat is flowing are not bound by plane parallel isothermal faces, there is no justification for assuming (as we just did) that the temperature gradient dT/dr is constant; hence in such cases dT/dr must be treated

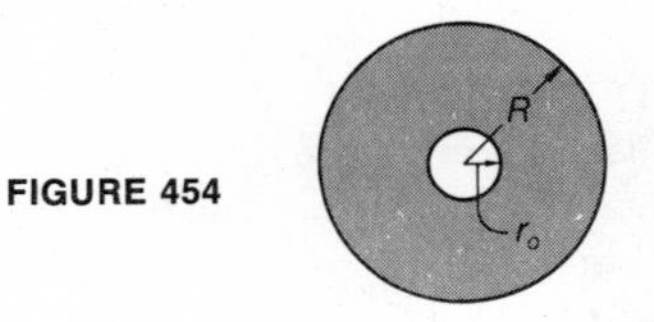

FIGURE 454

as a differential. As an example, consider a steam pipe of outer radius r_0 tightly surrounded by cylindrical insulation of inner radius r_0 and outer radius R (Figure 454). In this case, if u is the thermal energy leaking from the pipe per second per unit length, then $u = -k2\pi r(dT/dr)$ or $dr/r = -(2\pi k/u)\, dT$. Integrating across the insulation we get

$$T_0 - T_R = \frac{u}{2\pi k} \ln \frac{R}{r_0}$$

or

$$u = \frac{2\pi k(T_0 - T_R)}{\ln \dfrac{R}{r_0}}.$$

Electrostatic Analogies. In electrostatics we started with a flux vector $\vec{E}$ and from it obtained a scalar potential V, defined by $\vec{E} = -\nabla V$ or $E_r = -dV/dr$ in the one-dimensional case. In heat conduction we now have a scalar temperature T, and we can reverse the former process and obtain a heat flux vector $\vec{h}$ defined by $\vec{h} = -\nabla T$, or $h_r = -dT/dr$ for the one-dimensional cases we have been considering. We can then replace $-A\, dT/dr$ in equation 5-50 by the more general statement $\int \vec{h} \cdot d\vec{A}$, so that equation 5-50 now reads

$$u = k \int \vec{h} \cdot d\vec{A}.$$

This should have a familiar look—it is Gauss' Law now applied to heat conduction. Since solutions based upon the same fundamental relationships should have the same form regardless of particular symbols used, we should be able to write down solutions to heat conduction problems simply by finding an analogous electrostatic problem and making the following substitutions in the answers: $V \rightarrow T$,ϵ_0 (or $K\epsilon_0$) $\rightarrow k$, $\Sigma q \rightarrow u$. With the above substitutions in mind, note that the solution to the steam pipe insulation problem just discussed could simply have been copied from the answer to problem 280.

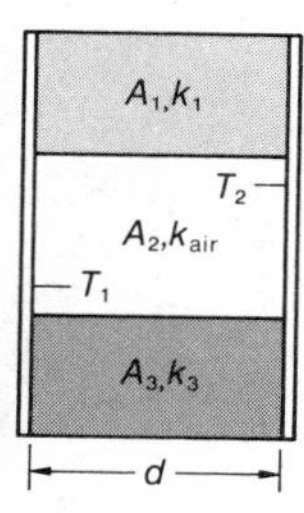

FIGURE 455

PROBLEM 773

On page 361 we considered the flow of heat through materials in series. Illustrated in Figure 455 is a typical wall: facing layers separated d apart by blocks in parallel of cross-sectional areas A_1 and A_3, with an air space of area A_2; the respective thermal conductivities are k_1, k_{air}, and k_3. If the temperatures of the inner surfaces of the facing layers are T_1 and T_2, $T_1 > T_2$, how much heat is moving through the wall?

PROBLEM 774

The average thermal conductivity of the earth's crust is 4.0×10^{-3} cal sec^{-1} cm^{-1} (°C)$^{-1}$, and the temperature gradient in the crust is about -30°C km^{-1} (minus because the temperature increases with decreasing radius.) At what average rate is heat flowing out of the surface of the earth? *Suggestion:* The depth of a thin crust layer is so small compared to the radius of curvature of the layer that the layer can be treated as a plane slab.

PROBLEM 775

A refrigerator with an effective wall area A, insulated with a thickness L of fiberglass whose thermal conductivity is k, is cooled by a mechanism operated by an electric motor of maximum power P. If the outside room temperature is T_0 and the mechanism is 75% efficient, what is the minimum temperature attainable inside the refrigerator?

PROBLEM 776

How long would it take to form a layer of ice 5.0 cm thick on top of a quiet pond if the air above the pond is at $-10°C$, the water below the ice is at 0°C? The heat of fusion of ice is 80 cal gm^{-1}, the thermal conductivity of ice is 5.0×10^{-3} cal sec^{-1} cm^{-1} $(°C)^{-1}$, and the density of ice at 0°C is 0.92 gm cm^{-3}. Neglect any further cooling of the ice after freezing, but assume that the top of the ice layer is in thermal equilibrium with the air above it. *Hint:* If the ice is y thick and this thickness is increasing at a rate dy/dt, then the amount of heat leaving the lower surface of the ice must equal $80\rho\ dy/dt$ per unit area per second.

PROBLEM 777

The walls of a "cold room" in a resort hotel are to be made up of a sheet of porcelainized steel 1.0 mm thick backed by a slab of cork held in place by a wall of wood 3.0 cm thick (Figure 456). The interior of the cold room is to be maintained at 5°C and the maximum ambient temperature expected outside the room is 30°C. If the heat leak into the room is not to exceed 10 watts m^{-2} of wall area at any time, how thick should the cork slab be? Assume complete thermal contact at all boundaries. The conductivity coefficients, in watts cm^{-1} $(°C)^{-1}$, are: porcelainized steel 0.083, cork 3.5×10^{-4}, wood 1.5×10^{-3}.

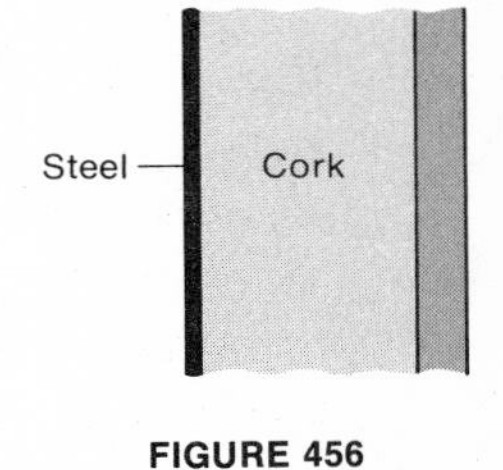

FIGURE 456

PROBLEM 778

When the builder of the cold room of problem 777 started to construct the final wall of the room, he found he had no more cork slabs. Rather than wait for a new supply he completed the room just as designed, but in place of the 8.0 cm of cork (see answer to problem 777) he left an 8.0 cm dead air space between steel and wood. If $k_{air} = 2.3 \times 10^{-4}$ watts cm^{-1} $(°C)^{-1}$, then, at the maximum expected temperature difference, how much heat m^{-2} leaked through the wall with the air space? What does your answer imply about the usefulness of dead air space for heat insulation?

PROBLEM 779

A laboratory glass cylinder of inner radius 7.0 cm, outer radius 8.0 cm, and height 10 cm is cemented to an insulating base. The cylinder is completely filled with water and then an insulating cover, through which a stirring rod has been frictionlessly mounted, is placed on the cylinder. When stirring has been continued long enough for steady state conditions to be reached, it is observed that the stirring motor, which is 90% efficient, is drawing 15 watts. $k_{\text{glass}} = 7.0 \times 10^{-3}$ watts cm^{-1} $(°\text{C})^{-1}$.

a. What must be the temperature difference between the inside and outside wall surfaces of the cylinder?
b. At what radius inside the glass wall would half of this temperature difference be found?

PROBLEM 780

A copper ball of radius r, density ρ, specific heat C, and initial temperature T_0 is held below the water surface of a swimming pool maintained at a temperature T_w, $T_0 > T_w$. The ball is covered with a very thin layer of insulating material of thickness d, $d \ll r$; the layer has a thermal conductivity k which is much lower than the conductivities of either copper or water.

a. What is the ball's temperature as a function of time?
b. How is the rate of cooling of the ball related to the instantaneous temperature difference between ball and water?
c. What reasonable assumption did you make in solving question (a)?

PROBLEM 781

Studies of seismic waves suggest that the earth possesses a distinct core with a radius of about 3.5×10^3 km. At one time it was thought this core was molten and was the source of the heat flux of 2.6×10^{10} kw observed at the surface of the earth (see problem 774). Assume the conductivity given in problem 774 to be relatively constant throughout the portion of the earth outside this core; then, as a consequence of the theory stated, what must be the temperature in the earth at the surface of this core?

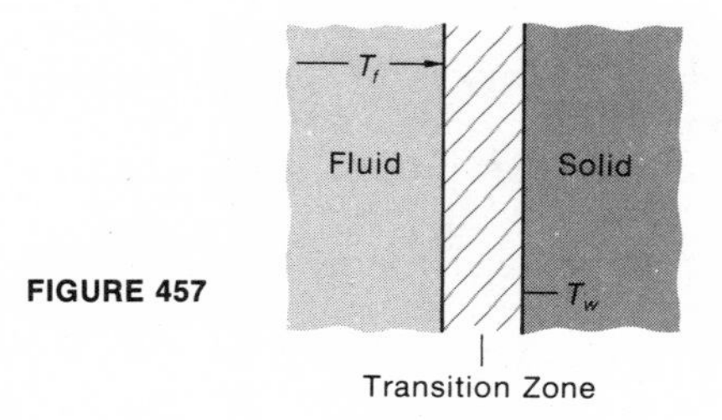

FIGURE 457

The assumption made in previous problems on heat flow—that a common boundary temperature exists at the interface between two substances—is valid for solids in thermal contact. It is seldom valid, however, when a fluid is in contact with a solid; in this case there is a temperature transition zone between the temperature T_f as measured in the fluid and the temperature T_w at the face of the confining solid (Figure 457). Neither the temperature gradient nor the zone width are usually known, so that in practice a heat transfer is measured experimentally and a heat transfer coefficient h is calculated by

$$h = \frac{u}{A(T_f - T_w)} \qquad T_f > T_w. \tag{5-53}$$

The coefficient h varies for different fluids, and is less for fluids at rest relative to the solid surface than for fluids moving. In computing a heat flow, h would appear in place of k/d in an inventory across boundaries that included a fluid-solid boundary.

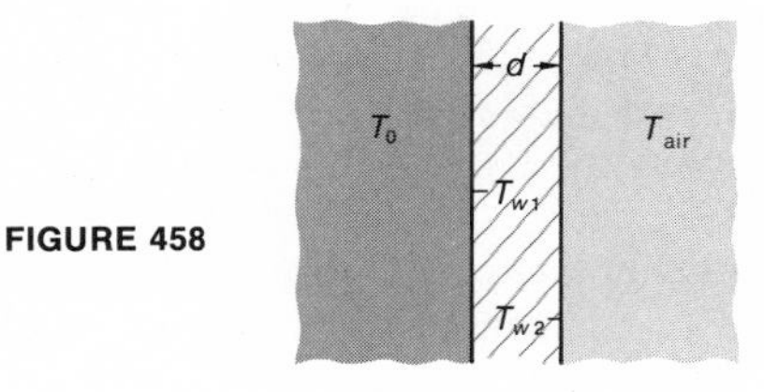

FIGURE 458

Thus in the case of a fluid confined by a solid which at its outer face is in contact with the ambient air (Figure 458), the fluid temperature T_0 and the air temperature T_{air} are known and the temperatures T_{w1} and T_{w2} at the faces of the solid are not known. For the heat flow we have

$$u = h_f A(T_0 - T_{\omega 1}) = kA\frac{(T_{\omega 1} - T_{\omega 2})}{d} = h_{\text{air}} A(T_{\omega 2} - T_{\text{air}}),$$

from which

$$u = \frac{A(T_0 - T_{\text{air}})}{(1/h_f) + (d/k) + (1/h_{\text{air}})}.$$

PROBLEM 782

a. If h for quiet air $= 5.0 \times 10^{-4}$ watts cm^{-2}(°C)$^{-1}$, what is the corrected answer to problem 778? Make the reasonable assumption that the inside surface of the steel wall is actually at 5°C.

b. If the heat transfer coefficient for quiet water is h_w, what is the corrected answer to problem 780?

PROBLEM 783

The essential part of a heat exchanger you are designing is to be a spaced array of thin-walled cylindrical tubes through which hot water at a temperature T_w will flow. Air at T_0 is to be blown over the tubes, $T_w > T_0$. The heat transfer coefficients for blowing air and flowing water are respectively 6×10^{-3} watts cm^{-2}(°C)$^{-1}$ and 0.5 watts cm^{-2}(°C)$^{-1}$. If you are concerned about maximum heat exchange per dollar cost of the array, in making a choice between copper tubes ($k_{\text{cu}} = 3.4$ watts cm^{-1}(°C)$^{-1}$) or aluminum tubes ($k_{\text{al}} = 1.6$ watts cm^{-1} (°C)$^{-1}$) how much weight should you give to the differences in thermal conductivity compared to the weight you should give to any differences in cost of the tubes?

PROBLEM 784

In view of what has been said about heat transfer between a fluid and a solid, does the linear temperature gradient specified for the bar of problem 739 seem realistic? What is a more probable temperature distribution down the bar?

SOLUTION

Unless the bar is completely surrounded by and in contact with a substance which is maintained at every contact point at a temperature exactly equal to the temperature of the bar at that point—an extremely improbable arrangement—a linear temperature gradient cannot be maintained in the bar.

Let us assume the bar is surrounded by a medium at T_0, choosing T_0 rather than the 0°C specified in the earlier problem in order to be more general. The bar could represent a cooling fin attached to a hot cylinder, and the medium could be a current of air or flowing water for which the heat transfer coefficient is h. At $x = 0$ the bar is in thermal contact with a heat source at T_s.

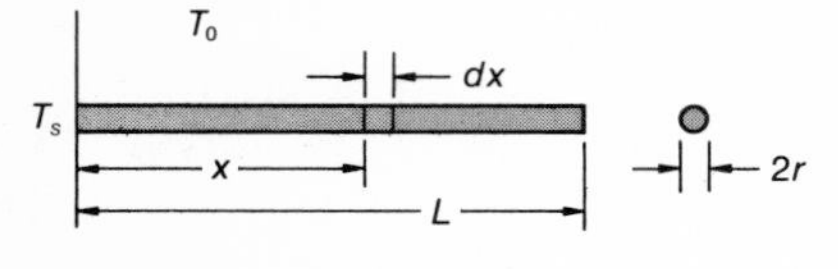

FIGURE 459

A small element of length dx (Figure 459) receives heat at a rate $-k\pi r^2 \, (dT/dx)|_x$; it delivers heat at a rate $h \; 2\pi r \; dx \; (T - T_0)$ to the surrounding medium and $-k\pi r^2 \, (dT/dx)|_{x+dx}$ to the bar beyond. For heat balance we have

$$\pi r^2 k \left[\left.\frac{dT}{dx}\right|_{x+dx} - \left.\frac{dT}{dx}\right|_x \right] = 2\pi r h (T - T_0) \; dx.$$

From elementary calculus we know that

$$\frac{(dT/dx)|_{x+dx} - (dT/dx)|_x}{dx} = \frac{d^2T}{dx^2}.$$

Rearranging our heat balance equation and making the above substitution we get

$$\frac{d^2T}{dx^2} - \frac{2h}{kr}(T - T_0) = 0.$$

Assuming $(T - T_0) = Ae^{\alpha x} + Be^{-\alpha x}$ and performing the indicated differentiations, we find that it is a solution provided $\alpha = \sqrt{2h/kr}$. When we evaluate the constants A and B by using the limits $T = T_s$ at $x = 0$ and $T = T_0$ at $x = L$, we end up with

$$T = T_0 + (T_s - T_0)(\cosh \alpha x - \sinh \alpha x \coth \alpha L).*$$

If you think the boundary condition $T = T_0$ at $x = L$ is questionable, it might appear more reasonable from the physical viewpoint to assume that under steady-state conditions of flow the temperature gradient approaches zero at the outer end of the bar, i.e., $dT/dx = 0$ at $x = L$. If we reevaluate the constants in the light of this boundary condition, we get

$$T = T_0 + (T_s - T_0) \; (\cosh \alpha x - \sinh \alpha x \tanh \alpha L).$$

---

* The hyperbolic functions are defined by $\sinh x = \frac{1}{2}(e^x - e^{-x})$ and $\cosh x = \frac{1}{2}(e^x + e^{-x})$.

For usual values of h and k, and for a bar for which $L \gg r$, αL comes out > 10. For high values such as this $\tanh \alpha L = \coth \alpha L = 1$. Then, both of the solutions reduce to $T = T_0 + (T_s - T_0)(\cosh \alpha x - \sinh \alpha x) = T_0 + (T_s - T_0)\, e^{-\alpha x}$.

PROBLEM 785

Under steady-state conditions of heat flow in the bar of problem 784,

a. at what rate is heat entering the bar at the hot end?
b. How much thermal energy has been stored in the bar?

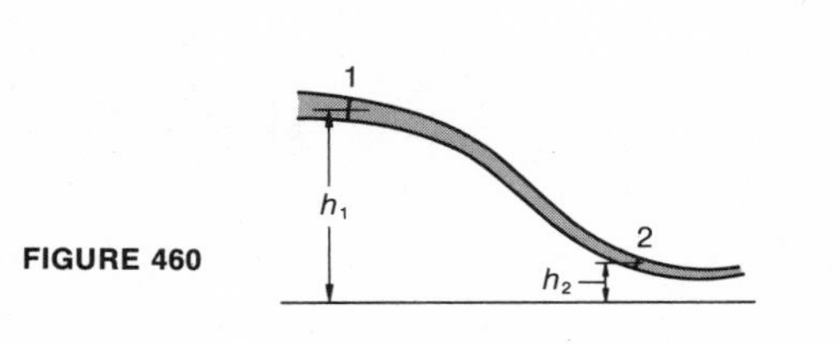

FIGURE 460

THE TRANSPORT OF ENERGY IN FLUID DYNAMICS

For steady, streamlined, isothermal flow of an incompressible, nonviscous fluid (Figure 460), an energy inventory per unit volume can be written for two different points in the stream:

$$P_1 + \tfrac{1}{2}\rho u_1^2 + \rho g h_1 = P_2 + \tfrac{1}{2}\rho u_2^2 + \rho g h_2. \qquad (5\text{-}54)$$

P is the static pressure in the fluid, u is the speed of the unit volume we are watching, and h is measured in reference to some selected gravitational potential energy datum. This is the well-known integral form of Bernoulli's equation. When calculations are made with the help of this equation, the data used must be expressed in absolute units.

An auxiliary equation to be used in conjunction with Bernoulli's equation is

$$\rho_1 u_1 A_1 = \rho_2 u_2 A_2. \qquad (5\text{-}55)$$

This statement requires that mass shall be conserved between 1 and 2 (or it gives us the comforting assurance that there are no sources or sinks between 1 and 2). With ρ constant it also tells us that the volume passing any cross section per unit time has a constant value.

If we divide equation 5-54 by ρg we get Bernoulli's equation in terms of "heads":

$$\frac{P_1}{\rho g} + \frac{u_1^2}{2g} + h_1 = \frac{P_2}{\rho g} + \frac{u_2^2}{2g} + h_2. \qquad (5\text{-}56)$$

$P/\rho g$ is called the pressure head or static head, $u^2/2g$ is the velocity head, and h is the potential head. This form of the Bernoulli is particularly useful in the laboratory, where often the various flow data are measured by manometers.

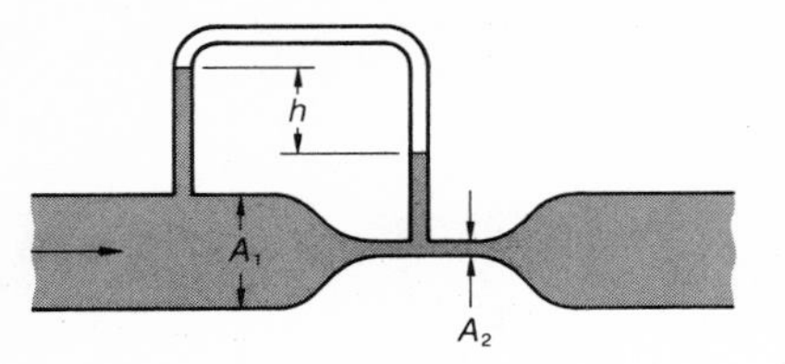

FIGURE 461

PROBLEM 786

One form of Venturi meter, in which only the variable h needs to be read, is as shown in Figure 461. Develop the formula for Q, the volume rate of flow in the pipe, in terms of h and the cross-sectional areas A_1 and A_2. Assume streamlined nonviscous flow.

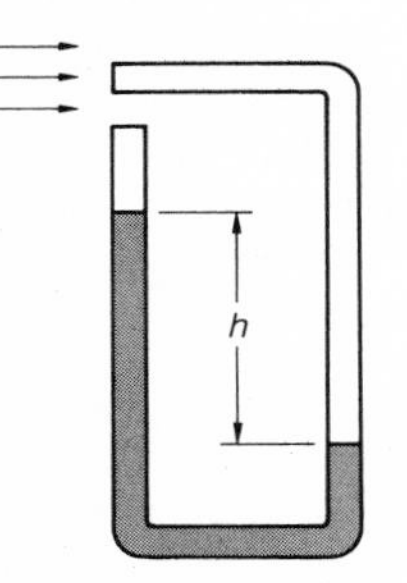

FIGURE 462

PROBLEM 787

The essential elements of a Pitot tube are as shown in Figure 462; one nozzle faces directly into the flow of a fluid of density ρ, the other is perpendicular to the stream flow. If the density of the manometer liquid is ρ_m, what is the local speed of flow? *Hint:* The nozzle facing the flow is exposed to both the static head and the velocity head: the perpendicular nozzle is exposed only to the static head.

PROBLEM 788

Old Faithful geyser in Yellowstone Park shoots water into the air to an average height of 38 m above ground.

a. What must have been the exit speed at the ground level?
b. What must have been the underground steam pressure at the geyser source 115 m below ground level?

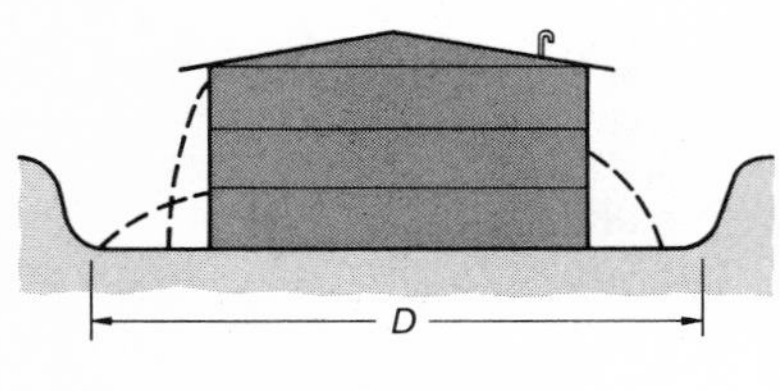

FIGURE 463

PROBLEM 789

For protection in case of fire or collapse, round steel storage tanks containing flammable oils and solvents are placed inside basins formed by earth walls (Figure 463). Such tanks usually have a radius equal to their wall height (see *Note* at end of problem). If such a tank is filled to the top of the side wall what should be the value of D, the floor diameter of the earth basin, so that liquid spurting from any hole in the wall falls inside the basin without hitting an earth wall? *Note:* If the slope of the roof is small, as it usually is, so that the roof can be treated as a flat plate, the ratio of contained volume to surface area is greatest, and hence the cost per unit volume of storage is least, when the height equals the radius. This is not difficult to prove; go ahead and do it.

Hint: You need first to determine the height on the wall of a hole for which an escaping stream will have a maximum range out from the foot of the wall. With this height determined you can then calculate the maximum range, and this in turn will lead to D.

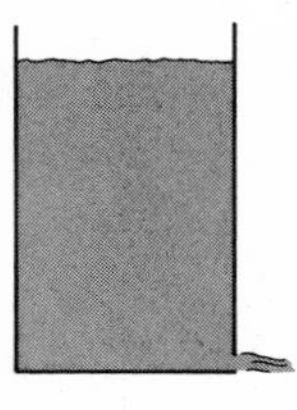
FIGURE 464

PROBLEM 790

An open-topped cylindrical barrel of cross-sectional area A is filled to a height H with a liquid of negligible viscosity. At $t = 0$ a small bung-hole of cross-sectional area a is opened at the very bottom of the barrel, as shown in Figure 464.

a. Assuming ideal flow, how long does it take the barrel to empty?
b. Because of the contraction of the exiting jet the actual discharge rate is only about 60% of the ideal rate. Taking this into account, what is the real time for emptying?

START OF SOLUTION
When the liquid stands y deep in the barrel, the theoretical speed of efflux at the orifice is $\sqrt{2gy}$ (from an application of equation 5-54). Also from conservation of flow we have $A(-dy/dt) = a\sqrt{2gy}$. Separating variables, $dy/\sqrt{y} = -(a/A)\sqrt{2g}\,dt$. When this is integrated between the limits of $y = H$ when $t = 0$ and $y = 0$ when $t = T$, a value for T is obtained.

PROBLEM 791
An irrigation standpipe, which discharges through a pipe of 6.0 in internal diameter as shown in Figure 465, is being filled at the rate of 150 $ft^3\ min^{-1}$. The diameter of the standpipe $\gg$ diameter of the bottom discharge line. Some time after the valve in the discharge line is opened a steady-state condition is reached in which the height of the water in the standpipe remains constant. What is this height? Neglect any energy of arrival of the water being delivered to the standpipe. *Note:* In problem 790 the discharge coefficient was given as 0.60; in situations like the present one, where the discharge is into a long pipe, the discharge coefficient drops to about 0.50.

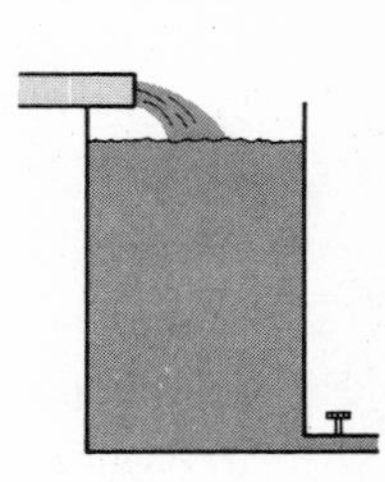

FIGURE 465

PROBLEM 792
Water flows vertically downward out of a round nozzle of radius r_0 with a speed u_0 (Figure 466). Since u_0 is not great, for some distance below the nozzle the water moves in a solid laminar stream. Over this distance what is the formula for the stream radius r as a function of the distance y below the nozzle. Over the distance considered the energy lost to viscosity is negligible.

FIGURE 466

PROBLEM 793
At a certain section of a thin airplane wing the shape is so designed that the air flowing over the flat lower surface has just 0.70 of the speed of the air flowing over the corresponding upper surface. In normal flight the lower surface is horizontal.

a. When the plane is flying at a speed of 480 mi hr^{-1} through air of density 0.068 lb ft^{-3}, what is the "lift" on the wing section mentioned?
b. When the plane rises to an altitude where the air density has dropped to 0.050 lb ft^{-3}, how fast must the plane fly to maintain the same lift as before? Assume the density of air is the same on both sides of the wing. This assumption is accurate enough for a first-order solution.

PROBLEM 794
Before diesel and electric traction, trains were pulled by steam engines. Due to leaks and auxiliary uses of steam, water was not completely conserved and hence had to be replaced occasionally. On long runs of express trains this replacement was often made "on the fly." From an engine passing over a shallow water trough set between the rails a scoop pipe was lowered into the water, as shown in Figure 467. If the pipe had a cross-sectional area of 18 in^2 and its discharge opening was 8.0 ft above the water surface, what volume of water per

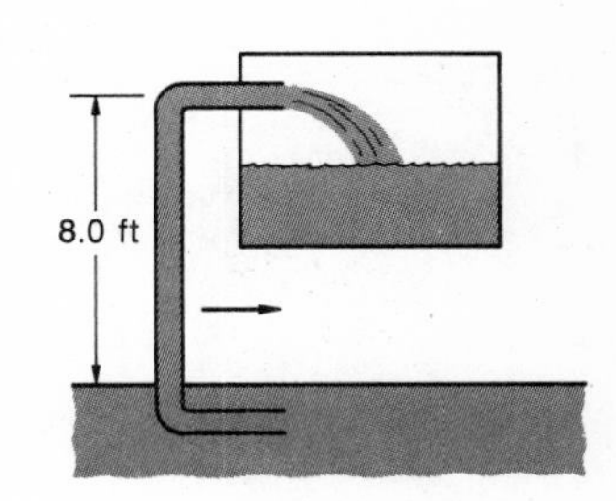

FIGURE 467

second was being placed in the holding tank when the train's speed was 30 mi hr^{-1}? Neglect any head losses due to bends in the pipe. *Suggestion:* If you have any difficulty getting started on the solution, try looking at the situation from the point of view of a coordinate system moving with the train.

Although the introduction of Bernoulli's equation was hedged by a number of ideal restrictions, in many cases equations 5-54 to 5-56 can be used for real liquids just as they stand. Except for heat exchangers and similar systems designed for transfer of thermal energy, most streamlined liquid flows can be considered as isothermal relative to the environment. Liquids are also relatively incompressible. The only significant departure from the idealized restrictions mentioned earlier develops from the fact of liquid viscosity; any accurate energy inventory must take into account the energy loss from this source. Adjusting Bernoulli's equation to real liquids we then have

$$P_1 + \tfrac{1}{2}\rho u_1^2 + \rho g h_1 = P_2 + \tfrac{1}{2}\rho u_2^2 + \rho g h_2 + W_v, \qquad (5\text{-}57)$$

where W_v is the energy lost to viscous friction per unit volume between points 1 and 2. You will recall from our earlier discussion of viscosity that the liquid layer adjacent to the container wall is motionless, so there is no friction between liquid and the confining wall; all of the energy lost to viscous friction is retained in the liquid itself. This means that W_v transforms to Q_v, where Q_v is the increase in the heat content of the liquid per unit volume during the passage of the liquid from point 1 to point 2. Hence Bernoulli's equation for real liquids could also be written

$$P_1 + \tfrac{1}{2}\rho u_1^2 + \rho g h_1 = P_2 + \rho\tfrac{1}{2}\rho u_2^2 + \rho g h_2 + Q_v. \qquad (5\text{-}58)$$

(Although the heading for this section refers to fluids, the discussion has been restricted to liquids. Gases are also fluids, but they are highly compressible. Streamlined flow of gases can be described by the differential form of Bernoulli's equation, $dP/\rho + u\,du + g\,dh = 0$, but the manipulation of this equation is beyond the intended scope of this workbook.)

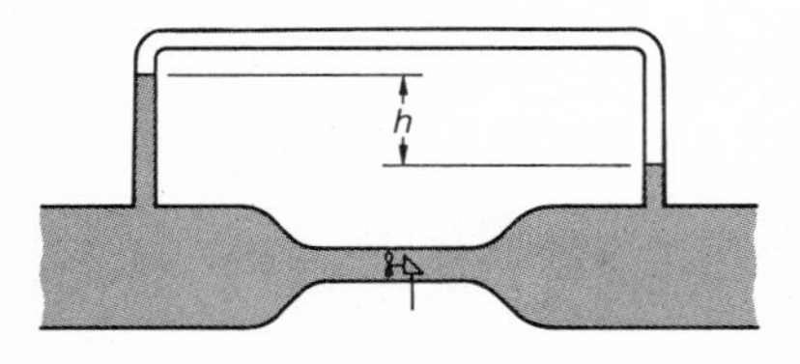

FIGURE 468

PROBLEM 795

Inserted into a horizontal pipe is a streamlined constriction carrying a turbine of 50% efficiency (Figure 468). A steady flow of Q liters sec^{-1} is maintained in the pipe. If a manometer attached to the upstream and downstream sections of the pipe as shown displays a pressure head difference of h meters, what is the power in watts being delivered by the turbine? Assume streamlined flow is maintained throughout the system.

PROBLEM 796

Carbon tetrachloride flows steadily in a uniform horizontal pipe. Between two points in the pipe 1.0 m apart the pressure difference is

30 cm of CCl_4, as shown by the manometer (Figure 469). If the specific heat of CCl_4 is 0.20 cal gm^{-1} $(°C)^{-1}$ and the pipe is made of thermally insulating material, what is the rise in the temperature of the CCl_4 per meter of travel in the pipe?

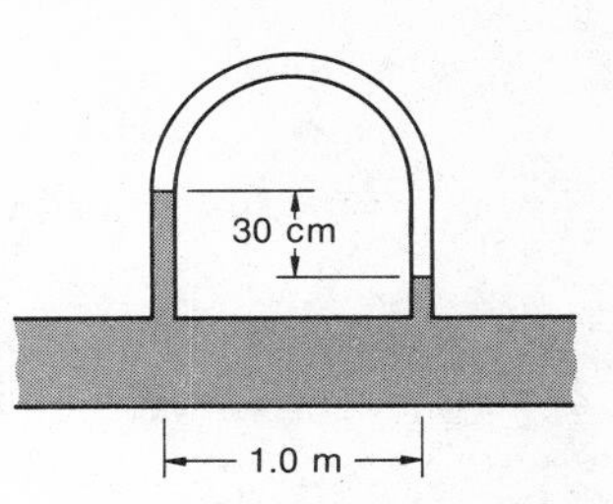

FIGURE 469

PROBLEM 797

A standard apparatus for demonstrating viscosity energy loss in pipe flow consists of a pipe of length L and uniform cross-sectional area a set horizontally in the side of a large vertical cylinder whose cross-sectional area $\gg a$ (Figure 470). The entry from the cylinder to the pipe is rounded so that no energy loss occurs there. Open-end manometers are mounted on the pipe. At the instant the water surface in the cylinder is 25 cm above the pipe the manometers read 15 cm and 11 cm respectively.

a. What is the length L of the pipe?

b. At the instant mentioned, how much water is flowing out of the pipe?

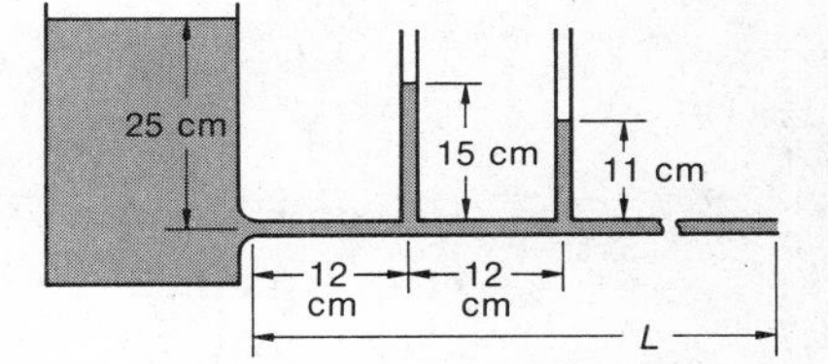

FIGURE 470

PROBLEM 798

An elevated tank of water whose drain valve is corroded and inoperative is to be drained by a siphon. A pipe is bent and placed over the tank edge (Figure 471). The greater the distance H the faster the tank will drain; however, the siphon will fail to operate if the static pressure in the pipe at the top of the bend drops to $1/40\ B_0$, because water at normal temperatures boils at that pressure. What limit does this boundary condition place on the length H? *Hint:* There are three points for which a Bernoulli equation inventory can be made—the water surface in the tank, the inside of the pipe at the top of the bend, and the outlet at the end of the pipe. These three inventories establish two independent equations, from which H can be calculated.

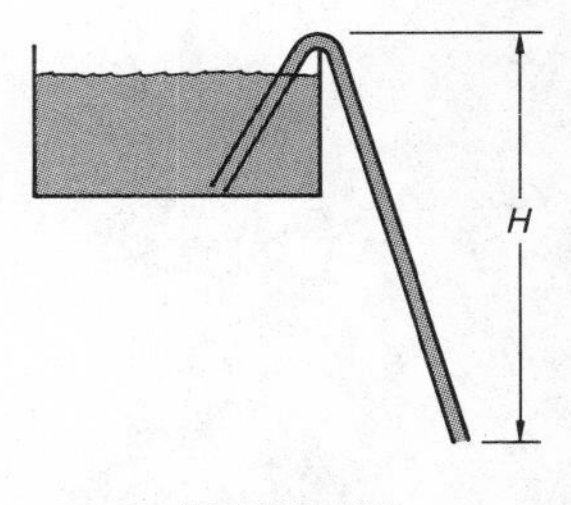

FIGURE 471

THE TRANSPORT OF ENERGY BY WAVES IN ELASTIC MEDIA

In an undisturbed elastic medium the individual mass elements are in equilibrium. If one of these elements is moved from its equilibrium position the forces in its neighborhood become unbalanced, and the disturbance is transmitted to nearby elements. These in turn move away from equilibrium; by this process the disturbance ultimately spreads through the medium with a speed of propagation that is characteristic of that medium. Such disturbances can be simple pulses such as a gun shot, but of greater interest are those that consist of a series of uniform pulses occurring at a constant rate.

In such disturbances each mass element of the elastic medium becomes a simple harmonic oscillator, with the equilibrium position of the oscillator remaining fixed in space. For the motion of such an oscillator we already have the description $y = y_0 \sin \omega t$, where y is the departure from the equilibrium position at time t, and y_0 is the maximum departure, or amplitude. This is an accurate description of the motion of one mass element; however, it reveals no relationship to the motion of other elements or to the orderly propagation of the disturbance that produces these motions. We can cure this deficiency by describing the motion not in terms of the time t on a clock at the mass element itself, but in terms of the time on a clock at some chosen reference element.* If the disturbance is moving through the medium with a speed c_m, the disturbance that reaches an element that is x away from the reference element is the same disturbance that was at the reference element x/c_m earlier. Thus the disturbance at x at time t there can be described in terms of the disturbance at the reference element at time $t-(x/c_m)$. Thus the description of the harmonic oscillation of a mass element at x in an elastic medium becomes

$$y = y_0 \sin \omega\left(t - \frac{x}{c_m}\right). \tag{5-59}$$

Equation 5-59 was developed to describe as a function of time the motion of an element fixed at x. However, we can get more use out of this equation than that. If we keep t constant and let y be a function of x, then equation 5-59 describes the condition of a disturbance travelling in the x-direction at a particular time t. Equation 5-59 thus serves two purposes: it describes the displacement motion of the individual mass element oscillators, and it describes the disturbance wave travelling through the medium at a speed c_m.

(In many textbooks you will find the angle in equation 5-59 reversed, i.e., $y = y_0 \sin \omega(x/c_m - t)$. There is no physical difference between the two forms of the wave description; the only difference is that the clocks used are out of phase by half a cycle.)

The sine function repeats for every 2π change in angle; thus

$$y_{x'} = y_0 \sin \omega\left(t - \frac{x'}{c_m}\right) = y_x = y_0 \sin\left[\omega\left(t - \frac{x}{c_m}\right) - 2\pi\right].$$

From this we get $(x' - x) = 2\pi c_m/\omega$. The distance $(x' - x)$ is the shortest distance in the direction of propagation between mass elements whose displacements are identical; this distance is called the wavelength λ. Putting this together with the fact that the angular frequency ω is related to the cyclic frequency ν and the period T by $\omega = 2\pi\nu = 2\pi/T$, we obtain the useful relationships

$$\lambda = \frac{c_m}{\nu} = c_m T. \tag{5-60}$$

With the aid of equation 5-60 we can transform equation 5-59 into other forms that are frequently found in textbooks:

$$\begin{aligned} y &= y_0 \sin 2\pi\left(\frac{t}{T} - \frac{x}{\lambda}\right) \\ &= y_0 \sin \frac{2\pi}{\lambda}(c_m t - x). \end{aligned} \tag{5-61}$$

---

* Since we assume no relative speed between elements, these clocks run at the same rate.

(Equations 5-59 and 5-61 are the working descriptions integrated from a more general wave equation $\delta^2\chi/\delta x^2 = (1/c_m^2)\delta^2\chi/\delta t^2$, where χ is some parameter of a disturbance moving in the x-direction through an elastic medium with a speed c_m. χ could represent the departure from equilibrium y, a variation from equilibrium pressure p, a variation from equilibrium density $d\rho$, etc.)

Longitudinal Waves

All movable mass elements in an elastic medium respond to a force or a pressure difference with an acceleration in the direction of the force. When the harmonic motions resulting from such a disturbance and the disturbance itself move in the same direction in the medium, the oscillations are called longitudinal and their description represents a longitudinal wave.

For the relation between a longitudinal pressure wave and the displacement it causes (a sound wave in air, for example), consider a small volume element of the medium A in cross section and dx long (Figure 472), dx small compared to the wavelength of the disturbance. In addition to the ambient pressure on the element a small pressure ripple of amplitude p_0 is moving through the medium in the x-direction; at a time t this ripple produces a small pressure difference dp on the element, as shown.

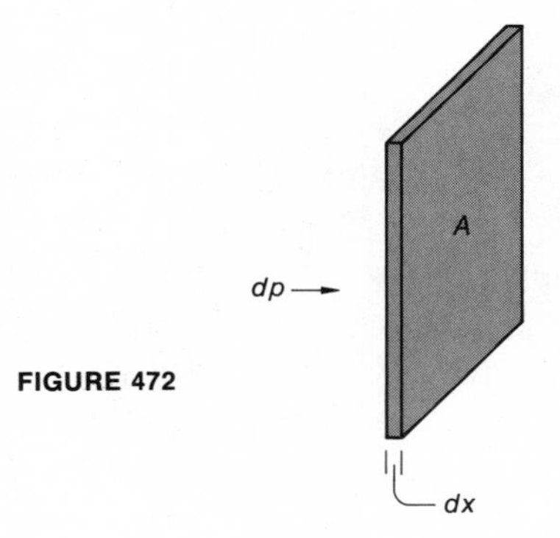

FIGURE 472

The mass of the element is $\rho_0 A\ dx$; the force on the element is $A\ dp$. By Newton's Second Law, $A\ dp = \rho_0 A\ dx\ \ddot{y}$.* Then $dp/dx = \rho_0\ddot{y} = -\rho_0\omega^2 y = -\rho_0\omega^2 y_0 \sin\omega\,(t - x/c_m)$. Integrating, we get

$$p = -p_0 \cos \omega\left(t - \frac{x}{c_m}\right),$$

where $p_0 = \rho_0 \omega y_0 c_m$. Note that since the pressure disturbance is a cosine function of time and the displacement is a sine function, in a longitudinal wave the disturbance and its resulting displacement are 90° out of phase.

* Remember, in keeping with equations 5-59 to 5-61, y is here not one of the standard Cartesian coordinates but a departure from equilibrium.

For an isotropic fluid medium of large extent the speed of propagation of longitudinal waves is given by

$$c_m = \sqrt{\frac{B}{\rho}}, \tag{5-62}$$

where B is the adiabatic bulk modulus of the fluid (cf. page 253) and ρ is the equilibrium density. If the fluid is a gas, $B = \gamma P$, so that for a gas

$$c_m = \sqrt{\frac{\gamma P}{\rho}} = \sqrt{\frac{\gamma RT}{MW}}, \tag{5-63}$$

where MW is the molecular weight of the gas.

In a solid the elastic moduli are so related that in any large volume a disturbance will induce not only longitudinal waves but waves in other directions as well; the resulting propagation is too complicated for consideration here. However, if a solid is restricted to one significant direction, such as a long bar of small diameter, a simple longitudinal wave can be established in the direction of the length of the bar. The speed of propagation in such a bar is given by

$$c_m = \sqrt{\frac{Y}{\rho}}, \tag{5-64}$$

where Y is Young's modulus for the material of the bar.

TRANSVERSE WAVES

Solids possess shear capability, and can thus sustain motions that are perpendicular to the direction of propagation of a disturbing wave. In general this motion is also too complex for consideration here. However, there is one transverse motion in a solid that is susceptible of simple analysis—the motion of mass elements in a stretched string or wire. If such a wire is disturbed perpendicular to its length, the resulting motion of the elements of the wire is perpendicular to the wire but the disturbance propagates along the wire with a speed given by

$$c_m = \sqrt{\frac{F}{\lambda}}, \tag{5-65}$$

where F is the tension in the wire and λ is its linear mass density.

PROBLEM 799

The working length of the A string (fundamental frequency 435 Hz) of a violin is about 32 cm and it has a linear density of about 2.5 $\times 10^{-2}$ gm cm^{-1}. To tune it to correct pitch what tension is required?

PROBLEM 800

A traveling wave in a wire under tension is described by $y = 0.40 \sin 20\pi\,(3.0\,t - 5.0 \times 10^{-4}\,x)$ where y and x are measured in cm perpendicular to the wire and in the direction of the wire respectively and t is in sec.

a. What are the values of the amplitude, the frequency, the speed of propagation of the wave down the wire, and the wavelength?
b. What is the maximum transverse speed of any point on the wire?
c. If the linear density of the wire was 2.0 gm cm^{-1}, what was the tension in the wire?

PROBLEM 801
In reporting the data describing the wave in problem 800, the second term in the parenthesis was incorrectly given as $+5 \times 10^{-4}x$. What effect would this error have on the computation of the values called for in the above problem?

PROBLEM 802
Most common metals under tension fail when the relative elongation $\Delta \ell / \ell$ exceeds about 10^{-3}. For a long, narrow metal rod under tension, what is the probable maximum of the ratio $c_{m,\text{trans}}/c_{m,\text{long}}$?

PROBLEM 803
A uniform endless belt of total length 25 ft and weight 10 lbwt is stretched between two pulleys. The tension in the belt (assumed everywhere equal for simplicity) is 80 lbwt.

a. At what linear speed will the belt just begin to leave contact with the pulleys?
b. What parameters could you change to increase this critical belt speed?
Comment: There is a hard way and an easy way to do this problem. The easy way is to recognize that the motion of the belt curving around a pulley is equivalent to a pulse travelling along the belt at the speed of the belt.

PROBLEM 804
A uniform chain of length L hanging vertically from a fixed support is shaken transversely at its bottom end with a frequency f such that waves are induced in the chain. How does the resulting wavelength vary with distance from the top of the chain, i.e., what is $d\lambda/dy$, where y is measured from the point of suspension? *Hint*: λ is a function of the tension T_y at some y, and the tension at any y is equal to the weight of the chain below that point.

PROBLEM 805
A uniform flexible cable of length 32 ft and total mass 16 lb hangs vertically from a firm attachment at its upper end. A 16-lb mass is attached to the lower end of the cable. A sharp horizontal blow is struck against this hanging mass. How long does it take for the resulting disturbance to reach the upper end of the cable?

PROBLEM 806

A very long horizontal wire under tension is attached at one end to a vibrator which causes vertical oscillations in the wire; the other end of the wire is fastened rigidly. The vibrator produces a travelling wave in the wire which is described by $y = y_0 \sin \pi(\alpha t - \beta x)$, where x is measured along the wire from the vibrator and α and β are known. If the wire has a linear density of σ* gm cm^{-1},

a. what is the tension in the wire?
b. What power is being supplied by the vibrator?

SOLUTION TO SECOND PART

The easy way to start is to remember, from problem 515, that the total energy of a simple harmonic oscillator is $E = \frac{1}{2}m\,\omega^2 A^2$, where m is the mass of the oscillator, ω is the angular frequency of the oscillator, and A is its maximum departure from the equilibrium position. A mass element of our vibrating wire is $\sigma\,dx$, $\omega = \pi\alpha$, and y_o is the amplitude of its motion; hence the total energy of a dx length of the wire is $E_{dx} = \frac{1}{2}\sigma\,dx\,\pi^2\alpha^2 y_o^2$. In one second the energy added to the wire is the energy contained in a length c_m of the wire, hence

$$P = \tfrac{1}{2}\sigma\pi^2\alpha^2 y_o^2 \int_x^{x+c_m} dx = \frac{\sigma\pi^2\alpha^3 y_o}{2\beta},$$

If you didn't remember that for an oscillator $E = \frac{1}{2}m\,\omega^2 A^2$, you have more work to do. We start with $\dot{y} = \pi\alpha y_o \cos \pi\alpha(t - \beta x/\alpha)$ and $\ddot{y} = -\pi^2\alpha^2 y$. Then

$$E_{dx} = (T + U)_{dx} = \tfrac{1}{2}\sigma\,dx\,\dot{y}^2 - \int_o^y F_y\,d_y$$

$$= \tfrac{1}{2}\sigma\,dx\,\dot{y}^2 - \int_o^y \sigma\,dx\,\ddot{y}\,dy = \sigma\,dx\,(\tfrac{1}{2}\dot{y}^2 + \pi^2\alpha^2 \int_o^y y\,dy)$$

$$= \tfrac{1}{2}\sigma\,dx\,\pi^2\alpha^2 y_o^2\,(\sin^2\underline{\quad\quad} + \cos^2\underline{\quad\quad}) = \tfrac{1}{2}\sigma\,dx\,\pi^2\alpha^2 y_o^2.$$

From here proceed as above.

* σ rather than λ is used here for linear density to avoid confusion with λ as wavelength.

PROBLEM 807

A very long flexible rope of mass σ per unit length is stretched horizontally under a tension T. A man takes hold of the middle of the rope and shakes it up and down with simple vertical harmonic motion $A \cos \omega t$, sending transverse waves down the rope in both directions. Before the motion becomes complicated by reflection from the distant ends,

a. what power must the man supply as a function of the time t?
b. what must be his average power supply?

Suggestion: To move the middle of the rope from its equilibrium position the man must supply a vertical force $2T \sin \phi^* \simeq 2T \tan \phi = (2T)dy/dx = (2T)(dy/dt)dt/dx$ for relatively small vertical oscillations. Also it is worth remembering that P can be calculated from $F \times v$.

* ϕ is the angle the stretched rope makes with the horizontal.

PROBLEM 808

Acoustic wave energy from a small source of sound moves equally in all directions from the source through an isotropic medium of large extent.

a. If at a distance r from the source the energy density u_r is known, what is the energy density in the medium at R from the source, $R > r$?

b. If the amplitude of the acoustic pressure variation in the medium at unit distance from the source is $p_{o,1}$, what is the amplitude of the acoustic pressure variation at r from the source?

Waves in Media of Limited Extent

Reflection at a Boundary

The discussion leading to equations 5-59, 5-60, and 5-61 implied an unbounded medium. Sound waves in the atmosphere and earthquake waves in the earth do move in media whose dimensions are many orders of magnitude greater than the wavelength. However, most of the waves we will be interested in move in more restricted media.

Let us look at two stretched wires fastened end-to-end at a junction J:

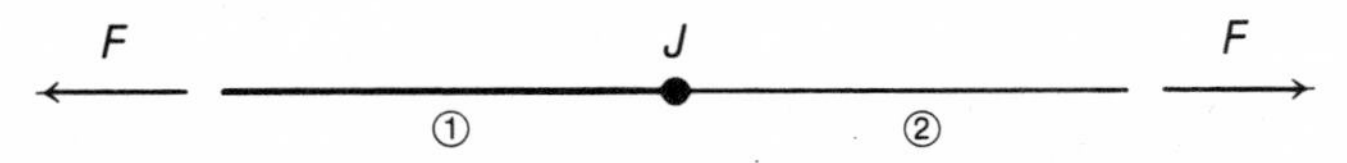

The wires are under a common tension F. A transverse wave moving to the right produces displacements in wire 1 described by $y_1 = A_1 \sin \omega_1(t - x/c_1)$ and in wire 2 by $y_2 = A_2\ \omega_2(t - x/c_2)$. These cannot be independent motions; at the junction y_1 must equal y_2 and the x and y components of F on both sides of the junction must be in equilibrium. In the general case it is not possible to satisfy these boundary conditions with the two displacement equations given. To evade this difficulty we postulate a third equation $y_{1,r} = A_r \sin \omega_r(t + x/c_1)$ to describe a displacement caused by a wave moving to the left in wire 1 as a result of a partial reflection at the junction. With the addition of this reflected wave our boundary conditions can be satisfied when the following relationships hold:

$$\omega_1 = \omega_2 = \omega_r$$

$$A_1 + A_r = A_2$$

$$\frac{\omega_1}{c_1} A_1 - \frac{\omega_r}{c_1} A_r = \frac{\omega_2}{c_2} A_2.$$

From these we can determine that

$$A_r = \frac{c_2 - c_1}{c_1 + c_2} A_1$$
$$A_2 = \frac{2c_2}{c_1 + c_2} A_1. \tag{5-66}$$

Thus the total displacement in wire 1 resulting from both the initial wave and the reflected wave is given by

$$y_{1,\text{total}} = A_1\left[\sin \omega_1\left(t - \frac{x}{c_1}\right) + \frac{c_2 - c_1}{c_1 + c_2} \sin \omega_1\left(t + \frac{x}{c_1}\right)\right] \tag{5-67}$$

and the displacement in wire 2 is described by

$$y_2 = \frac{2c_2}{c_1 + c_2} A_1 \sin \omega_1\left(t - \frac{x}{c_2}\right). \tag{5-68}$$

Standing Waves

There are two interesting special cases to which we can apply the above equations:

Fixed Termination. If wire 1 terminates in a rigid support at J, which physically is equivalent to saying that σ_2 is infinite and hence $c_2 = 0$, then $A_2 = 0$ and $A_r = -A_1$. Hence in this case the incident wave is completely reflected, and the displacement in wire 1 is given by

$$y_{1,\text{total}} = (-)\ 2A_1 \sin \frac{\omega_1 x}{c_1} \cos \omega_1 t$$
$$= (-)\ 2A_1 \sin \frac{2\pi x}{\lambda} \cos 2\pi \nu t. \tag{5-69}$$

The maximum amplitude $2A_1 \sin \omega_1 x/c_1$ or $2A_1 \sin 2\pi x/\lambda$ is a function of x, so that the motion of the elements of the cord has an envelope, as shown in Figure 473. From the diagram it is obvious why the motion described by equation 5-69 is called a standing wave. At all times it has positions of zero displacement, called nodes, at $\omega_1 x/c_1 = (n-1)\pi$ or at $x = (n-1)\lambda/2$; J is at one of these nodes. Midway between the nodes are positions of maximum displacement called loops or antinodes.

FIGURE 473

Stringed musical instruments illustrate a special application of equation 5-69 in which a vibrating string is fastened at both ends. Both ends are thus at nodal positions, so the length of the string must equal an integral number of half wavelengths, or $L = N\lambda/2$, N being 1, 2, 3, ... It is customary to speak of N as the number of the harmonic; when $N = 1$ the string is vibrating in its first harmonic, etc.

(In music the first harmonic is called the fundamental tone or the fundamental mode, the second harmonic is called the first overtone, etc.)

Longitudinal displacements behave at boundaries in the same way, and are described by the same equations developed above for transverse displacements. In particular longitudinal displacements reflect from fixed, solid boundaries with the same standing wave as described by equation 5-69. Thus a sound wave travelling down a closed-end organ pipe exhibits displacement nodes at the closed end and at $n\lambda/2$ from the end.

(It was consistent with the previous development to speak of displacements in discussing the wave in the organ pipe. However, what we hear in sound is not a displacement but a variation in pressure. The fundamental relation between a pressure variation dp and a displacement y is derived from the fact dp is proportional to dy/dx; thus a pressure variation dp and its resulting displacement y are 90° out of phase. Thus in a standing wave for pressure differences in a closed pipe there would be loop or antinode at the closed end.)

Free Termination. If the point J is free, which is physically equivalent to saying that $\sigma_2 = 0$ and hence $c_2 = \infty$, then from equation 5-66, $A_r = A_1$ and $A_2 = 2A_1$. Since $\sigma_2 = 0$, there is no transmitted wave, so that again the incident wave is reflected only. In this case we have

$$y_{1,total} = 2A_1 \cos \frac{\omega_1 x}{c_1} \sin \omega_1 t$$

$$= 2A_1 \cos \frac{2\pi x}{\lambda} \sin 2\pi\nu t.$$

The standing wave is shown in Figure 474. There are loops at the free end and at $n\lambda/2$ from the free end. (This diagram would also apply to the pressure standing wave in a closed pipe that was discussed above.)

FIGURE 474

PROBLEM 809

A plane sound wave is travelling down a long cylindrical tube of cross-sectional area A; the tube contains air of density ρ. The displacement of the air molecules from their equilibrium positions is described by $y = y_o \sin \omega(t - x/c_m)$, where x is measured from the sound source at the mouth of the tube.

a. How much energy per wave length is moving down the tube?
b. How much energy is being fed into the tube per second?

PROBLEM 810

The wave in the tube of problem 809 is reflected from the end of the tube in such a way that standing waves are established. What is then the average energy density in the tube due to wave motion?

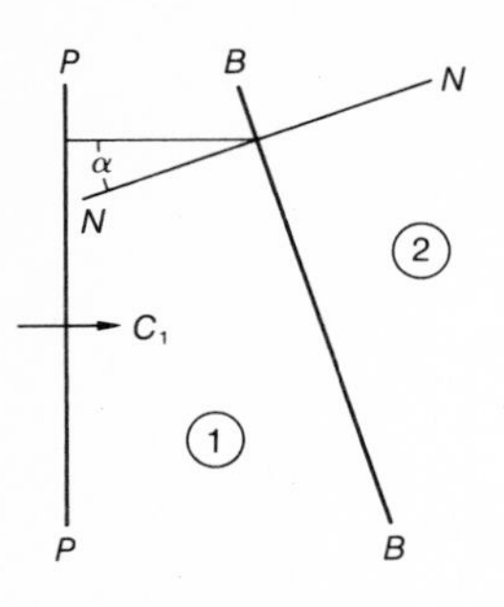

FIGURE 475

PROBLEM 811

A plane wave front of sound *P-P* moves in region 1 through quiet air where the speed of propagation is c_1 (Figure 475). It reaches a plane boundary *B-B* beyond which is region 2 with a polluted atmosphere where the speed of sound waves is $c_2 = \frac{3}{4}\, c_1$. *N-N*, the normal to the boundary *B-B*, makes an angle α_1 with the perpendicular to the wave front *P-P*.

a. What angle α_t will the perpendicular to the transmitted wave front P_t-P_t make with *N-N*?
b. What angle α_r will the perpendicular to the reflected wave front P_r-P_r make with *N-N*?

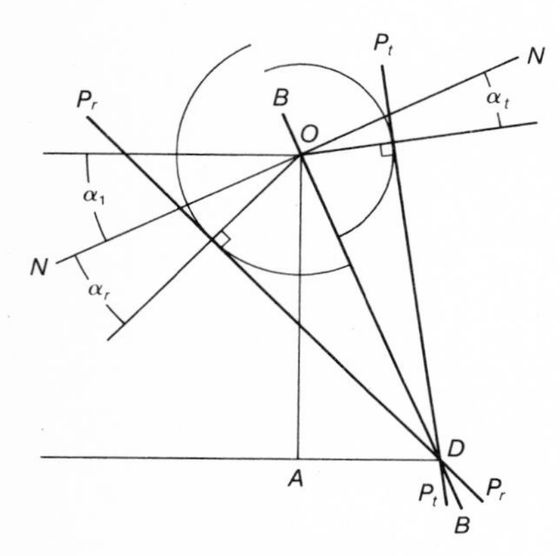

FIGURE 476

START OF SOLUTION

When *P-P* meets *B-B* at *O* a lower part of the wave front is still *AD* away from the boundary (Figure 476). In the time AD/c_1 it takes this part of the wave front to reach the boundary the part of the wave front that was at *O* earlier has been partly transmitted into region 2 a distance $c_2(AD/c_1)$, and partly reflected back into region 1 a distance $c_1(AD/c_1) = AD$. If semicircles are drawn about *O* of radii $(c_2/c_1)AD$ and *AD*, tangents drawn to these circles from *D* will represent the new wave fronts P_t-P_t and P_r-P_r. Then from the geometry it is easy to calculate that $\sin \alpha_t = (c_2/c_1) \sin \alpha_1$.

Comment: If you let *W-W* be an electromagnetic wavefront and *B-B* a boundary between dielectric media, the treatment given above can be applied with equal accuracy to the reflection and transmission of electromagnetic radiation incident on a dielectric boundary. In electromagnetic notation $c_1 = c/n_1$ and $c_2 = c/n_2$, where n_1 is called the index of refraction of medium 1, etc. Then the equation at the end of the solution above transforms to the familiar Snell's Law for the refraction of a light ray: $\sin \alpha_2 = (n_1/n_2) \sin \alpha_1$.

PROBLEM 812

A uniform string of mass *M*, length *L*, and under tension *T* is vibrating in a vertical plane in its fundamental mode. The ends of the string are fixed at the same horizontal level. A small object of mass *m* (*m* small enough not to affect the motion of the string) is fastened to the string at midpoint with glue that provides a force of attachment *f* between mass and string. What is the maximum amplitude that vibration may have without dislodging the mass if it is glued

a. on top of the string,
b. on the under side of the string?

PROBLEM 813

A horizontal wire of length *L* is clamped at both ends after being stretched to a tension *T*. It is set vibrating vertically in its first harmonic with an amplitude *A*, $A \ll L$. What is the maximum vertical force felt by either of the clamps? *Hint:* The solution is very simple if you take advantage of the fact that $A \ll L$.

PROBLEM 814

In a demonstration lecture on sound, a cello string was continuously bowed at such a point on the string that practically all of the energy in the string was in the third harmonic, which had an amplitude of 4.0 mm. From previous experiments with the time of decay of the vibration of that string in the third harmonic it was estimated that 1 percent of the energy of each cycle was delivered to the air as sound. The effective length of the string was 60 cm, its linear density was 10 gm m^{-1}, and it was under a tension of 20 kgwt.

a. How much power was being delivered to the air in sound?
b. What was the frequency of the sound?

Intensity

The most useful measure of energy being transported by wave motion is called the intensity I, defined by

$$I = \frac{\bar{P}}{A}, \tag{5-70}$$

where $\bar{P}$ is the average energy moving per sec through an area A in a direction perpendicular to A. When MKS units are used I is measured in watts m^{-2}. An equivalent definition of I is

$$I = \bar{u}\, c_m, \tag{5-71}$$

where u is the average energy density in the wave motion and c_m is the speed of propagation of the wave in the medium.

This energy becomes available through the induced motions of the mass elements of the medium. Since these motions are harmonic, the average energy available is proportional to the average square of the displacements—to $\overline{y^2}$ for transverse motions, to $\overline{y^2}$ directly or to $\overline{p^2}$ indirectly for longitudinal motion. (See, for example, problems 807, 809, 810, 814.) Since for SHM $\overline{y^2} = y_o^2/2$, in evaluating I we usually find that we must first determine y_o^2 or p_o^2.

PROBLEM 815

What is the intensity of the sound wave of problem 809 in terms of the maximum overpressure p_o? *Suggestion*: You can take the easy way and substitute y_o in terms of p_o from page 373 in the answer to question (b) of problem 809. However, just to learn a different approach why don't you evaluate $\bar{P}$ from the work done per unit time, thus:

$$\bar{P} = [\int_x^{x+\lambda} A\, p\, dy]v.$$

PROBLEM 816

Sound at the level of a whisper has an intensity of the order of 10^{-11} watts m^{-2}. If the principal tone in a whisper you hear has a frequency of 10^3 Hz,

a. what is the maximum amplitude of the motion of the air molecules in your ear?
b. What is the amplitude of the pressure variations in the air in your ear?
c. How much power is entering your ear from the whisper if the ear opening has an area approximately equal to 0.4 cm^2?

PROBLEM 817

The minimum intensity of sound that can be detected by an average normal ear is the order of 10^{-12} watts m^{-2}. You are at a party and you can just barely detect that someone in a group 20 ft from you is talking. How much closer must you get to the group to hear his speech at whisper level? (See problem 816.) *Hint*: As a first approximation you can neglect reflections and absorptions and sound energy from other sources in the room, and you can assume that all the power delivered to the air by the speaker you are interested in is distributed uniformly over any spherical surface centered on the speaker. Thus conserving energy and using equation 5-70, $I_1A_1 = I_2A_2$, where A_1 and A_2 are the areas of the appropriate spherical surfaces surrounding the speaker. (If you have difficulty finding a satisfactory solution maybe you should learn to lip-read.)

The Doppler Effect

So far in this section it has been tacitly assumed that only the wave moved; the source, the medium, any observer, or any energy measuring device were all at rest in an inertial system. If a source of disturbance moves relative to the medium, in comparison with the "at rest" case the nodes and antinodes of the wave will be closer together if the source moves in the direction of propagation, or farther apart if the source moves in the opposite direction. An observer at rest relative to the medium thus receives from a source moving toward him a shorter wavelength or a higher frequency than if the source had been stationary. If the source is moving away from the observer the observed wavelength is increased and the frequency is lowered. This is called the Doppler effect, and it has many practical applications for the measurement of relative speeds. This effect is described by

$$\nu' = \frac{1}{1 \mp (v_s/c_m)}\, \nu_o,$$

where ν_o is the frequency of the disturbance at the source, ν' is the frequency the observer measures, and v_s is the speed of the source relative to the medium in a line toward the observer. The upper sign in the denominator (−) is used for relative motion toward, the lower sign (+) is used for motion away from, the observer.

If the source is stationary but the observer moves toward the source, again the observer records a shorter wavelength or a higher frequency than at rest. This effect is described by

$$\nu' = \left(1 \pm \frac{v_o}{c_m}\right)\nu_o.$$

v_o is the speed of the observer relative to the medium in the direction toward the source. The upper sign (+) is used for relative motion toward the source, and the lower sign (−) for relative motion away from the source.

When both source and observer are moving relative to the medium we have

$$\nu' = \frac{1 \pm \frac{v_o}{c_m}}{1 \mp \frac{v_s}{c_m}}\nu_o.$$

It must be emphasized that in all three of these Doppler equations, v_o and v_s measure motions relative to the medium and in the direction between source and observer.

PROBLEM 818

You are standing on the curb of a narrow street as an ambulance drives by at 60 mi hr^{-1}, continuously sounding a siren whose frequency is 1242 Hz.

a. What were the upper and lower limits of the frequencies you heard?

b. If you plotted the frequencies you heard against the time you heard them, what would the graph look like?

PROBLEM 819

The first American tracking of the first space satellite Sputnik I was made by a group of early rising Caltech faculty members on the morning of October 19, 1957. The graph of their observational data is shown in Figure 477. The frequencies on the ordinate axis are

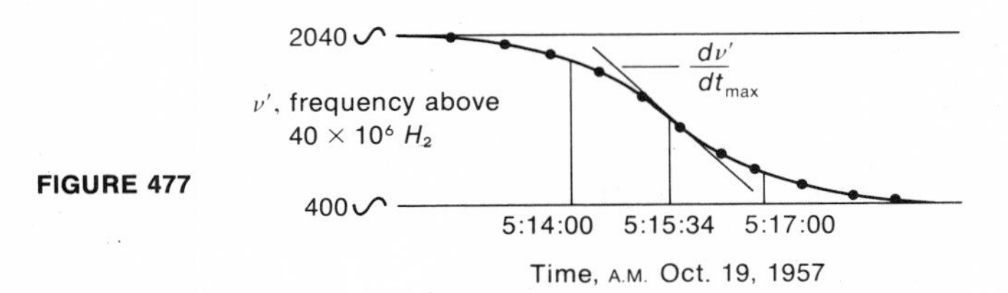

FIGURE 477

values above the local oscillator frequency 40 Mc, i.e., the actual frequencies observed ranged from $40 \times 10^6 + 400$ to $40 \times 10^6 + 2040$ cycles sec^{-1}. The maximum slope of the curve at frequency 1220 Hz and time 5:15:34 AM was measured to be 5.76 cycles sec^{-2}.

a. At what frequency was Sputnik I transmitting?

b. What was the speed of Sputnik I about 5:15 AM on that morning?

c. How far from the Caltech campus was Sputnik I at the time?

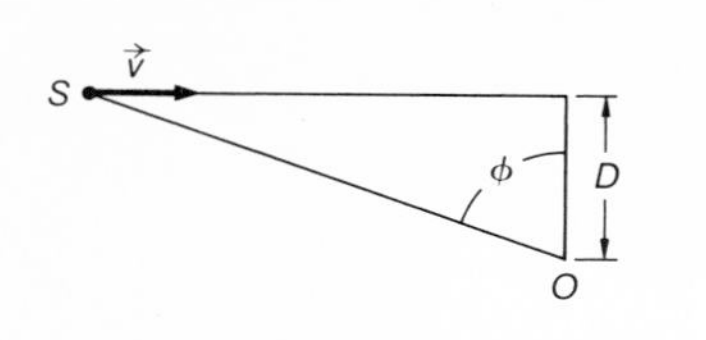

FIGURE 478

SOLUTION

Starting with the space diagram (Figure 478), at some time t a disturbance received at O was emitted by a source when it was at S moving with a velocity $\vec{v}$. The component of $\vec{v}$ in the direction from S to O was $v \sin \phi$; then the frequency heard at O at time t was

$$\nu' = \frac{\nu_o}{1 - \dfrac{v \sin \phi}{c_m}}.$$

Solving for the unknown ν_o emitted by Sputnik I we get

$$\nu_o = \nu'\left(1 - \frac{v \sin \phi}{c_m}\right)$$

and from this

$$\nu' - \nu_o = \frac{v \sin \phi}{c_m}\,\nu'. \tag{1}$$

In particular this becomes

$$\nu'_{\text{max}} - \nu_o = \frac{v \sin \phi}{c_m}\,\nu'_{\text{max}} = \frac{v}{c_m}\,\nu'_{\text{max}} \quad \text{for } \phi \simeq \frac{\pi}{2} \tag{2}$$

and

$$\nu'_{\text{min}} - \nu_o = \frac{v \sin \phi}{c_m}\,\nu'_{\text{min}} = \frac{-v}{c_m}\,\nu'_{\text{min}} \quad \text{for } \phi \simeq -\frac{\pi}{2}. \tag{3}$$

Dividing (2) by (3),

$$\frac{\nu'_{\text{max}} - \nu_0}{\nu'_{\text{min}} - \nu_0} = -\frac{\nu'_{\text{max}}}{\nu_{\text{min}}} \simeq -1. \tag{4}$$

(ν'_{max} and ν'_{min} differ by only 1640 cycles sec^{-1} in 40 Mc, hence for all reasonable accuracy $\nu'_{\text{max}}/\nu'_{\text{min}} = 1$ and the Doppler data curve is essentially symmetrical between ν'_{max} and ν'_{min}.)

a. Solving (4) for ν_o, we get

$$\nu_o = \frac{\nu'_{\text{max}} + \nu'_{\text{min}}}{2} = (40 \times 10^6 + 1220)\ \text{Hz}.$$

b. Rearranging (2) and (3) and adding,

$$v = \frac{(\nu'_{\text{max}} - \nu'_{\text{min}})\, c_m}{\nu'_{\text{max}} + \nu'_{\text{min}}} = \frac{1640 \times 3 \times 10^8}{(2 \times 40 \times 10^6 + 2440)} = 6.1\ \text{km sec}^{-1}.$$

c. $d\nu'/dt$, the slope of the Doppler data curve, is a maximum at the point of inflection; because of the symmetry of the graph this point occurs at ν'_{av}, or at $(1220 + 40 \times 10^6)$Hz. From (1),

$$\frac{d\nu'}{dt} = \frac{v\,\dot{\phi}\cos\phi}{c_m}\,\nu' + \frac{v \sin \phi}{c_m}\,\frac{d\nu'}{dt}.$$

Then

$$\frac{dv'}{dt} = \frac{v\,\dot{\phi}\cos\phi\, v'}{c_m\left(1 - \dfrac{v\sin\phi}{c_m}\right)} \simeq \frac{v\,\dot{\phi}\cos\phi}{c_m}\,v'$$

(assuming $v\sin\phi \ll c_m$). From the space diagram we get

$$\tan\phi = \frac{vt}{D}, \quad \frac{d(\tan\phi)}{dt} = \frac{-\dot{\phi}}{\cos^2\phi} = \frac{v}{D}$$

and this yields

$$\dot{\phi} = \frac{-v\cos^2\phi}{D}.$$

Substituting in (5)

$$\frac{dv'}{dt} = \frac{vv'\cos\phi}{c_m} \times \frac{-v\cos^2\phi}{D}.$$

and

$$\left.\frac{dv'}{dt}\right|_{\max} = \frac{-v^2\,v'_{\text{av}}}{D\,c_m} = -5.76 \text{ at } \phi = 0.$$

Solving for D, $D = 860$ km.

PROBLEM 820

A narrow beam of frequency ν emitted by a police radar unit is reflected from a car moving away from the unit with a ground speed V_c. When the reflected beam is received by the radar unit, determine the relation between the speed V_c and the observed relative frequency change $d\nu/\nu$

a. when the radar unit is at rest, and
b. when the radar unit is contained in a police vehicle that is pursuing the car with a speed V_p.
c. What does the answer to part (b) imply about the accuracy of radar used in pursuit?

PROBLEM 821

In a discharge tube an electric discharge forms a cloud of He ions (Figure 479); being positively charged, this cloud moves down the tube toward the cathode with a speed V_{cloud}. In addition, because of the high temperature of the cloud each ion possesses a random thermal motion with a speed V_{ion}; for simplicity assume the scalar value V_{ion} is the same for all ions. Each ion also emits light of frequency ν_o. What is the breadth of the spectral line for ν_o, i.e., what is $\nu_{o,\max} - \nu_{o,\min}$?

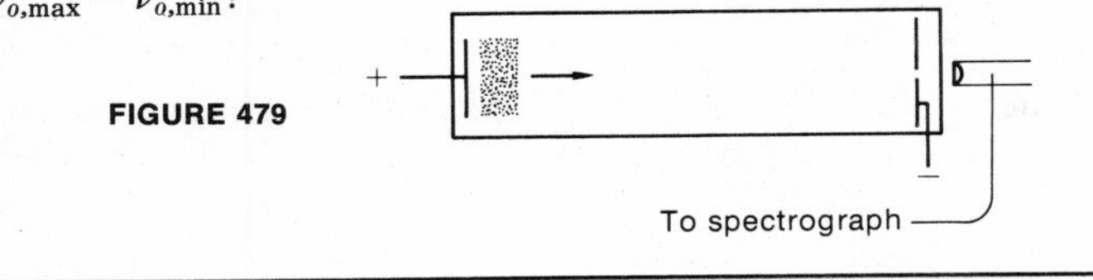

FIGURE 479

PROBLEM 822

A binary star system rotates in a plane that is edge-on to observers on Earth. The observed period for the system is 1.0 yr and the Doppler shift for the light from the two stars produces a maximum spread $d\lambda$ of 10 Å for the 6560 Å line of hydrogen. Assuming the stars have equal masses, what is the mass of either?

PROBLEM 823

An anchored submarine detector buoy is equipped with sonar; the direction of the received echo, as well as the echo frequency, can be determined by the detector and transmitted by radio to a search ship's plotting room. The pulses sent out by the detector have a frequency of 16×10^3 Hz. At a time 1.20 sec after a pulse is sent out an echo comes in from 30° E of N at a frequency of 16.08×10^3 Hz; a short time later the bearing of the echo has shifted to 29° E of N. If the watch officer on the search ship assumes the echo source is a submarine moving on a steady course at its normal 12-knot cruising speed,

a. at the time it made the first echo how far was the submarine from the buoy?
b. What was the submarine's course? (The speed of sound in sea water = 4400 ft sec^{-1}.)

In the problems just given there was little probability that either v_s or v_o would be a significant fraction of the speed of light, hence the clocks timing ν' and ν_o could be taken as running at the same rate. However, whenever either v_o/c or v_s/c is not negligible the relation between ν' and ν_o is affected not only by relative speeds but also by differences in timing. This leads to the following equation for the Doppler effect when speeds are relativistic:

$$\nu' = \sqrt{\frac{1 \pm \beta}{1 \mp \beta}}\, \nu_o \tag{5-72}$$

where the v in $\beta = v/c$ is a relative speed between source and observer and as before the upper signs are for motion toward and the lower signs for motion away.

PROBLEM 824

In most scientific work Doppler measurements are reduced to values of $d\lambda/\lambda$. When speeds are (a) nonrelativistic, (b) relativistic, how does the proper formula for $d\lambda/\lambda$ read for i. a moving source, ii. a moving observer?

PROBLEM 825

Measurements of the light from quasars give values for $d\lambda/\lambda$ as high as +2. Assuming this can be considered entirely due to the Doppler effect, what is the maximum speed relative to the earth observed so far for a quasar?

Addition of Independent Waves

Beats

In some of the Doppler problems the data for $d\nu$ or $d\lambda$ was obtained by counting "beats." Waves from different sources can move through a medium completely independent of one another; at any point in space their disturbance effects can be added vectorally, or if the displacements are in the same direction, can be added algebraically. Under certain conditions this summation may produce what are called beats.

For simplicity of illustration the point in space where we will make this summation is at $x=0$, and we will limit ourselves to displacements in the y direction caused by two waves of equal amplitude but different frequencies. Then at any time t

$$\begin{aligned} y_t &= y_{1,t} + y_{2,t} \\ &= A \sin \omega_1 t + A \sin \omega_2 t \\ &= 2A \cos \frac{\omega_1 - \omega_2}{2} t \sin \frac{\omega_1 + \omega_2}{2} t\,* \\ &= 2A \cos 2\pi \frac{\nu_1 - \nu_2}{2} t \sin 2\pi \frac{\nu_1 + \nu_2}{2} t. \end{aligned}$$

This is a particularly useful formula when ν_1 differs from ν_2 by only a small amount. Then $(\nu_1 + \nu_2)/2 = \nu_{av} \simeq \nu_1 \simeq \nu_2$, and the frequency of oscillation is very nearly that of either wave. However, the amplitude term $2A \cos 2\pi(\nu_1 - \nu_2)t/2$ is now governed by a very small term $\delta = (\nu_1 - \nu_2)$, so that it takes a long time $2/\delta$ for the amplitude to run through a 2π cycle. The graph of the summation of displacements due to the two waves is as shown in Figure 480. For an envelope two maxima

FIGURE 480

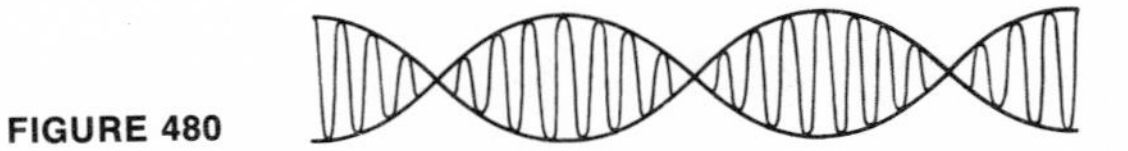

occur every cycle, hence the frequency of these maxima is $\nu_1 - \nu_2$. This is called the beat frequency; it is easily detected both audibly and by electronic methods.

*One of the standard formulas in the trigonometric section of any handbook is the sum $\sin A + \sin B = 2 \sin [(A+B)/2]$ and $[(A-B)/2]$.

PROBLEM 826

A merry-go-round revolves about a vertical axis at a speed of 8.0 rpm. A boy riding at 20 ft from the axis is blowing a horn that emits a note of frequency 121 Hz. Not far from the merry-go-round is a popcorn stand where a whistle of frequency 125 Hz is blowing. What is the range of the number of beats sec^{-1} heard by

a. the attendant standing at the center of the merry-go-round,
b. the boy,
c. a man standing on the ground beyond the merry-go-round on a line from the whistle that passes within 20 ft of the center of the merry-go-round?

PROBLEM 827

A low-flying twin-engined airplane flies directly over a stationary observer. A beat note between the two engines, which are running at constant but slightly different speeds, is observed to change from a maximum of 4 beats sec^{-1} during the plane's approach to 2 beats sec^{-1} during the plane's departure. If there was no wind what was the speed of the plane?

PROBLEM 828

A switchman in a signal tower mounted between two closely parallel railroad tracks hears no beats between a whistle on a train approaching from one direction and a whistle on the train approaching in the other direction. A brakeman on the first train, who knows his train is moving at a speed v_1, hears B beats sec^{-1} between the two whistles. When he learns by radiotelephone from the switchman that the latter heard no beats and that there is no wind, the brakeman computes the frequency of his train's whistle. What was that frequency?

Fourier Analysis of Repetitive Patterns

A further interesting addition of waves occurs in the analysis or synthesis of complex patterns that repeat at a regular rate. A complex wave can be broken down into a summation of simple harmonic waves by an appropriate Fourier series described by

$$y_t = a_0 + \Sigma a_n \cos \frac{2\pi n}{T} t + \Sigma\, b_n \sin \frac{2\pi n}{T} t.$$

The coefficients in the above expression are calculated as follows:

$$a_0 = \frac{1}{T} \int_0^T y_t \, dt$$

$$a_n = \frac{2}{T} \int_0^T y_t \cos \frac{2\pi n}{T} t \, dt$$

$$b_n = \frac{2}{T} \int_0^T y_t \sin \frac{2\pi n}{T} t \, dt$$

where T is the period and $n = 1,2,3, \ldots .$

In actual use these equations are much simpler than they may appear. As an example, let us consider a saw-toothed pattern. The pattern repeats at t_1, t_2, etc. (Figure 481). These values can be used, but it will prove more convenient to transform the horizontal time scale to 2π, 4π, etc.

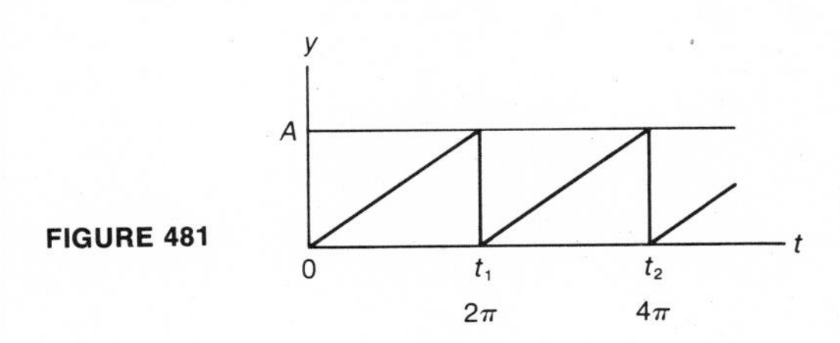

FIGURE 481

Thus the period $T = 2\pi$, the function has a slope $A/2\pi$, and $y_t = A/2\pi\, t$. Substituting in the above equations,

$$a_0 = \frac{1}{2\pi}\int_0^{2\pi}\frac{A}{2\pi}\,t\,dt = \frac{A}{2}$$

$$a_n = \frac{2}{2\pi}\int_0^{2\pi}\frac{A}{2\pi}\,t\cos nt\,dt = 0 \quad \text{for all } n;$$

$$b_n = \frac{2}{2\pi}\int_0^{2\pi}\frac{A}{2\pi}\,t\sin nt\,dt = -\frac{A}{n\pi} \quad \text{for all } n.$$

Then

$$y_t = \frac{A}{2} - \frac{A}{\pi}\sin t - \frac{A}{2\pi}\sin 2t - \ldots = \frac{A}{2} - \frac{A}{\pi}\sum_{n=1}^{\infty}\frac{1}{n}\sin nt. \qquad (5\text{-}73)$$

It is very easy to demonstrate on an oscilloscope that this summation does produce a saw-toothed wave that matches the diagram well except at 0, 2π, 4π, etc. It is not possible to provide an amplitude equal to A and one equal to zero at the same instant in time. As you can determine from the equation $y_t = A/2$, at $t = 2n$ and the summation rapidly changes sign as it passes through zero.

PROBLEM 829
By using the solution for the saw-toothed wave given in problem 828, determine the value of the sum $(1 - 1/3 + 1/5 - 1/7 \ldots)$. *Hint:* When $t = 3/2\pi$, $y_t = 3/4\,A$. Evaluate the right-hand side of equation 5-73 at $t = 3/2\pi$ and equate the result to $3/4\,A$.

PROBLEM 830
a. What combination of simple harmonic waves will produce the wave form shown in Figure 482. *Suggestion:* The function $y_{t,1}$ for the period $0 \leq t \leq \pi$ is not the same as the function $y_{t,2}$ applicable for the

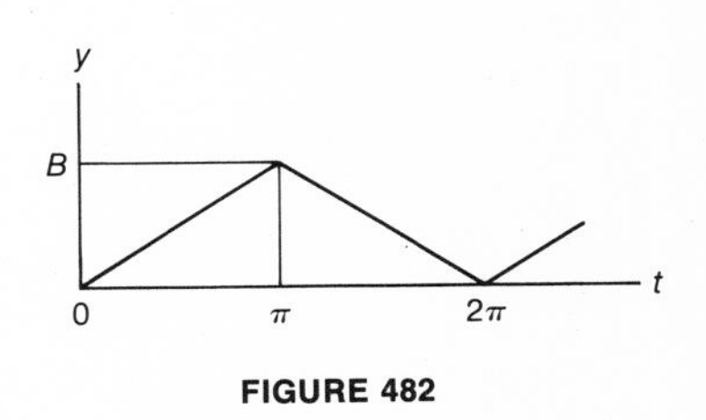

FIGURE 482

period $\pi \leqslant t \leqslant 2\pi$. Hence in solving for the Fourier coefficients the integral $\int_0^T y_t\,(\;)\,dt$ must be broken into $\int_0^{T/2} y_{t,1}\,(\;)\,dt + \int_{T/2}^{T} y_{t,2}\,(\;)\,dt$. From the diagram you should get $y_{t,1} = (B/\pi)t$ and $y_{t,2} = 2B - (B/\pi)t$.
b. If a string of a bass viol is plucked at its center, i. what harmonics of the string should be heard, and ii. how are the energies in these harmonics related, i.e., what are the ratios $E_a{:}E_b{:}E_c$ . . . ?
c. By evaluating y_t at $t = \pi$ determine the value of the sum $(1 + (1/3)^2 + (1/5)^2 + (1/7)^2$ . . .).

d. Using the result of part (c) determine the value of the sum $\sum_{n=1}^{\infty} 1/n^2$.
Hint:

$$\sum_{n=1}^{\infty} \frac{1}{n^2} = \left[1 + \left(\frac{1}{3}\right)^2 + \left(\frac{1}{5}\right)^2 \cdots\right] + \left[\left(\frac{1}{2}\right)^2 + \left(\frac{1}{4}\right)^2 \cdots\right].$$

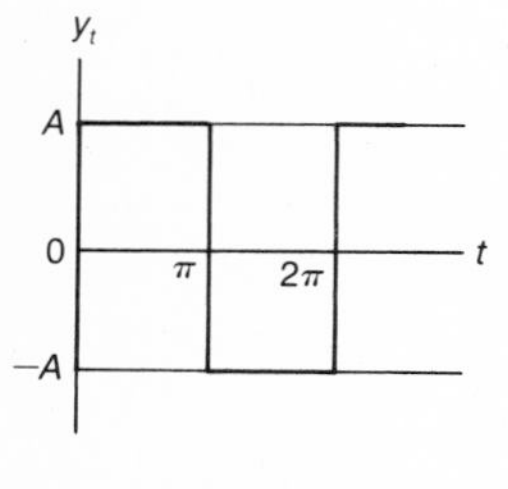

FIGURE 483

PROBLEM 831
A square wave like the one shown in Figure 483 is often used in electronic analysis with an oscilloscope. What combination of simple harmonic waves will produce such a square wave?

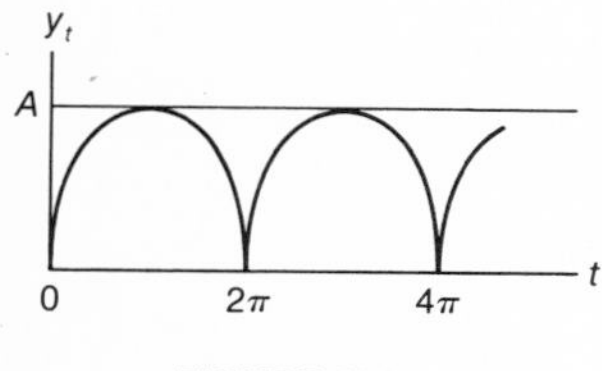

FIGURE 484

PROBLEM 832
Two half-wave rectifiers can be connected 180° out of phase to produce the sine wave pattern shown in Figure 484; with the coordinates chosen the equation of this pattern is $y_t = A \sin \frac{1}{2} t$ per cycle.
a. What combination of simple harmonic waves will reproduce this pattern?
b. By evaluating y_t for the Fourier sum at $t = \pi$, determine the value of

$$\left[1 - 2 \sum_{n=1}^{\infty} \frac{(-1)^n}{(2n-1)(2n+1)}\right].$$

THE TRANSPORT OF ENERGY BY ELECTROMAGNETIC RADIATION

The Electromagnetic Field

In the presence of an electric field a free electric charge is accelerated; if the electric field changes harmonically with time the acceleration of the charge also changes harmonically, and the charge becomes a simple harmonic oscillator. Thus a harmonically varying electric field propagating through space becomes a travelling disturbance that induces SHM in every charge it encounters, just as a wave in an elastic medium induced molecular accelerations. Since an electric field originates from charge, and a harmonic field originates from an oscillating charge, electromagnetic waves provide a cause-and-effect link between oscillating charges.

With one exception we can transfer to electromagnetic wave phenomena all we have learned from a study of elastic waves. The exception arises out of the fact that the travelling electric field, because of its vector orientation, does not possess the spherical symmetry we could assume for a travelling longitudinal disturbance in a homogeneous elastic medium. For the latter phenomenon a disturbance originating at O (Figure 485) that produces an amplitude p_O at unit distance from O would deliver r/c_m sec later a disturbance of amplitude p_O/r anywhere on the spherical surface

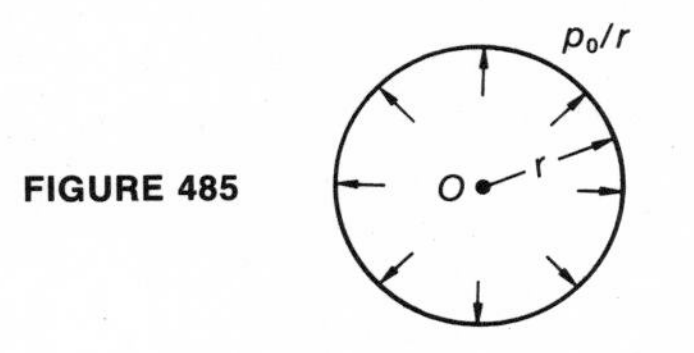

FIGURE 485

of radius r centered on O (see problem 808). However in the electromagnetic case an electron acceleration $\ddot{y}$ at O at time $(t - r/c)$ would produce at P, r/c sec later or at time t there, an electric field that was related to the vector acceleration of the charge at O, to $\vec{r}$, and to the angle θ in the following way:

$$E_{P,t} = \frac{-q\ddot{y}\left(\text{at } t - \dfrac{r}{c}\right) \sin\theta}{4\pi\epsilon_0 c^2 r}$$

$$= \frac{q\omega^2\, y_0 \sin\omega\left(t - \dfrac{r}{c}\right) \sin\theta}{4\pi\epsilon_0 c^2 r} \qquad (5\text{-}74)$$

$$= \frac{E_o \sin\omega\left(t - \dfrac{r}{c}\right)}{r} \sin\theta. \qquad (5\text{-}75)$$

$\vec{E_O}$ is the field tangent to a unit circle about O at the end of a diameter that is perpendicular to $\vec{y}$; E_O has the value $q\omega^2 y_0/4\pi\epsilon_0 c^2$.

The vector $\vec{E}_{P,t}$ is perpendicular to $\vec{r}$ and lies in the plane determined by $\vec{\ddot{y}}$ and $\vec{r}$ (Figure 486). It is this harmonically varying field $E_{P,t}$ that produces harmonic mo-

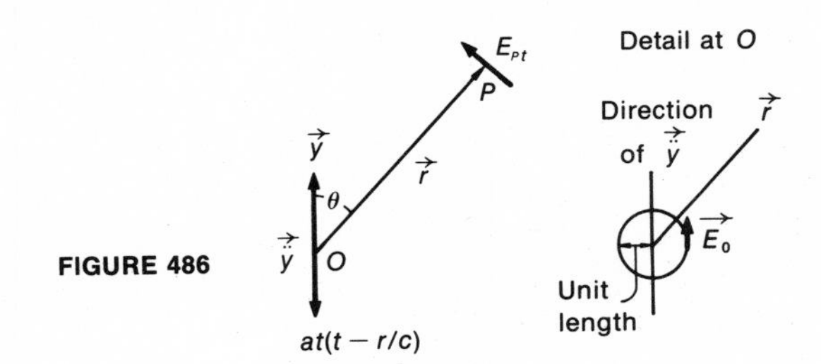

FIGURE 486

tions of charges at P: if a dipole at P is connected to a resonant circuit, a harmonically varying emf is generated in that circuit; if an eye is at P the resulting motions of the electrons in the receptors of the eye induce electrical signals to the brain; if the charges at P are not connected to a circuit their motions can become thermal energy or re-radiated energy.

Polarization. If the direction of $\vec{y}$ is fixed, as it would be in the case of an actual radiating dipole at O, then the field resulting from $\ddot{y}$ is linear in that direction also. Thus no matter how $\vec{r}$ is oriented in azimuth about $\vec{y}$, the field at P is polarized in the plane determined by $\vec{y}$ and $\vec{r}$; such radiation is called plane polarized. Radio, TV, and radar all depend on plane-polarized electromagnetic waves.

Electromagnetic field waves are also initiated by energy changes inside atoms. A planetary electron dropping from one energy level to a lower energy level inside the atom undergoes acceleration, and the field from that particular acceleration is polarized. However, such radiation has a maximum lifetime of less than 10^{-8} sec. Any observable field at P would be a repetition of an enormous number of such fields, all oriented randomly; hence the total field at P would be naturally unpolarized. Such a condition is characteristic of natural light.

Energy Density and Intensity

Since $E_{P,t}$ varies with θ, the energy flux through the surface of a sphere centered on O will not be uniform over the sphere. An electromagnetic wave originating at O and travelling in free space will produce at P an average energy density $u_{\text{em},P}$ whose value is

$$u_{\text{em},P} = \epsilon_O \overline{E_P^2}.* \qquad (5\text{-}76)$$

The average energy passing in unit time through a unit area at P that is perpendicular to $\vec{r}$, what we have learned to call intensity I, is then

$$\begin{aligned} I_P &= \epsilon_O c \, \overline{E_P^2} \\ &= \epsilon_O c \, \frac{E_O^2 \sin^2 \theta}{2r^2}. \end{aligned} \qquad (5\text{-}77)$$

From equation 5-75, with $\sin^2 \omega[t - (r/c)]$ averaging to $\frac{1}{2}$.

* Since $\overline{E_P^2} = \frac{1}{2}E_{P,\max}^2$, equation 5-76 has the same form as equation 5-6 for the energy density in an electrostatic field. However, the two formulae are not identical. They were derived from quite different backgrounds, one applying to an electrostatic field and the other, to a moving electromagnetic field.

PROBLEM 833

Integrate equation 5-77 for the power flowing through unit area at P over the entire spherical surface of radius r.

a. Determine the value of E_O^2 in terms of the power P_O being supplied at O.

b. Evaluate the intensity at P in terms of P_O.

SOLUTION

By conservation of energy the power entering the system at O must equal the power flowing out through any spherical surface centered on O; hence

$$P_O = \int_S I\, dS = \frac{\epsilon_O c\, E_O^2}{2} \int_0^{\pi} \frac{\sin^2\theta}{r^2}\, r\, d\theta\, 2\pi r \sin\theta = \frac{4}{3}\pi\epsilon_O c E_O^2$$

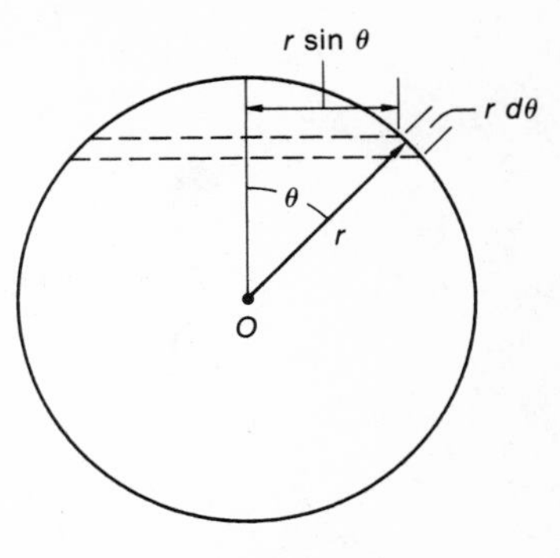

FIGURE 487

(Figure 487). Hence for part (a)

$$E_O^2 = \frac{3P_O}{4\pi\epsilon_O c}.$$

Substituting this value in equation 5-77,

$$I_P = \frac{3P_O}{8\pi}\frac{\sin^2\theta}{r^2}.$$

PROBLEM 834

A balloon carries an instrument package that measures cosmic ray intensities at various altitudes; data from the instruments is transmitted to the ground station by a radio transmitter sending at a power level of P_O watts. When the balloon is at a slant distance S from the ground station and at an elevation θ above the horizon,

a. what is the intensity of the signal strength received at the station if the sending antenna at the balloon is a dipole oriented vertically?
b. What is the intensity if the orientation is horizontal?
c. Considering only the effect on intensity of signal, if you are in charge of the experiment, which orientation would you specify for the sending antenna?

PROBLEM 835

The intensity of the sun's radiation at 1.0 A.U. from the center of the sun is approximately 1.4 kw m^{-2}. If the sun can be considered a uniformly luminous sphere of radius 7.0×10^5 km,

a. what must be the value of $E_\odot$ at the surface of this sphere?
b. What power is being radiated by the sun?

PROBLEM 836

On the average how much energy per sec reaches a square meter of the earth's surface at a point where the sun is at an elevation of θ above the horizon, when the sun time at that point is

a. noon,
b. 8:00 AM?

For simplicity in calculation, ignore the tilt of the earth's axis.

PROBLEM 837

a. Evaluate the energy radiated per cycle by an electron oscillating linearly in SHM with an angular frequency ω and a maximum excursion y_O.

The answer represents a depletion of the total energy of the oscillator. (See problem 515 if you have forgotten the value of this energy.) This depletion must be replaced every cycle if the oscillations are to be maintained.

b. What fraction of the oscillator's total energy is radiated? This fraction is called the radiation damping constant γ_R; to see why this name is appropriate, review problem 532. *Suggestion*: From part (a) of problem 833 you have the value of the energy radiated per sec in terms of E_O; from the text immediately following equation 5-75 you have E_O in terms of the oscillator parameters. Then if the energy per sec is divided by the frequency $\nu = \omega/2\pi$, you obtain the value of the energy radiated per cycle.

Radiation Pressure

Even though radiation is massless, since it transports energy it must also transport momentum (see problem 612). For electromagnetic radiation we have

$$p = \frac{E}{c}. \tag{5-78}$$

(Here E is energy, not electric field strength—another unfortunate but inescapable duplication in symbols.) For an area A perpendicular to $\vec{r}$ at P the energy E that arrives in time t is given by

$$E = I_P A t. \tag{5-79}$$

Then the linear momentum arriving at P during time t will be

$$p = \frac{I_P A t}{c}. \tag{5-80}$$

Since $F = dp/dt$, if a surface at A absorbs this momentum it must feel a perpendicular force

$$F_{\perp A} = \frac{I_P A}{c}. \tag{5-81}$$

PROBLEM 838

a. An instrument satellite exploring the region between the earth and the moon is equipped with a flat power panel of area A which always faces the sun and which absorbs all the solar radiation incident on it. What is the force on this panel resulting from the radiation pressure?

b. To ensure that there shall be no torque on the satellite from unbalanced radiation pressure, a flat panel of perfectly reflecting surface and of area B is mounted facing the sun symmetrically on the opposite side of the c.m. of the satellite from the power panel. What should be the value of B relative to A for no torque?

PROBLEM 839

Assuming the earth absorbs all of the solar radiation it receives, what is the ratio of the centrifugal to the centripetal forces on the earth from the sun?

PROBLEM 840

Tails of comets in the solar system point away from the sun. Assume this is due to the fact that the force on the grains of material in a comet's tail from the pressure of solar radiation must be greater than the gravitational force of attraction from the sun. What must be the upper limit for the value of the product $r\rho$, where r is the radius and ρ is the average density of a grain of comet-tail "dust" in MKS units?

PROBLEM 841

Mounted outside a satellite that is on a solar system mission are two spheres of identical size and D apart (Figure 488). One of the spheres is black and absorbs completely all radiation incident on it; the other sphere is perfectly reflecting for all angles of incidence. What is the maximum possible torque about the axis A-A midway between the two spheres that can develop from solar radiation? Ignore any reflected radiation that may fall on the black sphere. *Caution:* You can probably write down the answer for the force on the black sphere by analogy with problems you have already worked. However, this will not work for the bright sphere; better work that one out by integration.

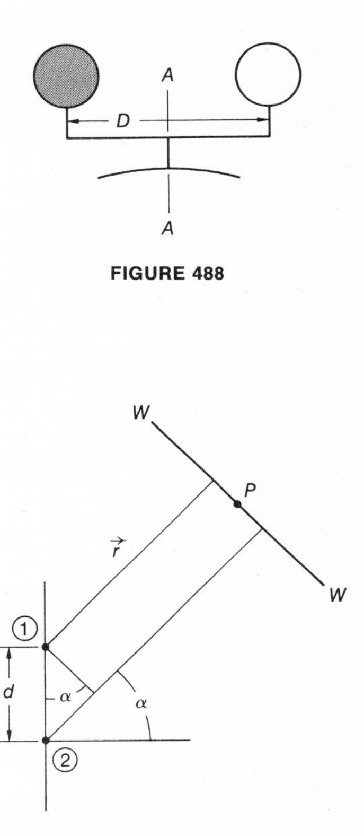

FIGURE 488

FIGURE 489

Addition of Electromagnetic Fields from Multiple Sources

In most practical applications dipole radiation is produced by an array of dipoles rather than by the single dipole discussed on page 391. The resulting field at some point P is then a summation of the individual fields that arrive at P simultaneously from a number of different radiators.

Let us consider a very simple situation: dipoles ① and ② oriented perpendicular to a plane determined by the point P and the centers of the dipoles (Figure 489). Thus the angle θ in equation 5-75 is made equal to $\pi/2$. The wavefront W-W passing P at any time t is a composite of two wavefronts,* one leaving dipole ① at time

*If $r \gg d$, the wavefronts at P are essentially plane and perpendicular to $\vec{r}$, and the $\vec{E}$ vectors at P lie in the plane of the wavefronts.

$(t-r/c)$ and one leaving dipole ② earlier at $[t-r/c-(d \sin \alpha)/c]$. If we consider the simplest possible case—the dipoles radiating in phase with $E_{O,1}=E_{O,2}$—then

$$E_{P,t}=\frac{E_O}{r}\sin \omega\left(t-\frac{r}{c}\right)+\frac{E_O}{r}\sin \omega\left(t-\frac{r}{c}-\frac{d \sin \alpha}{c}\right)$$

$$=\frac{2\,E_O}{r}\cos\frac{\omega d \sin \alpha}{2c}\sin\left[\omega\left(t-\frac{r}{c}\right)-\frac{\omega d \sin \alpha}{2c}\right]^*$$

$$=\frac{2\,E_O}{r}\cos\frac{\delta}{2}\sin\left[\omega\left(t-\frac{r}{c}\right)-\frac{\delta}{2}\right], \tag{5-82}$$

where

$$\delta=\frac{\omega d \sin \alpha}{c}=2\pi\,\frac{d \sin \alpha}{\lambda}$$

is the phase lag caused by the extra travel distance $d \sin \alpha$.

(A wave may be described either by $A \sin \omega t$ or by $A \cos \omega t$; the only difference in the two descriptions is that the clocks used differ by a quarter of a period. In many texts you will see the development start with a cosine description and end with

$$E_{P,t}=\frac{2\,E_O}{r}\cos\frac{\delta}{2}\cos\left[\omega\left(t-\frac{r}{c}\right)-\frac{\delta}{2}\right];$$

there is no physical difference between the two descriptions.)

We can increase generality in the use of equation 5-82 if we remove the stricture that both dipoles must operate in phase. If we allow the operating phase of dipole ② to lag dipole ① by a phase angle β, then the total phase lag of ② relative to ① is

$$\phi=\beta+\delta=\beta+2\pi\,\frac{d \sin \alpha}{\lambda}.$$

Equation 5-82 then becomes

$$E_{P,t}=\frac{2\,E_O}{r}\cos\frac{\phi}{2}\sin\left[\omega\left(t-\frac{r}{c}\right)-\frac{\phi}{2}\right]. \tag{5-83}$$

The amplitude of the composite field at P is $(2\,E_O/r)\cos(\phi/2)$ and the intensity will be

$$I_P=\epsilon_O c\,\overline{E_{P,t}^2}=\frac{\epsilon_O c\;4E_O^2\cos^2(\phi/2)}{2\,r^2}=4\,I_O\cos^2\frac{\phi}{2}, \tag{5-84}$$

where I_O is the intensity at P from one dipole alone radiating at the same rate.

*From the trig formula $\sin A+\sin B=2\sin[(A+B)/2]\cos[(A-B)/2]$.

PROBLEM 842

Two identical parallel dipoles are separated a distance equal to the wavelength λ of their radiation. What is the pattern of their radiation intensity in a plane perpendicular to the dipoles and through their centers if the dipoles are operating

a. in phase, and
b. 180° out of phase?

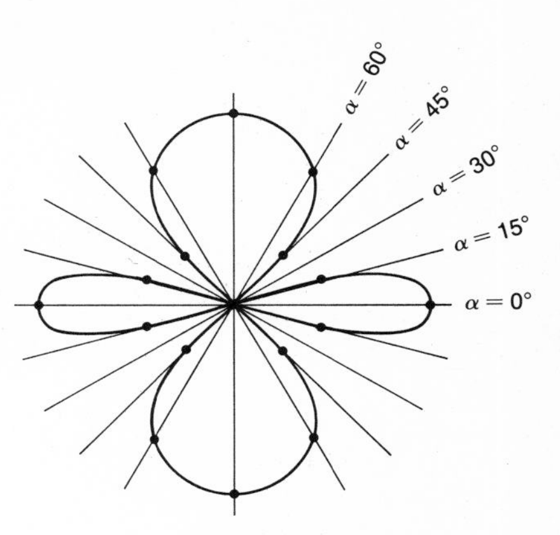

FIGURE 490

SOLUTION TO THE FIRST PART

For any dipole array we can hold $\epsilon_0 c/r^2$ constant and vary α; I_P then reduces to I_α, a function of α alone. For a pair of equal dipoles, equation 5-84 tells us that $I_\alpha = 4\ I_0 \cos^2 \phi/2$, and for a radiation pattern (Figure 490) we need to evaluate $\cos^2 \phi/2$ for various values of α. For question (a) $\beta = 0$, so that

$$\phi = 0 + 2\pi(\lambda \sin \alpha)/\lambda = 2\pi \sin \alpha, \text{ and } \phi/2 = \pi \sin \alpha.$$

The next step is a tabulation:

α	$\frac{\phi}{2}$	$\cos^2 \frac{\phi}{2}$	Relative Intensity
0°, 180°	0	1	$4\ I_0$
±30°, 150°	$\pm\frac{\pi}{2}$	0	0
±45°, 135°	$\pm\frac{\pi}{\sqrt{2}}$	0.36	$1.4\ I_0$
±60°, 120°	$\frac{\pm\sqrt{3}\pi}{2}$	0.83	$3.3\ I_0$
±90°	$\pm\pi$	1	$4\ I_0$
±15°, 165°	0.26π	0.47	$1.9\ I_0$

Figure 490 shows the radiation pattern—a plot of the data. The data for ±15° and ±165° were added after the other data were plotted and it was discovered there was an uncertainty in the region close to the $\alpha = 0$ axis.

PROBLEM 843

Two identical dipoles in a N-S line are separated a distance equal to $\lambda/2$.

a. If a maximum intensity is wanted in a direction 30°N of E of the dipoles, how should the south dipole be driven relative to the north dipole, i.e., what should β be for maximum I at $\alpha = 30°$?
b. If the dipoles are driven so as to satisfy the condition in question (a) what does the entire radiation pattern in a horizontal plane look like?

Hint: For maximum intensity $\cos^2 (\phi/2)$ must equal 1, hence $\phi/2$ must equal $n\pi$, where $n = 0, 1, 2 \ldots.$ Then $\beta + 2\pi(\lambda/2 \sin 30°)/\lambda$ must equal $n2\pi$. For $n = 0$, the south dipole must be driven in phase ahead of the north dipole just enough to offset the time delay from travelling the extra distance $(\lambda/2) \sin 30°$.

PROBLEM 844

A field engineer testing radiation intensity patterns is flying in a horizontal plane just above the ground in a helicopter that is making a ground speed of 120 mi hr^{-1} in a circular path of radius 2.0 mi, centered on the midpoint between two vertical transmitting dipole antenna that lie in a N-S line. For the frequency being used for the test the antennae are a half wavelength apart. Normally the antennae are operated in phase. However, the transmitter operator decides to play a joke on the field engineer by changing the phase relation between the two dipoles at such a rate that no change in radiation intensity can be observed aboard the helicopter. If he started making this change when the helicopter was due east of the antenna, at what rate was he changing the phase relation when the helicopter was $\theta°$ N of E from the antenna?

An increase in generality beyond that provided by equation 5-84 is needed when we have more than two adjacent dipoles radiating at the same frequency, or when we have only two dipoles radiating at different power levels. For these situations a vector summation of fields at P is much simpler than a summation by trigonometric formulae.

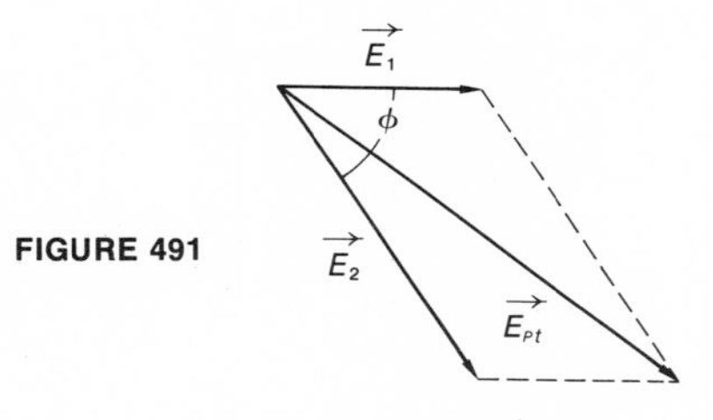

FIGURE 491

If two adjacent dipoles radiate at the same frequency but at different power levels, in a plane perpendicular to the dipoles and through their centers the composite field at P will be $\vec{E}_{P,t} = \vec{E}_1 + \vec{E}_2$. If $\vec{E}_1$ and $\vec{E}_2$ are in phase at P, then $E_{P,t} = E_1 + E_2 = (1/r)(E_{o,1} + E_{o,2}) \sin \omega(t - r/c)$. If there is a phase difference ϕ between E_1 and E_2 at P (Figure 491) then

$$E_{P,t}^2 = E_1^2 + E_2^2 + 2E_1E_2 \cos \phi$$
$$= \frac{1}{r^2}\left(E_{o,1}^2 + E_{o,2}^2 + 2E_{o,1}E_{o,2}\cos \phi\right)\sin^2 \omega\left(t - \frac{r}{c}\right).$$

By setting $\phi = 0$ this equation contains the special case of radiation from two dipoles being in phase at P. The general equation for the intensity at P from two radiating dipoles is

$$I_\alpha = \epsilon_0 c\, \overline{E_{P,t}^2} = \frac{\epsilon_0 c}{2r^2}\left(E_{o,1}^2 + E_{o,2}^2 + 2E_{o,1}E_{o,2} \cos \phi\right) \tag{5-85}$$
$$= I_{o,1} + I_{o,2} + 2\sqrt{I_{o,1}I_{o,2}} \cos \phi.$$

If the relation between $E_{o,1}$ and $E_{o,2}$ is known to be $E_{o,2} = aE_{o,1}$(or $I_{o,2} = a^2 I_{o,1}$) then

$$I_\alpha = I_{o,1}(1 + a^2 + 2a \cos \phi). \tag{5-86}$$

Note that for the special case of $E_{o,1} = E_{o,2}$, equation 5-85 reduces to equation 5-84 as it should. Note also that for $E_{o,1} = E_{o,2}$ Figure 491 shows clearly why the composite field $E_{P,t}$ lags the reference field E_1 by $\phi/2$.

The same vector addition method serves in the case of more than two radiators. Let us consider the case of n dipoles in line, separated by distances d from each other, all operating in phase and at the same power level. Then the composite field at P is a summation of n equal vectors, each lagging the one ahead of it by the same phase angle ϕ. (Figures 492 and 493 illustrate this summation for $n = 5$.) Then the

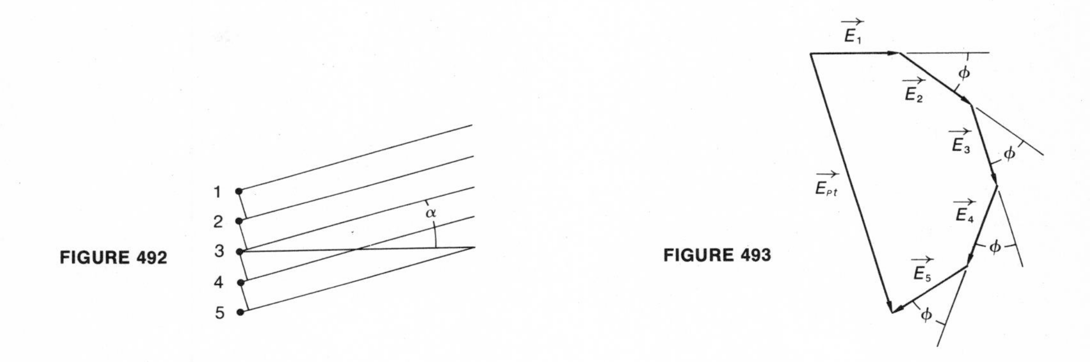

FIGURE 492

FIGURE 493

heads and tails of these vectors lie on the circumference of a circle, and the angle subtended at the center of the circle by the resultant $\vec{E}_{P,t}$ must $= n\phi$. By simple trigonometric formulae it can then be found that the scalar value of

$$E_{P,t} = E \frac{\sin n\phi/2}{\sin \phi/2}$$

and that $\vec{E}_{P,t}$ lags the reference vector $\vec{E_1}$ by $(n - 1)\phi/2$.

Then

$$I_P = \epsilon_o c \overline{E^2_{P,t}} = \frac{\epsilon_o c}{2r^2} E_o^2 \frac{\sin^2 n\phi/2}{\sin^2 \phi/2} = I_o \frac{\sin^2 n\phi/2}{\sin^2 \phi/2}. \tag{5-87}$$

Obviously when $\phi = 0$ the $\vec{E}$s add in a straight line, $E_{P,t} = nE$, and the intensity $I_P = n^2 I_o$. (This same result could have been obtained from equation 5-87 by evaluating

$$\frac{\sin^2 n\phi/2}{\sin^2 \phi/2} \text{ at } \phi = 0.)$$

When $\phi \neq 0$ it varies with α. When the angular displacement α is small $\sin \alpha$ will be small, and the value of $\phi = 2\pi(d \sin \alpha)/\lambda$ will depend on the ratio d/λ. Assuming for the moment that $d \simeq \lambda$ (so as to emphasize the dependence of ϕ on α),

ϕ will be small when α is small, and $E_{P,t}$ will then be the chord of a circle of large radius. $E_{P,t}$ is clearly less than nE, so that I_α will be less than $n^2 I_0$. As we increase α the value of ϕ also increases, until at some α, $n\phi = 2\pi$, or $\phi = 2\pi/n$. At this point the vector addition has become a closed polygon, and $E_{P,t}$ and $I_\alpha = 0$. This occurs at $\alpha = \sin^{-1}(\lambda/nd)$, which is called the position of the first intensity minimum. As we increase α still more, ϕ continues to increase; $n\phi$ is now $> 2\pi$, and the vector summation comprises more than one circle. At $n\phi \simeq 3\pi$ (the larger the value of n and the smaller the phase difference ϕ the more accurate this approximation becomes) $E_{P,t}$ reaches a secondary maximum, so that I_α at that point is also a secondary maximum. At $n\phi = 4\pi$ the vector polygon is again closed and the position of another minimum has been reached. We can summarize this process by saying that we reach successive minima when $\phi = (p/n)2\pi$, or at the positions $\alpha = \sin^{-1}(p\lambda/nd)$, except when $p = mn(p = 1, 2, 3, \ldots; m$ can have the values $0, 1, 2, \ldots)$. Successive secondary maxima occur when $\phi \simeq (2p + 1)\pi/n$ or at $\alpha \simeq \sin^{-1} [(2p + 1)\lambda/2nd]$; principal maxima repeat when $\phi = m2\pi$, or at $\alpha = \sin^{-1}(m\lambda/d)$. Note that the location of these principal maxima does not depend on n.

The process can be continued only as long as $\sin \alpha = m\lambda/d \leqslant 1$; thus the extent of the repetitive pattern of maxima and minima is highly sensitive to the ratio λ/d. If $d < \lambda$, the principal maxima appear only at $\alpha = 0$ and π; if $d = \lambda$ principal maxima appear at $\alpha = 0, \pm \pi/2$, and π. If $d > \lambda$, additional principal maxima may appear. Although the location of these principal maxima does not depend on n, the sharpness of their pattern, i.e., the angular distance between adjacent minima, does depend on n.

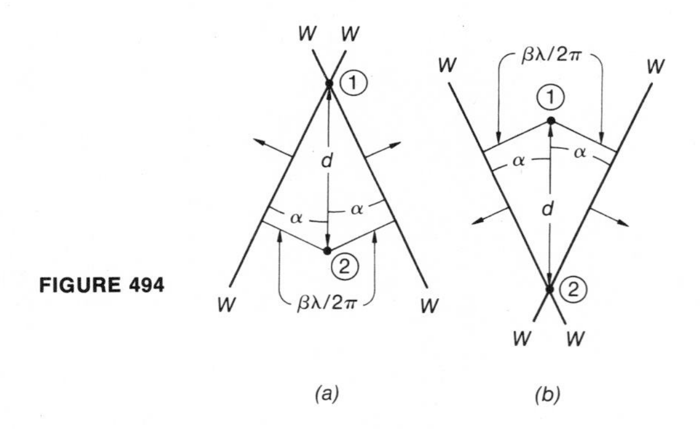

FIGURE 494

The direction of the entire pattern of maxima and minima can be shifted by introducing a phase shift β* between adjacent radiators. For the principal maxima we would then have $\phi = m2\pi = \beta + 2\pi(d \sin \alpha)/\lambda$ and these maxima would be located at $\alpha = \sin^{-1}[(m - \beta/2\pi)\lambda/d]$. The effect of this phase shift is illustrated in Figure 494 for the wavefront produced by two adjacent radiators (for simplicity we

*As it originally appeared in $\phi = \beta + 2\pi(d \sin \alpha)/\lambda$, β with a plus sign was an angle of lag (page 396). A lead angle is a negative lag angle; if you are told that the radiation from 2 leads the radiation from 1 by a phase difference Δ, then in computing ϕ you would have $\phi = -\Delta + 2\pi(d \sin \alpha)/\lambda$. Theoretically β can have any value: $\pi/3$, 1.5π, 29.4π, etc.; there is no reason one radiator cannot be made to lead or lag an adjacent radiator by 150 periods if that kind of a time delay serves a useful purpose. However, a realistic value for β is still controlled by the fact that $\sin \alpha \leqslant 1$.

will assume $m = 0$ and thus consider only the wavefront of the central maximum). In both (*a*) and (*b*) W-W represents an in-phase wavefront from both radiators; thus the advancing normal to the wavefront is in the direction of the central maximum. In (*a*) the radiation from 2 leads the radiation from 1 by β, i.e., 2's clock is running ahead of 1's clock by $\beta/2\pi$ of a cycle, and the central maximum occurs in the first or second quadrants. In (*b*), 2 lags 1 by β, and the central maximum occurs in the third or fourth quadrants. This phase shift of β between adjacent dipoles has the effect of making the line of real radiators appear from a direction α to be a line of apparent radiators all operating in phase, hence $I_\alpha = n^2 I_o$. Figure 495 shows a lead phase for a dipole relative to the one above it.

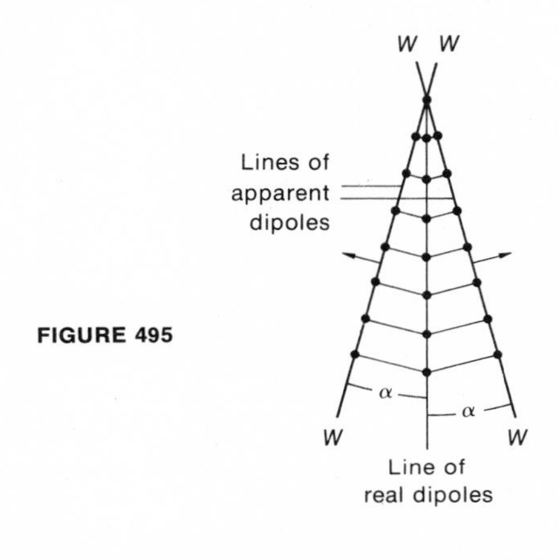

FIGURE 495

PROBLEM 845

Two vertical dipoles are in a N-S line and are separated by a distance equal to $\lambda/2$, where λ is the operating wavelength for both dipoles. The north dipole radiates twice as much power as the south dipole. What does the radiation intensity pattern in a horizontal plane through the center of the dipoles look like

a. when the dipoles are operated in phase with each other, and
b. when the south dipole lags the north dipole by 180°?

Suggestion: Before making any detailed tabulation of values of I_α vs. α, can you tell just by sketching the vector diagram for $E_{P,t}$ for either (a) or (b) whether there is any direction for which $I_\alpha = 0$?

PROBLEM 846

In a field training exercise a command post C is established on an E-W road. To communicate with units in the field two jeeps, each carrying shortwave equipment and a dipole antenna, are parked next to each other at C so that their antenna are in a N-S line (Figure 496). The officer had read somewhere that two dipoles operating in parallel in phase produced four times the intensity of one dipole, so he ordered one antenna disconnected from its own set and connected in parallel to the other antenna. A scout column S waiting 1.0 mi east of C on a N-S road is ordered to start north on that road and to keep listening for instructions as it moves. The scout column first

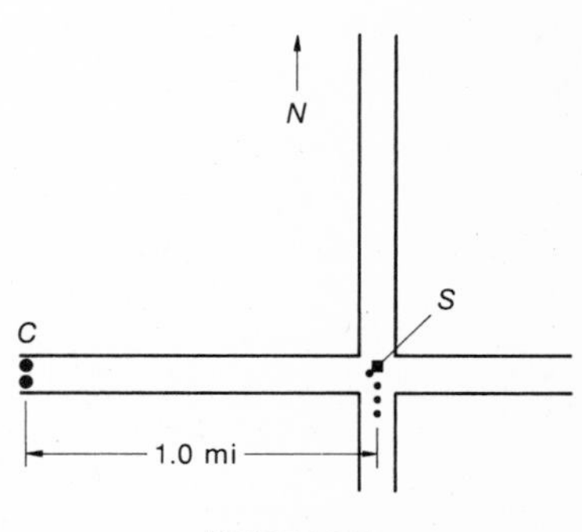

FIGURE 496

loses radio contact with C when it has gone 0.23 mi. How far has the column moved from its starting point

a. when its received intensity peaks again, and
b. when it loses radio contact the second time?

PROBLEM 847
For a N-S linear array of n parallel dipoles λ apart, all being operated at the same power level,

a. what relative phase shift β between adjacent dipoles is required to throw a maximum intensity of $n^2 I_0$ in the azimuthal direction i. 30° N of E, ii. 45° S of E, iii. 30° S of W, iv. due north?
b. When the central maximum beams of part (a) are established, are there companion central maximum beams in other directions and if so, what are these directions?

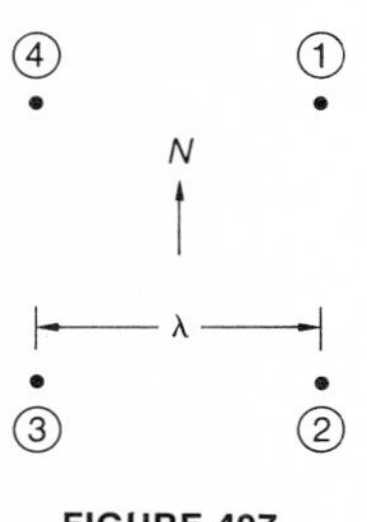

FIGURE 497

PROBLEM 848
Four vertical dipoles, arranged in a square that measures λ on a side, radiate at equal power. The square is oriented, and the dipoles identified, as shown in Figure 497. How should the radiation from the other dipoles be phased relative to the radiation from dipole 1 so as to throw maximum intensity in a direction 30° E of N? *Suggestion*: Don't try to work this out by formula or long computation; just make a sketch to show what the apparent array of dipoles is required to look like from the desired direction. The computations can then be done in your head.

PROBLEM 849
You have four equal dipoles, and you would like to arrange them in space and in phase so as to concentrate as much of the radiated energy as possible in one direction.

a. What is the simplest combination of spacing d and phasing β that you can think of to accomplish the desired result?
b. What is the maximum intensity possible, and what is its direction, for the arrangement you chose?
c. Make a rough sketch of what the radiation pattern of your chosen arrangement would look like.

SUGGESTED SOLUTION
There are only two simple arrangements for four dipoles—in a line or on the corners of a rectangle (or square). But at great distances radiators at the corners of a rectangle can produce maximum intensities in more than one direction; note the direction 30° W of N in problem 848, for example. It is therefore improbable that you can find a rectangular arrangement that will concentrate radiation in one direction only. Thus a linear arrangement seems more probable.

We have already learned that a linear array in phase produces equal maximum intensities in the direction $\alpha = 0$ and $\alpha = \pi$. We can eliminate these intensities by phasing the radiators with a lead or lag of $\pi/2$ between radiators; we obtain this value for β from the vector summation shown (Figure 498) or from the calculation $n\phi = 2\pi$. With β chosen at $\pi/2$, we then look to see if we can so space the dipoles so that from one end of the line, say from the direction $\alpha = \pi/2$, $\phi = 0$ and from the other direction, where $\alpha = 3\pi/2$, $\phi = \pi$. It should not be difficult to determine that this is accomplished by setting $d = \lambda/4$.

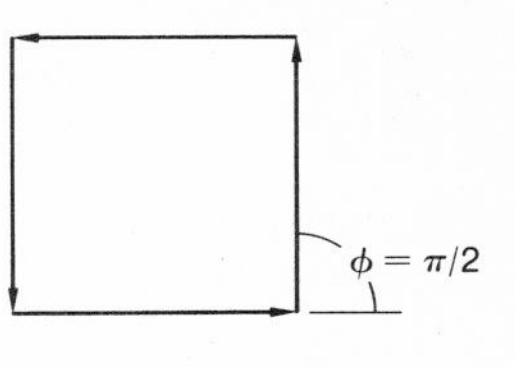

FIGURE 498

PROBLEM 850

You are at a remote radar tracking station and the motor driving your rotating parabolic radiation reflector burns out, so that you are unable to "sweep" the radar beam around the horizon by mechanical means. Being ingenious electronically, you hurriedly set up a linear array of n vertical, parallel transmitting dipoles λ apart, all to be operated in parallel on the same frequency and at the same power level.

a. If you plan to "sweep" the beam electronically at an azimuthal rate of $\omega = d\alpha/dt$, what phase shifting rate $d\beta/dt$ would you design for?
b. Are there any theoretical advantages to this electronic sweep system over a sweep carried out by mechanical rotation of the transmitter?
Note: This problem concerns itself only with the outgoing beam; you are to assume that a suitable arrangement for detecting the incoming reflected beam has also been arranged.

Physical Optics

So far we have discussed the addition of electromagnetic radiations from separate sources for the purpose of determining the energy intensity at some point or in some direction in space. Since all of the equations we have developed for this purpose are functions of the wavelength λ, we could just as easily have used the same equations for the measurement of λ. This use of the equations makes up the topic of physical optics. For this topic we need to learn nothing fundamentally new; by changing our description of an electromagnetic radiator from an actual oscillating dipole to a region of infinitesimal area in a radiation wavefront, we can apply to these simulated dipoles all we have learned about the radiation from real dipoles.

A plane wavefront W-W of radiation of wavelength λ falls perpendicularly on an opaque wall in which two extremely narrow linear slits S_1 and S_2 have been cut. The E vectors riding in the wavefront are parallel to the slits. These slits will act as if they were linear dipole radiators and their radiation will propagate in the region behind the wall just as would the radiation from real radiators. Thus radiation from both slits will be received at P on an observing screen (or photographic plate) that is placed D behind the wall and parallel to it (Figure 499). These radiations will be in phase when ϕ, the phase angle between radiations from S_1 and S_2 equals $0, 2\pi, 4\pi, \ldots$ just as before for two dipoles; from the geometry of Figure 499 this is the same as saying that $S_2P - S_1P = \Delta$ must equal $m\lambda (m = 0, 1, 2, \ldots)$.* For $\phi = \pi, 3\pi,$

*In physical optics it is customary to call m the "order" of the pattern. Thus the central maximum at O is the zero order maximum, etc.

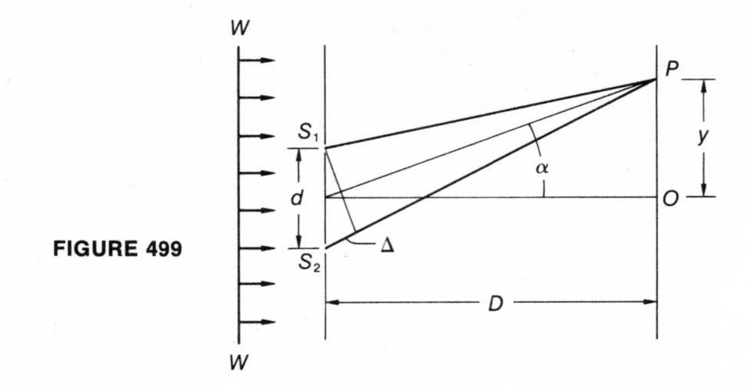

FIGURE 499

. . . the radiations from S_1 and S_2 are 180° out of phase and $\Delta = (m + 1/2)\lambda$. Thus maximum intensity occurs at any P for which $\sin \alpha = m\lambda/d$; minimum intensity occurs when $\sin \alpha = (m + \frac{1}{2})\lambda/d$.

Usually d is very large compared to $m\lambda$ so that $\sin \alpha \simeq \tan \alpha = y/D$. For this approximation the maximum and minimum intensities from a double slit occur at

$$y_{\max} = \frac{mD\lambda}{d} \tag{5-88}$$

and

$$y_{\min} = \frac{(m + \frac{1}{2})D\lambda}{d}. \tag{5-89}$$

In this development the two narrow slits have been treated as physically equivalent to the two in-phase dipoles discussed earlier; hence the intensity at P is still given by equation 5-84. In view of $\phi = 2\pi(d \sin \alpha)/\lambda$ this equation can be rewritten for a double slit as

$$I_P = 4\, I_o \cos^2 \frac{\pi\, d \sin \alpha}{\lambda}. \tag{5-90}$$

Note that for a double slit I_P is not a function of m, which tells us that $I_{\max}$ repeats for every $y_{\max}$. Thus the interference pattern at the screen from two slits will be as shown in Figure 500.

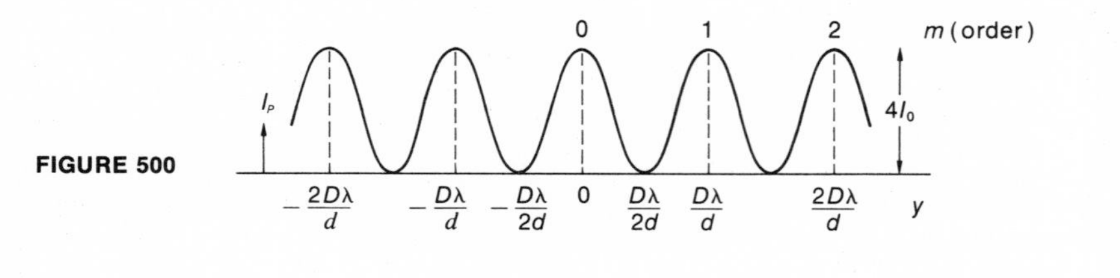

FIGURE 500

PROBLEM 851

In a standard double-slit arrangement for the measurement of monochromatic radiation, the two slits are 1.00 mm apart and 3.00 m in front of a photographic plate placed parallel to the plane of the slits.

a. If a measurement of the intensity pattern on the plate shows a

fifth-order maximum at a distance of 9.00 mm from the central maximum, what was the wavelength of the light?
b. If the measurement of the positions of higher order maxima relative to the central maximum is only accurate to 0.01 mm, what is the uncertainty in the wavelength determination when the measurement is made to the i. first-order maximum, and ii. the fifth-order maximum?

PROBLEM 852

A line source L emitting light of wavelength 5625 Å is placed 1.00 m directly back of slit S_1 in an opaque wall (Figure 501). A second slit S_2 is parallel to S_1 and d from it; both slits are parallel to L. As a training exercise in the measurement of microphotometer tracings, the intensity of the photographic image at a point equidistant from both slits is compared for three different exposures all of the same time duration: for slit S_2 covered, for slit S_1 covered, for both slits open. It is found that the intensity for both slits open equals the sum of the intensities from each slit by itself. How far apart were the slits?

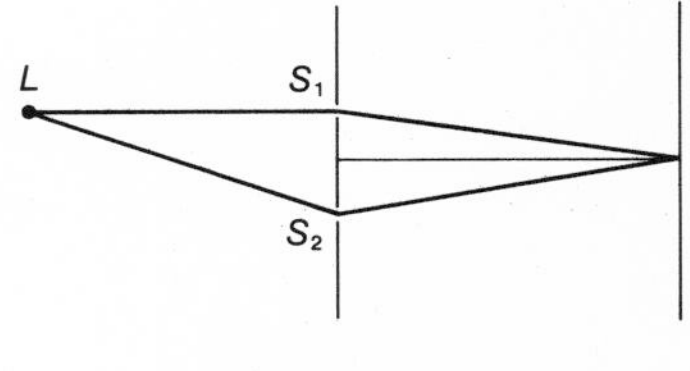

FIGURE 501

PROBLEM 853

If two adjacent maxima can be distinguished visually on the observing screen of Figure 499 only when their separation is at least 1.00 mm, what must be the double-slit spacing d to resolve the fourth- and fifth-order maxima of light of wavelength 5600 Å when $D = 2.00$ m?

PROBLEM 854

A small motor boat under way after sunset must carry aft and high up a white light that can be seen from all directions. A large cabin cruiser at night must in addition carry a forward-facing white light mounted near the bow and lower than the aft light. (These two white lights on the larger boat serve as range lights; when they appear one directly below the other the boat is being seen head-on.) Returning one evening to Seattle from a weekend cruise to Victoria you see a boat's light ahead of you; actually it is a large yacht heading toward you with its aft light mounted 4.0 ft higher than the bow light.

a. How close to this yacht are you before you are able to resolve the two lights if you depend on your eyes which have a pupil diameter of 5.0 mm?
b. Would it have been safer to use your binoculars, which have an object lens of 40 mm diameter?
Use a wavelength of 5600 Å as representative of the most intense part of the lights' spectrum.

Hint: This is exactly similar to Problem 853; the eye or the object lens serves as the observing screen and the two lights represent the double slit. Do you think any allowance should be made for the fact that the eye or the object lens receive a round pattern instead of a pattern spread over a rectangular space as in the other double-slit problems?

PROBLEM 855
In a double-slit diffraction pattern we assume each slit is transmitting energy at a rate per unit time that is proportional to E_o^2, or for a total of $2E_o^2$ coming through both slits. Yet at maxima our equations show an intensity proportional to $4E_o^2$; is there an inconsistency here? If not, what is the explanation? *Suggestion*: Graph the intensity pattern for a double slit, and show on this graph the intensity corresponding to $2E_o^2$.

Single-slit Diffraction

In the double-slit interference pattern of problem 855 it was assumed that the slits were so narrow they could be considered the equivalent of single dipole radiators. In practice this assumption is not valid; even a scratch only 0.001 mm wide (that's 25,000 scratches per inch) in an opaque film is approximately 2λ in width for visible light. Thus the radiation coming from any real slit must be considered more equivalent to the output from a linear array of a large number of dipoles arranged uniformly across the slit opening.

For n dipoles in line and in phase we have already discovered (page 400) that we find minima in directions $\alpha = \sin^{-1} p\lambda/nd$ and secondary maxima at $\alpha = \sin^{-1} (2p+1)\lambda/2nd$; $p = 1, 2, \ldots$. By making n large and d small (we are free to adjust these values to any extremes we consider necessary) we can simulate a wavefront by a continuous array of radiators; the product nd can be taken as the length of this array and thus equals the width of the slit w. Then for a single slit on which light of wavelength λ is incident in phase,

$$\alpha_{\min} = \sin^{-1} \frac{p\lambda}{w}, \tag{5-91}$$

and

$$\alpha_{\text{sec max}} \simeq \sin^{-1} \frac{(2p+1)\lambda}{2w}. \tag{5-92}$$

Note from page 400 that for principal maxima $\sin \alpha = m\lambda/d$. Since d is very small $m\lambda/d$ will be < 1 only for $m = 0$; hence a single slit can produce only a central or zero-order principal maximum.

Since this development was derived from the study of multiple dipole radiation in an in-phase wavefront moving in the direction α, strictly speaking equations 5-91 and 5-92 apply to a similar situation illustrated in Figure 502; this is called Fraunhofer diffraction.

In practice the wavefront is brought to a focus on a screen or photographic plate by the insertion of a simple converging lens of focal length D, as shown in Figure 503. In practice we measure y_α instead of α; since α is small $\sin \alpha \simeq \tan \alpha$

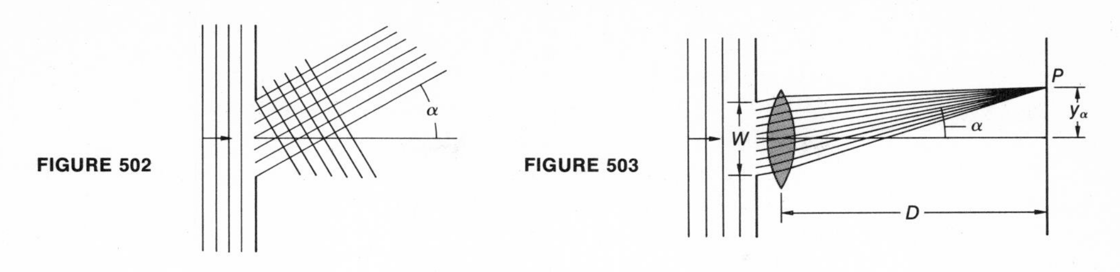

FIGURE 502

FIGURE 503

$= y_\alpha/D$. Thus, from equations 5-91 and 5-92 we find minima and secondary maxima for a single slit at

$$y_{\min} = \frac{pD\lambda}{w} \tag{5-93}$$

and

$$y_{\text{sec max}} = \frac{(2p+1)D\lambda}{2w}. \tag{5-94}$$

(We have now reached a point where we can check the validity of our assumption that a single slit can be simulated by a linear array of dipoles. Measurements of actual single-slit diffraction patterns match exactly the predictions of equations 5-93 and 5-94, confirming the assumption.)

The total phase change across the slit for any angle α is $n\phi = 2\pi(nd \sin\alpha)/\lambda = 2\pi(w \sin\alpha)/\lambda$. Substituting this in equation 5-87, for the intensity in the diffraction pattern of a single slit, we get

$$I_\alpha = I_o \frac{\sin^2\left(\dfrac{\pi\, w \sin\alpha}{\lambda}\right)}{\sin^2\left(\dfrac{\pi\, w \sin\alpha}{n\lambda}\right)}.$$

Because of the n, $(\pi\, w \sin\alpha)/n\lambda$ is very small, so that the sine of the angle can be replaced by the angle itself. Also at $\alpha = 0$, $I_\alpha = n^2 I_o = I_{\max}$. Making these changes we finally have for the single slit

$$I_\alpha = I_{\max} \frac{\sin^2\left(\dfrac{\pi\, w \sin\alpha}{\lambda}\right)}{\left(\dfrac{\pi\, w \sin\alpha}{\lambda}\right)^2}. \tag{5-95}$$

The central portion of a single slit diffraction pattern would be as shown in Figure 504.

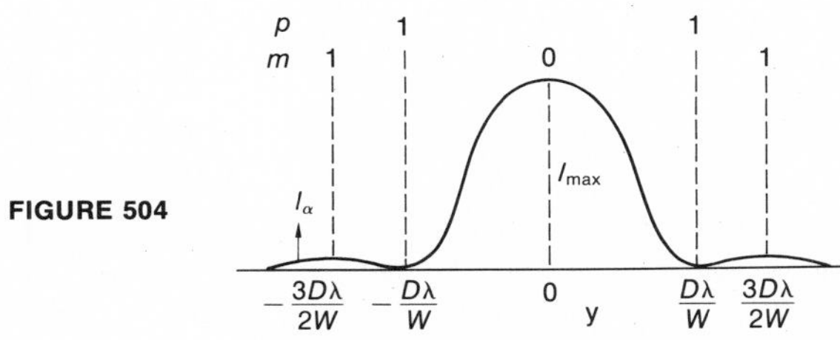

FIGURE 504

PROBLEM 856

If you take the distance between adjacent minima as the width of a maximum,

a. what change takes place in the width of the principal maximum of the diffraction pattern of a single slit when the slit width is doubled?
b. For $D = 1.00$ m and incident monochromatic light of wavelength 5800 Å, what width of single slit will produce a principal maximum whose width is exactly equal to that of the slit?
c. With the same parameters as in question (b) what width of principal maximum will be produced by a slit of width 1.00×10^{-2} mm?

PROBLEM 857

In problem 856 the width of the principal maximum of the diffraction pattern of a single slit was taken as the distance between the zero-order minima. In practice, what is often used as the width of a maximum is the width at half intensity, i.e., the width of the pattern where $I_\alpha = \frac{1}{2}I_{max}$. For the principal maximum of a single slit what is this width in terms of D, λ, and w?

START OF SOLUTION

Let $(\pi w \sin \alpha)/\lambda = \beta$. Then combining the requirements of the problem with equation 5-95, you get $\beta/\sqrt{2} = \sin \beta$. This must be solved for the value of β, and there are three obvious methods for this. (1) A neat but not very accurate way is to plot $y_1 = \beta/\sqrt{2}$ and $y_2 = \sin \beta$ on the same graph; the intersection of the two plots will determine the wanted value of β. (2) You can expand $\sin \beta$ in terms of β, which then makes the governing relationship an algebraic equation in β alone, with an infinite number of terms. Keeping only enough terms to constitute an easily solvable quadratic equation should yield at least two-figure accuracy for β. (3) A simple and practical way (although probably considered nonelegant by mathematical purists) is to look at $\beta/(2)^{1/2} = \sin \beta$ and recognize that for values of β near 1.41 rad both sides of the equation very nearly equal 1. Then with a slide-rule with trig scales or with a handbook, it should take you less than 30 sec to find out that the equation balances for $\beta = 1.39$.

PROBLEM 858

Light containing two separate wavelengths—λ_1 known to $= 5850$ Å and λ_2 unknown—is incident on a standard single-slit arrangement. On examining the diffraction pattern on a photographic plate it is determined that the sixth secondary maximum of λ_2 coincided with the position of the seventh minimum for λ_1. What was the wavelength λ_2?

PROBLEM 859

Monochromatic light of wavelength λ is incident on a standard single-slit arrangement. Back of the slit is a converging lens of 1.00 m focal length; a photographic plate is placed in the focal plane of the lens. For the separate cases of the slit width equalling i. 100 λ and ii. 1000 λ,

a. What is the angular width of the central maximum?
b. What is the linear width of the fifth-order secondary maximum?
c. How does the linear width of the central maximum compare with the widths of the secondary maxima?

PROBLEM 860
What is the general formula for the intensity of the successive secondary maxima in the diffraction pattern of a single slit? *Hint*: Combine equations 5-92 and 5-95.

PROBLEM 861
The accuracy of a trace from a particular microphotometer is 0.1 percent. When the sensitivity response of that microphotometer is adjusted to give full-scale deflection for the central maximum of a single-slit diffraction pattern you want to analyze, what is the highest order secondary maximum you expect to locate accurately?

Multiple-slit Diffraction

The discovery that a real single slit has a diffraction pattern of its own and is too wide to be considered a single dipole radiator should make us uneasy about the work we did in developing equation 5-90 and Figure 500 for a double slit. There are two ways of curing this uneasiness.

The harder way would be to treat each slit as we did the single slit, adding the contributions from two separate linear arrays of radiators to determine the intensity at some point P or in some direction α. (Note that α will be different for each slit because of the separation d of the slits.) However, there is a simpler way.

For this simpler way you start by comparing Figures 500 and 504. You will note that in Figure 504 there is only one principal maximum and that this maximum is $2D\lambda/w$ wide; in Figure 500 the principal maxima are evenly spaced and are $D\lambda/d$ wide. The quantity d will always be greater than w, so that several maxima of Figure 500 will be included between the minima on either side of the central maximum of Figure 504. Thus the intensity at any maximum of the double-slit interference pattern will be modulated by the intensity envelope from the single-slit diffraction pattern. Thus, for real slits the I_o of equation 5-90 must be replaced by

$$I_{\max} \frac{\sin^2\left(\dfrac{\pi\, w \sin\alpha}{\lambda}\right)}{\left(\dfrac{\pi\, w \sin\alpha}{\lambda}\right)^2}$$

from equation 5-95. Hence the intensity for the interference pattern for a real double slit is given by

$$I_\alpha = 4\, I_{\max} \frac{\sin^2\left(\dfrac{\pi\, w \sin\alpha}{\lambda}\right)}{\left(\dfrac{\pi\, w \sin\alpha}{\lambda}\right)^2} \cos^2 \frac{\phi}{2}. \tag{5-96}$$

If the radiation incident on the slits is all in phase, then $\phi = 2\pi(d \sin\alpha)/\lambda$ and equation 5-96 reduces to

$$I_\alpha = 4\, I_{\max} \frac{\sin^2\left(\dfrac{\pi\, w \sin\alpha}{\lambda}\right)}{\left(\dfrac{\pi\, w \sin\alpha}{\lambda}\right)} \cos^2 \frac{\pi\, d \sin\alpha}{\lambda}. \tag{5-97}$$

The effect of this modulation of the interference pattern of two slits by the diffraction pattern of one slit is to sharpen significantly the maxima by concentrating the energy (thereby increasing the intensity) in narrower regions on the screen or photographic plate. This effect is illustrated in Figure 505, which clearly shows how the finite width of real slits changes the real energy distribution from that shown in Figure 500 for infinitely narrow slits.

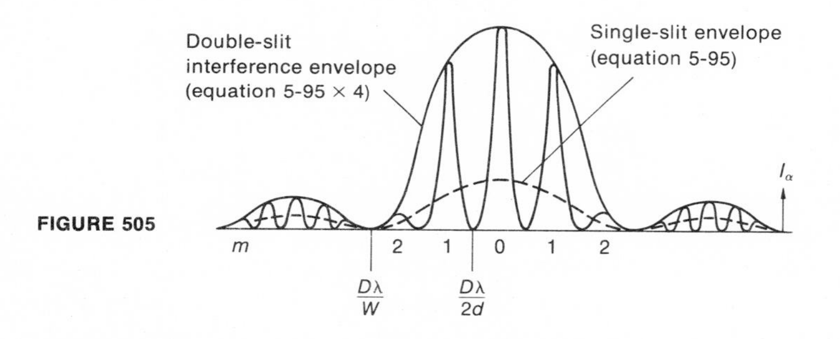

FIGURE 505

The Diffraction Grating. We found earlier that n dipoles produced more radiant energy and in more narrow concentrations than two dipoles; it would seem logical to expect this same enhanced result from n slits.

For n slits each of width w we can combine equation 5-87 with the replacement of

$$I_o \text{ by } I_{\max} \frac{\sin^2\left(\dfrac{\pi\, w \sin\alpha}{\lambda}\right)}{\left(\dfrac{\pi\, w \sin\alpha}{\lambda}\right)^2}$$

(from equation 5-95 for a real slit); we then get

$$I_\alpha = I_{\max} \frac{\sin^2\left(\dfrac{\pi\, w \sin\alpha}{\lambda}\right)}{\left(\dfrac{\pi\, w \sin\alpha}{\lambda}\right)^2} \frac{\sin^2\left(\dfrac{n\phi}{2}\right)}{\sin^2\left(\dfrac{\phi}{2}\right)}. \tag{5-98}$$

The final term in equation 5-98 represents the interference pattern for n slits. As in the case of n dipoles, this term produces principal maxima at $\phi = m2\pi$, secondary maxima at $n\phi \simeq (2p + 1)\pi$, and minima at $n\phi = p2\pi$, $p \neq mn$.

The intensity of principal maxima is

$$I_{\text{prin max}} = n^2 I_{\text{max}} \frac{\sin^2\left(\dfrac{\pi\, w \sin\alpha}{\lambda}\right)}{\left(\dfrac{\pi\, w \sin\alpha}{\lambda}\right)^2}. \tag{5-99}$$

For the geometrical arrangement in which the observing screen or photographic plate and the wavefront of incident radiation are all parallel to the plane of the slit assembly, these principal maxima occur at $\sin\alpha = m\lambda/d$, and for α small, at

$$y_{\text{prin max}} = \frac{mD\lambda}{d}. \tag{5-100}$$

In between these principal maxima there are $(n - 1)$ minima and $(n - 2)$ secondary maxima, located at

$$y_{\text{min}} = \frac{pD\lambda}{nd} = \frac{pD\lambda}{W}$$

and

$$y_{\text{sec max}} = \frac{(2p + 1)D\lambda}{2W},$$

where $W = nd =$ the width of the entire slit assembly.

Such a slit assembly is called a diffraction grating. It provides two important advantages for precision measurements of wavelength: the large concentration of energy in the $n^2 I_{\text{max}}$ intensity of the principal maxima, and the very narrow pattern of these maxima. Because of this narrowness the principal maxima are usually spoken of as spectral "lines"; the number m is the "order" of the spectrum.

The precision capability of a grating can be appreciated by evaluating its power to resolve small percentage differences in wavelength. A standard criterion for discriminating between two wavelengths that are nearly equal, and hence whose principal maxima are closely adjacent, requires that a principal maximum of one wavelength, say $\lambda + d\lambda$, shall coincide with the minimum adjacent to a principal maximum of the other wavelength λ. This requirement is met by

$$y_{\text{max},\lambda + d\lambda} = \frac{mD(\lambda + d\lambda)}{d} = y_{\text{min},\lambda} = \frac{(mn + 1)D\lambda}{nd}$$

from which we get for a diffraction grating

$$\frac{d\lambda}{\lambda} = \frac{1}{mn}. \tag{5-101}$$

So far we have assumed that the incident wavefront was parallel to the plane of the grating, hence β (see page 400) equalled 0. However, if the incoming radiation is

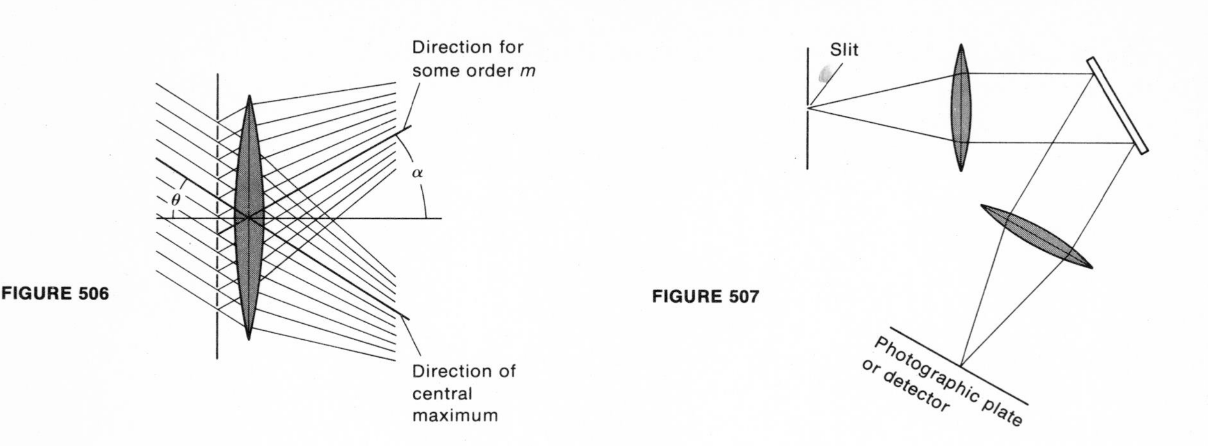

FIGURE 506

FIGURE 507

at an angle θ with the grating (Figure 506), $\beta = (2\pi/\lambda)\Delta = 2\pi(d \sin \theta)/\lambda$. Then $\phi = 2\pi d (\sin \theta + \sin \alpha)/\lambda$ and the principal maxima of the interference pattern occur when $(\sin \theta + \sin \alpha) = m\lambda/d$. For the location of these principal maxima we have

$$\sin \alpha_m = \left(\frac{m\lambda}{d} - \sin \theta\right) \tag{5-102}$$

and

$$y_{\text{prin max}} = D \tan \alpha_m. \tag{5-103}$$

Note that for the central principal maximum, for which $m = 0$, $\sin \alpha = -\sin \theta$. (Note also that α may be large, so that we are denied the use of the approximation $\sin \alpha \simeq \tan \alpha$.)

Diffraction gratings can be transmission gratings, such as shown in Figure 506, or reflection gratings. Reflection gratings may be plane, in which case they are combined with converging lenses as shown in Figure 507 or they made concave with a focal length of their own in which case the lenses can be dispensed with.

PROBLEM 862

The central maximum of the diffraction pattern of a single slit is contained within the limits $y = \pm D\lambda/w$ (see Figure 504). What must be the ratio d/w when the interference pattern of two slits has

a. a minimum,

b. a maximum, coinciding with these limits?

c. How many principal maxima of the interference patterns are contained within the diffraction central maximum in case (a), and in case (b)?

Hint: Put together equation 5-93 and either equation 5-88 or 5-89.

PROBLEM 863

Light from a mercury arc that contains two strong spectral lines at 5769.6 Å and 5790.6 Å is incident normally on a grating that is ruled with 5000 lines cm^{-1}. If at the observing position you want to obtain an angular separation between these two Hg lines of at least $0°10'$ of arc, what spectrum order would you use?

PROBLEM 864

From theoretical considerations you expect to find spectral lines of Neon at 4827.35 Å and 4827.59 Å. How wide a grating of 4900 lines cm^{-1} would you need to resolve these lines in

a. the first order,
b. the fourth order?

PROBLEM 865

If you want a diffraction grating that at normal incidence will suppress all the even orders of the spectrum, what ratio of slit spacing d to slit width w would you specify?

PROBLEM 866

Can you, by any manipulation of the parameters of a diffraction grating, suppress the zero-order spectrum when the incident radiation is

a. normal to the plane of the grating,
b. at an angle θ to the normal? *Hint*: A judicious use of equation 5-102 in connection with equation 5-99 should help.

PROBLEM 867

Light is incident on a transmission grating of 9000 lines cm^{-1} at an angle that is not known. The diffracted light is focussed by a lens of 1.00 m focal length on a screen 2.00 m long that is parallel to the grating; the center of the screen is at the foot of a perpendicular from the center of the grating. Spectral lines are observed on the screen at +12.81 cm and +74.81 cm from its center. Assume that the light was monochromatic.

a. What was its wavelength?
b. What other lines could have been observed on the screen?
c. From the data given can you determine the angle of incident light? If not, what additional information would you need?

PROBLEM 868

A compact form of grating spectrograph is the Littrow mounting shown in Figure 508; parallel light from a slit S in the focal plane of a converging lens is reflected back through the lens to a focus on a photographic plate placed alongside the slit. The angle θ can be adjusted to place any desired order of spectrum in line with the slit; thus by rotating the grating a spectrum of several orders can be swept past the slit location.

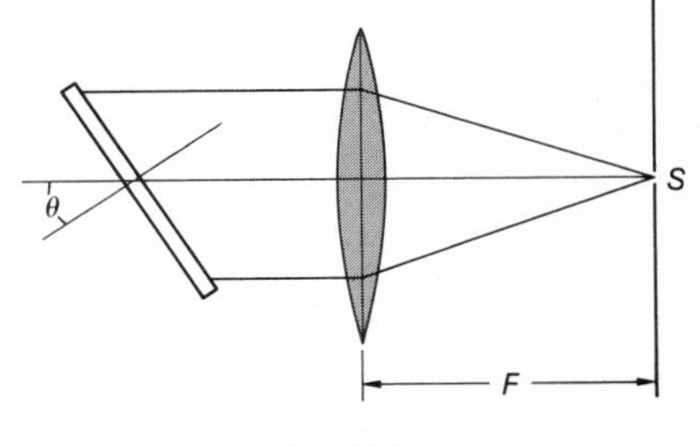

FIGURE 508

a. What should be the angle θ for a particular order m at the slit location, in terms of the other parameters of the grating?

b. What is this angle for the fourth order of the D line of Na at 5890 Å if the grating is ruled with 5200 lines cm^{-1}?
c. In order to suppress other orders of other wavelengths between 3800 Å and 7200 Å that might fall in the same position as the fourth order of the D line, what screening filters should you put in the light incident on the slit?
d. The D line of Na is actually a doublet D_1 at 5895.92Å and D_2 at 5889.95 Å. In the fourth order how far apart are these doublet components on the photographic plate if $F = 1.20$ meters?
e. How much improvement in the separation of the doublet would you gain by going to the fifth order?

SUPPLEMENTARY PROBLEMS

PROBLEM 869

A space vehicle of mass $2m$ and its final booster stage of mass m are together in a circular parking orbit of radius R from the center of the earth. The two are separated by an explosion whose impulse is tangent to the orbit.

a. If the impulse was the minimum possible for the vehicle to escape from the earth, what was the velocity of the final stage immediately after the explosion?
b. What was the semimajor axis of the orbit for the final stage resulting from the explosion?
c. What was the lower limit for the amount of energy that had to be delivered by the explosion?

PROBLEM 870

A solid, uniformly charged sphere, with radius R and total charge Q, splits to form two equal uniformly charged spheres, density remaining unchanged.

a. What is the potential energy of one of the smaller spheres due to its charge alone?
b. What is the potential energy of the two-sphere system immediately after the split?
c. If the original sphere was a U^{238} nucleus with a radius of about 7×10^{-15} m, what was the relative speed of the two smaller spheres immediately after the split?
d. What was the speed of either sphere when they were a very large distance apart?

PROBLEM 871

In designing a camera that is to be equipped with a simple lens of 50 mm focal length, how much axial movement of the lens must you allow for if the lens is to give sharply focussed images for object distances of 1.0 m to ∞?

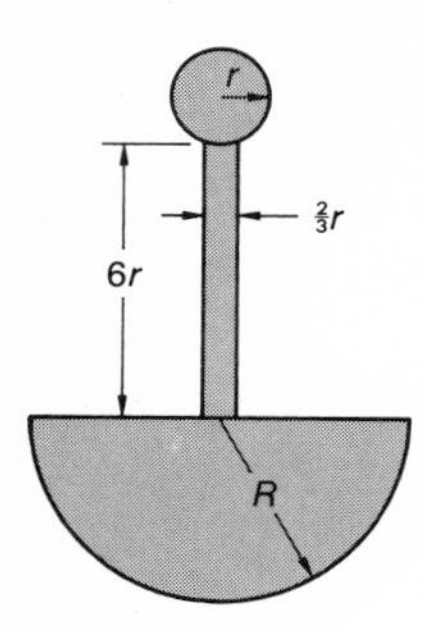

FIGURE 509

PROBLEM 872

The basic structure of a child's toy consists of a solid sphere mounted on a solid cylindrical rod which in turn is mounted on a solid hemisphere (Figure 509). The structure is made of the same uniform material throughout. For the toy to stay in equilibrium when the rod is at any angle above the horizontal, what should R be, in terms of r?

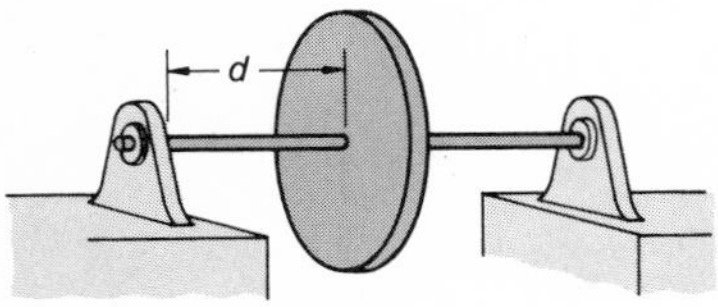

FIGURE 510

PROBLEM 873

A solid cylindrical flywheel, of mass M and radius R, is firmly mounted at the middle of a steel shaft of radius r. The shaft is supported by bearings, as shown in Figure 510. When the flywheel is freely rotating at an angular speed of ω_0, one of the bearings runs out of lubricant and begins to bind, slowing the wheel at a uniform rate. At a time t sec after the binding starts the wheel is rotating with a speed ω_t; at that instant the bearing "seizes" so that the shaft in the bearing cannot move.

a. How much heat was developed in the bearing to cause it to "seize"?

b. What was the twist in the shaft between the flywheel and the faulty bearing during the first t sec?

c. What was the maximum twist in the strained shaft?

Suggestion: To simplify the calculations, make the reasonable assumption that the rotational energy of the shaft is negligible compared to that of the flywheel.

PROBLEM 874

In problem 873, the bearing failure produced a twist in the shaft greater than was considered tolerable. The assembly was therefore scheduled for a redesign that would allow for a lesser twist in the shaft in case of bearing failure. If you were handed the problem of redesign, would you shorten the shaft, increase the diameter of the shaft, or decrease the radius of the flywheel?

PROBLEM 875

An object of mass M_1, moving with a linear speed u_1 on a frictionless horizontal surface, strikes a stationary object of mass M_2. At the point of contact M_2 is covered with a powder that explodes when touched. As a result of the impact plus explosion, M_1 rebounds backward at an angle of 150° with its original path and with a speed v_1.

a. With what velocity did M_2 move after the impact?

b. What was the lower limit for the energy released in the explosion?

PROBLEM 876

A small ball carrying a charge $+q$ hangs from a suspension point by an insulating thread that is d long (Figure 511). When a large plane sheet uniformly charged with $+\sigma$ C m^{-2} is held vertically a distance d from the suspension point, the thread swings away from the sheet at an angle θ with the vertical. Then the charged sheet is removed and a large plane grounded conducting sheet is held vertically a distance d from the suspension point; the thread swings toward the sheet at an angle θ with the vertical. What was the value of q?

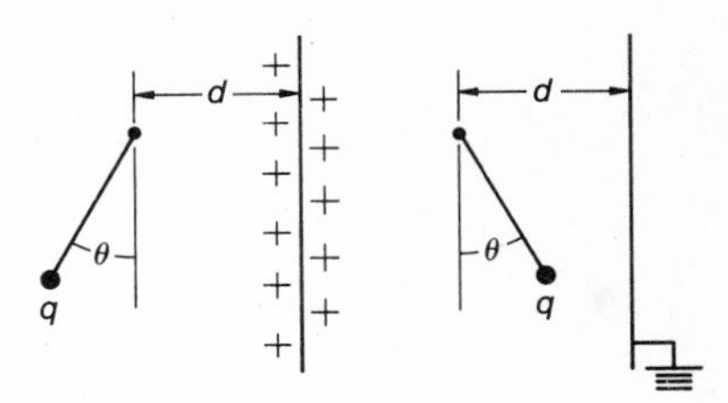

FIGURE 511

PROBLEM 877

A mass m is hung by a cord from a fixed support. A mass 2 m is hung from mass m by an ideal spring of unstretched length L and spring constant k (Figure 512). At $t=0$ the cord is burned. What is the subsequent equation of motion for the 2 m mass, i.e., what is y_{2m} as a function of t, with $y=0$ as shown? Assume dimensions of the masses negligible compared to L.

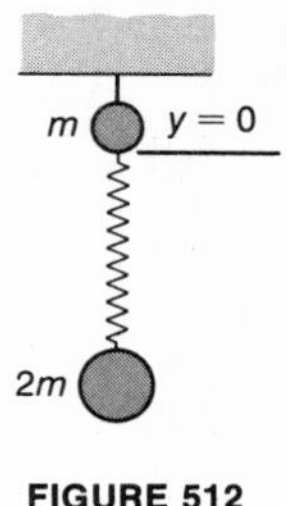

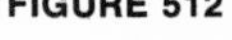
FIGURE 512

PROBLEM 878

A satellite is in circular orbit about the earth at an altitude of 290 km. The ratio m/A of the satellite's mass to its cross-sectional area equals 50. It is observed that the period of the satellite decreases at the rate of 1.0 sec every 100 revolutions about the earth. If you assume that this decrease is due entirely to inelastic impact with the atoms in the atmosphere, what must be the density of the atmosphere at an altitude of 290 km? *Warning:* There is a considerable amount of work to do in this problem in establishing relationships between ΔT and ΔE, and then between ΔE and ρ. Unless you heed the warning given on page 5 and do all your work in algebraic terms, you can get hopelessly involved in a vast mess of numbers. Insert numerical data only at the very end, after you have established a direct relationship between ρ and ΔT.

PROBLEM 879

A mass M of uniform rod is bent into four adjacent semicircles all lying in the same plane (Figure 513). What is the moment of inertia of the bent rod about an axis through one end A and perpendicular to the plane of the rod? *Hint:* A semicircle is half of a full circle.

FIGURE 513

PROBLEM 880

A partly filled bottle with straight sides and cross-sectional area A, fitted with a small pouring tube, sits upright with the tube uncapped. When a finger is held over the tube and the bottle inverted, the liquid surface stands half-way down in the bottle, as shown in Figure 514. A short time after the finger is removed and liquid starts pouring out the flow stops, and the bottle "breathes" with the intake of a bubble of air. If the density of the liquid is such that $\rho g L = B_0/40$, what is the lower limit for the amount of liquid dispensed before the first

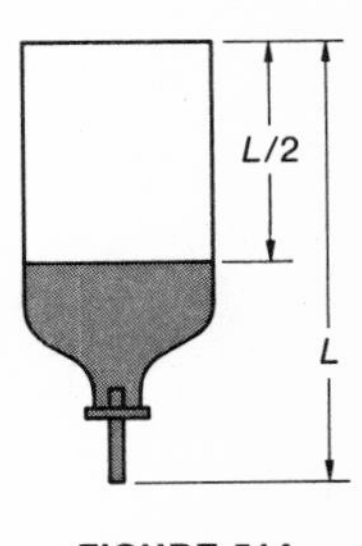

FIGURE 514

"breath"? Would this arrangement form the basis for a useful patent for an automatic pouring spout?

PROBLEM 881

Two gases of different molecular weights MW_1 and MW_2 are in thermal equilibrium at a temperature T in an enclosure. When a small hole of area A develops in the wall of the enclosure, what is the ratio of the rates of escape of the two gases? *Hint:* Appendix H may help you. This problem is related to the method of gaseous separation by thermal diffusion.

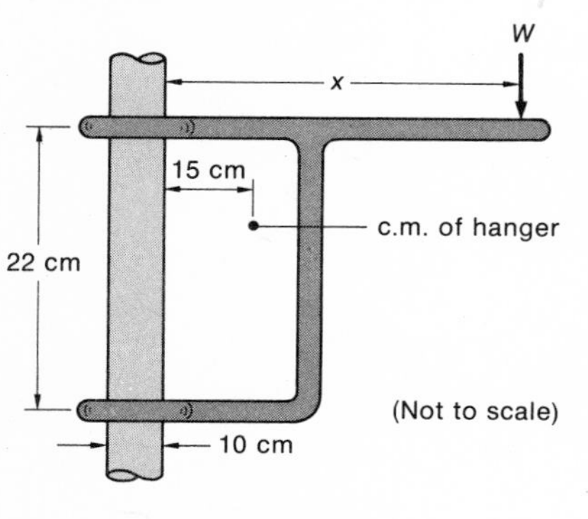

FIGURE 515

PROBLEM 882

Supports that can be moved by sliding up and down vertical posts are useful in many applications. Such a support is shown in Figure 515, with pertinent dimensions. If the coefficient of static friction between post and support is 0.30, and if a load W that is 50 times the weight of the hanger is placed on the hanger, how far out on the hanger must it be if no slippage is to occur?

PROBLEM 883

An electric cable consists of a copper wire 0.10 cm in diameter surrounded by a cylinder of plastic insulation 1.0 cm in external diameter. The coefficient of thermal conductivity of the plastic is 2.1×10^{-3} watts $\text{cm}^{-1}(°\text{C})^{-1}$ and is relatively independent of temperature over normal ranges. The plastic starts to decompose at 200°C. If the outer surface of the insulation is kept at 20°C,

a. what is the maximum steady rate of I^2R heat loss per cm length along the wire that can be allowed without failure of the insulation?
b. What maximum current can be carried by the wire if its resistance in the 20°C–200°C range is $2.1 \times 10^{-2}\ \Omega\ \text{m}^{-1}$?

PROBLEM 884

If you decide that the relativistic range has been reached when the kinetic energy exceeds $10^{-4}\ mc^2$ for any particle,

a. what is the temperature at which a plasma of protons must be treated relativistically?
b. At a relativistic temperature such as found in question (a) it would not be reasonable to expect that the normal gas law $PV = NRT$ would apply. Develop a relativistic counterpart to this gas law, in terms of $\beta = v/c$ and E, the total energy per molecule. *Hint:* Carry through the development exactly as was done in deriving $PV = NRT$, but remember to define p relativistically by $p = \gamma mv = \gamma mc\beta$.

PROBLEM 885

You can shift a linear beam of monochromatic light in air parallel to itself by interposing in the beam a parallel-sided slab of transparent material whose index of refraction is n and whose thickness is d

(Figure 516). What is Δ, the distance the beam is shifted, in terms of n, d, and the angle θ between the beam and the normal to the side of the slab?

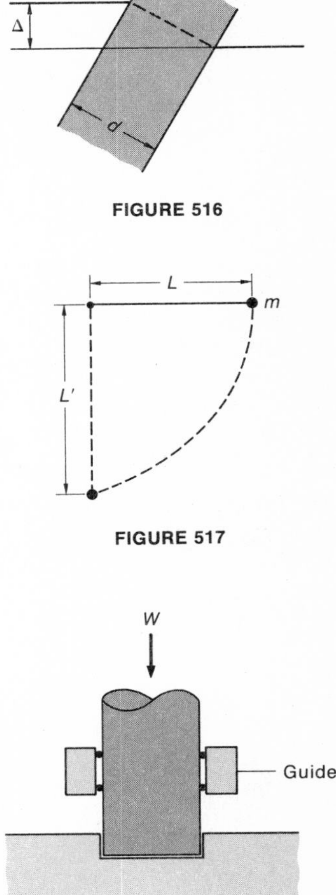

FIGURE 516

FIGURE 517

PROBLEM 886
A small ball of mass m is suspended from a fixed support by a rubber cord of unstretched length L and elastic constant k (Figure 517). The mass of the cord is negligible compared to m. When the ball is released from a point level with the support and L from it, what is L', the vertical distance from the support to the bottom of the swing?

PROBLEM 887
A load W is transmitted axially to a thrust bearing by a round shaft with a flat end face (Figure 518). If the coefficient of sliding friction between shaft end and bearing surface is μ,

a. what is the relation between the frictional torque at the bearing and the radius of the shaft?
b. If you had a choice of sizes for the shaft, would you choose a large shaft so as to reduce the pressure at the bearing, or would you choose the smallest shaft that would carry the load?
c. If the radius of the shaft is R and it rotates with an angular speed ω, what power is being lost in the bearing?

FIGURE 518

PROBLEM 888
With switches S_2 and S_3 open, S_1 is closed and C_1 is fully charged (Figure 519). Then after S_1 is opened S_2 is closed. Then S_3 is closed.

a. How much energy was lost in the spark that developed across S_2 as it was closing?
b. How much energy was lost in the spark that developed across S_3 as it was closing?

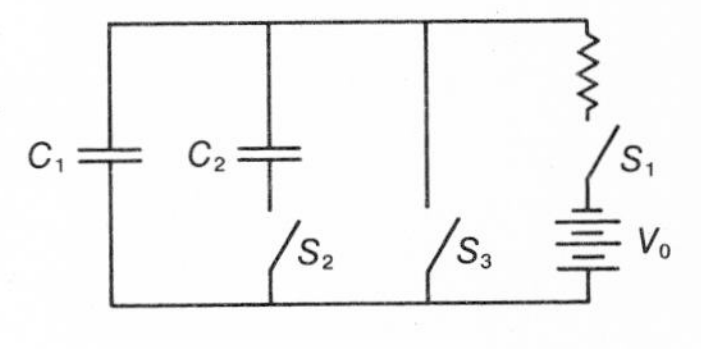

FIGURE 519

PROBLEM 889
In "western" films seen on television it is common to see an actor who has just been shot fall backward as if he had been hit a very heavy blow. Is this just exaggerated histrionics, or is it realistic?

Assume that a .45 calibre bullet weighs 0.033 lbwt and is moving at a speed of 800 ft sec^{-1} when it hits and remains embedded in a dummy weighing 180 lbwt. What is the approximate force of the blow felt by the dummy?

PROBLEM 890
Slow positrons are short-lived; they cannot exist for any length of time without coming in contact with one of the slow electrons that are so abundant in nature. The result of such contact is pair annihilation, in which both electron and positron disappear. What could be the products of this annihilation?

PROBLEM 891

What is the maximum possible wavelength for photons produced by the electron-positron annihilation discussed in problem 890?

PROBLEM 892

An inflated free passenger balloon, weighing 600 lbwt with basket and equipment, and carrying a sand bin holding 200 lb of sand, is on display at the Kansas State Fair. It is kept from rising by a vertical anchor rope under 160 lbwt tension. A boy weighing 140 lbwt climbs into the basket to inspect it at close range, and accidentally releases the anchor rope. Misunderstanding instructions shouted to him as the balloon starts to rise, he opens the sand release valve, allowing sand to leave the basket at a rate of 20 lb sec^{-1}. How fast was the balloon rising as the sand bin emptied?

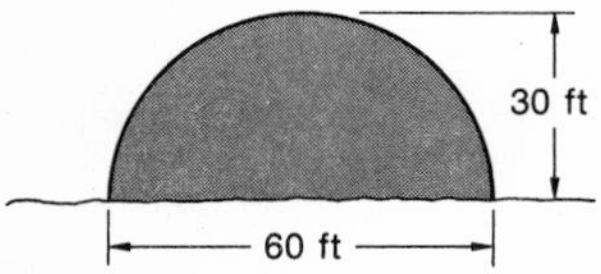

30 ft

60 ft

FIGURE 520

PROBLEM 893

A workman inspecting a vent at the top of the hemispherical dome of a nuclear energy generating plant (Figure 520) loses his hold and starts sliding. His work clothes are so oil-stained that he slides essentially without friction. Where on the ground does he land?

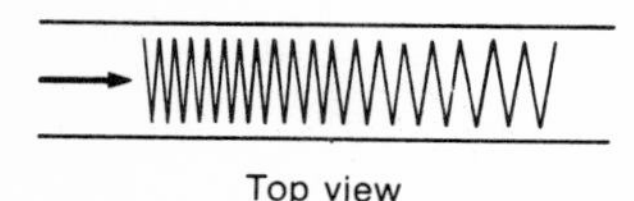

Top view

FIGURE 521

PROBLEM 894

A uniform spring of mass M, unstretched length L, and force constant k rests in a smooth, horizontal channel as shown in Figure 521. A horizontal force in line with the spring's axis is gradually applied at one end until the spring acquires a horizontal acceleration a along its entire length.

a. What is the length of the accelerated spring?

b. How much energy, not including its kinetic energy of motion, is stored in the accelerated spring? *Caution:* The strain is not uniform along the accelerated spring.

PROBLEM 895

A one-eyed observer stands with his eye 54 cm above a plane mirror lying horizontally on a table. He holds a transparent centimeter scale horizontally at such a height that when he looks vertically downward he sees 2.0 cm in the image of the scale in the mirror for every 1.0 cm of the scale he sees directly. After placing a 15 cm thick block of plastic on the mirror he repeats the experiment by again holding the scale so that he sees 2.0 cm in the mirror coinciding with 1.0 cm on the scale.

a. At what height did he hold the scale during the first observation?

b. If the scale was 22 cm above the mirror during the second observation, what was the index of refraction of the plastic?

PROBLEM 896

A helicopter flying a horizontal path at a ground speed v has mounted on a tip of the main rotor a smoke generator with which it makes smoke patterns in the air. There is no wind. The smoke generator is R distant from the axis of the rotor. When the rotor is revolving at a certain angular speed ω_c, the smoke pattern as seen from the ground is shown below:

a. What is ω_c?

At different rotational speeds of the rotor, two other patterns can be observed from the ground:

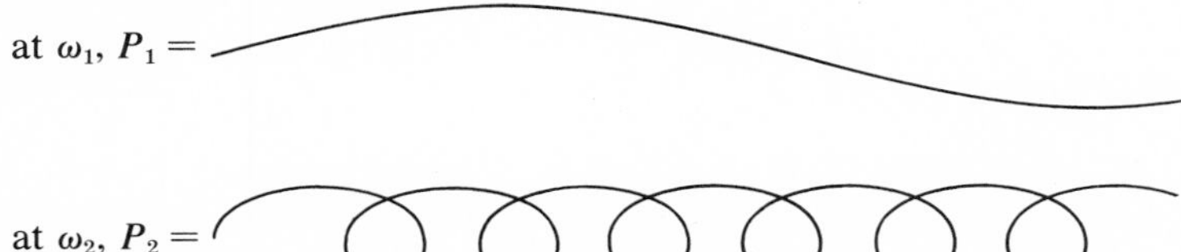

b. What can you say about ω_1?
c. What can you say about ω_2?
d. From the point of view of an observer at rest on the ground, what would be the parametric equations for the pattern P_1? (Disregard any change in v due to a change in ω, i.e., assume v remains constant throughout the problem.)

PROBLEM 897

Two spherical bodies, one of mass m_1 and radius r_1 and the other of mass m_2 and radius r_2, attract each other gravitationally. They are released from rest at a very great distance apart. What is their relative speed at the instant of first collision?

PROBLEM 898

A copper rod of negligible electrical resistance is bent into a long narrow plane U as shown in Figure 522. In a region where there is a uniform vertical magnetic field B the plane of the U is set at an angle θ with the horizontal, and a metal rod of length $\frac{3}{2}d$ is placed crosswise on the U. The cross rod has a linear mass density λ and an electrical resistance of R per unit length.

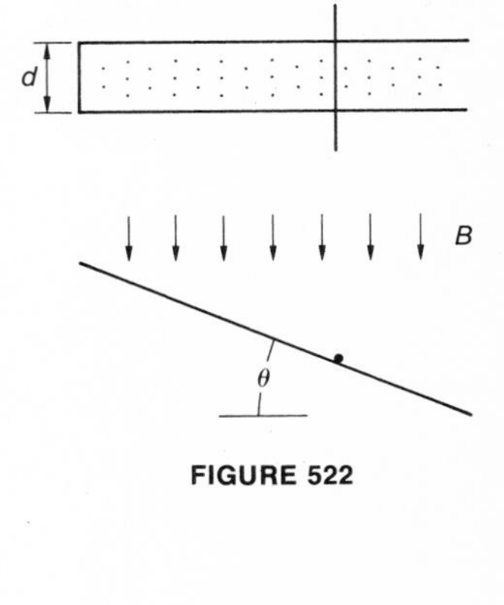

FIGURE 522

a. When the rod is released to roll down the U, what terminal speed does it attain?
b. At terminal speed, what power is being expended?

PROBLEM 899

A cylinder of cross-sectional area 1.0×10^2 cm^2 and closed at one end is fitted with a tight but freely sliding piston (Figure 523). The space between the piston and the closed end is filled with 1.0 liter of an ideal

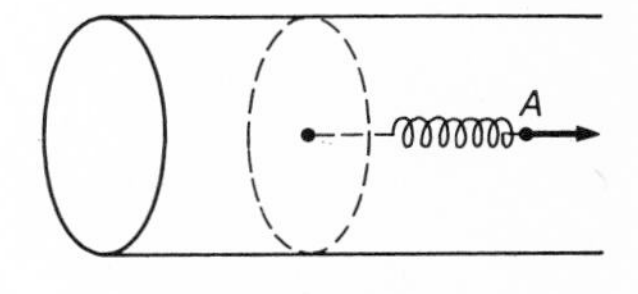

FIGURE 523

gas in equilibrium with STP conditions outside. An axial force applied at A is communicated to the piston through a spring whose force constant is 5.1×10^7 dynes cm^{-1}. The force is slowly increased from zero (temperature remaining constant) until the volume of the gas has doubled. How far has A moved?

PROBLEM 900

A mass m is connected to a linear spring of spring constant k; the mass is also subjected to a frictional force $\gamma m v$ resisting motion. The mass is released from rest when the spring is extended a distance A beyond its unstrained length; exactly 10 cycles later the amplitude of m's motion has decreased to A/e.

a. Assuming that γ remained constant, what was its value?
b. What was the average loss of energy per cycle?
c. How does this average value compare to the energy lost during the first cycle?

PROBLEM 901

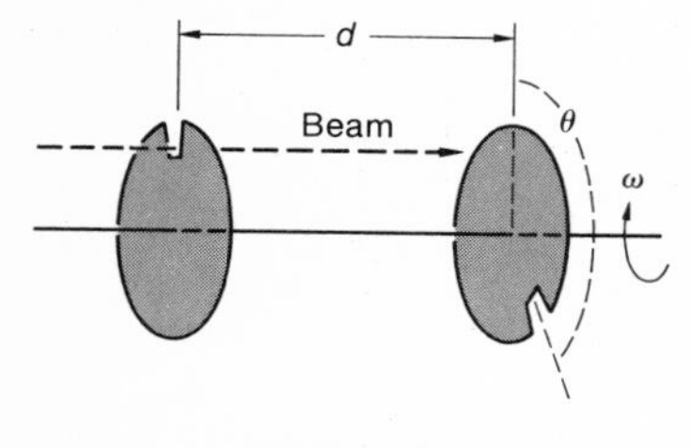

FIGURE 524

Another form of velocity selector (cf. problem 79), used in both molecular and particle experiments, consists essentially of two slotted discs mounted on a common axis at a distance d apart. The slots are displaced relative to each other by an angle θ, as shown in Figure 524; the axis is driven at an angular speed ω. Particles with all speeds in a collimated beam will get through the slot in the first disc for a short time interval; to get through the slot in the second disc as well, a particle must travel the distance d in just the time it takes for the second slot to line up with the beam. (Note that for this line-up the second slot could have moved θ, $\theta + 2\pi$, $\theta + 4\pi$, etc.).

a. If $d = 1.0$ m, $\omega = 24000$ rpm, and $\theta = 60°$, what speeds will get through the velocity selector?
b. Because of the finite width of the slots, the selected speeds are not sharply defined. If the slots have an angular width at the beam of 2°, what is the spread of values for the highest speed found in question (a)?

PROBLEM 902

The speed of sound in a confined gas is measured and found to be 0.632 times the $v_{\rm rms}$ of the gas molecules as determined from the temperature.

a. What must be the value of c_v for the gas?
b. How many degrees of freedom of the gas are activated?
c. By considering the effect of a small change in the measurement of the speed of sound—say from 0.632 $v_{\rm rms}$ to 0.640 $v_{\rm rms}$—comment on whether a speed of sound measurement represents a good experimental method for determining c_v.

PROBLEM 903

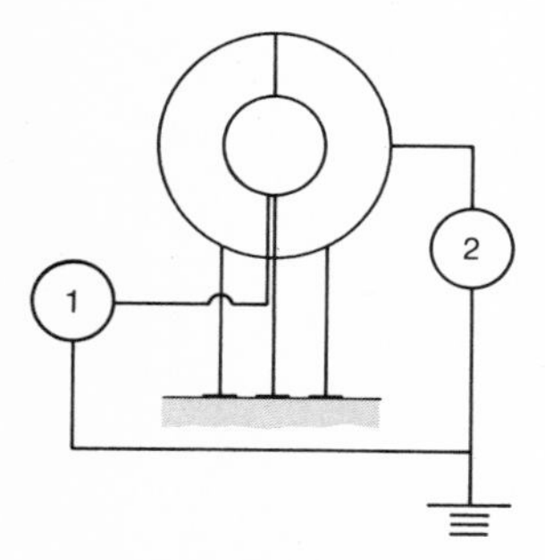

FIGURE 525

A metal sphere of radius r on an insulated stand is connected to electrostatic voltmeter No. 1 as shown in Figure 525. The sphere is charged so that the voltmeter reads V_0, and the charging source is disconnected. Then two metal hemispherical thin shells of radius R, $R > r$, are brought up on insulated stands to surround the charged sphere concentrically. The outside of this shell is connected to electrostatic voltmeter No. 2 as shown in Figure 525.

a. What are the readings of voltmeter No. 1 and voltmeter No. 2? The two metal hemispherical shells are then removed temporarily so that two solid hemispherical shells, inner radius r and outer radius R and made of material of dielectric constant K, can be fitted around the inner sphere (care being taken to avoid discharging the inner sphere); the outer metal shells are then placed back around the dielectric (Figure 526).
b. What are now the readings of voltmeter No. 1 and voltmeter No. 2?

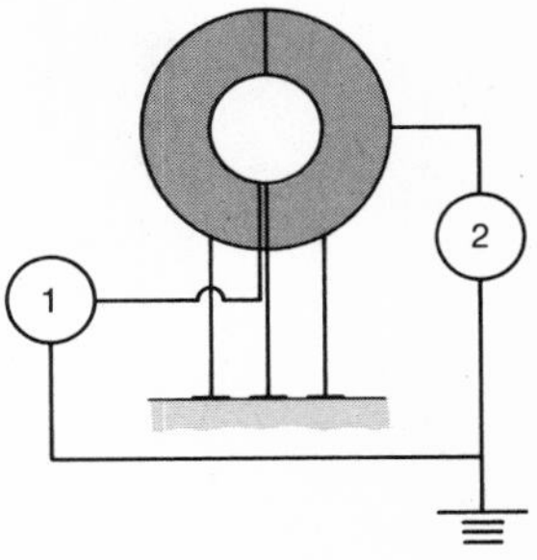

FIGURE 526

PROBLEM 904
During a lull in a road machinery demonstration the drivers of two rival scoop loaders engage in a race. The scoop of one of the loaders runs from tangent to the ground up in a 45° circular arc of radius R (Figure 527). The scoop is new and made of smoothly polished metal. During the race this scoop picks up an empty coke bottle lying on the ground. The bottle flies up into the air and lands in the lap of the driver, which is on a level with the upper lip of the scoop and exactly R behind it. How fast was the loader being driven?

FIGURE 527

PROBLEM 905
A number n of identical balls, each of mass m and radius r, are strung like beads at random and at rest along a smooth, rigid horizontal rod of length L mounted between immovable supports (Figure 528); r/L is small but not negligible. Collisions between balls, or between balls and supports, are perfectly elastic. If one of the balls is struck horizontally so as to acquire a speed v, what is the resulting average outward force felt by the supports?

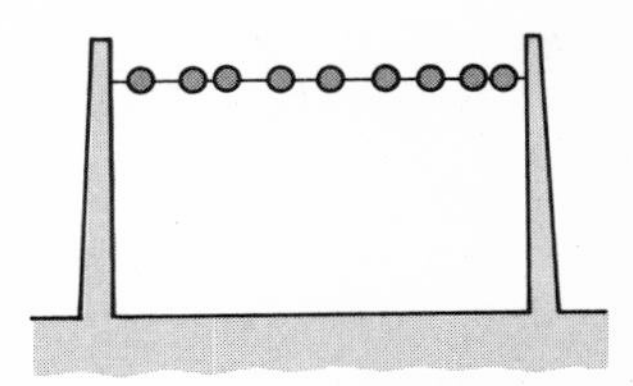

FIGURE 528

PROBLEM 906
A passenger sitting next to the window on the north side of an eastbound jet plane, which is flying on a level course 6.0 mi above ground, looks down as the plane is passing over the flat Kansas plains, and notes the regular grid of north-south and east-west roads 1.0 mi apart.* He also notes that his plane and its shadow are simultaneously in line with the same north-south road, and that the plane's shadow crosses a north-south road every 6.0 sec. He then observes a small plane at a lower altitude; when its shadow is 3.0 mi east of his plane's shadow the small plane itself seems in line with the north-south road 1.0 mi east of its shadow. Twenty-four sec later the small plane does not seem to have moved relative to the ground, but it and its shadow are now in a north-south line with the jet and its shadow. What was the small plane's course and altitude?

* In the early settlement of the central plains by homesteaders, the roads were run along section lines wherever possible.

PROBLEM 907

A hydrogen atom in its ground state ($n = 1$, see problem 625) is excited by the absorption of 12.7 ev of energy. What spectral lines could be produced during the atom's return to the ground state?

PROBLEM 908

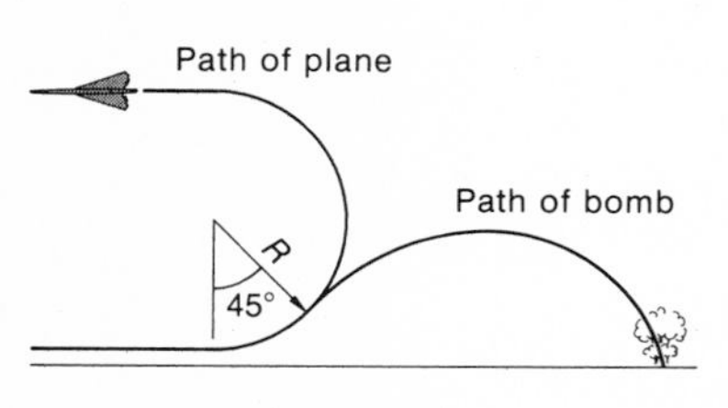

FIGURE 529

For delivery of bombs with nuclear warheads a method of "toss bombing" has been developed: a plane on a bombing run, flying at tree-top level to avoid radar detection, starts into a vertical, circular loop and releases a bomb at the instant its course is 45° above the horizontal (Figure 529). After bomb release, the plane continues in the loop until it is again flying horizontally directly away from the bombing target. If the plane's constant air speed was 600 mi hr^{-1} and the radius of the circular loop was such as to produce a constant centripetal acceleration of 2 g, how far horizontally was the plane from the impact point when the bomb arrived there? (For simplicity, assume level ground and no wind.)

PROBLEM 909

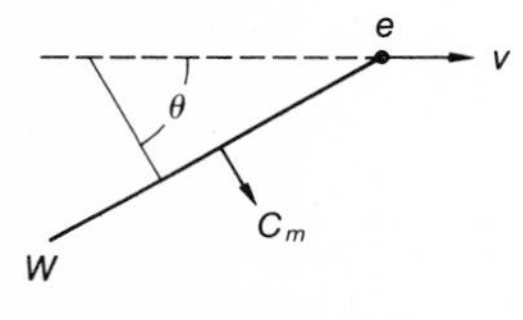

FIGURE 530

An electron moving through a medium at a speed v greater than the speed of light c_m in that medium creates a wavefront of electromagnetic radiation (called Cerenkov radiation) that propagates through the medium in the same manner as a bow wave propagates from the bow of a boat moving through water (Figure 530). A detector placed at the appropriate angle θ with the direction of the motion of the electron can observe this wavefront and thus provide data for the measurement of the speed of the electron. If such radiation is observed at 35° with the known direction of a fast electron moving through a liquid whose index of refraction is 1.36 ($c_m = c/1.36$), what must have been the kinetic energy of the electron?

PROBLEM 910

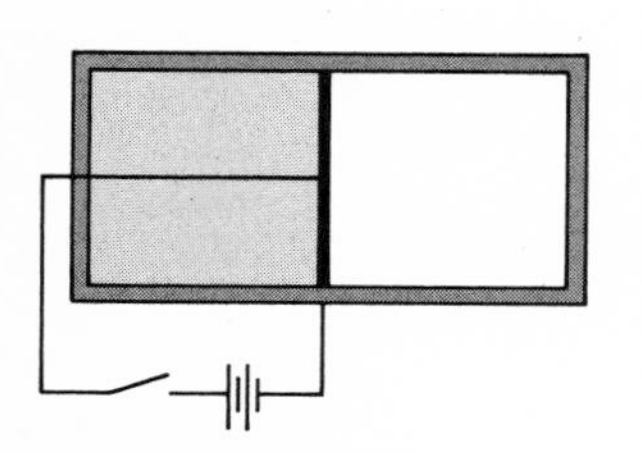

FIGURE 531

A closed, insulated cylinder of negligible heat capacity is fitted with a freely sliding piston of mass m, also of negligible heat capacity (Figure 531). The volume on one side of the piston is evacuated; the volume on the other side is filled with N moles of an ideal diatomic gas. The piston is held at the center of the cylinder by a restraining wire; by means of an external circuit the wire can be used to heat the gas. When the temperature of the gas reaches T the wire breaks. As the piston hits the end of the cylinder,

a. what is the temperature of the gas;
b. what is the impact speed of the piston?
c. After a long time, what is the final temperature of the gas?

PROBLEM 911

If a projector mounted at the back of a lecture hall is to project a 35 mm slide so as to fill a 2.0 m screen at the front of the hall 11 m from the projector, what focal length is needed for the lens of the projector?

PROBLEM 912

The ascent of a space vehicle at a Cape Kennedy launch could be described, for the first 400 seconds, by $y = 2.1 \times 10^{-3}\, t^2 - 3.8 \times 10^{-6}\, t^3$, where y is the altitude above the launch platform in nautical miles and t is seconds after lift-off.

a. When did the maximum vertical speed occur?
b. What was this maximum vertical speed?
c. What was the average vertical speed during the period from $t = 100$ sec to $t = 300$ sec?
d. What was it from lift-off to $t = 400$ sec?

PROBLEM 913

The hydroelectric power house at the McNary Dam on the Columbia River is designed for a minimum river flow of 8.0×10^4 ft^3 sec^{-1} and a minimum head of water above the turbine outlets of 51 ft. If the exit speed of the water at these outlets is 8.0 ft sec^{-1} and the overall efficiency of the generator-turbine system is 88 percent, what is the minimum power output at McNary Dam?

PROBLEM 914

A star under continuous observation displays the following changes in its spectrogram: At $t = 0$, its spectrogram shows a sharp line spectrum similar to that of a laboratory comparison spectrum. At $t = 40$ days the lines have doubled, the doublet components moving to either side of the normal line position and with one component brighter than the other. At $t = 80$ days the lines are again sharp and single. At $t = 120$ days the lines are again doublets, but this time the fainter and brighter components have exchanged positions in the spectrum. At $t = 160$ days the spectrum is the same as at $t = 0$. This pattern repeats every 160 days. For the 4860 Å line of hydrogen, the maximum doublet separation was 8.0 Å.

a. What was the mass of the observed star system?
b. What can you suggest in explanation of the exchange of the doublet positions every 80 days?

PROBLEM 915

If the maximum allowable tensile stress parallel to the grain for structural grade Douglas Fir lumber is 1650 lbwt in^{-2}, what should be the depth of a rectangular Douglas Fir beam 4 inches wide to be used to support the sun deck of the beach house of Problem 382? (Consider the weight of the beam itself to be included in the floor load given in that problem.)

PROBLEM 916

A pile driver consists essentially of a weight to be dropped on to the upper end of a pile, and a mechanism for lifting and releasing the weight. In driving piles for a dock, a weight of 1500 lbwt is dropped from a height of 4.0 ft onto the top of a pile weighing 1.0 tonwt. The falling weight is observed to stay on top of the pile after impact, and the pile is observed to sink 15 inches. What was the average force resisting penetration of the pile?

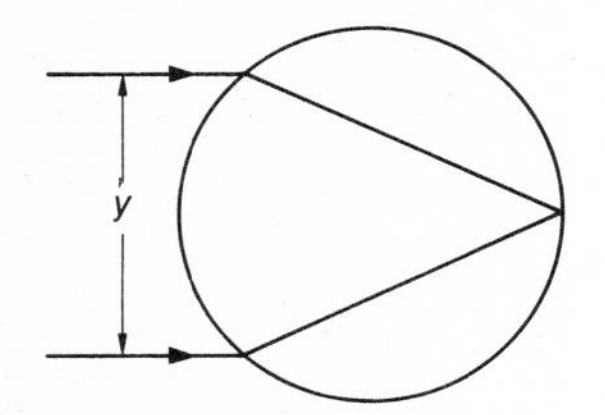

FIGURE 532

PROBLEM 917

In the calibration of a photometric instrument it is necessary to determine that two parallel light rays are exactly y apart. This is done by letting both rays fall normally and symmetrically on a glass rod of radius R and index of refraction 1.60, and then adjusting the separation of the rays until they come exactly to a focus at the opposite circumference of the rod (Figure 532). In terms of R, what was y?

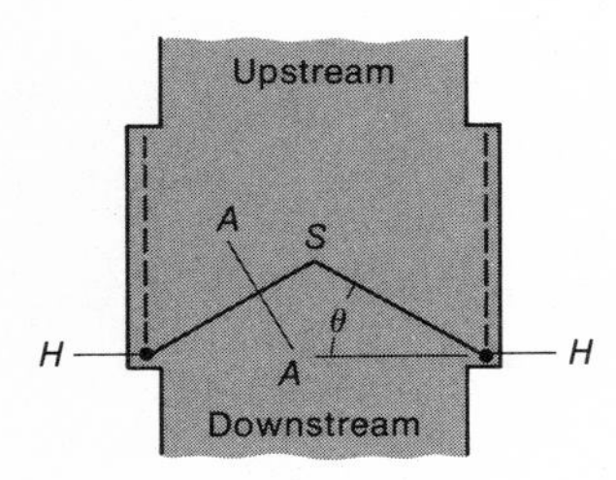

FIGURE 533

PROBLEM 918

A standard form of lock gate used at large canal locks (the Panama Canal, for example) is the so-called mitre gate, as shown in Figure 533. Each of the two leaves of the gate is W wide, is hinged at H, and the vertical edges of the leaves meet in a seal at S. When the upstream lock is full of water (Figure 534),

a. what is the force on the seal?
b. What is the hinge reaction, both in magnitude and direction?

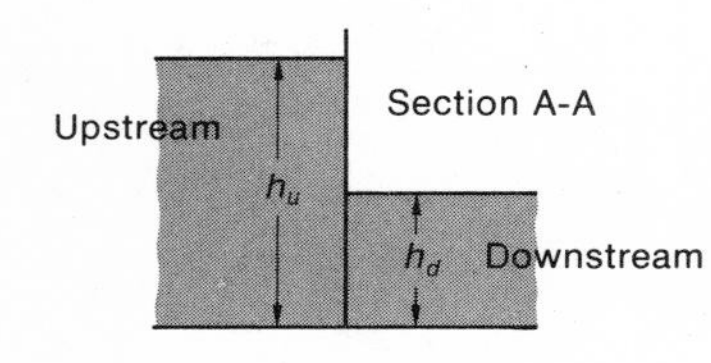

FIGURE 534

PROBLEM 919

The horizontal motion of a particular point on a reciprocating lever is shown in the x-t diagram in Figure 535. Sketch the corresponding $\dot{x}$-t and $\ddot{x}$-t diagrams.

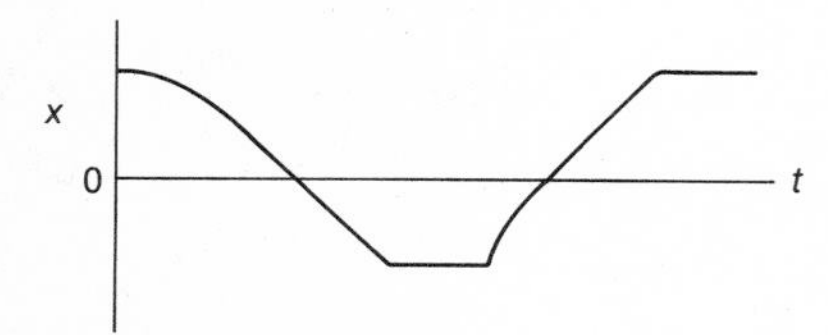

FIGURE 535

PROBLEM 920

A soft rubber tube lying limp and empty on a horizontal surface has an unstretched diameter of 8.0 cm and a wall thickness of 0.10 cm. The stretch modulus for the rubber $Y = 1.0 \times 10^7$ dynes cm^{-2}. When the atmospheric pressure is 1.0×10^6 dynes cm^{-2},

a. how many moles of an ideal gas at 30°C can be pumped into the tube per unit length to fill it out to a cylindrical shape but without stretching the rubber?
b. What will be the diameter of the tube if a total of 0.02 moles of an ideal gas at 30°C are pumped into the tube per unit length?

PROBLEM 921

A large spherical vase, filled with water of index of refraction 1.33 and whose glass walls have relatively negligible thickness, is resting on a window sill on a clear, sunny day. The radius of the sphere of the vase is 12 cm. How far from the center of the vase would it be inadvisable to leave anything that is highly flammable?

PROBLEM 922

An object projected straight up an incline set at angle θ with the horizontal comes to a stop and then slides back down the incline.

a. If the down trip takes exactly twice as long as the up trip, what is the coefficient of sliding friction between the object and the surface of the incline?
b. What percentage of the original energy of projection was lost to friction during the round trip?

PROBLEM 923

When cartons of merchandise are unloaded from trucks at a warehouse or supermarket, a roller track conveyor is often used. Such a conveyor consists of a strong steel framework supporting a series of parallel rollers mounted on ball-bearing axles (Figure 536). The rollers are spaced 6.0 inches apart, are 3.0 inches in diameter, and each has a moment of inertia about its axis of symmetry of 0.20 lb ft^2. If the rollers are motionless when a 30-lb carton starts down the conveyor, and the conveyor is set at an angle of 15° with the horizontal, what is the speed of the carton down the conveyor? Assume no slipping between carton and rollers.

FIGURE 536

PROBLEM 924

A uniform rectangular plate, $a \times 2a$ and of mass M, is suspended at P from a frictionless, horizontal axis that is perpendicular to the plate (Figure 537). From what other points on the plate could the plate be suspended from a horizontal axis so as to swing in its own plane with the same period as from P?

FIGURE 537

PROBLEM 925

A barrel filled with liquid stands on end. A frictional force F is required to hold a cork in the bung-hole against the internal pressure (Figure 538). If the cork blows out, what force is necessary to push the cork back into the bung-hole if this is attempted immediately after outflow starts?

In your solution you may assume that the barrel is open at the top to the ambient pressure, or you may assume that the barrel is closed and the liquid is under pressure greater than the ambient pressure. How does the difference between these two assumptions affect your answer?

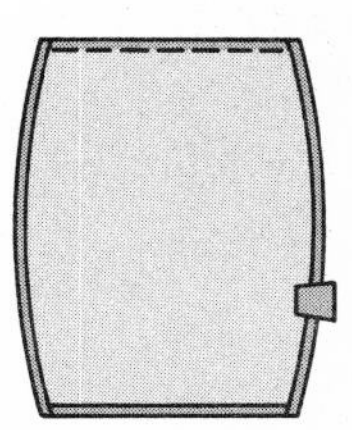

FIGURE 538

PROBLEM 926

In a demonstration experiment a ball is shot from A with an initial speed v at an angle θ with an inclined plane AB (Figure 539). At the same instant, a cup is released from A and slides frictionlessly down the plane, which is at an angle α with the horizontal. If θ is chosen

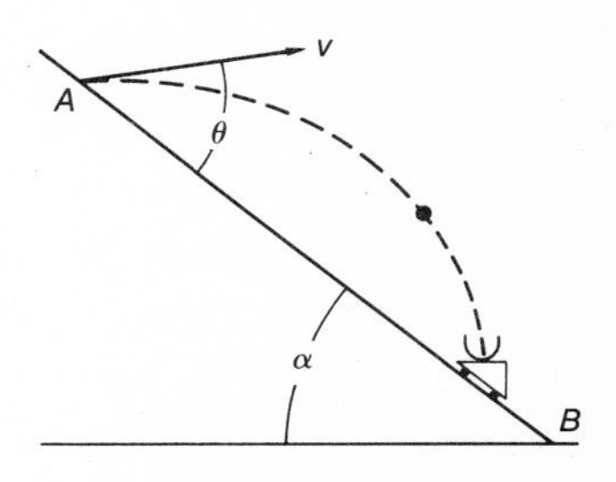

FIGURE 539

properly, the ball will always land in the cup regardless of the value of v or of α, providing $\alpha < 90°$. What is the proper value of θ? *Suggestion:* A rectangular coordinate system chosen with its axes parallel to and perpendicular to the inclined plane should make the correct answer obvious.

PROBLEM 927

The inner walls of a closed and dark container are completely lined with perfectly reflecting surfaces. A small window in the container wall is opened to admit monochromatic radiation of wavelength 5890 Å; when the average photon density in the container reaches 10^{13} photons m^{-3} (about the intensity of normal daylight) the window is closed.

a. What is the pressure of the photon gas in the container?
b. If the volume of the container is now reduced by a factor of two, what is the new wavelength of light within the container? (Disregard any low-energy radiation that might have been in the container originally.)

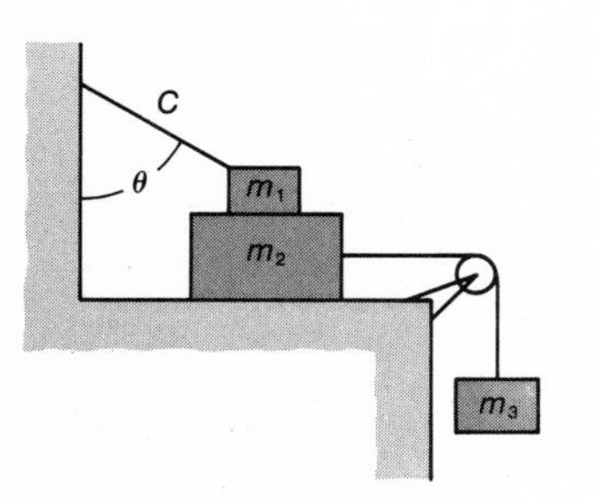

FIGURE 540

PROBLEM 928

There is a coefficient of friction μ between m_2 and m_1 and between m_2 and the horizontal surface it rests on (Figure 540). The pulley is of negligible mass and frictionless.

a. What must be the minimum value of m_3 for m_2 to move?
b. If m_3 has at least this value, what is its acceleration.
c. What is the tension in the cord C?

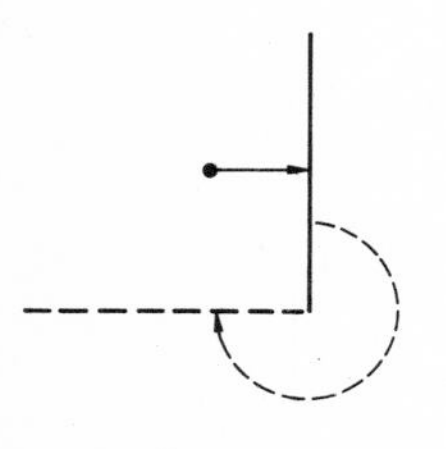

FIGURE 541

PROBLEM 929

A uniform straight stiff bar of mass $6m$ and length L is attached to a horizontal axle at one end (Figure 541). When the bar is standing vertically at rest above the axle it is struck at its midpoint by a solid ball of mass m moving horizontally with a speed u; the collision is perfectly elastic. After the impact the bar swings through 270°, comes to rest horizontally *and stays there.* On the assumption that the frictional torque of the axle bearing is constant,

a. what is the necessary minimum speed u of the ball?
b. When the ball had this minimum speed before impact, what was its speed after impact?

PROBLEM 930

X-rays are produced when electrons emitted essentially at rest by a cathode are accelerated by an electric potential difference V between cathode and metal target. The energy of the electrons at the target may be absorbed by the molecules of the target material, thus raising the target temperature; or it may cause ejection of lower energy electrons from the target; or it may cause the production of one or more (usually more) high-energy photons called X-rays or gamma rays, this result being the most probable for high-energy electrons. If the X-rays ejected per second are counted according to their frequency (or wavelength) and the number per second per frequency is plotted against frequency, the resulting curve can be interpreted as showing the distribution of photon energies as a function of frequency in much the same way that the Maxwell distribution curve depicted the distribution of speeds (and therefore of kinetic energies) as a function of molecular speeds. However, there is one significant difference between these two curves: the photon distribution curve drops sharply to zero number of photons at a cut-off frequency ν_c, whereas the Maxwell curve shows a continuous distribution of molecules out to $\nu \rightarrow \infty$.

a. What important deduction can you make to explain this difference between the two curves?
b. If a particular photon distribution curve shows a cut-off frequency $\nu_c = 4.82 \times 10^{18}$ sec^{-1} for an accelerating voltage $V = 20{,}000$ volts, what must be the value of Planck's constant h?

PROBLEM 931
A smooth tether ball, of radius 10 cm and weight 600 gmwt, hangs from a pole in a school playground (Figure 542). What is the speed of a horizontal wind blowing across the playground when the tether is at an angle of 39° with the vertical?

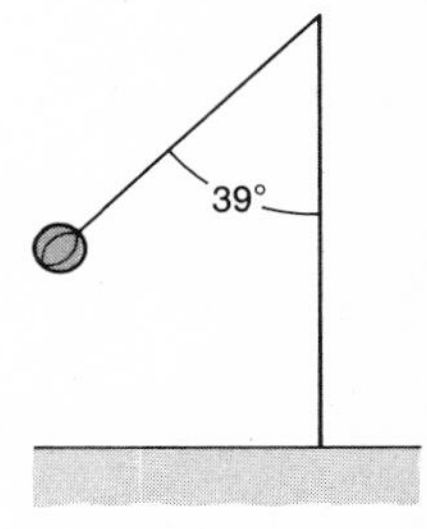

FIGURE 542

PROBLEM 932
The tide is flowing out of San Francisco Bay at 6.0 knots due west. A motor boat is moving into the bay from the Pacific Ocean. A compass on board shows its true heading to be 30° south of east; the Kenyon meter shows the speed of the boat in the water to be 12 knots. The helmsman notices that the smoke from a fire on nearby Alcatraz Island is blowing due east; at the same time he notices that a small pennant on his boat is being blown due north.

a. What is the course being made good, i.e., the velocity of the boat relative to the ground?
b. What is the true wind velocity?
c. What is the wind speed as measured by an anemometer on the boat?

PROBLEM 933
A jai alai court is 176 ft long between vertical end walls 40 ft high. The balls used are so made that their coefficient of restitution at the end walls, which are faced with polished granite, is essentially 1.0. A shot a player would like to make, because of the difficulty for the opposing player in judging its rebound, is off the front wall and back to where the floor and rear wall meet, as shown in Figure 543. If the shot also reaches maximum height, it is less likely to be intercepted before reaching the rear wall; this height is limited by a horizontal ceiling screen which is usually mounted 50 ft above the floor. If such

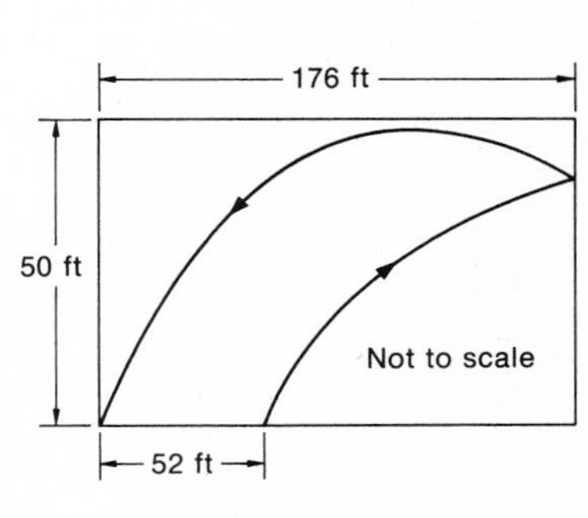

FIGURE 543

a shot is made starting practically at floor level (not realistic but it makes the math easier) and at a point 52 ft in front of the rear wall, at what velocity does it leave the player who makes it?

PROBLEM 934

Approximately 2.3×10^3 joules (or about 540 calories) are required to evaporate 1.0 cm^3 of water at STP. Part of this energy goes into the work of expanding the water vapor against the ambient pressure. It seems reasonable to assume that the rest of the energy goes into the work of bringing the water molecules up through the surface against the field of molecular attraction.

a. How much was this "rest of the energy"?
b. How much energy was expended per molecule?
c. If the range of the force of molecular attraction is of the order of 1.0 Å, what was the average force on each evaporating molecule?

PROBLEM 935

You want to take a broadside picture of an automobile that is about to pass 20 m away from you. You know that the focal length of your camera lens is 5.0 cm, and you guess that the speed of the automobile is 40 km hr^{-1}. If you want to limit any blur in the image in the negative due to the car's movement to less than 0.1 mm, what is the minimum shutter speed you must use?

Anode
$+V$
δ
Cathode
L

FIGURE 544

PROBLEM 936

A beam of electrons entering a small hole on the axis of an accelerating anode is distributed over a small cone angle δ; the anode is at a $+V$ potential relative to the cathode from which the electrons came (Figure 544). What magnetic field B should be provided parallel to the axis of the anode so that the enitre beam will pass through an equal exit hole on the axis of the anode a distance L from the entrance hole?

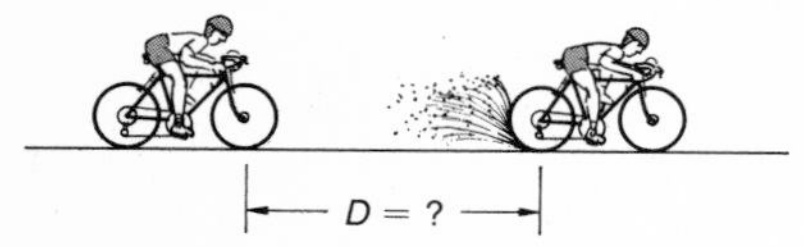

FIGURE 545

PROBLEM 937

Two racing cyclists in the Tour de France are moving in single file at speed v along a muddy road in Brittany. How far back (from wheel axle to wheel axle) should the rear cyclist stay to avoid mud being thrown by the rear wheel of the front cyclist (Figure 545)? *Suggestion:* Consider this problem in a coordinate system moving with the cycles. If in this system you sketch the mud trajectory from the front cycle, and then use a little judgment, you can get an answer by inspection that will differ from an accurately computed answer by less than 0.01 percent. (Which is certainly accurate enough for the rear cyclist!)

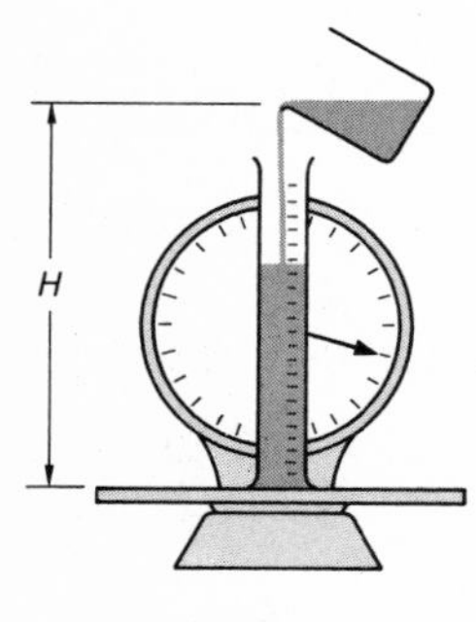

FIGURE 546

PROBLEM 938

Liquid is slowly poured from a beaker into a tall graduate that sits on the platform of a weighing scale, as shown in Figure 546. The cross-sectional area of the graduate is A cm², its weight is W gmwt, the density of the liquid is ρ gm cm^{-3}, and the liquid is being poured at a rate of v cm^{-3} sec^{-1}.

a. What was the height of the liquid in the graduate when the scale reading was a maximum?
b. What was this maximum reading?
c. By how much did this maximum reading differ from the weight of the graduate and the liquid it contained at the instant of maximum reading?

PROBLEM 939

A blob of sticky putty is placed on the inside top surface of a glass and metal case, in preparation for a demonstration experiment. The case is then hung from a spring, of spring constant k (Figure 547). Shortly after the system has come to rest the putty comes unstuck and falls. The putty hit the inside bottom of the case at the exact instant the case was momentarily at rest at the top of its first half oscillation. The case had a mass M, the putty a mass m, and the putty's vertical dimension was negligible compared to h, the inside height of the case.

FIGURE 547

a. What was h?
b. What was the amplitude of oscillations with the putty stuck in the bottom of the case?
c. Since these later oscillations took place about the former rest point for the system, they indicate an increase in kinetic energy; where did this increase come from?

PROBLEM 940

In the circuit shown in Figure 548, $V = 24$ volts DC, $R_1 = 4 \times 10^4\ \Omega$, $C = 100\ \mu$F, $R_2 = 2\Omega$, and $L = 4$H. With the capacitor uncharged at $t = 0$ the switch is placed at A; 9.2 sec later it is moved to B.

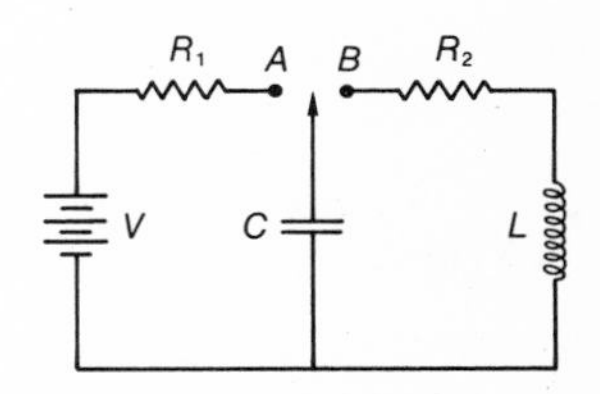

FIGURE 548

a. What was the voltage across C at $t = 9.2$ sec?
b. After $t = 9.2$ sec what was the frequency of oscillation of charge in the C-R_2-L circuit?
c. After 30 such oscillations how much energy had been dissipated in R_2?

PROBLEM 941

For pistol practice on a police shooting range an outline of a man is cut from plywood and mounted on a base, as shown in Figure 549. The back edge of the base is hinged, so that the dummy cannot slide but it can tip over. The overall weight of dummy and base is W lbwt. and its c.m. is 2.0 ft vertically above, and 2.0 ft horizontally forward of, the hinge. The assembly has a radius of gyration of 2.1 ft relative to a horizontal axis through its c.m.

The bullets used for practice have a mass m lb and a horizontal speed v ft sec^{-1}; they go *through* the dummy, losing 0.75 of their kinetic energy in the passage. If the dummy is to tip over only when a bullet goes through its head 6.0 ft above the hinge, what must W be?

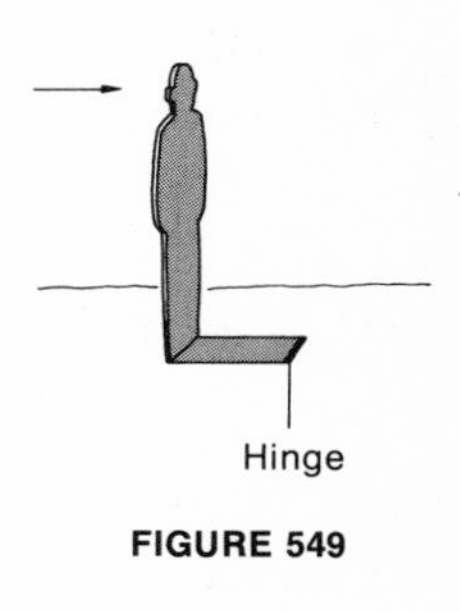

FIGURE 549

PROBLEM 942

A large insulated container is fitted at the top with a vertical glass tube which is accurately ground and polished on the inside. The container is filled with a gas whose γ is to be determined, and then is "corked" by insertion into the tube of a close-fitting polished ball of mass m. The fit of the ball in the tube is such that no gas can leak by the ball, but friction between the ball and tube is negligible. After equilibrium has been reached, i.e., when the gas pressure in the tube is just sufficient to support the ball, the ball is given a small inward displacement, and the period T of its subsequent oscillations is timed. The cross-sectional area of the tube is A, the volume of the container is V, and the barometric pressure is B; assuming the gas behaves like an ideal gas, what is γ? (This is a standard method, devised by Ruchhardt, for determining γ.)

PROBLEM 943

Two raindrops, each of radius r and each carrying an electric charge Q, bump together in the storm and coalesce to form a larger drop.

a. What would have to be the relation between Q and r for the change to take place isothermally, i.e., for the internal energy of the two-drop system to remain unchanged?

For $r = 1.0$ mm, what was the potential

b. of an original drop;

c. of the final drop?

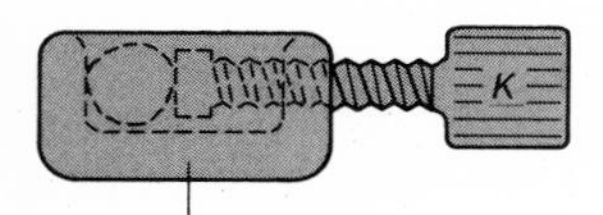

FIGURE 550

PROBLEM 944

A force of 5×10^2 newtons must be applied directly across a diameter to crack a macadamia nut by compression (Figure 550). (Macadamia nuts are even harder to crack than hickory nuts.) For such nuts a screw type nutcracker, such as can be found in handcraft shops in Yugoslavia, should be used. If the pitch of the screw is such that the screw advances p cm for every full turn of the screw, what torque must be applied to the knob k to crack a macadamia nut?

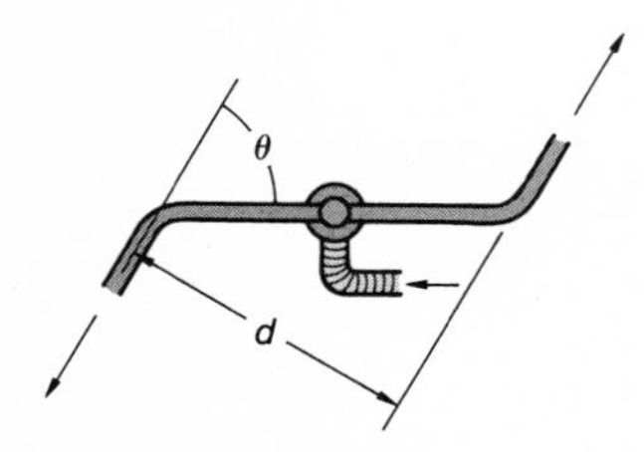

FIGURE 551

PROBLEM 945

A rotary lawn sprinkler has nozzles set at θ with the main arm of the sprinkler, as shown in Figure 551; the perpendicular distance between the nozzles is d, and the nozzle openings have a cross-sectional area A. Water is supplied to the sprinkler from a large hose where the gauge pressure is P. If the sprinkler is observed to rotate at a constant N rev sec^{-1}, what is the frictional torque at the central bearing at that speed?

PROBLEM 946

A sailboat is on a course due east; the windvane at the masthead indicates that the wind relative to the boat is coming from $\theta°$ S of E. The boom carrying the foot of the mainsail is set at an angle α with the fore-and-aft centerline of the boat, $\alpha < \theta$. Assuming that the mainsail is a vertical plane containing the boom (a bad assumption), and assuming that the drive of the boat comes entirely from the component of the relative wind perpendicular to the sail (also a bad assumption), what should α be for maximum forward driving force? (In spite of the bad assumptions, this problem produces a reasonably good answer.)

PROBLEM 947

Theoretically it should be possible to maintain a charged soap bubble at a radius R, even though the interior and exterior pressures are equal. (Practically, the difficulties of demonstrating this are so frustrating that one is tempted to forswear physics and take up folksinging instead.) With the inside of the bubble connected to the outside atmosphere, so that the pressures stay equalized, it should be possible to vary the size of the bubble by changing its electric potential (Figure 552).

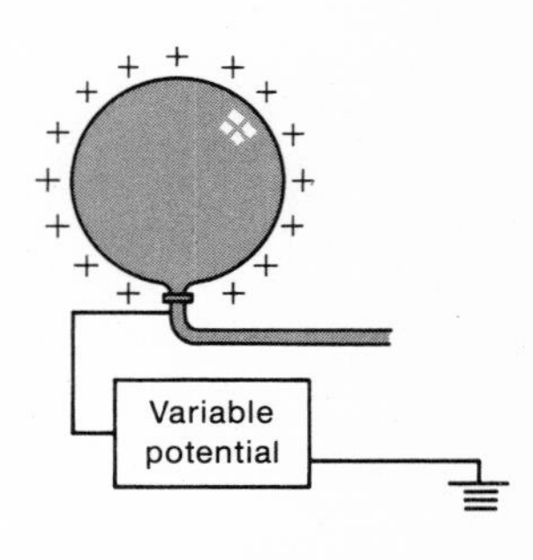

FIGURE 552

a. What is the theoretical relation between the electric potential of such a charged bubble and its radius? Take $\sigma_{\text{surface}} = 7.0 \times 10^{-2}$ j m^{-2}.
b. Is there any upper or lower limit to the size of the bubble that can be manipulated in this way?
c. At any radius R, what is the relation between the electrical energy stored on the bubble and the surface energy stored in the skin of the bubble?
d. In changing the bubble from a radius R to one of $2R$, how much work had to be done by the source of electric potential?

PROBLEM 948

A simple harmonic oscillator of mass m and natural frequency ω_0 is being driven in forced oscillations by a force $F = F_0 \cos \omega t$, $\omega > \omega_0$. At an instant that m has reached maximum amplitude this driving force is disconnected, and the oscillator resumes natural oscillations.
a. What was the maximum kinetic energy of the oscillator under forced vibrations?
b. How does this compare with the maximum kinetic energy of the oscillator after it returned to natural motion?
c. If these two maximum energies are not equal, how do you account for the difference?

PROBLEM 949

You want to design a synchrotron that will provide particles of rest mass m and charge q with a total relativistic energy E. What must be the relation between the required magnetic field B and the maximum track radius R for such an accelerator?

PROBLEM 950

Two small particles, of mass $m_1 = 10$ gm and $m_2 = 20$ gm, are held at rest in space 20 cm apart. An attractive force exists between them. When released, they move toward each other, collide head-on, and then rebound. Their speeds immediately after the collision are measured at $v_1 = 200$ cm sec^{-1} and $v_2 = 100$ cm sec^{-1}. The coefficient of restitution at collision was $e = 0.80$.

a. How much work was done by the attractive force prior to the collision?
b. If the attractive force is assumed to be directly proportional to the distance of separation, i.e., if $F = -ks$, what was the value of k?

PROBLEM 951

A long, straight, insulated antenna wire, of radius r and parallel to the ground d above it, is given such a positive charge that an electrostatic voltmeter between wire and ground reads V.

a. What is λ, the charge per unit length on the wire?
b. What is σ_x, the charge per unit area induced on the ground, as a function of x, the distance from directly beneath the wire and perpendicular to it?
c. What force, in addition to its weight, must be supplied to support the central portion of the wire?

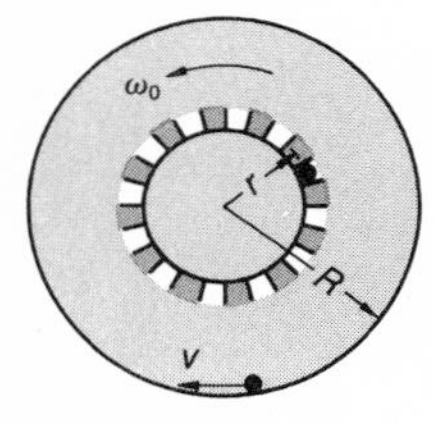

FIGURE 553

PROBLEM 952

A roulette wheel consists of a large, shallow horizontal dish with raised sides, and an array of open, numbered slots arranged in the dish in a circle of radius r, as shown in Figure 553. The wheel has a moment of inertia I_0 about its vertical axis of symmetry. The wheel is set spinning freely about this axis with an initial angular speed ω_0; then a small solid ball of mass m is thrown on to the wheel with a horizontal speed v tangential to a circle of radius R, and in the opposite direction to the motion of the wheel.

a. When the ball stops bouncing around and stays in one of the slots at rest relative to the wheel, how much energy has been lost?
b. What is your guess as to where this energy has gone?

PROBLEM 953

A viscous liquid of density ρ is flowing at a steady rate down a vertical, cylindrical tube of radius R, under the action of gravity alone. What is the rate of flow, in terms of η?

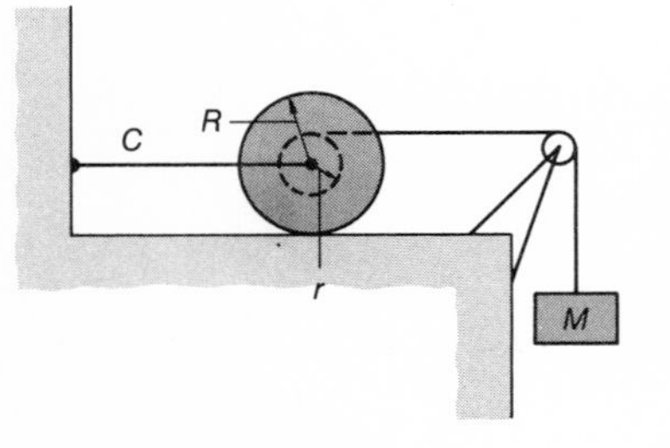

FIGURE 554

PROBLEM 954

A spool has a mass m and a radius of gyration k; there is a coefficient of sliding friction μ between spool and the horizontal surface it rests on (Figure 554). A horizontal cord C runs from the frictionless axle of the spool back to an anchored point. When the system is released,

a. what is the acceleration of M;
b. what is the tension in the cord C;
c. what is the limit of this tension as M is made very large compared to m?

PROBLEM 955

As a police car and a car believed to be carrying an escaped criminal come abreast and parallel while travelling in the same direction on a highway, a police officer fires horizontally at the other car, with his rifle pointed at right angles to the car's line of travel. The muzzle speed of the bullet is 500 m sec^{-1}, and the police car is travelling 80 km hr^{-1} at the time of the shot. Two bullet holes are subsequently discovered on opposite sides of the suspect car; the one on the side that was nearest the police car is found to be 2.2 cm further forward in the car than the hole on the opposite side; the car is 2.0 m wide at the holes. If the police bullet lost 19% of its energy in making the first hole, and if it is assumed both holes were made by the same bullet, a. how fast was the suspect car travelling at the time of the shot? b. What reasonable assumptions could you make so that the solution to question (a) would be relatively simple?

PROBLEM 956

The frequency of radiation from a receding radiant object in space as measured by observers on earth is given by

$$\nu = \nu_0 \sqrt{\frac{1-\beta}{1+\beta}}$$

(see equation 5-72), where ν_0 is the "proper" frequency emitted by the object; the relativistic form of the Doppler equation is used for generality. Then the energy received on earth per radiant atom

$$h\nu = h\nu_0 \sqrt{\frac{1-\beta}{1+\beta}},$$

where $h\nu_0$ is the proper energy released per atom at the source. What happened to the difference in energy $h\nu_0 - h\nu$?

PROBLEM 957

A metal wire of linear density 0.03 gm cm^{-1} is stretched between rigid supports under a tension of 1.5 kgwt. The wire is set vibrating in its fundamental mode with an amplitude of 3.0 cm. When a uniform magnetic field of strength 0.20 webers m^{-2} is established perpendicular to the plane of vibration of the wire,

a. what is the maximum potential difference developed between the ends of the wire?
b. If the wire supports are 70 cm apart, what is the frequency of the voltage developed between the ends of the wire? Is this a true "sine wave" voltage?

PROBLEM 958

In simple harmonic motion calculations for mechanical vibrators it is customary to neglect the mass m of the driving spring or the moment of inertia I_r of the driving torsion rod. If you include these neglected factors,

a. what is the corrected formula for the period of a simple, linear, spring driven undamped oscillator?
b. What is it for the period of a simple, undamped rotational oscillator driven by a torsion rod?
c. If you want to use the uncorrected formula for the period of a simple, spring-driven undamped oscillator, but at the same time don't want to incur an error greater than 1 percent, what is the upper limit for the value of m in terms of M, the driven mass?
Suggestion: The simplest approach is by energies. For example, consider the spring. When the spring is treated as massless the kinetic energy of the system is just the kinetic energy $\frac{1}{2}M\dot{x}^2$ of the driven mass. When the mass m of the spring is included, the kinetic energy of the system will include not only $\frac{1}{2}M\dot{x}^2$ but the integrated kinetic energy of each element dm of the spring moving with a speed which would be a linear fraction of $\dot{x}$. This integration should evaluate the system energy as $T = \frac{1}{2}\mathcal{M}\dot{x}^2$, where $\mathcal{M} = M$ plus some function of m. $\mathcal{M}$ would then be the effective mass of the oscillating system.

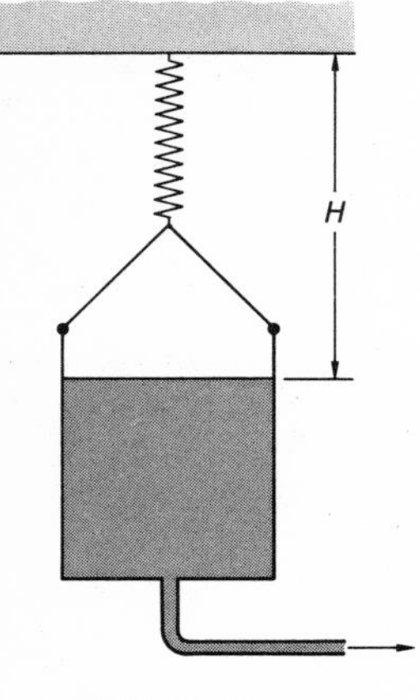

FIGURE 555

PROBLEM 959
One method of providing a constant hydrostatic head for a flow measurement is to place the liquid to be used in a container hung from a spring, as shown in Figure 555; as the liquid flows out of the container the spring shortens just enough to keep H constant. If you intend using water as your liquid, and your container is a cylindrical pail of internal radius 18 cm, you need a spring of what force constant?

PROBLEM 960
Small masses m_1 and m_2 are placed at opposite ends of a long, thin rigid rod of length L and mass M. The rod is mounted horizontally in a clamp at the end of a vertical axis of a rotator.

a. How far from m_1 should the rod be clamped so that a minimum amount of energy is necessary to set the rod to rotating in a horizontal plane with an angular speed ω?
b. Does this clamping point possess any special significance other than meeting the requirement of minimum energy?
c. What is the minimum energy at an angular speed ω?

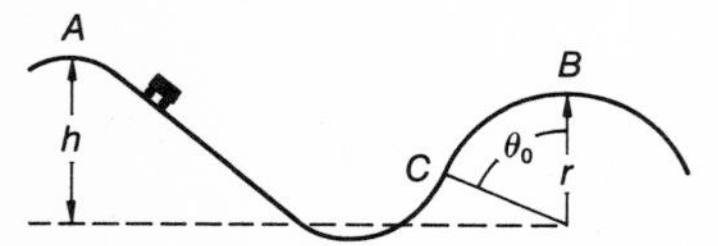

FIGURE 556

PROBLEM 961
The essential geometry of a roller-coaster is as shown in Figure 556; the circular convex curvature of the B section begins at C. If a small car rolling freely barely gets over the hump at A,

a. where on the B section is the car most likely to leave the track?
b. What is the maximum value of h such that the car will not leave the track at the critical point determined in question (a)?

PROBLEM 962
It is estimated that a mountain climber attached to a safety rope in the normal manner can withstand a jerk from the rope of only about 1500

lbwt before severe pain and possibly serious injury, ensue. If a 160-lb climber is at the end of 30 ft of a standard $\frac{7}{16}$ inch nylon climbing rope that is held at the other end by a static belay, i.e., the held end is not movable,* through what distance could he free-fall before being restrained by the rope without risking severe pain or injury? See Figure 557 for data on his climbing rope.

*Actually, in good mountaineering practice a dynamic belay, which allows for slippage of the rope during stress, would be used. This slippage absorbs a portion of the energy generated by the fall.

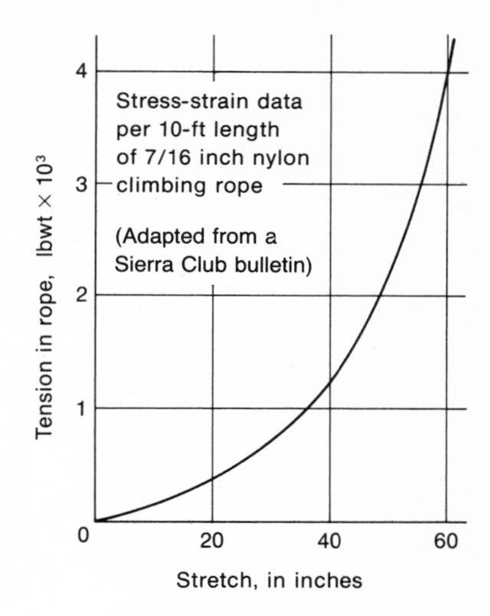

FIGURE 557

PROBLEM 963

A uniform slender rod (a meter stick, perhaps) stands vertically on a smooth floor with its lower end against a fixed peg, as shown in Figure 558. When the top of the rod is given an infinitesimal displacement in a direction away from the peg, it is observed that as the rod falls its lower end loses contact with the peg when the rod is at an angle θ with the horizontal.

a. What is the value of θ?
b. When the rod is horizontal, with what horizontal speed is it moving?

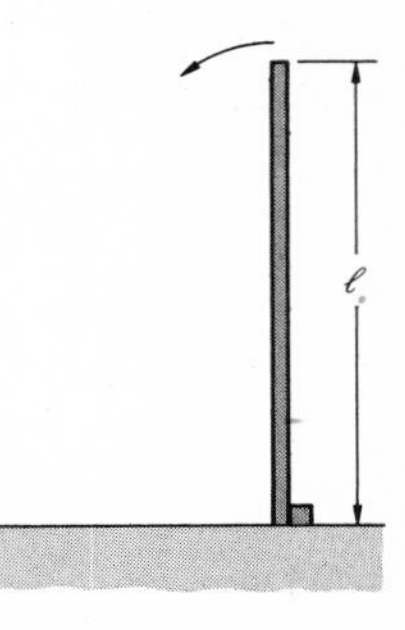

FIGURE 558

PROBLEM 964

In a particular cyclone the wind velocities are essentially horizontal and tangent to circles whose centers all lie on a common vertical axis (this axis being at the "eye" of the cyclone). These tangential speeds are observed to be a function of the distance R from this axis as follows:

$$\text{for } R < 10^3\text{m},\ v_R = 0.050\ R\ \text{m sec}^{-1}$$

$$\text{for } R > 10^3\text{m},\ v_R = \frac{5 \times 10^4}{R}\ \text{m sec}^{-1}$$

Assuming that the density remains constant throughout the cyclone at 1.3 kg m^{-3} and that at $R \gg 10^3$m the air pressure $= B_0 = 1.0\,A$s, what is the air pressure at the center of the cyclone? *Suggestion:* Integrate the centripetal force provided by the pressure gradient on an infinitesimal volume element of air.

PROBLEM 965

What do you think happened to the potential energy lost by the unrolling tape of problem 366?

PROBLEM 966

During the First Planetary War, a rear gunner in a spacecraft, traveling in essentially field-free space at a velocity of 20.0 km sec^{-1} due south galactic, observed an enemy war capsule at a distance of 1.30 km due north of him; the scanner at the same instant indicated the enemy capsule was traveling with a velocity v_W due west galactic and slowing down with a constant acceleration a_W due east galactic. 2.00 sec after the observation, the gunner launched a Paralyzo rocket,

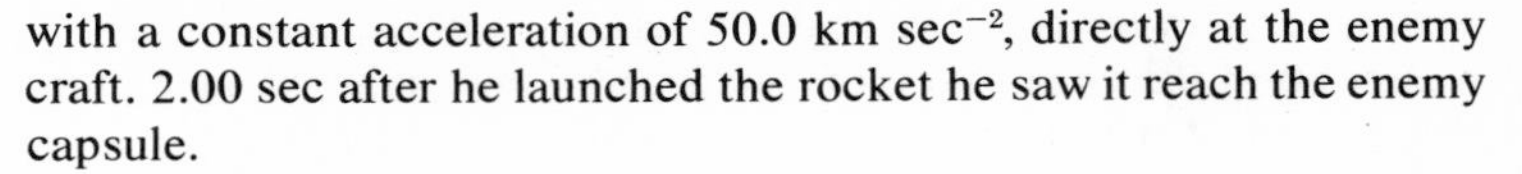
with a constant acceleration of 50.0 km sec^{-2}, directly at the enemy craft. 2.00 sec after he launched the rocket he saw it reach the enemy capsule.

a. In what galactic compass direction did the gunner aim the rocket at its launch?
b. What was v_W?
c. What was a_W?

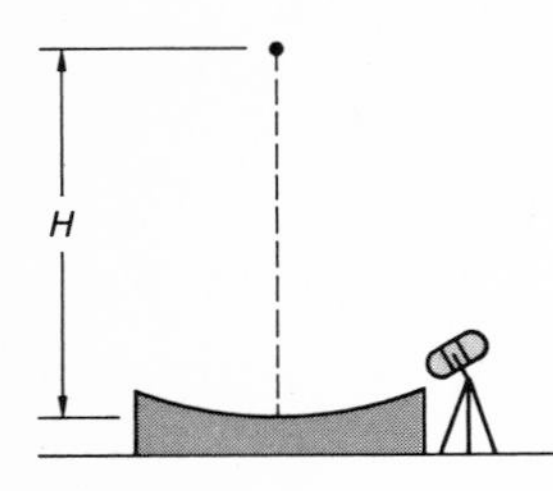

FIGURE 559

PROBLEM 967

In a lecture demonstration a small steel ball is dropped from a height H onto a hardened steel plate (Figure 559). It continues to bounce with increasing frequency until all its energy is dissipated. A microphone was set up next to the plate to detect the sound of the bounces. If the ball is dropped at $t = 0$ and T sec later all sound has ceased, indicating the ball has come to rest, what is the value of the coefficient of restitution? *Hint:* See Appendix D for the value of various summations.

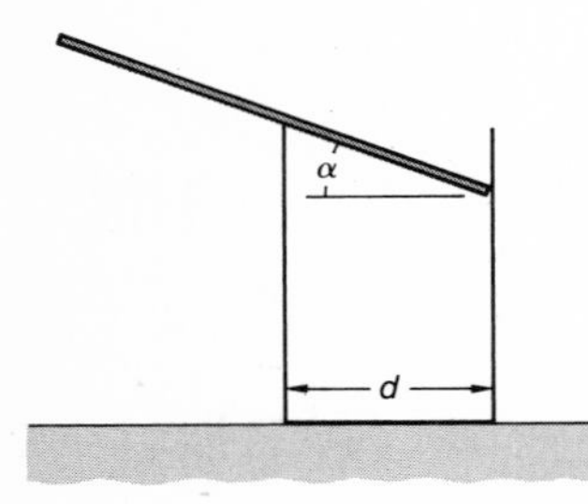

FIGURE 560

PROBLEM 968

A cylindrical drinking glass, with smooth sides and a smooth, rounded drinking edge, has a uniform soda straw of length L balanced on the edge of the glass, as shown in Figure 560. For a given diameter of glass d, what must α be for this trick to work?

PROBLEM 969

With good tires and dry, level pavement, the coefficient of friction between tires and pavement is about 0.7. A car's normal weight with driver is usually distributed about 55 percent on the rear wheels and 45 percent in front. A car is started from rest and driven with constant acceleration until it reaches a certain speed v; at that instant all four brakes are jammed full on and the car skids to a stop.

a. How does the time of positive acceleration compare with the time of skidding?
b. How does the distance traversed during positive acceleration compare with the distance skidded?
c. If the total distance covered was 630 ft, what was v?

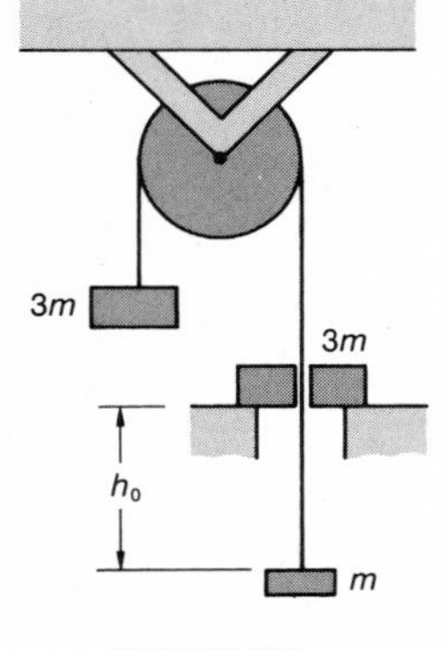

FIGURE 561

PROBLEM 970

An Atwood's machine is as shown in Figure 561: the slotted weight of mass $3m$ rests on a solid shelf, with a central hole in the shelf large enough for the hanging weight of mass $1m$ to pass through and pick up the $3m$ weight. When the $1m$ mass is released from rest h_0 below the slotted mass, the system goes through a cyclic motion of diminishing amplitude. Assume moment of inertia of pulley and friction at pulley axle are negligible.

a. When the system finally comes to rest, how much mechanical energy has disappeared from the system?

b. Where did this energy go?
c. How long after release did it take for the system to come to rest?

PROBLEM 971
Relative to the earth the moon does not rotate—the same moonscape faces the earth at all times. What is the moon's angular momentum about its own axis?

PROBLEM 972
The aerator nozzle 34 cm below the surface of the water in a tropical fish tank forms air bubbles of 3.0 mm diameter. For bubbles of this size the viscous drag of the water $= 0.2\rho_{H_2O}\pi\, r^2v^2$, r the radius of the bubble and v its speed.

a. What is the terminal speed of the rising bubbles?
b. How long after formation does it take the bubbles to reach 99 percent of this terminal speed?
c. How long does it take the bubbles to rise to the surface?
d. As the bubbles rise they encounter decreasing hydrostatic pressure. Should the resulting change in the size of the bubbles be taken into account in your calculations?

PROBLEM 973
A space rocket, of total mass M, sits on its launching pad. At $t=0$ the rocket motors are ignited. They eject mass vertically downward at a constant rate $\dot{M}$ and with a nozzle speed of v. As $v\dot{M} < Mg$, the rocket does not rise immediately at ignition.

a. How long after ignition did lift-off begin?
b. What was the total mass of the rocket at lift-off?

PROBLEM 974
You can make a fairly accurate determination of the friction between a lead screw and its chasing nut by mounting the lead screw vertically and letting the nut run down the screw (Figure 562). The nut has a mass M and a radius of gyration k, and the screw has a pitch of p turns per unit length.

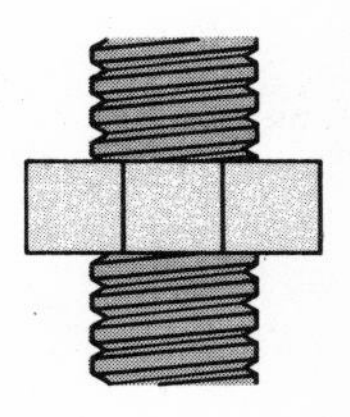
FIGURE 562

a. If the nut is released from rest, how much time should it take to run down a length L of the screw, in the ideal case of no friction?
b. In an actual test, the nut only got down to 0.9 L in the time obtained in question (a) for the ideal case. What was the frictional torque (assumed constant) between nut and screw?
c. In a similar test on a different nut-screw combination, the nut was given an initial angular speed, and it was observed that the nut ran down the screw without any change in angular speed. What was the frictional torque in this case?

APPENDIXES

APPENDIX **A**

USEFUL DATA

B_0 standard atmospheric pressure – 1.013×10^5 N m^{-2}

c speed of light in free space – 2.9979×10^8 m sec^{-1}
value for normal use – 3.00×10^8 m sec^{-1}

c_m speed of sound in dry air at STP – 331 m sec^{-1} = 1086 ft sec^{-1}

ϵ_0 permittivity of free space - 8.85×10^{-12} farads m^{-1}

$$\frac{1}{4\pi\epsilon_0} = 10^{-7} c^2$$

G universal gravitational constant – 6.67×10^{-11} N m^2 kg^{-2}

g_s standard value of gravitational acceleration at earth's surface – 980.665 cm sec^{-2} = 32.174 ft sec^{-2}

g value of g_s for normal use – 9.8 m sec^{-2} = 32 ft sec^{-2}

h Planck's constant – 6.63×10^{-34} joule sec

k Boltzmann's constant – 1.38×10^{-23} joule (°K)$^{-1}$

m_{el} mass of the electron – 9.11×10^{-31} kg

m_{pr} mass of the proton – 1.67×10^{-27} kg

$M_\oplus$ mass of the earth – 5.98×10^{24} kg

$M_\odot$ mass of the sun – 1.97×10^{30} kg

$M_☽$ mass of the earth's moon – 7.34×10^{22} kg

N_0 Avogadro's number – 6.02×10^{23} molecules mole^{-1}

q_{el} charge of the electron – 1.60×10^{-19} coulomb

R universal gas constant – 8.31 joules mole^{-1} (°K)$^{-1}$

$R_\oplus$ radius of the earth (average) – 6.37×10^6 m

$R_\odot$ radius of the sun – 6.95×10^8 m

$R_☽$ radius of the earth's moon – 1.74×10^6 m

Radius of moon's orbit about the earth – 3.8×10^8 m

Radius of earth's orbit about the sun – 1 A.U. = 1.49×10^{11} m

Densities at Standard Conditions

$\rho_{air} = 1.293$ kg m^{-3}

ρ_{water} $\begin{cases} = 998.2 \text{ kg m}^{-3} \text{ at } 20°\text{C} \\ = 1.00 \text{ gm cm}^{-3} \text{ for normal calculations} \end{cases}$

$\rho_{Hg} = 13.6$ gm cm^{-3}

	Elastic Constants		Thermal Coefficients		Specific Heats
	Y in N m^{-2} $\times 10^{10}$	n in N m^{-2} $\times 10^{10}$	λ—(°C)$^{-1}$ $\times 10^{-5}$	α—(°C)$^{-1}$ $\times 10^{-5}$	cal gm^{-1} (°C)$^{-1}$
Brass	9.0	3.5	1.8		0.089
Glass	6.0	2.4		2.5	0.17
Lead			2.9		0.031
Mercury				18.2	0.033
Steel	20	8.2	1.1		
Water (average for range 0–20°C)					1.00

Selected Values for the Vapor Pressure and Density of Saturated Water Vapor

Temperature in °C	Vapor Pressure in mm of Hg	Density gm cm^{-3} $\times 10^{-6}$
−5	3.2	3.3
20	17.6	17.3
21	18.7	18.3
21.5	19.2	18.8
22	19.8	19.4
25	23.8	23.0
30	31.9	30.4
50	92.6	83.2
91	546.3	439.5

APPENDIX B

CONVERSION FACTORS

Mass	1.0 kg = 1.0×10^3 gm = 2.2 lb 1.0 ton = 2.0×10^3 lb 1.0 metric ton = 1.0×10^3 kg 1.0 amu = 1.66×10^{-27} kg = 931.5 Mev in mass energy units
Length	1.0 m = 1.0×10^2 cm = 1.0×10^{-3} km = 39.37 in = 3.28 ft = 6.21×10^{-4} mi 1.0 statute (land) mile = 5280 ft = 1.61 km 1.0 nautical mile = 6076 ft = 1.15 statute mile 1.0 fathom = 6.0 ft 1.0 A.U. = 1.49×10^{11} m = 92.9×10^6 mi 1.0 Å = 1.0×10^{-10} m 1.0 ly = 9.46×10^{15} m
Time	60 sec = 1.0 min 3.6×10^3 sec = 1.0 hr } mean solar time units 8.64×10^4 sec = 1.0 day } 1.0 sidereal year = 366.256 sidereal days = 365.256 mean solar days
Area	1.0 m^2 = 1.0×10^4 cm^2 = 1.55×10^3 in^2 = 10.76 ft^2 1.0 in^2 = 6.45 cm^2 1.0 barn = 1.0×10^{-28} m^2
Volume	1.0 m^3 = 1.0×10^6 cm^3 = 6.1×10^4 in^3 = 35.3 ft^3 1.0 L = 1.0×10^{-3} m^3 = 1.0×10^3 cm^3

Speed	1.0 m sec^{-1} = 3.6 km hr^{-1} = 2.24 mi hr^{-1} 60 mi hr^{-1} = 88 ft sec^{-1} 1.0 knot = 1.0 nautical mi hr^{-1} = 1.85 km hr^{-1}
Force	1.0 N = 1.0 × 10^5 dynes = 0.225 lbwt 1.0 lbwt = 32 lbal = 4.45 N 1.0 kgwt = 9.8 N
Pressure	1.0 N m^{-2} = 10 dynes cm^{-2} = 1.45 × 10^{-4} lbwt in^{-2} 1.0 bar = 1.0 × 10^5 N m^{-2} = 1.0 × 10^6 dynes cm^{-2} 1.0 As = 760 mm Hg
Energy	1.0 j = 1.0 × 10^7 ergs = 1.0 watt sec = 2.78 × 10^{-7} kwh = 3.73 × 10^{-7} HPh = 9.48 × 10^{-4} Btu 1.0 ev = 1.60 × 10^{-19} j 1.0 Mev = 1.60 × 10^{-13} j 1.0 cal = 4.186 j 1.0 amu = 1.49 × 10^{-10} j
Power	1.0 watt = 1.0 × 10^{-3} kw = 1.0 j sec^{-1} 1.0 HP = 0.746 kw
Frequently used prefixes	μ (micro) = 10^{-6} m (milli) = 10^{-3} c (centi) = 10^{-2} k (kilo) = 10^3 M (mega) = 10^6 G (giga) = 10^9

APPENDIX C

ABBREVIATIONS

A	ampere
Å	angstrom
AC	alternating current
amu	atomic mass unit
As	atmosphere
A.U.	astronomical unit
Btu	British thermal unit
C	coulomb
°C	degrees Centigrade
cm	centimeter
c.m.	center of mass
DC	direct current
emf	electromotive force
ev	electron volt
°F	degrees Fahrenheit
ft	foot
gm	gram
HP	horsepower
HPh	horsepower hour
hr	hour
Hz	(Hertz) – cycles sec^{-1}
j	joule
°K	degrees Kelvin
kg	kilogram
kgwt	kilogramweight
km	kilometer
kw	kilowatt
kwh	kilowatt hour
L	liter
lb	pound
lbal	poundal
lbwt	poundweight
ly	light year
m	meter
Mev	million electron volts
mi	mile
min	minute
mm	millimeter
MW	molecular weight
N	newton
rad	radian
rev	revolution
rpm	revolutions min^{-1}
rps	revolutions sec^{-1}
sec	second
SHM	simple harmonic oscillations
STP	standard temperature and pressure
w	watt

APPENDIX D

SYMBOLS

From the English Alphabet

A	area
a	acceleration, linear
B	ambient pressure, bulk modulus, magnetic field strength
C	capacity, dimensional symbol for current
c	speed of light in free space, specific heat for liquids and solids
c_m	speed relative to the medium
c_p	specific heat of a gas at constant pressure
c_v	specific heat of a gas at constant volume
E	total energy, electric field strength
$\mathscr{E}$	electromotive force
e	base of natural logarithms, coefficient of restitution, reduced electronic charge
F or f	force
G	universal gravitational constant
g	acceleration due to gravitational force
h	heat transfer coefficient, Planck's constant
I	current, intensity, moment of inertia
i	$\sqrt{-1}$
J	impulse
K	dielectric constant
k	Boltzmann's constant, force constant, general constant of proportionality, coefficient of thermal conductivity
L	angular momentum, self inductance, dimensional symbol for length

ℓ	length
M	dimensional symbol for mass, mutual inductance, mass
m	mass
N	number of moles
N_0	Avogadro's number (molecules mole^{-1})
n	shear modulus
n_0	number of molecules per unit volume
P	power, pressure
p	linear momentum, overpressure
Q	dimensional symbol for charge, thermal energy
q	electric charge
R	electrical resistance, universal gas constant
R or r	radius
S	entropy
S or s	distance
T	dimensional symbol for time, absolute temperature, kinetic energy, period of oscillation, tension
t	time
U	potential energy
u	energy density, heat or energy flow, relative speed
V	electric or gravitational potential, potential difference, volume
v	speed
W	work
Y	elastic (Young's) modulus
Z	impedance

From the Greek Alphabet

α (alpha)	angle, angular acceleration, coefficient of volume thermal expansion
β (beta)	angle, the ratio v/c
γ (gamma)	coefficient of drag, damping constant, ratio of specific heats, symbol for $1/\sqrt{1-\beta^2}$
Δ (Delta)	an increment or change
δ (delta)	angle, phase angle, a small change
ϵ (epsilon)	a very small quantity
ϵ_0	permittivity of free space
η (eta)	coefficient of viscosity
Θ (Theta)	angle, dimensional symbol for temperature
λ (lambda)	coefficient of linear thermal expansion, linear mass or charge density, mean free path, wavelength
μ (mu)	reduced mass, prefix for micro
μ_0	permeability constant
μ_k	coefficient of sliding friction
μ_{st}	coefficient of static friction

ν (nu)	frequency
ρ (rho)	mass or charge density
Σ (Sigma)	indicates summation
σ (sigma)	coefficient of surface tension, collision cross-section, surface density of mass or charge
τ (tau)	torque
τ_0	torsional constant
Φ (Phi)	magnetic flux
ϕ (phi)	angle, phase angle
Ω (Omega)	angular rate of precession, solid angle, ohm
ω (omega)	angular frequency, angular speed

Mathematical Symbols and Useful Approximations

$\propto$	proportional to
$\simeq$	approximately equals
$a > b$	means a is greater than b
$a \geqslant b$	means a is greater than or equal to b
$a < b$	means a is less than b
$a \gg b$	means a is very much greater than b, i.e., b/a is very small
$\rightarrow$	approaches as a limit
$\bar{x}$	the average of x over a specified range
$\vec{p}$	the vector p

For θ very small:

$$\sin\theta \simeq \tan\theta \simeq \theta$$
$$1 - \cos\theta = 2\sin^2(\theta/2) \simeq \theta^2/2$$

For β very small:

$$\sqrt{1-\beta^2} \simeq 1 - \beta^2/2$$
$$\frac{1}{\sqrt{1-\beta^2}} \simeq 1 + \beta^2/2$$

Summations

$$\left.\begin{aligned}\sum_{n=1}^{\infty} x^n &= \frac{x}{1-x} \\ \sum_{n=1}^{\infty} (-x)^n &= \frac{-x}{1+x}\end{aligned}\right\} x < 1$$

APPENDIX 

STANDARD COORDINATE SYSTEMS

The Cartesian System in Three Dimensions

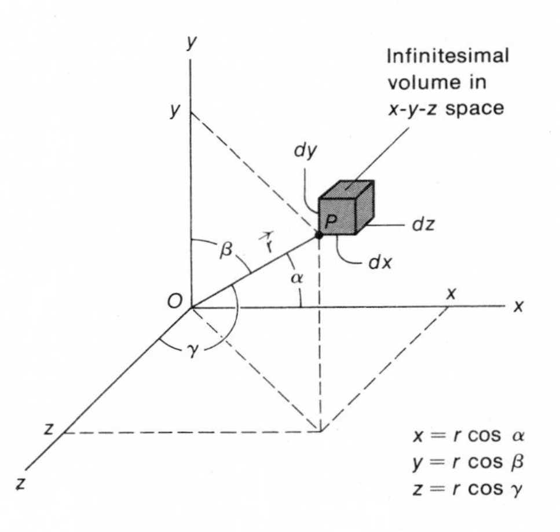

If A is a scalar function of position x, y, and z, then the rate of change of A (the gradient of A) in the x direction at any point P is found by evaluating $\delta A/\delta x$ at P, etc.

Polar or Rotational Coordinates

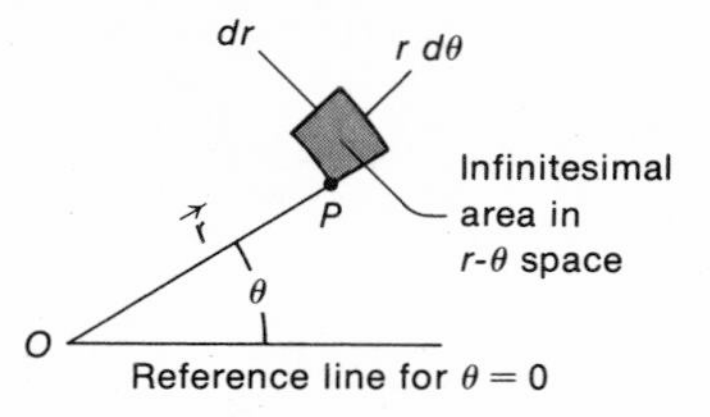

If B is a scalar function of position r and θ, then the gradient of B in the r direction is evaluated by $\delta B/\delta r$ and the gradient in a direction perpendicular to r in the plane of the diagram is evaluated from $(1/r)\ (\delta B/\delta\theta)$.

Spherical Coordinates

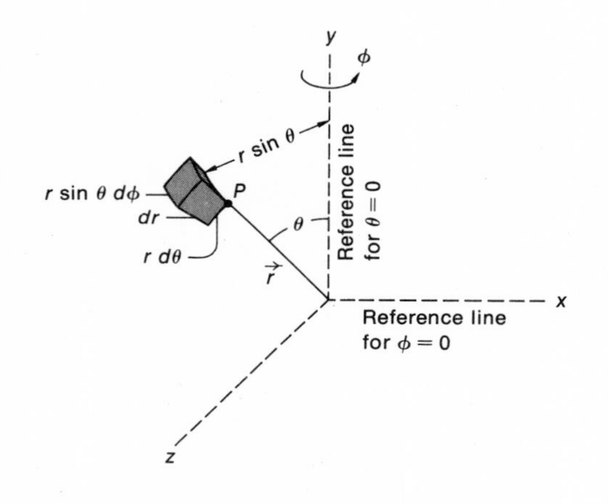

If C is a scalar function of position r, θ, and ϕ, the gradient of C in the direction $\vec{r}$ is $\delta C/\delta r$, the gradient of C in a plane determined by $\vec{r}$ and the y-axis and in a direction perpendicular to $\vec{r}$ is $(1/r)\ (\delta C/\delta\theta)$, and the gradient of C in a plane perpendicular to the y-axis and in a direction perpendicular to $\vec{r}$ is $[1/(r \sin\theta)]\ (\delta C/\delta\phi)$.

APPENDIX F

VERY SHORT OUTLINE OF VECTOR ANALYSIS

1. $\vec{V}_1 + \vec{V}_2 = \vec{V}_2 + \vec{V}_1$; vectors added in any order will produce the same result.

2. In Cartesian coordinates $\vec{V} = V_x\vec{i} + V_y\vec{j} + V_z\vec{k}$, where $\vec{i}$, $\vec{j}$, and $\vec{k}$ are unit vectors along the x, y, and z axes. $V_x = V\cos\alpha$, $V_y = V\cos\beta$, and $V_z = V\cos\gamma$. α is the angle between V and the x-axis, measured in the plane containing V and the x-axis; β and γ are similarly defined. α, β, and γ are often referred to as the cosine direction angles.

3. Dot Product or Scalar Product.

$\vec{F}\cdot\vec{s} = \vec{s}\cdot\vec{F} = Fs\cos(F, s)$ where (F, s) is the angle between $\vec{F}$ and $\vec{s}$. In the dot product the order of the vectors does not affect the result. In terms of components $\vec{F}\cdot\vec{s} = (F_x\vec{i} + F_y\vec{j} + F_x\vec{k})\cdot(s_x\vec{i} + s_y\vec{j} + s_z\vec{k}) = F_xs_x + F_ys_y + F_zs_z$.

4. Vector or Cross Product.

$\vec{v}\times\vec{B} = -\vec{B}\times\vec{v} = vB\sin(v, B)\,\vec{p}$, where $\vec{p}$ is a unit vector perpendicular to the plane containing v and B and positive in the direction a right-handed screw would advance if turned from $\vec{v}$ to $\vec{B}$. In the cross product, reversing the order of the vectors reverses the direction of the product. Determinants are useful in expanding the cross product:

$$\vec{v}\times\vec{B} = \begin{vmatrix} \vec{i} & \vec{j} & \vec{k} \\ v_x & v_y & v_z \\ B_x & B_y & B_z \end{vmatrix} = (v_yB_z - B_yv_z)\vec{i} - (v_xB_z - B_xv_z)\vec{j} + (v_xB_y - B_xv_y)\vec{k}.$$

5. The Triple Vector Product.

$\vec{a}\times(\vec{b}\times\vec{c}) \neq (\vec{a}\times\vec{b})\times\vec{c}$ except that when $\vec{a}$, $\vec{b}$, and $\vec{c}$ are mutually perpendicular both products equal zero. When $\vec{a}$ and $\vec{b}$ are parallel and $\vec{c}$ is perpendicular to both, $\vec{a}\times(\vec{b}\times\vec{c}) = -ab\,\vec{c}$.

6. The vector operator ∇ (called del) is defined by

$$\nabla = \frac{\delta}{\delta x}\vec{i} + \frac{\delta}{\delta y}\vec{j} + \frac{\delta}{\delta z}\vec{k}.$$

If ϕ is a scalar function of position, the gradient of ϕ is computed from

$$\text{grad } \phi = \nabla \phi = \frac{\delta \phi}{\delta x}\vec{i} + \frac{\delta \phi}{\delta y}\vec{j} + \frac{\delta \phi}{\delta z}\vec{k}.$$

If $\vec{V}$ is a vector the divergence of $\vec{V}$ is a scalar and is computed from

$$\text{div } \vec{V} = \nabla \cdot \vec{V} = \frac{\delta V_x}{\delta x} + \frac{\delta V_y}{\delta y} + \frac{\delta V_z}{\delta z}.$$

APPENDIX G

RELATIVE MOTION IN ROTATION (NONRELATIVISTIC)

A primed coordinate system rotates with a constant angular speed ω relative to an at-rest unprimed system; the z and z' axes coincide along the axis of relative rotation.

Relative Speeds

As seen in the unprimed system, at the instant an object is at P it is moving with velocity $\vec{v}$, which has components v_x and v_y. See Figure G-1.

At the same instant, as seen in the primed system the object is at P' and is moving with a velocity $\vec{v}'$, with components $v'_{x'}$ and $v'_{y'}$. See Figure G-2.

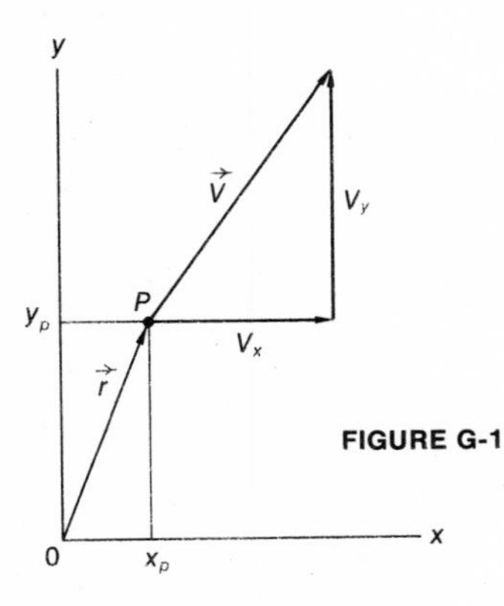

FIGURE G-1

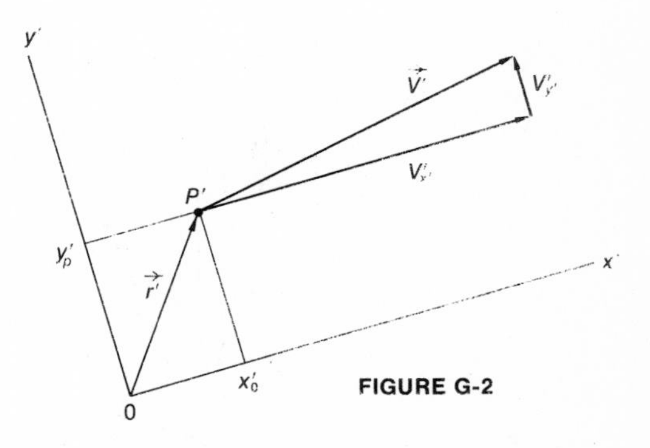

FIGURE G-2

When we put the two diagrams together, as in Figure G-3, we see that $\vec{v}'$ is compounded of $\vec{v}$ and the apparent rotational motion $-\omega \times r$ of P as seen from the moving system. (Note that $\vec{r}' = \vec{r}$.) By following the dashed construction lines it can be seen that

$$v'_{x'} = (v_x + \omega y) \cos \omega t + (v_y - \omega x) \sin \omega t$$

$$v'_{y'} = (v_y - \omega x) \cos \omega t - (v_x + \omega y) \sin \omega t$$

which confirm the equations near the bottom page 82.

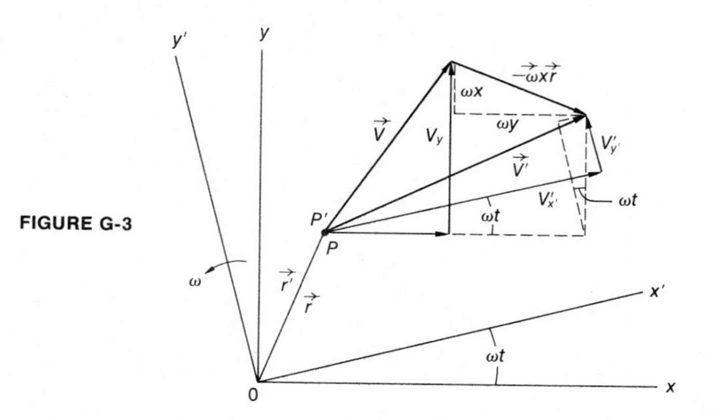

FIGURE G-3

Relative Accelerations

By the same process we can construct the diagram in Figure G-4, which illustrates the equation

$$\vec{a}' = \vec{a} - 2\vec{\omega} \times \vec{v} - \omega^2\vec{r}.$$

By following the dashed lines one can confirm the values for $a'_{x'}$ and $a'_{y'}$ which were also given on page 82.

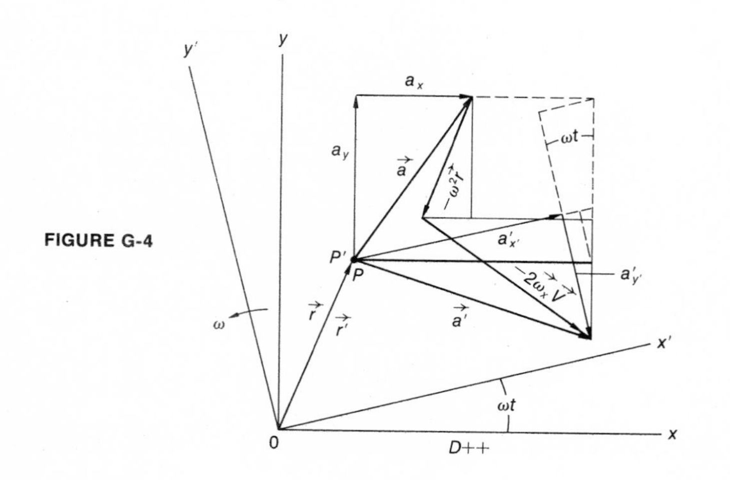

FIGURE G-4

APPENDIX H

NUMBER OF THERMAL MOLECULES HITTING A SMALL AREA ON WALL OF CONTAINER

For molecules in a gas at temperature T and with an average speed $\bar{v}$ – on the average at any instant, the number of molecules passing through the volume element shown in Figure H-1 is $n_0 r^2 \, dr \, \sin\theta \, d\theta \, d\phi$; they emerge from this volume element in all directions with equal probability. The number of those emerging that will head for

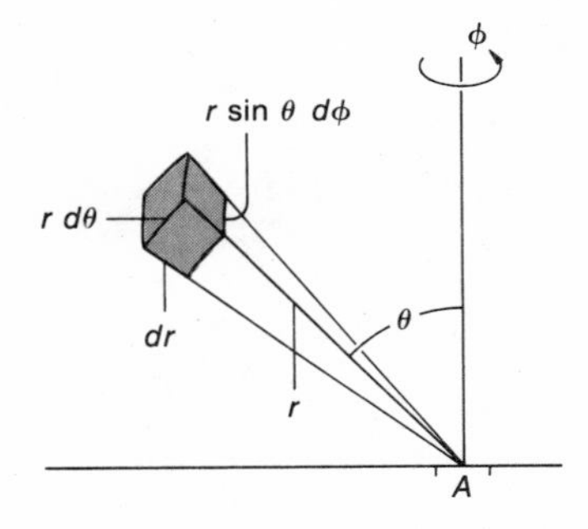

the area A will be $[(A \cos\theta/4\pi r^2)] \, n_0 r^2 \, dr \, \sin\theta \, d\theta \, d\phi$. Even if some of these molecules suffer a collision on the way toward A momentum toward A will be preserved; on the average any molecule from the volume element that is diverted from its path toward A will be matched by a molecule not from the volume element that is diverted toward A. Hence on the average

$$\frac{A \cos\theta \, n_0 r^2 \sin\theta \, d\theta \, d\phi}{4\pi r^2}$$

molecules per volume element will arrive at A. In one second the number arriving may come from as far away as $r = \bar{v}$. Thus the total number of molecules arriving per second at A

$$= \frac{An_0}{4\pi} \int_0^{\bar{v}} dr \int_0^{\pi/2} \sin\theta \cos\theta \, d\theta \int_0^{2\pi} d\phi$$

$$= \frac{An_0\bar{v}}{4}.$$

If A is a small opening in the wall of the container, then $An_0\bar{v}/4$ is the number of molecules escaping per second.

This result is needed in problem 719.

AVERAGE ENERGY LOSS DURING A LARGE NUMBER OF RANDOM ELASTIC COLLISIONS

In order to apply the results to molecular collisions we will assume that the collisions are perfectly elastic, that any particular molecule makes a large number of collisions per unit time, and that these collisions take place over a random distribution of impact angles. It will also be assumed that the energies involved are low enough so that the collisions can be treated nonrelativistically.

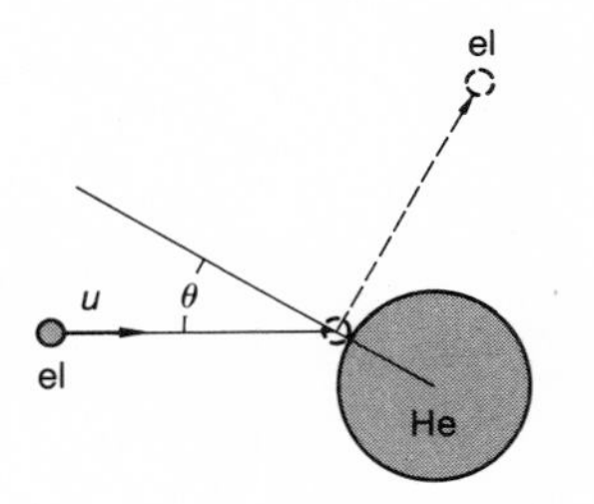

A molecule of mass m moving with a speed u collides at an impact angle θ with a molecule of mass M essentially at rest (Figure I-1). If the collision is perfectly elastic, from the solution to problem 598 we have for the post-collision energy of m

$$E_{m,\text{final}} = \frac{1}{2} m\, u^2 \left[\sin^2\theta + \left(\frac{M-m}{M+m}\right)^2 \cos^2\theta \right].$$

Then the energy lost by m in the collision is

$$\Delta E_m = \frac{1}{2} \frac{4Mm}{(M+m)^2} mu^2 \cos^2\theta$$

and the percentage loss is

$$\frac{\Delta E_m}{E_m} = \frac{4Mm}{(M+m)^2} \cos^2\theta \simeq \frac{4m}{M} \cos^2\theta \text{ for } m \ll M.$$

During a very large number of random collisions it is reasonable to expect that the impact angles θ will be uniformly distributed over the range from $-\pi/2$ to $+\pi/2$. Thus over the duration of these collisions the average energy percentage loss per collision should be

$$\left(\frac{E}{E}\right)_{av} = \frac{4m}{M}\left(\cos^2\theta\right)_{av} = \frac{4m}{M}\left[\frac{1}{\pi}\int_{-\pi/2}^{\pi/2}\cos^2\theta\, d\theta\right] = \frac{2m}{M}$$

A normal atom would appear electrically neutral to a charged particle outside the atom. Thus, for an electron with energy below any excitation level of He, a collision with a He atom should, at least to a first approximation, be like the particle collision discussed above. Hence the above average percentage loss of energy is applicable to Problem 718 (d).

ANSWERS

2. (a) 9.5×10^{12} km (b) 6.3×10^4 A.U.

3. 51 ft

4. 10.4 sec

5. $x \simeq t/5$; you had to assume that the sound and the flash originated at the same time and the same place—an assumption that is not always valid

6. (a) 13.5 hr (b) 135 nautical miles

7. $R = 0.081\,t$ for R in nautical miles and t in μsec

8. (a) 30 mi hr^{-1}
   (b) approximately 30 mi hr^{-1}
   (c) about 60 mi hr^{-1} (d) 0

9. $\Delta t \simeq 0.1$ sec

10. 50 mi hr^{-1}

11. 55 ft

12. *B* finishes first by 2.7 min; see diagram

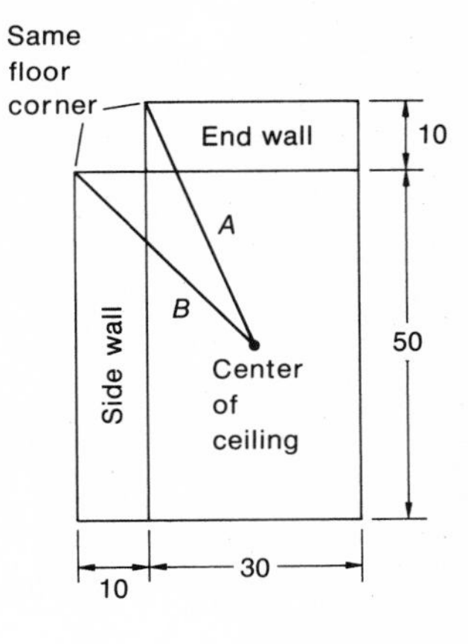

13. (a) 15 mi hr^{-1} (b) nearly 11 min

14. approximately 13 min longer

15. 6 blocks east of Central Avenue

16. 54 mi hr^{-1}

17. 18 min 41.2 sec

18. 18 min 47.0 sec

19. This is the time at which $v = 0$—the time when the object was released from rest

20. (a) This is the speed acquired at the time the clock was started
(b) 3.9 m sec^{-1}; it is the same as v_0 in Figure 5

21. The acceleration decreases for increasing speed, because of air friction; the linear relationship breaks down at about 41 m sec^{-1}

22. This area represents the distance the object fell from rest before the timing was started

23. $S_{1-2} = v_0(t_2 - t_1) + \frac{1}{2}a(t_2^2 - t_1^2)$

24. −16 ft sec^{-2}

25. (a) 4.6 mi hr^{-1} sec^{-1} (b) 43.5 mi hr^{-1}

26. (a) about 12 sec (b) about 10 sec
(c) about −8 ft sec^{-2} during braking
(d) about 0.5 mi

29. −3.8 ft sec^{-2}

30. 5.5 ft sec^{-2}; the v-t graph should look like this:

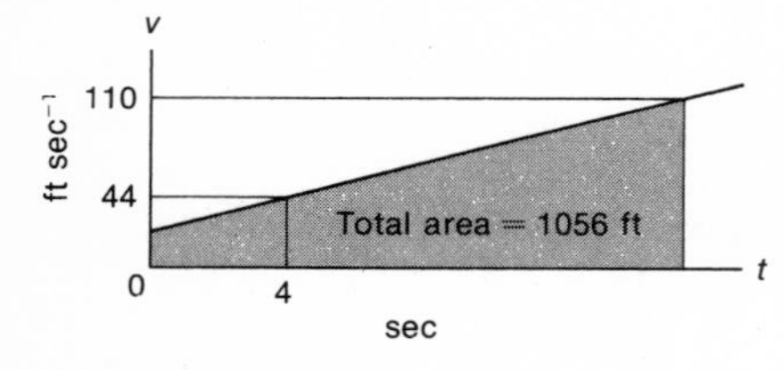

31. (a) 52.5 mi hr^{-1} (b) 11/4 ft sec^{-2}

32. 7.9 ft sec^{-2}

33. 8/9; the v-t graph should look like this:

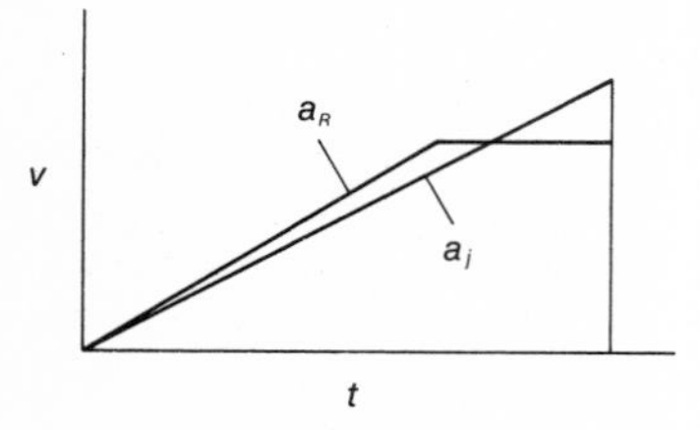

34. (a) 2.8 sec (b) 1.5×10^2 ft (c) 25 sec

35. $\frac{1}{2}a_cT^2 - [v_m - (a_c - a_p)t]T + \frac{1}{2}(a_c - a_p)t^2 = 0$

36. (a) 4 (b) −1.5 ft per 100 ft (c) the cars touched each other and moved with constant acceleration (freight cars are 50–60 ft long)

37. less than 1.1 sec

38. Theoretical: (a) 14.3 sec (b) 140 m sec^{-1}
By estimation from graph: (a) 14.5 sec
(b) 127 m sec^{-1}

39. $t_{reaction} = 0.07\ d$

40. 256 ft, 144 ft, 64 ft, 16 ft

41. (a) 40 ft sec^{-1} (b) the same—mass does not affect the answer.

42. 77 m

43. 54 ft if you assume that the stern of the scow was 5 ft above the water and that the drop point for the orange was 3 ft above the walkway

44. 51 ft with the same assumptions as for problem 43

46. (a)

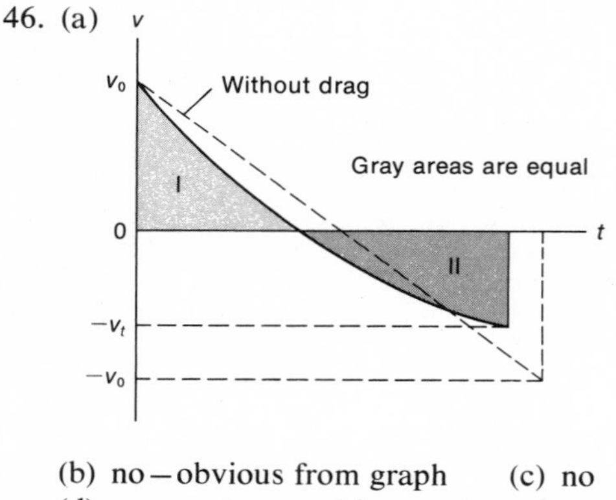

(b) no—obvious from graph (c) no
(d) no—$v_t < v_0$ (e) $t_{rise} < t_{fall}$

47. 4.9 m; the balls must be 0.5 sec apart in flight, with one arriving at one hand just as another leaves his other hand

48. 3 mi

49. 7 cars

50. (a) (1) $v_{max} = \sqrt{2aL}$ (2) there is no unique v_{max}
(b) (1) $Flow_{max} = 1/(t_0 + \sqrt{2L/a})$
(2) there is no unique $Flow_{max}$

51. (a)

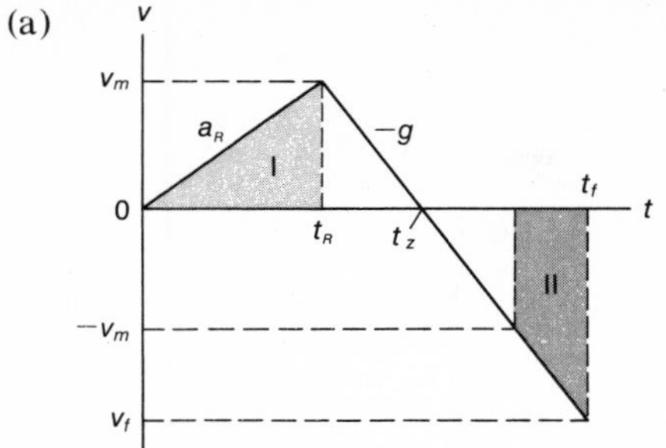

The quick way from the graph: area I = area II, the altitude of area II is greater than v_m, the altitude of area I, hence the base of area II must be less than t_R, the base of area I. Then the time of fall $t_f - t_z$ must be less than the time of rise t_z. Or by evaluating areas and equating, you can work it out algebraically that $t_f - t_z = t_z/\sqrt{1 + g/a_R}$.

(b) $v_m = a_R t_R$

(c) $\frac{1}{2} a_R t_R^2\ (1 + a_R/g)$

(d) $v_f = -a_R t_R \sqrt{1 + g/a_R}$

(e) $\Delta t = t_R(1 + a_R/g)(1 - 1/\sqrt{1 + g/a_R})$

54.

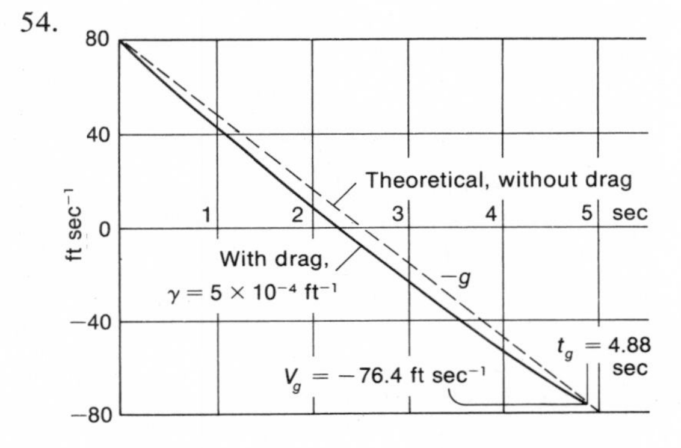

55. (a) 15 sec (b) 121 m sec^{-1}

56. (a) 9.3 m sec^{-1}
(b) 49 m
(c) 98 m sec^{-1}; actually the drag coefficient probably increased long before the balloon reached terminal speed
(d) T^{-1}

57. (a) 53 ft (b) 24 sec

59. (a) $\bar{v} = 28$ m sec^{-1}

60. (a) $\bar{I}_{0\text{-}\pi/\omega} = 0.64\ I_0$
(b) $\bar{I}_{0-2\pi/\omega} = 0$ by inspection

61. (a) $\bar{U} = A_0^2/2$
(b) Since $U_{max} = A_0^2$, $\bar{U} = \frac{1}{2} U_{max}$

62. $I_{eff} = (1/\sqrt{2}) I_0 = 0.71\ I_0$; $V_{eff} = 0.71\ V_0$

63. $\overline{KE}_t = \frac{1}{4} m\omega^2 x_0^2 = \overline{PE}_t$

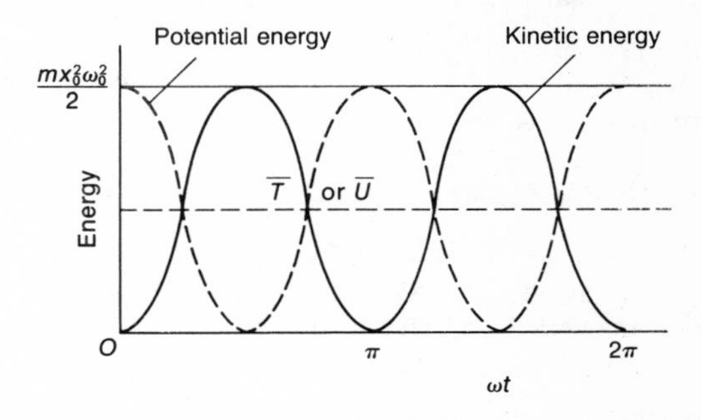

64. (a) $\overline{PE} = kA^2/6$ (b) $\overline{KE} = kA^2/3$
(c) $\overline{KE} = 2\ \overline{PE}$ (d) No; as can be seen from the diagrams, the averages in problem 63 and in problem 64 refer to quite different functions

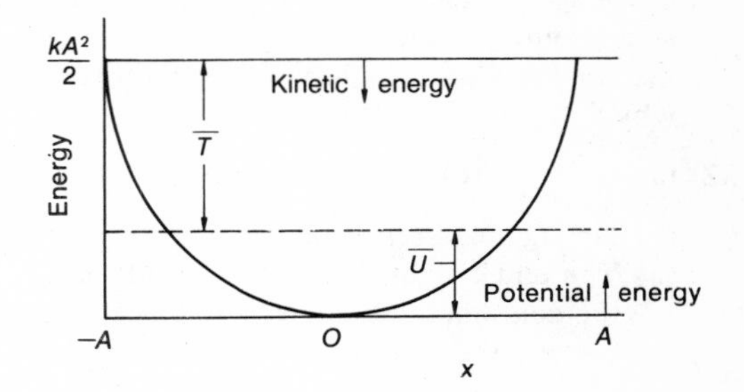

66. 20°.9, 69°.1

67. $\theta_1 + \theta_2 = \pi/2$

68. $\theta = \tan^{-1} v\sqrt{2/gH}$

69. (a) 43°.4 (b) $t = 0.125$ sec late

70. $V = A\sqrt{H/2g}$

71. (a) $y = (Eq_e/2mv^2)x^2$
(b) $y = (Eq_e\ d^2)/2mv^2$; $v_{y,d} = (Eq_e\ d)/mv$
(c) $y = [(Eq_e\ d)/mv^2](x - d/2)$

72. (a) 24 cm (b) Negligibly

73. (a) If E is doubled, y is doubled
(b) Yes; since y is linear with E, this provides the basis for using an oscilloscope as a field strength meter. Later, when we learn that E is linear with the voltage V, we will realize that an oscilloscope can be used as a voltmeter.

75. $x = 2v \sin\theta\,(v\cos\theta - u)/g$

76. 0.56 mi. if you didn't get this answer perhaps you forgot that the metal chunk had the upward speed of the plane at the instant it became separated from the plane

77. $D = 2\sqrt{HS}\,\sin\theta$

78. (a) $\theta = \sin^{-1}\sqrt{\dfrac{1-\sqrt{1-g^2s^2/v_0^4}}{2}}$

(b) $t_{\min} = (\sqrt{2}\,v_0/g)\,\sqrt{1-\sqrt{1-g^2s^2/v_0^4}}$

79. (a) $d_{AB} = (mv^2 \sin 2\theta)/Eq_e$ (b) v
It may be instructive to compare these answers with those for problem 65

80. $\theta = \tan^{-1}\dfrac{v_0^2 \pm \sqrt{v_0^4-(g^2S^2+2gHv_0^2)}}{gS}$

81. $D = [(v\cos\theta)/g][v\sin\theta \pm \sqrt{v^2\sin^2\theta - 2gH}$
This answer can take several forms, depending on how you solve the quadratic equation. The form given here is probably the simplest. The larger value for this answer would produce a successful shot, but the smaller value would not. Can you see why?

82. (a) 1°39′ (b) 1°45′

83. (a) He signalled a touchdown
(b) The end caught the ball 3.4 yds from the goal line, but carried the ball 0.6 yds beyond the goal line

84. $v_0 \sin\theta > \sqrt{gh/2}$

85. $v_0 \cos\theta > \sqrt{gD^2/2h}$

87. (a) 18.5 mi sec^{-1} (b) 0.0195 ft sec^{-2}

88. (a) −7.4 rad sec^{-2} (b) 680 rev

89. (a) 6.0 rev (b) −3.0 rad sec^{-2}

90. (a) 3.07 km sec^{-1}
(b) 0.224 m sec^{-2} = 2.3% of g at surface of the earth

91. 9% high

92. (a) 39.2 ± 0.4 rpm
(b) The shutter was about 18% slow at 1/25 sec; a nominal exposure of 0.04 sec actually was 0.047 ± 0.005 sec

93. 3.9 rps (rev sec^{-1})

94. $\theta = 54°.7$

95.

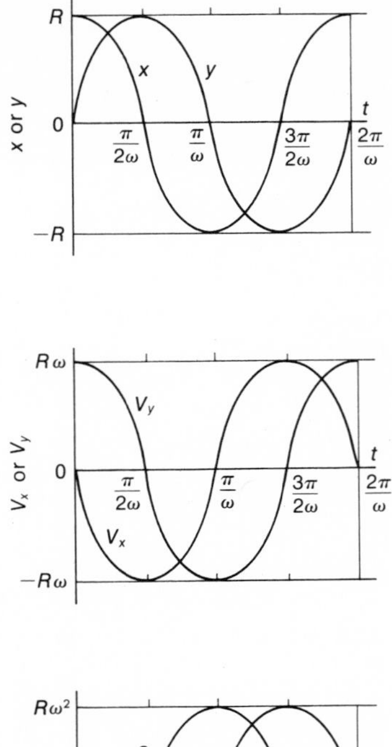

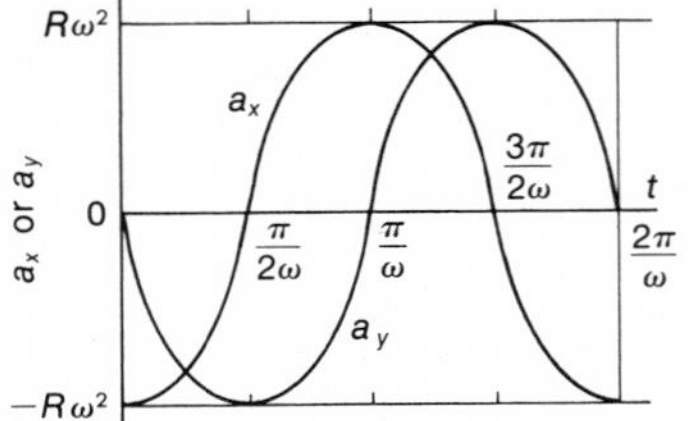

96. (a) i. 1040 mi hr^{-1}; for comparison, see answer to problem 87 ii. 0.11 ft sec^{-2} radially inward
(b) In Wainwright, Alaska—latitude 70°.5

97. 800 mi hr^{-1}

98. (a) 3 rad sec^{-1} clockwise
(b) 6 rad sec^{-2} counter-clockwise
(c) 12 cm sec^{-1} to the right
(d) 20 cm sec^{-2} to the left

99. (a) $\dfrac{x^2}{L^2} + \dfrac{y^2}{(L/3)^2} = 1$, the equation of an ellipse of semimajor axis L and semiminor axis $L/3$

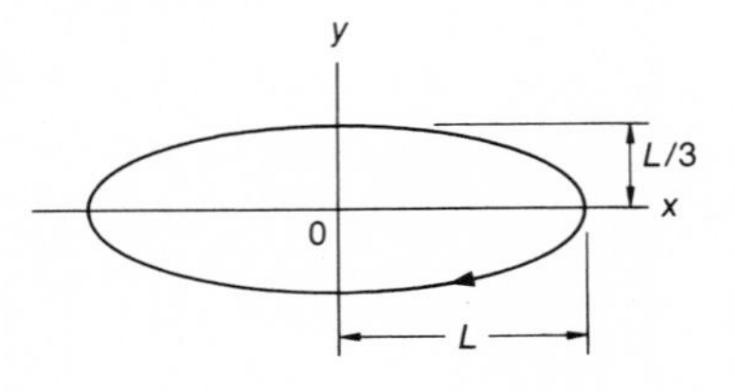

(b) $v_t = \omega L\sqrt{\sin^2 \omega t + 1/9 \cos^2 \omega t}$
(c) $a_t = \omega^2 L\sqrt{\cos^2 \omega t + 1/9 \sin^2 \omega t}$
(d) Since $a_x = -\omega^2 x$ and $a_y = -\omega^2 y$, $a = -\omega^2\sqrt{x^2 + y^2} = -\omega^2 r$, hence acceleration is centripetal
(e) At the ends of the major axis, and nowhere else

101. (a) 3800 mi hr^{-1} backward
(b) 2900 mi hr^{-1} backward

102. 3900 mi hr^{-1}

103. $V_{W-G} = 6\sqrt{2}$ knots; vector diagram

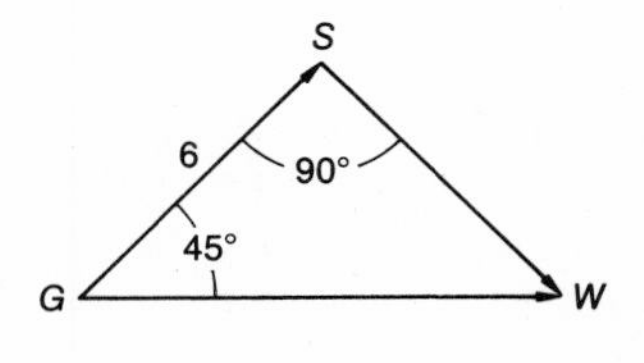

104. Wind 28 knots from 22° S of W

105. Wind must have an easterly component of 53 mi hr^{-1}

106. $v_{\text{wind}} = 62$ mi hr^{-1}

107. (a) 0.72 sec
(b) 0.83 sec
(c) 0.77 sec

108. 40 mi hr^{-1}

Note pertaining to problem 109 and those that follow: Ground measurement is usually made in statute miles or kilometers; navigation is done with distance in nautical miles and speed in knots.

109. (a) 11°33′ E of N
(b) 1 hr 46 min

110. (a) at 36°52′ S of E
(b) 2 hr 54 min

111. (a) 207 knots
(b) 39° S of E

112. (a) 1 hr 27 min
(b) 35 knots from 31°.4 S of E
(c) 11°.5 E of S
(d) 12.4 min

113. 20 mi hr^{-1}

114. (a) Due north
(b) 25 nautical miles

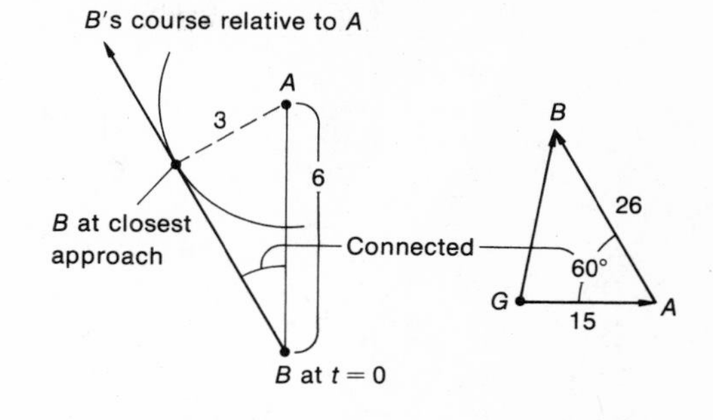

115. 28 knots due south

116. (a) $x_{I,t} = -A + v_s t$; $y_{II,t} = B - v_y t$
(b) $Av_y = Bv_x$
(c) $\Delta S = \sqrt{A^2 + B^2 + (v_x^2 + v_y^2)\, t^2 - 2t(Av_x + Bv_y)}$

117. (a) $t_{\Delta_{\min}} = (Av_x + Bv_y)/(v_x^2 + v_y^2)$
(b) $\Delta_{\min} = \sqrt{A^2 + B^2 - [(Av_x + Bv_y)^2/v_x^2 + v_y^2]}$

118. 38°.3 E of N. Probably the simplest way to do this problem is by a coordinate system based on the bomber. As a preliminary for that, we find the fighter's velocity relative to the bomber by this diagram:

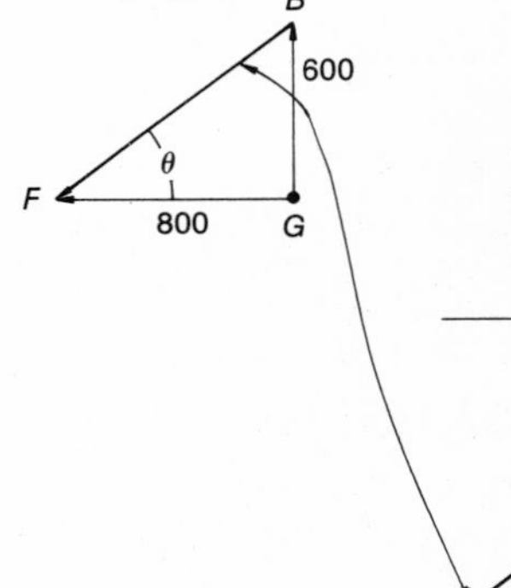

Then we may draw the space diagram in the bomber's coordinate system:

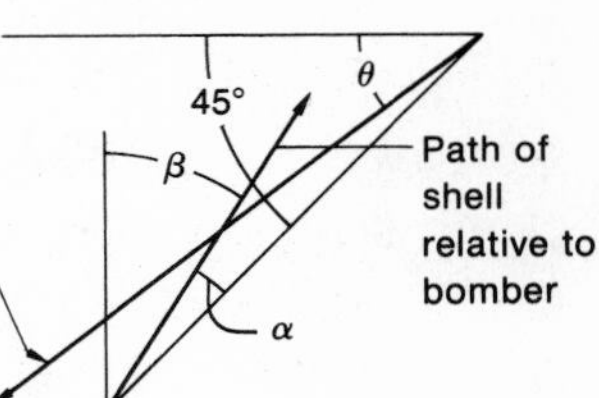

$\sin \alpha/(V_{F-B}t) = [\sin (45 - \theta)]/V_{\text{shell}}\, t$, which can be solved for α. Then $\beta = 45° - \alpha$.

119. (a) $600\sqrt{3}$ mi hr^{-1}
(b) 30° N of E

120. This problem was designed to produce the classical answer, $\tan\theta = V/c$, from

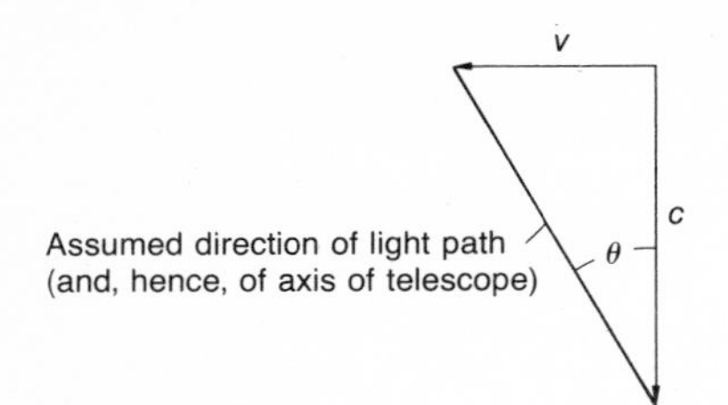

For the correct relativistic answer see problem 153.

121. (a) 10 ft sec^{-1} (b) 20° W of N

122. Method 2 is shorter by 4 min

123. $\sin^{-1} v_B/(v_M + v_R)$ N of E

124. $\cos^{-1} (2v_c + v_b)/3v$

125. (a) $\vec{a} = 3.45 \times 10^5$ m sec^{-2} at 105° with the x-axis
(b) $\vec{v} = 1.34 \times 10^4$ m sec^{-1} at 48° with the x-axis

126. $t_A/t_B = (v^2 - m^2)/(v^2 - 4m^2) > 1$, hence $t_A > t_B$, hence driver B wins; v = speed of either boat relative to water

127. (a) 1.2 nautical miles (b) 7.2 min

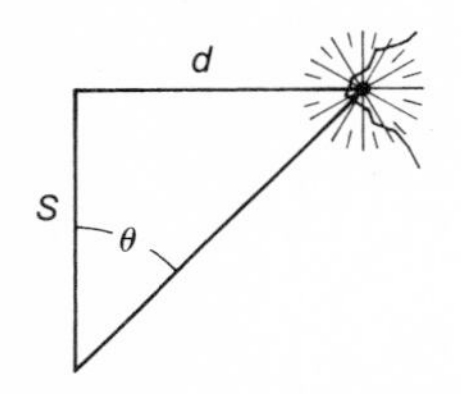

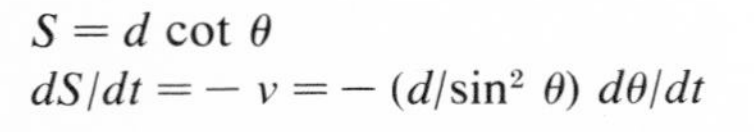
$S = d \cot\theta$
$dS/dt = -v = -(d/\sin^2\theta)\, d\theta/dt$

128. (a) 8.4 ft sec^{-1} horizontally forward
(b) 4.4 ft sec^{-2} downward

129. (a) $v = 2.6$ m sec^{-1} at $(5.2t + \pi/2)$ radians clockwise from the vertical
(b) $v = (2.6 - 0.17t)$ m sec^{-1} at $[(5.2 - 0.35t)t + \pi/2]$ radians clockwise from the vertical

130. 1.5 ft sec^{-1} at $3.5t$ radians clockwise from north

131. (a) i. 0 ii. 2v horizontally forward
iii. $\sqrt{2}$v forward at 45° below the horizontal
(b) i. v^2/R toward center of wheel
ii. same iii. same
(c) All points except the one instantaneously in contact with track

133. (a) 39° backward
(b) 1.3 sec
(c) 36° forward

134. 1.5 ft sec^{-1} at $3.5t$ radians clockwise from south; this is the same problem of the rotating observer that must be solved by an astronomer

135. $H_{max} = R + (gR^2/2v_o^2) + (v_o^2/2g)$. When you begin this problem you do not know what point on the wheel rim is the starting point for the highest flying mud; your first step is to locate this point by a maxima-minima operation. Selecting any point P on the rim,

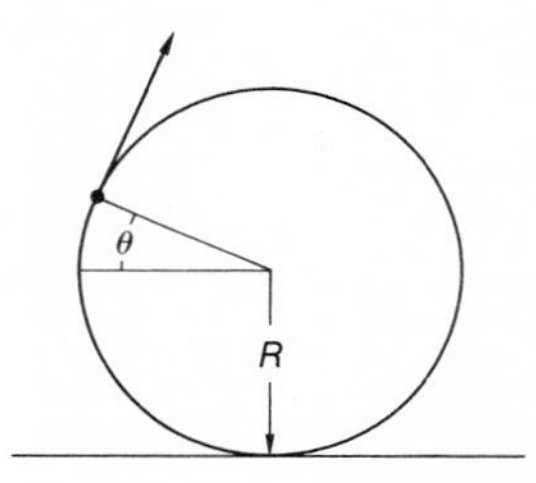

the upward component of the speed of a glob of mud leaving P is $v_o \cos\theta$, and the total height above the road reached by a glob leaving P is $H = R + R\sin\theta + (v_0 \cos\theta)^2/2g$. Differentiating this with respect to θ and setting the result = 0, $dH/d\theta = R\cos\theta - (2v_0^2 \sin\theta\cos\theta)/2g = 0$. From this, for H_{max} we have $\sin\theta = gR/v_0^2$ and $\cos^2\theta = 1 - \sin^2\theta = (v_0^4 - g^2R^2)/v_0^4$. These values can be substituted back into the above equation for H, to determine H_{max}.

136. $v_W = (r_2 - r_1)v_p/2r_2$

137. $L = L'$ for (a), (b), and (c)

138. (a) L
(b) $v_{bar} = \sqrt{v'^2 + u^2 + \sqrt{2}\, v'u}$
at $\tan^{-1} 1/(1 + \sqrt{2}u/v')$ above x-axis

140. 453 mi hr^{-1} on a course 36.°6 E of N

141.

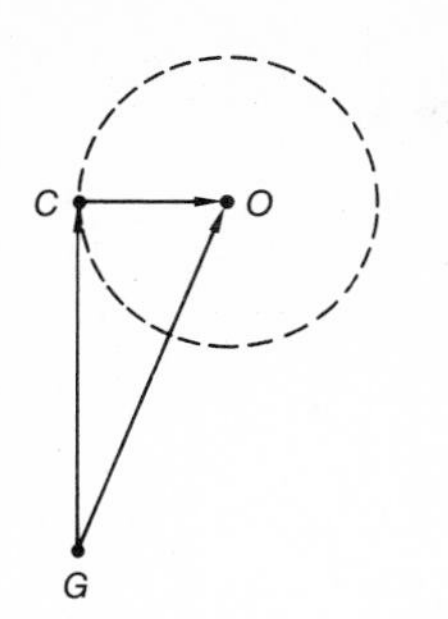

All directions between due north and 30° N of E. To meet the required condition the scalar length of the vector $\overrightarrow{OC}$ must always equal 20 m sec^{-1}; $\overrightarrow{CG}$ is the motion vector of the new coordinate system relative to ground, $\overrightarrow{OG}$ is the original object motion vector. Thus the point C must lie on a circle of radius 20 centered on O; all directions for $\overrightarrow{CG}$ between $\pm \sin^{-1} OC/OG$ will permit this.

142. (a) (b)

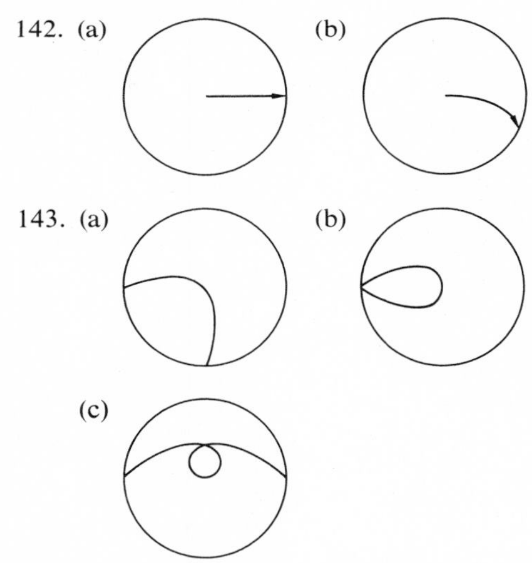

143. (a) (b)

(c)

(d) v' is not constant, because of the centrifugal acceleration
(e) No; v' increases with r, and so does $\omega^2 r$, hence both $2\omega v'$ and $\omega^2 r$ accelerations increase to make increased acceleration perpendicular to v'
(f) The greatest radius of curvature is at edge of disc

144. (a) counterclockwise (b) clockwise

145. (a) toward the northeast
(b) toward the southwest

147. 979.81 cm sec^{-2} 982.02 cm sec^{-2}

148. Even now the earth is not rigid; earlier it was more plastic than it is now. The surface adjusts at every point so as to be perpendicular to $\vec{g}_\theta$ at that point; since the deviation toward the axis between $\vec{g}_\theta$ and the radius vector increases with increasing θ, the earth's surface flattens at an increasing rate near the poles.

149. At the north and south poles, and at all points on the equator

150. (a) At $\theta = 0$, $\Delta x = 69.6$ cm eastward
(b) At $\theta = 45°$, $\Delta x = 49.2$ cm eastward

151. (a) 98.4 cm westward
(b) 0.103 m sec^{-1} westward

152. 1.96 m westward and a negligible amount (7×10^{-5}m) northward

154. (a) $L = \sqrt{1-\beta^2}\, L'$ (b) $L = 86.6$ cm

155. (a) $\theta = \tan^{-1} \gamma \tan \theta'$
(b) $L = \sqrt{1-\beta^2 \cos^2\theta'}\, L'$

156. $u = 0.82\, c$

157. $$v = \frac{\sqrt{v'^2(1-u^2/2c^2) + u^2 + \sqrt{2}\, uv'}}{1 + uv'/(\sqrt{2}\, c^2)}$$

at $\theta = \tan^{-1} \dfrac{\sqrt{1-u^2/c^2}}{1+\sqrt{2}u/v'}$

158. 0.74 c

159. −0.91 c

160. 0.70 c

161. 1 hr 02 sec

162. 0.95 c

166. 11×10^2 lbwt

167. $dF = d(mv)/dt = v\, dm/dt = mv_0/60$ lbal

168. No effect; the boy must continue to supply only the frictional force, which remains constant—see statement of problem 167

171. A force equal to his weight

172. (a) 200 lbwt (b) 0

173. 0.50 sec

174. (a) 5.5 ft sec^{-2}
(b) 17 lb

175. (a) $g/4 = 2.45$ m sec^{-2} (b) 4.9 N
(c) 0.98 N (d) 0.245 N

176. 1.4 cm sec^{-2}

177. (a) $a_{el} = \dfrac{F - (m + M)g}{m + M}$
(b) $T = m\left[g + \dfrac{F - (m + M)g}{m + M}\right]$
(c) $a'_{el} = \dfrac{F - Mg}{M}$
(d) $t = \sqrt{\dfrac{2s}{g + (F - Mg)/M}}$

178. 6

179. $\dfrac{x^2}{x_o^2} + \dfrac{y^2}{mv_o^2/k} = 1$; an ellipse with center at the origin and major axis $2x_o$ and minor axis $2v_o\sqrt{m/k}$

180. $S = 70.5$ m

181. (a) 7.0 m sec^{-2} (b) 2.5 m sec^{-2}
(c) 44 sec (d) 56 m

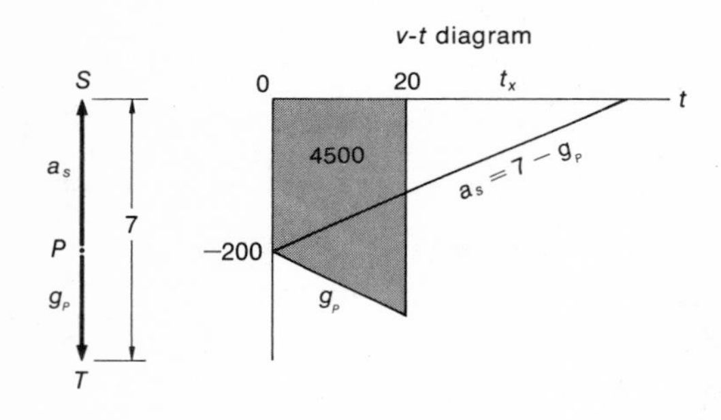

182. 2.5 m sec^{-2}

183. (a) $a_1 = \dfrac{4m_2m_3 - m_1(m_2 + m_3)}{4m_2m_3 + m_1(m_2 + m_3)}g$
(b) $T = \dfrac{16m_1m_2m_3g}{4m_2m_3 + m_1(m_2 + m_3)}$
(c) If $\dfrac{4m_2m_3}{m_2 + m_3} = m_1$, then $a_1 = 0$

184. (a) $\frac{1}{3}g$ (b) 280 lbwt

185. (a) $m_4 = \dfrac{4m_1m_2 + m_1m_3 - m_2m_3}{3m_2 - m_1}$
(b) $\dfrac{m_1}{m_2} < 3$

186. 1.0 sec

187. (a) m_2; the net force on m_2 is $T + m_2g$ downward, where T is the force in the cord connecting all the masses
(b) $a_1 = -\dfrac{m_1(m_2 + m_3)g}{m_1(m_2 + m_3) + 4m_2m_3}$; for the constraint equation you should have obtained $a_1 = (a_2 - a_3)/2$
(c) $m_1m_2 > m_3(m_1 + 4m_2)$

188. 120 lbwt

189. (a) $g \sin\theta$
(b) $a_x = -g \sin\theta \cos\theta$, $a_y = -g \sin^2\theta$
(c) $mg \cos\theta$

190. 21.9 N

191. (a) $a_0 = g \tan\theta$
(b) $F = (M + m)g \tan\theta$

192. $\vec{a}_m = -\dfrac{(M + m)g}{m + M/\sin^2\theta}\vec{j}$, where $\vec{j}$ is a unit vector in the $+y$ direction
It should be obvious that m feels no force, and hence has no acceleration, in the x-direction

193. $2g \tan\theta$

194. $F = g \tan\theta\ [2M + m(1 + \sin^2\theta)]$

195. (a) $M = m$; for the constraint equation you should have $a_{m,y} = 2a_{m,x} \tan\theta$
(b) 28°

196. (a) 0.80 m (= 3 tan 15°)
(b) −1.1 m sec^{-1}
(c) 1.5 sec

197. (a) $y = (\omega^2/2g)(x^2 - R^2/2) + h$
(b) $\omega_{max} = \dfrac{2}{R}\sqrt{gh}$

198. (a) 7 radians sec^{-1}, independent of x
(b) $\omega^2 = \dfrac{2kg}{1 - (2kr/\sqrt{1 + 4k^2x^2})}$
not independent of x

199. 724 days

200. (a) 360 N (b) due west

202. (a) $\pi G\sigma$, perpendicular to base plane of shell, toward shell
(b) 0

203. $M = 5.98 \times 10^{24}$ kg

204. (a) 2.7×10^{-3} m sec^{-2}
(b) 2.7×10^{-3} m sec^{-2}

205. (a) at 3.42×10^5 km from center of the earth

(b) 0—or the radius of curvature is infinite

(c) $\dfrac{1}{1.49 \times 10^{11}}$ m^{-1}; at that point the astronaut is on the same trajectory as the c.m. of the earth-moon system around the sun

206. $g_{alt} = 0.224$ m sec^{-2}; science *is* consistent

207. (a) 10.9 km sec^{-1}
(b) 11 km sec^{-1}
(c) Negligibly

208. 0.039 kg m^{-2}

209. $(mM/\omega)(R/d)^2$

210. $m = \Delta W/4v\omega$; did you realize that ΔW arose only from the product of the mass times the difference in the Coriolis accelerations?

211. (a) $\vec{g_r} = -(4/3)\,\pi G\rho\vec{r}$, where ρ is the density of the earth, assumed to be uniform
(b) $g_r = (r/R)\,g_R$
(c)

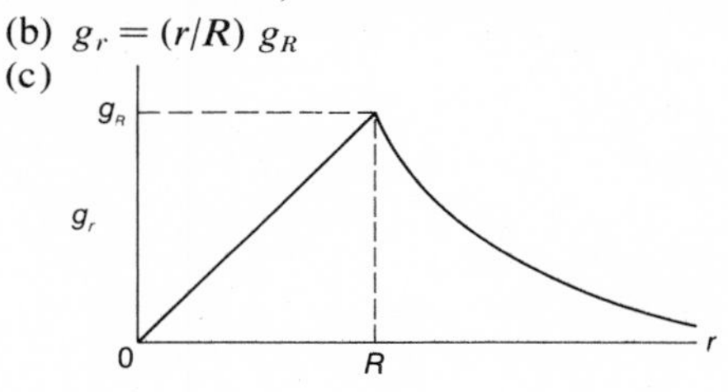

212. $\Delta W/W = 1/R = 0.025\%$

213. (a) For center of coordinates placed at C,

$$x_{c.m.} = \frac{L^2 - W(2L + D)}{2(L + D)}$$

$$y_{c.m.} = \frac{D^2 - LW}{2(L + D)}$$

(b) $L = 1.62\,D$

214. With coordinates shown, $x_{c.m.} = 0$, $y_{c.m.} = 4R/3\pi$

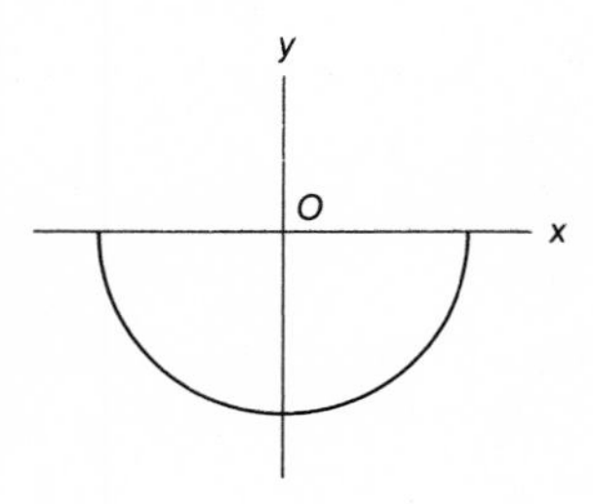

215. For coordinates as shown, $x_{c.m.} = -W/3$, $y_{c.m.} = -h/3$

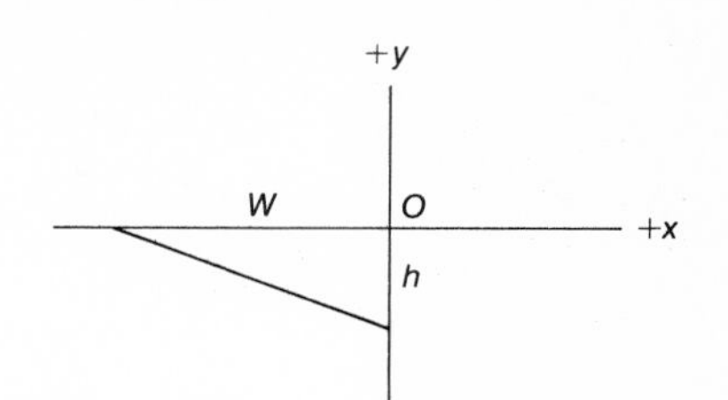

216. For coordinates shown,

$$x_{c.m.} = \frac{R\sin(\alpha/2)}{\alpha/2}$$

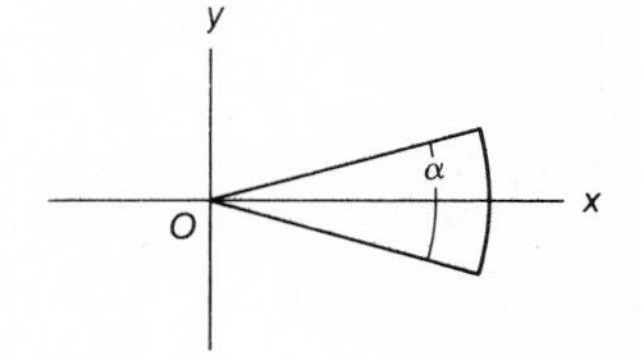

217. $x_{c.m.} = r^3/(R^2 - r^2)$, $y_{c.m.} = 0$

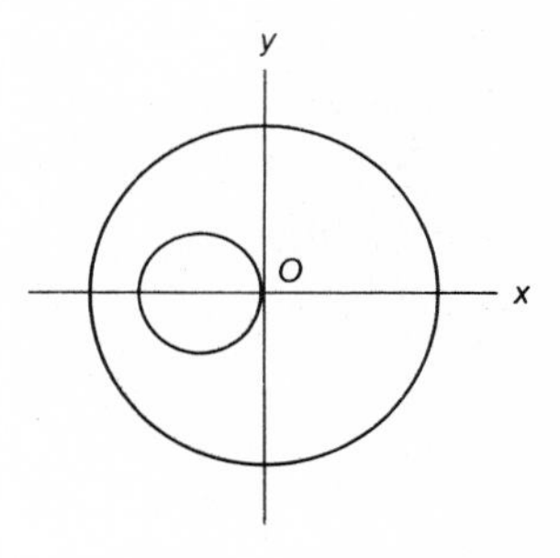

218. (a) $dg/g \simeq 1 \times 10^{-7}$ (b) i. no effect
ii. will increase weight by about 1×10^{-7}

219. (a) $M_{☽} = \dfrac{M_{\oplus}}{81.5}$
(b) $M_{☽} = 7.35 \times 10^{22}$ kg
(c) $g_{☽} = 1.63$ m sec^{-2}

220. c.m. of $\oplus$–$\odot$ system is 282 mi from center of sun; including the mass of the moon would make a difference of about 1.2% in the result

221. (a) $M_{\odot} = 3.29 \times 10^5\,M_{\oplus}$
(b) $M_{\odot} = 1.97 \times 10^{30}$ kg

222. $m_{♂} = 0.104\,M_{\oplus}$

223. 4.78 $M_{\odot}$; the data is applicable to β Aurigae

224. (a) Integrating, $\int_{v_0}^{\dot r} \dot r\, d\dot r = \int_R^r -(GM/r^2)\, dr$, $\dot r^2 = v_0^2 - 2GM(1/R - 1/r)$
(b) $r_{(\dot r = 0)} = \dfrac{R}{1 - Rv_0^2/2GM}$
(c) $v_0^2 = 2GM/R = 2\, g_R\, R$

225. $F_{el} = 1.2 \times 10^{36}\, F_{grav}$

226. 6.7×10^{15} cycles sec^{-1}

227. $v = 0.73 \times 10^{-2} c$, hence nonrelativistic

228. The maximum movement of the proton is 5.8×10^{-4} Å; from that point of view it could be considered at rest. However, the orbital speed of the proton would be 1.23×10^3 m sec^{-1}, which could hardly be called "at rest."

229. 37 N; since the nucleus does not burst apart under such forces between protons, there must be even stronger nuclear forces of attraction

230. 2.7×10^{21} N C^{-1} radially outward

231. (a) $E = Qx/[4\pi\epsilon_0(x^2 + R^2)^{3/2}]$ along the axis $= (Q/4\pi\epsilon_0 R^2) \cos\theta \sin^2\theta$, where 2θ is the angle subtended at x by the loop
(b) For $x = R/\sqrt{2}$ or $\theta = \tan^{-1}\sqrt{2}$

232. (a) $E = (Q/2\pi\epsilon_0 R^2)(1 - x/\sqrt{x^2 + R^2})$ outward along the axis $= (Q/2\pi\epsilon_0 R^2)(1 - \cos\theta)$, where 2θ is the angle subtended at x by the disc
(b) $E_{max} \simeq Q/2\pi\epsilon_0 R^2$ very close to the plane surface of the disc, where this surface appears as an infinite flat sheet. Note that on the other side of the disc the field has reversed direction. Hence the field must have passed through zero between the two plane surfaces of the disc.

233. (a) $6.1 \times 10^{10}\sqrt{Q/R}$ m sec^{-1} for Q and R in MKS units
(b) Q/R must be less than about 2×10^{-7} (for $x < 0.1\, c$)

234. (a) $\vec E = mv^2/qd$, radial
(b) $\vec E = mv^2/(\sqrt{2}\, qd)$ at 135° with original v

235. If the current is measured in the fixed coordinate system in which a point P at r from the beam is at rest (which would be the normal procedure) then $\lambda = (1 \times 10^{-6})/v$, where λ is the linear charge density and v is the speed of motion of this charge. $E_r = \lambda/2\pi\epsilon_0 r = (1.8 \times 10^4)/rv$ for MKSC units.

236. (a) 1.15×10^{-9} C m^{-2}
(b) 3.7×10^{24} electrons

237. 5.9×10^{-13} C m^{-3}

238. (a) i. 0 ii. σ/ϵ_0 (b) i. σ/ϵ_0 ii. 0

239. (a) $E_x = \dfrac{\sigma x}{2\epsilon_0}\left[\dfrac{1}{\sqrt{r^2 + x^2}} - \dfrac{1}{\sqrt{R^2 + x^2}}\right]$
(b) $E_x \to 0$ for $x >> R$

240. $V_{y=0} = 0$

241. (a) r_x must be $> 20d$ (b) r_y must be $> 23d$

242. If you try it you will find it much easier

243. (a) $E_r = \dfrac{V}{r \ln R/r}$ radially outward
(b) $E_R = \dfrac{V}{R \ln R/r}$ radially outward

244. (a) $E = (\lambda/2\pi\epsilon_0 d)(1 - a/d)$ downward
(b) $V_Y - V_P = (\lambda/2\pi\epsilon_0)[\ln y/d + a(1/y - 1/d)]$

245.

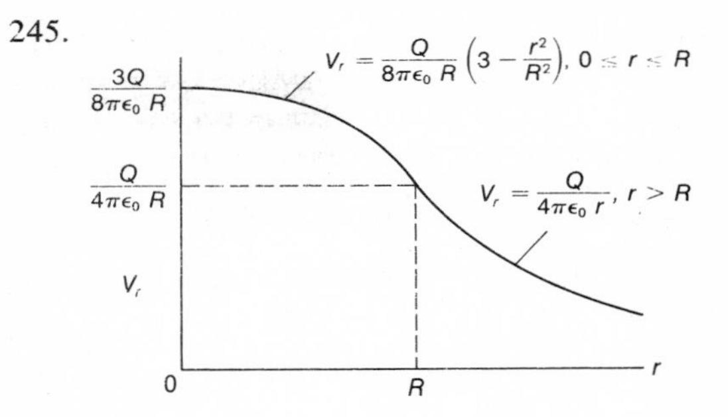

246. (a) i. $\vec E_r = \dfrac{q_e}{4\pi\epsilon_0 r^2}\left(1 - \dfrac{r^3}{R^3}\right)$ radially outward, $0 \leq r \leq R$
ii. $\vec E_r = 0$, $r > R$
(b) i. $V_r = \dfrac{q_e}{4\pi\epsilon_0}\left(\dfrac{1}{r} - \dfrac{1}{R} - \dfrac{R^2 - r^2}{2R^3}\right)$
$0 < r < R$
ii. $V_r = 0$, $r \geqslant R$

247. (a) $Np/2\epsilon_0$ (b) 0

248. From discussion on page 132 note that α, the direction angle for $\vec E$, is constant for any particular θ. Thus at all points on the cone the field crosses the surface at the same angle.

249. $\sigma d/\epsilon_0$

250. 51 volts

251. (a) $V = (mdu^2/q_e L) \tan \theta$ (b) $\frac{1}{2}L \tan \theta$

252. $n = mgd/Vq_e$

253. (a) $V_A = -0.65q/4\pi\epsilon_0 a$
(b) $\vec{F} = 0.92q^2/16\pi\epsilon_0 a^2$ pointing outward from center of square

254. (a) $q/4\pi\epsilon_0 s$
(b) $2q/4\pi\epsilon_0 s$
(c) $3\sqrt{3}\, q/4\pi\epsilon_0 s$
(d) $(-2/\sqrt{3})\, q$

255. (a) $-g_R R^2/r$
(b) $-g_R R$
(c) $-3g_R R/2$

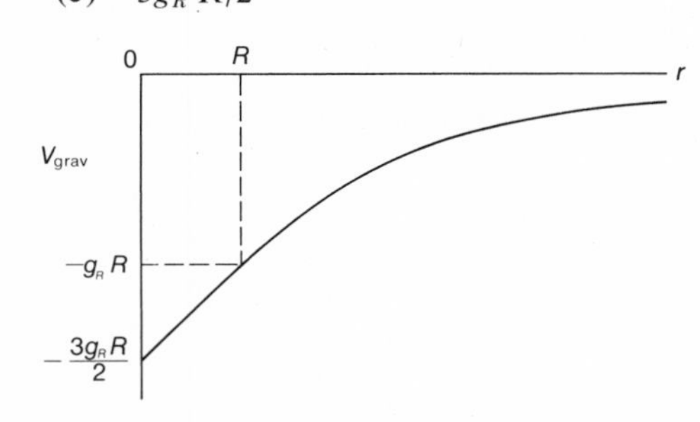

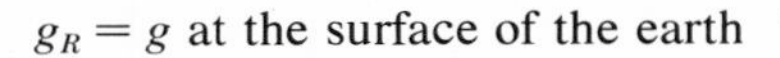
$g_R = g$ at the surface of the earth

256. $V_R = -9.5 \times 10^8$ N m kg^{-1}; about 93% of this is contributed by the sun, and the moon's contribution is negligible

257. σ/ϵ_0

258. No

259. $4\, V_0$

260. Since equilibrium was not disturbed by the shorting wire, no charge moved, so spheres must have been at same potential. Then $V_1 = V_2$, which leads to $q_2 = 2q_1$. From $F = q_1 q_2/4\pi\epsilon_0 d^2$ we obtain $q_1 = \sqrt{2\pi\epsilon_0 F}\, d$ and $q_2 = 2\sqrt{2\pi\epsilon_0 F}\, d$.

261. (a) $\frac{8}{11}V$ (b) $\frac{3}{11}q_0$

262. (a) 1.8×10^6 volts
(b) $V_{max} = 3 \times 10^6\, r$, if r is in meters

263. 30 volts

264. (a) $V_R = V_0 - (\lambda/2\pi\epsilon_0) \ln R/r$
(b) $\lambda/2\pi\epsilon_0 R$
(c) No

265. (a) $E_x = -\frac{q}{4\pi\epsilon_0}\left\{\frac{(d+x)}{[(d+x)^2+y^2]^{3/2}} + \frac{(d-x)}{[(d-x)^2+y^2]^{3/2}}\right\}$

$E_y = \frac{qy}{4\pi\epsilon_0}\left\{\frac{1}{[(d-x)^2+y^2]^{3/2}} - \frac{1}{[(d+x)^2+y^2]^{3/2}}\right\}$

(b) For $x = 0$, $E_y = 0$ for all y. Thus a plane surface perpendicular to the x-axis and passing through $x = 0$ is an equipotential surface. A thin conducting sheet could be placed in this plane without disturbing the environment.
(c) By computing V from the charges for $x = 0$ or better, by noting that no work is required to move a charge from infinity to anywhere on this surface, you find that $V_{(x=0)} = 0$.

266. (a) The real charge $+q$ and the image charge $-q$, which are $2d$ apart, make this problem an exact duplicate of problem 265, with the grounded conducting sheet acting as the equipotential surface at $x = 0$ in that problem. Then from that problem we have at the surface that $E_r = -2qd/[4\pi\epsilon_0(d^2 + r^2)^{3/2}]$. Since E_r must arise from σ_r/ϵ_0, $\sigma_r = -2qd/[4\pi(d^2 + r^2)^{3/2}]$.
(b) $R = \sqrt{99}\, d = 9.95\, d$
(c) $E_{(r=R)} = 10^{-3}\, E_{(r=0)}$
(d) $V = -q/8\pi\epsilon_0 d$; note that this value is the same, whether calculated from the charges on the sheet or from the image

267. (a) The original charge $+q$ establishes the two $-q$ image charges, and these in turn require the $+q$ image. This array of charges is exactly the same as for problem 253. From that problem we have $V_A = -0.65\, q/4\pi\epsilon_0 d$.

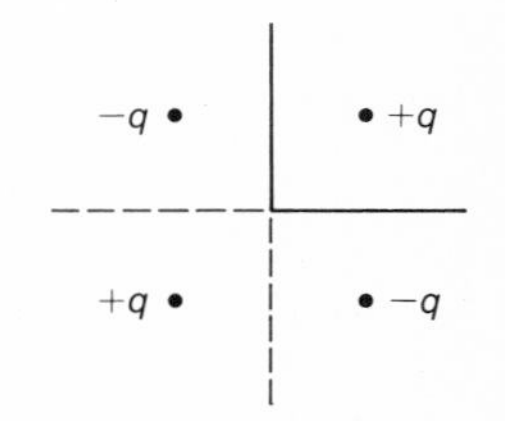

(b) $F_A = 0.92\, q^2/16\pi\epsilon_0 d^2$ out from corner

268. 4×10^{-15} N per electron

269. $V = 2.7 \times 10^5 \sqrt{W}$ volts *Comment:* For no breakdown in air, V must be $< 3 \times 10^6 R$ volts; if available maximum V is 6×10^4 volts, then for no breakdown R must be $>$ 2 cm. From $V \leqslant 6 \times 10^4$ volts and $V \geqslant 2.7 \times 10^5 \sqrt{W}$ volts for the experiment to work, the mass of the hemisphere must be about 5 grams. For a mass of 5 gm and $R > 2$ cm, hemisphere must be made of metal foil.

270. (a) $Q^2/2\epsilon_0 A$
(b) None

271. $\epsilon_0 A/d$

272. (a) $C_{\text{sphere}} = 4\pi\epsilon_0 R$
(b) $C_{\oplus} \simeq 7 \times 10^{-4}$ farad $= 7 \times 10^2\ \mu f$

273. (a) 3×10^3 volts
(b) 2.7×10^{-5} C m^{-2} $= 1.7 \times 10^{10}$ electron charges cm^{-2}—about 2.3×10^4 times the earth's surface charge in normal fair weather (cf. problem 236)

274. Since $1/C_{\text{comb}} = 1/C_1 + 1/C_2 + 1/C_3 + \ldots$, $1/C_{\text{comb}} > 1/C_n$, where C_n is the capacity of any one of the individual capacitors; hence $C_{\text{comb}} < C_n$

275. (a) $C_3 = C_1^2/(C_1 + C_2)$
(b) $C_3 = [(C_1 + C_2)C_1]/C_2$

277. $V_{\text{max}} = \epsilon_0 A E_{cr}/C_0$

278. 5.2×10^3 volts

279. (a) $\dfrac{dC}{dt} = \dfrac{(K-1)uC_0}{W}$

(b) $\dfrac{dV}{dt} = -\dfrac{(K-1)uV_0}{W[1 + (ut/W)(K-1)]^2}$

(c) $V_t = \dfrac{V_0}{K}$

280. (a) $-\dfrac{\lambda}{2\pi K\epsilon_0} \ln \dfrac{R}{r}$

(b) $-\dfrac{\lambda}{2\pi\epsilon_0}\left[\dfrac{1}{K} \ln \dfrac{R}{r} + \ln 2\right]$

281. $\vec{B} = mv/qd$ perpendicular to the plane of $\vec{v}_0$ and $\vec{v}_f$

282. A uniform horizontal field $B \perp$ to the beam will turn the beam in a vertical circle of radius R. To deflect the beam through an angle θ in the horizontal distance L, $L = R \sin\theta$, or R must $= L/\sin\theta$. Then $\vec{B} = mu/q_e R = (mu \sin\theta)/q_e L$, horizontal and perpendicular to the beam. Now the vertical displacement at L is $L(1 - \cos\theta)/\sin\theta$, which is not the same as the vertical displacement in the electric case in problem 251. It is not possible by a circular arc to achieve both a given displacement and a given deflection over a given horizontal distance. You might try a variation of this problem, in which it is required that a vertical displacement y_0 be provided in a horizontal distance L. The answer is $B = 2muy_0/[q_e(L^2 + y_0^2)]$ and the angular deflection is $\sin^{-1} 2Ly_0/(L^2 + y_0^2)$.

283. (a) $\omega = qB/m$, which is independent of r
(b) no adjustment necessary, because independent of r
(c) not a spiral, but a series of semicircles

284. (a) $B_r = \dfrac{m\omega}{q\sqrt{1 - r^2\omega^2/c^2}}$

(b) $\omega = \dfrac{qB}{m\sqrt{1 + (qB/mc)^2 r^2}}$

285. (a) minus
(b) $\tan^{-1} 2\pi rp$
(c) It is not possible to write an equation for v that does not contain the term q/m, or vice versa

(d) $v = \dfrac{q}{m}\,\dfrac{B\sqrt{1 + (2\pi rp)^2}}{2\pi p}$

286. (a) A charged particle emerging at r with a radial speed $v \sin\alpha$ will be bent by a B parallel to the axis into a circle. To fit inside R, this circle must have a radius r (draw the cross-sectional geometry to see this). Then $B = (mV \sin\alpha)/qr$.
(b) B must be parallel to the axis, in *either* direction

287. (a) $q/m = 8V/(dB)^2$
(b) For any ion entering S at an angle its d' will be a chord of a circle of diameter d, and hence $d' < d$, and all such ions will fall inside d. Thus the outer edge of the image on P is sharp.

288. (a) $v = E/B_0$
(b) $q/m = 2EA/B_0^2$
(c) The particles are charged, otherwise they would not have responded to E as they did. The ratio q/m is the same for

all, otherwise the trace would not have been narrow. The particles all have the same speed, otherwise they would not all have responded in the same way to B. However, there is nothing in the data that would discriminate among particles with q/m, $2q/2m$ . . . , nq/nm, etc.; hence it is not possible to say that the particles are all alike.

289. By passing through the wire a current $I = I_0 \sin \omega t$; the wire will vibrate with a frequency $\nu = 2\pi/\omega$

290. $I = F/aB$, where F is the force needed to keep the beam in balance. (Note that with electrons moving as shown, the net torque on the coil is zero, so that its plane will stay perpendicular to B.) The equation itself does not provide any answer to the question about the suitability of this method for measuring I or B. However, there are more independent methods for measuring I than there are for B, hence this method is more useful for measuring B.

291. (a) The loop is in a region containing a magnetic field with an upward component (and probably an outward component also if the wire seems to rest firmly on the table)
(b) $B_y = 2\pi F_t/IL$, where F_t is the tension in the wire

292. 22 m

293. (a) $B_x = (I/2\pi\epsilon_0 c^2)[d/x(d-x)]$, coordinates as shown

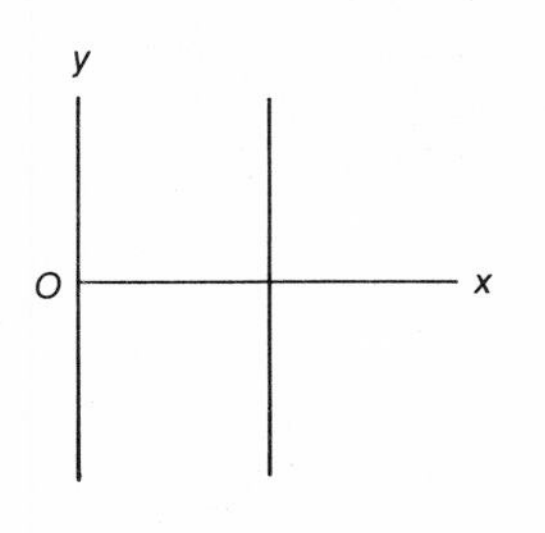

(b) $I^2/2\pi\epsilon_0 c^2 d$ outward

294.

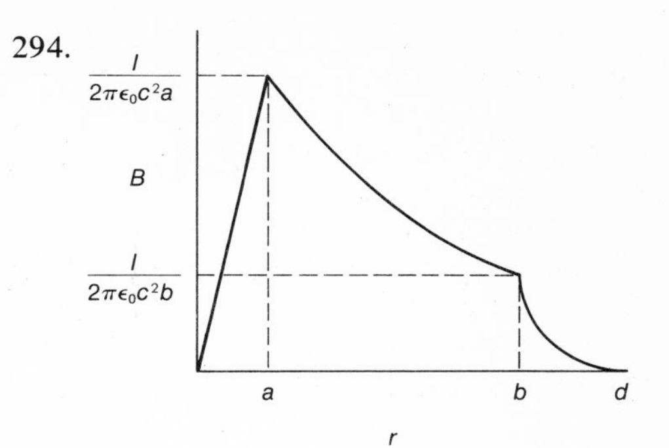

For $r < a$, $B_r = \dfrac{I}{2\pi\epsilon_0 c^2} \dfrac{r}{a^2}$

For $a < r < b$, $B_r = \dfrac{I}{2\pi\epsilon_0 c^2 r}$

For $b < r < d$, $B_r = \dfrac{I}{2\pi\epsilon_0 c^2 r} \dfrac{(d^2 - r^2)}{(d^2 - b^2)}$

295. (a) $$B_x = \frac{NIR^2}{2\pi\epsilon_0 c^2} \left\{ \frac{1}{(x^2 + R^2)^{3/2}} + \frac{1}{[(d-x)^2 + R^2]^{3/2}} \right\}$$
$B_y = B_z = 0$
(b) $$\frac{dB}{dx} = \frac{NIR^2}{2\pi\epsilon_0 c^2} \left\{ \frac{-3x}{(x^2 + R^2)^{5/2}} + \frac{3(d-x)}{[(d-x)^2 + R^2]^{5/2}} \right\},$$
which $= 0$ for $(d - x) = x$ or for $x = d/2$
(c) $\left.\dfrac{d^2B}{dx^2}\right|_{x=d/2} = 0$ when $d = R$
(d)

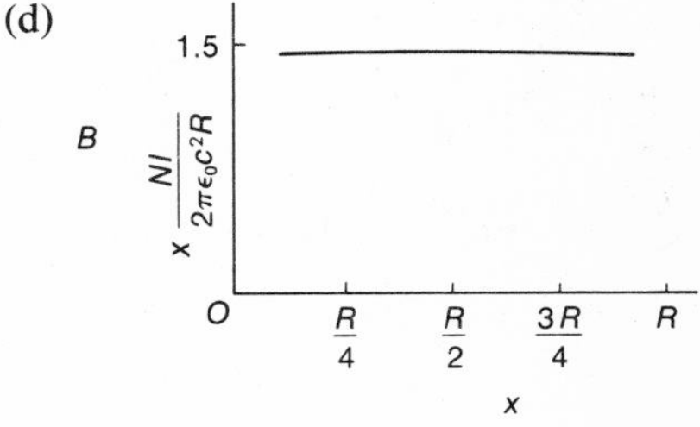

296. $(I/2\pi\epsilon_0 c^2 R)(\cot \theta - 1)$

297. (a) $$B_x = \frac{Q\omega}{2\pi\epsilon_0 c^2 R^2} \left[\frac{2x^2 + R^2}{\sqrt{x^2 + R^2}} - \frac{2x^2}{\sqrt{x^2}} \right],$$
where Q is the total charge on the disc
(b) $B_{x=0} = \dfrac{Q\omega}{2\pi\epsilon_0 c^2 R}$

298. 36 millivolts

299. (a) $\mathscr{E}_{\text{max}} = \pi RBL_0 D/2$
(b) no difference, the product NA is the same for both forms

300. (a) $\mathscr{E} = \omega NAB \sin \omega t$ if $t = 0$ when plane of coil and B are perpendicular
(b) $\mathscr{E}_\theta = \omega NAB \sin \omega t \sin \theta$
(c) Yes; when the axis of rotation is exactly lined up with the direction of the field the induced voltage (or the induced current in a detection circuit) is zero. (Note, however, that this null method does not discriminate between θ and $\theta + \pi$.) This is the basis principle of the earth inductor compass.

301. $\mathscr{E} = \omega NAB \sin \omega t$

302. (a) $\mathscr{E} = -\omega NAB_0 \cos 2\omega t$
(b) $\mathscr{E} = \omega NAB_0 \sin 2\omega t$
Note that the induced emfs have a frequency double that of the basic frequencies. The difference in signs indicates that one emf lags (or leads) the other by 90°.

303. The magnetic field resulting from this circuit current would be in opposition to the original field responsible for the emf. The secondary emf generated by this opposing field is in the opposite direction to the original open-circuit emf, and must be subtracted from it to find the net emf available to the circuit. For this reason this opposing emf is called the "back emf."

304. $L = n^2\pi r^2/\epsilon_0 c^2$

305. $M = \text{pan}/\epsilon_0 c^2$

306. $L = \dfrac{\ln b/a}{2\pi\epsilon_0 c^2}$ per unit length

307. $L_{\text{unit length}} = \dfrac{\ln (d - r)/r}{\pi\epsilon_0 c^2}$

308. $\mu_k = (\sin \theta_1 - \sin \theta_2)/\cos \theta_1$

309. $\mu_k = (\sin^2 \theta_3 - \sin^2 \theta_4)/\sin \theta_3 \cos \theta_3$

310. (a) 70.4 sec
(b) 102 mi hr^{-1}—which must seem unrealistically high. If a more realistic value for the coefficient of friction had been given, the answer for the speed would have been even more unrealistic. The explanation is that an actual v-t curve would look like this

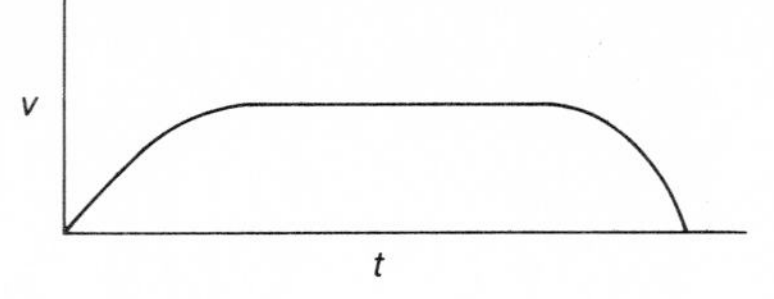

instead of like this

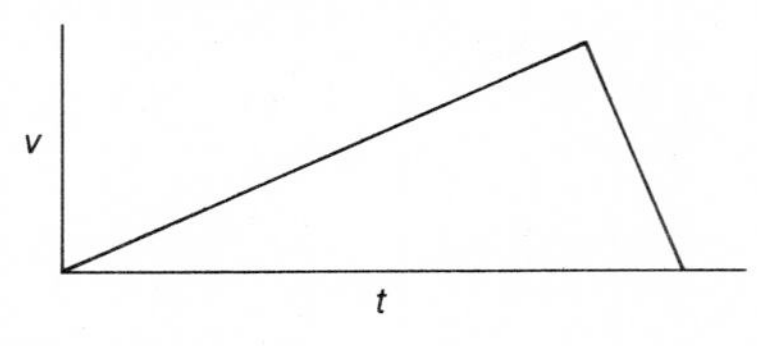

311. $10\sqrt{3}$ gmwt tension

312. $F = [W(\sin \theta + \mu \cos \theta)]/(\cos \theta - \mu \sin \theta)$, F and W in compatible units

313. (a) $\alpha = \theta + \tan^{-1} \mu$ for maximum acceleration
(b) $a_{\max} = [(B/W)\sqrt{1 + \mu^2} - (\sin \theta + \mu \cos \theta)]g$, B and W in practical units

314. $\mu = 0.61$; solution simplest by use of coordinate system moving with the car

315. (a) $-N_1 = ma_{m,x}$
$\mu N_1 - mg = ma_{m,y}$
$N_1 - N_2 \sin \theta = Ma_{M,x}$
$-\mu N_1 + N_2 \cos \theta - Mg = Ma_{M,y}$
$a_{m,x} = a_{M,x} = a_{M,y} \cot \theta$

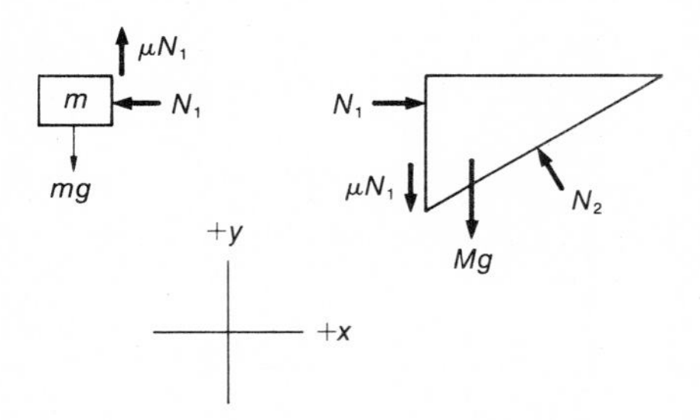

(b) For $a_{m,x} = a_{m,y} \tan \theta$,
$\mu = (m + M \tan^2 \theta)/[(m + M) \tan \theta]$

316. $\dfrac{m_1 - \mu_2 m_2}{m_2 + \mu_1 m_1} g \leqslant a \leqslant \dfrac{m_1 + \mu_2 m_2}{m_2 - \mu_1 m_1} g$

317. (a) $\mu_{crate} < \mu_{tires}$

(b)

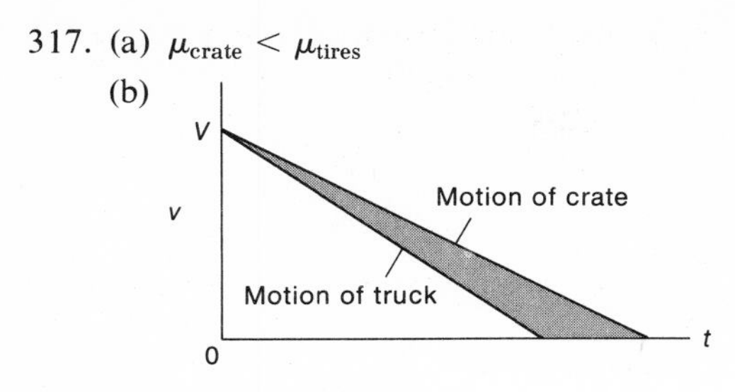

(c) the hatched area
(d) at the instant the truck stopped
(e) 0.67

318. $v_0 \geqslant \sqrt{v^2 + 10\ \mu g d}$

320. 4.4 m

321. (a) $v_r = (\Delta P/4\eta)(R^2 - r^2)$

(b)

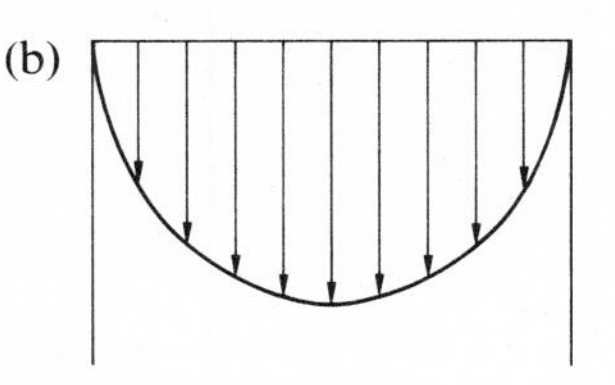

322. (a) $Q = \Delta P \pi R^4/8\eta$ (b) $v_{av} = \frac{1}{2} v_{max}$

323. 7.7×10^2 dynes

324. 5.4×10^2 cm dynes

325. (a) $v_{term} = 2r^2 \rho_w g/9\eta$
(b) $8.5 \times 10^3\ r^2$ sec, if r is in cm
(c) $1.2 \times 10^{10}\ r^4$ cm if r is in cm
(d) From the answer to part (a) it is clear that an increase in r will cause an increase in speed as long as Stoke's Law holds.

326. $r_{max} = 4.9 \times 10^{-3}$ cm; see limitation on applicability of Stoke's Law following problem 324

327. (a) 282 sec (b) 7.1×10^2 cm sec^{-1}
(c) 1.8 sec (d) 1.4×10^{-3} sec

328. Percentage error $= (v_{term,\, accurate} - v_{term,\, approx})/v_{term,\, accurate} = -\rho_{air}/(\rho_w - \rho_{air}) = -0.13\%$

329. 4×10^2 gmwt

330. approximately $RB_0/2w$ if $w << R$

331. (a) 44% (b) 138

332. $\rho_\ell = h_w/h_\ell$ gm cm^{-3}

333. 1.7 ft maximum depth

334. To the extreme top of the tube if there is any pressure in the capsule greater than the vapor pressure of mercury

335. 157 lbwt

336. (a) 2.3 inches below drain channel
(b) Only change is due to evaporation, hence new water level is 3.3 inches below drain channel

337. $\Delta w = \rho_a[V - (w/\rho_w)]g$ in absolute units
$= \rho_a[V - (w/\rho_w)]$ in practical units

338. (a) 4 wg (b) $4\rho_\ell$ (c) 3 wg

339. 8.0×10^3 kgwt m^{-2}

340. Weight of water displaced by barge plus weight of water left in box equals total weight in box; for balance these weights must be equal. If boxes are equal in cross-section, then for balance water heights in boxes must be equal.

341. At rest, $(P - p)A - T_0 - V\rho_s g = 0$ and $(P - p)\,A - V\rho_\ell g = 0$

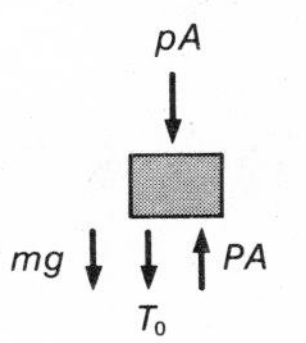

Accelerated, $(P' - p')A - T - V\rho_s g = V\rho_s a$ and $(P' - p')A - V\rho_\ell g = V\rho_\ell a$

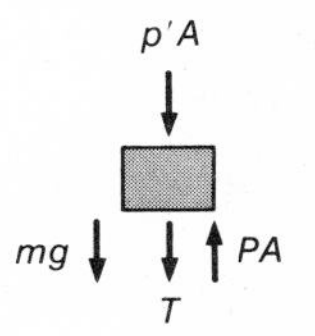

Solving for T, $T = T_0(1 + a/g)$

342. (a) 22 gmwt (b) T_{min} (at top) = 2.5 gmwt T_{max} (at bottom) = 46.5 gmwt

343. For $T = 0$, $\omega = 2\sqrt{5}$ rad sec^{-1} = 42.7 rpm

344. $T = 40.5$ gmwt at 33° below the horizontal. In solving for θ you should get $5.45 \tan\theta - \sin\theta = 4.02$. This can be solved explicitly for θ, but a lot of work is in-

volved. It is much simpler and faster to look at the trig tables and make a stab at a cut-and-try answer; you should hit the answer on the nose by the third try.

345. $\eta = [2r^2g(\rho_{ball} - \rho_{liq})]t/9y$, with r = radius of ball

346. (a) 12.9 sec (b) 22.5 m

348. 16.4 m
(b) $\ddot{y}_{down} = g/9 + D\dot{y}$, $\ddot{y}_{up} = g/9 - D\dot{y}$; $\ddot{y}_{down} > \ddot{y}_{up}$
(c) For $\dot{y}_{up}$ = constant, $\ddot{y}_{up} = 0$ or $g/9 = Dv$; then $v_{up} = 11.8$ m sec^{-1}

350. (a) $\frac{7}{5}MR^2$ (b) $\frac{1}{3}ML^2$

351. $R\sqrt{0.3}$

352. (a) $\frac{289}{288}MR^2$ (b) i. yes; error = $\frac{1}{288}$, which is less than 1% ii. no; error = $\frac{73}{937}$, which is much greater than 1%

353. (a) $\frac{1}{3}mw^2$ (b) $\frac{1}{12}mh^2$

354. (a) Truck dips in front from torque of decelerating force about axis through c.m.
(b) Actual acceleration $< -\mu g$, hence crate slides farther

355. (a) i. if surface is smooth cylinder could roll and slip at the same time, even when moving at constant speed
ii. if surface has friction, constant speed means no horizontal force, hence no relative motion between surface of cylinder and horizontal surface; hence cylinder must be rolling without slipping
(b) 0, for reason given in (ii)

356. g

357. (a) $mg \sin \theta$ (b) $\alpha_0 = (2g \sin \theta)/r$
(c) $2g \sin \theta$

358. (a) $\frac{1}{2}mg \sin \theta$ (b) $\frac{1}{2}mg \sin \theta$
(c) upward ∥ to incline

359. (a) assuming rolling without slipping, $a_0 = \frac{2}{3}g \sin \theta$
(b) $\frac{1}{3}mg \sin \theta$ Since this is less than the $\frac{1}{2}mg \sin \theta$ available—see (a) of problem 358—assumption of rolling without slipping is justified

360. (a) 6.3×10^2 lbwt ft sec^{-1}
(b) 1.9×10^3 lbwt ft

361. From four dynamical equations and two constraint equations you get $a_{pulley} = [F(m_1 + m_2 + m_p)]/[(2m_1 + m_p)(2m_2 + m_p)] - \mu g = 0.39$ m sec^{-2}

362. (a) $a_0 = \frac{5}{7}g \sin \theta = \frac{5}{7}a_{block}$
(b) $\frac{2}{7}mg \sin \theta$
(c) upward ∥ to plane

363. (a) $\frac{5}{6}g \sin \theta$
(b) $a = [(M + m)g \sin \theta]/(m + \frac{7}{5}M)$

364. Relative to the sphere alone of problem 362, there are now two new forces acting on the sphere, a normal force and a tangential frictional force. For these to have no effect their torques about P must cancel; $\Sigma\tau_P = \mu Nr - Nr = 0$, hence

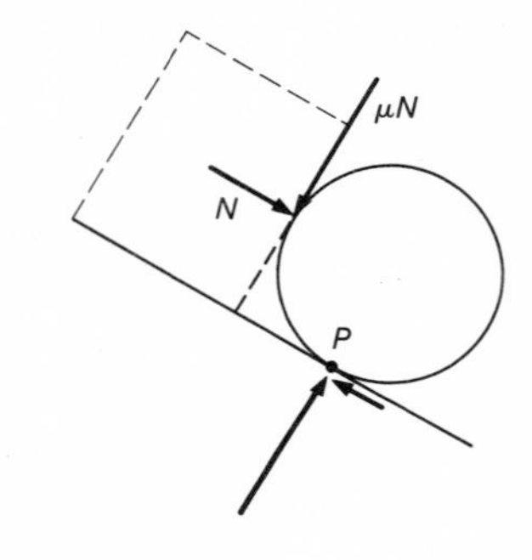

(a) $\mu = 1$
(b) $\mu_{plane} = 4/(2 + 7 \cot \theta)$

365. 27°.5

366. $\sqrt{3L/(g \sin \theta)}$; you should have found that the linear acceleration of the center of the roll at any instant was independent of the radius of the roll at that instant, which is another way of finding out that the linear acceleration of the center of the roll was constant at $\frac{2}{3}g \sin \theta$

367. (a) $a_M = -2(m + 2M)g/(m + 6M)$
(b) $-\frac{2}{3}g$

368. (a) $(M + 2m_1)g$
(b) $\dfrac{M^2 - 2M(m_1 + m_2) - 12\, m_1m_2}{M - 3(m_1 + m_2)}g$
The scale reading is the sum of the forces *on* the beam; to get at these forces it is necessary to construct separate free body diagrams for the beam and the masses, with separate dynamical equations

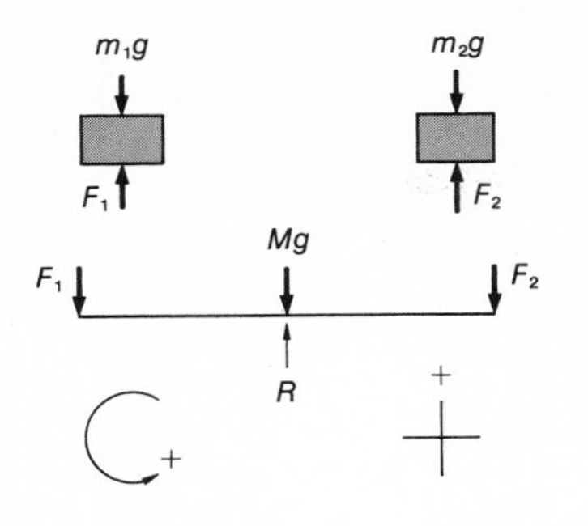

$F_1 - m_1g = m_1a_1$
$F_2 - m_2g = m_2a_2$
$(F_2 - F_1)\ L/2 = 1/12\ ML^2\alpha$
$R = F_1 + F_2 + Mg$
Constraints: $a_1 = -a_2 = -\ L/2\ \alpha$

369. (a) $15R\omega^2/2\pi g$ (b) 18° (c) $\mu = 0.45$, a reasonable value for ribbed vinyl on a napped surface

370. (a) at the bottom of the rim
(b) $\sin^{-1} 2R\alpha/5g$

371. $\tau = (\frac{5}{2}M + m)\ gR \sin\theta$

372. $\alpha_{max} = (2g \cos\theta)/R$

373. $r^2\omega_0^2/18\mu g$

374. (a) $v_0 = \frac{2}{5}r\omega_0$, r being the radius of the marble
(b) $v_0^2/2\mu g$ (c) $v_0 = \frac{1}{5}r\omega_0$

375. (a) $\dfrac{v_{pend}}{v_{ball}} = \sqrt{\dfrac{7\ell^2}{2r^2 + 5\ell^2}}$
(b) $\dfrac{T_{pend}}{T_{ball}} = \sqrt{\dfrac{2r^2 + 5\ell^2}{7\ell^2}}$

376. $\eta = \dfrac{FR_0^2(R_2^2 - R_1^2)}{4\pi hvR_1^2R_2^2}$

378. $F_J = (WL \tan\theta)/4d$

379. $\tan\theta_{max} = \mu_1/[a - (1 - a)\ \mu_1\mu_2]$

381. (a) $60 \tan\theta$ lbwt
(b) $60 \sec\theta$ lbwt
(c) $60 \cos\theta$ lbwt
(d) at bottom of swinging arc; ($T_{max} = 60(3 - 2\cos\theta)$ lbwt

382. (a) burn the thread to the 500-gm weight
(b) initial acceleration of 100-gm mass = $1.1g$ at $\tan^{-1}\frac{15}{16}$ with the original direction of the thread to the 500-gm mass

383. $\mu_{min} = \tan(\theta/2)$

384. total number of bricks = a

385. $W/\sqrt{3}$

386. $w_{max} = \frac{3}{2}W$

387. (a) nothing—tube still stands upright
(b) tube falls over toward the side the top ball rests against

388. $0.63a$

389. $\tan^{-1} 2$

390. (a) $\theta = \sin^{-1}(\pi \sin\alpha)/2$
(b) $\sin\alpha < 2/\pi$ or $\alpha < 39.5°$

391. For a log to float upright in stable equilibrium its center of buoyancy must lie above its center of gravity. This requirement cannot be met by logs of reasonably uniform density.

392. $W_{casting} > \frac{1}{5}W_{log}$

393. $\mu_{min} = 1/(9 - \cot\theta)$

394. $F_{min} = (1 + \mu)mg/(1 - \mu)$

395. 250 lbwt (obviously he has been well fed and is heavily clothed against the cold)

396. (a) $\rho gV/\pi rT$ (b) $2w\rho gV/T$.
Note the interesting fact that the answer is independent of r; it applies equally validly to tall, slender tanks and to short tanks of large diameter—even to a tank only one band high. What does this suggest about the comparison of material costs for tall tanks vs. short tanks that hold equal volumes? For another question about material costs vs. contained volume, see Note at end of Problem 789.

397. $\dfrac{Wh}{(1/\sqrt{3}) + (2h/L)}$

398. (a) 0 (b) IBA, where A is the area of the loop

399. No; any plane loop enclosing an area A of any shape that is immersed in a uniform magnetic field B and carrying a current I is subject to a torque $IBA \sin\theta$, where θ is the angle between the normal to the plane of the loop and the direction of the field B

400. (a) $I = \dfrac{\tau_0 \theta}{NBA \sin \theta}$

(b) $\dfrac{dI}{d\theta} = \dfrac{\tau_0}{NBA}\left(\dfrac{\sin \theta - \theta \cos \theta}{\sin^2 \theta}\right)$

(c) Mount the coil in a radial magnetic field thus:

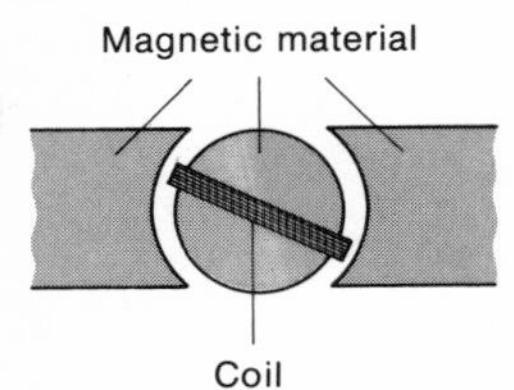

401. (a) $\dfrac{abd(a^2+b^2)I^2 \sin \phi}{\pi \epsilon_0 c^2[(a^2-b^2)^2+4a^2b^2 \sin^2 \phi]}$

(b) $\phi_{\tau\max} = \sin^{-1} \dfrac{a^2-b^2}{2ab}$

(c) $\tau_{\max} = \dfrac{dI^2(a^2+b^2)}{4\pi\epsilon_0 c^2(a^2-b^2)}$

402. (a) 20 lbwt (b) 47 lb ft^{-3}

403. $0.013\ b^3\rho g$, where ρ is the density of the water

404. (a) $2W/3$ (b) $\sin^{-1} 1/\sqrt{3}$ (c) $R/\sqrt{3}$ from O on a line at $(90° - \sin^{-1} 1/\sqrt{3})$ from OB. From the diagrams and text you have the following information: from (*a*) c.m. must be located on a line at $(90° - \theta)$ from OB; from (*b*) $PR/2 = Wr \sin \theta$, where r is distance of c.m. from O; from (*c*) $(PR/2) \sin \theta = Wr \sin (90° - 2\theta)$; from (*d*) $r = R \sin \theta$.

405. $\mu = 0.80$; diagram for vector solution:

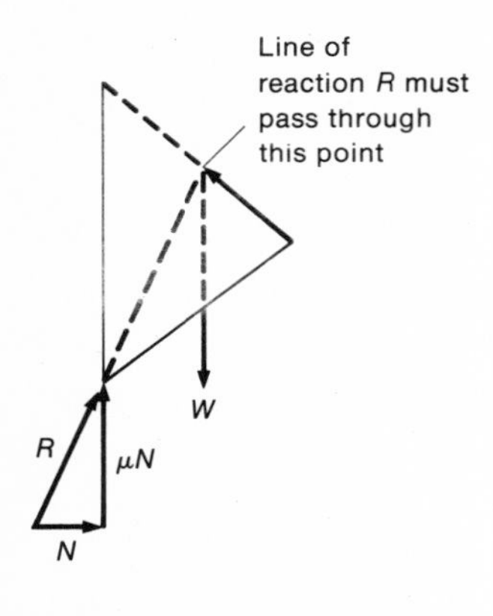

406. $W(r_2 - r_1)/2$

407. 3.1 lbwt

408. (a) $F_{\min} = \mu W/\sqrt{1+\mu^2}$

(b) $\theta = \tan^{-1} \mu$ above horizontal

409. $F = (\mu W\sqrt{x^2+h^2})/(x+\mu h)$

410. $\theta = \tan^{-1} 1/(3\sqrt{3}) = 10.°9$

411. 2.0 tonwt

412. (a) 0; CG only comes into use when you include the weight of members HG and GF, BH provides a force only when you include the weight of AB and BC or when there is a load between A and C

(b) at C, making the abutment reactions at A and E equal

413. $W/(2\sqrt{3})$ tension

414. $FD = W/2$ compression, $CD = 3W/2$ tension, $GD = \sqrt{5}W/8$ compression

415.

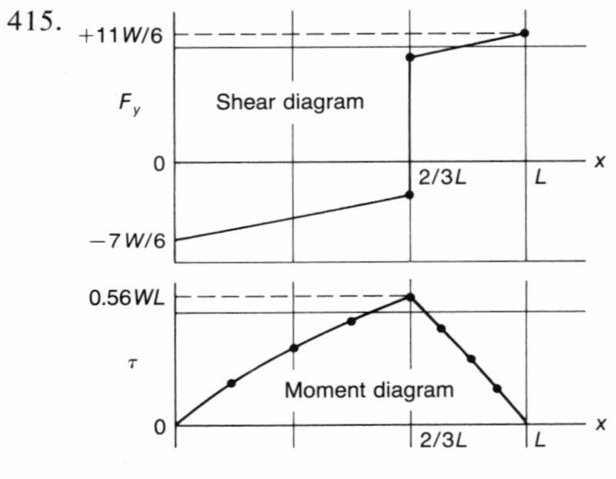

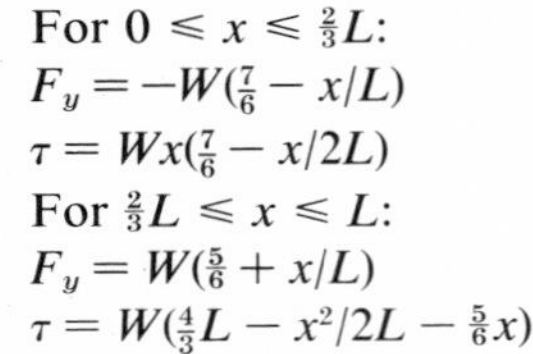

For $0 \leq x \leq \frac{2}{3}L$:
$F_y = -W(\frac{7}{6} - x/L)$
$\tau = Wx(\frac{7}{6} - x/2L)$
For $\frac{2}{3}L \leq x \leq L$:
$F_y = W(\frac{5}{6} + x/L)$
$\tau = W(\frac{4}{3}L - x^2/2L - \frac{5}{6}x)$

416.

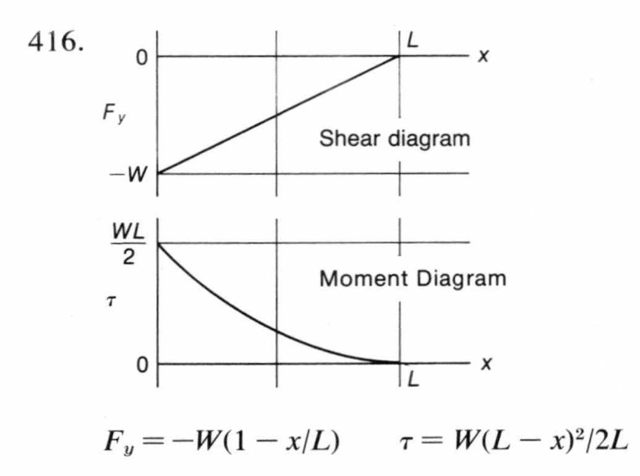

$F_y = -W(1 - x/L)$ $\tau = W(L-x)^2/2L$

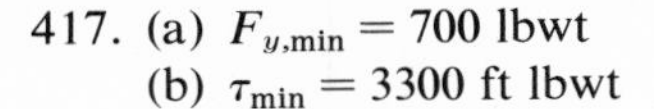

417. (a) $F_{y,\min} = 700$ lbwt
(b) $\tau_{\min} = 3300$ ft lbwt

418.

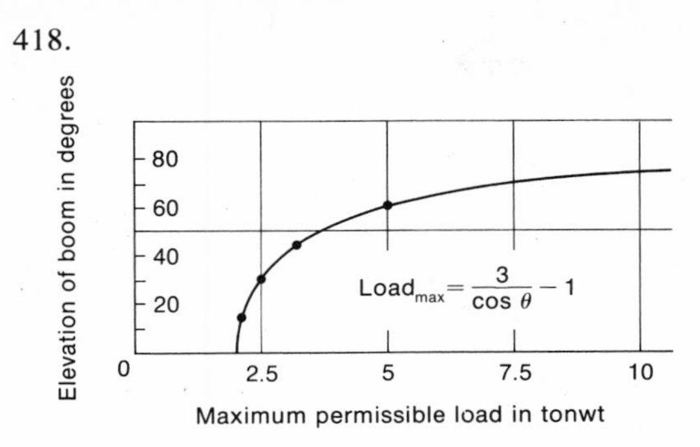

419. $\mu \geqslant \tan\theta$

420. 380 lbwt

421. (a) 265 gmwt (b) 79°1

422. The entire weight of the backstop is carried by the upper triangular framework

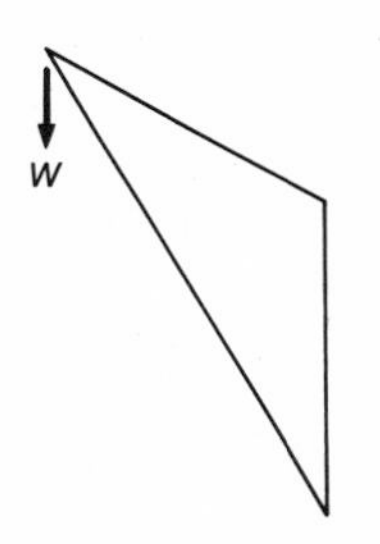

the lower member only serves to steady the backstop during impact. In spite of this the design shown is probably the simplest for maximum stability.

423.

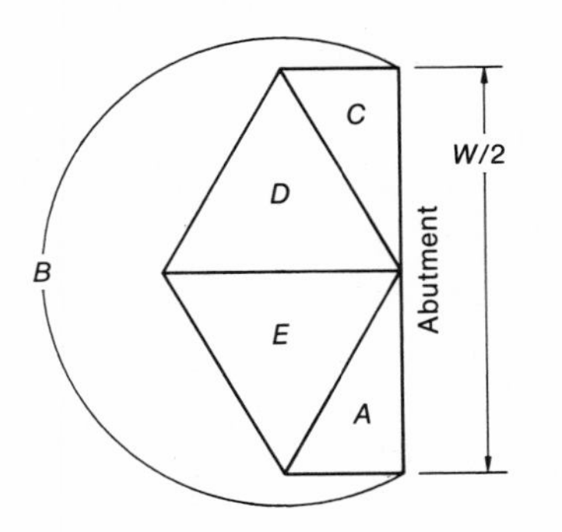

424. Note that all of the forces in the structural members are exactly double those in the previous problem; only the abutment reactions are asymmetrical

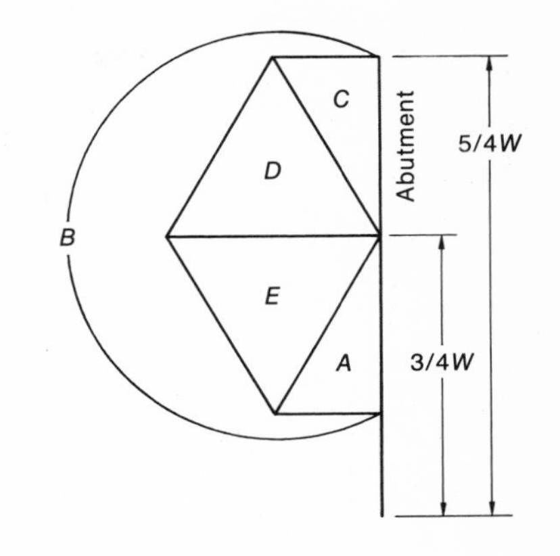

425. 42.3 min

426. (a) 7.9×10^3 m sec^{-1} (b) no longer—it took the same time

427. $T = (2\pi/3)\sqrt{2b/g}$

428. $\omega = 10.8/\sqrt{r}$; ω in rad sec^{-1} if r is in meters

429. $(L - R) \geqslant 13.1\,R$

430. $T = 2\pi\sqrt{2L/3g}$

431. (a) for $T_{\min}$, $d = L/(2\sqrt{3})$
(b) $T_{\min} = 2\pi\sqrt{L/(\sqrt{3}\,g)}$
(c) 0.43 m

432. $T = 2\pi\sqrt{\dfrac{R^2 - L^2/6}{g\sqrt{R^2 - L^2/4}}}$

433. $g = \dfrac{4\pi^2(d_1 + d_2)}{T^2}$

434. $T = 2\pi\sqrt{\dfrac{7(R-r)}{5g}}$

435. (a) $T = 2\pi\sqrt{5L/6g}$ (b) same as (a); for each element of mass in the rod the motion is the same in both cases

436. $I_f = mr(r + gT^2/4\pi^2)$

437. (a) for $\nu_{\max}$, $d = R/\sqrt{2}$
(b) $\nu_{\max} = \dfrac{1}{2\pi}\sqrt{\dfrac{g}{R\sqrt{2}}}$

438. $E = \dfrac{4\pi^2\nu^2 m(a^2 + b^2)}{3\sqrt{(Qa)^2 + (qb)^2}}$

439. (a) $\nu = (1/6\pi)\sqrt{10g/L}$ (b) $A_m = 0.09\,L$

440. (a) Since the apparent "g" field increases with upward acceleration, the period should decrease, or the frequency should increase
(b) $\nu_{\text{accel}} = 2\nu_{\text{normal}}$

442. (a) $4\pi^2 M/T^2$ (b) $(L/2) - (gT^2/4\pi^2)$
(c) $\dot{y} = (\pi L/T) - (gT/2\pi)$ (d) 0

443. $2\pi \sqrt{3M/2k}$

444. (a) $F_x = -8kx(x^2/L^2 - 3x^4/L^4 \ldots)$, not SHM
(b) $F_y = -4ky$, hence SHM
(c) $T = 2\pi \sqrt{M/4k}$ for system (b)

445. (a) $2\pi \sqrt{2M/k}$ (b) $2\pi \sqrt{M/2k}$
(c) $2\pi \sqrt{M/2k}$ for oscillations in the line of the springs

446. (a) $T = 2\pi \sqrt{M(k_1 + k_2)/k_1 k_2}$
(b) no, it would be the same

447. (a) $m = kR/2g$ (b) $m = 2kR/3g$

448. (a) $T = 2\pi \sqrt{2ML/(24kL + 3Mg)}$
(b) $T \simeq 2\pi \sqrt{2L/3g}$

449. (a) $T = 2\pi \sqrt{2ML/(24kL - 3Mg)}$
(b) $k > Mg/8L$

450. $T_{\text{torsion}} = T_{\text{spring}} < T_{\text{simple pendulum}}$

451. $T_{\text{cir}}/T_{\text{sq}} = \sqrt{3/\pi}$

452. $I_{\text{irreg}} = 2MR^2$

453. $r > L/55$ before the moment of inertia of the rod needs to be included

454. $T = P \sqrt{\dfrac{1}{[(\sin \phi)/\phi] + 1}}$, assuming that the moment of inertia of R about its axis is negligible compared to the moment of inertia of C about the same axis. The sequence of development would go thus: $P = 2\pi \sqrt{I_C/Mgd}$, where I_C is the unknown moment of inertia of C about the axis of R, and d is the distance from the center of mass M of C to the axis of R; $Mgd \sin \phi = 4\tau_0 \phi$, where τ_0 is the torsion constant of the entire rod R (remember that the torsion constant of a half rod is twice the torsion constant of the whole rod; $T = 2\pi \sqrt{I_C/(4\tau_0 + Mgd)}$.

456. Using the same coordinate system as shown in problem 455.

$$x_1 = vt + \frac{m_2 L}{2(m_1 + m_2)} \cos \sqrt{\frac{k(m_1 + m_2)}{m_1 m_2}}\, t$$
$$x_2 = L + vt - \frac{m_1 L}{2(m_1 + m_2)} \cos \sqrt{\frac{k(m_1 + m_2)}{m_1 m_2}}\, t$$

457. (a) $k = 5.6 \times 10^2$ N m^{-1}
(b) For the linear portion of the curve at x_0, $dF/dx = \text{constant} = k$

458. $\nu = 6.3 \times 10^{13}$ vib sec^{-1}

459. (a) computed value for ratio = 1.92 (actual value from experiment, 1.76) (b) computed value = 1.14 (actual value from experiment, 1.08)

460. $x_m = A \cos \sqrt{(k/m)(2 \mp \sqrt{3})}\, t$
$x_{2m} = (A/2)(1 \pm \sqrt{3}) \cos \sqrt{(k/m)(2 \mp \sqrt{3})}\, t$
$\nu_1 = (1/2\pi) \sqrt{(k/m)(2 - \sqrt{3})}$
$\nu_2 = (1/2\pi) \sqrt{(k/m)(2 + \sqrt{3})}$

461. None; the equilibrium positions relative to the point of attachment are changed from those in the above problem, but the equations of motion from these equilibrium positions are the same as before

462. (a) $M_1/M_2 = (k_1 + 2k_2)/2k_2$
(b) $\nu = (1/2\pi) \sqrt{(k_1 + 2k_2)/M_1}$
$= (1/2\pi) \sqrt{2k_2/M_2}$

463. (a) 39.1 cm from anchor for mode 1, 52.9 cm from anchor for mode 2
(b) $T_1 = 1.6$ sec, $T_2 = 0.83$ sec

465. $kx_0 \geqslant 41\mu mg$

467. $t_{\text{rel}} = 2/\gamma$

468. (a) 0.099 sec^{-1}
(b) Because γ is small,
$\omega_0^2 \simeq \omega^2 = 4\pi^2/T^2 = 1.26$;
also, $\omega^2 = k/m = A\rho_w g/A\rho_L L = 1.2g/L$, from data in problem 392.
Then, $L = 1.2g/1.26 \simeq 31$ ft

469. $\eta = 4.6 \times 10^{-3}\, d\, \sigma$, where σ is surface density of metal sheet

470. (a) $x = \dfrac{R \cos \omega t}{1 - \omega^2/\omega_0^2}$
(b) $\ddot{x}_{470} = \dfrac{1}{1 - \omega^2/\omega_0^2}\, \ddot{x}_{428}$
Note that as $\omega \rightarrow \omega_0$, acceleration is greatly increased

471. $y_{\text{total}} = 2.8\ y_0$; this is essentially a straight substitution in equation 3-68

472. For effective damping ω^2 must be $>> \omega_0^2 = 4k/M$, so that $k << M\omega^2/4$; Suitable damping could probably obtained with $k = M\omega^2/400$

473. $(1/2\pi)\ \sqrt{g/L}$

474. (a) μWS (b) $\mu WS/(1 - \mu/\sqrt{3})$

475. $\dfrac{W(\mu \cos \theta + \sin \theta)}{\cos \theta - \mu \sin \theta}$

476. $\frac{5}{8}$ the weight of his own bike

478. $0.16\ WL$

479. 3.5×10^3 joules

480. (a) $\frac{1}{2}kx_0^2$ (b) $mgL(1 - \cos \theta_0)$, m the mass of the pendulum bob

481. $\frac{1}{2}WL(1 - \sin \theta)$, W the weight of the ladder

482. $\frac{1}{2}WL(1 - \sin \theta)$

483. $WL[\frac{1}{2}(1 - \sin \theta) + \mu \cos \theta]$

484. (a) $v_0/3$ (b) $\frac{1}{6}mv_0^2$

485. (a) $\sqrt{3}/2\ W$ (b) In part (a) of problem 397 horizontal force applied at the ground provided a counter-clockwise torque about the point of contact with the curb. In this problem a horizontal force applied in line with the axle provided a clockwise torque about the same point of contact.

486. $F \cot^2 \theta$

487. $F_{DE} = (\frac{1}{2}\sqrt{3})\ W$ compression

489. $0.3\ W$

490. 54 mi hr^{-1}

493. (a) $(F_0/ML)^{1/2}$ (b) $\frac{2}{3}F_0L$

494. (a) $\omega = [(F_0/ML)^2 - (\mu g/L)^2]^{1/4}$
(b) $(L/2\mu g)[(F_0/M)^2 + 3(\mu g)^2]^{1/2}$

495. (a) $v_{\max} = 0.11\ c = 3.3 \times 10^7$ m sec^{-1}
(b) 3.4×10^3 ev

496. $\Delta m/m = gh/c^2$

497. (a) $\Delta U = -7.8 \times 10^4$ joules
(b) Work done by the brakes

498. $W = \Delta U = \frac{1}{2}\ \rho_{\text{water}}\ Ah^2g$; note that the water and the cork simply exchange places

499. $U_R = Q^2/8\pi\epsilon_0 R = 4.5 \times 10^9\ Q^2/R$ joules, Q in C and R in m

500. $U_R = 3Q^2/20\pi\epsilon_0 R = 5.4 \times 10^9\ Q^2/R$ joules if Q in C and R in m

501. (a) $\Delta U = 4.5 \times 10^9\ Q^2/R - 5.4 \times 10^9\ Q^2/R = -1.6 \times 10^9\ \pi^2\rho^2R^5$ joules if ρ and R in MKSC units
(b) Increase in the heat content of the sphere

502. (a)

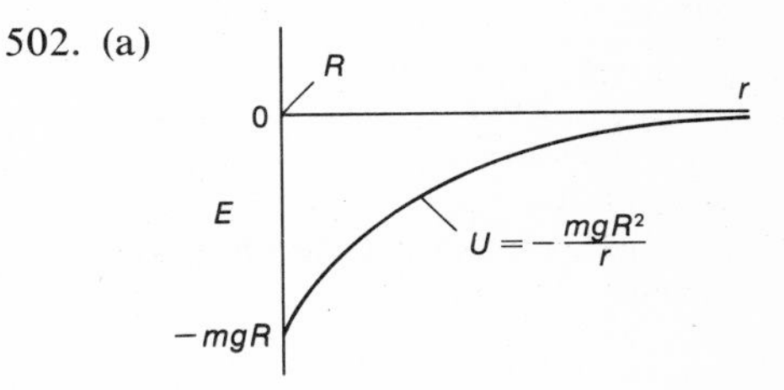

(b) Masses whose E_r lies on the curve are momentarily at rest just prior to falling back toward the earth
(c) Masses whose E_r lies between the curve and $E = 0$ are in a closed orbit about the earth
(d) Masses whose E_r lies above the $E = 0$ axis are in open orbit and will escape the earth's gravitational field

503. (a) $\Delta U = 48\pi\epsilon_0 rV_0^2$, where r is the radius of one of the original drops and V_0 its original electric potential
(b) The ΔU came from kinetic energy acquired during the storm plus a contribution from surface energy

504. (a) $U_0 = \frac{8}{3}\pi\epsilon_0 RV^2$
(b) $\Delta U/U = \frac{3}{11}$
(c) I^2R loss in the wire during the passage of charge plus energy lost in sparks that developed when the wire first touched the spheres

505. (a) $\rho_0 = 6.2 \times 10^{25}$ C m^{-3}
(b) $U = 8.4 \times 10^{-12}$ joules,
(c) For uniformly charged nucleus $U = 5.6 \times 10^{-12}$ joules, hence U for nonuniform charge is about 50% greater than U for a uniform charge

506. (a) 20.02 ft (b) 40.06 ft

508. $L_f = L(1 + 4F_0/3AY)$

509. (a) $A_{tube}/A_{rod} = 0.32$; in terms of metal used the tube is about 3 times as effective as the rod
(b) $r_{tube} = 1.26 r_{rod}$

510. (a) 1:5 (b) $\frac{1}{2}$

511. $(5.4 \times 10^2 \pi R^4 \theta)/L$ m N

512. 1043 kg m^{-3}

513. (a) $B = 7.8 \times 10^7$ N m^{-2}
(b) No. Plotting equation 4-15 for both sea water and ring material, it is clear that a change in density per change in hydrostatic pressure is greater for the ring material. Hence if the ring rises above the equilibrium level, the buoyant force will increase and the ring will accelerate upward; the opposite effect will result from a motion below the equilibrium level.

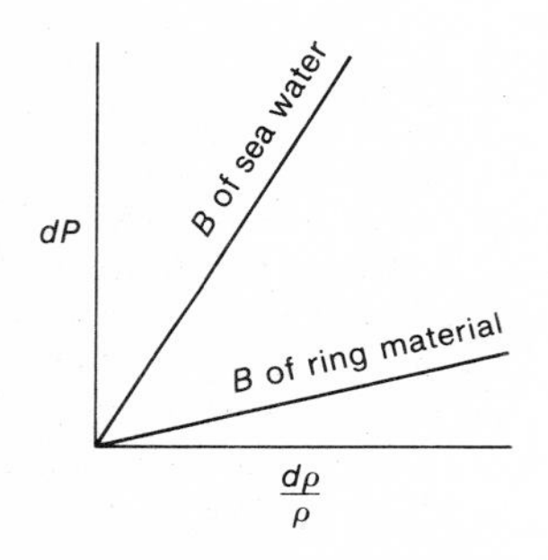

(c) No; suggestion valid only for the special case of an object whose density is very nearly that of sea water and whose bulk modulus was greater than $B_{sea\ water}$

514. 4×10^6 atmospheres

515. (a)

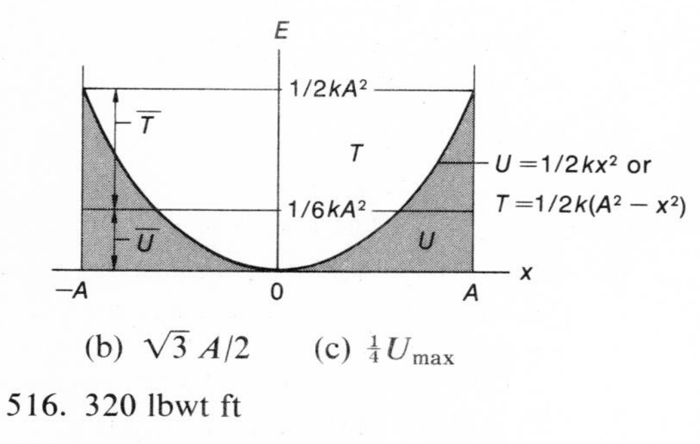

(b) $\sqrt{3}\, A/2$ (c) $\frac{1}{4} U_{max}$

516. 320 lbwt ft

517. $d + (mgh/k)^{1/2}$

518. 7.0 ft

519. (a) 1.21 m (b) 0.53 m
(c) $A = 0.34$ m, $T = 0.52$ sec

520. 1.45 joules, the sum of 1.20 j of kinetic energy of the c.m. and 0.25 j of internal kinetic energy of the spring-mass system

521. (a) $W_{brass} = 0.35\, mgh$, $W_{steel} = 0.15\, mgh$
(b) This part of the problem draws attention to the question of how the mass m reached equilibrium: was it lowered infinitely slowly by an outside agency, or was it released and allowed to oscillate, the oscillations finally damping out? In the first case $\frac{1}{2}mgh$ of work would be done against the outside agency; in the second case $\frac{1}{2}mgh$ of energy would be absorbed in damping.
(c) $\ell_{brass}/\ell_{steel} = \frac{9}{20}$
(d) $\Delta\ell_{brass} = \Delta\ell_{steel} = 0.43\, h$

522. 118 lbwt ft

523. $W_\theta / W_Y = \mu^2 Y/n$

524. $8MgR/9\pi^3$. The solution is based on the comparison between B and C. If the datum for potential energy (gravitational) is taken at the center of the rod, in B the system possesses a gravitational potential energy of $-4MgR/3\pi$ and an elastic potential energy stored in two half-rods twisted through π, in C the system possesses zero gravitational potential energy plus the elastic potential energy stored in two half-rods twisted through $\pi/2$.

525. 1.0 joule

526. 2.6×10^4 lbwt

527. 0.5 kw

528. 188 watts

529. 3×10^4 HP

530. 7.7×10^2 HP

531. (a)

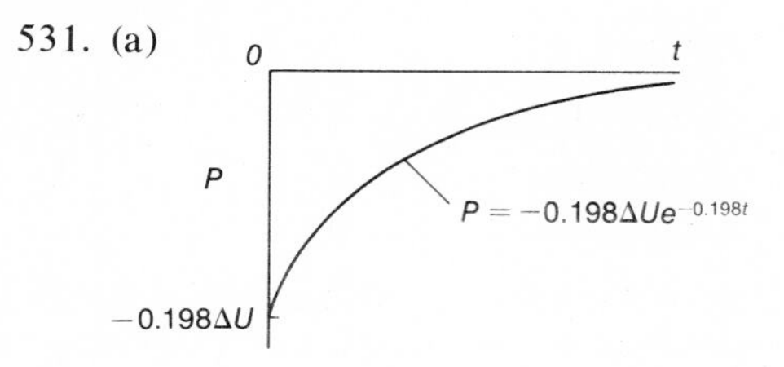

(b) Since the exponent of e must be dimensionless, the factor 0.198 must have the dimension T^{-1}. Then the right-hand side of the equation for power has the dimensions of $\Delta U/T$.
(c) $P_{2,1} = -0.13\Delta U$

532. $P_t = -\gamma m e^{-\gamma t} X_0^2 (\omega_0^4/\omega_\gamma^2) \sin^2 \omega_\gamma t$

533. (a) $dT/dt = mg^2 t(\sin\theta - \mu\cos\theta)^2$
(b) $dU/dt = -mg^2 t \sin\theta (\sin\theta - \mu\cos\theta)$
(c) $\Delta P = -mg^2 t\mu \cos\theta (\sin\theta - \mu\cos\theta)$— the rate of work being done against friction

534. 80%

535. There is no gain in energy from the work done in braking, hence the efficiency of a braking operation is zero

536. 50%

537. (a) $(1 - \mu\tan\theta)/(1 + \mu\cot\theta)$
(b) For $Eff_{\max}$, $\theta = \frac{1}{2}\tan^{-1} 1/\mu$
(c) $Eff_{\max} = (\sqrt{1+\mu^2} - \mu)/(\sqrt{1+\mu^2} + \mu)$

538. $\frac{3}{2}(L - D)$

539. (a) 16 ft $\sec^{-1}$ with the assumption that the c.m. moves from about 3.5 ft above ground to about 7.5 ft above ground
(b) 3.5 HP

540. (a) Anywhere on the $E = 0$ axis
(b) $v_{esc} = 11.2$ km $\sec^{-1}$ for either a space probe or a hydrogen molecule

541. eV

542. $\dot{y} = \sqrt{g(L^2 - y_0^2)}/L$

543. $34/\sqrt{a}$ m $\sec^{-1}$ if a is in m

544. 733 watts. The answer contains more significant figures than the data justifies; this was done in order to show the necessary inclusion of the power required to change the momentum of the sand.

545. 3.7 m $\sec^{-1}$

546. 168 lbwt

547. (a) 2.2×10^{-14} m (b) 2.5×10^6 m $\sec^{-1}$
(c) 2.8×10^{-14} m, 1.0×10^6 m $\sec^{-1}$

548. $dr = mV/qB^2 r$

549. (a) 3.1×10^2 HP (b) 2.6×10^2 HP
(c) 0.42 HPh (d) 0.50 kwh

550. (a) ω_0. The loop alternately accelerates and decelerates every half revolution, coming back to ω_0 whenever its plane is parallel to the field.
(b) Energy is gained in the first and third quarter-cycles and lost in the other quarter-cycles. These energy changes appear at the source of the constant current I.

551. $\dfrac{WvN(\sin\theta + \mu\cos\theta)}{550E(1 + \mu\tan\phi)}$ HP

553. (a) 3.1×10^7 m $\sec^{-1}$
(b) 5.2×10^{-20} kg m $\sec^{-1}$

554. $\Delta E/E = \frac{1}{6}$

555. $\vec{J} = 127m$ at $\tan^{-1} 0.817$ above horizontal

556. $\vec{J} = m\sqrt{2g}(\sqrt{h_0} + \sqrt{h})$ vertically upward

557. $10F$

558. 0.03 sec

560. 94 ft $\sec^{-1}$

561. (a) 3.3×10^2 kgwt. If a fluid of density ρ is being ejected from an opening of cross-sectional area A at a speed v, the mass leaving the opening per unit time = $\rho A v$. Then the momentum leaving the opening per unit time = $\rho A v^2$.
(b) 2.2×10^2 kgwt (It pays to flee!)

562. 50% more than the proper charge

563. (a) 2π rad $\sec^{-1}$
(b) $v_{c.m.} = (r + L/2)\, 2\pi$
(c) $\pi m L^2$
(d) $2\pi m(r + L)$
(e) $2\pi r/L$

564. (a) $F = A/\omega$, where $A = \pi I\ddot{\theta}/50r$ = constant if $\ddot{\theta}$ is constant
(b) $\ddot{\theta} = B\omega$, where $B = 50Fr/\pi I$ = constant if F is constant

565. 42 watts

566. (a) $P_y = (Mv/L)(gy + v^2)$
(b) $\Delta P = Mv^3/2L$
(c) In any small interval of time, an element of the rope has its momentum changed from 0 to mv, where m is the mass of the element. The impact that provides this sudden change in momentum is essentially inelastic.

567. The lower limit could be about 0.5: since for any size of area there would be some slippage around the edge, not all of the wind from the direction in which the area faces would reach it; as the size of the area decreases this slippage factor would increase. The upper limit could be about 1.2: it is reasonable to expect that a number of arriving air molecules will rebound from the surface and not hit other arriving molecules; this would increase momentum transfer. Thus for a reasonable estimate, $0.5 < k < 1.2$

569. $\frac{2}{3}L$ below suspension point

570. $\pi L/6$

571. $v_{AB}/v_{AC} = -\frac{2}{7}$. The only simple way to solve this problem is to separate the two bars and consider them individually at the moment of impact, thus:

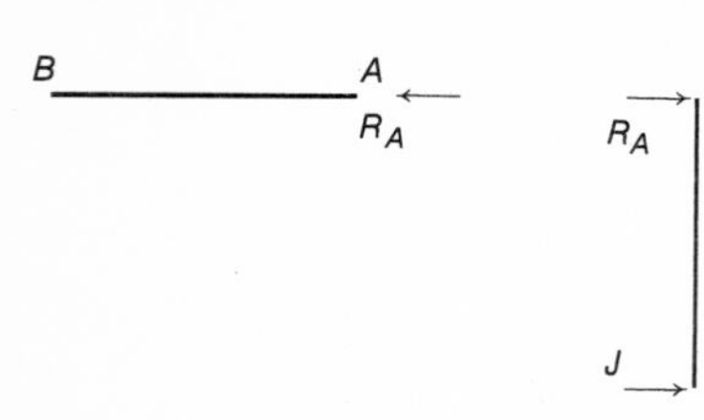

572. 0.7 D, where D is the diameter of the ball

573. For elastic impact the incident angle with the normal must equal the rebound angle; for the present case this means that the ball must rebound at 2θ with the vertical. But for R_{max} this angle must = 45°. Hence $\theta = 22.5°$.

574. $v^2 = 7gd/(3 \sin 2\theta)$

575. $v = \dfrac{u(e \cos \theta \tan \phi - \sin \theta)}{(1 + e) \tan \phi}$

576. (a) $e_0^2 h_0/8$ (b) $\frac{1}{2}h_0$

577. $\mu = [(1 - e)/(1 + e)] \cot \theta$

578. (a) $\theta = \phi$; the inbound and rebound paths at a right-angled corner are parallel
(b) 5

579. 0.25

580. (a) $\Delta E = (1 - e^{2n}) E_0$ (b) $e = 0.9$

581. (a) $e = \tan^2 \theta$ (b) $\Delta E/E = 1 - \tan^2 \theta$

582. (a) 0.71 (b) 1.7 ft sec^{-1} horizontal

583. 64 cm sec^{-1}

584. If direction of u_1 is taken as +,
(a) $v_1 = \dfrac{m_1u_1 - m_2u_2 - em_2(u_1 + u_2)}{m_1 + m_2}$

$v_2 = \dfrac{m_1u_1 - m_2u_2 + em_1(u_1 + u_2)}{m_1 + m_2}$

(b) $\Delta E = \dfrac{m_1m_2(1 - e^2)(u_1 + u_2)^2}{2(m_1 + m_2)}$

585. 1.56 kg

586. 1.1 kg

587. 54° from original path of ball carrier

588. 120°. The two original momentum vectors and their vector sum all have the same scalar values. Then the vectors must form an equilateral triangle.

589. (a) $\Delta T = \frac{1}{3}m(u_2 - u_1)^2$ with m the mass of 1 freight car; note that the impact was inelastic, for which $e = 0$
(b) In this case $\Delta T = \frac{1}{4}m(u_2 - u_1)^2$—less energy is lost
(c) Same loss as in (a); the reduced mass $m_1m_2/(m_1 + m_2)$ is symmetrical in m_1 and m_2, hence they can be exchanged without altering the value of the reduced mass
(d) ΔT would be a minimum for $m_1 = m_2$
(e) No difference; the energy loss is a function of relative speed, which is the same in both cases cited

590. You cited the Chevrolet driver for (1) going 43 mi hr^{-1} in a 35 mi hr^{-1} zone and (2) failure to yield the right of way

591. 2.1 ft

592. (a) $\frac{11}{14}u$ in the same direction as the original u
(b) $\Delta T/T = 12\%$
(c) Largely spent in work against friction during the transfer of mass

593. (a) $\dfrac{\Delta E_m}{E_0} = \dfrac{4\ m/M}{(1 + m/M)^2}$
(b) $\Delta E_m/E_0$ is largest for $m/M = 1$ or $m = M$. When the masses are equal and one is at rest, in an elastic collision all of the kinetic energy of the moving mass is transferred to the at-rest mass.
(c) The percentage energy loss is the same

in either case. If $m/M = f$, where f is a small fraction, note that the percentage loss given in (a) has the same value for f or $1/f$.

(d) Choose Be; the loss per collision is greater for $m/M = \frac{1}{9}$ than for $m/M = \frac{1}{12}$.

594. 0.71

595. 0.22

596. $v_D = 3.6 \times 10^5$ m sec^{-1}; $v_{He} = 1.9 \times 10^7$ m sec^{-1}, about the upper limit for nonrelativistic three-figure accuracy

597. (a) $m_{fuel}/M_0 = 63\%$
(b) When $v = v_0$; at that instant the ejected fuel is motionless relative to the coordinate system in which v is measured and hence possesses no kinetic energy in that system. Thus at that speed all of the net energy of the fuel is being used to increase the speed of the rocket.

599. (a) $m_1/m_2 = 1$ (b) $m_1/m_2 = e$

600. $\Delta E/E = (1 - e) \cos^2 \theta$

601. (a) $m_1/m_2 = \cos 2\theta$
(b) $e = (1 + 2 \sin^2 \theta)/2 \cos^2 \theta$
(c) $\Delta E/E = 1 - e$

602. Two facts of physics should govern your decision: (1) a ball at rest will depart from a collision along the line of centers at impact, (2) when a moving mass collides head-on and perfectly elastically with an equal mass at rest, all momentum is transferred to the second mass. In the light of these facts you should choose a line of approach near the right limit of the permissible angle and aim head-on for the nearer X ball. This ball when struck will then strike the second X ball, removing both from the vicinity of the jack. Your ball should stop when it hits the X ball; however, some remaining angular momentum may make it creep even closer to the jack.

603. 67%

604. $M = [\sin (2\theta + \phi)/\sin \phi]\ m_{el}$

605. $v_n = u/\sqrt{3}$ due north from inspection of appropriate circle diagram

606. (a) $T_{c.m.} = \dfrac{m_1 m_2 (u_1 - u_2)^2}{2(m_1 + m_2)}$

(b) The kinetic energy of the c.m.
$= \dfrac{(m_1 u_1 + m_2 u_2)^2}{2(m_1 + m_2)} = \dfrac{p_0^2}{2(m_1 + m_2)}$

(c) $T'_{c.m.} = \dfrac{e^2 m_1 m_2 (u_1 - u_2)^2}{2(m_1 + m_2)}$

(d) $\Delta T_{c.m.}/T_{c.m.} = (1 - e^2)$

(e) $\dfrac{\Delta T}{T_0} = \dfrac{m_1 m_2 (1 - e^2)(u_1 - u_2)^2}{(m_1 + m_2)(m_1 u_1^2 + m_2 u_2^2)}$

607. 54°.7

608. (a)

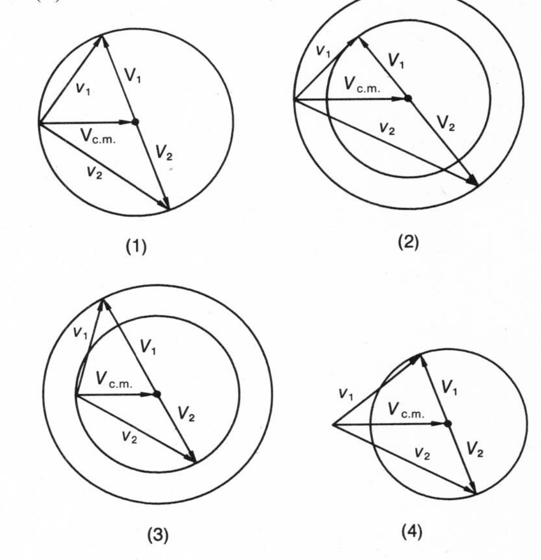

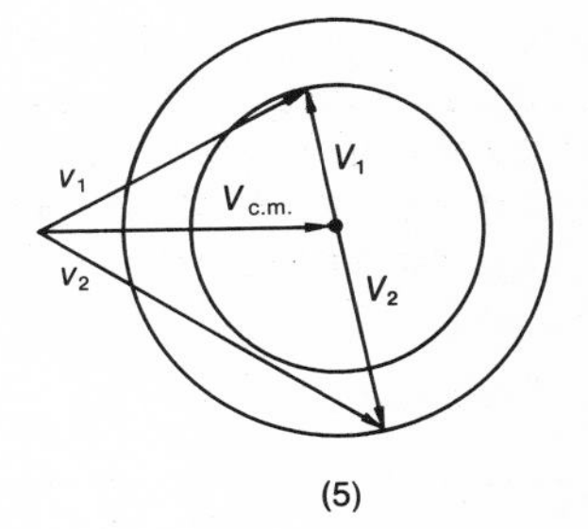

(b) (2), (4), and (5)

609. (a) $V_1/V_{c.m.} = m_2/m_1$ (b) $m_{gas}/m_p = V_p/V_{c.m.} = \sin 15° \simeq \frac{1}{4}$; $m_p/m_{gas} \simeq 4$

610. $T_2 = [(M - m_2)^2 - m_1^2]c^2/2M$

611. 279.5 Mev

612. E/c $(= h\nu/c)$

613. 2.04 Mev

614. (a) $m_{\text{comp}} = m\sqrt{2(1+\gamma)}$
(b) $v_{\text{comp}} = v/(1 + 1/\gamma)$

615. $E_\nu = \dfrac{\Delta E(\Delta E + 2m_H c^2)}{2(\Delta E + m_H c^2)}$

616. (a) negligible (b) $m_D = 2.01414$ amu

617. $J_{\min} = m\sqrt{2gL/3}$

618. $2mT/q_{el}B$

620. (a) $L_{el} = 1.05 \times 10^{-34}$ MKSC units, which is about half of L_{el} for a real hydrogen atom
(b) $\omega_p = 1.6 \times 10^{23}$ rad sec^{-1}
(c) The situation is stable. Since the proton in the Bohr model is centrally located in the electromagnetic field of the revolving electron, electromagnetic forces on the distributed charge of the proton will be inward.

621. (a) $q_{el}/m = 8VR^2/[B^2(R^2 - a^2)^2]$
(b) $(R^2 - a^2)/2R$

622. 11° W of S

623. (a) at $L/(2\sqrt{3})$ from center of the rod
(b) J^2/m
(c) at $L/(2\sqrt{3})$ from center of rod, on the opposite side of the center from where J struck

624. $E = \frac{13}{32}md^2\omega_0^2$

625. (a) $E_r = -\, q_{el}^2/8\pi\epsilon_0 r$
(b) $L_r = -2mr^2E_r$
(c) $E_n = -\, mq_{el}^4/32\pi^2\epsilon_0^2 n^2\hbar^2$
(d) $\Delta E = hcR\ [1/n^2 - 1/(n+1)^2]$, where $R = mq_{el}^4/8\epsilon_0^2 ch^3 = 1.097 \times 10^7$ m^{-1} is called the Rydberg constant

626. (a) $37.3\ R^2\omega_0^2$
(b) $v_{\text{girl}} = 1.06\ R\omega_0$
(c) The moment of inertia of each actor about a vertical axis through his or her individual c.m.

628. (a) The plane nosed downward
(b) He pulled back on the elevator controls to lift the nose of the plane, whereupon it precessed to the left

630. $\tau = \frac{1}{2}M\omega^2[r^4 \cos^2\delta + \frac{1}{4}(r^2 + L^2/3)^2 \sin^2\delta]^{1/2} \times \sin\left[\tan^{-1}\left(\dfrac{r^2 + L^2/3}{2r^2}\tan\delta\right) - \delta\right]$

631. (a) $Mg + I\omega^2/r$
(b) No change; both $\vec{\Omega}$ and $\vec{\omega}$ change on reversal, so that $\vec{\Omega} \times \vec{\omega}$ stays constant

632. No; there is no mechanism by which the original angular momentum mvr of the electron about Q can be changed, hence this angular momentum must be maintained throughout the passage of the electron. Hence a finite distance must always exist between Q and the electron.

633. $v = [2GMR/(D^2 - R^2)]^{1/2}$, where M is the mass of the sun

634. ΔE must be $> GmM/2R$, m the mass of the body, M the mass of the earth

635. $d_{\min} = R_0\{[(GM)^2 + R_0^2v_0^4 \sin^2\theta - 2GMR_0v_0^2 \sin^2\theta]^{1/2} - GM\}/(R_0v_0^2 - 2GM)$

636. $v_0^2 > 2GM/R_0$

637. $v_0^2 > 2GMd/b^2$

638. $d = K[1 + \csc(\phi/2)]/2T$, K the same as that appearing in equation 4-31. Because in this case K is a very small gravitational term, and T cannot usually be measured too accurately, this value of d would be an unacceptable measure of the size of the nucleus.

639. (a) It is a pure pipedream; it would be totally impossible to control the direction of the particles to the accuracy of nuclear dimensions.
(b) No; there are too many other variables in the experiment of at least the same order of magnitude as the relative size of the nucleus. These macroscopic objections do not include the quantum mechanical objections.

640. 0.01 PA

641. (a) $T = -E$ (b) $T = -\frac{1}{2}U$

642. Per kg of mass, 4.4×10^8 j or about 120 kwh

643. $\Delta E = -\frac{3}{4}E_{3h}$ (note that since E_{3h} is $-$, ΔE is $+$)

644. (a) 1.06×10^4 km
(b) $\Delta E = 1.4 \times 10^7$ j

645. (a) 360 km (b) i. 7.84×10^3 m sec^{-1}
ii. 7.66×10^3 m sec^{-1}

646. 2.56×10^6 j kg^{-1} from the back side of the moon 2.57×10^6 j kg^{-1} from the side of the moon facing the earth

647. (a) 13.6 ev

648. (a) $\dfrac{M-m}{M-2m} \times 6.57 \times 10^6$ m

(b) $\left[\dfrac{M-m}{M-2m}\right]^{3/2} \times 1.47$ hours

Note: After several stages separated in accordance with this agreement had crashed to the earth without vaporizing, the agreement was cancelled.

649. $T = 12.5 \times 10^{10}\ m$, $U = -12.0 \times 10^{10}\ m$; $T + U$ is positive, hence the object is in an open orbit around the sun

650. (a) $T = 11.90 \times 10^{10}\ m$, $U = -11.95 \times 10^{10}\ m$; $T + U$ is negative, hence object is in a closed orbit
(b) 1.31×10^8 km
(c) $e = 0.9916$
(d) 1.70×10^7 km
(e) 0.82 yr

651.

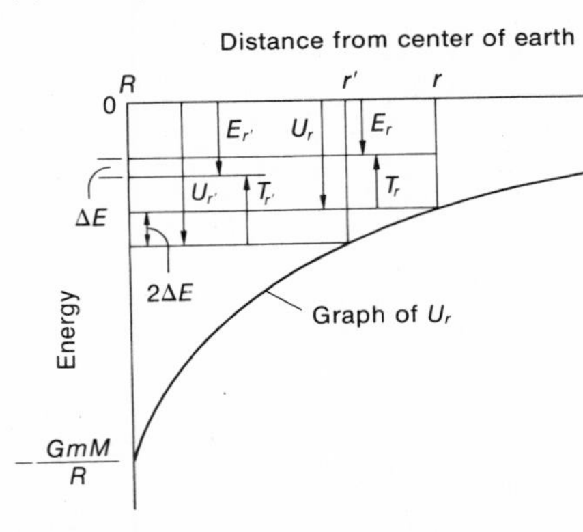

652. $F = Q^2/2KA\epsilon_0$

653. (a) $V_0 = Q/C_{\max}$ (b) $10V_0$
(c) $4.5\ Q^2/C_{\max}$

654. (a) $2R$
(b) Is the same—$2R$; the equality of the two answers should not be surprising, since the difference in potential between the two locations is just the work required to move a unit charge between the two locations

655. (a) $0.1Q$
(b) $0.45\ Q^2/C_{\max}$ work done *by* the capacitor in moving $0.9Q$ back to the source
(c) $0.45\ Q_0^2/C_{\max}$

656. (a) $[2/(1+f)]\ V_0$
(b) $[(1-f)/(1+f)]CV_0^2$

657. (b) $\frac{1}{2}CV_0^2L/[L + x(K-1)]$

658. $v = \left[\dfrac{(K-1)CV_0^2}{mK(K+1)}\right]^{1/2}$

659. $L_1 + L_2$

660. Per unit length $L = (\ln b/a)/2\pi\epsilon_0 c^2$

661. $N^2 d(\ln b/a)/2\pi\epsilon_0 c^2$

662. $L_{\text{neg}} = 1/8\pi\epsilon_0 c^2$ per unit length

663. (a) 22.4×10^{-3} m^3 = 22.4 liters
(b) 2.69×10^{19} molecules cm^{-3}
(c) 99.98%

664. (a) $v_{\text{rms}} = 1.58 \times 10^2\sqrt{T/MW}$ m sec^{-1}
(b) i. 1×10^4°K, ii. 2×10^4°K

665. (a) 4 (b) 1

666. (b) $\frac{1}{4}$

667. 4.1×10^{-2} m^3

668. (a) i. 0.98×10^5 N m^{-2} = 0.97 As
ii. 0.95×10^5 N m^{-2} = 0.94 As
(b) 11
(c) 11

669. 20 cm Hg

670. 842 mm Hg

671. 5.0 cm

673. (a) i. $PT^{\gamma/(1-\gamma)}$ = constant
ii. $VT^{1/(\gamma-1)}$ = constant
(b) $W_{\text{ad}} = [NR/(\gamma - 1)](T_0 - T_f)$

674. $V_a/V_d = V_b/V_c$

676 (a) $V_0/V_f = 15.6$
(b) 46.8 As = 687 lbwt in^{-2}

677. (a) 142°C
(b) $P_{\text{Ne}} = 1.4$ As $P_{\text{H}} = 4.3$ As

678. (a) 1740°K (b) 5.8

679. (a) 0.9 J (b) 2 (c) No; different isotherms pass through the point P_0, V_0 for the different gases

680. (a) $dQ/dt = -PAv$
(b) $dT/dt = 2PAv/5NR$

681. (a) Starting from the same point of state the adiabatic curve is steeper than the isothermal curve. Hence the area under the curve between a given P_0 and P_f is less under the adiabatic than under the

isothermal. Hence less work is done in an adiabatic cycle.

(b) $[V_0/(\gamma - 1)]\{P_0 - P_f - \gamma P_f[(P_0/P_f)^{1/\gamma} - 1]\}$

(c) $Eff = \dfrac{P_0 - P_f - \gamma P_f[(P_0/P_f)^{1/\gamma} - 1]}{P_0 - P_f}$

(d) For $P_0/P_f = e^2$, $Eff_{\text{ad}} = 30.5\%$, $Eff_{\text{iso}} = 27.2\%$. In the adiabatic cycle energy had to be supplied only along the constant volume path, whereas in the isothermal cycle energy had to be supplied not only along the same constant volume path but along the isothermal path as well. Even though more work is done by the isothermal cycle, you could reasonably suspect that its efficiency was less.

682. (a) 

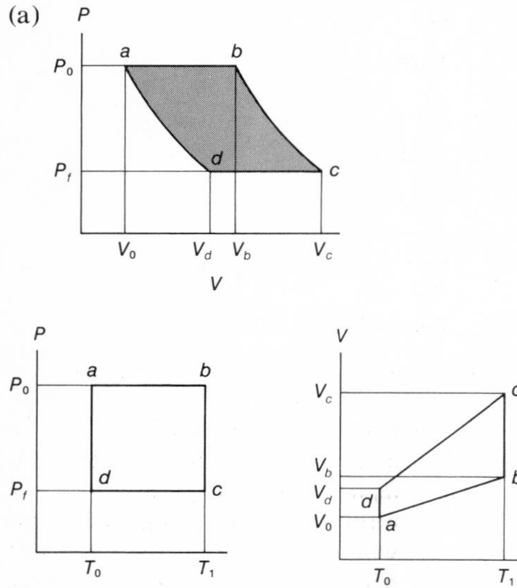

(b) The enclosed area on the P-V diagram only

(c) i. along paths ab and bc
ii. along paths cd and da

683. (a) $Eff = 1 - (V_b/V_a)\,(T_a/T_b)$, where T_b and T_a are the highest and lowest temperatures available to the cycle

(b) The above efficiency is not a function of γ, and hence is not a function of the complexity of the molecules

(c) With P_a, V_a (the starting point) and P_f fixed, increase T_b

684. (a) $Eff = 1 - (V_2/V_1)^{\gamma-1}$

(b) Increase the compression ratio V_1/V_2

685. (a) $Q_0/Q_1 = T_0/T_1$

(b) $W_{\text{net}} = NR(T_0 - T_1)\ln(V_b/V_a)$

(c) $Eff = (T_0 - T_1)/T_0$

686. (a) $Q_{DC} = NRT_1 \ln V_b/V_a = P_fV_c \ln V_b/V_a$

(b) $Q_{ba} = NRT_0 \ln V_b/V_a = P_0V_a \ln V_b/V_a$

(c) $W = NR(T_0 - T_1)\ln V_b/V_a$

(d) $T_1/(T_0 - T_1)$

(e) Decreases with increasing $(T_0 - T_1)$

687. (a) 13.9 (b) 4.96

688. (a) $Q_{ab} = W_{ab} + \Delta U_{ab} = P_1(V_2 - V_1) + [P_1/(\gamma - 1)](V_2 - V_1) = [\gamma/(\gamma - 1)]P_1(V_2 - V_1)$

(b) $Q_{acb} = Q_{ac} = W_{ac} = P_1V_1 \ln(V_c/V_1) = [\gamma(\gamma - 1)]\ P_1V_1 \ln(V_2/V_1)$
$\Delta U_{cb} = [P_1/(\gamma - 1)](V_2 - V_1)$
This equals the area under the adiabatic from c to $b = W_{cb}$, the work done on the gas. Then net work $W_{acb} = W_{ac} - \Delta U_{cb} = [P_1/(\gamma - 1)]\ [\gamma V_1 \ln(V_2/V_1) - (V_2 - V_1)]$

(c) $Q_{ab} - Q_{abc} = [\gamma P_1/(\gamma - 1)][(V_2 - V_1) - V_1 \ln(V_2/V_1)]$
$W_{ab} - W_{acb,\text{net}} = [\gamma P_1/(\gamma - 1)][(V_2 - V_1) - V_1 \ln(V_2/V_1)]$

689. (a) $V_{b,1}/V_{b,2} = e$ (b) $W_1/W_2 = 3$

(c) Since both engines operate between the same temperatures, the efficiencies are equal

690. (a) Equally divided, the temperature difference across each engine = 300°K

(b) $T_2 = 756$°K, $T_3 = 478$°K

(c) 900 $NR \ln V_b/V_a$ in MKS units for either arrangement

(d) Best efficiency = 37% when arranged for equal efficiency

692. (a) $\Delta S_{ad} = [NR/(\gamma - 1)] \ln T_2/T_1$

(b) $\Delta S_{ad} = (1/\gamma)\,\Delta S_{ab}$ from problem 691

693. (a) $\Delta S_{ab} = [\gamma/(\gamma - 1)]\ NR \ln V_b/V_a$
$\Delta S_{bc} = 0$ $\Delta S_{cd} = -\Delta S_{ab}$ $\Delta S_{da} = 0$

(b) $\Delta S = 0$

694. (a) 0

(b) $\Delta S > 0$, entropy of the universe increases

(c) Since the system returns to its original state there is no change in its entropy

695. (a) 0.71 j $(°\text{K})^{-1}$

(b) −0.71 j $(°\text{K})^{-1}$

696. (a) 8.6×10^2 j (°K)$^{-1}$ (b) 0
(c) $\Delta S_{univ} > 0$

697. (a) $c_p = 37$ j mole^{-1} (°K)$^{-1}$, $\gamma = 1.29$
(b) 28.7 j

699. (a) $c_v = 29.4$ j mole^{-1} (°K)$^{-1}$ $c_p = 38.5$ j mole^{-1} (°K)$^{-1}$
(b) $\gamma \simeq \frac{9}{7}$, hence probably 7 degrees of freedom, hence probably polyatomic
(c) $c_p - c_v = 9.1$ j mole^{-1} (°K)$^{-1}$, which is greater than $R = 8.3$ j mole^{-1} (°K)$^{-1}$. Hence the internal energy of the gas during expansion must increase at a greater rate than is required to increase the temperature. This probably means that an internal potential energy is developed against intermolecular forces, which in turn means the gas cannot be "ideal."

700. (a) 1.396 (b) 29.3 j mole^{-1} (°K)$^{-1}$
(c) 106 j

701. If P_f is the pressure necessary for static equilibrium of the piston, on its first down trip the piston went past this point in the same way a weight dropped on a spring originally sinks below the equilibrium position. Thereafter the piston moved up and down in damped SHM, carrying the pressure through a sequence of very short adiabatic zig-zags. The state diagram would look something like this:

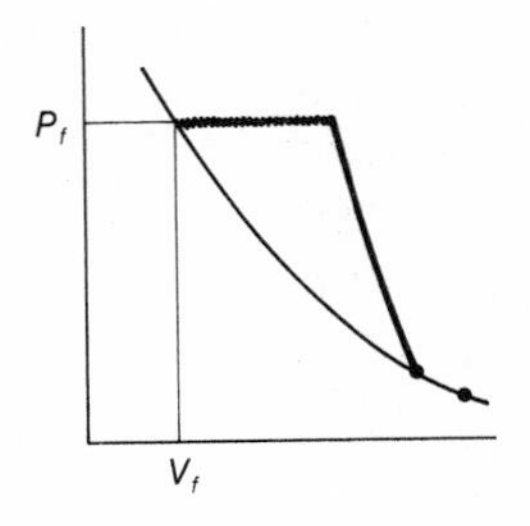

702. (b) 4.7×10^2 m sec^{-1}

703. (a) $\overline{v^2} = 3kT/m$
(b) $v_{rms} = \sqrt{3kT/m}$
(c) 5.1×10^2 m sec^{-1}

704. (a) $v_{mp} = \sqrt{2kT/m}$
(b) 4.2×10^2 m sec^{-1}

705. (a) i. 1.83% ii. 1.66%
(b) i. 1.79% ii. 1.97%
From the definition of v_{mp}, or from its position on the speed distribution curve, you would expect a larger fraction of molecules near v_{mp} than near any other speed. Note that this expectation is realized when you deal with actual speeds but not when you deal in percentages.

706. $n_{(E)}/N = (2/\sqrt{\pi})(1/kT)^{3/2}E^{1/2}e^{-E/kT}\,dE$

707. (a) $\bar{E} = \frac{3}{2}kT$, which agrees with $E_{av} = \frac{3}{2}kT$ developed earlier (remember, $\bar{E} = \int n_{(E)}E/N$)
(b) $E_{mp} = \frac{1}{2}kT$ (remember, E_{mp} occurs when $[d\,f(E)]/dE = 0$)

708. $\frac{1}{2}T_1$

709. $N_2/N_1 = 1.11$

710. $f/(1 + \sqrt{2}e^{-a})$

711. (a) $0.31/kT$ (b) 0.92%

713. $10^{-10^{12}}$

714. $\left.\dfrac{\rho_{He}}{\rho_{N_2}}\right|_h = \frac{1}{7}\exp\,(6m_{He}\,gh/kT)$, where m_{He} is the mass of a helium atom

715. (a) $n_{n+1}/n = e^{-\hbar\omega/kT}$
(b) i. e^{-24} ii. $e^{-0.72}$
(c) It is obvious from the answers that very many more energy states are utilized at high temperatures than at low temperatures. This would explain qualitatively why c_v, which is a measure of the capacity for storing energy, increases with significant increases in temperature.

716. (a) 1.05 N m^{-2} $\simeq 1 \times 10^{-5}$ As
(b) 1.05×10^{-2} N m^{-2} $\simeq 1 \times 10^{-7}$ As

717. (a) $\eta = m\bar{v}/(3\sqrt{2}\,\sigma)$; since m, v, and σ are not functions of pressure, η is independent of pressure
(b) No change, for the reason given in (a)

718. (a) 3.4×10^9 sec^{-1} (did you check first to see whether the speed of the electron was nonrelativistic?)
(b) $v_{He} = 1.26 \times 10^3$ m sec^{-1}, $v_{el} = 2.65 \times 10^6$ m sec^{-1}, hence helium molecules are essentially at rest relative to the electron
(c) A reasonable limit for the assumption

would be $v_{el} = 10^2\ v_{He}$, or $v_{el} = 1.26 \times 10^5$ m sec^{-1}
(d) For above limiting speed, $n = 2.2 \times 10^4$ collisions

719. 4.5×10^9 barns

720. 5.6×10^9 molecules sec^{-1}

721. (a) $t = 0.42\ V/\bar{v}A$
(b) 6.7×10^{12} sec $= 2.1 \times 10^5$ years

722. $\left.\dfrac{n_{He}}{n_{N_2}}\right|_{24\text{ hr}} = 0.985$

723. $B_{\text{correct}} = B_{\text{measured}} + 23.8$, if B in mm Hg

724. Dewpoint inside = 21.1°C, dewpoint outside is at 21.3°C; hence dew forms first on outside

725. (a) 21°C (b) 14.9 gm

726. (a) 733.2 mm Hg (b) 0.832 gm
(c) 50°C (d) 882 mm Hg
(e) 1020 mm Hg

728. (b) 730 mm Hg

729. $c = 3\ R = 6$ cal mole^{-1} (°C)$^{-1}$—from 6 degrees of freedom for a three-dimensional harmonic oscillator and $\frac{1}{2}R$ per degree

730. Mix together $\frac{1}{2}$ cup of boiling water and $\frac{1}{2}$ cup of water from an ice-water mixture

731. 15.6°C; tower height not necessary since *mgh* negligible compared to thermal energy for any reasonable *h*

732. $c_{\text{kettle}} = 0.18$ cal gm^{-1} (°C)$^{-1}$

733. (a) 125 cal (b) 28°C

734. (a) $\alpha \simeq 3\lambda$ (b) $A_t \simeq A_0[1 + 2\lambda(t - t_0)]$

735. 4.8 j

736. $r_t = \dfrac{d[2 + (\lambda_A + \lambda_B)(t - t_0)]}{2(\lambda_A - \lambda_B)(t - t_0)}$

737. 1.23 cm

738. $V = [\alpha_{gl}/(\alpha_{Hg} + \alpha_{oil})]\ V_0$

739. (a)

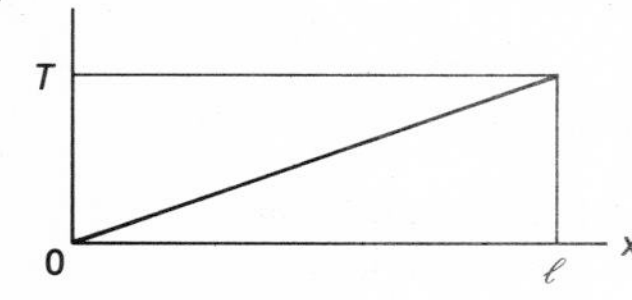

(b) $d(dx) = dx\ \lambda t$ (c) $\Delta\ell = \ell_0\lambda T/(2 - \lambda T)$

740. $dy/dt = 0.074$ mm sec^{-1} (the cone angle does not change with temperature)

742. $P_{\text{inside}} = B_0 + (4\sigma/R)$

743. (a) $R_{\text{int}} = rR/(R - r)$
(b) Concave toward smaller bubble

744. $16\pi R^2\sigma$

745. 300 ergs

746. $E_{\text{film}}/E_{\text{gas}} = 1$

747. (a) $V_{\max} = 2\sqrt{r\sigma_{\text{water}}/\epsilon_0}$
(b) $U_{el} = 2\ U_{\text{surf}}$

748. (a) 0.78 ohms (b) $\Delta P = 12.5$ watt

750. $[R_1(E_1 - E_2)]/E_2$

751. no change; the equation for balance is the same for either arrangement

752. $\Delta R_2/R_{2,t_0} = \alpha(t - t_0)$

753. $\lambda_x = 2.1 \times 10^{-4}$ (°C)$^{-1}$

754. 8.0 mi (this is known as the Murray loop test)

755. (a) $\dfrac{P_L}{P_0} = \dfrac{1 + [1 - (4RP_L/V^2)]^{1/2}}{2}$
(b) Interpreting the above equation, the higher the voltage the higher the efficiency. The upper limit to a practical transmission voltage is determined by the increased cost of insulators and rectifiers for higher voltages.

756. (a) 4.5 volts (b) 1.0 volts
(c) 1.0 watt

758. (a) $I_t = (V/R)\ e^{-t/RC}$
(b) $V_{R,t} = V\ e^{-t/RC}$, $V_{C,t} = V(1 - e^{-t/RC})$
(c) $P_R = (V^2/R)\ e^{-2t/RC}$
(d) $P_C = (V^2/R)\ e^{-t/RC}\ (1 - e^{-t/RC})$
(e) $P_{\text{bat}} = (V^2/R)\ e^{-t/RC}$
(f) At $t = 0$, $V_R = V$

759. (a) $I_0 = V/R$
(b) $I_t = [V/(R + R_1)](1 + (R_1/R)\ e^{-(R+R_1)t/RR_1C}$
(c) $V_{R,t\text{large}} = VR/(R + R_1)$
(d) $U_C = CV^2R^2/[2(R + R_1)^2]$

760. (a) $V_1 = V_2 = C_1V/(C_1 + C_2)$
(b) $\Delta E = C_1C_2V^2/[2(C_1 + C_2)]$

761. 11 megohms

762. (b) $q = CV(1 - e^{-Rt/2L})$
(c) $V_C = V(1 - e^{-Rt/2L})$ $V_R = 2Ve^{-Rt/2L}$
$V_L = -Ve^{-Rt/2L}$

763. (a) $V_C = V \cos \omega t$ $V_L = V \cos (\omega t - \pi) = -V \cos \omega t$
(b) $\omega = 1/\sqrt{LC}$ or $\nu = 1/(2\pi\sqrt{LC})$
(c) $I_{max} = \sqrt{C/L}\ V$

764. (a) For P_{max}, $C = 1/(\omega^2 L)$
(b) $P_{max} = V_0^2/2R$
(c) $1/[\omega^2 L + (\omega R/\sqrt{3})]$

765. (a) $V_2/V_1 = [1 + (\omega L/R)^2]^{1/2}$
(b) $P = I^2R$ in both cases
(c) $\tan^{-1} \omega L/R$

766. $R_x = R_S$ $L_x = 1/[(2\pi\nu)^2 C_s]$

767. $\nu_{crit} = (1/2\pi)(C_1C_2L_1L_2)^{-1/4}$

768. (a) 4.5 amp (b) 0.28 sec (c) 5.5 amp

769. (a) For P_{max}, $R_L = R_S$ (b) 50%

770. For P_{Lmax}, $L_SC_L = 1/\omega^2$ and $R_L = R_S$

771. (a) 10 kc (b) 30 watts
(c) The actual average power does not change with the frequency in the circuit shown; the current, the power factor, and the efficiency of operation do change with frequency

772. $V_{AB} = V_D + 0.053\ V_R$

773. $u = (k_1A_1 + k_{air}A_2 + k_3A_3)[(T_1 - T_2)/d]$ per unit time

774. 2.6×10^{10} kw

775. $T_0 - 3PL/4kA$

776. 5.1 hr

777. 8.0 cm

778. 6.8 watts m^{-2}; since this is less than it would have been with the cork, dead air is a better insulation than cork

779. (a) 4.0°C (b) 7.47 cm

780. (a) $T = T_w + (T_0 - T_w)e^{-3kt/Cr\rho d}$
(b) $dT/dt = -(3k/Cr\rho d)\ \Delta T$ — a linear relationship
(c) In computing dQ/dT you had to assume the entire ball was at the same temperature. Because the conductivity of copper is large compared to k, this was a reasonable assumption.

781. 1.6×10^5°C. It is now believed the theory mentioned is incorrect; the heat flux at the surface of the earth is now thought to originate in radioactivity outside of the central core.

782. (a) 5.8 watts m^{-2}
(b) $T = T_w + (T_0 - T_w)e^{-3t/[Cr\rho(d/k+1/h_w)]}$

783. Even without computing the heat flow exactly you know that the equation will contain a $(1/h_w + d/k + 1/h_{air})$ term in the denominator. The d/k term will be negligible compared to the transfer terms, hence differences in the thermal conductivity will have little effect on the heat exchange. Thus economic factors of tube and installation costs should govern the decision.

785. (a) $u_{(x=0)} = \pi\sqrt{2r^3hk}\,(T_S - T_0)$ (b) $Q = \pi r^2\rho c[T_0L + \sqrt{kr/2h}\,(T_S - T_0)(1 - e^{-\sqrt{2h/kr}\,L})]$

786. $$Q = A_1\left[\frac{2gh}{(A_1/A_2)^2 - 1}\right]^{1/2}$$

787. $u = \sqrt{2\rho_m gh/\rho}$

788. (a) 27.3 m sec^{-1} (b) 16 As

789. $D = 4R$

790. (a) $T = [(A/a)^2 - 1]^{1/2}\sqrt{2H/g}$
$\simeq (A/a)\sqrt{2H/g}$ for $A \gg a$
(b) $T_{act} \simeq 1.7\ (A/a)\sqrt{2H/g}$

791. 10.1 ft

792. $r_y = r_0[1 + (2gy/u_0^2)]^{-1/4}$

793. (a) 540 lbwt ft^{-2} (b) 560 mi hr^{-1}

794. 4.7 ft^3 sec^{-1}

795. $P = 4.9Qh$ watts if all data are in MKS units

796. 3.5×10^{-3}°C m^{-1}

797. (a) 57 cm (b) 65 cm^3 sec^{-1}

798. $H_{max} = 10$ m

799. 20 kgwt

800. (a) $y_0 = 0.40$ cm $\nu = 30$ sec^{-1}
$v = 6 \times 10^3$ cm sec^{-1} $\lambda = 200$ cm
(b) $\dot{y}_{max} = 75$ cm sec^{-1}
(c) 7.2×10^7 dynes

801. No change in values; only change is in direction of propagation

802. 3.2×10^{-2}

803. (a) 80 ft sec^{-1} (b) Increase tension in the belt or choose a belt with less mass per unit length

804. $d\lambda/dy = -(\sigma/2f)\sqrt{g/(L-y)}$

805. 0.83 sec

806. (a) $\alpha^2\sigma/\beta^2$

807. (a) $P_t = 2A^2\omega^2\sqrt{T\sigma}\sin^2\omega t$
(b) $\bar{P} = A^2\omega^2\sqrt{T\sigma}$

808. (a) $u_R = (r/R)^2 u_r$ (b) $p_r = (1/r)\, p_{0,1}$

809. (a) $E_\lambda = \pi\rho A\omega c_m y_0^2$
(b) $P = \frac{1}{2}\rho A\omega^2 c_m y_0^2$

810. $\bar{u} = \rho\omega^2 y_0^2$

811. (b) $\alpha_r = \alpha_1$

812. (a) $[(f+mg)ML]/\pi^2 mT$
(b) $[(f-mg)ML]/\pi^2 mT$

813. $F_{y,\max} = \pi Td/L$

814. (a) 0.41 watt (b) 350 Hz

815. $I = p_0^2/2\rho c_m$

816. (a) 3.4×10^{-11} m; note that this is less than a molecular diameter
(b) 9×10^{-10} As
(c) 0.4×10^{-15} watts

817. 6 ft

818. (a) 1350 Hz 1150 Hz
(b)

ν — 1350 Hz, 1242 Hz, 1150 Hz; t

820. (a) $d\nu/\nu \simeq 2V_c/c_m$
(b) $d\nu/\nu \simeq 2(V_p - V_c)/c_m$
(c) Since in pursuit situations $(V_p - V_c)$ will probably be small compared to V_c, $d\nu/\nu$ will be relatively inaccurate compared to the "at rest" measurement

821. $\Delta\nu_0 \simeq (2\, V_{\text{ion}}\nu_0)/c$

822. 3.6×10^{33} kg

823. (a) 0.43 nautical mile (b) 2°.8 S of W

824. (a) i. $\dfrac{d\lambda}{\lambda} = \mp\dfrac{V_S}{c_m}$ ii. $\dfrac{d\lambda}{\lambda} = \dfrac{-1}{1 \pm c_m/V_0}$
(b) i. and ii. $d\lambda/\lambda = \sqrt{(1 \mp \beta)/(1 \pm \beta)} - 1$, with $\beta = V_{\text{rel}}/c$
As usual, the upper signs apply to relative motion toward, the lower to relative motion away

825. 0.8 c, away from the earth

826. (a) 4 sec^{-1} (b) 2 sec^{-1}–6 sec^{-1}
(c) 2.2 sec^{-1}–5.8 sec^{-1}

827. 250 mi hr^{-1}

828. $B(740 - V_1)/2V_1$ if V_1 is in mi hr^{-1}

829. $\pi/4$

830. (a) $y_t = B/2 - (4B/\pi^2)\,\Sigma\,(1/n^2)\cos nt$, for n odd only
(b) i. only the odd harmonics 1, 3, 5, . . .
ii. $E_1{:}E_3{:}E_5{:} \ldots {:}E_n = 1{:}\frac{1}{81}{:}\frac{1}{625}{:} \ldots : 1/n^4$
(c) $\pi^2/8$ (d) $\pi^2/6$

831. $y_t = (4A/\pi)(\sin t + \frac{1}{3}\sin 3t + \frac{1}{5}\sin 5t + \ldots + 1/n \sin nt + \ldots)$

832. (a) $y_t = \dfrac{2A}{\pi}\left[1 - 2\sum_{n=1}^{\infty}\dfrac{\cos nt}{(2n-1)(2n+1)}\right]$
(b) $\pi/2$

834. (a) $I_{S,\theta\uparrow} = (3P_0\cos^2\theta)/(8\pi S^2)$
(b) $I_{S,\theta\rightarrow} = (3\,P_0\sin^2\theta)/(4\pi S^2)$
(c) $\left.\dfrac{I\uparrow}{I\rightarrow}\right|_{S,\theta} = \cos^2\theta/(2\sin^2\theta)$. This ratio < 1 for elevation angles $\theta > 35°$. Hence unless you expect strong winds that may cause the horizontal movement of the balloon to be greater than the vertical movement, a horizontal antenna would give the receiving station the larger signal.

835. (a) 2.2×10^5 volts m^{-1}
(b) 3.9×10^{23} kw

836. (a) 1.4 cos θ kw m^{-2}
(b) 0.7 cos θ kw m^{-2}

837. (a) $P_\nu = (q^2\omega^3 y_0^2)/(6\epsilon_0 c^3)$
(b) $\gamma_R = (q^2\omega)/(3m\epsilon_0 c^3)$

838. (a) $4.7 \times 10^{-6}\,A$ newtons if A is in m^2
(b) $B = \frac{1}{2}A$

839. 1.7×10^{-14}

840. $r\rho < 3I_\odot R_\odot^2/(4Gc\,M_\odot) = 5.9 \times 10^{-4}$ kg m^{-2}

841. $\tau_{A\text{-}A} = 0$. You should have found that the force on the reflecting sphere $= \pi r^2 I/c$, which is exactly the same force felt by the absorbing sphere.

842. (b) $\phi/2 = (\pi/2)(1 + 2 \sin \alpha)$

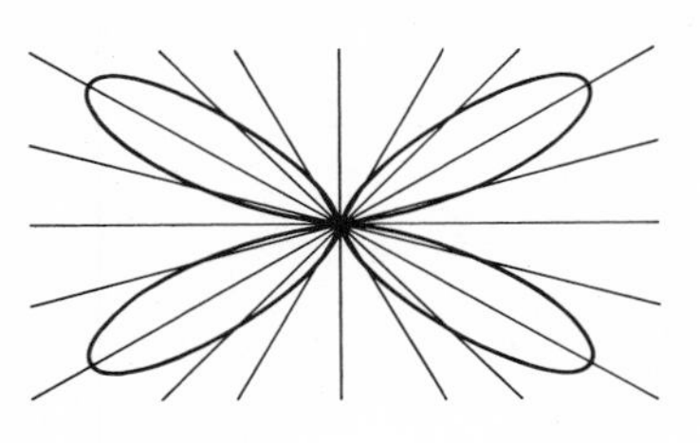

843. (a) $\beta = -\pi/2$
(b)

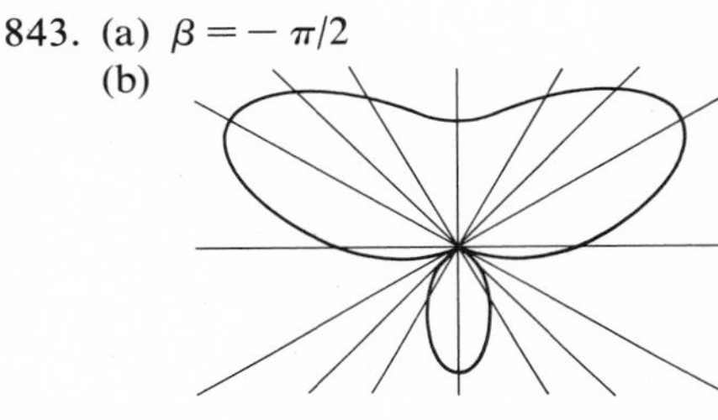

844. $d\beta/dt = -$ (cos α/120 cycles sec^{-1}

845. (a)

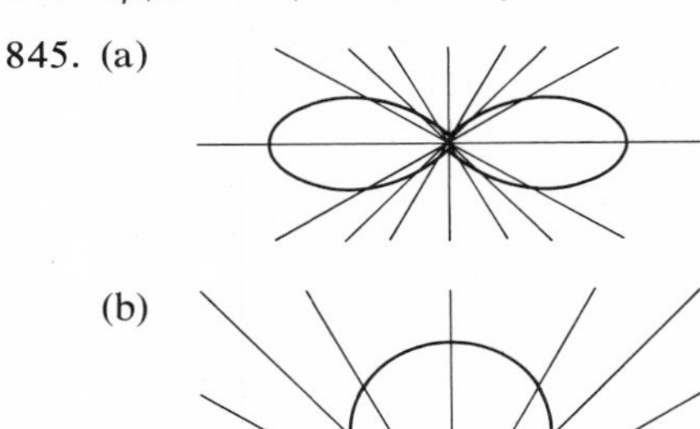

(b)

846. (a) 0.50 mi (b) 0.92 mi

847. (a) i. $\pm\pi$, lead or lag ii. $\sqrt{2}\,\pi$ lag
iii. π lag iv. 0, dipoles in phase
(b) i. at 30° N of W ii. at 45° S of W
iii. at 30° S of E iv. at due E, W and S

848. $\beta_2 = \sqrt{3}\,\pi$ lead $\beta_3 = 0.7\,\pi$ lead
$\beta_4 = \pi$ lead

849. (a) linear array with $d = \lambda/4$ and $\beta = \pi/2$
(b) 16 I_0 in line with array
(c)

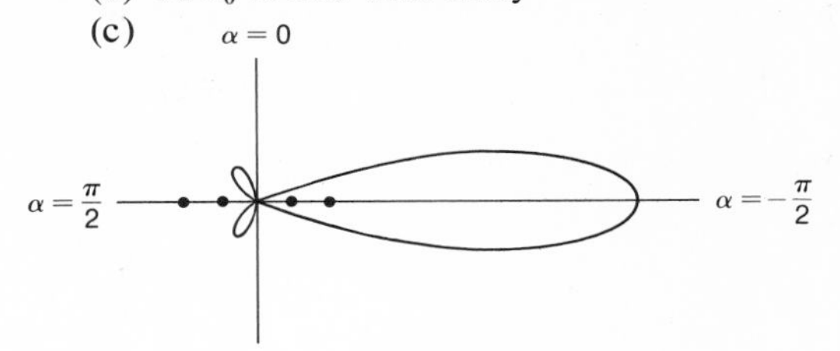

850. (a) $d\beta/dt = -2\pi\omega \cos \alpha$
(b) As can be seen from the diagrams on page 400, beams are being swept simultaneously and symmetrically in both hemispheres. Thus the electronic sweep for the same ω gives information at twice the rate provided by the mechanical sweep.

851. (a) 6000 Å (b) i. 33 Å ii. 6.6 Å

852. 0.375 mm

853. 1.12 mm

854. (a) 10.9 km for a rectangular detector. Since the eye is a round detector a reduction factor of 1/1.22 must be applied, reducing the distance to 8.9 km = 5.5 mi.
(b) For $d = 40$ mm, distance = 44 mi. However, the light will probably be below the horizon at that distance.

855. No inconsistency; intensity varies from being proportional to $4E_0^2$ at maximum to 0 at minimum, with an average of $2E_0^2$. The energy per unit time is accounted for but is distributed unevenly in space.

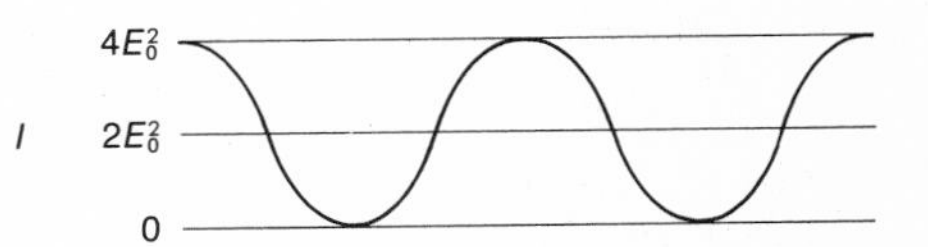

856. (a) The width of the principal maximum is halved
(b) $w = \sqrt{2D\lambda} = 1.08$ mm
(c) $W = 11.6$ cm

857. $W_{1/2\,\text{max}} = 0.884\, D\lambda/w$

858. 6240 Å

859. (a) i. $\theta_{\text{cen max}} = 1°.2$ ii. $\theta_{\text{cen max}} = 0°.11$
(b) i. 1.0 cm ii. 1.0 mm
(c) $W_{\text{cen max}}/W_{\text{sec max}} = 2$

860. $I_{\text{sec max}} = 4\, I_{\max}\, [\pi^2(2p+1)^2]; p = 1, 2, 3, \ldots$

861. ninth order

862. (a) $d/w = m + \frac{1}{2}$ (b) $d/w = m$
(c) case (a), $2m + 1$ case (b), $2m - 1$

863. the third or greater order

864. (a) 3.94 cm (b) 0.98 cm

865. $d = 2w$

866. (a) Not possible to suppress the central maximum at normal incidence
(b) Zero order suppressed when $\sin\theta = q\lambda/w$, $q = 1, 2, 3, \ldots$

867. (a) 5244 Å (b) A third line could be observed at −36.79 cm
(c) From the information given all we can reasonably assume is that the three lines represent successive orders, the value of the order m not known. If, however, the central line at +12.81 cm happens to be brighter than the others it would be a good guess that it was the central maximum. In that case $\theta = -7^\circ\!.3$.

868. (a) $\theta = \sin^{-1}(m\lambda/2d)$, d is the grating spacing (b) $37^\circ\!.8$ (c) Filters to absorb lines of 3927 Å and 4712 Å
(d) 0.94 mm (e) 25%

869. (a) $0.172\sqrt{GM_\oplus/R}$ (b) $0.51\, R$
(c) $\Delta E_{\min} = 0.515\, GmM_\oplus/R$

870. (a) $U = 1.7 \times 10^9\, Q^2/R$ if Q is in coulombs and R is in meters
(b) $4.8 \times 10^9\, Q^2/R$
(c) $v_{\text{rel}} = 1.94 \times 10^7$ m sec^{-1}
(d) $v_f = 1.77 \times 10^7$ m sec^{-1}
Did you obtain the values for (c) and (d) by relativistic methods?

871. $\Delta = 2.6$ mm

872. $R = 2.6\, r$

873. (a) $\Delta Q = (MR^2/4)\,(\omega_0^2 - \omega_f^2)$
(b) $\theta_{\text{twist}} = [MR^2(\omega_0 - \omega_f)d]/(2\pi r^4 nt)$
(c) $\theta_{\max} = \left[\dfrac{MR^2 d}{\pi r^4 n}\left\{\omega_t^2 + \dfrac{MR^2(\omega_0 - \omega_t)^2 d}{4\pi r^4 n t^2}\right\}\right]^{1/2}$

874. Because of the $1/r^2$ influence on $\theta_{\max}$, increasing the radius of the shaft is an obvious way to reduce the twist; however, this increases the size and cost of the bearings. Decreasing R would reduce the flywheel inertia; decreasing d would increase stress.

875. (a) $\vec{v}_2 = (M_1/M_2)[u_1^2 + v_1^2 + \sqrt{3}\, u_1 v_1]^{1/2}$ at $\tan^{-1} - 1/[2(u_1/v_1) + \sqrt{3}]$ with direction of u_1
(b) $\Delta E_{\min} = \frac{1}{2}(M_1/M_2)\,[(M_1 - M_2)u_1^2 + (M_1 + M_2)v_1^2 + \sqrt{3}\, M_1 u_1 v_1]$

876. $8\pi d^2\sigma(1 - \sin\theta)^2$

877. $y_{2m,t} = -[\frac{2}{3}(L + 2mg/k) + \frac{1}{2}gt^2]$
$2mg/3k \cos\sqrt{3k/2m}\, t$

878. 7.4×10^{-16} gm cm^{-3}. Actually at that altitude both the satellite and the atmosphere will be electrically charged, and the drag from electrical forces may be of the same order of magnitude as the drag from mass impact.

879. $22\, Mr^2$, with r the radius of each semicircle.

880. $Q_{\min} = LA/160$. This arrangement would not make a satisfactory automatic pouring spout; the amount delivered depends too much on the amount still left in the bottle.

881. $N_1/N_2 = (\rho_1/\rho_2)(MW_2/MW_1)^{3/2}$. Note that the ratio increases with increasing density ratio, hence if several enclosures are connected in series with connecting small holes, the rate of separation increases with each succeeding enclosure. This is the principle of the cascade method of gaseous separation.

882. 32 cm

883. (a) 1.03 watts cm^{-1} (b) 70 amperes

884. (a) 7.1×10^{8}°K (b) $PV = \frac{1}{3}N\, N_0\beta^2 E$

885. $\Delta = d\sin\theta\,[1 - \cos\theta/\sqrt{n^2 - \sin^2\theta}]$

886. $L' = \dfrac{3\left(\dfrac{mg}{k} + L\right) + \sqrt{\dfrac{9\, mg}{k}\left(\dfrac{mg}{k} + 2L\right) + L^2}}{4}$

887. (a) $2\mu WR$ (b) Choose smallest R that will support the load (c) $P = 2\mu WR\omega$

888. (a) $\Delta E_{S_2} = -\frac{1}{2}\, C_1 C_2 V_0^2/(C_1 + C_2)$
(b) $\Delta E_{S_3} = -\frac{1}{2}\, C_1^2 V_0^2/(C_1 + C_2)$

889. Assuming a penetration of 0.5 ft at an average speed of 400 ft sec^{-1}, $F = dp/dt =$ 660 lbwt, a fairly heavy blow. Hence it is realistic for an actor to fall backwards when "shot."

890. Annihilation products must meet the following requirements:

1. charge must be conserved, and since $q_{el+} + q_{el-} = 0$, the total final charge must $= 0$;
2. momentum must be conserved, and since for slow particles the linear momentum is essentially zero and since for particles the collision occurred at the c.m., both the final linear and angular momenta must $= 0$;
3. mass energy must be conserved, and since for slow particles $T_0 \simeq 0$, E_f must $= E_0 = 2\ m_{el}c^2$.

These requirements could be met by creation of the following:

(a) a single uncharged "neutrett" at rest and of mass $2m_{el}$ or a number of alike uncharged neutretts moving with equal speeds symmetrically from the c.m. point;

(b) n alike photons, each with energy $h\nu = 2m_{el}c^2/n$, symmetrically distributed in space.

Particles described in (a) are not found in nature; particles described in (b) are found, usually with $n = 2$.

891. 0.0243 Å

892. 46.7 ft $\sec^{-1}$

893. On the ground 3.7 ft out from the base of the dome

894. (a) $L - (Ma/2k)$ (b) $M^2a^2/6k$

895. (a) 18 cm above the mirror (b) 1.67

896. (a) For the trace to have points like a cycloid, V must $= R\omega_c$; hence $\omega_c = V/R$
(b) V must be $> R\omega_1$, hence $\omega_1 < \omega_c$
(c) This is the reverse of (b) — $\omega_2 > \omega_c$
(d) $x = Vt + R\cos\omega_1 t$, $y = R\sin\omega_1 t$

897. $v_{rel} = [2G(m_1 + m_2)/(r_1 + r_2)]^{1/2}$

898. (a) $v_{term} = (3\lambda Rg\sin\theta)/(2B^2\cos^2\theta)$
(b) $P_{term} = (9\lambda^2 Rdg^2\tan^2\theta)/4B^2$

889. 11 cm

900. (a) $\gamma \simeq \omega_0/10\pi = (1/10\pi)\sqrt{k/m}$
(b) $\overline{\Delta E_\nu} = 0.043\ kA^2$
(c) $\Delta E_1 = 0.091\ kA^2$

901. (a) $v = 2400/(1 + 6n)$, $n = 0, 1, 2, \ldots, v$ in m $\sec^{-1}$
(b) $v_{max} = 2400 \pm 80$ m $\sec^{-1}$

902. (a) $c_v = 5R$
(b) With $R/2$ per degree of freedom, 10 degrees of freedom must exist in the gas
(c) Obviously this is a very rough way of determining c_v

903. (a) V_0 $(r/R)\ V_0$
(b) $(V_0/K)\ [1 + r(K-1)/R]$ $(r/R)\ V_0$

904. $v = \sqrt{gR(3 - \sqrt{2})} = \sqrt{1.59gR}$

905. $\bar{f} = mv^2/(L - 2nr)$; this is essentially a one-dimensional model of a perfect gas in an enclosure

906. The small plane is at an altitude of 1.5 mi flying at a speed of 150 mi hr^{-1} eastbound

907. From $\Delta E = hcR[1 - (1/n^2)]$, $n = 4$; then possible lines are $\lambda_{4\text{-}3} = 18760$ Å $\lambda_{4\text{-}2} = 4860$ Å $\lambda_{4\text{-}1} = 972$ Å $\lambda_{3\text{-}2} = 6560$ Å $\lambda_{3\text{-}1} = 1026$ Å $\lambda_{2\text{-}1} = 1215$ Å

908. 8.7 mi

909. $T_{el} = 0.66$ Mev

910. (a) $T_{imp} = 0.76\ T$
(b) $v_{imp} = \sqrt{1.2\ NRT/m}$ (c) T

911. 19 cm

912. (a) 184 sec after lift-off
(b) $\dot{y}_{max} = 0.39$ nautical mi $\sec^{-1}$
(c) 0.35 nautical mi $\sec^{-1}$
(d) 0.23 nautical mi $\sec^{-1}$

913. 3.0×10^5 kw

914. (a) 2.0×10^{33} kg (b) The star system must be an eclipsing binary, both stars alike

915. 6 inches

916. 5600 lbwt. Did you take into consideration the decrease in gravitational potential energy of the pile and weight?

917. $y = 1.92\ R$

918. (a) $(\rho gW/4\sin\theta)(h_u^2 - h_d^2)$
(b) same value as (a), at $(\pi/2 - 2\theta)$ with wall

919.

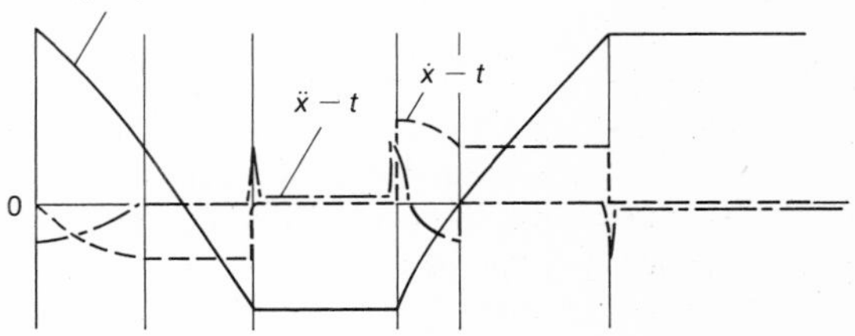

920. (a) 2.0×10^{-3} moles cm^{-1} (b) 24.6 cm

921. 24 cm

922. (a) $0.6 \tan \theta$ (b) 75%

923. 4.4 ft sec^{-1}

924. At a point on the other short edge directly opposite to P, and at all points on a circle of radius $\frac{5}{12}a$ with its center at the center of the plate

925. $2F$; the answer is not affected by the pressure on the surface of the liquid

926. θ must $= \pi/2$; if the ball is shot in a direction perpendicular to the plane, then relative to the plane both the ball and the cup have zero initial speed and the same acceleration

927. (a) 1.12×10^{-6} N m^{-2}
(b) 4670 Å; γ for photons $= \frac{4}{3}$

928. (a) $m_3 > \mu[2m_1/(1 + \mu \cot \theta) + m_2]$
(b) $a_3 = -g\left\{\dfrac{m_3 - \mu[2m_1/(1 + \mu \cot \theta) + m_2]}{m_2 + m_3}\right\}$
(c) $F_C = \mu m_1 g/(\sin \theta + \mu \cos \theta)$

929. (a) $u_{\min} = 7.53 \sqrt{gL}$
(b) $v = -5.85 \sqrt{gL}$ horizontally

930. (a) The Maxwell distribution assumes a continuous distribution of energies among a large number of particles, with zero probability that one particle possesses all the available energy. The photon distribution curve exhibits a finite possibility of one photon possessing all the available energy, and the cut-off point establishes the value of that one-particle energy. By extension this also suggests the possibility of energy distribution by discrete quanta.
(b) 6.63×10^{-34} j sec

931. 24 mi hr^{-1}

932. (a) 7.4 knots at 53°.8 S of E
(b) 4.4 knots due east
(c) 6.0 knots

933. 70 mi hr^{-1} at 33°.6 above the horizontal

934. (a) 2.1×10^3 j $\simeq$ 500 calories
(b) 6.3×10^{-20} j molecule^{-1}
(c) 6.3×10^{-10} N

935. $\frac{1}{300}$ sec

936. $B = (2\pi/L) \sqrt{2mV/q}$

937. $D = (V^2/g) + \sqrt{2}\, r$, r the radius of a bicycle tire

938. (a) $y_{\max} = H - (v^2/2gA^2)$
(b) $W + \rho AH + (\rho v^2/2gA)$ gmwt
(c) $\rho v^2/gA$

939. (a) $h = (g/k)[2m + (\pi^2 M/2)]$
(b) $(mg/k)\sqrt{4 + \pi^2 M/(M + m)}$
(c) loss of potential energy of m, which is below where it originally was at rest

940. (a) 21.6 volts
(b) 7.95 Hz
(c) $\Delta E = 1.98 \times 10^{-2}$ j

941. $W = mv/5$

942. $\gamma = 4\pi^2 mV/[T^2 A^2(B + mg/A)]$

943. (a) $Q^2/r^3 = 7.3 \times 10^{-11}$ if data are in MKSC units
(b) 2.4×10^3 volts
(c) 3.9×10^3 volts

944. 0.8 N m

945. $2\, PA\, d$

946. $\alpha = \theta/2$

947. (a) $V = 2.5 \times 10^5 \sqrt{r}$ volts
(b) $r_{\max} = 10$ cm with $V = 8 \times 10^4$ volts, about the maximum potential that can be obtained at a lecture demonstration $r_{\min} = 0.7$ cm, limited by the breakdown field in air of 3×10^6 volts m^{-1}
(c) $U_{\text{elec}} = 2\, U_{\text{surf}}$
(d) $\Delta U = 72\pi R^2 \sigma_{\text{surf}} = 16\, R^2$ j if R is in m

948. (a) $T_{\max} = \omega^2 F_0^2/[2m(\omega_0^2 - \omega^2)^2]$
(b) $T_{\text{max, forced}}/T_{\text{max, natural}} = \omega^2/\omega_0^2$ which is > 1
(c) Difference arises from the work done by $F_0 \cos \omega t$ on the inbound quarter-cycles

949. $BR = \sqrt{E^2 - m^2c^4}/qc$

950. (a) 4.7×10^5 ergs
(b) 2.3×10^3 dyne cm^{-1}

951. (a) $\lambda = \dfrac{2\pi\epsilon_0 V}{\ln [(2d - r)/r]}$
(b) $\sigma_x = -\dfrac{2d\epsilon_0 V}{(d^2 + x^2) \ln [(2d - r)/r]}$
(c) $F_{\text{unit length}} = \pi\epsilon_0 V^2/d\{\ln [(2d - r)/r]\}^2$

952. (a) $\Delta E = [I_0 m(V^2 + 2VR\omega_0 + r^2\omega_0^2) - m^2V^2(R^2 - r^2)]/2(I_0 + mr^2)$

(b) Probably most of the energy went in wear on the wheel and slots, with a small amount in heat and sound

953. $Q = \pi\rho g R^4/8\eta$

954. (a) $a_M = -g\,\dfrac{(r^2 - \mu m r R/M)}{r^2 + mk^2/M}$

(b) $F_C = mg\left[\dfrac{k^2(1 + \mu m/M) + \mu r(R + r)}{r^2 + mk^2/M}\right]$

(c) For $M \gg m$,
$F_C \rightarrow (mg/r^2)[k^2 + \mu r(R + r)]$

955. (a) 98 km hr^{-1} (b) 1. the forward motion of the bullet was not affected by going through the first hole 2. the energy lost at the first hole can be entirely deducted from the transverse motion

956. The "proper energy" $h\nu_0$ is the difference in the total *internal* energies of the radiant object before and after radiating, i.e., $h\nu_0 = (M - M')c^2$. The $h\nu$ delivered to an external observer is the difference in the total energies, internal and kinetic, of the radiant object before and after radiating, i.e., $h\nu = \gamma(M - M')c^2$.

957. (a) $\text{emf}_{max} = 0.84$ volts
(b) $\nu = 50$ Hz; the voltage is a true sine wave

958. (a) $T = 2\pi\sqrt{(M + \frac{1}{3}m)/k}$
(b) $T = 2\pi\sqrt{(I + \frac{1}{3}I_r)/\tau_0}$
(c) $m \leqslant 0.06\,M$

959. $k = 1.0 \times 10^3$ N m^{-1}

960. (a) $x = (m_2 L + \frac{1}{2}ML)/(m_1 + m_2 + M)$
(b) It is the location of the c.m. of the rod-masses system
(c) $W_{min} = \dfrac{L^2\omega^2[(M/4)(m_1 + m_2) + m_1 m_2]}{2(m_1 + m_2 + M)}$

961. (a) By normal dynamical methods you find that the normal force N between car and track along the curved section beginning at C is described by $N = 3mg \cos\theta - 2mgh/r$. For any h it is clear that N is least for $\cos\theta$ least, or for θ as large as possible. Thus if N is to be made equal to zero, which is the precondition for the car to leave the track, it can only happen at $\theta = \theta_0$.
(b) $h_{max} = (3r \cos\theta_0)/2$

962. Maximum distance for free-fall before rope begins to stretch is about 24 ft. *Solution:* the maximum allowable tension in the rope occurs when the rope has been stretched 43 inches in each 10 ft, for a total stretch of 10.8 ft for a 30 ft rope. By counting squares you can determine that the area under the curve up to the 43-inch limit is approximately 1880 lbwt ft. Thus the permissible energy to be stored in the 30-ft rope equals about 5640 lbwt ft, and this amount of energy is then the entire permissible loss of gravitational energy of the climber. This includes both the loss during free-fall and the further loss as the rope stretches.

963. (a) $\theta = \sin^{-1}\frac{2}{3} = 42°$ (b) $\frac{1}{3}\sqrt{g\ell}$

964. $B_C = 0.968$ As

965. There may be a small amount of energy lost in impact when the end of the tape hits the slope; however, most of the energy is stored elastically in the tape. At any time t there is a force stretching the tape of $\frac{1}{3}m_t g \sin\theta$.

966. (a) 35°.6 W of N galactic
(b) 15 km sec^{-1} (c) 0.23 km sec^{-2}

967. $e = \dfrac{T\sqrt{g/2H} - 1}{T\sqrt{g/2H} + 1}$

968. $\alpha = \cos^{-1}\sqrt[3]{2d/L}$

969. (a) $t_+/t_- = 1.8$ (b) $S_+/S_- = 1.8$
(c) 68 mi hr^{-1}

970. (a) 2 mgh$_0$
(b) in impacts between $3m$ and the shelf and between m and $3m$; also, in a practical case, in stretching of the cord
(c) $26\sqrt{h_0/g}$

971. 2.36×10^{29} kg m^2 sec^{-1}

972. (a) 31 cm sec^{-1} (b) 1.1×10^{-4} sec
(c) 1.1 sec (d) Assuming isothermal conditions and ignoring change in surface energy as size of bubble changes, $dr \simeq r/100$, which is negligible

973. (a) $t_{\text{lift-off}} = M/\dot{M} - V/g$ (b) $\dot{M}V/g$

974. (a) $\sqrt{(2L/g)(1 + 4\pi^2k^2p^2}$ (b) $0.05Mg/\pi p$
(c) $Mg/2\pi p$

INDEX